Enantioselective Organocatalysis

Edited by
Peter I. Dalko

1807–2007 Knowledge for Generations

Each generation has its unique needs and aspirations. When Charles Wiley first opened his small printing shop in lower Manhattan in 1807, it was a generation of boundless potential searching for an identity. And we were there, helping to define a new American literary tradition. Over half a century later, in the midst of the Second Industrial Revolution, it was a generation focused on building the future. Once again, we were there, supplying the critical scientific, technical, and engineering knowledge that helped frame the world. Throughout the 20th Century, and into the new millennium, nations began to reach out beyond their own borders and a new international community was born. Wiley was there, expanding its operations around the world to enable a global exchange of ideas, opinions, and know-how.

For 200 years, Wiley has been an integral part of each generation's journey, enabling the flow of information and understanding necessary to meet their needs and fulfill their aspirations. Today, bold new technologies are changing the way we live and learn. Wiley will be there, providing you the must-have knowledge you need to imagine new worlds, new possibilities, and new opportunities.

Generations come and go, but you can always count on Wiley to provide you the knowledge you need, when and where you need it!

William J. Pesce
President and Chief Executive Officer

Peter Booth Wiley
Chairman of the Board

Enantioselective Organocatalysis

Reactions and Experimental Procedures

Edited by
Peter I. Dalko

WILEY-VCH Verlag GmbH & Co. KGaA

The Editor

Dr. Peter I. Dalko
Université René Descartes, Paris V
CNRS, Laboratoire de Chimie et Biochimie
Pharmacologiques et Toxicologiques
45, rue des Saints-Pères
75270 Paris, Cedex 06
France

■ All books published by Wiley-VCH are carefully produced. Nevertheless, authors, editors, and publisher do not warrant the information contained in these books, including this book, to be free of errors. Readers are advised to keep in mind that statements, data, illustrations, procedural details or other items may inadvertently be inaccurate.

Library of Congress Card No.: applied for

British Library Cataloguing-in-Publication Data
A catalogue record for this book is available from the British Library.

Bibliographic information published by the Deutsche Nationalbibliothek
Die Deutsche Nationalbibliothek lists this publication in the Deutsche Nationalbibliografie; detailed bibliographic data are available in the Internet at http://dnb.d-nb.de

© 2007 WILEY-VCH Verlag GmbH & Co. KGaA, Weinheim

All rights reserved (including those of translation into other languages). No part of this book may be reproduced in any form – by photoprinting, microfilm, or any other means – nor transmitted or translated into a machine language without written permission from the publishers. Registered names, trademarks, etc. used in this book, even when not specifically marked as such, are not to be considered unprotected by law.

Printed in the Federal Republic of Germany
Printed on acid-free paper

Cover design   Schulz Grafik-Design, Fußgönheim
Typesetting   Asco Typesetters, Hong Kong
Printing   betz-druck GmbH, Darmstadt
Bookbinding   Litges & Dopf Buchbinderei GmbH, Heppenheim
Wiley Bicentennial Logo   Richard J. Pacifico

ISBN 978-3-527-31522-2

Contents

Preface XV

List of Contributors XVII

List of Abbreviations XXI

1 Asymmetric Organocatalysis: A New Stream in Organic Synthesis 1
Peter I. Dalko

1.1   Introduction 1
1.2   Historical Background 2
1.3   Catalysts 6
1.3.1   Privileged Catalysts 9
1.3.1.1   Proline 9
1.3.1.2   Cinchona Alkaloids 10
1.3.1.3   TADDOLs and Derivatives 11
1.3.1.4   Binaphthol Derivatives 11
1.4   Reaction Types 12
1.4.1   Covalent Catalysis 12
1.4.2   Non-Covalent Organocatalysis 12
1.5   How This Book is Organized 13
   References and Notes 13

2 Enamine Catalysis 19

2.1 Aldol and Mannich-Type Reactions 19
Fujie Tanaka and Carlos F. Barbas, III

2.1.1   Introduction 19
2.1.2   Aldol Reactions 20
2.1.2.1   Aldol Reactions of Alkyl Ketone Donors 20
2.1.2.2   Aldol Reactions of α-Oxyketone Donors 23
2.1.2.3   Aldol Reactions of Aldehyde Donors 25
2.1.2.4   Aldol Reactions with Ketone Acceptors 30

2.1.2.5	Intramolecular Aldol Reactions 31
2.1.2.6	Mechanism and Transition States of Aldol Reactions and Effects of Water on Aldol Reactions 31
2.1.2.7	Catalyst Recycling 38
2.1.2.8	Catalyst Development Strategies 38
2.1.3	Mannich Reactions 38
2.1.3.1	Mannich-Type Reactions of Aldehyde Donors with Glyoxylate Imines 38
2.1.3.2	Mannich-Type Reactions of Aldehyde Donors with Other Preformed Imines 45
2.1.3.3	Three-Component Mannich Reactions using Aldehyde Donors 45
2.1.3.4	Mannich-Type Reactions of Ketone Donors 47
	References 51

2.2 α-Heteroatom Functionalization 56
Mauro Marigo and Karl Anker Jørgensen

2.2.1	Introduction 56
2.2.2	Direct α-Amination of Aldehydes and Ketones 59
2.2.3	Direct α-Amination of α-Cyanoacetates and β-Dicarbonyl Compounds 64
2.2.4	Direct α-Oxygenation Reactions of Aldehydes and Ketones 64
2.2.5	Direct α-Oxygenation Reactions of β-Ketoesters 67
2.2.6	Direct α-Halogenation Reactions of Aldehydes and Ketones 68
2.2.6.1	Direct α-Fluorination of Aldehydes 68
2.2.6.2	Direct α-Fluorination of β-Ketoesters 69
2.2.6.3	Direct α-Chlorination of Aldehydes and Ketones 69
2.2.6.4	Direct α-Chlorination of β-Ketoesters 72
2.2.6.5	Direct α-Bromination of Aldehydes and Ketones 72
2.2.7	Direct α-Sulfenylation of Aldehydes 73
2.2.8	Direct α-Sulfenylation of Lactones, Lactams, and β-Diketones 74
2.2.9	Direct α-Selenation of Aldehydes and Ketones 74
	References 75

2.3 Direct Conjugate Additions via Enamine Activation 77
Cyril Bressy and Peter I. Dalko

2.3.1	Introduction 77
2.3.2	Factors Determining the Stereoselectivity of the Organocatalytic Conjugate Additions 77
2.3.3	Addition of Ketones to Nitroolefins and Alkylidene Malonates 79
2.3.3.1	Proline 79
2.3.3.2	Pyrrolidine Amines and Pyrrolidine Amine Salts as Catalysts for Michael-Type Addition of Ketones to Activated Olefins 79
2.3.3.3	Chiral Primary Amines 82

2.3.3.4	Amine/Thiourea Catalysts	82
2.3.4	Addition of Aldehydes to Nitroolefins and Alkylidene Malonates	84
2.3.4.1	Aminopyrrolidine Catalysts	84
2.3.4.2	Addition of Aldehydes and Ketones to Enones	87
2.4	Conclusions	92
	References	93

3 Iminium Catalysis 95
Gérald Lelais and David W. C. MacMillan

3.1	Introduction	95
3.2	The Catalysis Concept of Iminium Activation	95
3.3	Development of the "First-Generation" Imidazolidinone Catalysts	96
3.4	Development of the "Second-Generation" Imidazolidinone Catalysts	98
3.5	Cyloaddition Reactions	98
3.5.1	Diels–Alder Reactions	98
3.5.2	[3+2]-Cycloadditions	102
3.5.3	Cyclopropanations	103
3.5.4	Epoxidations	104
3.5.5	[4+3]-Cycloadditions	105
3.6	1,4-Addition Reactions	106
3.6.1	Friedel–Crafts Alkylations	106
3.6.2	Mukaiyama–Michael Reactions	108
3.6.3	Michael Reactions of α,β-Unsaturated Ketones	108
3.7	Transfer Hydrogenation	110
3.8	Organocatalytic Cascade Reactions	111
3.8.1	Cascade Addition–Cyclization Reactions	111
3.8.2	Cascade Catalysis: Merging Iminium and Enamine Activations	113
3.9	Conclusions	115
	References	116

4 Ammonium Ions as Chiral Templates 121
Takashi Ooi and Keiji Maruoka

4.1	Introduction	121
4.2	Homogeneous Catalysis with Chiral Quaternary Ammonium Fluorides	121
4.2.1	Aldol and Nitroaldol Reactions (Preparation of Chiral Quaternary Ammonium Fluorides)	121
4.2.2	Trifluoromethylation	124
4.2.3	Hydrosilylation	124
4.3	Homogeneous Catalysis with Chiral Quaternary Ammonium Bifluorides	125
4.3.1	Aldol and Nitroaldol Reactions	125
4.3.2	Michael Reaction	127

4.4	Homogeneous Catalysis with Chiral Quaternary Ammonium Phenoxides *129*	
4.5	Heterogeneous Catalysis: Chiral Phase-Transfer Catalysis *129*	
4.5.1	Pioneering Study *129*	
4.5.2	Monoalkylation of Glycinate Schiff Base: Asymmetric Synthesis of α-Amino Acids *130*	
4.5.3	Dialkylation of Schiff Bases Derived from α-Alkyl-α-Amino Acids *138*	
4.5.4	Michael Reaction of Glycinate Benzophenone Schiff Bases *141*	
4.5.5	Aldol and Mannich Reactions *142*	
4.6	Conclusions *147*	
	References *147*	
5	**Organocatalytic Enantioselective Morita–Baylis–Hillman (MBH) Reactions** *151*	
	Candice Menozzi and Peter I. Dalko	
5.1	Addition of Ketones and Aldehydes to Activated Olefins *151*	
5.1.1	Reaction Mechanism *154*	
5.1.2	Diastereoselectivity *156*	
5.1.3	Chiral Amine Catalysts *157*	
5.1.3.1	Chiral (Thio)Urea/Amine Catalysts Systems *167*	
5.1.4	Chiral Phosphine-Catalyzed MBH Reactions *169*	
5.1.4.1	Phosphine/Binaphthol-Derived Catalysts *172*	
5.2	Asymmetric Aza-MBH Reactions *174*	
5.2.1	Reaction Mechanism *175*	
5.2.2	Chiral Amine Catalysts *175*	
5.2.2.1	Cinchona-Derived Catalysts *175*	
5.2.2.2	Non-Natural Tertiary Amine/Phenol Catalysts *178*	
5.2.2.3	Chiral Acid/Achiral Amine *179*	
5.2.3	Phosphine-Mediated Aza-MBH Reactions *180*	
5.3	Conclusions *183*	
	References and Notes *184*	
6	**Asymmetric Proton Catalysis** *189*	
	Jeff D. McGilvra, Vijaya Bhasker Gondi, and Viresh H. Rawal	
6.1	Introduction *189*	
6.2	Conjugate Addition Reactions *193*	
6.2.1	Conjugate Addition Reactions of α,β-Unsaturated Carbonyl Substrates *193*	
6.2.2	Conjugate Addition Reactions of Nitroolefins *197*	
6.2.3	Conjugate Addition Reactions of Vinyl Sulfones *206*	
6.3	Hydrocyanation Reactions *207*	
6.3.1	Hydrocyanation Reactions of Aldehydes *207*	

6.3.2	The Strecker Reaction 209
6.3.3	Hydrocyanation Reactions of Ketones 214
6.4	Mannich Reactions 215
6.5	Aza-Henry Reactions of Aldimines 220
6.6	Acyl Pictet–Spengler Reactions of Iminium Ions 222
6.7	Aza-Friedel–Crafts Reactions of Aldimines 224
6.8	Hydrophosphonylation Reactions of Aldimines 225
6.9	Direct Alkylation Reactions of α-Diazoesters 227
6.10	Imine Amidation Reactions 228
6.11	Transfer Hydrogenation Reactions of Imines 229
6.12	Morita–Baylis–Hillman Reactions 231
6.13	Cycloaddition Reactions 234
6.13.1	Hetero Diels–Alder (HDA) Reactions 235
6.13.2	Diels–Alder (DA) Reactions 241
6.14	Aldol and Related Reactions 244
6.15	Conclusion and Prospects 248
6.16	Addendum 249
	References 250

7   Chiral Lewis Bases as Catalysts   255
Pavel Kočovský and Andrei V. Malkov

7.1	Introduction 255
7.2	Allylation Reactions 256
7.2.1	Catalytic Allylation of Aldehydes 256
7.2.2	Stoichiometric Allylation of Aldehydes and Ketones 264
7.2.3	Allylation of Imines 264
7.3	Propargylation, Allenylation, and Addition of Acetylenes 265
7.3.1	Addition to Aldehydes 265
7.3.2	Addition to Imines 266
7.4	Aldol-Type Reactions 267
7.5	Hydrocyanation and Isonitrile Addition 273
7.5.1	Cyanation of Aldehydes 273
7.5.2	Cyanation of Imines (Strecker Reaction) 274
7.6	Reduction of Imines 275
7.7	Epoxide Opening 278
7.8	Conclusions and Outlook 281
	References 281

8   Asymmetric Acyl Transfer Reactions   287
Alan C. Spivey and Paul McDaid

8.1	Introduction 287
8.2	Type I Acyl Transfer Processes 290
8.2.1	Type I Acylative KR of Racemic Alcohols and Amines 291

8.2.1.1	Aryl Alkyl sec-Alcohols 291
8.2.1.2	Allylic sec-Alcohols 299
8.2.1.3	Propargylic sec-Alcohols 302
8.2.1.4	1,2-Amino Alcohols and Mono-Protected 1,2-Diols 302
8.2.1.5	α-Chiral Primary Amines 306
8.2.2	Type I Acylative ASD of Achiral/*meso*-Diols 307
8.2.3	Type I Asymmetric Acyl Addition to π-Nucleophiles: Steglich and Related Rearrangements and Additions to Silyl Ketene Acetals/Imines 309
8.3	Type II Acyl Transfer Processes 311
8.3.1	Type II Alcoholative ASD of Achiral/*meso*-Cyclic Anhydrides 312
8.3.2	Type II Alcoholative KR of Racemic Anhydrides, Azlactones, N-Carboxyanhydrides, Dioxolanediones and N-Acyloxazolodinethiones 316
8.3.3	Type II Asymmetric Alcohol, Phenol, Enol, and Amine Addition to Ketenes 321
8.4	Concluding Remarks 324
	References 324

9   Nucleophilic N-Heterocyclic Carbenes in Asymmetric Organocatalysis 331
Dieter Enders, Tim Balensiefer, Oliver Niemeier, and Mathias Christmann

9.1	The Benzoin Condensation 331
9.2	The Stetter Reaction 338
9.3	Further Applications 344
9.3.1	a^3 to d^3 Umpolung 344
9.3.2	Transesterifications and Polymerizations 349
	References 353

10   Ylide-Based Reactions 357
Eoghan M. McGarrigle and Varinder K. Aggarwal

10.1	Introduction 357
10.2	Epoxidation 357
10.2.1	Sulfur Ylide-Catalyzed Epoxidation 358
10.2.1.1	Catalysis via Sulfide Alkylation/Deprotonation 359
10.2.1.2	Conditions and Results for Alkylation/Deprotonation Cycle 359
10.2.1.3	Catalysis via Ylide Formation from a Carbene Source 360
10.2.1.4	Carbenes from *In Situ* Diazo Compound Generation 361
10.2.1.5	Scope of Carbene-Based Methodologies 362
10.2.1.6	Catalytic Cycle Using the Simmons–Smith Reagent 364
10.2.1.7	Synthesis of Glycidic Amides 364
10.2.1.8	Optimum Sulfide in Carbene-Based Cycle 364
10.2.1.9	Diastereoselectivity in S-Ylide Epoxidations 365
10.2.1.10	Enantioselectivity in S-Ylide Epoxidations 366

10.2.1.11	Applications of S-Ylide Epoxidation in Synthesis	367
10.2.2	Te-Ylide-Catalyzed Epoxidation	369
10.2.3	Se-Ylide-Catalyzed Epoxidation	369
10.2.4	Summary of Ylide-Mediated Epoxidation	370
10.3	Asymmetric Aziridination	370
10.3.1	Sulfur Ylide-Catalyzed Aziridination	370
10.3.1.1	Ylide Regeneration via Carbenes	370
10.3.1.2	Scope and Limitations of the Carbene-Based Cycle	371
10.3.1.3	Catalysis via Sulfonium Salt Formation and Deprotonation	373
10.3.1.4	Selectivity in S-Ylide Aziridinations	374
10.3.1.5	Use of S-Ylide Aziridinations in Synthesis	375
10.3.1.6	Summary of Ylide-Catalyzed Aziridination	376
10.4	Asymmetric Cyclopropanation	377
10.4.1	S-Ylide-Mediated Cyclopropanation	377
10.4.1.1	Catalytic Cycle Based on Carbene Generation of Ylide	377
10.4.1.2	Catalysis via *In Situ* Generation of Diazo Compounds	378
10.4.1.3	Control of Selectivity Using Carbene-Derived Ylides	379
10.4.1.4	Catalytic Cycle Based on Sulfide Alkylation and Sulfonium Salt Deprotonation	379
10.4.1.5	Scope and Limitations of the Alkylation/Deprotonation Method	380
10.4.1.6	Control of Selectivity in the Alkylation/Deprotonation Method	381
10.4.1.7	Use of S-Ylide-Catalyzed Cyclopropanations in Synthesis	381
10.4.2	Tellurium Ylide-Mediated Cyclopropanation	382
10.4.3	Nitrogen Ylide-Mediated Cyclopropanation	382
10.4.3.1	Development of a Catalytic Method	382
10.4.3.2	Enantioselective Cyclopropanation	383
10.4.3.3	Mechanism of N-Ylide-Catalyzed Cyclopropanation	385
10.4.4	Summary of Ylide-Catalyzed Cyclopropanation	385
10.5	Summary of Ylide-Catalyzed Asymmetric Reactions	386
	References 386	

11	**Organocatalytic Enantioselective Reduction of Olefins, Ketones, and Imines**	**391**
	Henri B. Kagan	
11.1	Introduction	391
11.2	The General Concept	391
11.3	Asymmetric Organocatalytic Reduction of Olefins	393
11.4	Reduction of Ketones or Ketimines	397
11.4.1	Hantzsch Ester/Chiral Brønsted Acids	397
11.4.2	Silanes/Phase-Transfer Catalysts	398
11.4.3	Silanes/Formamide Catalysts	399
11.5	Conclusions	401
	References 401	

12 Oxidation Reactions 403
Alan Armstrong

12.1	Epoxidation of Alkenes 403
12.1.1	Ketone-Mediated Epoxidation 404
12.1.2	Iminium Salt-Catalyzed Epoxidation 410
12.1.3	Amine-Catalyzed Epoxidation 412
12.2	Epoxidation of Enoates, Enones, and Enals 413
12.2.1	Epoxidation of Enoates 413
12.2.2	Epoxidation of Enones 414
12.2.2.1	Phase-Transfer Catalysis 414
12.2.2.2	Polyleucine-Mediated Epoxidation 416
12.2.3	Epoxidation of Enals 417
12.3	Asymmetric Baeyer–Villiger Reaction 418
12.4	Oxidation of Thioethers 419
12.5	Resolution of Alcohols by Oxidation 420
12.6	Conclusions 421
	References 421

13 Shape- and Site-Selective Asymmetric Reactions 425
Nicolas Bogliotti and Peter I. Dalko

13.1	Introduction 425
13.2	Acylation Reactions with Oligopeptide-Based Catalysts 426
13.2.1	Regioselective Acetylation of Glucose Derivatives 426
13.2.2	Regioselective Phosphorylation 427
13.3	Sequence-Specific RNA-Cleaving Agents 428
13.3.1	RNA Hydrolysis by DNA-DETA Adduct 429
13.3.2	RNA Hydrolysis by PNA-DETA Adduct 430
13.3.3	Methyl Phosphonate Oligonucleotides as RNA-Cleaving Agents 431
13.4	Chiral Cavitands: Calixarenes and Cyclodextrins 432
13.4.1	Calixarenes 433
13.4.1.1	Specific Esterification of *N*-Acetyl-L-Amino Acids 433
13.4.2	Cyclodextrins 434
13.4.2.1	Aryl Glycoside Hydrolysis 435
13.4.2.2	Phosphate Hydrolysis 435
13.4.2.3	Transamination Reactions 437
13.4.2.4	Amidation Reactions 437
13.4.2.5	Epoxidation of Alkenes 438
13.5	Molecular Imprinting 442
13.5.1	Enantioselective Ester Hydrolysis 443
13.6	Conclusion 446
	References 446

14	**Appendix I: Reaction Procedures** *451*	
14.1	Aldol Reactions *451*	
14.2	Mannich Reaction *457*	
14.3	Pictet–Spengler Reaction *459*	
14.4	Nitroaldol (Henry) Reaction *460*	
14.5	Hydrocyanation and Cyanosilylation Reactions *461*	
14.6	Alkylation of α-Diazoesters *464*	
14.7	Heteroatom Addition to Imines *464*	
14.8	α-Heteroatom Functionalization *466*	
14.9	1,4-Additions *468*	
14.10	Morita–Baylis–Hillman (MBH) Reactions *476*	
14.11	Cyclopropanation *478*	
14.12	Aziridination *480*	
14.13	Epoxide Formation *481*	
14.14	Epoxide Opening *488*	
14.15	Nucleophilic Substitution Reactions *488*	
14.16	Allylation Reactions *490*	
14.17	Acylation *491*	
14.18	[4+2] Cycloadditions *500*	
14.19	[3+2]-Cycloadditions *503*	
14.20	Acyloin and Benzoin Condensations *504*	
14.21	Reduction of Ketones and Imines *505*	
14.22	Hydrogenation of Olefins *506*	
14.23	Multicomponent Domino Reactions *508*	
	References 510	
15	**Appendix II: Catalysts** *513*	
	Subject Index *531*	

Preface

Asymmetric synthesis, the ability of controlling the three dimensional structure of the molecular architecture has revolutionized chemistry in the second half of the XXth century. This concept continues to influence the development of basically all fields of science. Amongst the various ways of creating enantiomerically enriched products, catalytic methods (i.e. when chemical transformations are controlled by a small amount of chiral compounds) are considered as the most appealing. It is difficult to conceive that the impressive knowledge accumulated in this field was gained in a relatively short period of time. New concepts and methods are emerging continuously, allowing more selective, economically more appealing and environmentally friendlier transformations. In this context, asymmetric organocatalysis is a «fast lane» of the chemical highway: the progress in the last decade has been simply spectacular.

Performing chemical transformations with a small amount of organic molecules is not a novel concept: enantioselective organocatalytic transformations were developed prior to organometallic ones. The relatively narrow scope of these transformations, however, did not stir particular interest in the past. Nowadays the situation is changing. The renewed interest is due to the serendipitous discovery of a number of selective transformations and also to the realization of the tremendous potential which is inherent to these novel forms of activations, which are also complementary to the existing ones. After the milestone book of Berkessel and Gröger (*Asymmetric Organocatalysis, From Biomimetic Concepts to Applications in Asymmetric Synthesis* VCH, Weinheim, 2005), this multiauthor book is the state of the art of this rapidly evolving field. The chapters are written by organic chemists, leaders at the forefront of research and able to provide an insider's view. I am grateful to all colleagues who agreed to contribute to this project, despite their many other obligations and busy schedules: the result is more than impressive.

It is the aim of this book to provide a concise and comprehensive treatment of this rapidly evolving field, focusing on the preparative aspect of this chemistry. In fact, the use of organocatalytic transformations in a multistep synthesis remains scare. This book wishes to promote the application of these reactions, giving solid synthetic evidence. Additionally, a collection of sample procedures of typical

Enantioselective Organocatalysis: Reactions and Experimental Procedures
Edited by Peter I. Dalko
Copyright © 2007 WILEY-VCH Verlag GmbH & Co. KGaA, Weinheim
ISBN: 978-3-527-31522-2

organocatalytic transformations is given in Appendix I. Despite the spectacular advancement, there is room for further development, and it is the wish of the Editor that this manual should be rapidly updated.

This book is suggested for graduate students as well as all organic chemists.

Paris, January 2007

Peter I. Dalko

List of Contributors

Varinder K. Aggarwal
University of Bristol
School of Chemistry
Cantock's Close
Bristol, BS8 1TS
United Kingdom

Alan Armstrong
Imperial College London
Department of Chemistry
South Kensington Campus
London, SW7 2AZ
United Kingdom

Tim Balensiefer
RWTH Aachen University
Institute of Organic Chemistry
Landoltweg 1
52074 Aachen
Germany

Carlos F. Barbas, III
The Scripps Research Institute
The Skaggs Institute for
Chemical Biology
Departments of Chemistry and
Molecular Biology
10550 North Torrey Pines Road
La Jolla, CA 92037
USA

Nicolas Bogliotti
ESPCI
Laboratoire de Chimie Organique
10, rue Vauquelin
75231 Paris Cedex 05
France

Cyril Bressy
Université Paul Cezanne
d'Aix-Marseille-CNRS
Faculté des Sciences de St Jerôme
Boîte D12
SymBio, Laboratoire RESO
13397 Marseille Cedex 20
France

Mathias Christmann
RWTH Aachen University
Institute of Organic Chemistry
Landoltweg 1
52074 Aachen
Germany

Peter I. Dalko
Université René Descartes, Paris V
Laboratoire de Chimie et Biochimie
Pharmacologiques et Toxicologiques
45, rue des Saints-Pères
75270 Paris Cedex 06
France

Enantioselective Organocatalysis: Reactions and Experimental Procedures
Edited by Peter I. Dalko
Copyright © 2007 WILEY-VCH Verlag GmbH & Co. KGaA, Weinheim
ISBN: 978-3-527-31522-2

List of Contributors

Dieter Enders
RWTH Aachen University
Institute of Organic Chemistry
Landoltweg 1
52074 Aachen
Germany

Vijaya Bhasker Gondi
University of Chicago
Department of Chemistry
5735 S. Ellis Avenue
Chicago, JL 60637
USA

Karl Anker Jørgensen
The Danish National Research
Foundation: Center for Catalysis
Department of Chemistry
Aarhus University
8000 Aarhus C
Denmark

Henri B. Kagan
Université Paris-Sud
Institut de Chimie Moléculaire
et des Matériaux d'Orsay
CNRS UHR 8182
91405 Orsay
France

Pavel Kočovský
University of Glasgow
Department of Chemistry
Joseph Black Building
Glasgow, G12 8QQ
United Kingdom

Gérald Lelais
Genomic Institute of the Novetis
Foundation
Medicinal Chemistry
10675 John Jay Hopkins Drive
San Diego, CA 92121
USA

David W. C. MacMillan
Princeton University
Frick Laboratories
Washington Road
Princeton, NJ 08544
USA

Andrei V. Malkov
University of Glasgow
Department of Chemistry
Joseph Black Building
Glasgow, G12 8QQ
United Kingdom

Mauro Marigo
The Danish National Research
Foundation: Center for Catalysis
Department of Chemistry
Aarhus University
8000 Aarhus C
Denmark

Keiji Maruoka
Kyoto University
Graduate School of Science
Department of Chemistry
Sakyo
Kyoto 606-8502
Japan

Paul McDaid
Pfizer Process Development Center
Loughbed API Plant
Ringaskiddy
Co. Cork
Ireland

Eoghan M. McGarrigle
University of Bristol
School of Chemistry
Cantock's Close
Bristol, BS8 1TS
United Kingdom

Jeff D. McGilvra
University of Chicago
Department of Chemistry
5735 S. Ellis Avenue
Chicago, JL 60637
USA

Candice Menozzi
ESPCI
Laboratoire de Chimie Organique
10, rue Vauquelin
75231 Paris Cedex 05
France

Oliver Niemeier
RWTH Aachen University
Institute of Organic Chemistry
Landoltweg 1
52074 Aachen
Germany

Takashi Ooi
Kyoto University
Graduate School of Science
Department of Chemistry
Sakyo
Kyoto 606-8502
Japan

Viresh H. Rawal
University of Chicago
Department of Chemistry
5735 S. Ellis Avenue
Chicago, JL 60637
USA

Alan C. Spivey
Imperial College London
Department of Chemistry
South Kensington Campus
London, SW7 2AZ
United Kingdom

Fujie Tanaka
The Scripps Research Institute
Department of Molecular Biology
10550 North Torrey Pines Road
La Jolla, CA 92037
USA

List of Abbreviations

Ac	acetyl
AIBN	2,2-azobis(isobutyronitrile)
Alloc	allyloxycarbonyl
ASD	asymmetric desymmetrization
BAMOL	1,1′-biaryl-2,2′-dimethanol
BEMP	2-*tert*-butylimino-2-diethylamino-1,3-dimethyl-perhydro-1,3,2-diazaphosphorine
[bmim]BF$_4$	1-butyl-3-methylimidazonium tetrafluoroborate
BINOL	1,1-bi(2-naphthol)
Bn	benzyl (CH$_2$Ph)
Boc	*tert*-butoxycarbonyl
C	conversion
CAN	ceric ammonium nitrate
cat	catalyst
Cbz	benzyloxycarbonyl
CD	cyclodextrin
CPME	cyclopentyl methyl ether
CVAM	catalytic asymmetric vinylogous Mukaiyama (reaction)
CX$_n$	calix [n]arene
Cy	cyclohexyl
d	day
DA	Diels-Alder (reaction)
DABCO	diazabicyclo[2.2.2]octane
DBU	1,8-diazabicyclo[5.4.0]undec-7-ene
DEA	direct electrostatic activation
DETA	diethylenetriamide
(DHQ)$_2$AQN	hydroquinine anthraquinone-1,4-diyl diether
(DHQD)$_2$AQN	hydroquinidine anthraquinone-1,4-diyl diether
DIPEA	diisopropylethylamine
DKR	dynamic kinetic resolution
DMAP	4-(dimethylamino)pyridine
DMM	dimethoxymethane
E	electrophile

Enantioselective Organocatalysis: Reactions and Experimental Procedures
Edited by Peter I. Dalko
Copyright © 2007 WILEY-VCH Verlag GmbH & Co. KGaA, Weinheim
ISBN: 978-3-527-31522-2

EDTA	ethylenediaminetetraacetic acid
ee	enantiomeric excess
ent	enantiomeric
er	enantiomeric ratio
EWG	electron withdrawing group
FADH	dihydroflavin adenin dinucleotide
Fmoc	9-fluorenylmethyloxylacrbonyl
GABA	γ-aminobutyric acid
GC	gas chromatography
GLC	gas-liquid chromatography
GTLC	gradient thin layer chromatography
HDA	hetero-Diels-Alder (reaction)
3-HDQ	3-hydroxy quinuclidine
HEPES	4-(2-hydroxyethyl)-1-piperazineethanesulfonic acid (widely used buffering agent to maintain physiological pH)
HFIPA	1,1,1,3,3,3-hexafluoroisopropyl acryl ester
HMPA	hexamethylphosphoramide
HOMO	highest occupied molecular orbital
HPLC	high pressure liquid chromatography
i	iso
ICD	isocupreidine
KR	kinetic resolution
LDA	lithium diisopropylamide
LUMO	lowest unoccupied molecular orbital
MBH	Morita-Baylis-Hillman (reaction)
MIP	molecular imprinted polymer
MS	molecular sieve
MVK	methyl vinyl ketone
n	normal
N/A	not available
NADH	dihydronicotinamide adenine dinucleotide
Nap	naphthyl
NCS	N-chlorosuccinimide
NFSI	N-fluorobenzenesulfonimide
NHC	N-heterocyclic carbenes
NMP	N-methylpyrrolidine
NMR	nuclear magnetic resonance
NOBIN	2-amino-2'-hydroxy-1,1'-binaphthyl
Nu	nucleophile
Oxone	$2KHSO_5 \cdot KHSO_4 \cdot K_2SO_4$
PBO	P-aryl-2-phosphabicyclo[3.3.0]octane
Phe	(S)-phenylalanyl
PIP	2-phenyl-2,3-dihydroimidazo[1,2a]pyridine
PIQ	2-phenyl-1,2-dihydroimidazo[1,2a]quinoline
PKR	parallel kinetic resolution

PMHS	polymethylhydrosiloxane
PMP	*p*-methoxyphenyl
PNA	peptide nucleic acid
PNP	*p*-nitrophenyl
PPY	4-(pyrrolidino)pyridine
Pyr	pyridine
RDS	rate-determining step
rec SM	recovered starting material
ROMP	ring opening metathesis polymerisation
rt	room temperature
s	(enantio)selectivity value
sec	secondary
SES	(trimethylsilyl)ethansulfonyl
TADDOL	*trans*-4,5-bis-(diphenyl-hydroxymethyl)-2,2-dimethyl-1,3-dioxolane
TADMAP	3-(2,2-triphenyl-1-acetoxyethyl)-4-dimethylamino)pyridine
TBAF	tetra-*n*-butylammonium fluoride
TBDPS	*tert*-butyldiphenylsilyl
TBS	*tert*-butyldimethylsilyl
TCA	trichloroacetic acid
tert	tertiary
TES	triethylsilyl
TFA	trifluoroacetic acid
TFAA	trifluoroacetic anhydride
TIPS	triisopropylsilyl
TLC	tin layer chromatography
TMG	1,1,3,3-tetramethylguanidine
TMP	thymidine monophosphate
TMSCN	trimethylsilyl cyanide
TOF	turnover frequency
TON	turnover number
Troc	2,2,2-trichloroethyl
Trt	trityl (triphenylmethyl)
Ts	tosyl, *p*-toluenesulfonyl
TS	transition state
UNCA	urethane-protected α-amino acid *N*-carboxy anhydride
VAPOL	vaulted biphenantrol
VMA	vinilogous Mukaiyama aldol (reaction)

1
Asymmetric Organocatalysis: A New Stream in Organic Synthesis

Peter I. Dalko

1.1
Introduction

In common with metal complexes and enzymes, small organic molecules may promote chemical transformations. Organocatalysis provides a means of accelerating chemical reactions with a substoichiometric amount of organic molecules, which do not contain a metal element [1, 2].

Despite this rich historical past, the use of small organic molecules as chiral catalysts has only recently been recognized as a valuable addition and/or alternative to existing, well-established, often metal-based methodologies in asymmetric synthesis. Driven both by distinguished scientific interest, which usually accompanies emerging fields, and the recognition of the huge potential of this new area, organocatalysis has finally developed into a practical synthetic paradigm [3–15]. The question must be asked, however, as to why it has taken so long for chemists to appreciate and exploit the potential of small organic molecules as chiral catalysts. Why was not the imagination of the vast majority of the chemical community captured by the perspectives of asymmetric organocatalysis, when metal complex-derived catalysis underwent steady development for enantioselective reactions?

Principally, asymmetric organocatalytic reactions were, for a long time, considered to be inefficient and limited in scope. In parallel, organometallic catalysts provided a flexible ground for all types of reaction, and thus received disproportionate emphasis. Although today the vast majority of reactions in asymmetric catalysis continue to rely on organometallic complexes, this picture is changing, and organic catalysis is becoming an increasingly important segment of organic chemistry, offering a number of advantages over metal-based and bioorganic methods.

Today, reactions can be performed under an aerobic atmosphere, with wet solvents; indeed, the presence of water is often beneficial to the rate and selectivity of the reaction. The operational simplicity and ready availability of these mostly inexpensive bench-stable catalysts – which are incomparably more robust than

Enantioselective Organocatalysis: Reactions and Experimental Procedures
Edited by Peter I. Dalko
Copyright © 2007 WILEY-VCH Verlag GmbH & Co. KGaA, Weinheim
ISBN: 978-3-527-31522-2

enzymes or other bioorganic catalysts – makes organocatalysis an attractive method for the synthesis of complex structures. Unlike any earlier developed system, organocatalytic reactions provide a rich platform for multicomponent, tandem, or domino-type multistep reactions [16], allowing increases in the structural complexity of products in a highly stereocontrolled manner. In addition, fewer toxicity issues are often associated with organocatalysis, although this applies only when utilizing the more notorious metals. It should also be pointed out that little is currently known regarding the toxicity of many organic catalysts; moreover, there is no risk of metal leakage, and no expensive recovery process is required for waste treatment. Nowadays, increasing numbers of industrial applications are based on asymmetric organocatalytic reactions, and the environmentally friendly, "green" aspect of this chemistry – coupled with the sustainability of the catalysts – is considered widely for replacing standard, metal-based reactions [17, 18].

1.2
Historical Background

The history of organocatalytic reactions has a rich past, there being evidence that such catalysis has in the past played a determinant role in the formation of prebiotic key building blocks such as sugars. In this way, the reactions have led to the introduction and widespread use of homochirality in the living word [19]. Enantiomerically enriched amino acids such as L-alanine and L-isovaline, which may be present in up to 15% enantiomeric excess (*ee*) in carbonaceous meteorites, were able to catalyze the aldol-type dimerization of glycolaldehyde, as well as the reaction between glycolaldehyde and formaldehyde producing sugar derivatives. For example, Pizzarello and Weber were able to demonstrate that L-isovaline, which was found in the Murchison meteorite, promotes the self-aldol reaction of glycolaldehyde in water, generating aldol products such as L-threose and D-erythrose with up to $10.7 \pm 1.2\%$ and 4.8 ± 0.9 *ee*, respectively [19]. Proline, the most efficient natural amino acid catalyst in aldol-type condensations is scarcely present in meteorites. Asymmetric photolysis in interstellar clouds may produce optically active proline, however, indicating that proline may also have been transported to Earth [20]. The formation of sugars under prebiotic conditions was amplified in a number of elegant *de-novo* constructions of complex, differentiated carbohydrates by chemical synthesis [21]. It is likely, therefore, that these aldol products were the precursor of complex molecules such as RNA and DNA. Prebiotic RNA most likely played a central role in orchestrating a number of key biochemical transformations necessary for life, in which sugars served as chiral templates [22]. For example, it is considered, that amino acid homochirality in proteins was determined during asymmetric aminoacylation, which is the first step in protein synthesis and thus was critical for the transition from the putative RNA world to the theater of proteins [23]. According to this concept, the selectivity (L or D) of amino acids was determined in large part by the preestablished homochirality of RNA.

Organic molecules have been used as catalysts from the early age of synthetic chemistry. Indeed, the discovery of the first organocatalytic reaction is attributed to J. von Liebig, who found – accidentally – that dicyan is transformed into oxamide in the presence of an aqueous solution of acetaldehyde (Scheme 1.1). Subsequently, this efficient reaction found industrial application by forming the basis of the Degussa oxamide synthesis.

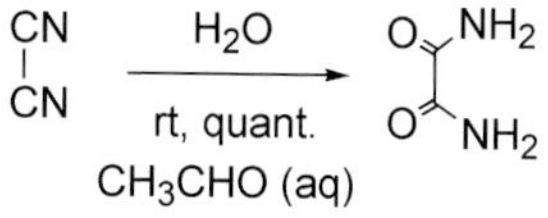

Scheme 1.1 von Liebig's oxamide synthesis.

Undoubtedly, the discovery of enzymes and enzyme functions had an important impact on the development of asymmetric catalytic reactions. The first asymmetric reaction – a decarboxylative kinetic resolution – was discovered by Pasteur [24], who observed that the organism *Penicillium glauca* destroyed more rapidly one of the enantiomers (d) from a racemic solution of ammonium tartrate. Asymmetric decarboxylation reactions were re-examined under non-enzymatic conditions by Georg Breding during the early 1900s. Breding, who had a remarkably wide interdisciplinary interest, was motivated to find the chemical origin of enzyme activity observed in living organisms. In his early experiments he showed enantiomerical enrichment in the thermal decarboxylation of optically active camphorcarboxylic acid in d and l limonenes, respectively [25]. As an extension of this work he studied this decarboxylation reaction in the presence of chiral alkaloids, such as nicotine or quinidine, and established the basic kinetic equations of this kinetic resolution [26]. The first asymmetric C–C bond forming reaction is attributed also to his name. This milestone achievement is related to Rosenthaler's work, who was able to prepare mandelonitrile by the addition of HCN to benzaldehyde in the presence of an isolated enzyme, emulsin [27]. Breding was also able to perform this reaction in the presence of alkaloids as catalysts, such as the pseudoenantiomeric quinine and quinidine (Scheme 1.2) [28]. It should be noted

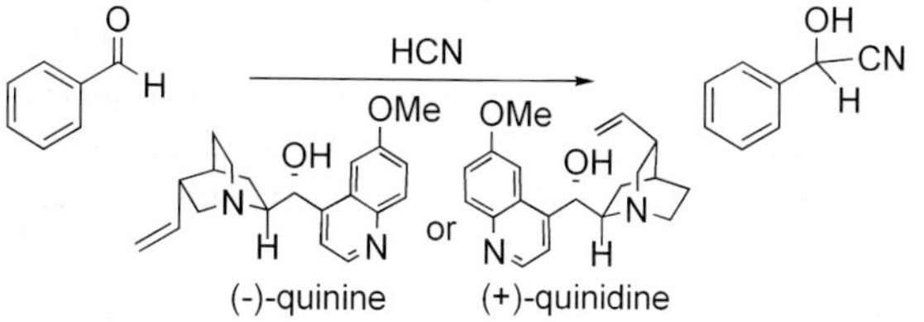

Scheme 1.2

that, although these studies were considered as conceptually groundbreaking, the enantioselectivity of the reaction was less than 10%.

Although catalytic transformations gained increasing importance after the First World War, asymmetric reactions were considered at the time to be an academic curiosity. Of note, the determination of enantioselectivity was hampered by a lack of methods to achieve not only efficient purification but also reliable analyses. Hence, the presence of a chiral impurity – which often arose from the catalyst – spoiled the determination of the correct, optical rotation-based *ee*-values.

Nitrogen-containing natural products such as alkaloids (in particular strychnine, brucine and cinchona alkaloids) and amino acids (including short oligopeptides) were among the first organic catalysts to be tested. The acylative kinetic resolution of racemic secondary alcohols was initiated during the late 1920s by Vavon and Peignier in France [29], and, independently, also by Wegler in Germany [30]. These authors showed that brucine and strychnine were able to induce enantiomeric enrichment either in the esterification of *meso* dicarboxylic acids or in the kinetic resolution of secondary alcohols, albeit with low *ee*-values.

Also, Wolfgang Langenbeck's contribution should be remembered, who developed reactions, which were promoted by simple amino acids, or, by small oligopeptides [31]. A major part of these studies were dedicated to reactions which emulated enzyme functions by using simple amino acids or small peptides. Not surprisingly, enamine-type reactions were among the first to be discovered. This finding was initiated by the studies of Dakin who, in 1909, noted that in a Knoevenagel-type condensation between aldehydes and carboxylic acids or esters with active methylene groups, the amine catalysts could be mediated by amino acids [32]. The reaction was extended to aldol and related transformations, and studied systematically from the early 1930s onwards with notable success, essentially with non-asymmetric systems.

The reinvestigation of Breding's asymmetric cyanohydrin synthesis by Prelog during the mid-1950s [33] undoubtedly promoted the concept of asymmetric synthesis, and led the way to more efficient reactions. The advent of synthetically useful levels of enantioselectivity can be dated to the late 1950s, when Pracejus reported that methyl phenyl ketene could be converted to $(-)$-α-phenyl methylpropionate in 74% *ee* by using O-acetylquinine as catalyst [34].

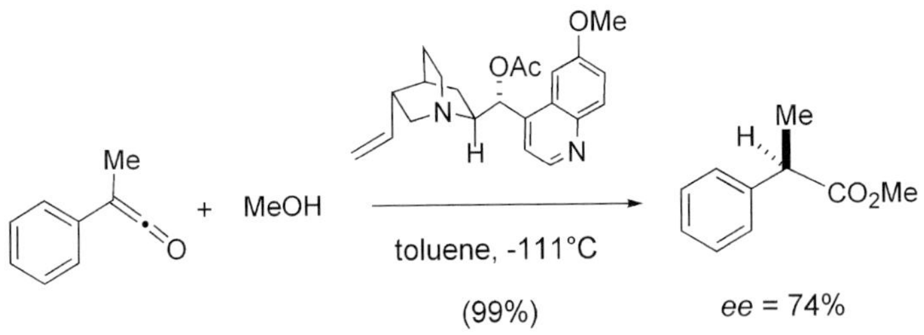

Scheme 1.3 Pracejus' enantioselective ester synthesis from phenyl methyl ketene.

This quite impressive result inspired the reinvestigation of other possible reaction manifolds for the cinchona catalyst system. Bergson and Långström reported the first Michael addition of β-keto esters to acrolein using 2-(hydroxymethyl)-quinuclidine as catalyst [35]. Although they never determined the enantiomeric excess, these authors noted the optical activity of their products. Wynberg and co-workers carried out extensive studies of the use of cinchona alkaloids as chiral Lewis-base/nucleophilic catalysts [36], and showed this class of alkaloid to be a versatile catalyst, promoting a variety of 1,2- and 1,4-additions of a wide range of nucleophiles to carbonyl compounds. Noteworthy, in these early studies it was often observed that natural cinchona alkaloids were superior, in terms of both catalytic activity and selectivity, to modified cinchona alkaloids derived from modification of the C-9 hydroxyl group. In order to rationalize this phenomenon, Wynberg proposed that the natural cinchona alkaloids were bifunctional catalysts utilizing both the tertiary amine and hydroxyl group to activate and orient the nucleophile and electrophile, respectively, thus achieving optimum asymmetric catalysis [36].

Another key event in the history of organocatalytic reaction was the discovery of efficient L-proline-mediated asymmetric Robinson annulation reported during the early 1970s. The so-called Hajos–Parrish–Eder–Sauer–Wiechert reaction (an intramolecular aldol reaction) allowed access to some of the key intermediates for the synthesis of natural products (Scheme 1.4) [37, 38], and offered a practical and enantioselective route to the Wieland–Miescher ketone [39]. It is pertinent to note, that this chemistry is rooted in the early studies of Langenbeck and in the extensive investigations work of Stork and co-workers on enamine chemistry [40].

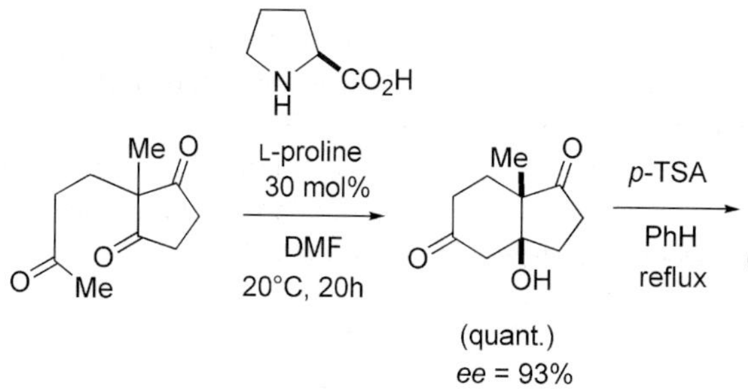

Scheme 1.4 The L-proline-mediated Robinson annulation.

This L-proline-mediated annulation received a considerable synthetic and mechanistic interest [41]. It was demonstrated that other amino acids, such as (R)-phenylalanine, could replace in some cases advantageously the L-proline [42]. Earlier applications in total syntheses appeared, however, as singular events, as in Woodward's synthesis of erythromycin (Scheme 1.5) [27]. Remarkably, in this synthesis a racemic keto aldehyde 5 could be used for aldolization with D-proline

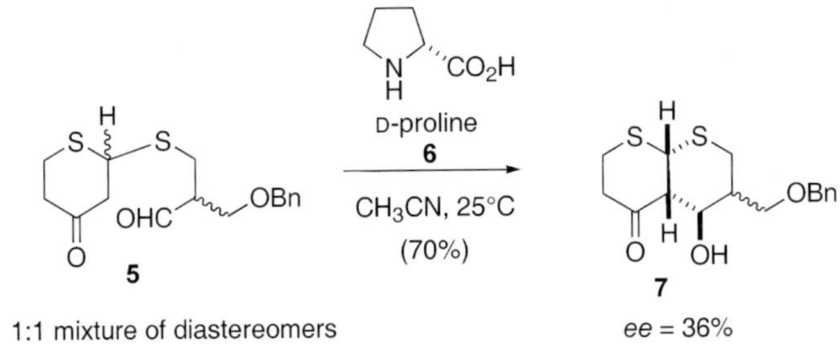

Scheme 1.5 The D-proline-mediated intramolecular aldol reaction in Woodward's erythromycin synthesis.

(**6**) as catalyst. All of the chiral centers of the erythronolide backbone are derived either directly or indirectly from this rather poor reaction, which was only of 36% ee. However, an optically pure downstream product of **7** was separated by simple recrystallization, which made the process eminently practical (Scheme 1.5).

The late 1970s and early 1980s marked a clear turning point, with the advent not only of more general efficient asymmetric organocatalysts but also of organocatalytic reactions. During this period a number of reactions which proceeded via ion-pairing mechanisms (i.e., similar of that noted with the propos of cinchona alkaloids) were uncovered. In addition, chiral diketopiperazines were developed by Inoue as chiral Brønsted acids for the asymmetric hydrocyanation reactions [43], this reaction paving the way for the efficient hydrocyanation reactions of aldimines developed some years later by the groups of Lipton and Jacobsen [44, 45]. The advent of efficient phase-transfer reactions dates back to the mid-1980s, when researchers at Merck reported that substituted 2-phenyl-1-indanone systems could be alkylated with remarkably high enantioselectivity (up to 94%) in the presence of catalytic amounts of substituted N-benzylcinchoninium halides (50% NaOH/toluene) [46]. Mention should be made here of the chiral amine-mediated cycloaddition reactions, which were pioneered by Kagan [47], as well as the earliest examples of the enantioselective oxidation of chalcones using polyamino- or resin-bonded polyamino acid under tri- and biphasic conditions, the so-called Julià reaction [48]. Reinvestigation of the Hajos–Parrish–Eder–Sauer–Wiechert reaction by List and Barbas during the late 1990s also opened an avenue for a number of related transformations such as the enantioselective intermolecular cross-aldol reactions, as well as Mannich, Michael and Diels–Alder-type transformations, and the application of these transformations in multistep (domino) reactions [3, 16].

1.3
Catalysts

As metals easily form Lewis acids, organic catalysts are more prone to form heteroatom-centered Lewis bases. Among these catalysts, the N- and P-based forms are the most studied, with amine catalysts being more easily available than their phosphorus-containing counterparts, due mainly to the natural abundance. There is no natural P-containing chiral substrate for catalytic use, and consequently all of these catalysts are man-made [49]. One particular advantage of phosphorus-based catalysts is their ability to act as both a nucleophilic *and* stereogenic reaction center. The difference in Brønsted basicity of the phosphorus atom compared to the amine function may also be advantageous in avoiding base-mediated secondary reactions.

It should be noted that not only the Lewis base but also typical Lewis acid roles can be emulated by organocatalytic systems. The proton is arguably the most common Lewis acid found in Nature, and these exist in two forms classified by the nature of the hydrogen bond: polar covalent (RX–H) and polar ionic ($RX^+H \cdots Y^-$). In the former case, in asymmetric transformations the chiral information is dictated by the chiral anion, whilst in the latter case the anion is nonchiral and the enantioselectivity is introduced by a chiral ligand (usually an amine base), which complexates the proton. This activation is discussed more extensively in Chapter 7.

Another class of activation is related to the particular reactivity of the nitrogen, and is referred to as *aminocatalysis*. Amine catalysts may give rise either to enamine or iminium intermediates [50]; the former activation results in an increased electron density at the reaction center(s), while the latter activation corresponds to a decrease in electron density at the reaction center(s). One peculiarity of this type of chemistry is the facile equilibrium between these two electron-rich and electron-deficient states (i.e., the acid–base form) of the same center. It is easy to conceive this equilibrium simply by considering protonation-deprotonation, which on the one hand may activate the reagent and, on the other hand may contribute to the kinetic lability of the ligand. The peculiarity of this activation is the fact that, due to this equilibrium process, the same center may act as either a Lewis acid or Lewis base, depending on the reaction conditions (Scheme 1.6). While both intermediates are formed in the same mixture, the relative concentrations of these structures is determined by the reaction conditions, leading to chemical transformations which follow entirely different mechanistic pathways and usually result in different products. More importantly, the same catalyst may promote complementary nucleophile/electrophile activation (i.e., promoting reactions via

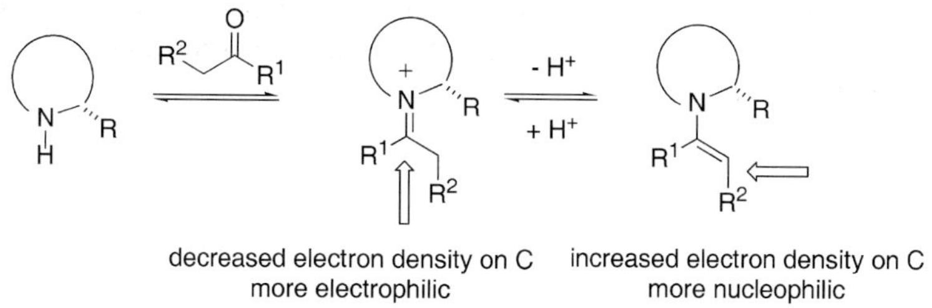

decreased electron density on C
more electrophilic

increased electron density on C
more nucleophilic

Scheme 1.6

enamine and iminium intermediates, respectively) in the same reaction pot, in a domino sequence [51].

Another particular area of organic catalysis is that of Lewis acid activation by a Lewis base. This catalysis represents a powerful means of modulating the electron density of weakly electron-withdrawing centers. Such an interaction operates under well-defined circumstances between donor and acceptor entity, and results in a decreased electron density on the central atom in question. This chemistry is described in greater detail in Chapter 10.

In contrast to most organometallic catalysts – which achieve catalytic activity via a single (usually Lewis acid) function [52] – most of the currently used efficient organocatalysts have more than one active center. The vast majority of these catalysts are bifunctional catalysts, having commonly a Brønsted acid and Lewis base center [17]. Such catalysts are able to activate both the donor and acceptor, respectively, and this results not only in a considerable acceleration in reaction rate but also in increased selectivity due to the highly organized transition state (TS). Moreover, the fact that the reaction occurs in a confined reaction space means that the catalyst functions as an "entropy trap", in the sense originally proposed by Westheimer ("... by overcoming the unfavorable entropy of activation usually inherent in a chemical reaction.") [53]. In the transition states, hydrogen bond interactions constitute a major driving force in the formation of specific molecular and complex geometries. Thus, protein and nucleic acid secondary and tertiary structural elements – as well as many natural and artificial host–guest complexes – are partly based on the directive power of intra- and intermolecular hydrogen bond formation [54]. It should be noted that Brønsted acids may also participate actively in the chemical transformation; indeed, in many cases the chiral proton transfer determines both the rate and selectivity of the global process.

Although the first catalysts to be identified were naturally occurring molecules having a rigid backbone, organocatalytic reactions have essentially evolved from the ligand chemistry of organometallic reactions. Today, the large array of ligands developed for metal-mediated reactions are still among the best-performing organocatalysts. Paradoxically, these ligands were considered originally exclusively to be chiral manifolds, without realizing the benefits of the presence of the catalytically active functionalities. The main advantages of synthetic molecules over their natural counterparts are their readily available enantiomers and easily tunable structures. Moreover, compound classes having no naturally occurring analogues can also be obtained. While there is no naturally occurring source of phosphorus-containing chiral compounds for catalytic use, there is understandably an intense synthetic activity to close the gap, and in this respect there are two complementary strategies for catalyst development:

- variation of the structure of an efficient catalyst family, usually that of a privileged class [55]; and
- the generation and testing of a large number of catalysts (library), and selection of the one(s) having the best kinetic/selectivity profile.

The massive involvement of automatization and computation, both in the generation of novel catalyst structures and in the evaluation of the reactions, facilitates the emergence of new catalyst structures. This approach is particularly useful when preparing peptide-based catalysts, which fold into defined secondary structures in either organic or aqueous solution. The "oligopeptide approach" has some advantages. First, the efficiency of the catalyst can be improved by varying the nature of the amino acids using combinatorial synthetic methods. Second, the structural simplicity of the oligopeptides contrasts with the complexity of the enzymes and thus renders easier the mechanistic investigations. Third, the flexibility of the method is of great use. It is possible to prepare a peptide sequence that can produce, eventually, the opposite enantiomer or its diastereoisomer, a process barely amenable with enzymes. Moreover, this oligopeptide approach may provide the solution for the reactivity versus selectivity problem, notably when the steric hindrance of the chiral appendage compromises the reactivity of the catalyst. Here, the strategy consists of building a simplified version of a complex chiral environment around the catalytic site, very much like that found in enzymes, where the chiral handle is thus distant from the active site. Such artificial enzymes may comprise a short oligopeptide sequence that includes an active site (e.g., imidazole), and a basic secondary structure, for example an α/β-turn or α/β-hairpin. It is interesting to contemplate, that with the spectacular increase in molecular weight and complexity of many catalyst structures, it is not only the selectivity but also the kinetic profile of the catalyst that is sharply ameliorated.

1.3.1
Privileged Catalysts

Some catalysts may have the extraordinary capacity to mediate efficiently not only one but rather a variety of seemingly unrelated chemical transformations. The term "privileged" chiral catalysts was coined in analogy to pharmaceutical compound classes that are active against a number of different biological targets [55]. Indeed, there is a steadily growing number of such organic compounds included in this list, the details for some of which are outlined in the following sections.

1.3.1.1 Proline [7k, 56]

L-Proline is perhaps the most well-known organocatalyst. Although the natural L-form is normally used, proline is available in both enantiomeric forms [57], this being somewhat of an asset when compared to enzymatic catalysis [58]. Proline is the only natural amino acid to exhibit genuine secondary amine functionality; thus, the nitrogen atom has a higher pK_a than other amino acids and so features an enhanced nucleophilicity compared to the other amino acids. Hence, proline is able to act as a nucleophile, in particular with carbonyl compounds or Michael acceptors, to form either an iminium ion or enamine. In these reactions, the carboxylic function of the amino acid acts as a Brønsted acid, rendering the proline a bifunctional catalyst.

The high, and often exceptional, enantioselectivity of proline-mediated reactions can be rationalized by the capacity of the molecule to orchestrate highly organized transition states by an extensive hydrogen-bonding network. In all proline-mediated reactions, proton-transfer from the amine or the carboxylic acid group of proline to the forming alkoxide or imide is essential for charge stabilization and to facilitate C–C bond formation in the transition state [59]. While most of the partial steps in aminocatalytic reactions are in equilibrium, the enhanced nucleophilicity of the catalyst can entail a number of equilibrated reactions with electrophiles present in the medium, resulting in a low turnover number. However, this drawback can be remedied by upsetting the equilibrium by higher catalyst loading, whilst the catalyst is of low cost.

Synthetic shortcomings related to proline are persistent, however. For example, in the dimerization or oligomerization of α-unbranched aldehydes, it is difficult to avoid competing reactions. Reactions with acetaldehyde or acetophenone afford generally low yields and selectivity in aldol reactions. Although proline continues to play a central role in aminocatalysis, its supremacy is being challenged either by new synthetic analogues [60], or by more complex oligopeptides. Structural analogues or derivatives offer better rates and selectivity in a number of reactions.

1.3.1.2 Cinchona Alkaloids [61]

The readily available and inexpensive cinchona alkaloids having pseudoenantiomeric forms such as quinine and quinidine, or cinchonine and cinchonidine, are among the most efficient catalysts (Scheme 1.7). The key structural feature responsible for their synthetic utility is the presence of the tertiary quinuclidine nitrogen, which complements with the proximal polar hydroxyl function of the natural compound. The presence of these Lewis acidic (H-bonding) and Lewis basic (quinuclidine nitrogen) sites makes them bifunctionally catalytic. The range

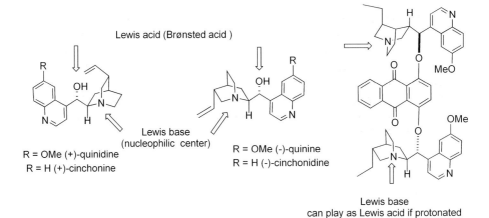

Scheme 1.7 Some cinchona alkaloids and cinchona-derived catalysts.

of reaction types over which the cinchona alkaloids impart high enantioselectivity is astonishing. In the past, modification of the cinchona backbone has resulted in notorious decreases or even losses of selectivity, and consequently such derivatives were disregarded as catalysts. The major event causing modified cinchonas to become the center of attention was the development of dimeric cinchona alkaloid ligands for the asymmetric dihydroxylation of simple olefins [62]. In fact, a very large number and variety of derivatives offer very high levels of selectivity over a wide diversity of reactions.

1.3.1.3 TADDOL and Derivatives [63]

TADDOL is one of the oldest, and most extraordinarily versatile, chiral auxiliaries (Scheme 1.8). The initial design of TADDOL was driven by practical considerations, mainly because it is derived from tartaric acid – the least-expensive chiral starting material with twofold symmetry available from natural sources. The two hydroxyl functions of the genuine molecule can act as a double hydrogen-bond donor, allowing the formation of bidentate complexes. Moreover, these functions can be easily substituted, giving access to a variety of derivatives.

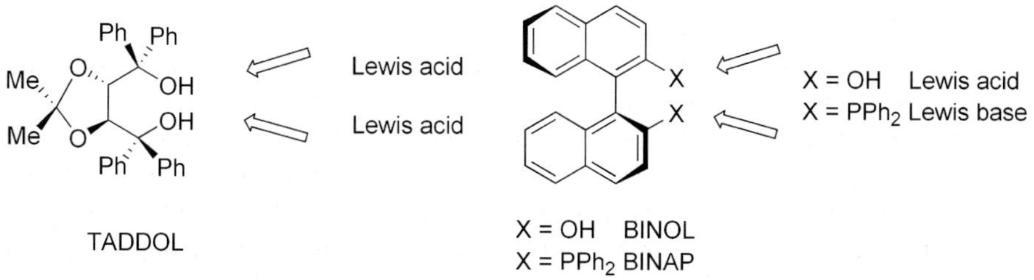

Scheme 1.8 Taddol and binaphthol-derived catalysts.

1.3.1.4 Binaphthol Derivatives

The enantiomeric atropoisomers of 1,1′-binaphthyl-2,2′-diol (BINOL) and bis-diphenylphosphonate derivatives (BINAP) are completely synthetic molecules that have been developed to exploit the axial dissymmetry induced by the restricted rotation about the biaryl bond (Scheme 1.8) [64]. During the past 15 years, these compounds have become the most widely used ligands for both stoichiometric and catalytic asymmetric reactions, with many analogues and derivatives having been developed recently.

1.4
Reaction Types

Normally, organocatalytic reactions proceed either by a much "tighter" or a much "looser" transition structure than chiral metal complex-mediated reactions. The former class involves compounds, which are acting as covalently (truly) bonded reagents, with the bonding energy between catalyst and substrate exceeding 15 kcal. The latter class includes reactions via non-covalent complexes, and usually via ion pairing as dominant interactions, and encompasses interactions lower than 4 kcal mol^{-1} [4].

1.4.1
Covalent Catalysis

The vast majority of organocatalytic reactions proceeds via covalent formation of the catalyst–substrate adduct to form an activated complex. Amine-based reactions are typical examples, in which amino acids, peptides, alkaloids and synthetic nitrogen-containing molecules are used as chiral catalysts. The main body of reactions includes reactions of the so-called generalized enamine cycle and charge accelerated reactions via the formation of iminium intermediates (see Chapters 2 and 3). Also, Morita–Baylis–Hillman reactions (see Chapter 5), carbene-mediated reactions (see Chapter 9), as well as asymmetric ylide reactions including epoxidation, cyclopropanation, and aziridination (see Chapter 10), and oxidation with the in situ generation of chiral dioxirane or oxaziridine catalysts (see Chapter 12), are typical examples.

1.4.2
Non-Covalent Organocatalysis

There are a growing number of asymmetric organocatalytic reactions, which are accelerated by weak interactions. This type of catalysis includes neutral host–guest complexation, or acid–base associations between catalyst and substrate. The former case is highly reminiscent of the way that many enzymes effect reactions, by bringing together reactants at an active site and without the formation of covalent bonds. The chemistry of this organocatalysis is discussed in Chapter 13.

Weak acid–base chiral complex formation represents hydrogen bond catalysis (see Chapter 9) and deprotonation followed by cation/anion association under homogeneous, and also under phase-transfer conditions (see Chapter 4) [14, 65].

1.5
How This Book is Organized

The goal of this handbook is to bring together all important aspects of the rapidly growing field of asymmetric organocatalysis. The authors have attempted this difficult task in order to provide some practical guidelines for all of those who wish to familiarize themselves with this new domain, and also to provide useful information to those who are contributing actively to the extraordinary evolution of this field. The book is divided into two complementary parts:

- "Reactions" (Chapters 2 to 13), which discuss the most currently used organocatalytic transformations and provide a scholar *mechanism-based* treatment of the major reaction types.
- "Experimental Procedures" (Chapter 14), which gathers critically selected how-to-do-it protocols, classified according to *transformation* types.

References and Notes

1 Metalloids are usually included, as non-metallic elements. This book will consider arbitrarily Si and discard B.
2 The exclusion of metal elements from the catalyst structure is not always strictly considered. Catalysts where the metal arguably is not the reaction center ("spectator") such as in metallocenes are discussed, under certain conditions. For an excellent review of this chemistry see: G.C. Fu, *Acc. Chem. Res.* **2006**, *34*, 853–860.
3 A. Berkessel, H. Gröger, *Asymmetric Organocatalysis, From Biomimetic Concepts to Applications in Asymmetric Synthesis.* VCH, Weinheim, **2005**.
4 General reviews on asymmetric organocatalysis: (a) P.I. Dalko, L. Moisan, *Angew. Chem. Sonderausgabe* **2005**, A147–A184; (b) Y. Hayashi, *J. Synth. Org. Chem. Jpn.* **2005**, *63*, 464–477; (c) P.I. Dalko, L. Moisan, *Angew. Chem., Int. Ed.* **2004**, *43*, 5138–5175; (d) H.F. Jiang, Y.G. Wang, H.L. Liu, et al. *Chin. J. Org. Chem.* **2004**, *24*, 1513–1531; (e) P.I. Dalko, in: *The McGraw-Hill 2003 Yearbook*. McGraw-Hill, New York, **2003**, pp. 312–315; (f) P.I. Dalko, L. Moisan, *Angew. Chem. Int. Ed. Engl.* **2001**, *40*, 3726–3748.
5 Special issues on asymmetric organocatalysis. B. List, C. Bolm (Eds.), *Adv. Synth. Catal.* **2004**, *346*, 1007–1249; K.N. Houk, B. List (Eds.), *Acc. Chem. Res.* **2004**, *37* (8).
6 Reviews on organometallic/organocatalytic interface: (a) B. Plietker, *Angew. Chem. Int. Ed.* **2006**, *45*, 190–192; (b) P. Kočovský, A.V. Malkov, *Russ. Chem. Bull.* **2004**, *53*, 1806–1812.
7 Reviews on aminocatalysis: (a) B. List, *Chem. Commun.*, **2006**, 819–824; (b) G. Lelais, D.W.C. MacMillan, *Aldrichimica Acta*, **2006**, *39*, 79–87; (c) M. Marigo, K.A. Jørgensen, *Chem. Commun.* **2006**, 2001–2011; (d) J. Seayad, B. List, *Org. Biomol. Chem.* **2005**, *3*, 719–724; (e) A.J.A. Cobb, D.M. Shaw, D.A. Longbottom, J.B. Gold, S.V. Ley, *Org. Biomol. Chem.* **2005**, *3*, 84–96; (f) X.F. Wei, *Chin. J. Org. Chem.* **2005**, *25*, 1619–1625; (g) S.B. Tsogoeva, *Lett. Org. Chem.* **2005**, *2*, 208–213; (h) B. List, *Acc. Chem. Res.*, **2004**, *37*, 548–557; (i) S. France, D.J. Guerin, S.J. Miller, T. Lectka, *Chem. Rev.* **2003**, *103*, 2985–3012; (j) B. List, *Tetrahedron* **2002**, *58*, 5572–5590; (k) M. Movassaghi, E.N. Jacobsen, *Science* **2002**, *298*, 1904; (l) E.R. Jarvo, S.J. Miller, *Tetrahedron* **2002**, *58*, 2481–2495; (m) H. Gröger, J. Wilken, *Angew. Chem. Int. Ed. Engl.* **2001**, *40*, 529–532; (i) B. List, *Synlett*, **2001**, 1675–1686. See also Ref. [61].
8 Morita–Baylis–Hillman reactions: (a) K.Y. Lee, S. Gowrisankar, J.N. Kim, *Bull. Korean Chem. Soc.* **2005**, *26*, 1481–1490; (b) D. Basavaiah, A.J. Rao, T. Satyanarayana, *Chem Rev.* **2003**, *103*, 811–891; (c) P. Langer, *Angew. Chem. Int. Ed.* **2000**, *39*, 3049–3052.
9 Reviews on asymmetric catalysis by chiral hydrogen-bond donors: (a) T. Akiyama, J. Itoh, K. Fuchibe, *Adv. Synth. Catal.* **2006**, *348*, 999–1010; (b) M.S. Taylor, E.N. Jacobsen, *Angew. Chem. Int. Ed.* **2006**, *45*, 1520–1543;

(c) C. Bolm, T. Rantanen, I. Schiffers, L. Zani, *Angew. Chem. Int. Ed.* **2005**, *44*, 1758–1763; (d) H. Yamamoto, K. Futatsugi, *Angew. Chem. Int. Ed.* **2005**, *44*, 1924–1942; (e) C.A. Hunter, *Angew. Chem. Int. Ed.* **2004**, *43*, 5310–5324; (f) P.M. Pihko, *Angew. Chem. Int. Ed.* **2004**, *43*, 2062–2064; (g) M. Oestreich, *Nachr. Chem.* **2004**, *52*, 35–38; (h) P.R. Schreiner, *Chem. Soc. Rev.* **2003**, *32*, 289–296; (i) S.J. Connon, *Chem. Eur. J.* **2006**, *12*, 5418–5427.

10 Reviews on asymmetric acylation: (a) E. Vedejs, M. Jure, *Angew. Chem. Int. Ed.* **2005**, *44*, 3974–4001; (b) D.E.J.E. Robinson, S.D. Bull, *Tetrahedron: Asymm.* **2003**, *14*, 1407–1446; (c) J.R. Dehli, V. Gotor, *Chem. Soc. Rev.* **2002**, *31*, 365–370; (d) F.F. Huerta, A.B.E. Minidis, J.-E. Bäckvall, *Chem. Soc. Rev.* **2001**, *30*, 321–331; (e) J. Eames, *Angew. Chem. Int. Ed.* **2000**, *39*, 885–888.

11 Reviews on asymmetric reactions via carbenes: D. Enders, T. Balensiefer, *Acc. Chem. Res.* **2004**, *37*, 534–541.

12 Reviews on asymmetric oxidation reactions: (a) A.K. Yudin, *Aziridines and Epoxides in Organic Synthesis*, Wiley-VCH, Weinheim, 2006; (b) E.M. McGarrigle, D.G. Gilheany, *Chem. Rev.* **2005**, *105*, 1563–1602; (c) Q.-H. Xia, H.-Q. Ge, C.-P. Ye, Z.-M. Liu, K.-X. Su, *Chem. Rev.* **2005**, *105*, 1603–1662; (d) Y. Shi, *Acc. Chem. Res.* **2004**, *37*, 488–496; (e) D. Yang, *Acc. Chem. Res.* **2004**, *37*, 497–505; (f) W. Adam, C.R. Saha-Moller, P.A. Ganeshpure, *Chem. Rev.* **2001**, *101*, 3499–3548.

13 Reviews on Lewis base-catalyzed transformations: (a) S.E. Denmark, J. Fu, *Chem. Commun.* **2003**, 167; (b) S.E. Denmark, J. Fu, *Chem. Rev.* **2003**, *103*, 2763; (c) J.W.J. Kennedy, D.G. Hall, *Angew. Chem., Int. Ed.* **2003**, *42*, 4732; (d) S.E. Denmark, R.A. Stavenger, *Acc. Chem. Res.* **2000**, *33*, 432.

14 Reviews on asymmetric phase-transfer reactions: (a) B. Lygo, B.I. Andrews, *Acc. Chem. Res.* **2004**, *37*, 518–525; (b) K. Maruoka, T. Ooi, *Chem. Rev.* **2003**, *103*, 3013–3028; (c) C. Najera, *Synlett* **2002**, 1388–1403; (d) M.J. O'Donnell, *Aldrichimica Acta* **2001**, *34*, 3–15.

15 Reviews on asymmetric shape and site selective organocatalytic reactions: (a) R. Breslow (Ed.), *Artificial Enzymes*, Wiley-VCH, Weinheim, 2005; (b) C.W. Lim, B.J. Ravoo, D.N. Reinhoudt, *Chem. Commun.* **2005**, 5627–5629; (c) C. Alexander, L. Davidson, W. Hayes, *Tetrahedron* **2003**, *59*, 2025–2056; (d) G. Wulff, *Chem. Rev.* **2002**, *102*, 1–28; (e) B. Clapham, T.S. Reger, K.D. Janda, *Tetrahedron* **2001**, *57*, 4637–4662; (f) B. Sellergren, *Angew. Chem. Int. Ed.* **2000**, *39*, 1031–1037.

16 (a) J. Seayad, B. List, Catalytic asymmetric multi-component reactions. In: J. Zhu, H. Bienayme (Eds.), *Multi-Component Reactions*. Wiley-VCH: Weinheim, Germany, 2004; (b) D.B. Ramachary, M. Kishor, G.B. Reddy, *Org. Biomol. Chem.* **2006**, *4*, 1641–1646; (c) H.-C. Guo, J.-A. Ma, *Angew. Chem. Int. Ed.* **2006**, *45*, 354–366.

17 (a) H. Bernard, G. Bulow, U.E.W. Lange, H. Mack, T. Pfeiffer, B. Schäfer, W. Seitz, T. Zierke, *Synthesis* **2004**, 2367–2375; (b) M. Breuer, K. Ditrich, T. Habicher, B. Hauer, M. Kesseler, R. Sturmer, T. Zelinski, *Angew. Chem. Int. Ed.* **2004**, *43*, 788–824; (c) H. Tye, P.J. Comina, *J. Chem. Soc., Perkin 1* **2001**, 1729–1747; (d) H.J. Federsel, *Nature Rev. Drug Discov.* **2005**, *4*, 685–697; (e) H.U. Blaser, B. Pugin, F. Spindler, *J. Mol. Catal., Chemical* **2005**, *231*, 1–20.

18 C. Thirsk, D. Jay, *Chem. Ind.* **2004**, *16*, 15–17.

19 S. Pizzarello, A.L. Weber, *Science* **2004**, *303*, 1151.

20 G.M.M. Caro, U.J. Meierhenrich, W.A. Schutte, B. Barbier, A.A. Segovia, H. Rosenbauer, W.H.-P. Thiemann, A. Brack, J.M. Greenberg, *Nature* **2002**, *416*, 403.

21 (a) M. Klussmann, H. Iwamura, S.P. Mathew, D.H. Wells, Jr., U. Pandya, A. Armstrong, D.G. Blackmond,

Nature **2006**, *441*, 621; (b) Y. Hayashi, M. Matsuzawa, J. Yamaguchi, S. Yonehara, Y. Matsumoto, M. Shoji, D. Hashizume, H. Koshino, *Angew. Chem. Int. Ed.* **2006**, *45*, 4593–4597; (c) A.B. Northrup, D.W.C. MacMillan, *Science* **2005**, 305, 1752; (d) A. Córdova, M. Engqvist, I. Ibrahem, J. Casas, H. Sunden, *Cheminform*, **2005**, *36*, (32); (e) A. Córdova, I. Ibrahem, J. Casas, H. Sundén, M. Engqvist, E. Reyes, *Chem. Eur. J.* **2005**, 11, 4772–4784; (f) A. Córdova, M. Engqvist, I. Ibrahem, J. Casas, H. Sundén, *Chem. Commun.* **2005**, 2047–2049; (g) U. Kazmaier, *Angew. Chem. Int. Ed.* **2005**, *44*, 2186–2188. (h) See also: G. Carrea, S. Colonna, D.R. Kelly, A. Lazcano, G. Ottolina, S.M. Roberts, *Trends Biotechnol.* **2005**, *23*, 507–513.

22 (a) G.F. Joyce, G.M. Visser, C.A.A. van Boeckel, J.H. van Boom, L.E. Orgel, J. van Westerenen, *Nature* **1984**, *310*, 602–604; (b) M. Bolli, R. Micura, A. Eschenmoser, *Chem. Biol.* **1997**, *4*, 309–320.

23 K. Tamura, P. Schimmel, *Science* **2004**, *305*, 1253.

24 H.B. Kagan. In: E.N. Jacobsen, A. Pfaltz, H. Yamamoto (Eds.), *Comprehensive Asymmetric Catalysis*, Springer-Verlag, Berlin, **1999**, Vol. 1, pp. 101–118.

25 G. Breding, R.W. Balcom, *Ber. Deutsch. Chem. Ger.* **1908**, *41*, 740–751.

26 G. Breding, K. Fajans, *Ber. Deutsch. Chem. Ger.* **1908**, *41*, 752–763.

27 L. Rosentahler, *Biochem. Z.* **1908**, *14*, 232.

28 G. Breding, P.S. Fiske, *Biochem. Z.* **1912**, *46*, 7.

29 M.M. Vavon, P. Peignier, *Bull Soc. Fr.* **1929**, *45*, 293.

30 R. Wegler, *Liebigs Ann. Chem.* **1932**, *498*, 62.

31 See W. Langenbeck, *Die Organische Katalysatoren und ihre Beziehungen zu den Fermenten* 2. Aufl. Springer-Verlag, **1949**; (b) W. Langenbeck, Fortschritte der Chemischen Forschung, Vol. 6, Springer Verlag, Berlin, p. 301, **1966**; see also in (c) B. Lukowczyk, *J. Pract. Chem* **1959**, *8*, 372–378.

32 H.D. Dakin, *J. Biol. Chem.* **1909**, *7*, 49.

33 V. Prelog, M. Wilhelm, *Helv. Chim. Acta*, **1954**, *37*, 1634.

34 H. Pracejus, *Justus Liebigs Ann. Chem.* **1960**, *634*, 9–22.

35 B. Långström, G. Bergson, *Acta Chem. Scand.* **1973**, *27*, 3118–3119.

36 For a review, see: H. Wynberg, *Top. Stereochem.* **1986**, *16*, 87–129.

37 Z.G. Hajos, D.R. Parrish, *J. Org. Chem.* **1974**, *39*, 1615–1621.

38 U. Eder, G. Sauer, R. Wiechert, *Angew. Chem. Int. Ed. Engl.* **1971**, *10*, 496–497.

39 See for example: S.J. Danishefsky, J.J. Masters, W.B. Young, J.T. Link, L.B. Snyder, T.V. Magee, D.K. Jung, R.C.A. Isaacs, W.G. Bornmann, C.A. Alaimo, C.A. Coburn, M.J. Di Grandi, *J. Am. Chem. Soc.* **1996**, *118*, 2843–2859.

40 G. Stork, R. Terrell, J. Szmuszkovicz, *J. Am. Chem. Soc.* **1954**, *76*, 2029–2030.

41 For a pioneering review dealing with proline as catalyst, see: J. Martens, *Top. Curr. Chem.* **1984**, *125*, 165–246.

42 T.P. Brady, S.H. Kim, K. Wen, E.A. Theodorakis, *Angew. Chem.* **2004**, *116*, 757–760; *Angew. Chem. Int. Ed.* **2004**, *43*, 739–742 and references cited therein.

43 A. Mori, S. Inoue. In: E.N. Jacobsen, A. Pfalz, H. Yamamoto (Eds.), *Comprehensive Asymmetric Catalysis*. Springer, Heidelberg, **1999**, pp. 983–994 and references therein.

44 M.S. Iyer, K.M. Gigstad, N.D. Namdev, M. Lipton, *J. Am. Chem. Soc.* **1996**, *118*, 4910–4911.

45 M.S. Sigman, E.N. Jacobsen, *J. Am. Chem. Soc.* **1998**, *120*, 4901–4902.

46 (a) U.-H. Dolling, P. Davis, E.J.J. Grabowski, *J. Am. Chem. Soc.* **1987**, *106*, 446–447; (b) D.L. Hughes, U.-H. Dolling, K.M. Ryan, E.F. Schoenewaldt, E.J.J. Grabowski, *J. Org. Chem.* **1987**, *52*, 4745–4752.

47 For a review on catalytic Diels–Alder reactions, see: (a) H.B. Kagan, O. Riant, *Chem. Rev.* **1992**, *92*, 1007–1019.

48 (a) S. Juliá, J. Masana, J.C. Vega, *Angew. Chem. Int. Ed.* **1980**, *19*, 929–931; (b) S. Juliá, J. Guixer, J. Masana, J. Rocas, S. Colonna, R. Annuziata, H. Molinari, *J. Chem. Soc., Perkin Trans. 1* **1982**, 1317. For a mechanistic discussion of this reaction, see: (c) A. Berkessel, N. Gasch, K. Glaubitz, C. Koch, *Org. Lett.* **2001**, *3*, 3839–3842.

49 For a review on phosphines, see: J.L. Methot, W.R. Roush, *Adv. Synth. Catal.* **2004**, *346*, 1035–1050.

50 As defined in Ref. [7i]: "There are two aminocatalytic pathways. Iminium catalysis directly utilizes the higher reactivity of the iminium ion in comparison to the carbonyl species and facilitates Knoevenagel-type condensations, cyclo- and nucleophilic additions, and cleavage of σ-bonds adjacent to the α-carbon. Enamine catalysis on the other hand involves catalytically generated enamine intermediates that are formed *via* deprotonation of an iminium ion, and react with various electrophiles or undergo pericyclic reactions."

51 (a) J.W. Yang, M.T. Hechavarria Fonseca, B. List, *J. Am. Chem. Soc.* **2005**, *127*, 15036–15037; (b) Y. Huang, A. Walji, C.H. Larsen, D.W.C. MacMillan, *J. Am. Chem. Soc.* **2005**, *127*, 15051–15053.

52 For reviews on bifunctional catalysts: (a) M. Shibasaki, H. Sasai, T. Arai, *Angew. Chem. Int. Ed. Engl.* **1997**, *36*, 1236–1256; (b) H. Gröger, *Chem. Eur. J.* **2001**, *7*, 5246–5251; (c) G.J. Rawlands, *Tetrahedron* **2001**, *57*, 1865–1882; (d) M. Shibasaki, N. Yoshikawa, *Chem. Rev.* **2002**, *102*, 2187–2209; (e) J.-A. Ma, D. Cahard, *Angew. Chem. Int. Ed.* **2004**, *43*, 4566–4583; (f) M. Kanai, N. Kato, E. Ichikawa, M. Shibasaki, *Synlett* **2005**, *10*, 1491–1508.

53 (a) F.H. Westheimer, *Adv. Enzymol.* **1962**, *24*, 441–482; (b) F.H. Westheimer, *Adv. Phys. Org. Chem.* **1985**, *21*, 1–35.

54 (a) G.A. Jeffrey, W. Saenger, *Hydrogen Bonding in Biological Structures*. Springer, Berlin, **1994**; (b) D.G. Lonergan, G. Deslongchamps, S. Tomás, *Tetrahedron Lett.* **1998**, *39*, 7861–7864.

55 T.P. Yoon, E.N. Jacobsen, *Science* **2003**, *299*, 1691–1693.

56 W.D. Allen, E. Czinki, A.G. Csaszar, *Chem. Eur. J.* **2004**, *10*, 4512–4517.

57 The natural compound is extracted essentially from chicken feathers.

58 J. Sukumaran, U. Hanefeld, *Chem. Soc. Rev.* **2005**, *34*, 530–542.

59 (a) S. Bahmanyar, K.N. Houk, *Org. Lett.* **2003**, *5*, 1249–1251; (b) S. Bahmanyar, K.N. Houk, H.J. Martin, B. List, *J. Am. Chem. Soc.* **2003**, *125*, 2475–2479; (c) L. Hoang, S. Bahmanyar, K.N. Houk, B. List, *J. Am. Chem. Soc.* **2003**, *125*, 16–17; (d) S. Bahmanyar, K.N. Houk, *J. Am. Chem. Soc.* **2001**, *123*, 12911–12912; (e) S. Bahmanyar, K.N. Houk, *J. Am. Chem. Soc.* **2001**, *123*, 11273–11283.

60 For related derivatives, see: (a) M. Nakadai, S. Saito, H. Yamamoto, *Tetrahedron* **2002**, *58*, 8167–8177; (b) Z. Tang, F. Jiang, L.-T. Yu, X. Cui, L.-Z. Gong, A.-Q. Mi, Y.-Z. Jiang, Y.-D. Wu, *J. Am. Chem. Soc.* **2003**, *125*, 5262–5263; (c) F. Fache, O. Piva, *Tetrahedron: Asymmetry* **2003**, *14*, 139–143; (d) T.J. Dickerson, K.D. Janda, *J. Am. Chem. Soc.* **2002**, *124*, 3220–3221; (e) A.J.A. Cobb, D.M. Shaw, S.V. Ley, *Synlett* **2004**, 558–560.

61 (a) S.-K. Tian, Y. Chen, J. Hang, L. Tang, P. McDaid, L. Deng, *Acc. Chem. Res.* **2004**, *37*, 621–631; (b) C. Ó'Dálaigh, *Synlett* **2005**, 875–876.

62 (a) H.C. Kolb, M.S. VanNieuwenhze, K.B. Sharpless, *Chem. Rev.* **1994**, *94*, 2483–2547; (b) I.E. Marko, J.S. Svendsen. In: E.N. Jacobsen, A. Pfaltz, H. Yamamoto (Eds.), *Comprehensive Asymmetric Catalysis*. Springer, New York, **1999**, Vols. I–III, Chapter 20.

63 D. Seebach, A.K. Beck, A. Heckel, *Angew. Chem. Int. Ed. Engl.* **2001**, *40*, 92–138.

64 See for example: (a) M. Shibasaki, S. Matsunaga, *Chem. Soc. Rev.* **2006**, *35*, 269–279; (b) J.M. Brunel, *Chem. Rev.*, **2005**, *105*, 857–898; (c) A.K. Unni, N. Takenaka, H. Yamamoto, V.H. Rawal, *J. Am. Chem. Soc.* **2005**, *127*, 1336–1337.

65 M. Makosza, M. Fedorynski, *Catal. Rev., Sci. Eng.* **2003**, *45*, 321–367.

2
Enamine Catalysis

2.1
Aldol and Mannich-Type Reactions

Fujie Tanaka and Carlos F. Barbas, III

2.1.1
Introduction

Chiral amines (both primary and secondary amines) and amino acids have been used as catalysts for aldol reactions, Mannich-type reactions, and other reactions that proceed through enamine intermediates. An enamine-based catalytic cycle is shown in Scheme 2.1. The catalytic cycle includes formation of an iminium intermediate between a donor carbonyl compound and the amine-containing catalyst, the formation of an enamine intermediate from the iminium, C–C bond forma-

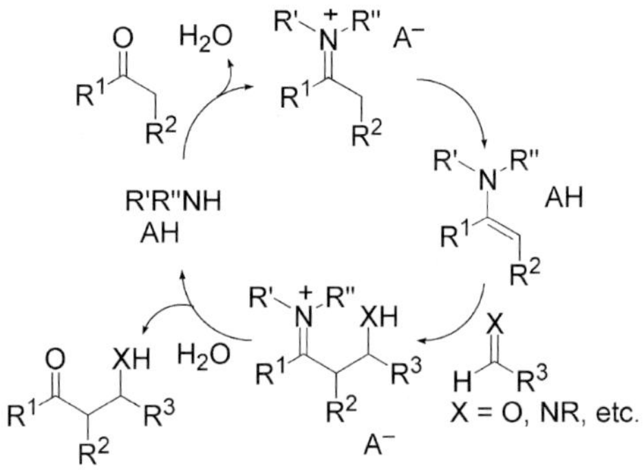

Scheme 2.1 The enamine catalytic cycle. An enamine derived from an amine- or amino acid-catalyst can react with a variety of electrophiles. The aldehyde and ketone reactants that form enamines and act as nucleophiles are often described as "donors". Aldehyde and imine reactants that serve as electrophiles are described as "acceptors" for aldol and Mannich reactions, respectively. Ketones also serve as acceptors for aldol reactions.

Enantioselective Organocatalysis: Reactions and Experimental Procedures
Edited by Peter I. Dalko
Copyright © 2007 WILEY-VCH Verlag GmbH & Co. KGaA, Weinheim
ISBN: 978-3-527-31522-2

tion between the enamine and acceptor, and hydrolysis of the resulting iminium to afford the product. Historically, (S)-proline derivatives have been used in stoichiometric asymmetric enamine chemistry since the 1960s [1]. Unlike the catalytic reactions that were to follow, these early reactions involved the formation and isolation of enamines that were then reacted. Subsequently, the early 1970s witnessed the development of the first asymmetric intramolecular aldol reactions catalyzed by amino acids containing either primary amines, for example with phenylalanine and alanine, or the secondary amine provided by proline [2, 3]. Although direct catalytic asymmetric aldol reactions of the intramolecular "Hajos–Parish" variety have been used continuously since their discovery, somewhat surprisingly the scope of this type of amino acid catalysis remained largely unexplored until recently. In 1997, we initiated comparative studies between proline catalysis and the aldolase antibody catalysts that we had created to catalyze intermolecular aldol reactions using the enamine mechanism common to enzymes [4a–e]. These studies led us to determine that proline-catalyzed reactions were analogous in many ways to the reactions catalyzed by aldolase antibodies, and that aldolase antibodies and proline could catalyze many of the same reactions, both using enamine catalysis. This new insight led us to study proline and other amino acids and their derivatives as catalysts for intermolecular aldol reactions [5, 6], and tandem Michael-aldol [43] reactions that had been well studied with aldolase antibodies [4c–e]. With this success, in 2000 we began to view proline as the "elusive" open-active site aldolase capable of generating an enamine and reacting it with a wide variety of electrophiles, lacking the limitations of scope that are imposed by the steric constraints of an enzyme's active site. Enamine-based direct asymmetric amine- or amino acid-catalyzed intermolecular reactions via *in-situ*-generated enamine intermediates has since undergone significant development [6–8]. An attractive feature of this chemistry is that it can afford highly enantiomerically enriched products under mild conditions, without the requirement of preformation or isolation of enamines, or preactivation of carbonyl compounds; this is in contrast to approaches based on preformed enolate equivalents such as silyl enol ethers. The amine- or amino acid-based reactions can be performed simply by mixing reactants and a catalyst in an appropriate solvent under air at room temperature. The simplicity of these reaction conditions stand in stark contrast to reactions that use lithium amides to form enolate intermediates and control the stereochemical outcome of reactions through the use of a stoichiometric chiral auxiliary. Such approaches lack atom economy and typically require low temperatures, absolute solvents, and an atmosphere of inert gas, such as argon or nitrogen.

2.1.2
Aldol Reactions

2.1.2.1 Aldol Reactions of Alkyl Ketone Donors
(S)-Proline (**1**), its derivatives, and related molecules catalyze aldol reactions of ketone donors [6]. For example, (S)-proline, (2S,4R)-4-hydroxyproline (**2**), and

Table 2.1 Aldol reactions of acetone and aldehydes [6].

Entry	R	Catalyst	Yield [%]	ee [%]
1	4-NO$_2$C$_6$H$_4$	1	68	76
2		2	85	78
3		3	60	86
4	i-Pr	1	97	96
5	c-C$_6$H$_{11}$	1	60	85

5,5-dimethylthiazolinium-4-carboxylic acid (3) all similarly catalyzed aldol reactions (Table 2.1) [6]. These reactions were performed by mixing a large excess of acetone (20 vol%), an aldehyde, and the catalyst (0.2 equiv to the aldehyde) in organic solvent, such as DMSO, at room temperature. The typical enantioselectivities of these reactions were 60–90% enantiomeric excess (*ee*) for arylaldehyde acceptors (aldol electrophiles), and up to 96% *ee* for α,α-disubstituted aldehyde acceptors. Although the enantioselectivities were not perfect in these reactions, these results indicate that simple chiral amino acids can catalyze intermolecular direct asymmetric aldol reactions via *in-situ*-generated enamines, a chemistry shared with much more complex enzyme catalysts.

(S)-Proline is a natural amino acid that is non-toxic, safe, and available at low cost. Indeed, (S)-proline is one of the few edible catalysts for asymmetric synthesis. As described later in this chapter, (S)-proline is an excellent catalyst that provides high enantioselectivities for aldol reactions involving certain reactants, and is also an effective catalyst of other reactions. Although reaction conditions using (S)-proline can be optimized, proline is not always the best catalyst for many reactions. Other amino acids and small amine derivatives have also been used to catalyze reactions in order to afford desired products in good yields with high regio-, diastereo-, and enantioselectivities within a reasonable reaction time. The selection of reaction conditions – for example, solvents, temperatures, concentrations of reactants and of catalyst, and catalyst loading amounts – is often important to obtain the desired results. For example, proline amide derivative 4 was used for aldol reactions at −25 °C to afford high enantioselectivities (Table 2.2) [9]. Other proline amide derivatives [10, 11] and proline hydrazide [12] derivatives

Table 2.2 Aldol reactions of acetone and aldehydes catalyzed by proline amide **4** [9].

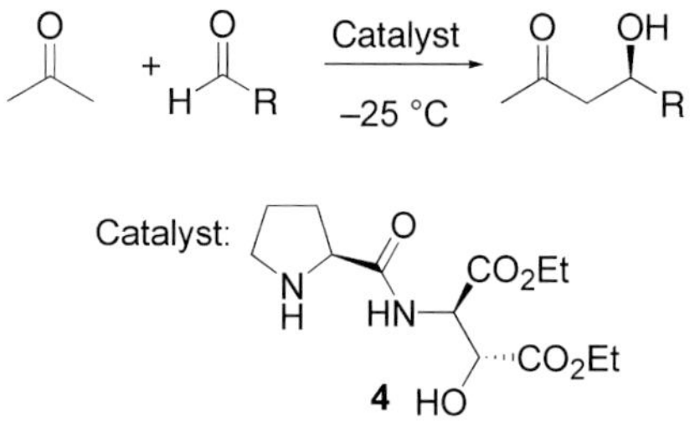

Entry	R	Yield [%]	ee [%]
1	4-$NO_2C_6H_4$	62	99
2	Ph	68	98
3	i-Pr	75	>99
4	c-C_6H_{11}	80	99

have also been used at $-35\,°C$ and $0\,°C$, respectively, for the aldol reaction of acetone and aldehydes to afford products with high enantioselectivities.

(S)-Proline-catalyzed aldol reactions involving 2-butanone afforded the products of C–C bond formation at the methyl group, the less substituted α-position of the ketone as the major regioisomers (Fig. 2.1) [6, 9]. The regioselectivity of the aldol reaction of 2-butanone was reversed using a proline amide derivative as the catalyst, as shown in Scheme 2.2 [13]. The (S)-proline-catalyzed aldol reactions of cyclohexanone and of cyclopentanone afforded both *anti*- and *syn*-products (*anti:syn* ∼ 2:1) with moderate enantioselectivities (63–89% ee) [6]. The selectivity

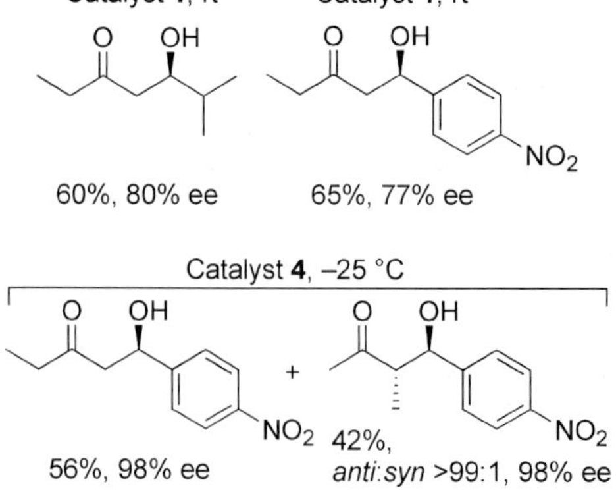

Fig. 2.1 Products of (S)-proline **1**- and **4**-catalyzed aldol reactions of 2-butanone [6, 9].

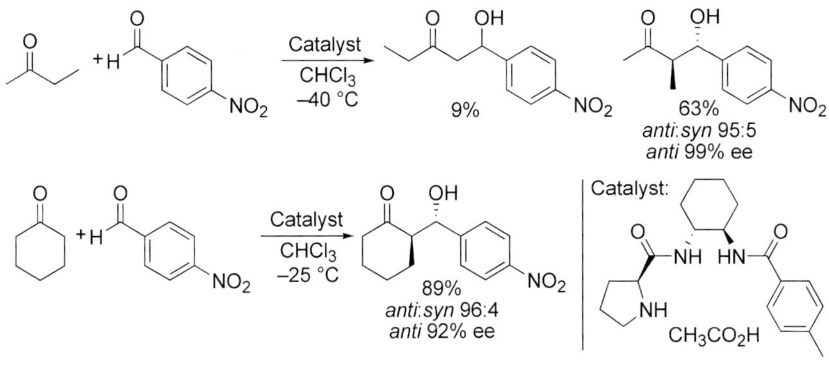

Scheme 2.2 Aldol reactions using a proline amide catalyst [13].

was improved using a proline amide derivative (Scheme 2.2) [13]. Proline-derived diamine-acid combination catalysts and O-protected-4-hydroxyproline derivatives also afforded good diastereo- and enantioselectivities for the aldol reactions of cyclopentanone in water [14, 15], as discussed later in this section. The aldol reactions of a series of ketone donors with chloral using a proline-derived tetrazole catalyst afforded the products with high enantioselectivities [16] (see below, later in this section).

2.1.2.2 Aldol Reactions of α-Oxyketone Donors

Direct asymmetric aldol reactions of α-oxyketones provide concise routes to carbohydrate derivatives. Guided by studies initially performed with aldolase antibodies, hydroxyacetone was first studied as a donor under proline catalysis [4b,d]. (S)-Proline catalyzed the aldol reactions of hydroxyacetone [6, 17] and of 2,2-dimethyl-1,3-dioxan-5-one (5) [18], as shown in Tables 2.3 and 2.4, respectively. Reaction with hydroxyacetone provided products with C–C bond formation at the hydroxy group-substituted α-position, with exclusive regioselectivity. In reactions with these oxyketones, the *anti*-aldol products were obtained with high diastereo- and enantioselectivities in many cases. The configurations of the newly generated stereocenters depended on the catalyst used for the aldol reactions, and not on the stereochemistries of the acceptor aldehydes (Scheme 2.3). That is, (S)-proline and (R)-proline provided either *anti*-diastereomer in reactions with a chiral aldehyde acceptor, indicative of catalyst control in the stereochemistry-defining step. Deprotection and reduction of the ketone group of these products afforded expedient access to carbohydrates (Scheme 2.4) [18]. (For *syn*-selective cross-aldol reactions of ketone donors and aldehyde acceptors see Chapter 14.1.1). The (S)-proline-catalyzed reaction between 5 and pentadecenal was used for the synthesis of phytosphingosines [19].

The regioselectivities of the aldol reactions of hydroxyacetone were reversed from those of the proline-catalyzed reactions when small peptide catalyst 6 or 7 containing (S)-proline at the N-terminal was used, as shown in Table 2.5 [20].

Table 2.3 (S)-Proline-catalyzed aldol reactions of hydroxyacetone [6, 7].

Entry	R	Yield [%]	anti:syn	anti ee [%]
1	i-Pr	62	>20:1	>99
2	c-C$_6$H$_{11}$	60	>20:1	>99
3	(dioxolane)	40	2:1	>97
4	Ph	83	1:1	80

Table 2.4 (S)-Proline-catalyzed aldol reactions of 2,2-dimethyl-1,3-dioxan-5-one [18].

Entry	R	Conditions[a]	Yield [%]	anti:syn	anti ee [%]	Ref.
1	i-Pr	A	97	>98:2	94	18a
2	i-Bu	B	75	10:1	98	18c
3	CH$_2$OBn	A	40	>98:2	97	18a
4	CH$_2$(OMe)$_2$	A	69	94:6	93	18a
5	CH$_2$OAc	B	60	>15:1	98	18c

[a] Conditions: A, Ketone **5** (2.3 mmol, 1 equiv.), aldehyde (2.3 mmol, 1 equiv.), (S)-proline (0.69 mmol, 0.3 equiv.), DMF (1.2 mL), 2 °C, 6 days; B, ketone **5** (0.5 mmol, 5 equiv.), aldehyde (0.1 mmol, 1 equiv.), (S)-proline (0.02 mmol, 0.2 equiv.), DMF (0.2 mL), 4 °C, 72 h.

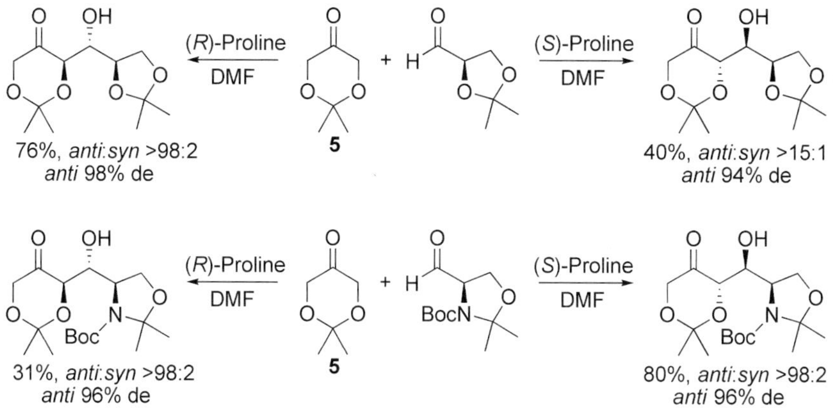

Scheme 2.3 (S)- and (R)-Proline-catalyzed aldol reactions of **5** [18].

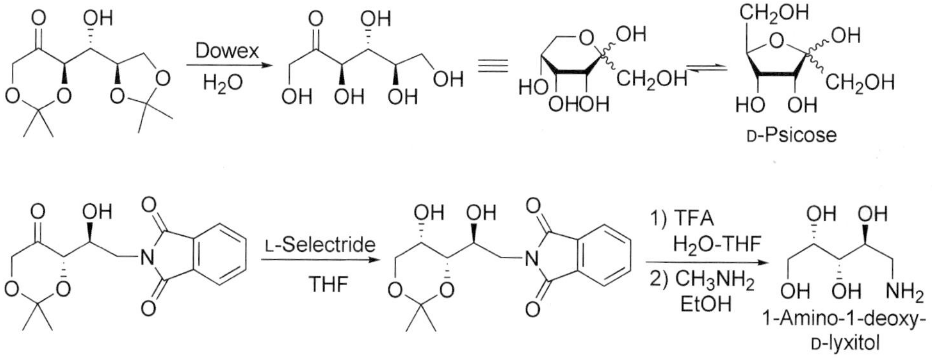

Scheme 2.4 Expedient synthesis of carbohydrates enabled by simple manipulation of proline aldol products [18].

2.1.2.3 Aldol Reactions of Aldehyde Donors

(S)-Proline-catalyzed aldehyde donor reactions were first studied in Michael [21] and Mannich reactions (see below), and later in self-aldol and in cross-aldol reactions. (S)-Proline-catalyzed self-aldol and cross-aldol reactions of aldehydes are listed in Table 2.6 [22–24]. In self-aldol reactions, the reactant aldehyde serves as both the aldol donor and the acceptor; whereas in cross-aldol reactions, the donor aldehyde and acceptor aldehyde are different.

Self-aldolization of acetaldehyde provided one-step enantioselective syntheses of (5R)- and (5S)-Hydroxy-(2E)-hexenal using either (R)- or (S)-proline as catalysts of the homologation of three acetaldehyde units; ee-values of up to 90% were obtained [25]. Using this methodology, a series of triketides was prepared by slow addition of propionaldehyde into acceptor aldehyde and (S)-proline in DMF, and the isolated lactols were converted to the corresponding δ-lactones [26]. The product enantiomeric purity was typically moderate because of the isomerization

Table 2.5 Peptide **6**- and **7**-catalyzed aldol reaction of hydroxyacetone [20].

Catalyst:

Pro-Phe-Phe-Phe-OMe (**6**)

Pro-Phe-Phe-Phe-Phe-OMe (**7**)

Entry	R	Catalyst	Yield of A [%]	ee of A [%]	Yield of B [%]
1[a]	4-NO$_2$C$_6$H$_4$	6	82	82	18
2[b]		7	76	87	21
3[a]	2,6-Cl$_2$C$_6$H$_3$	6	84	96	–[c]
4[b]		7	73	96	–[c]

[a] Catalyst **6** (0.2 equiv.), 4 days.
[b] Catalyst **7** (0.1 equiv.), 6 days.
[c] Not reported.

during the second aldol step, but these reactions are significant in terms of asymmetric formation of carbohydrates from aldehydes in one-step and were suggested to be potential pre-biotic routes to carbohydrates.

Significant for cross-aldol reactions, when an aldehyde was mixed with (S)-proline in a reaction solvent, the dimer (the self-aldol product) was the predominant initial product. Formation of the trimer typically requires extended reaction time (as described above). Thus, it is possible to perform controlled cross-aldol reactions, wherein the donor aldehyde and the acceptor aldehyde are different. In order to obtain a cross-aldol product in good yield, it was often required that the donor aldehyde be slowly added into the mixture of the acceptor aldehyde and (S)-proline in a solvent to prevent the formation of the self-aldol product of the donor aldehyde. The outcome of these reactions depends on the aldehydes used for the reactions. Slow addition conditions can sometimes be avoided through the use of excess equivalents of donor or acceptor aldehyde – that is, the use of 5–10 equiv. of acceptor aldehyde or donor aldehyde. In general, aldehydes that easily form self-aldol products cannot be used as the acceptor aldehydes in

Table 2.6 (S)-Proline-catalyzed cross-aldol reactions of aldehyde donors.[a]

Entry	R¹	R²	Yield [%]	anti:syn	anti ee [%]	Ref.
1	Me	Et	80	4:1	99	22
2	Me	i-Bu	88	3:1	97	22
3	n-Bu	i-Pr	80	24:1	98	22
4	OBn	CH$_2$OCH$_2$Ph	73	4:1	98	23
5	Me	CH$_2$OTIPS	75	4:1	99	23
6[b]	phthalimido-N–	i-Pr	87	>100:1	>99.5	24
7[b]	i-Pr	CH$_2$-N-phthalimide	89	6:1	97	24

[a] TIPS = triisopropylsilyl.
[b] The reaction was performed in NMP instead of in DMF.

Conditions: Entry 2: A solution of donor aldehyde (2.0 mmol, 2 equiv.) in DMF (500 µL) was added slowly over 2.5 h to a stirring mixture of acceptor aldehyde (1.0 mmol, 1 equiv.) and (S)-proline (0.10 mmol, 0.1 equiv.) in DMF (500 µL) at 4 °C. The resulting mixture was stirred for 16 h at the same temperature. Entry 6: A mixture of donor aldehyde (2 mmol, 1 equiv.), acceptor aldehyde (20 mmol, 10 equiv.), and (S)-proline (0.6 mmol, 0.3 equiv.) in N-methylpyrrolidone (NMP) (1.0 ml) was stirred at 4 °C for 36 h.

cross-aldol reactions. Certain features of aldehydes typically determine whether an aldehyde can be a donor or an acceptor in a cross-aldol reaction. For example, α,α-disubstituted aldehydes (such as isobutyraldehyde), aldehydes possessing a bulky substituent, and non-enolizable aldehydes (such as arylaldehydes), can serve as acceptors and linear alkyl aldehydes can act as donors. The aldol products of the reactions of aldehyde donors have an aldehyde group, and these types of product often easily isomerize at the position alpha to the aldehyde. In order to obtain high enantioselectivities for these reactions, carefully controlled reaction conditions and work-up and purification procedures must be used.

(S)-Proline-catalyzed cross-aldol reaction of aldehydes followed by Mukaiyama aldol reaction sequence was used for the synthesis of prelactone B [27]. The products of the aldol reactions of O-protected α-oxyaldehydes are protected carbohydrates, and were also transformed to highly enantiomerically enriched hexose derivatives, again through a second Mukaiyama aldol reaction (Scheme 2.5) [28]. The products of the aldol reactions of N-protected α-aminoaldehyde donor were easily converted to the corresponding highly enantiomerically enriched β-hydroxy-α-amino acids and their derivatives (Scheme 2.6) [24]. (For experimental details see Chapter 14.1.1).

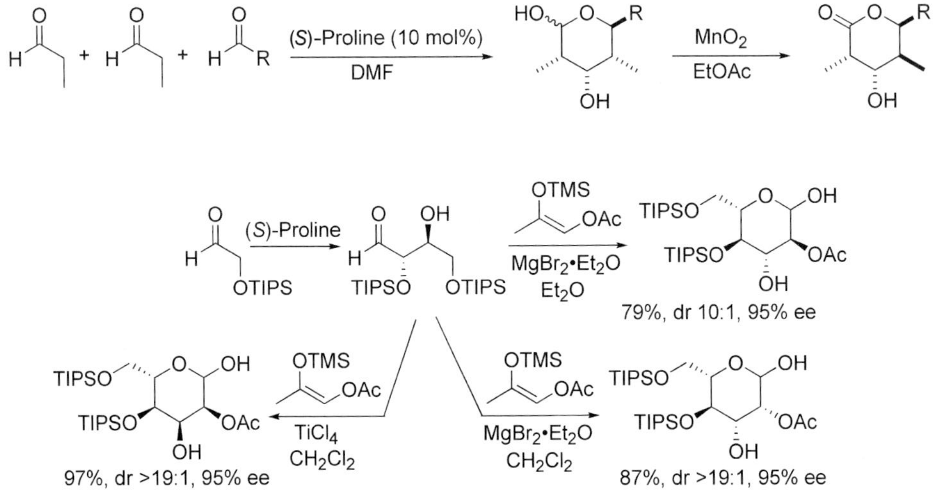

Scheme 2.5 Syntheses of carbohydrate derivatives using the (S)-proline-catalyzed aldol reactions [26, 28].

Although (S)-proline was an efficient catalyst for the aldol reactions of many aldehyde donors, the (S)-proline-catalyzed aldol reactions of α,α-disubstituted aldehyde donors were inefficient [29]. By using a fluorescence-based screening methodology [30], pyrrolidine-derived amine and acid combination catalysts were developed [29]. The combination catalyst 8-CF$_3$CO$_2$H efficiently catalyzed the aldol reactions of α,α-disubstituted aldehyde donors to afford products containing quaternary carbon centers (Table 2.7) that are typically difficult to synthesize [29]. The acid component of the combination catalyst is important for the reaction: the use of 8-CF$_3$SO$_3$H instead of 8-CF$_3$CO$_2$H was also efficient for the reaction, but the reaction with 8-CH$_3$CO$_2$H afforded the products with only low enantioselectivities. For the formation of the corresponding racemic aldol products, pyrrolidine-CH$_3$CO$_2$H was an efficient catalyst [29a]. A pyrrolidine-derived diamine-protonic acid combination was studied for the first time in the acetone aldol reaction, but was ineffective as compared to proline [6]. Later, a series of pyrrolidine-derived diamine-protonic acid combination catalysts were tested in

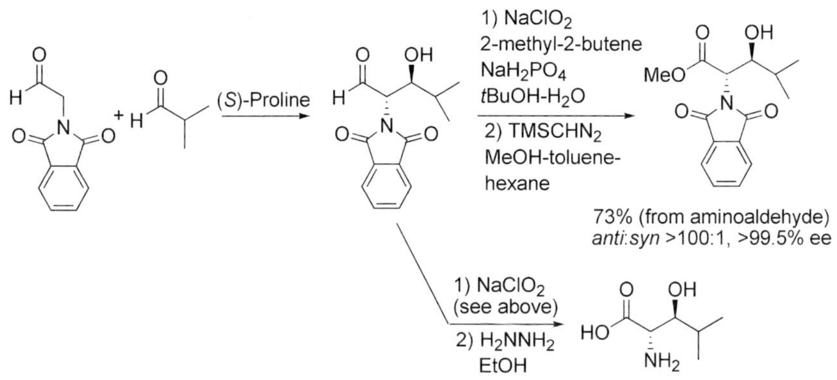

Scheme 2.6 Syntheses of β-hydroxy-α-amino acid derivatives using the (S)-proline-catalyzed aldol reactions [24].

Table 2.7 Diamine 8-CF$_3$CO$_2$H-catalyzed aldol reactions of α,α-disubstituted aldehyde donors to afford hydroxyaldehydes with a quaternary carbon atom [29].

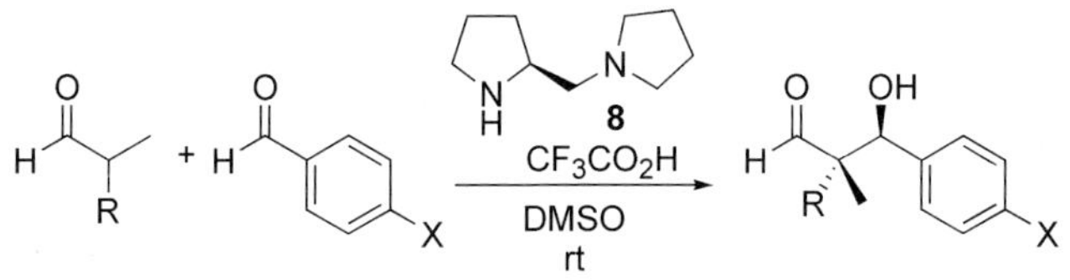

Entry	R	X	Yield [%]	anti:syn	anti ee [%]
1	Me	NO$_2$	94	–	95
2	Me	Br	80	–	95
3	Et	NO$_2$	96	62:38	91
4	(CH$_2$)$_8$CH$_3$	NO$_2$	93	69:31	91
5	CH$_2$C$_6$H$_4$-4-t-Bu	NO$_2$	91	85:15	96

Conditions: Diamine 8 (0.05 mmol, 0.1 equiv), CF$_3$CO$_2$H (0.05 mmol, 0.1 equiv), donor aldehyde (1.0 mmol, 2 equiv), acceptor aldehyde (0.5 mmol, 1 equiv), DMSO (0.5 mL). Entries 1 and 2, 12 h; entry 3, 72 h; entries 4–6, 48 h. For entry 6, donor aldehyde (5.0 mmol).

the aldol reactions of acetone; in these cases, α,β-unsaturated products (formally H_2O-eliminated products from the aldols) formed as byproducts with the desired aldol products (up to 93% ee) [31]. Pyrrolidine bearing a sulfonamide group also catalyzed the aldol reactions of α,α-disubstituted aldehyde donors with high enantioselectivities [32].

2.1.2.4 Aldol Reactions with Ketone Acceptors

Intermolecular aldol reactions with ketone acceptors that possess an electron-withdrawing group are also catalyzed by (S)-proline and diamine 8-CF_3CO_2H, as shown in Scheme 2.7 [33–36]. Reactions with unsymmetrical ketone acceptors provided concise syntheses of compounds containing a quaternary (tetrasubstituted) carbon center, but these reactions are significantly more rare than their aldehyde counterpart and present significant challenges in terms of reactivity and stereocontrol. Enantiomerically enriched α-hydroxy-phosphonates have also been also prepared by (S)-proline-catalyzed reactions [35]. Aldol reactions involving an α-ketoacid acceptor were accomplished by the use of pyrrolidine amide catalyst containing a pyridine moiety to selectively recognize the carboxylic acid of the

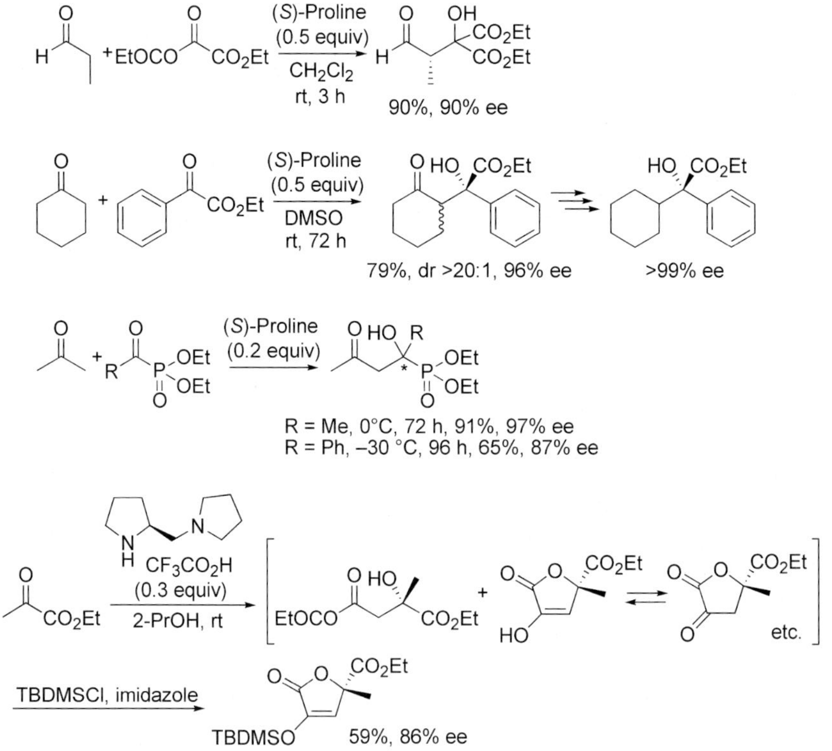

Scheme 2.7 Aldol reactions with ketone acceptors [33–36].

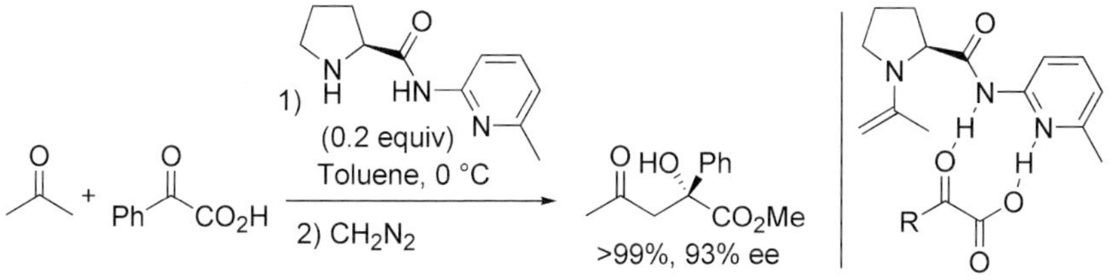

Scheme 2.8 Aldol reaction of a ketoacid acceptor using proline amide catalyst bearing a molecular recognition site for the carboxylic acid of the acceptor [37].

acceptors (Scheme 2.8) [37]. Dimerization of 2,2-dimethyl-1,3-dioxan-5-one (5) was also catalyzed by (S)-proline [18a].

2.1.2.5 Intramolecular Aldol Reactions

(S)-Proline, its derivatives, and other amino acids, such as (S)-phenylalanine, were used to catalyze asymmetric intramolecular aldol reactions and aldol condensations of *meso*-triketones (Scheme 2.9) [2, 3, 38–42], and these reactions were used as key steps for the synthesis of enantiomerically enriched steroids and other bioactive compounds. (For experimental details see Chapter 14.1.3). Empowered with mechanistic insight into these reactions gained through studies of aldolase antibodies [4], the classical Hajos–Parrish-type reaction was re-examined in 2000 [43]. This study demonstrated the potential of proline to catalyze reactions via both imine and enamine mechanisms, and that imine and enamine catalytic pathways could be linked in a tandem or cascade reaction processes. Interesting and different properties of pyrrolidine derivatives were demonstrated. For example, one family of molecules afforded the aldol addition product, while other pyrrolidine derivatives provided the aldol condensation products [43]. In the reaction between 2-methyl-cyclohexane-1,3-dione and methylvinyl ketone, some pyrrolidine derivatives catalyzed the synthesis of only the Michael product via iminium activation of methylvinyl ketone, while other pyrrolidine derivatives and amino acids catalyzed the Michael addition and aldol condensation reaction, providing the Robinson annulation product in one step via a tandem imine, enamine process [2, 43]. These results readily demonstrated that pyrrolidine derivatives could be tuned to perform different catalytic tasks. More recently, cyspentacin was also shown to catalyze the intramolecular aldol reactions [44]. However, cyspentacin-derived tetrazole afforded a different aldol product in racemic form. Other key intermediates of bioactive compounds have also been enantioselectively synthesized by intramolecular aldol reactions using proline derivatives as catalysts (Scheme 2.10) [45].

2.1.2.6 Mechanism and Transition States of Aldol Reactions and Effects of Water on Aldol Reactions

For the (S)-proline-catalyzed aldol reaction of acetone and 4-nitrobenzaldehyde in DMSO, involvement of a single molecule of (S)-proline in the C–C bond-

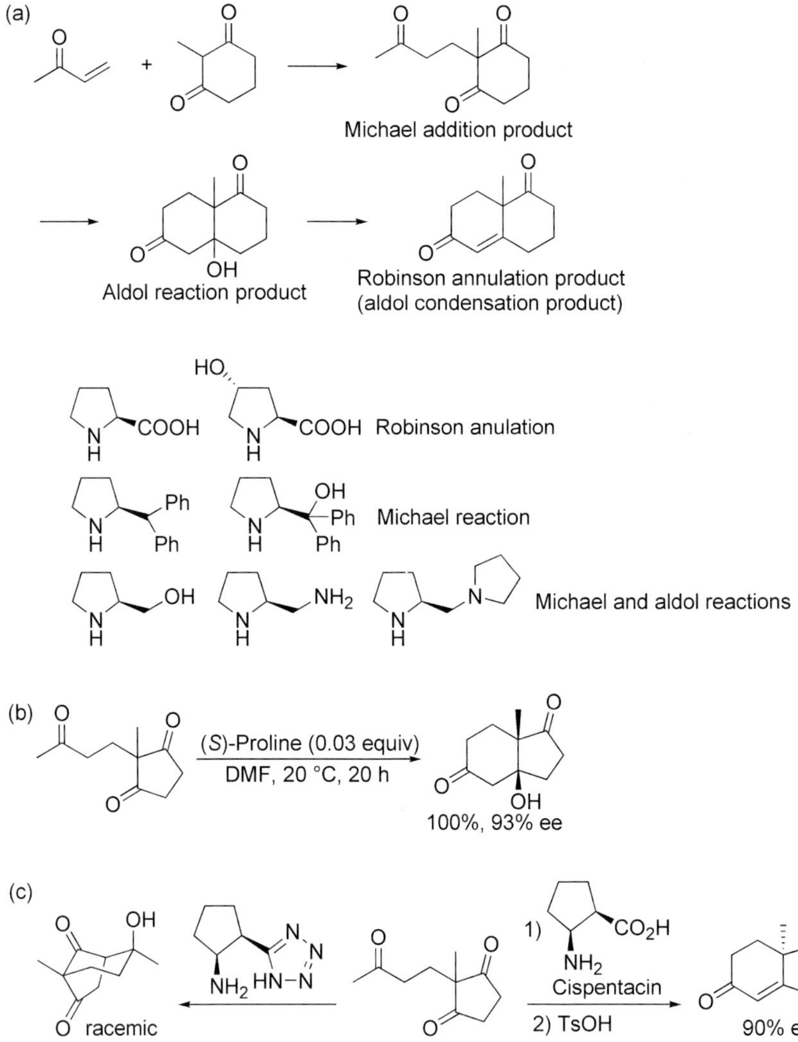

Scheme 2.9 Intramolecular aldol reactions: (a) [43]; (b) [2b]; (c) [44].

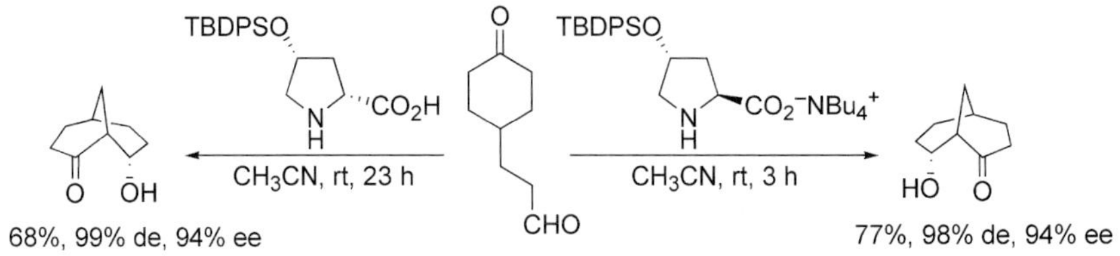

Scheme 2.10 Intramolecular aldol reactions to form bicyclo[3,3,1]alkanones [45].

formation step was first supported by studies that demonstrated a liner relationship between the *ee*-value of the catalyst (S)-proline and that of the aldol product [6]. The proposed pathway and the most suitable transition state of the (S)-proline-catalyzed aldol reaction of acetone and acetaldehyde are summarized in Scheme 2.11 [6, 46–49]. The pathway includes iminium formation, enamine formation, C–C bond formation, and hydrolysis of iminium. The *anti*-enamine (s-*trans*-enamine) is predominately used for the C–C bond formation over the *syn*-enamine (s-*cis*-enamine) in this proline-catalyzed reaction. The facial selectivities of the enamine and of the acceptor aldehyde are controlled by the carboxylic acid of the (S)-proline. The methyl group of acetaldehyde is at an equatorial position in the six-membered transition state of the C–C bond formation. In the (S)-proline-catalyzed intramolecular aldol reaction shown in Scheme 2.9(b), computational analysis has suggested that both the enamine formation and the C–C bond formation represent the rate-limiting steps [50]. In intermolecular aldol reactions, it has been demonstrated experimentally that faster enamine formation contributes to faster product formation [29a].

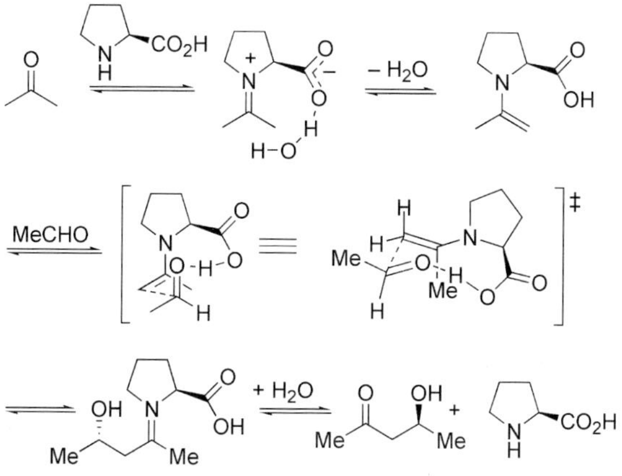

Scheme 2.11 The proposed pathway and most suitable transition state of the (S)-proline-catalyzed aldol reaction of acetone [6, 46–49].

For the aldol reactions of aldehyde donors using (S)-proline or diamine (S)-8-CF$_3$CO$_2$H, the major products and the proposed most suitable transition state that explains the stereochemistries of the products are also shown in Scheme 2.12 [8, 29a].

Although experimental support exists for one proline molecule in the transition state of proline-catalyzed intramolecular aldol reactions was later developed [6, 51], reaction rates and stereoselectivities are not fully explained by a simple transition-state model. For example, the insoluble portion of the catalyst cannot participate in the catalysis. However, when a catalyst with a low enantiopurity is used, the solubilized portion of the catalyst can have a higher enantiopurity (i.e., racemic

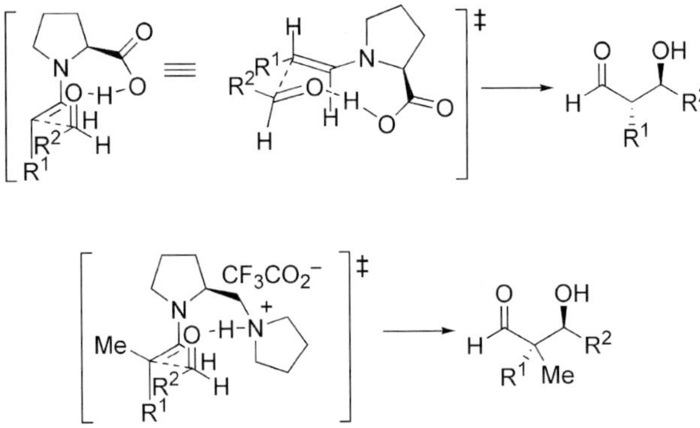

Scheme 2.12 The proposed most suitable transition states of the (S)-proline-catalyzed and diamine 8-CF$_3$CO$_2$H-catalyzed aldol reactions of aldehyde donors [29b].

catalyst precipitates) or lower enantiopurity (i.e., highly enantiopure catalyst precipitates). Thus, the enantiopurity of the catalyst added to the reaction mix may not be linearly related to the net enantioselectivity of the reaction as it is performed [52]. Indeed, this mechanism lends support to the idea that homochiral aldols and carbohydrates could have been formed prebiotically under amino acid catalysis [26]. A further complication is the fact that the product formed in the catalytic reaction can become involved in the reaction itself and affect the reaction rate and/or enantioselectivity of the net reaction [53].

The existence of the enamine intermediate of proline-catalyzed reaction with acetone as a donor was detected by mass analysis [54], but not by ^{1}H NMR. The formation of the presumed enamine intermediate generated from pyrrolidine-acetic acid and isobutyraldehyde was confirmed by ^{1}H NMR [29a]. In this study, the enamine formation in the presence of pyrrolidine-acetic acid was observed within 5 min, but the enamine was shown to form only very slowly in the absence of acid. In these pyrrolidine derivative–acid combination catalysts, the acid component was shown to be important both for faster enamine formation and for the stereocontrol in the C–C bond-forming step. These catalyst systems are essentially "split-proline" systems that allow for the contributions of the pyrrolidine and carboxylate functionalities of proline to be probed independently.

The use of different acid functionalities on pyrrolidine-derived catalysts has improved the reaction rate of some aldol reactions. For example, pyrrolidine-based tetrazole derivative **9** (Fig. 2.2) catalyzed many aldol reactions with rates faster than proline, with similar stereocontrol [16, 18b, 24, 55]. The faster reaction rates with tetrazole derivative **9** in DMSO as compared with proline were attributed to the lower pK_a of the tetrazole moiety as compared to the carboxylic acid group in DMSO (tetrazole p$K_{a(DMSO)}$ 8.2; acetic acid p$K_{a(DMSO)}$ 12.3) [55, 56]. In addition, tetrazole derivative **9** is more soluble than proline in many organic solvents. A higher actual concentration of the catalyst in the solution phase of a reaction mix-

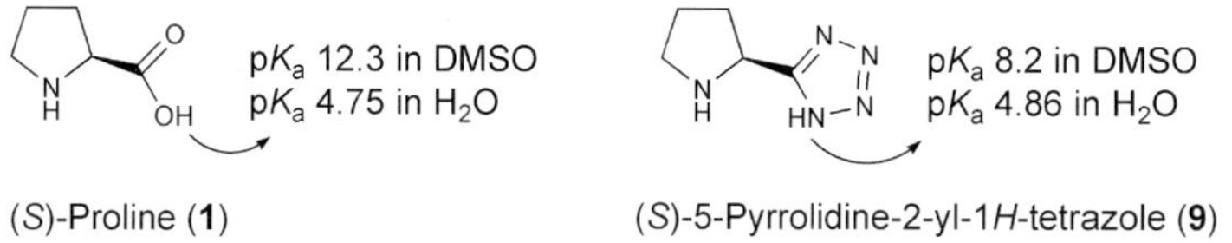

Fig. 2.2 Proline and its tetrazole derivative used for aldol reactions [55, 56].

(S)-Proline (**1**): pK_a 12.3 in DMSO, pK_a 4.75 in H$_2$O

(S)-5-Pyrrolidine-2-yl-1H-tetrazole (**9**): pK_a 8.2 in DMSO, pK_a 4.86 in H$_2$O

ture contributes to faster initial reaction rates. Note that catalyst **9** is not always better than proline. Rather, the outcome of a reaction with proline or with catalyst **9** depends on the reactants: in some cases, aldol reactions with proline afforded much higher enantioselectivity than those with tetrazole catalyst **9** [35]. In the case of diamine-acid catalysts, as described above, the acid component (both type and amount) also affected the outcome of the catalyzed aldol reactions [29, 31]. Phenol derivatives were also used as acid additives in pyrrolidine-catalyzed non-asymmetric aldol reactions [57]. Amides [9, 13, 20], sulfonamides [32, 58], thioamides [59], and hydroxy groups [9, 11, 60] on pyrrolidine-derived catalysts also function as acids for aldol reactions.

In the (S)-proline-catalyzed aldol reactions, the addition of a small amount of water did not affect the stereoselectivities [6]. However, a large amount of water often resulted in products with low enantiomeric excess; water molecules interrupt the hydrogen bonds and ionic interactions critical for the transition states that lead to the high stereocontrol. For example, in the (S)-proline-catalyzed aldol reaction of acetone and 4-nitrobenzaldehyde in DMSO, the addition of 10% (v/v) water to the reaction mixture reduced the *ee*-value from 76% (no water) to ∼30% [6]. Note that the addition of a small amount of water into (S)-proline-catalyzed reactions often accelerates the reaction rate, and the addition of water should be investigated when optimizing these reactions [61].

In the case of the **9**-catalyzed aldol reactions of ketones and aldehyde donors that have a high affinity for water (e.g., chloral, trifluoroacetaldehyde, aqueous formaldehyde; or the corresponding hydrates of the aldehydes), the addition of 100–500 mol% water to the reaction mixture accelerated the reaction rate and afforded the products with higher enantiomeric excess (Scheme 2.13) [16]. The presence of a catalytic amount of water (20 or 50 mol%) or no addition of water

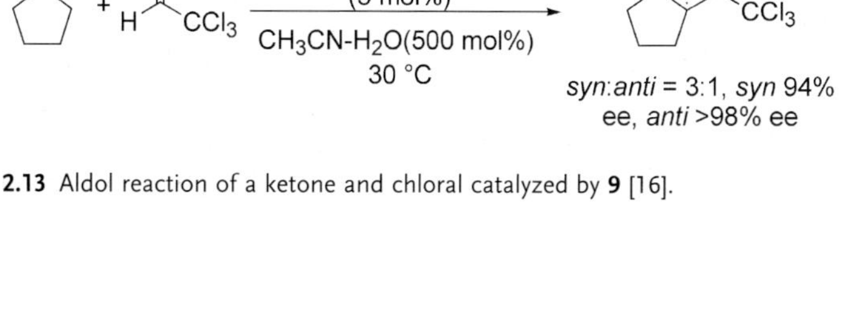

syn:anti = 3:1, syn 94% ee, anti >98% ee

Scheme 2.13 Aldol reaction of a ketone and chloral catalyzed by **9** [16].

resulted in only a low conversion. The major diastereomer of the reaction of cyclopentanone was *syn*; thus, different transition states from those shown in Schemes 2.11 and 2.12 were suggested for these reactions.

Many of the above-described pyrrolidine-based catalysts (e.g., proline) are useful for preparing aldol products with high diastereo- and enantioselectivities in organic solvents, or in organic solvents with a small amount of water. However, most of those catalysts typically either do not catalyze aldol reactions, or they catalyze but afford very low or no enantioselectivities, in water, in water with surfactants, or in organic solvents with a large amount of water [62]. Some proline amide derivatives or small peptides **6** and **7** also catalyzed aldol reactions with high enantioselectivity in aqueous media (see Table 2.5), as described above [20]. Aldol reactions that exhibit high stereocontrol in water without any organic solvent were performed using **10**-CF_3CO_2H (Table 2.8) [14]. Amine **10** has hydrophobic alkyl chains, and thus **10**-CF_3CO_2H assembles with hydrophobic reactants through hydrophobic interactions in water. As a result, it is suggested that the transition state for the stereodefining C–C bond-forming step is similar to that

Table 2.8 Direct asymmetric aldol reactions in water [14].

Entry	Catalyst	X	Time [h]	Yield [%]	anti:syn	anti ee [%]
1	10-CF_3CO_2H	NO_2	24	99	89:11	94
2	10-CF_3CO_2H	CO_2Me	48	89	91:10	91
3	10-CF_3CO_2H	H	72	46	90:10	99
4	10-CH_3CO_2H	NO_2	24	81	60:40	21
5	10	NO_2	5	99	94:6	1
6	11	NO_2	5	78	84:16	22
7	8	NO_2	5	99	77:23	0
8	8-CF_3CO_2H	NO_2	96	0	–	–
9	(S)-Proline	NO_2	96	0	–	–

Conditions: Amine or proline (0.05 mmol, 0.1 equiv.), acid (if used, 0.05 mmol, 0.1 equiv.), cyclohexanone (1.0 mmol, 2 equiv.), and aldehyde (0.5 mmol, 1 equiv.) in water (1.0 mL).

formed in neat organic solvents. With the designed catalyst, the reaction is believed to occur in an hydrophobic microenvironment within the catalyst. Such a feature allows this catalyst system to use water as a solvent and affords aldol product diastereoselectivity and enantioselectivity similar to those obtained from the same reaction performed in DMSO. In most cases of the aldol reactions catalyzed by **10**-CF_3CO_2H in water, use of only 1–2 equiv. of a ketone donor relative to the acceptor aldehyde efficiently afforded the products in good yield within 1 to 3 days. In this system, the acid component of the catalyst was also an important contributor to high enantioselectivities: Amines **10**, **11**, and **8** alone without acid additive, and **10**-CH_3CO_2H, also catalyzed the aldol reaction in water, though the enantioselectivities were poor (Table 2.8, entries 4–7). Note that amine **8**-CF_3CO_2H catalyzed the aldol reaction in DMSO [29b], but did not catalyze the reaction in water (Table 2.8, entry 8) [14]. Significantly, both ketone and aldehyde donors were shown to be effective in this system.

Aldol reactions using catalyst **12** were also performed with water without any organic solvent (Table 2.9) [15]. In these examples, it is assumed that the reaction occurred in the organic phase composed of reactants and the catalyst separated from the water phase. A large excess of ketone was used in these reactions. Catalysts possessing a *tert*-butyldimethylsilyl (TBS) or a triisopropylsilyl (TIPS) group instead of the *tert*-butyldiphenylsilyl (TBDPS) group on **12** also effectively catalyzed the reaction with water, but hydroxyproline **2** did not catalyze the reaction under the same conditions; this indicated that the hydrophobic substituents of

Table 2.9 Direct asymmetric aldol reactions with water catalyzed by **12** [15].

Entry	n	R	Time [h]	Yield [%]	anti:syn	anti ee [%]
1	1	Ph	18	78	13:1	>99
2	1	4-$NO_2C_6H_4$	5	86	20:1	>99
3	1	c-C_6H_{11}	24	76	>20:1	>99
4	0	Ph	18	74	9:1	>99

Conditions: Catalyst **12** (0.04 mmol, 0.1 equiv.), ketone (2.0 mmol, 5 equiv.), aldehyde (0.4 mmol, 1 equiv.), and water (0.13 mL).
TBDPS = *tert*-butyldiphenylsilyl.

the catalysts were important for the catalysis with water. The same reaction using **12**, but performed neat, provided lower diastereo- and enantioselectivities than those obtained in the reactions with water. The addition of water improved not only the selectivity of the reactions but also the rates. Similar results were also reported with the use of (S)-tryptophan as a catalyst of aqueous aldol reactions [63].

2.1.2.7 Catalyst Recycling

Systems have been developed that allow the recycling of catalysts. The first case study involved simple adsorption of proline onto silica gel [6], but the system suffered from a loss in enantioselectivity. More recently, promising results have been obtained with fluorous proline derivatives [64] used for aldol reactions; the recycling of fluorous catalysts has been demonstrated using fluorous solid–liquid extraction. Solid phase-supported catalysts through covalent bonds [65] and through noncovalent interactions [66] were also used for aldol reactions. Proline and other catalysts can be recycled when ionic liquids or polyethylene glycol (PEG) were used as reaction solvents [67].

2.1.2.8 Catalyst Development Strategies

In addition to rational design strategies, the optimization of catalysts and reaction conditions through the screening of catalyst libraries and reaction conditions is an important strategy for the development of reactions and their catalysts. For experiments involving libraries, the use of methods that can evaluate library members quickly and simultaneously is key. For example, using fluorogenic substrates allows progress to be detected at an early stage of the reaction on a small scale in a high-throughput format. Examples of reaction monitoring systems are shown in Scheme 2.14. Fluorescence-based reaction monitoring systems that use a fluorogenic maleimide have been used to rapidly and effectively evaluate enamine-based aldol reactions [29, 30, 68]. Fluorogenic aldehydes have also been developed, and these aldehyde reactants are useful in the development of aldol catalysts [69]. Better catalysts were also identified by supporting them on beads together with a reactant substrate. When the reactions were performed with another reactant that was conjugated to a dye and the remaining the dye-containing reactant removed was by washing, beads containing active catalysts were identified by the color of the dye [70].

2.1.3
Mannich Reactions

2.1.3.1 Mannich-Type Reactions of Aldehyde Donors with Glyoxylate Imines

(S)-Proline has been used to catalyze Mannich-type reactions of enolizable carbonyl donors. Reactions of unmodified aldehydes and N-p-methoxyphenyl (PMP)-protected glyoxylate imine in the presence of a catalytic amount of (S)-proline at room temperature afforded enantiomerically enriched β-aminoaldehydes, as

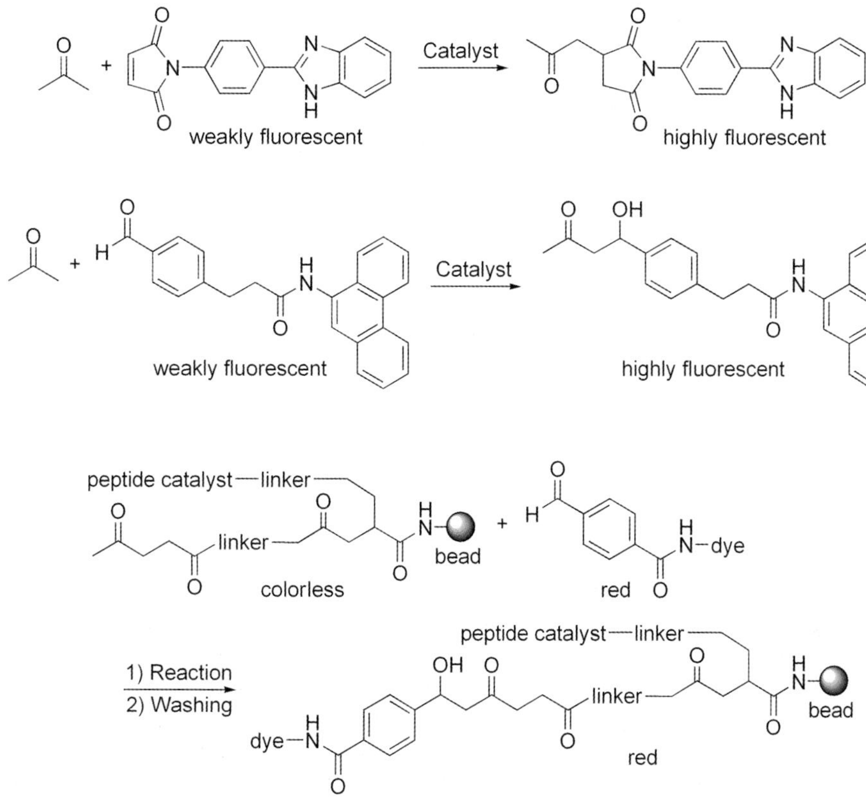

Scheme 2.14 Reaction monitoring systems for catalyst discovery and reaction optimization through screening.

shown in Table 2.10 [71]. (For experimental details see Chapter 14.2.1). The reactions afforded the (2S,3S)-syn-product as the major diastereomer, and the diastereoselectivities were excellent when aldehydes with a longer alkyl chain were used. The enantioselectivities of the syn-products were also excellent. The reactions also proceeded efficiently in a wide range of solvents, including DMSO, Et$_2$O, EtOAc, and THF, and afforded the desired products in good yields with high diastereo- and enantioselectivities. The addition of water (up to 20%, v/v) into the reactions did not affect the enantioselectivity, although the yield was moderate in the reaction with 20% (v/v) water. Lesser quantities of water facilitated the reaction and increased the yield. The use of ionic liquids [bmim]BF$_4$ and [bmim]PF$_6$ as solvents in these reactions provided facile product isolation, catalyst recycling, significantly improved reaction rates (approximately 4- to 50-fold), and afforded the products with excellent yields and diastereo- and enantioselectivities. (S)-Proline also catalyzed the Mannich reactions of α,α-disubstituted aldehyde donors (Table 2.11) [72].

Table 2.10 (S)-Proline-catalyzed Mannich-type reactions of aldehyde donors and N-PMP-protected glyoxylate imine [71a,b].

Entry	R	Yield [%]	syn:anti[a]	syn ee [%]
1	Me	72	1.1:1 (3:1)	99
2	i-Pr	81	>10:1 (19:1)	93
3	n-Bu	81	3:1 (>19:1)	99
4	n-Pent	81	>19:1 (>19:1)	>99
5	$CH_2CH=CH(CH_2)_4CH_3$	89	>19:1 (>19:1)	99

Conditions: Aldehyde (0.75 mmol, 1.5 equiv.), N-PMP-protected α-imino ethyl glyoxylate (0.5 mmol, 1 equiv.), (S)-proline (0.025 mmol, 0.05 equiv.), dioxane (5 mL), rt, 2–24 h. PMP = p-methoxyphenyl.

[a] Ratio after purification. Ratio of the crude product is indicated in parentheses. Epimerization at α-position of the aldehyde of the Mannich products occurred during work-up and silica gel column purification.

Table 2.11 (S)-Proline-catalyzed Mannich-type reactions of α,α-disubstituted aldehyde donors and N-PMP-protected glyoxylate imine [72].

Entry	R^1	R^2	Yield [%]	syn:anti	syn ee [%]
1	Me	Ph	66	85:15	86
2	$-(CH_2)_4-$		94	–	98
3	Me	2-Thienyl	80	83:17	92
4	Me	4-MeC_6H_4	82	75:25	99
5	Me	$4\text{-}t\text{-BuC}_6H_4CH_2$	80	60:40	>99

Table 2.12 anti-Mannich-type reactions of aldehyde donors and N-PMP-protected glyoxylate imine catalyzed by (3R,5R)-5-methyl-3-pyrrolidinecarboxylic acid (**13**) [73].

Entry	R	13 [equiv.]	Time [h]	Yield [%]	anti:syn	anti ee [%]
1	Me	0.05	1	70	94:6	>99
2	i-Pr	0.05	3	85	98:2	99
3	n-Bu	0.02	1	71	97:3	99
4	n-Pent	0.05	3	80	97:3	>99

Conditions of entries 1–2, and 4: Aldehyde (0.5 mmol, 2 equiv.), N-PMP-protected α-imino ethyl glyoxylate (0.25 mmol, 1 equiv.), **13** (0.0125 mmol, 0.05 equiv.), DMSO (2.5 mL).

Whereas the (S)-proline-catalyzed Mannich reactions afforded (2S,3S)-syn-isomers as the major products, (3R,5R)-5-methyl-3-pyrrolidinecarboxylic acid (**13**) catalyzed the reactions and afforded (2S,3R)-anti-products in good yield with high, almost perfect, diastereo- and enantioselectivities (Table 2.12) [73]. The reaction rates of the **13**-catalyzed Mannich reactions were approximately two- to threefold faster than the corresponding (S)-proline-catalyzed reactions that afford the syn-products. Thus, the reactions with only 0.01 or 0.02 equiv. of **13** afforded the desired products in reasonable yields within a few hours.

Note that catalyst **13** was designed for anti-selective Mannich reactions based on the study of proline-catalyzed Mannich reactions. Four considerations are key for the diastereo- and enantioselectivities observed in the (S)-proline-catalyzed reactions (Scheme 2.15a):

1. (E)-Enamine intermediates predominate.
2. The s-trans conformation of the (E)-enamine reacts in the C–C bond-forming transition state. The s-cis conformation results in steric interaction between the enamine and the substituent at the 2-position of the pyrrolidine ring.
3. C–C bond formation occurs at the re-face of the enamine intermediate. This facial selection is controlled by proton-transfer from the carboxylic acid to the imine nitrogen.
4. The enamine attacks the si-face of the (E)-imine. The facial selectivity of the imine is also controlled by this proton transfer that increases the electrophilicity of the imine.

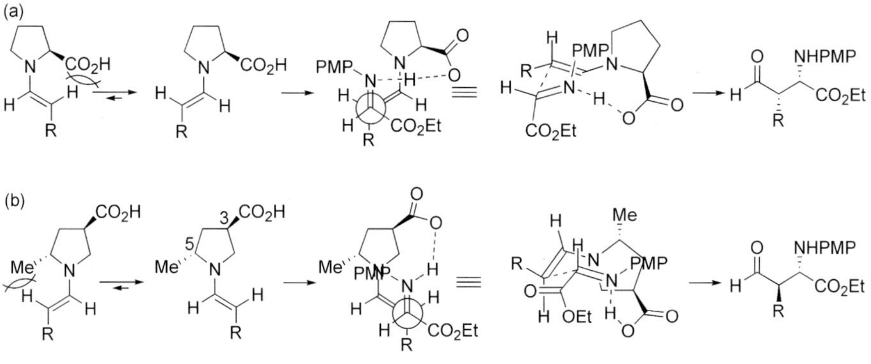

Scheme 2.15 The proposed, most suitable transition states of the (S)-proline- and **13**-catalyzed Mannich reactions [73].

In order to form the *anti*-products enantioselectively, the reaction face of either the enamine or the imine must be opposite that utilized in the proline-catalyzed reactions. In the reactions catalyzed by **13** (Scheme 2.15b), the methyl group at 5-position of the pyrrolidine ring acts to fix the conformation of the enamine and the acid functionality at the 3-position controls the enamine and imine face selection in the transition state (Scheme 2.15b). In order to avoid steric interactions between the substituent at the 5-position of this catalyst and the imine in the transition state, catalyst **13** has a *trans* configuration for substituents at the 3- and 5-positions. Although both the 3-carboxylic acid and 5-methyl groups of catalyst **13** were critical for excellent *anti*-selectivity and enantioselectivity, the reaction with 3-pyrrolidinecarboxylic acid showed that the 3-carboxylic acid group alone had a significant role in the stereoselection.

Whereas the (S)-proline- and **13**-catalyzed Mannich reactions afforded (2S,3S)-*syn*-products and (2S,3R)-*anti*-products, respectively, as shown in Scheme 2.15, with high diastereo- and enantioselectivities, the (S)-pipecolic acid (**14**)-catalyzed reaction afforded (2S,3S)-*syn*- and (2S,3R)-*anti*-products with moderate diastereoselectivities but high enantioselectivities for both the *syn*- and *anti*-products [74] (Scheme 2.16). This was explained by computational analyses indicating that (S)-pipecolic acid uses both the s-*trans* and s-*cis* conformations of the enamine similarly (the energy differences: 0.2 kcal mol^{-1} for pipecolic acid versus 1.0 kcal mol^{-1} for proline) in the C–C bond-forming transition state [74]. Note that (S)-pipecolic acid was not a catalyst for the aldol reaction of acetone and 4-nitrobenzaldehyde [6].

Proline derivatives, such as (2S,4R)-4-hydroxyproline (**2**), (2S,4R)-4-*tert*-butoxyproline, (2S,3S)-3-hydroxyproline [71b] and tetrazole-containing pyrrolidine **9** [75] also catalyzed the Mannich-type reactions using aldehydes as nucleophiles via enamine intermediates, and afforded the *syn*-isomer as the major diastereomer with high enantioselectivity at room temperature. On the other hand,

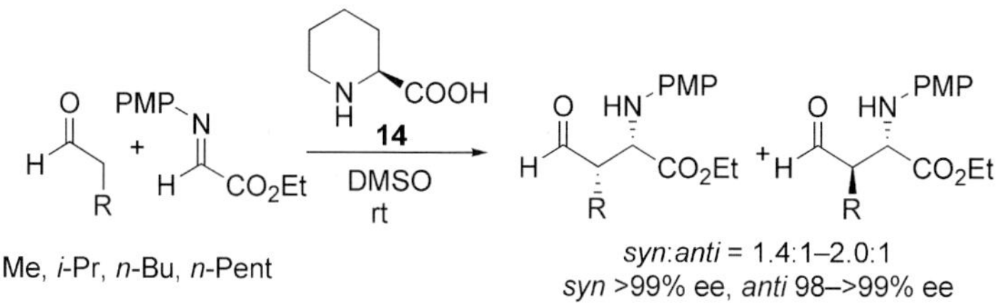

Scheme 2.16 (S)-Pipecolic acid (**14**)-catalyzed Mannich reactions afford both *syn-* and *anti*-products [74].

pyrrolidine derivatives possessing a 2-bulky substituent that does not have an acid functional group, such as **15** [76], and sulfonamide-containing binaphthyl derivative **16** [77] catalyzed the Mannich reactions and afforded the *anti*-isomer as the major diastereomer with high enantioselectivity at room temperature (Fig. 2.3). Early studies with 2-methoxymethylpyrrolidine demonstrated that it catalyzed Mannich reactions and afforded the *anti*-product as the major diastereomer, but the enantioselectivities of the *anti*-products are lower than those seen in the more recently developed catalyst systems [78]; this indicated that the methoxymethyl group is too small to provide the stereocontrol needed for high selectivities.

The (S)-proline-catalyzed Mannich-type reactions that afford enantiomerically enriched aminoaldehyde derivatives constitute efficient routes to access enantiomerically enriched α- and β-amino acid derivatives and β-lactams [71a,b] and enzyme inhibitors [79] (Scheme 2.17). The Mannich products obtained from the

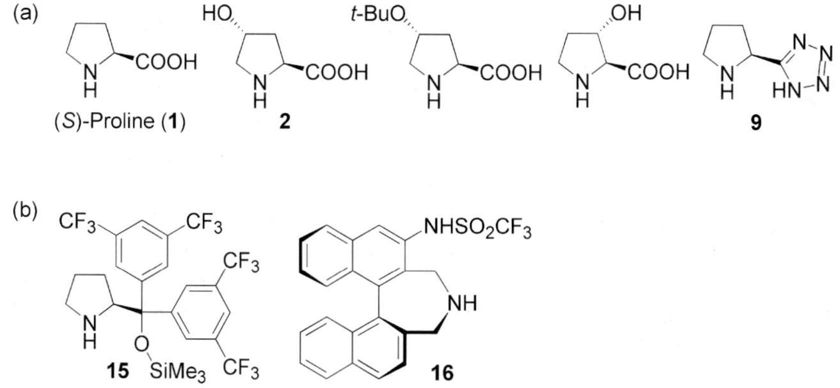

Fig. 2.3 Catalysts for the Mannich-type reactions of aldehydes and glyoxylate imines that use *in-situ*-generated enamine intermediates and that selectively afford (a) *syn*-products or (b) *anti*-products with high enantioselectivities.

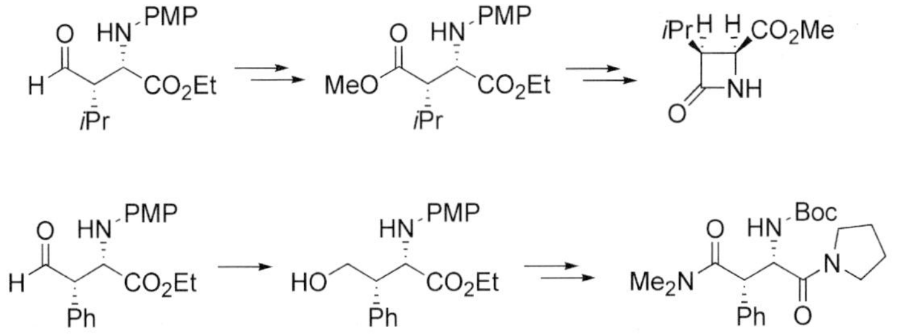

Scheme 2.17 Synthesis of important compounds through (S)-proline-catalyzed Mannich reactions: β-lactam (top) [71a,b] and serine protease inhibitor (bottom) [79].

(S)-proline-catalyzed reactions using unmodified aldehydes as nucleophiles retain the aldehyde group, and the aldehyde group of the products can be used for further transformations in the same reaction vessel. For example, one-pot Mannich-oxime formation [71b], Mannich-allylation [71c], and Mannich-cyanation [80] reactions have been demonstrated (Scheme 2.18). Mannich-type reaction products that possess an aldehyde functionality are easily epimerized during work-up and silica gel column purification. In the one-pot Mannich-cyanation reaction sequence, the cyanohydrin was obtained without epimerization at the α-position of the original aldehyde Mannich products. Thus, this one-pot sequence minimizes potential epimerization of the Mannich products.

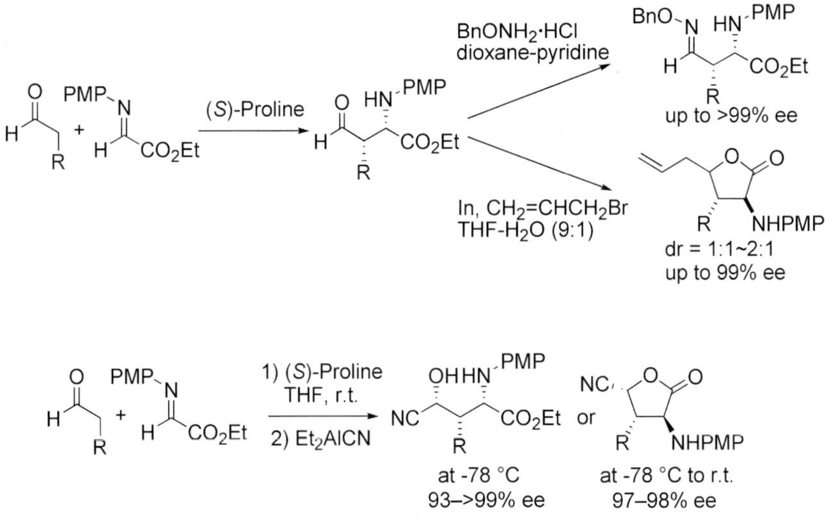

Scheme 2.18 One-pot Mannich-oxime formation [71b], Mannich-allylation [71c], and Mannich-cyanation [80] reactions.

Table 2.13 (S)-Proline-catalyzed Mannich-type reactions of aldehyde donors and preformed imines [71b].

Entry	R	Ar	Yield [%]	syn:anti	syn ee [%]
1	Me	4-NO$_2$C$_6$H$_4$	81	>10:1	99
2	Me	4-CNC$_6$H$_4$	72	7:1	98
3	Me	Ph	65	4:1	93
4	Me	3-BrC$_6$H$_4$	89	3:1	96
5	n-Pent	4-NO$_2$C$_6$H$_4$	60	>19:1	90

2.1.3.2 Mannich-Type Reactions of Aldehyde Donors with Other Preformed Imines

(S)-Proline also catalyzed the Mannich-type reactions of unmodified aldehydes and N-PMP-protected imines to afford the corresponding enantiomerically enriched β-aminoaldehydes at 4 °C (Table 2.13) [71b]. The products were isolated after reduction with NaBH$_4$, though oxidation to the β-amino acid is also possible. These reactions also provided the syn-isomer as the major diastereomer with high enantioselectivities, and proceeded well in other solvents (e.g., dioxane, THF, Et$_2$O). In the reaction of propionaldehyde and the N-PMP-imine of 4-nitrobenzaldehyde in DMF, the addition of water (up to 20%, v/v) did not affect the enantioselectivity. Similar results were obtained for the (S)-proline-catalyzed Mannich-type reactions with the glyoxylate imine where water did not reduce enantioselectivity [71b]. However, the enantioselectivity of the reaction of propionaldehyde and the N-PMP-imine of benzaldehyde in DMF was decreased by the addition of water or MeOH [71b].

The (S)-proline-catalyzed Mannich reactions of aldehyde donors and N-PMP-protected imines of fluorinated aldehyde, such as CF$_3$CHO, C$_2$F$_5$CHO, and PhCF$_2$CHO, were also used for the expedient synthesis of fluorinated aminoalcohols [81].

2.1.3.3 Three-Component Mannich Reactions using Aldehyde Donors

(S)-Proline also catalyzed Mannich reactions in a three-component (donor aldehyde, 4-methoxyaniline, arylaldehyde) protocol – that is, without preformation of imine (Table 2.14) [71b, 82]. (For experimental details see Chapter 14.2.2). This three-component format also afforded the syn-Mannich products in good yields with high diastereoselectivity and enantioselectivities when slow addition of donor aldehyde and/or formation of the imine prior to addition of donor aldehyde was used at a lower reaction temperature, such −20 °C. Reactivity of benzaldehyde and of N-PMP-imine of benzaldehyde as acceptors was compared in the

2 Enamine Catalysis

Table 2.14 (S)-Proline-catalyzed three-component Mannich reactions of aldehyde donors [71b, 82].

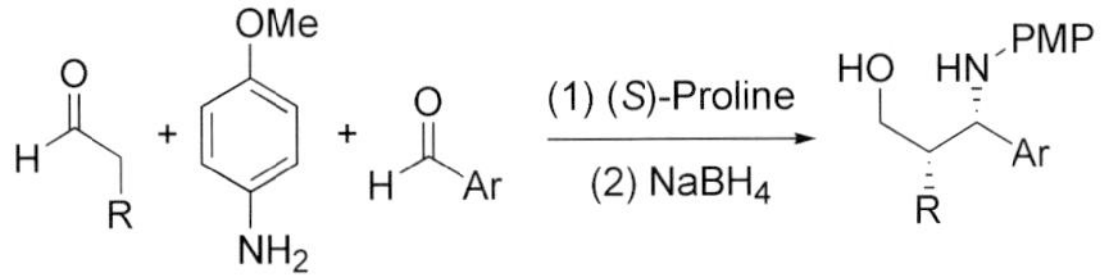

Entry	R	Ar	Yield [%]	syn:anti	syn ee [%]	Ref.
1	Me	4-NO$_2$C$_6$H$_4$	87	65:1	98	71b
2	Me	Ph	82	4:1	94	71b
3	n-Pent	4-NO$_2$C$_6$H$_4$	87	138:1	>99	71b
4	Me	4-ClC$_6$H$_4$	91	>95:5	98	82
5	Me	Ph	90	>95:5	98	82

Conditions: Entries 1–3: To a mixture of ArCHO (0.5 mmol), 4-methoxyaniline (0.5 mmol), and (S)-proline (0.15 mmol) in DMF (3 mL), donor aldehyde (5.0 mmol) in DMF (2 mL) was slowly added (0.2 mL min^{-1}) at −20 °C. The mixture was stirred at the same temperature for 4–10 h. The mixture was diluted with Et$_2$O and reduction performed by addition of NaBH$_4$ [71b]. Entries 4 and 5: After stirring a solution of ArCHO (1.0 mmol), 4-methoxyaniline (1.1 mmol), and (S)-proline (0.1 mmol) in N-methyl-2-pyrrolidinone (1.0 mL) for 2 h at rt, propionaldehyde (3.0 mmol) was added to the mixture at −20 °C, and stirring was continued for 20 h at the same temperature. The reaction was worked-up and reduction with NaBH$_4$ performed without purification [82].

(S)-proline-catalyzed reaction using propionaldehyde as donor and the results showed that the imine reactivity was approximately sevenfold higher than that of the aldehyde [83]. Under basic conditions, it is generally accepted that nucleophilic addition to an aldehyde is typically faster than addition to an aldimine, but nucleophilic addition to an aldimine is faster than addition to an aldehyde when protonation of the imine nitrogen occurs [83]. In the (S)-proline-catalyzed three-component Mannich reactions in the absence of arylaldehyde, self-Mannich products were obtained with moderate to high diastereo- and enantioselectivities (Scheme 2.19) [71b, 82].

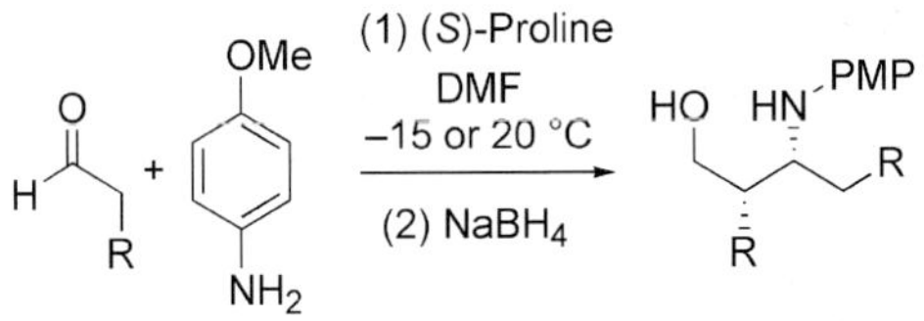

Scheme 2.19 (S)-Proline-catalyzed self-Mannich reactions.

2.1 Aldol and Mannich-Type Reactions

Table 2.15 (S)-Proline-catalyzed Mannich-type reactions of ketone donors and glyoxylate imines.

Entry	R¹	R²	Yield [%]	syn:anti	syn ee [%]	Ref.
1	H	H	82	–	95	84
2	H	Me	72	>19:1	>99	84
3	Me	Me	47	>19:1	>99	84
4	H	CH$_2$CH=CH$_2$	79	>19:1	>99	84
5	H	OH	62	>19:1	99	84
6	Me	OH	68	4:1	98	84
7	–(CH$_2$)$_3$–		81	>19:1	>99	84
8	–OCMe$_2$O–		72	97:3	99	85

Conditions: Entries 1–7: Ketone (1 mL), N-PMP-protected α-imino ethyl glyoxylate (0.5 mmol, 1 equiv.), (S)-proline (0.1 mmol, 0.2 equiv.), DMSO (4 mL), rt, 2–24 h. Entry 8: After stirring (S)-proline (0.3 mmol) and 2,2,2-trifluoroethanol (1 mL) for 3 min at rt, the ketone (1.0 mmol) was added. After 15 min, the imine (1.0 mmol) was added and the mixture stirred for 20 h at rt.

2.1.3.4 Mannich-Type Reactions of Ketone Donors

(S)-Proline was also a good catalyst for the Mannich-type reactions of ketones and glyoxylate imines [84, 85], as shown in Table 2.15. For these Mannich reactions of ketone donors, an excess amount of ketone donors was typically used. The reactions selectively afforded *syn*-products with high enantioselectivities at room temperature. In the reactions of unsymmetrical alkyl ketone donors, such as 2-butanone, the C–C bond formation occurred exclusively at the more substituted α-carbon (Table 2.15, entries 2 and 4). When the reaction was performed using α-hydroxyketone as a donor, C–C bond formation occurred exclusively at the hydroxy-substituted α-carbon (entries 5 and 6). For the reactions of α-aminoketones, azidoketone afforded 1,2-diamino products and phthalimidoacetone afforded the 1,4-diamino product (Scheme 2.20) [86]. The regioselectivity of the reaction of phthalimidoacetone is explained by exclusive enamine formation through deprotonation at the methyl group, whereas enamine formation by deprotonation at the side of the protected amino group-substituted methylene is disfavored due to steric hindrance.

The Mannich reactions of ketones were also performed using sulfonamide-containing pyrrolidine **17**, and afforded the products with similar regio-, diastereo-,

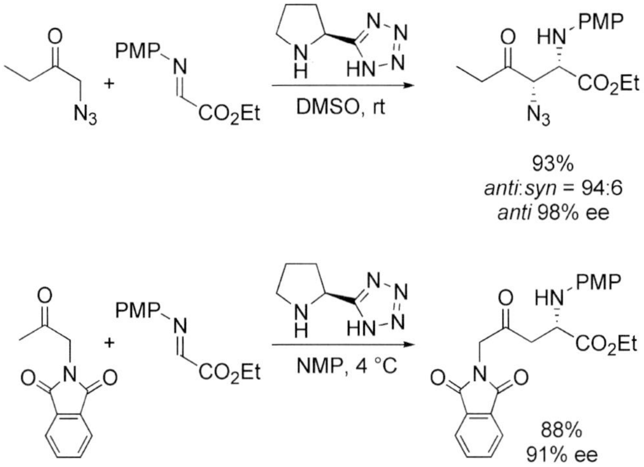

Scheme 2.20 Mannich-type reactions of azidoketones and of phthalimidoacetone.

and enantioselectivities (Scheme 2.21) to those obtained in the (S)-proline-catalyzed reactions [87]. Other pyrrolidine-based catalysts containing sulfonamide functionality were also used for the Mannich reactions [88]. Tetrazole-containing catalyst 9 catalyzed the same reactions with faster reaction rates in a lower catalyst loading (0.01–0.05 equiv.), whilst retaining high diastereo- and enantioselectivities [75].

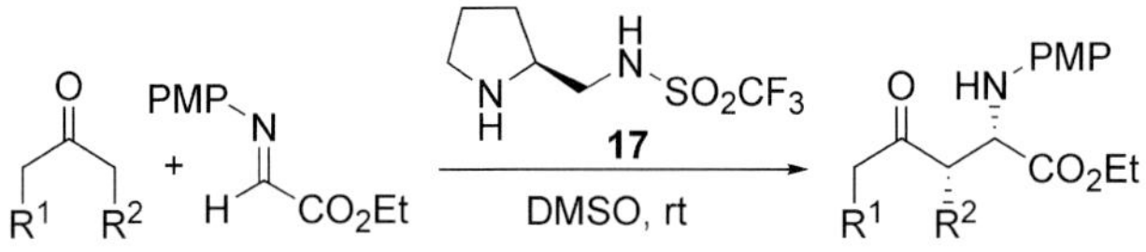

Scheme 2.21 Mannich-type reactions catalyzed by sulfonamide-containing pyrrolidine 17 [87].

(S)-Proline-catalyzed additions of ketones to a cyclic imine provide concise routes to access precursors for the synthesis of indole alkaloids (Scheme 2.22) [89].

(S)-Proline also catalyzed Mannich reactions of ketone donors in a three-component (donor ketone, 4-methoxyaniline, aryaldehyde) protocol, as shown in Table 2.16 [84b, 90, 91]. In these three component reactions, the C–C bond formation occurred at both α-positions of unsymmetrical alkyl ketones (entry 3), and the ratio of the regioisomers depended on the reactant ketones and aldehydes. When the reaction was performed using a ketone donor possessing an α-hydroxy or methoxy group, C–C bond formation occurred exclusively at the oxy-substituted α-carbon (entries 5–7); the major diastereomer was again the syn-product. The enantioselectivities of (S)-proline-catalyzed three-component

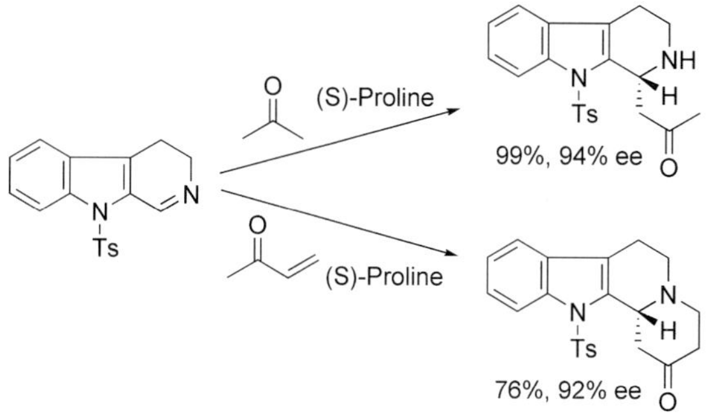

2.1 Aldol and Mannich-Type Reactions

Scheme 2.22 (S)-Proline-catalyzed addition of ketones to a cyclic imine.

Table 2.16 (S)-Proline-catalyzed three-component Mannich reactions of ketone donors.

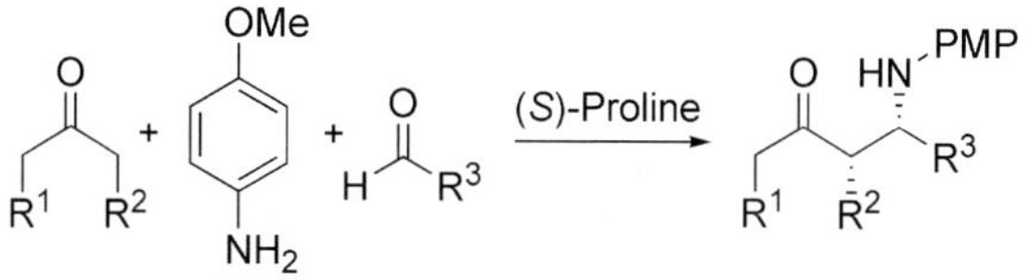

Entry	R^1	R^2	R^3	Yield [%]	syn:anti	ee [%]	Ref.
1	H	H	4-NO$_2$C$_6$H$_4$	50	–	94	90a,b
2	H	H	i-Bu	90	–	93	90a,b
3	H	Me	4-NO$_2$C$_6$H$_4$	66	>19:1	98	84b
	Me	H	4-NO$_2$C$_6$H$_4$	27	–	95	84b
4	Me	H	1-Naphthyl	73	–	96	84b
5	H	OMe	4-NO$_2$C$_6$H$_4$	93	>97.5:2.5	98	90b
6	H	OH	4-NO$_2$C$_6$H$_4$	92	20:1	>99	90b
7	H	OH	Ph	83	9:1	93	90b
8	–OCMe$_2$O–		CH(OMe)$_2$	91	>99.5:0.5	98	91
9	–OCMe$_2$O–		CO$_2$Et	91	>98:2	98	91
10	–OCMe$_2$O–		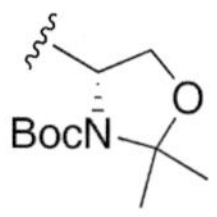 BocN-(tetrahydropyranyl)	67	>98:2	>96	91

Conditions: Entries 1 and 2: (S)-Proline (0.35 mmol), 4-methoxyaniline (1.1 mmol), aldehyde (1.0 mmol), ketone (2 mL), DMSO (8 mL), rt, 12 h. Entries 6 and 7: (S)-Proline (0.2 mmol), 4-methoxyaniline (1.1 mmol), aldehyde (1.0 mmol), hydroxyacetone (1 mL), DMSO (9 mL), rt, 3–16 h. Entries 8–10: Ketone 5 (0.77 mmol), aldehyde (0.38 mmol), 4-methoxyaniline (0.42 mmol), (S)-proline (0.11 mmol for entries 8 and 10, 0.04 mmol for entry 9), DMF (1 mL), H$_2$O (1.5 mmol for entry 8, 1.1 mmol for entry 9), 2 °C, 2–5 days.

Table 2.17 *anti*-Mannich-type reactions of ketone donors and N-PMP-protected glyoxylate imine catalyzed by (R)-3-pyrrolidine-carboxylic acid (**18**).

Entry	R^1	R^2	Yield [%]	anti:syn	anti ee [%]
1	Me	Me	91	97:3	97
2	H	$CH_2CH=CH_2$	85	>95:5	91
3	$-(CH_2)_3-$		96	>99:1	96
4	$-CH_2SCH_2-$		78	>99:1	99

Conditions: Ketone (5.0 mmol, 10 equiv. for entries 1 and 2, 1.0 mmol, 2 equiv. for entries 3 and 4), N-PMP-protected α-imino ethyl glyoxylate (0.5 mmol, 1 equiv.), **18** (0.05 mmol, 0.1 equiv.), 2-PrOH (1.0 mL).

Mannich reactions varied from excellent to low depending on the reactants; Table 2.16 includes the results of highly enantioselective reactions. Reactions using 2,2-dimethyl-1,3-dioxane-5-one (**5**) provide for concise syntheses of enantiomerically enriched protected amino sugars (entries 8–10) [91]. In reactions using **5**, the addition of 1 to 10 equiv. of H_2O increased the rate and stereoselectivity of the reaction.

To perform *anti*-selective Mannich-type reactions, (R)-3-pyrrolidinecarboxylic acid (**18**) was used as a catalyst (Table 2.17) [92]. In the case of Mannich-type reactions of aldehyde donors, (3R,5R)-5-methyl-3-pyrrolidinecarboxylic acid (**13**) catalyzed the *anti*-selective reactions and afforded high diastereo- and enantioselectivities, as described above (Table 2.12 and Scheme 2.15). Catalyst **13**, however, was ineffective in the Mannich-type reactions of ketones. The **13**-catalyzed Mannich-type reaction between 3-pentanone and N-PMP-α-imino ethyl glyoxylate was very slow. It was suggested that the low efficiency of catalyst **13** in the ketone reaction originated from relatively slow formation of the enamine intermediate due to steric interaction with the methyl group of the catalyst (Scheme 2.23a). In the case of reactions with catalyst **18**, *anti*-selectivity has been explained by the stereo-directing effect of the acid group at the 3-position (Scheme 2.23b): when proton transfer occurs from the acid at the 3-position of the catalyst to the imine nitrogen, the nucleophilic carbon of enamine **A** is properly positioned to react with the imine, whereas the nucleophilic carbon of enamine **B** is too far from

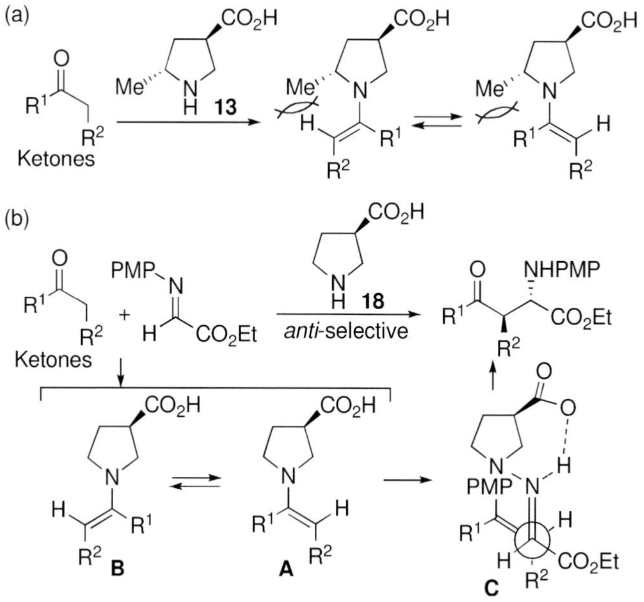

Scheme 2.23 (a) Steric interactions in the possible transition state of 13-catalyzed Mannich-type reactions. (b) The proposed, most suitable transition state of the *anti*-selective Mannich-type reactions catalyzed by (R)-3-pyrrolidinecarboxylic acid (**18**).

the imine carbon to form a bond. Thus, only **A** advances to C–C bond formation via transition state **C**, although enamine conformations **A** and **B** may have similar free energies. Since **18** does not have an α-substituent on the pyrrolidine, neither enamine **A** nor **B** has a disfavored steric interaction like that shown in Scheme 2.23a.

References

1 (a) S. Yamada, K. Hiroi, K. Achiwa, *Tetrahedron Lett.* **1969**, 4233–4236; (b) S. Yamada, G. Otani, *Tetrahedron Lett.* **1969**, 4237–4240.
2 (a) U. Eder, G. Sauer, R. Wiechert, *Angew. Chem., Int. Ed.* **1971**, *10*, 496; (b) Z.G. Hajos, D.R. Parrish, *J. Org. Chem.* **1974**, *39*, 1615–1621.
3 N. Cohen, *Acc. Chem. Res.* **1976**, *9*, 412–417, and references cited therein.
4 (a) J. Wagner, R.A. Lerner, C.F. Barbas, III, *Science* **1995**, *270*, 1797; (b) C.F. Barbas, III, A. Heine, G. Zhong, T. Hoffmann, S. Gramatikova, R. Bjornestsdt, B. List, J. Anderson,

E.A. Stura, I.A. Wilson, R.A. Lerner, *Science* **1997**, *278*, 2085; (c) G. Zhong, T. Hoffmann, R.A. Lerner, S. Danishefsky, C.F. Barbas, III, *J. Am. Chem. Soc.* **1997**, *119*, 8131–8132. (d) T. Hoffmann, G. Zhong, B. List, D. Shabat, J. Anderson, S. Gramatikova, R.A. Lerner, C.F. Barbas, III, *J. Am. Chem. Soc.* **1998**, *120*, 2768–2779; (e) B. List, R.A. Lerner, C.F. Barbas, III, *Org. Lett.* **1999**, *1*, 59–62.

5 B. List, R.A. Lerner, C.F. Barbas, III, *J. Am. Chem. Soc.* **2000**, *122*, 2395–2396.

6 K. Sakthivel, W. Notz, T. Bui, C.F. Barbas, III, *J. Am. Chem. Soc.* **2001**, *123*, 5260–5267.

7 B. List, R.A. Lerner, C.F. Barbas, III, *J. Am. Chem. Soc.* **2000**, *122*, 2395–2396.

8 W. Notz, F. Tanaka, C.F. Barbas, III, *Acc. Chem. Res.* **2004**, *37*, 580–591.

9 Z. Tang, Z.-H. Yang, X.-H. Chen, L.-F. Cun, A.-Q. Mi, Y.-Z. Jiang, L.-Z. Gong, *J. Am. Chem. Soc.* **2005**, *127*, 9285–9289.

10 S. Samanta, J. Liu, R. Dodda, C.-G. Zhao, *Org. Lett.* **2005**, *7*, 5321–5323.

11 H.-M. Guo, L.-F. Cun, L.-Z. Gong, A.-Q. Mi, Y.-Z. Jiang, *Chem. Commun.* **2005**, 1450–1452.

12 C. Cheng, J. Sun, C. Wang, Y. Zhang, S. Wei, F. Jiang, Y. Wu, *Chem. Commun.* **2006**, 215–217.

13 J.-R. Chen, H.-H. Lu, X.-Y. Li, L. Cheng, J. Wan, W.-J. Xiao, *Org. Lett.* **2005**, *7*, 4543–4545.

14 N. Mase, Y. Nakai, N. Ohara, H. Yoda, K. Takabe, F. Tanaka, C.F. Barbas, III, *J. Am. Chem. Soc.* **2006**, *128*, 734–735.

15 Y. Hayashi, T. Sumiya, J. Takahashi, H. Gotoh, T. Urushima, M. Shiji, *Angew. Chem., Int. Ed.* **2006**, *45*, 958–961.

16 H. Torii, M. Nakadai, K. Ishihara, S. Saito, H. Yamamoto, *Angew. Chem., Int. Ed.* **2004**, *43*, 1983–1986.

17 W. Notz, B. List, *J. Am. Chem. Soc.* **2000**, *122*, 7386–7387.

18 (a) D. Enders, C. Grondal, *Angew. Chem., Int. Ed.* **2005**, *44*, 1210–1212; (b) C. Grondel, D. Enders, *Tetrahedron* **2006**, *62*, 329–337; (c) J.T. Suri, D.B. Ramachary, C.F. Barbas, III, *Org. Lett.* **2005**, *7*, 1383–1385; (d) J.T. Suri, S. Mitsumori, K. Albertshofer, F. Tanaka, C.F. Barbas, III, *J. Org. Chem.* **2006**, *71*, 3822–3828.

19 D. Enders, J. Palecek, C. Grondal, *Chem. Commun.* **2006**, 655–657.

20 Z. Tang, Z.-H. Yang, L.-F. Cun, L.-Z. Gong, A.-Q. Mi, Y.-Z. Jiang, *Org. Lett.* **2004**, *6*, 2285–2287.

21 J.M. Betancort, C.F. Barbas, III, *Org. Lett.*, **2001**, *3*, 3737–3740.

22 A.B. Northrup, D.W.C. MacMillan, *J. Am. Chem. Soc.* **2002**, *124*, 6798–6799.

23 A.B. Northrup, I.K. Mangion, F. Hettche, D.W.C. MacMillan, *Angew. Chem., Int. Ed.* **2004**, *43*, 2152–2154.

24 R. Thayumanavan, F. Tanaka, C.F. Barbas, III, *Org. Lett.* **2004**, *6*, 3541–3544.

25 A. Cordova, W. Notz, C.F. Barbas, III, *J. Org. Chem.* **2002**, *67*, 301–303.

26 N.S. Chowdari, D.B. Ramachary, A. Cordova, C.F. Barbas, III, *Tetrahedron Lett.* **2002**, *43*, 9591–9595.

27 P.M. Pihko, A. Erkkila, *Tetrahedron Lett.* **2003**, *44*, 7607–7609.

28 A.B. Northrup, D.W.C. MacMillan, *Science* **2004**, *305*, 1752–1755.

29 (a) N. Mase, F. Tanaka, C.F. Barbas, III, *Org. Lett.* **2003**, *5*, 4369–4372; (b) N. Mase, F. Tanaka, C.F. Barbas, III, *Angew. Chem., Int. Ed.* **2004**, *43*, 2420–2423.

30 F. Tanaka, R. Thayumanavan, C.F. Barbas, III, *J. Am. Chem. Soc.* **2003**, *125*, 8523–8528.

31 M. Nakadai, S. Saito, H. Yamamoto, *Tetrahedron* **2002**, *58*, 8167–8177.

32 W. Wang, H. Li, J. Wang, *Tetrahedron* **2005**, *46*, 5077–5079.

33 A. Bogevig, N. Kumaragurubaran, K. Jørgensen, *Chem. Commun.* **2002**, 620–621.

34 O. Tokuda, T. Kano, W.-G. Gao, T. Ikemoto, K. Maruoka, *Org. Lett.* **2005**, *7*, 5103–5105.

35 S. Samanta, C.-G. Zhao, *J. Am. Chem. Soc.* **2006**, *128*, 7442–7443.

36 P. Dambruoso, A. Massi, A. Dondoni, *Org. Lett.* **2005**, *7*, 4657–4660.
37 Z. Tang, L.-F. Cun, X. Cui, A.-Q. Mi, Y.-Z. Jiang, L.-Z. Gong, *Org. Lett.* **2006**, *8*, 1263–1266.
38 S. Danishefsky, P. Cain, *J. Am. Chem. Soc.* **1975**, *97*, 5282–5284.
39 F.R. Clemente, K.N. Houk, *J. Am. Chem. Soc.* **2005**, *127*, 11294–11302, and references cited therein.
40 A. Przezdziecka, W. Stepanenko, J. Wicha, *Tetrahedron: Asymm.* **1999**, *10*, 1589–1598.
41 H. Shigehisa, T. Mazutani, S. Tosaki, T. Oshima, M. Shibasaki, *Tetrahedron* **2005**, *61*, 5057–5065.
42 K. Inomata, M. Barrague, L.A. Paquette, *J. Org. Chem.* **2005**, *70*, 533–539.
43 T. Bui, C.F. Barbas, III, *Tetrahedron Lett.* **2000**, *41*, 6951–6954.
44 S.G. Davis, R.L. Sheppard, A.D. Smith, J.E. Thomson, *Chem. Commun.* **2005**, 3802–3804.
45 (a) N. Itagaki, M. Kimura, T. Sugahara, Y. Iwabuchi, *Org. Lett.* **2005**, *7*, 4185–4188; (b) N. Itagaki, T. Sugahara, Y. Iwabuchi, *Org. Lett.* **2005**, *7*, 4181–4183.
46 K.N. Rankin, J.W. Gauld, R.J. Boyd, *J. Phys. Chem. A* **2002**, *106*, 5155–5159.
47 S. Bahmanyar, K.N. Houk, H.J. Martin, B. List, *J. Am. Chem. Soc.* **2003**, *125*, 2475–2479.
48 C. Allemann, R. Gordillo, F.R. Clemente, P.H.-Y. Cheong, K.N. Houk, *Acc. Chem. Res.* **2004**, *37*, 558–569.
49 S. Bahmanyar, K.N. Houk, *Org. Lett.* **2003**, *5*, 1249–1251.
50 F.R. Clemente, K.N. Houk, *Angew. Chem., Int. Ed.* **2004**, *43*, 5766–5768.
51 L. Hoang, B. Bahmanyar, K.N. Houk, B. List, *J. Am. Chem. Soc.* **2003**, *125*, 16–17.
52 M. Klussmann, H. Iwamura, S.P. Mathew, D.H. Wells, U. Pandya, A. Armstrong, D.G. Blackmond, *Nature* **2006**, *441*, 621–623.
53 (a) S.P. Mathew, H. Iwamura, D.G. Blackmond, *Angew. Chem., Int. Ed.* **2004**, *43*, 3317–3321; (b) H. Iwamura, D.H. Wells, Jr., S.P. Mathew, M. Klussmann, A. Armstrong, D.G. Blackmond, *J. Am. Chem. Soc.* **2004**, *126*, 16312–16313.
54 C. Marquez, J.O. Metzger, *Chem. Commun.* **2006**, 1539–1541.
55 A. Hartikka, P.I. Arvidsson, *Tetrahedron: Asymmetry* **2004**, *15*, 1831–1834.
56 F.G. Bordwell, *Acc. Chem. Res.* **1988**, *21*, 456–463.
57 C. Ji, Y. Peng, C. Huang, N. Wang, Y. Jiang, *Synlett* **2005**, 986–990.
58 A. Berkessel, B. Koch, J. Lex, *Adv. Synth. Catal.* **2004**, *346*, 1141–1146.
59 D. Gryko, R. Lipinski, *Adv. Synth. Catal.* **2005**, *347*, 1948–1952.
60 G. Zhong, J. Fan, C.F. Barbas, III, *Tetrahedron Lett.* **2004**, *45*, 5681–5684.
61 (a) D.E. Ward, V. Jheengut, *Tetrahedron Lett.* **2004**, *45*, 8347–8350; (b) A.I. Nyberg, A. Usano, P.M. Pihko, *Synlett* **2004**, 1891–1896; (c) P.M. Pihko, K.M. Laurikainen, A. Usano, A.I. Nyberg, J.A. Kaavi, *Tetrahedron* **2006**, *62*, 317–328.
62 (a) A. Cordova, W. Notz, C.F. Barbas, III, *Chem. Commun.* **2002**, 3024–3025; (b) T. Darbre, M. Machuqueiro, *Chem. Commun.* **2003**, 1090–1091; (c) Y.-Y. Peng, Q.-P. Ding, Z. Li, P.G. Wang, J.-P. Cheng, *Tetrahedron Lett.* **2003**, *44*, 3871–3875; (d) Y.-S. Wu, W.-Y. Shao, C.-Q. Zheng, Z.-L. Huang, J. Cai, Q.-Y. Deng, *Helv. Chim. Acta* **2004**, *87*, 1377–1384; (e) Y.-S. Wu, Y. Chen, D.-S. Deng, J. Cai, *Synlett* **2005**, 1627–1629; (f) S.S. Chimni, D. Mahajan, V.V.S. Babu, *Tetrahedron Lett.* **2005**, *46*, 5617–5619.
63 Z. Jiang, Z. Liang, X. Wu, Y. Lu, *Chem. Commun.* **2006**, 2801–2803.
64 F. Fache, O. Piva, *Tetrahedron: Asymm.* **2003**, *14*, 139–143.
65 (a) M. Benaglia, G. Celentano, F. Cozzi, *Chem. Commun.* **2001**, *343*, 171–173; (b) G. Szollosi, G. London, L. Balaspiri, C. Somlai, M. Bartok, *Chirality* **2003**, *15*, S90–S96;

(c) M.R.M. Andreae, A.P. Davis, *Tetrahedron: Asymm.* **2005**, *16*, 2487–2492; (d) F. Calderon, R. Fernandez, F. Sanchez, A. Fernandez-Mayoralas, *Adv Synth. Catal.* **2005**, *347*, 1395–1403; (e) J.D. Revell, D. Gantenbein, P. Krattiger, H. Wennemers, *Biopolymers* **2006**, *84*, 105–113.

66 (a) M. Gruttadauria, S. Riela, C. Aprile, P.L. Meo, F. D'Anna, R. Noto, *Adv. Synth. Catal.* **2006**, *348*, 82–92; (b) Z. Shen, J. Ma, Y. Liu, C. Jiao, M. Li, Y. Zhang, *Chirality* **2005**, *17*, 556–558.

67 (a) P. Kotrusz, I. Kmentova, B. Gotov, S. Toma, E. Solcaniova, *Chem. Commun.* **2002**, 2510–2511; (b) T.-P. Loh, L.-C. Feng, H.-Y. Yang, J.-Y. Yang, *Tetrahedron Lett.* **2002**, *43*, 8741–8743; (c) S. Chandrasekhar, N.R. Reddy, S.S. Sultana, C. Narsihmulu, K.V. Reddy, *Tetrahedron* **2006**, *62*, 338–345.

68 F. Tanaka, R. Thayumanavan, N. Mase, C.F. Barbas, III, *Tetrahedron Lett.* **2004**, *45*, 325–328.

69 F. Tanaka, N. Mase, C.F. Barbas, III, *J. Am. Chem. Soc.* **2004**, *126*, 3692–3693.

70 P. Krattiger, R. Kovasy, J.D. Revell, S. Ivan, H. Wennemers, *Org. Lett.* **2005**, *7*, 1101–1103.

71 (a) A. Cordova, S. Watanabe, F. Tanaka, W. Notz, C.F. Barbas, III, *J. Am. Chem. Soc.* **2002**, *124*, 1866–1867; (b) W. Notz, F. Tanaka, S. Watanabe, N.S. Chowdari, J.T. Turner, R. Thayumanavan, C.F. Barbas, III, *J. Org. Chem.* **2003**, *68*, 9624–9634; (c) A. Cordova, C.F. Barbas, III, *Tetrahedron Lett.* **2003**, *44*, 1923–1926; (d) N. Chowdari, D. Ramachary, A. Cordova, C.F. Barbas, III, *Synlett* **2003**, 1906–1909.

72 N.S. Chowdari, J.T. Suri, C.F. Barbas, III, *Org. Lett.* **2004**, *6*, 2507–2510.

73 S. Mitsumori, H. Zhang, P.-H.-Y. Choeng, K.N. Houk, F. Tanaka, C.F. Barbas, III, *J. Am. Chem. Soc.* **2006**, *128*, 1040–1041.

74 P.H.-C. Cheong, H. Zhang, R. Thayumanavan, F. Tanaka, K.N. Houk, C.F. Barbas, III, *Org. Lett.* **2006**, *8*, 811–814.

75 A.J.A. Cobb, D.M. Shaw, S.V. Ley, *Synlett* **2004**, 558–560.

76 J. Franzen, M. Marigo, D. Fielenbach, T.C. Wabnitz, A. Kjaersgaard, K.A. Jørgensen, *J. Am. Chem. Soc.* **2005**, *127*, 18296–18304.

77 T. Kano, Y. Yamaguchi, O. Tokuda, K. Maruoka, *J. Am. Chem. Soc.* **2005**, *127*, 16408–16409.

78 A. Cordova, C.F. Barbas, III, *Tetrahedron Lett.* **2002**, *43*, 7749–7752.

79 J.M. Janey, Y. Hsiao, J.D. Armstrong, III, *J. Org. Chem.* **2006**, *71*, 390–392.

80 S. Watanabe, A. Cordova, F. Tanaka, C.F. Barbas, III, *Org. Lett.* **2002**, *4*, 4519–4522.

81 S. Fustero, D. Jimenez, J.F. Sanz-Cervera, M. Sanchez-Roselle, E. Esteban, A. Simon-Fuentes, *Org. Lett.* **2005**, *7*, 3433–3436.

82 Y. Hayashi, W. Tsuboi, I. Ashimine, T. Urushima, M. Shoji, K. Sakai, *Angew. Chem. Int. Ed.* **2003**, *42*, 3677–3680.

83 Y. Hayashi, T. Urushima, M. Shoji, T. Uchiyama, I. Shiina, *Adv. Synth. Catal.* **2005**, *347*, 1595–1604.

84 (a) A. Cordova, W. Notz, G. Zhong, J.M. Betancort, C.F. Barbas, III, *J. Am. Chem. Soc.* **2002**, *124*, 1842–1843; (b) W. Notz, S. Watanabe, N.S. Chowdari, G. Zhong, J.M. Betancort, F. Tanaka, C.F. Barbas, III, *Adv. Synth. Catal.* **2004**, *346*, 1131–1140.

85 B. Westermann, C. Neuhaus, *Angew. Chem., Int. Ed.* **2005**, *44*, 4077–4079.

86 N.S. Chowdari, M. Ahmad, K. Albertshfer, F. Tanaka, C.F. Barbas, III, *Org. Lett.* **2006**, *8*, 2839–2842.

87 W. Wang, J. Wang, H. Li, *Tetrahedron Lett.* **2004**, *45*, 7243–7246.

88 A.J.A. Cobb, D.M. Shaw, D.A. Longbottom, J.B. Gold, S.V. Ley, *Org. Biomol. Chem.* **2005**, *3*, 84–96.

89 T. Itoh, M. Yokoya, K. Miyauchi, K. Nagata, A. Ohsawa, *Org. Lett.* **2003**, *5*, 4301–4304.

90 (a) B. List, *J. Am. Chem. Soc.* **2000**, *122*, 9336–9337; (b) B. List, P. Pojarliev, W.T. Biller, H.J. Martin, *J. Am. Chem. Soc.* **2002**, *124*, 827–833; (c) W. Notz, K. Sakthivel, T. Bui, G. Zhong, C.F. Barbas, III, *Tetrahedron Lett.* **2001**, *42*, 199–201.

91 D. Enders, C. Grondal, M. Vrettou, G. Raabe, *Angew. Chem., Int. Ed.* **2005**, *44*, 4079–4083.

92 H. Zhang, M. Mifsud, F. Tanaka, C.F. Barbas, III, *J. Am. Chem. Soc.* **2006**, *128*, 9630–9631.

2.2
α-Heteroatom Functionalization

Mauro Marigo and Karl Anker Jørgensen

2.2.1
Introduction

In this chapter, we will outline the application of organocatalysis for the enantioselective α-heteroatom functionalization of mainly aldehydes and ketones. Attention will be focused on enantioselective amination-, oxygenation-, fluorination-, chlorination-, bromination-, and sulfenylation reactions catalyzed by chiral amines. The scope, potential and application of these organocatalytic asymmetric reactions will be presented as the optically active products obtained are of significant importance, for example in the life-science industries.

Optically active molecules having a chiral carbon atom attached to a heteroatom adjacent to a carbonyl functionality (Fig. 2.4) are of fundamental use in a large number of fields in chemistry.

A variety of methods are available for the stereochemical construction of these optically active α-heteroatom-substituted carbonyl compounds, and in recent years a large number of new procedures have been developed applying asymmetric catalysis to carbonyl compounds, or their equivalents, as substrates [1]. The "trick" to these reactions is the direct C–H to C–Het transformation, as outlined in Eq. (1).

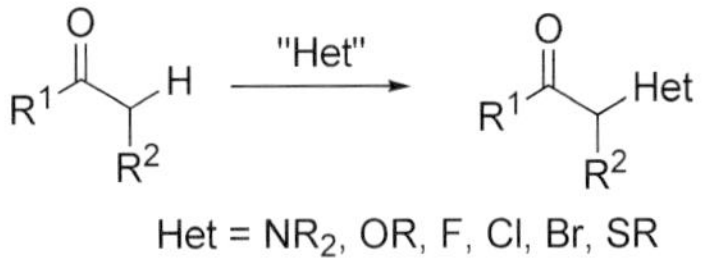

$$\text{Het} = NR_2, OR, F, Cl, Br, SR \tag{1}$$

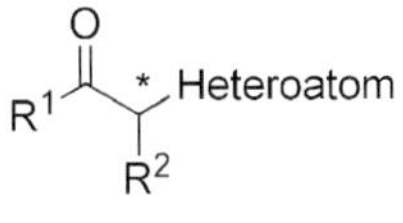

Fig. 2.4 Optically active compounds having a carbon–heteroatom bond adjacent to a carbonyl functionality.

Enantioselective Organocatalysis: Reactions and Experimental Procedures
Edited by Peter I. Dalko
Copyright © 2007 WILEY-VCH Verlag GmbH & Co. KGaA, Weinheim
ISBN: 978-3-527-31522-2

2.2 α-Heteroatom Functionalization

Several catalytic asymmetric approaches can be considered for the C–H to C–Het transformation demonstrated in Eq. (1), and the use of chiral Lewis acid- and chiral organic molecules has attracted considerable attention during recent years.

In the following sections we will describe recent developments and applications in the organocatalytic α-heteroatom functionalization of aldehydes and ketones catalyzed by chiral amines [1]. Notably, the C–H to C–Het transformations associated with amination, oxygenation, halogenation (fluorination, chlorination, bromination), and sulfenylation will be outlined (Scheme 2.24).

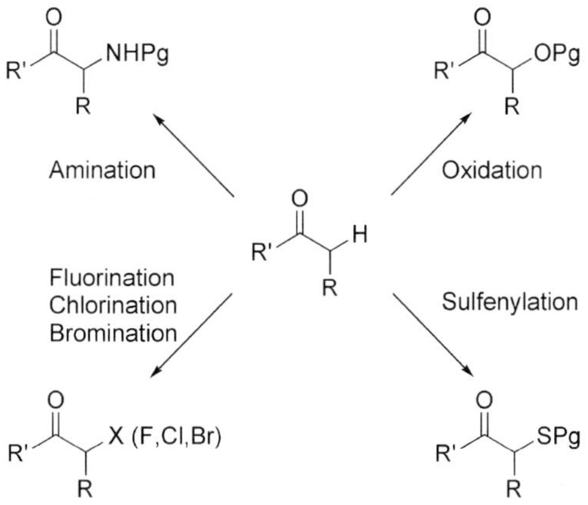

Scheme 2.24 Organocatalytic α-heteroatom functionalization of carbonyl compounds.

The direct activation and transformation of a C–H bond adjacent to a carbonyl group into a C–Het bond can take place via a variety of mechanisms, depending on the organocatalyst applied. When secondary amines are used as the catalyst, the first step is the formation of an enamine intermediate, as presented in the mechanism as outlined in Scheme 2.25. The enamine is formed by reaction of the carbonyl compound with the amine, leading to an iminium intermediate, which is then converted to the enamine intermediate by cleavage of the C–H bond. This enamine has a nucleophilic carbon atom which reacts with the electrophilic heteroatom, leading to formation of the new C–Het bond. The optically active product and the chiral amine are released after hydrolysis.

The stereochemical information on the chiral center formed by the mechanism outlined in Scheme 2.25 is determined by the R*-substituent in the chiral amine. Here, two types of interaction are operating, namely electronic and steric. The electronic interaction [2] outlined to the left in Figure 2.5 seems to take place when the R*-substituent has an acidic hydrogen atom/proton such as a carboxylic acid or tetrazole in the 2-position of a pyrrolidine ring. The acidic hydrogen atom in these R*-substituents interacts with a lone pair in the heteroatom Y of the incoming electrophile. The electrophilic addition of the heteroatom takes thus place

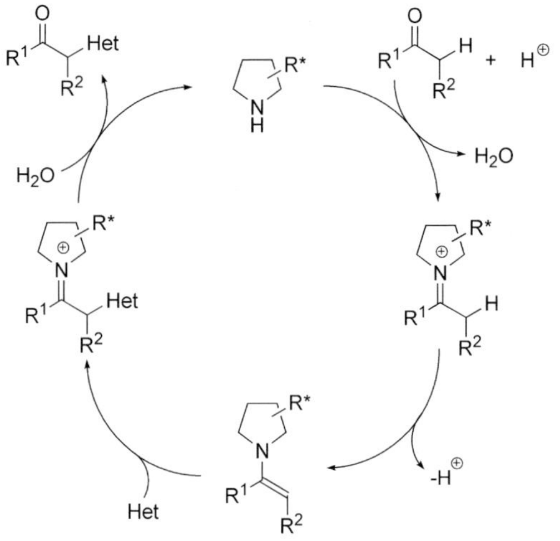

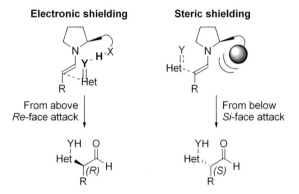

Scheme 2.25 Mechanism for the catalytic enantioselective α-heteroatom functionalization of aldehydes and ketones catalyzed by secondary chiral amines.

Fig. 2.5 Electronic and steric interactions in the approach of the electrophilic heteroatom (Het) to the nucleophilic carbon atom in the chiral enamine intermediate.

from the same face of the chiral substituent due to this hydrogen-bonding interaction, leading to an approach to the *Re*-face of the nucleophilic enamine-carbon atom.

The face-selectivity originating from a steric shielding is outlined to the right in Figure 2.5. The chiral substituent in e.g. the 2-position of the pyrrolidine ring shields the *Re*-face of the enamine carbon atom forcing an approach of the electrophile from the opposite site (*Si*-face approach) of the chiral substituent.

The configuration of the chiral carbon atom formed in the α-position of the carbonyl functionality is thus dependent on the type of interaction between the electrophilic heteroatom and the catalyst. This change in interaction will therefore give rise to the opposite absolute configuration in the optically active product formed if the catalyst has the same absolute configuration.

A different mechanism operates in the direct α-heteroatom functionalization of carbonyl compounds when chiral bases such as cinchona alkaloids are used as the catalysts. The mechanism is outlined in Scheme 2.26 for quinine as the chiral catalyst: quinine can deprotonate the substrate when the substituents have strong electron-withdrawing groups. This reaction generates a nucleophile in a "chiral pocket" (see Fig. 2.6), and the electrophile can thus approach only one of the enantiotopic faces.

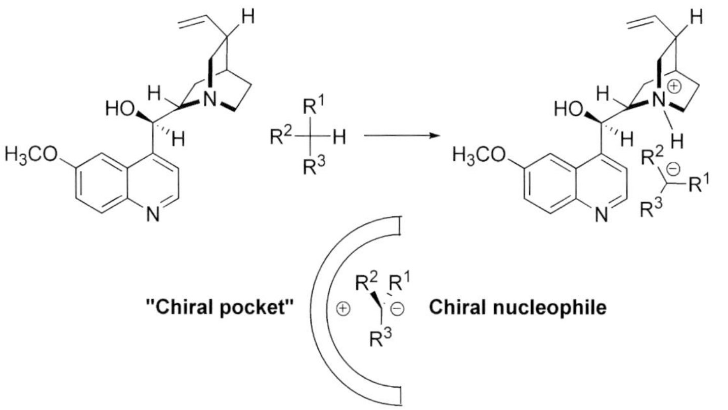

Scheme 2.26 Catalytic formation of a chiral nucleophile by an optically active cinchona alkaloid.

2.2.2
Direct α-Amination of Aldehydes and Ketones

The enantioselective introduction of a nitrogen atom functionality in the α-position of aldehydes and ketones leads to optically active synthetic targets such as α-amino acids and amino alcohols, which have great potential in many life-science molecules [3].

The direct α-amination of aldehydes by azodicarboxylates as the electrophilic nitrogen source can be catalyzed by, for example L-proline **3a**, to give the α-hydrazino aldehydes **4** having (R)-configuration in moderate to good yields and with excellent enantioselectivities (89–97% *ee*) (Scheme 2.27) [4]. The optically active α-hydrazino aldehydes **4** are prone to racemization, and it was found beneficial to reduce them directly with NaBH$_4$ to stereochemical stable compounds which, by treatment with NaOH, can cyclize to form the N-amino oxazolidinones **5** in a one-pot process. The N-amino group in **5** could be cleaved with Zn/acetone to give the oxazolidinone **6** (Scheme 2.27).

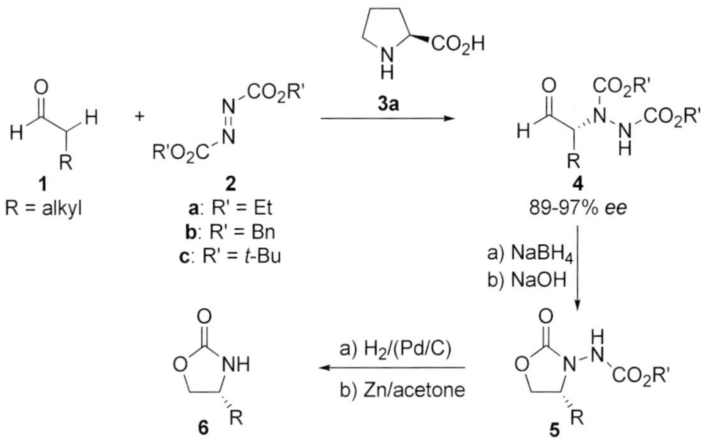

Scheme 2.27 Direct enantioselective α-amination of aldehydes catalyzed by L-proline, and further transformations to optically active oxazolidinones. (For experimental details see Chapter 14.8.1).

It has also been shown that the optically active α-hydrazino aldehydes **4** can be converted directly to the α-amino acid derivatives **7** in moderate yields by oxidation of the aldehyde functionality to the corresponding carboxylate with $KMnO_4$, followed by esterification and reductive hydrazine cleavage (Eq. 2) [4a].

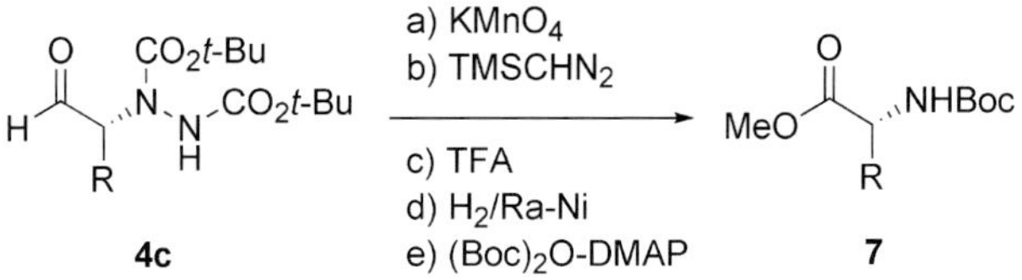

(For experimental details see Chapter 14.8.1). The L-proline-catalyzed direct amination of aldehydes has been extended to also include α,α-disubstituted aldehydes **1** using azodicarboxylates as the electrophilic nitrogen source (Scheme 2.28) [5]. The direct α-amination of α,α-disubstituted aldehydes **1** was especially well-suited for α-alkyl-α-aryl aldehydes, which gave the optically active products in moderate to good yields and up to 87% ee. The –N–N– bond in **5** was cleaved by hydrogenation using Pd/C in MeOH/AcOH, followed by treatment with $NaNO_2$ in AcOH/HCl.

The absolute configuration of the optically active hydrazino aldehydes **4** obtained in the organocatalytic direct α-amination of aldehydes can be changed to the formation of the S-enantiomer without changing the absolute configuration of the organocatalyst, due to a steric-shielding, rather than to an electronic-shielding mechanism (see Fig. 2.5). The application of (S)-2-[bis(3,5-bistrifluoromethylphenyl)trimethylsilanyloxymethyl]pyrrolidine **3b** as the organocatalyst for the α-amination of aldehydes **1** with diethyl azodicarboxylate **2a** takes place within

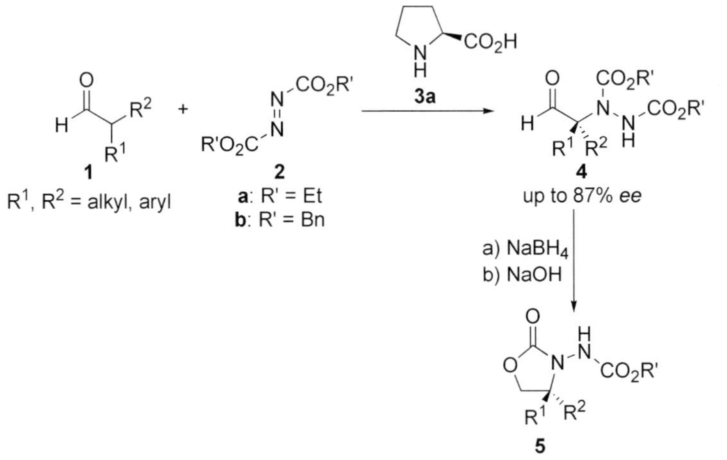

Scheme 2.28 The direct enantioselective α-amination of α,α-disubstituted aldehydes catalyzed by L-proline, and further transformations to optically active oxazolidinones.

minutes at room temperature to give **4** with *S*-configuration in good yields and with excellent enantioselectivities (90–97% *ee*) (Eq. 3) [6].

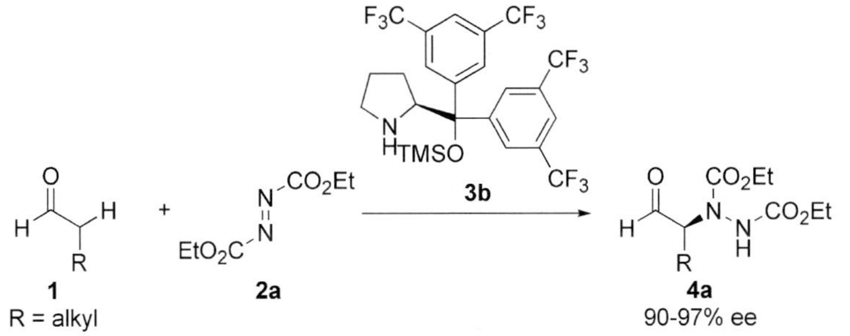

Barbas and colleagues have applied the organocatalytic direct amination of aldehydes in a series of reports [7]. By combining acetone, various aldehydes, dibenzyl azodicarboxylate and L-proline as the catalyst, a one-pot synthesis of functionalized β-amino alcohols was achieved [7a]. The scope of the reaction was found to be quite general for various aldehydes, and the optically active β-amino alcohols were obtained in high yields with low diastereoselective control. However, excellent enantioselectivity of especially the *anti*-adduct was obtained.

The same authors have shown that the direct α-amination reaction could also be used to construct the quaternary stereocenter in the enantioselective total synthesis of the cell-adhesion inhibitor BIRT-377 [7b]. The L-proline-derived tetrazole **3c** catalyzed the direct α-amination of 3-(4-bromophenyl)-2-methylpropanal **8** with dibenzyl azodicarboxylate **2b** to give the amino aldehyde **9** in 95% yield and with

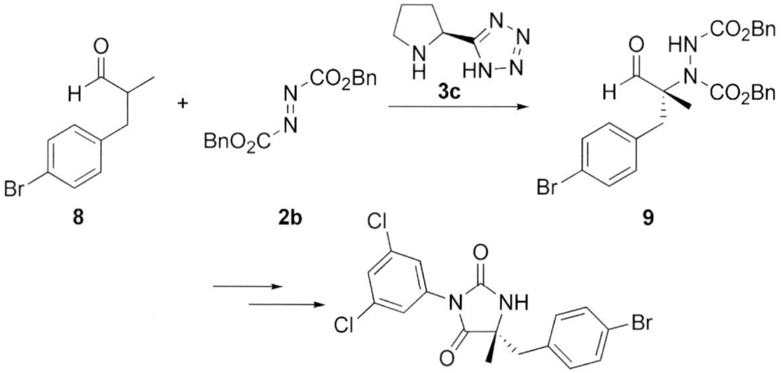

Scheme 2.29 The application of organocatalytic enantioselective α-amination reaction of 3-(4-bromophenyl)-2-methylpropanal **8** for the total synthesis of the optically active cell-adhesion inhibitor BIRT-377 [7b].

80% *ee*. From the optically active α-aminated aldehyde **9**, BIRT-377 was synthesized by standard transformations (Scheme 2.29).

Antagonists of metabotropic glutamate receptors and G-protein-coupled receptors associated with various neurodegenerate diseases have been prepared by proline-catalyzed direct α-amination reactions (Scheme 2.30) [7c]. Both, indane carbaldehyde **10** and analogous compounds having an ester functionality [leading to 1-aminoindan-1,5-dicarboxylic acid (AIDA)] and or a phosphonate substituent [(*RS*)-1-amino-5-phosphonoindan-1-carboxylic acid; APICA], all reacted with di-

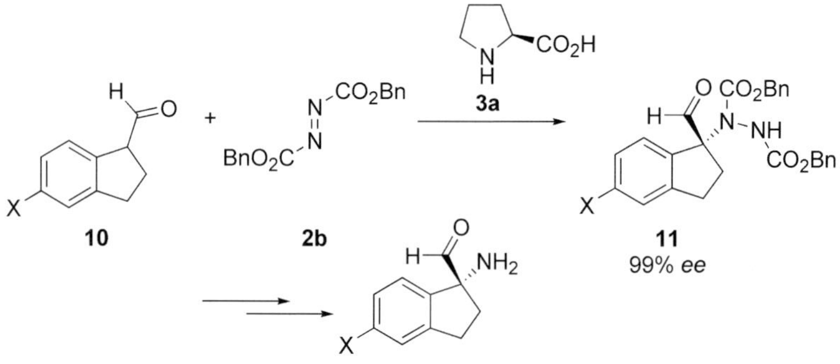

Scheme 2.30 Synthesis of 1-aminoindan-1,5-dicarboxylic acid (AIDA) and (*RS*)-1-amino-5-phosphonoindan-1-carboxylic acid (APICA) (known antagonists of metabotropic glutamate and G-protein-coupled receptors) using the L-proline-catalyzed enantioselective α-amination reaction.

benzyl azodicarboxylate **2b** in the presence of L-proline as the catalyst and products **11** having a quaternary stereocenter were obtained with excellent enantioselectivity (99% ee). It was observed that low yield in the cleavage of the –N–N– bond with the previously mentioned procedures was obtained. However, Barbas and colleagues [7c] developed an alternative route based on SmI$_2$ by first applying a one-pot trifluoroacetylation-selective benzyloxycarbonyl deprotection protocol giving the trifluorohydrazine, followed by –N–N– bond cleavage with SmI$_2$.

Ketones have also been the substrate for the direct α-amination of catalyzed by L-proline [8]. When the nitrogen source is diethyl azodicarboxylate **2a** and L-proline the catalyst, the amination reaction proceeded with excellent enantioselectivities – for acyclic ketones in the range of 94–98% ee, while cyclohexanone gave 84% ee. Furthermore, the reaction is highly regioselective and takes place at the more substituted carbon atom. The stereocenter formed in the α-amination of ketones was found to be less prone to racemization compared to the related aldehydes. We have demonstrated the synthetic utility of the optically active α-aminated ketones by the diastereoselective reduction of the carbonyl functionality with NaBH$_4$ to give the corresponding *syn*-α-amino alcohol, while the use of Et$_3$SiH-TiCl$_4$ provided the corresponding *anti*-α-amino alcohol [8].

In order to account for the absolute configuration of the α-aminated aldehydes and ketones using L-proline **3a** and (*S*)-2-[bis(3,5-bistrifluoro-methylphenyl)-trimethylsilanyloxymethyl]pyrrolidine **3b** (having identical absolute configuration) which leads to the formation of products with opposite absolute configuration at the α-carbon stereocenter, the two transition states presented in Figure 2.6 have been proposed [6].

In the case of proline-catalyzed α-amination of aldehydes, the generally accepted catalytic cycle presented in Scheme 2.25 does not seem detailed enough to explain some of the results obtained for this particular transformation by Blackmond and co-workers [9]. In fact, their studies revealed product acceleration, a positive nonlinear effect, and asymmetric amplification. These properties of the

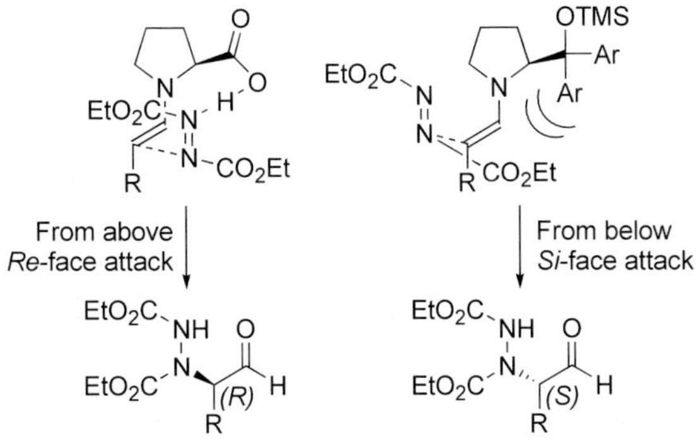

Fig. 2.6 Proposed transition-state models for L-proline **3a** (*S*)-2-[bis(3,5-bistrifluoro-methylphenyl)trimethylsilanyloxymethyl]pyrrolidine **3b**-catalyzed α-amination of aldehydes.

system were suggested to be the result of the involvement of the optical active products in a more complex catalytic cycle.

2.2.3
Direct α-Amination of α-Cyanoacetates and β-Dicarbonyl Compounds

The α-cyanoacetates **12** are optimal substrates for the approach outlined in Scheme 2.26 due to the low pK_a of the α-proton. It has been reported that the quinidine-derived alkaloid β-isocupridine (β-ICD) can catalyze the direct α-amination of α-cyanoacetates **12** (Eq. 4) and β-dicarbonyl compounds [10], probably by an enolate having a chiral β-ICD-H$^+$ counterion as the intermediate. The α-amination of α-cyanoacetates **12** with di-*tert*-butyl azodicarboxylate **2c** is an efficient process that proceeds with only 0.5 mol% of β-ICD. The expected products **13**, having a stereogenic quaternary carbon center, were isolated in excellent yields and with excellent levels of enantioselectivity independently by the nature of the aryl-substituent in the α-cyanoacetates, while the β-dicarbonyl compounds give slightly lower enantioselectivty (83–90% *ee*).

$$\text{Ar}\overset{\text{CN}}{\underset{\text{CO}_2\text{R}^1}{\bigg|}} + \overset{\text{Boc}}{\underset{\text{Boc}}{\text{N=N}}} \xrightarrow[\text{0.5 mol%}]{\text{[β-ICD]}} \text{Ar}\overset{\text{Pg}_{\diagdown}\text{N}^{\diagup}\text{NHPg}}{\underset{\text{CN}}{\bigg|\text{CO}_2\text{R}^1}} \quad (4)$$

12   **2c**   **(S)-13**
99% yield and 97-99% *ee*

Further developments by Deng et al. on the α-amination of α-cyanoacetates **12** (Eq. 4) showed that 6′-OH-modified cinchona alkaloids, which are accessible from either quinine or quinidine, were also effective catalysts for the reaction leading to optically active products in 71–99% yield and up to 99% *ee* [11].

2.2.4
Direct α-Oxygenation Reactions of Aldehydes and Ketones

The hydroxy group in the α-position to a carbonyl group is a common feature of many natural and biologically active compounds. Furthermore, this functionality is an obvious precursor in the synthesis of other important building blocks such as diols.

The first electrophilic source of oxygen introduced for the proline-catalyzed α-oxygenation of aldehydes and ketones was nitrosobenzene, based on the use of this reagent in the asymmetric metal-catalyzed oxidation of tin enolates [12]. A number of research groups, including those of Zhong [12a], MacMillan [13b], and Hayashi [13c], applied nitrosobenzene as the terminal oxidant reagent and independently reported the ability of L-proline to control both the O/N-selectivity, as well as the enantioselectivity in a variety of solvents and reaction conditions (Scheme 2.31).

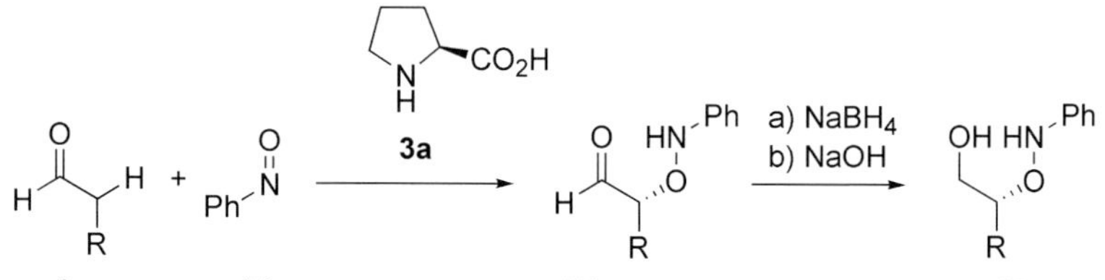

Scheme 2.31 Direct enantioselective α-oxidation of aldehydes using nitrosobenzene as the oxygen donor and L-proline as the catalyst.

It was found that the product **15** of the organocatalytic oxidation was relatively unstable, and was conveniently reduced *in situ* using NaBH$_4$. Nevertheless, other transformations can also be performed directly on the crude reaction mixture, thereby maintaining the high optical purity, as reported by MacMillan and colleagues [13b] and later by Zhong [14a] and the group of Ley [14b]. These reactions included the reductive amination and allylation of the aldehyde functionality, and the formation of compounds which are the formal [4+2]-cycloaddition products of nitrosobenzene with a diene.

Also for this transformation, Blackmond and co-workers [15] detected a rate acceleration and amplification of the enantiomeric excess.

The α-oxidation of aldehydes was later further extended to the use of ketones as nucleophiles. In order to develop this reaction into a useful process, a considerable effort was made to optimize the reaction conditions as several different problems arose; these included a lower reaction rate and yields because of the formation of the di-addition product at the two enolizable carbon atoms and lower O/N-selectivity. Hayashi et al. [16a] and Córdova et al. [16c] partly solved these problems by using a relatively large excess of ketone and by applying the slow addition method leading to good chemical yields (44–91%), with near-enantiopure products being obtained (96–99% ee).

The addition of nitrosobenzene to ketones catalyzed by proline has been applied by Ramachary and Barbas in the desymmetrization of *meso*-cyclohexanone derivatives [17].

Several research groups have tested a number of other secondary amines (**3c–f**) for this important transformation [18]. Derivative **3c** was found to be a very efficient catalyst for the α-oxidation of aldehydes and ketones, as the same high yields and enantioselectivities could be obtained in the case of both aldehydes and ketones, but with a lower catalyst loading and shorter reaction times.

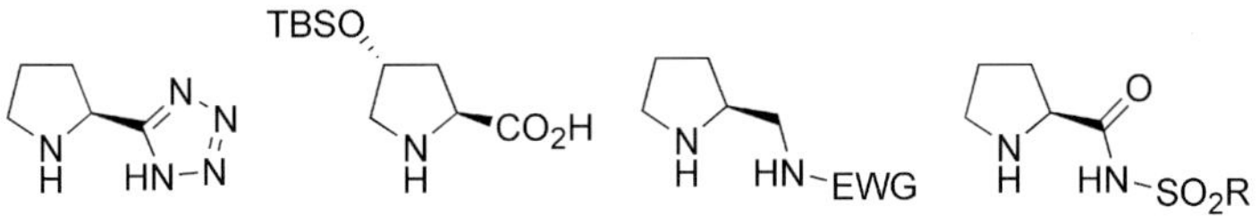

Fig. 2.7 Catalyst structures. The acidic proton in the group in the 2-position of the pyrrolidine ring is a common feature.

Catalysts **3a** and **3c–f** all have the acidic proton in the group in the 2-position of the pyrrolidine ring as the common feature (Fig. 2.7). It has been pointed out that the role of this functionality is not only to control the enantioselectivity as in other related proline-catalyzed reactions (see Fig. 2.5), but also to control the regioselectivity of the reaction. Both Córdova and colleagues [18d] and Cheong and Houk [19] have performed quantum mechanical computational studies in attempts to understand the mechanism and to explain the greater electrophilicity of the oxygen atom over the nitrogen atom of nitrosobenzene under these specific reaction conditions. The investigations by Cheong and Houk [19] showed that a higher energy for the N-anti transition state accounted for the excellent O/N-selectivity observed (Fig. 2.8). The preferential protonation of the nitrogen atom is a consequence of its higher basicity, and this fact leads to the electrophilic attack of the enamine at the oxygen atom.

One drawback for the direct organocatalytic α-oxidation of aldehydes and ketones is the use of nitrosobenzene, which is an "expensive oxygen source". This has led to further investigations in order be able to use other oxidants. Recently, Córdova et al. [20] reported that L-α-methyl proline could incorporate O_2 in the α-position of an aldehyde. The presence of tetraphenylporphyrin (TPP) as sensitizer was necessary to promote the formation of singlet O_2 as the electrophilic species. Although, the enantioselectivities obtained were only moderate (54–66% ee), this represents undoubtedly a very intriguing alternative to the use of nitrosobenzene in this type of reaction.

An interesting method for the α-oxidation of carbonyl compounds leading to optically active α-hydroxy carboxylic acid derivatives has recently been developed

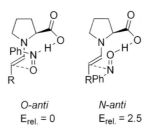

O-anti
$E_{rel.} = 0$

N-anti
$E_{rel.} = 2.5$

Fig. 2.8 Computational rationalization of the O/N-selectivity in L-proline-catalyzed α-oxidation of aldehydes by nitrosobenzene.

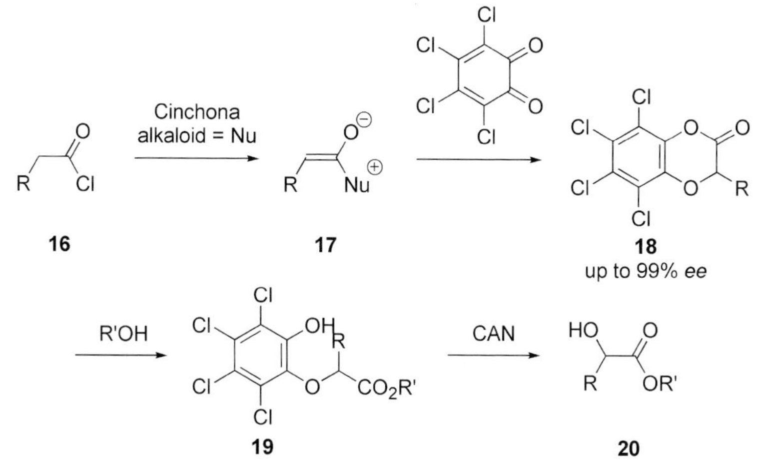

Scheme 2.32 Formation of optically active α-hydroxyesters by reaction of acid chlorides with benzoylquinidine acting a catalytic chiral nucleophile, followed by [4+2]-cycloaddition with o-chloranil.

by Lectka and co-workers, as outlined in Scheme 2.32 [21]. The method is based on the formation of chiral, catalytically derived zwitterionic "ketene" enolates **17** formed from acid chloride **16** and a cinchona alkaloid, benzoylquinidine. This intermediate undergoes a [4+2]-cycloaddition with o-chloranil to give **16** in good yields and up to 99% *ee*, which by methanolysis and ceric ammonium nitrate (CAN)-oxidation affords the optically active α-hydroxyesters **20**. In all the cases, the alcoholysis/oxidation sequence proceeds in high yield, under mild conditions, and with full preservation of the optically activity of compound **18**.

2.2.5
Direct α-Oxygenation Reactions of β-Ketoesters

The direct enantioselective α-hydroxylation of activated ketones [22], specifically cyclic β-dicarbonyl compounds, can be performed using dihydroquinine as the chiral catalyst and simple commercially available peroxides as the oxidant. The use of cumyl hydroperoxides led to the α-hydroxylation of β-ketoesters **21** in high yields and moderate to good enantioselectivities (66–80% *ee*) (Eq. 5). These optically active alcohols (**22**) undergo a diastereoselective reaction to *anti*-diols with excellent diastereoselectivity (99:1) using BH_3-4-ethylmorpholine as the reducing agent.

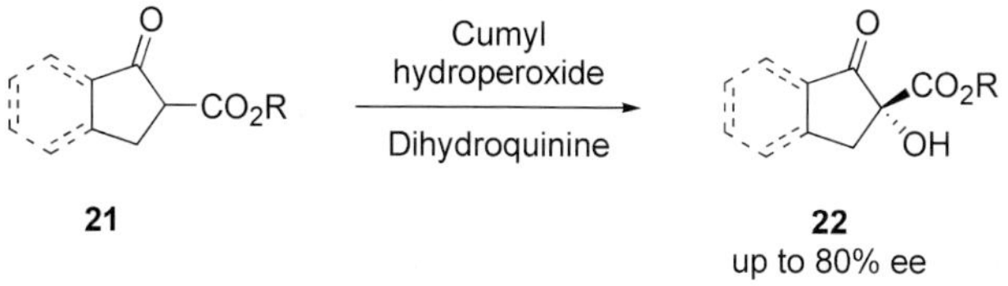

2.2.6
Direct α-Halogenation Reactions of Aldehydes and Ketones

Organic compounds having a stereogenic C–halogen (C–X) center are important in various fields of science, either for use in further manipulations or because the stereogenic C–X center has a unique property which is of specific importance for a given molecule. The involvement of these functional groups for further stereospecific manipulations, together with their increasing importance in medicinal chemistry and material sciences, have led to an increased search for catalytic asymmetric C–X bond-forming reactions [23].

Organocatalysis has led to the development of new methods for the asymmetric α-halogenation of carbonyl compounds leading to the formation of stereogenic C–X centers. Hence, details of direct enantioselective fluorination, chlorination, and bromination reactions will be presented in the following sections.

2.2.6.1 Direct α-Fluorination of Aldehydes

Various chiral amines can catalyze the direct enantioselective α-fluorination of aldehydes. Enders and Hüttl have focused on the use of Selectfluor for the α-fluorination of aldehydes and ketones [24a]. For the aldehydes, no enantiomeric excess was reported using L-proline as the catalyst. In an attempt to perform direct enantioselective α-fluorination of ketones, cyclohexanone was used as the model substrate and a number of chiral amines were tested for their enantioselective properties; however, the enantiomeric excess was rather low and in the range of 0 to 36% ee.

The other approaches to the direct enantioselective α-fluorination of aldehydes all used N-fluorobenzenesulfonimide (NFSI) **23** as the fluorinating reagent. When we applied (S)-2-[bis(3,5-bistrifluoro-methylphenyl)trimethyl-silanyloxymethyl]pyrrolidine **3b** as the catalyst (Scheme 2.33) [24b], a variety of aldehydes were α-fluorinated in a highly enantioselective manner (91–97% ee) in MTBE as the solvent, and the corresponding α-fluoro alcohols **25** were obtained in moderate to good yields after reduction with $NaBH_4$ [24c,d]. The α-fluorination reactions were based on the same catalytic concept, but using the chiral imidazolidinone **3g** (Scheme 2.33). Among investigations reported by the group of Barbas, a large number of catalysts were evaluated, and under catalytic conditions (30 mol%) up to 88% ee was obtained for linear aldehydes, though the conversion was rather low. The scope of the reaction was demonstrated for various linear and branched aldehydes, and an equimolar amount of catalyst was needed in order to obtain moderate to good yields of the optically active α-fluorinated compounds [24c]. The α-fluorination developed by the MacMillan group used a salt of the imidazolidinone compound (Scheme 2.33, **3g**) as catalyst, and the addition of 10% i-PrOH was found to significantly improve the properties of the catalytic system [24d]. The latter development led to a highly enantioselective α-fluorination of linear aldehydes in moderate to high yields, with enantioselectivities in the range of 91–99% ee.

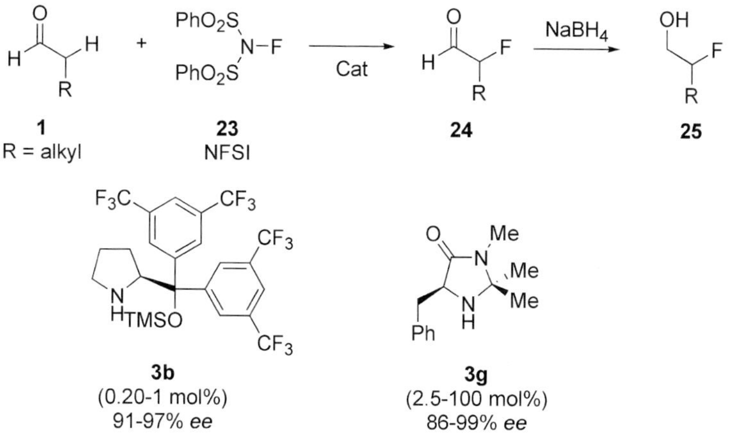

2.2 α-Heteroatom Functionalization

Scheme 2.33 Direct enantioselective α-fluorination of aldehydes catalyzed by (S)-2-[bis(3,5-bistrifluoro-methylphenyl)trimethyl-silanyloxymethyl]pyrrolidine **3b** and the salt of imidazolidinone compound **3g**.

2.2.6.2 Direct α-Fluorination of β-Ketoesters

The direct enantioselective organocatalytic α-fluorination can also be performed with cinchona alkaloid derivatives as catalyst under phase-transfer reaction conditions [25]. The fluorination reaction by NFSI of β-ketoesters **21**, readily enolizable substrates, generated a stereogenic quaternary C–F bond in high yields and with enantioselectivities up to 69% *ee* for the optically active products **26** (Eq. 6).

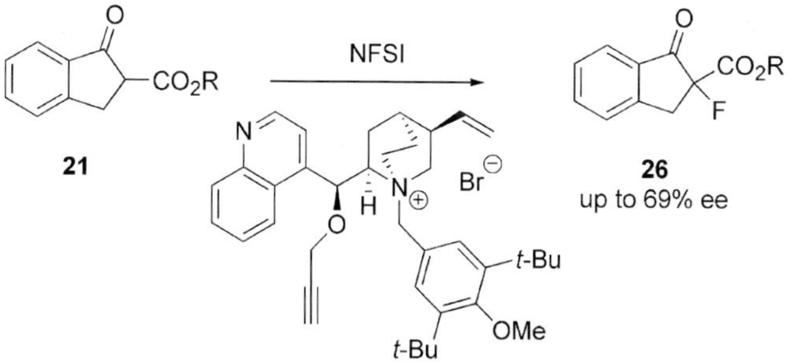

2.2.6.3 Direct α-Chlorination of Aldehydes and Ketones

The direct enantioselective α-chlorination of aldehydes leads to important intermediates **27** in organic synthesis (Scheme 2.34). This reaction was developed independently by two groups; MacMillan et al. applied the salt of the imidazolidinone **3g** as the catalyst, as L-proline was found to be a poor catalyst for this reaction (Scheme 2.34) [26a]. Various chlorinating reagents were tested, and the per-chlorinated quinone **28** was found to provide the best enantioselectivity. It was

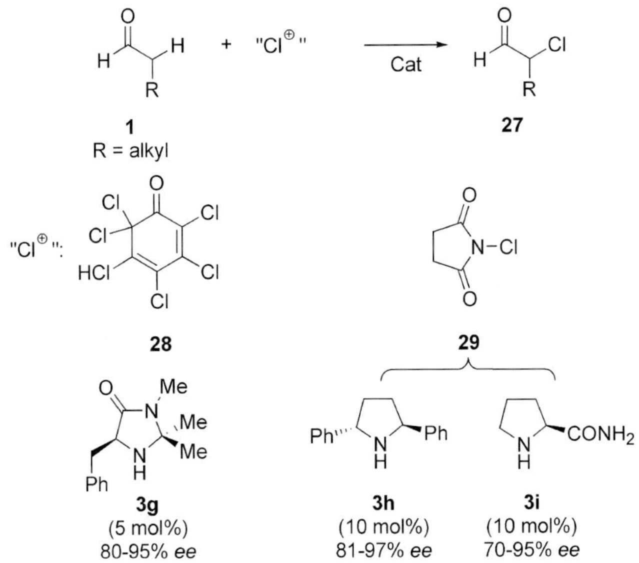

Scheme 2.34 Organocatalytic enantioselective α-chlorination of aldehydes. (For experimental details see Chapter 14.8.2).

also shown that a number of linear aldehydes could successfully be α-chlorinated in good to high yields, and with enantioselectivities in the range of 80–95% ee.

The other development for the direct enantioselective α-chlorination of aldehydes was reported by the present authors, and was seen to differ from the first approach in several ways. For example, the catalysts used for the α-chlorination reaction were the C_2-symmetric (2R,5R)-diphenylpyrrolidine **3h** and L-proline amide **3i**, while the electrophilic chlorinating reagent was N-chlorosuccinimide (NCS) **29** (Scheme 2.34) [26b]. Both chiral amines were effective catalysts for the α-chlorination reaction, and the optically active products **27** were obtained in high yields; the use of catalyst **3h** provided generally higher enantiomeric excess (81–97% ee) than **3i** (70–95% ee).

The synthetic application of the α-chloro aldehydes has been demonstrated by the preparation of a variety of important chiral building blocks (Scheme 2.35) [26b]. The α-chloro aldehydes could be reduced to the corresponding optically active α-chloro alcohols in more than 90% yield, maintaining the enantiomeric excess by using NaBH$_4$. It was also shown that optically active 2-aminobutanol – a key intermediate in the synthesis of the tuberculostatic, ethambutol – could be obtained in high yields by standard transformations from 2-chlorobutanol. Furthermore, the synthesis of an optically active terminal epoxide was demonstrated. The 2-chloro aldehydes could also be oxidized to α-chloro carboxylic acids in high yields without loss of optical purity, and further transformations were also presented.

The mechanism for the α-chlorination of aldehydes catalyzed by the C_2-symmetric (2R,5R)-diphenylpyrrolidine **3h** is probably different from that out-

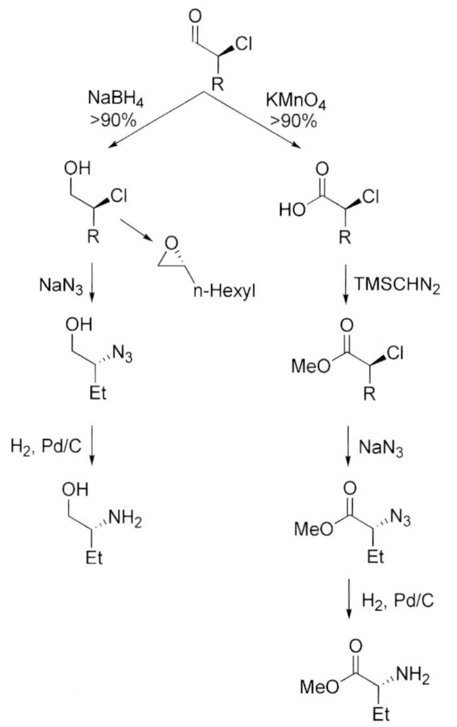

Scheme 2.35 Various transformations of optically active α-chloro aldehydes.

lined in Scheme 2.25, as computational studies showed no face-shielding for the density functional theory (DFT) – optimized chiral enamine intermediate [27]. Based on experimental and theoretical investigations, it was postulated that the reaction might proceed via an initial kinetically controlled N-chlorination of the enamine followed by a [1,3]-sigmatropic shift leading to the energetically favorable catalyst-iminium-C-chlorinated intermediate.

Ketones have also been enantioselectively α-chlorinated, although the catalysts used for the α-chlorination of aldehydes were found not to be optimal for the functionalization of these substrates. Poor yields and moderate enantiomeric excess were observed with, for example, (2R,5R)-diphenylpyrrolidine **3h** and L-proline amide **3i** due also to poly-chlorination of the starting material [28]. A catalyst screening revealed that the C_2-symmetric 4,5-diphenylimidazoline **3j** constitutes, in combination with 2-nitrobenzoic acid as a rate-accelerating additive, an efficient catalytic system for the α-chlorination of both cyclic and acyclic ketones **30** using NCS **29** as the chlorinating reagent (Eq. 7) [28]. The scope of the direct enantioselective α-chlorination showed that the optically active α-chloro ketones **31** were obtained in 86–98% ee, with the highest enantioselectivity for the cyclic ketones.

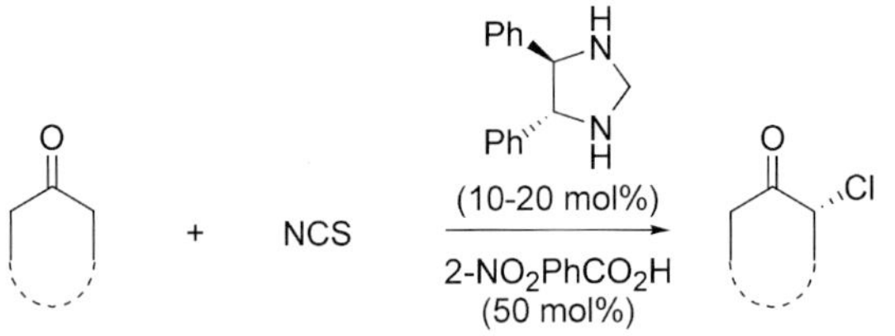

(7)

Various transformations of the optically active α-chloro ketones were also presented [28]: optically active 2-chloro cyclohexanone was oxidized by a Baeyer–Villiger oxidation to the corresponding lactone in 81% yield without reduction in enantiomeric excess, and the ketone functionality was selectively reduced to the corresponding syn-α-chloro alcohol with a high diastereomeric ratio.

2.2.6.4 Direct α-Chlorination of β-Ketoesters

Cinchona alkaloids have also been applied as catalysts for the direct α-chlorination of 1,3-dicarbonyl compounds **32** using trichloroquinolinone **33** as the chlorinating reagent for the direct approach (Eq. 8) [29]. A number of acyclic and cyclic β-dicarbonyl compounds could be α-chlorinated in moderate to good yield. The enantioselectivity was highly dependent on the substrate, with the cyclic β-ketoesters giving the highest enantioselectivity (90–96% ee), while the acyclic β-ketoesters and β-dicarbonyl compounds gave enantioselectivities in the range of 51 to 89% ee. The best catalyst was benzoylquinidine, and the reaction was further extended to include enantioselective bromination using tribromoquinolinone as the brominating reagent.

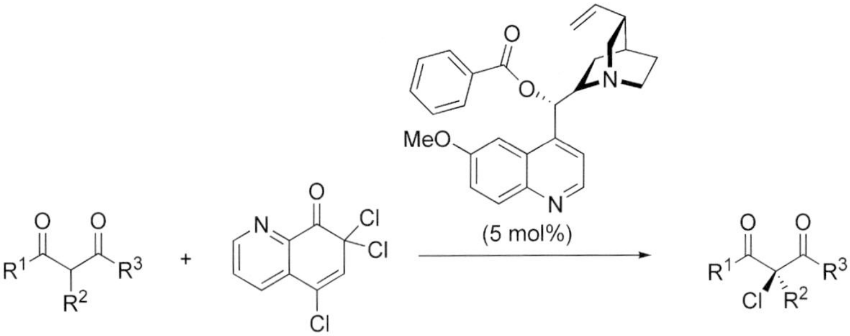

(8)

2.2.6.5 Direct α-Bromination of Aldehydes and Ketones

Aldehydes and ketones can also be directly α-brominated using the catalytic concepts presented in Scheme 2.34 and Eq. (7) [30]. The easily synthesized and air-stable 4,4-dibromo-2,6-di-tert-butyl-cyclohexa-2,5-dienone was found to be the best reagent for the functionalization of aldehydes (enantioselectivities in the range of 68–96% ee) and for the preparation of the optically active αbromoketones (73–94% ee). (S)-2-[Bis(3,5-bistrifluoro-methylphenyl)trimethyl-silanyloxymethyl]pyrrolidine **3b** was also found to be an efficient catalyst for the α-bromination of aldehydes using 4,4-dibromo-2,6-di-*tert*-butyl-cyclohexa-2,5-dienone as the bromination reagent [6]. Various aldehydes were α-brominated in good yield and in 94–95% ee.

2.2.7
Direct α-Sulfenylation of Aldehydes

Optically active compounds having a free thiol functionality are interesting molecules because they can serve as inhibitors of zinc-containing enzymes [31].

The direct catalytic approach to these synthetic targets was presented recently by Wang et al. using catalyst **3e**, which promoted the racemic sulfenylation of aldehydes and ketones using commercially available electrophilic sulfur sources [32]. Almost contemporarily, we presented the first enantioselective version of this transformation by using 1-benzylsulfanyl-1,2,4-triazole **35** as the sulfur source and (S)-2-[bis(3,5-bistrifluoro-methylphenyl)trimethyl-silanyloxymethyl]-pyrrolidine **3b** as the catalyst [33]. The application of these reagents for the direct α-sulfenylation of aldehydes **1** resulted in a highly enantioselective process (Scheme 2.36), and the absolute configuration of the products **36** was in agreement with the model discussed previously for the steric-shielding intermediate (see Fig. 2.5). The products of the reaction of **36** could be quantitatively transformed into compounds **37** by reduction with NaBH$_4$. The optically active amino thiol precursor **38** was instead isolated after reductive amination.

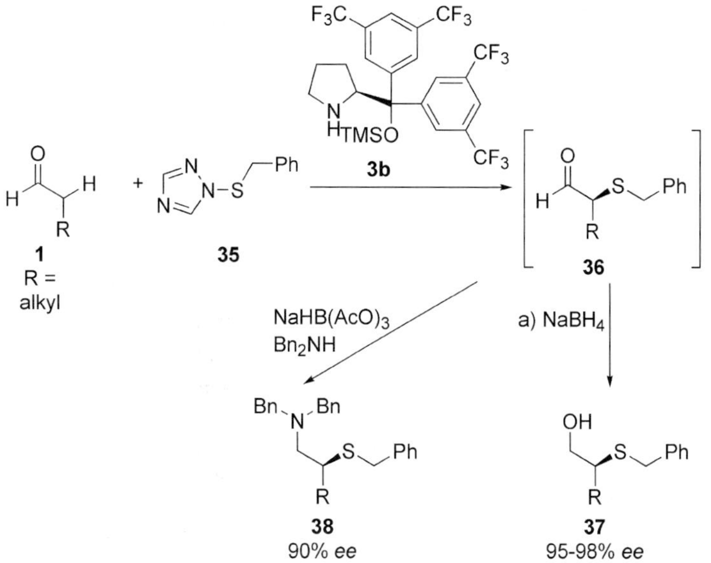

Scheme 2.36 Organocatalytic enantioselective α-sulfenylation of aldehydes.

The use of **36** was justified by the deprotection of the benzyl group that led to the free thiol functionality using Na/NH$_3$(l).

It should be noted that the more sterically demanding α,α-disubstituted aldehydes could be efficiently sulfenylated when the same catalyst was used, even if with lower enantiomeric excess (61% ee).

2.2.8
Direct α-Sulfenylation of Lactones, Lactams, and β-Diketones

Compound **35**, along with some other structurally related sulfenylating agents, has also been found to be useful for the functionalization of 1,3-dicarbonyl compounds **38** (Eq. 9). Lactones and β-dicarbonyl compounds were α-sulfenylated in the presence of (DHQD)$_2$PYR in high yields and moderate to good enantioselectivities (51–91% ee) [34].

$$\tag{9}$$

2.2.9
Direct α-Selenation of Aldehydes and Ketones

Wang and co-workers recently developed an efficient strategy for the α-selenation of aldehydes and ketones using the proline derivatives as catalysts; under mild conditions, the products were obtained in good yield (Eq. 10) [35].

$$\tag{10}$$

Even if chiral organocatalysts were used, the authors did not focus on the development of an asymmetric version of the reaction.

Acknowledgments

These studies were made possible by a grant from The Danish National Research Foundation. M.M. thanks EU: HMPT-CT-2001-00317 for financial support.

References

1. For recent reviews on organocatalysis including, for example, C–C and C–heteroatom bond formation, see: (a) P.L. Dalko, L. Moisan, *Angew. Chem. Int. Ed.* **2004**, *43*, 5138; (b) A. Berkessel, H. Gröger, *Asymmetric Organocatalysis*. VCH, Weinheim, Germany, **2004**; (c) J. Seayed, B. List, *Org. Biomol. Chem.* **2005**, *3*, 719; (d) E.R. Jarvo, J.M. Scott, *Tetrahedron* **2002**, *58*, 2481; (e) B. List, *Tetrahedron* **2002**, *58*, 5573; (f) B. List, *Synlett* **2001**, *11*, 1675; (g) See also *Acc. Chem. Res.* **2004**, *37*, number 8, special edition devoted to asymmetric organocatalysis; (e) M. Marigo, K.A. Jørgensen, *Chem. Commun.* **2006**, 2001.

2. See, for example: (a) C. Allemann, R. Gordillo, C.R. Clemente, P.H. Cheong, K.N. Houk, *Acc. Chem. Res.* **2004**, *37*, 558; (b) P.H.-Y. Cheong, K.N. Houk, *J. Am. Chem. Soc.* **2004**, *126*, 13912; (c) S. Bahmanyar, K.N. Houk, *Org. Lett.* **2003**, *5*, 1249; (d) S. Bahmanyar, K.N. Houk, *J. Am. Chem. Soc.* **2001**, *123*, 11273.

3. For recent accounts on amination reactions, see: (a) J.M. Janey, *Angew. Chem. Int. Ed.* **2005**, *44*, 4292; (b) R.O. Duthaler, *Angew. Chem. Int. Ed.* **2003**, *42*, 975.

4. (a) A. Bøgevig, K. Juhl, N. Kumaragurubaran, W. Zhuang, K.A. Jørgensen, *Angew. Chem. Int. Ed.* **2002**, *41*, 1790; (b) B. List, *J. Am. Chem. Soc.* **2002**, *124*, 5656.

5. H. Vogt, S. Vanderheiden, S. Brase, *Chem. Commun.* **2003**, 2448.

6. J. Franzén, M. Marigo, D. Fielenbach, T.C. Wabnitz, A. Kjærsgaard, K.A. Jørgensen, *J. Am. Chem. Soc.* **2005**, *127*, 18296.

7. (a) N.S. Chowdari, D.B. Ramachary, C.F. Barbas, III, *Org. Lett.* **2003**, *5*, 1685; (b) N.S. Chowdari, C.F. Barbas, III, *Org. Lett.* **2005**, *7*, 867; (c) J.T. Suri, D.D. Steiner, C.F. Barbas, III, *Org. Lett.* **2005**, *7*, 3885.

8. N. Kumaragurubaran, K. Juhl, W. Zhuang, A. Bøgevig, K.A. Jørgensen, *J. Am. Chem. Soc.* **2002**, *124*, 6254.

9. (a) H. Iwamura, S.P. Mathew, D.G. Blackmond, *J. Am. Chem. Soc.* **2005**, *127*, 11770; (b) H. Iwamura, D.H. Wells, S.P. Mathew, M. Klussmann, A. Armstrong, D.G. Blackmond, *J. Am. Chem. Soc.* **2005**, *127*, 16312.

10. S. Saaby, M. Bella, K.A. Jørgensen, *J. Am. Chem. Soc.* **2005**, *127*, 8120.

11. X. Li, H. Li, L. Deng, *Org. Lett.* **2005**, *7*, 167.

12. N. Momiyama, H. Yamamoto, *J. Am. Chem. Soc.* **2003**, *125*, 6038; see also: H. Yamamoto, N. Momiyama, *Chem. Commun.* **2005**, 3514.

13. (a) G. Zhong, *Angew. Chem. Int. Ed.* **2003**, *42*, 4247; (b) S.P. Brown, M.P. Brochu, C.J. Sinz, D.W.C. MacMillan, *J. Am. Chem. Soc.* **2003**, *125*, 10808; (c) Y. Hayashi, J. Yamaguchi, K. Hibino, M. Shoji, *Tetrahedron Lett.* **2003**, *44*, 8293.

14. (a) G. Zhong, *Chem. Commun.* **2004**, 606; (b) S. Kwumarn, D.M. Shaw, D.A. Longbottom, S.V. Ley, *Org. Lett.* **2005**, *7*, 4189.

15 S.P. Mathew, H. Iwamura, D.G. Blackmond, *Angew. Chem. Int. Ed.* **2004**, *43*, 3317.

16 (a) Y. Hayashi, J. Yamaguchi, T. Sumiya, M. Shoji, *Angew. Chem. Int. Ed.* **2004**, *43*, 1112; (b) Y. Hayashi, J. Yamaguchi, T. Sumiya, K. Hibino, M. Shoji, *J. Org. Chem.* **2004**, *69*, 5966; (c) A. Bøgevig, H. Sunden, A. Córdova, *Angew. Chem. Int. Ed.* **2004**, *43*, 1109; (d) A. Córdova, H. Sunden, A. Bøgevig, M. Johansson, F. Himo, *Chem. Eur. J.* **2005**, *11*, 3673.

17 D.B. Ramachary, C.F. Barbas, III, *Org. Lett.* **2005**, *7*, 1577.

18 (a) N. Momiyama, H. Torii, S. Saito, H. Yamamoto, *Proc. Natl. Acad. Sci. USA* **2004**, 5374; (b) Y. Hayashi, J. Yamaguchi, K. Hibino, T. Sumiya, T. Urushima, M. Shoji, D. Hashizume, H. Koshino, *Adv. Synth. Catal.* **2004**, *346*, 1435; (c) W. Wang, J. Wang, L. Liao, *Tetrahedron Lett.* **2004**, *45*, 7235; (d) H. Sunden, N. Dahlin, I. Ibrahem, H. Adolfsson, A. Córdova, *Tetrahedron Lett.* **2005**, *46*, 3385.

19 P.H.-Y. Cheong, K.N. Houk, *J. Am. Chem. Soc.* **2004**, *126*, 13912.

20 (a) A. Córdova, H. Sunden, M. Enqvist, I. Ibrahem, J. Casas, *J. Am. Chem. Soc.* **2005**, *127*, 8914; (b) H. Sunden, M. Enqvist, J. Casas, I. Ibrahem, A. Córdova, *Angew. Chem. Int. Ed.* **2004**, *43*, 6532.

21 T. Bekele, M.H. Shah, J. Wolfer, C.J. Abraham, A. Weatherwax, T. Lectka, *J. Am. Chem. Soc.* **2006**, *128*, 1810.

22 M.R. Acocella, O.G. Mancheno, M. Bella, K.A. Jørgensen, *J. Org. Chem.* **2004**, *69*, 8165 and references therein.

23 See, for example: (a) H. Ibrahim, A. Togni, *Chem. Commun.* **2004**, 1147; (b) M. Oestreich, *Angew. Chem. Int. Ed.* **2005**, *44*, 2324.

24 (a) D. Enders, M.R.M. Hüttl, *Synlett*, **2005**, 991; (b) M. Marigo, T.C. Wabnitz, D. Fielenbach, A. Braunton, A. Kjærsgaard, K.A. Jørgensen, *Angew. Chem. Int. Ed.* **2005**, *44*, 3703; (c) D.D. Steiner, N. Mase, C.F. Barbas, III, *Angew. Chem. Int. Ed.* **2005**, *44*, 3706; (d) T.D. Beeson, D.W.C. MacMillan, *J. Am. Chem. Soc.* **2005**, *127*, 8826.

25 D.Y. Kim, E.J. Park, *Org. Lett.* **2002**, *4*, 545.

26 (a) M.P. Brochu, S.P. Brown, D.W.C. MacMillan, *J. Am. Chem. Soc.* **2004**, *126*, 4108; (b) M. Marigo, A. Braunton, S. Bachmann, M. Marigo, K.A. Jørgensen, *J. Am. Chem. Soc.* **2004**, *126*, 4790.

27 N. Halland, M.A. Lie, A. Kjærsgaard, M. Marigo, B. Schiøtt, K.A. Jørgensen, *Chem. Eur. J.* **2005**, *11*, 7083.

28 M. Marigo, S. Bachmann, N. Halland, A. Braunton, K.A. Jørgensen, *Angew. Chem. Int. Ed.* **2004**, *43*, 5507.

29 G. Bartoli, M. Bosco, A. Carlone, M. Locatelli, P. Melchiorre, L. Sambri, *Angew. Chem. Int. Ed.* **2005**, *44*, 6219.

30 S. Bertelsen, N. Halland, S. Bachmann, M. Marigo, A. Braunton, K.A. Jørgensen, *Chem. Commun.* **2005**, 4821.

31 (a) W. Yuan, B. Munoz, C. Wong, *J. Med. Chem.* **1993**, *36*, 211; (b) E.M. Gordon, J.D. Godfrey, N.G. Delaney, M.M. Asad, D. Von Lagen, D.W. Cushman, *J. Med. Chem.* **1988**, *31*, 2199.

32 W. Wang, H. Li, L. Liao, *Tetrahedron Lett.* **2004**, *45*, 8229.

33 M. Marigo, T.C. Wabnitz, D. Fielenbach, K.A. Jørgensen, *Angew. Chem. Int. Ed.* **2005**, *44*, 794.

34 S. Sobhani, D. Fielenbach, M. Marigo, T.C. Wabnitz, K.A. Jørgensen, *Chem. Eur. J.* **2005**, *11*, 5689.

35 (a) W. Wang, J. Wang, H. Li, *Org. Lett.* **2004**, *6*, 2817; (b) J. Wang, H. Li, Y. Mei, B. Lou, D. Xu, D. Xie, H. Gou, W. Wang, *J. Org. Chem.* **2005**, *70*, 5678.

2.3
Direct Conjugate Additions via Enamine Activation

Cyril Bressy and Peter I. Dalko

2.3.1
Introduction

Small chiral organic molecules may catalyze the asymmetric addition of ketones, and aldehydes to electron-deficient olefins, such as vinylidene acetones, nitroolefins, enones, and vinyl sulfones. In this chapter we will describe the inter- and intramolecular reactions in which activation of the carbonyl compound takes place via enamine formation.

Although asymmetric conjugate additions have over the years been dominated by the application of chiral Lewis acids as catalysts [1, 2], more recently organocatalysts have been added as efficient tools [3, 4].

Chiral organic molecules may promote conjugate additions in numerous ways (see also Chapters 3, 4 and 6). An important part of this chemistry is related to aminocatalytic reactions [5], that is, when activation takes place either via iminium intermediates (see Chapter 3) or, by activating the donors by forming enamines (this Chapter).

2.3.2
Factors Determining the Stereoselectivity of the Organocatalytic Conjugate Additions

The partial steps of the conjugate addition in aminocatalytic reactions are in dynamic equilibrium, and thus products are formed under thermodynamic control. This fact is translated also in the geometry of the enamine intermediates, leading to the product, which can be either *E* or *Z* (Fig. 2.9). The geometry of the enamine depends on the catalyst structure and also on the substrate. Whilst proline-catalyzed reactions form preferentially, with α-alkyl substituted ketones, the *E*-isomer, enamines derived from pipecolic acid afford an approximate 1:1 mixture of the *E* and *Z* isomers [6]. In turn, small- and medium-sized cyclic ketones afford the *E* isomer.

Moreover, the relative stability and the reactivity of the rotamers such as **A** and **B** (Fig. 2.10) have a determinant role in the facial selectivity of the addition.

Enantioselective Organocatalysis: Reactions and Experimental Procedures
Edited by Peter I. Dalko
Copyright © 2007 WILEY-VCH Verlag GmbH & Co. KGaA, Weinheim
ISBN: 978-3-527-31522-2

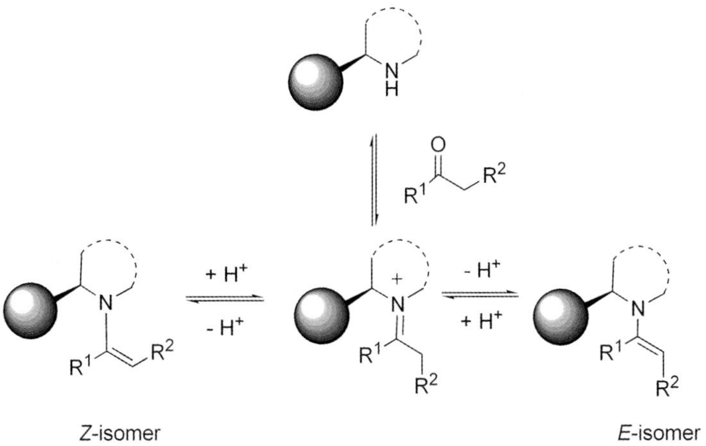

Fig. 2.9 Interconversion of the Z/E geometry of the enamines.

Finally, the acceptor (activated olefin) may approach the enamine by two different ways, as depicted in Figure 2.11. Stabilizing interactions such as hydrogen bonding, or electrostatic interactions between the catalyst and the olefin, may offset repulsive (steric) interactions preferring transition state **I**. It is considered, that the acceptor approaches the enamine via an acyclic synclinal transition state, as suggested by Seebach and Golinski [7].

The analysis of these simplified conditions reveals that, in conjugate additions, there is a number of viable alternative reaction paths leading to different transition states, and rendering possible the formation of a number regio- and stereo-isomers. Until very recently it was considered that organocatalytic reactions were unselective, with limited scope, but this picture is rapidly changing and this approach to chemistry is beginning to provide the first important results.

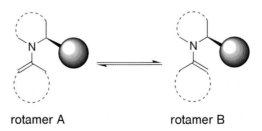

Fig. 2.10 Enamine-rotamers as factors of stereocontrol.

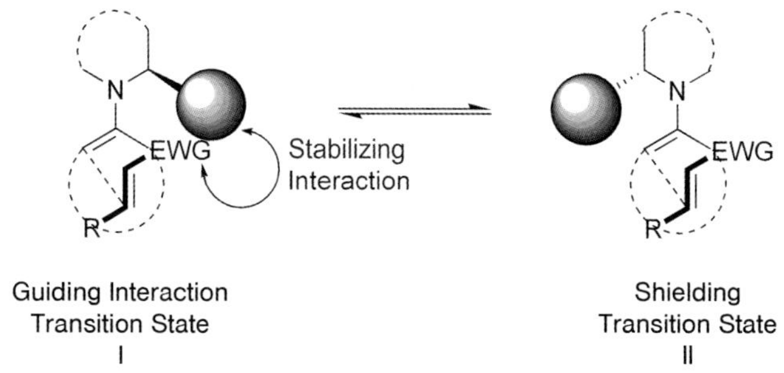

Figure showing two transition state structures with stabilizing interaction arrows between them.

Guiding Interaction Transition State
I

Shielding Transition State
II

Fig. 2.11 The guiding and shielding transition states.

2.3.3
Addition of Ketones to Nitroolefins and Alkylidene Malonates

2.3.3.1 Proline

Proline was among the first compounds to be tested in asymmetric conjugated reactions, both as a chiral ligand [8] and also as an organic catalyst [3]. The earliest asymmetric intermolecular Michael-type addition, in which proline catalyzed the reaction (arguably via enamine formation) was reported by Barbas and colleagues [9, 10] and by List and co-workers [11]. The reaction, which proceeded in high chemical yield (85–97%) and diastereoselectivity, albeit afforded near-racemic products in dimethyl sulfoxide (DMSO) [11] (Scheme 2.37). The enantioselectivity of the addition was later ameliorated by Enders, who demonstrated that a small amount of methanol rather than DMSO was beneficial to the enantioselectivity of the addition reaction [12].

Modification of the proline structure leads to more efficient chiral pyrrolidine catalysts, showing better selectivity and improved synthetic scope.

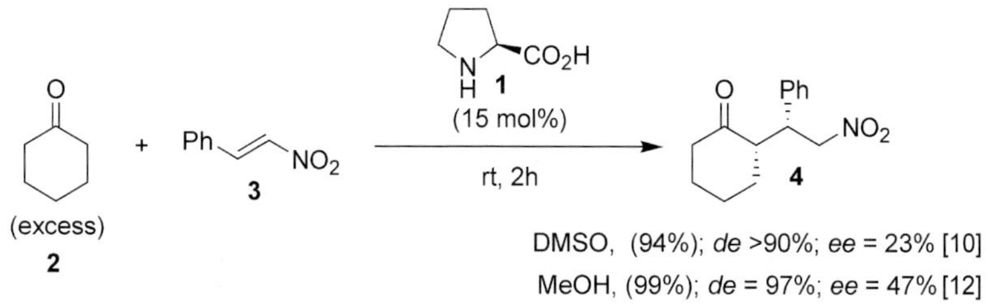

Scheme showing cyclohexanone (2, excess) + PhCH=CHNO$_2$ (3) with proline catalyst 1 (15 mol%), rt, 2h, giving product 4.

DMSO, (94%); de >90%; ee = 23% [10]
MeOH, (99%); de = 97%; ee = 47% [12]

Scheme 2.37 The (S)-proline-catalyzed addition of cyclohexanone to nitrostyrene 3.

2.3.3.2 Pyrrolidine Amines and Pyrrolidine Amine Salts as Catalysts for Michael-Type Addition of Ketones to Activated Olefins

A handful of catalysts having an N-containing side chain or heterocycle were developed (some are depicted in Scheme 2.38), whereupon either the free amine or the corresponding salts were shown to promote the highly *syn*-selective addition

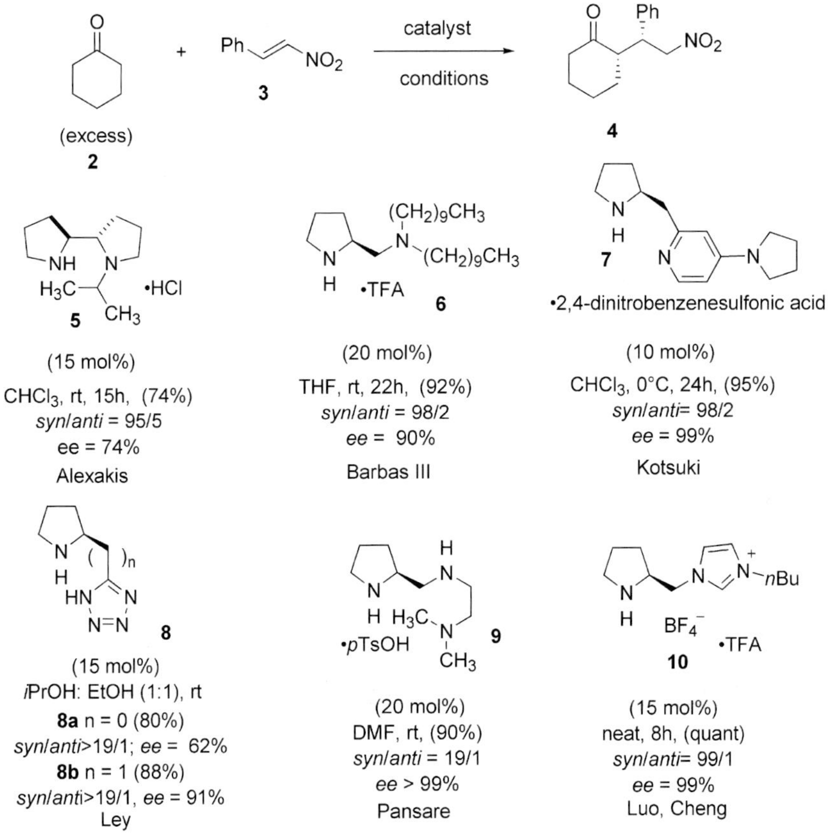

Scheme 2.38 Chiral pyrrolidine-derived catalysts for Michael-type addition of ketones to activated olefins.

of α-alkyl-substituted cyclic and acyclic ketones to nitrostyrenes (Schemes 2.38 and 2.39) [13–19]. (For experimental details see Chapter 14.9.2). The role of the acide co-catalyst is to increase ligand instability, and this results in an overall rate acceleration and increased conversion. The match between the catalyst and co-catalyst is the determinant factor both for the selectivity and for the rate of transformation. Whilst further functionalization of the pyrrolidine core may increase the selectivity of the alkylation, it considerably decreases the reactivity. Whilst the selectivity profile of the catalysts has been considerably improved during the past few years, the selectivity of the addition has remained highly substrate-dependent. Among ketone reagents, the best selectivity was notoriously obtained with cyclohexanone [enantiomeric excess (*ee*)-values up to 99%; diasteromeric ratio (*dr*) = 99:1], while a considerably lower enantioselectivity can be achieved with cyclopentanone, or with acyclic ketones. Two factors relating to the catalyst structure – namely, the position of the heteroatom or heterocycle on the side chain – are important in terms of the overall selectivity of the transformation. For example, **8b** affords better diastereoselectivity and enantioselectivity compared to **8a** (Scheme 2.43). Such variation can be attributed to the difference in rotational freedom generating increased steric volume of the tetrazole side chain and allowing a more efficient shielding of one of the reacting faces.

Whilst in some cases near-stoichiometric amounts of reagents can be used [16], the excess (5–10 equiv.) of ketone reagent is preconceived in order to ensure convenient kinetics and conversion. Reactions can be run typically at room temperature [20] in a polar aprotic solvent such as $CHCl_3$, or in THF or i-PrOH. The presence of water was noted to be beneficial in some cases [21]. In ionic liquids, such as in [bmim]BF_4, a low (5 mol%) catalyst concentration can be applied, while the enantioselectivity of the alkylation is modest in this solvent [22]. The ionic liquid-derived hybrid catalyst **10**, used neat with a small amount of trifluoroacetic acid (TFA) as co-catalyst, affords quantitatively **4**, though in a remarkably high dr ($syn/anti = 99/1$) and ee (99%). It should be noted that this catalyst can easily be recovered by extraction, and re-used without loss of activity.

The alkylation of asymmetric acyclic ketones takes place regioselectively on the most-substituted carbon, thus affording the *syn* isomers as major products. α-Hydroxyketones showed *anti* selective additions similar to that observed in related aldol, and Mannich-type additions (Scheme 2.39). Such selectivity is due to the preferred formation of the Z-enamine intermediate, stabilized by intramolecular hydrogen bonding between the hydroxy group and the tertiary amine of the catalyst [23].

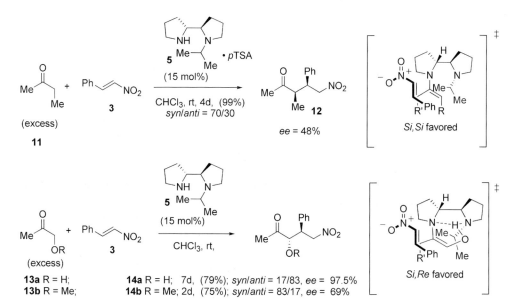

Scheme 2.39 The variation of *syn:anti* diastereoselectivity of the conjugate addition in the function of the substrate.

Pyrrolidine-amine-based catalysts such as **18** and **19** can also mediate the addition of ketones to alkylidene malonates (Scheme 2.40) [14]. Aldehydes (e.g., isovaleraldehyde) do not react with vinylidenemalonate **16** under these conditions.

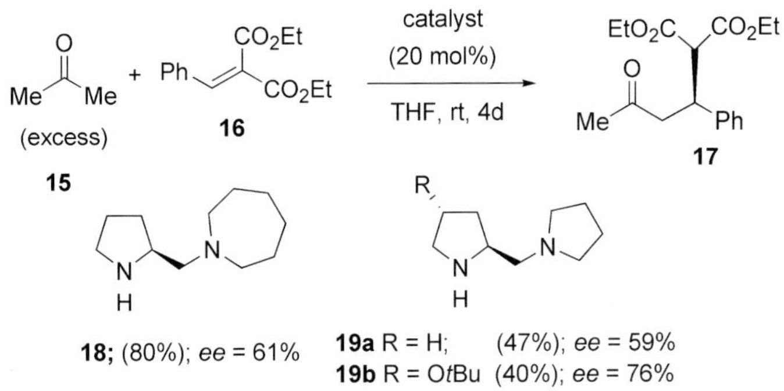

Scheme 2.40 The catalytic direct addition of acetone to alkylidene malonate **16**.

2.3.3.3 Chiral Primary Amines

Alanine **20** and alanine-containing small oligopeptides showed good stereocontrol in the addition of ketones to nitroolefins (Scheme 2.41) [24]. The L-ala-L-ala dipeptide **21** was more selective than the monomer **20**, while the hybrid derivative **22**, mediated the addition, with an *ee*-value of 93% [25].

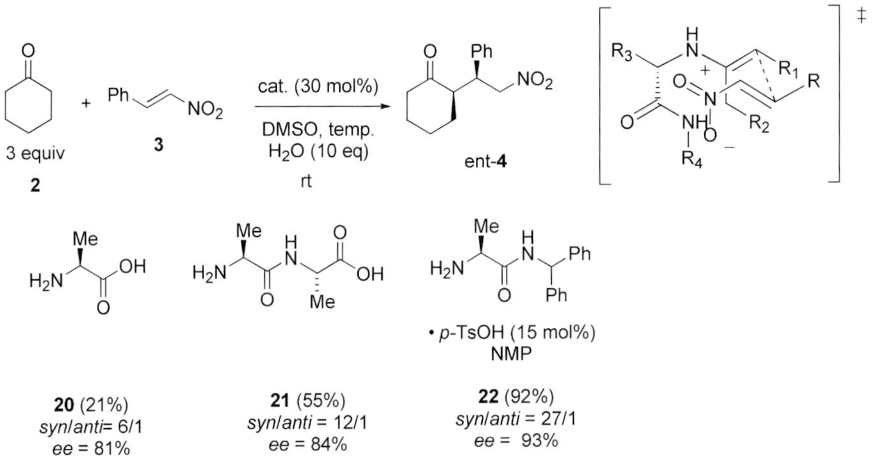

Scheme 2.41 The addition of cyclohexanone to nitrostyrene, mediated by Cordova's catalyst.

2.3.3.4 Amine/Thiourea Catalysts

The groups of Tsogoeva, Tang and Jacobsen each pioneered the development of the bifunctional catalyst, combining the nucleophilic amine and thiourea

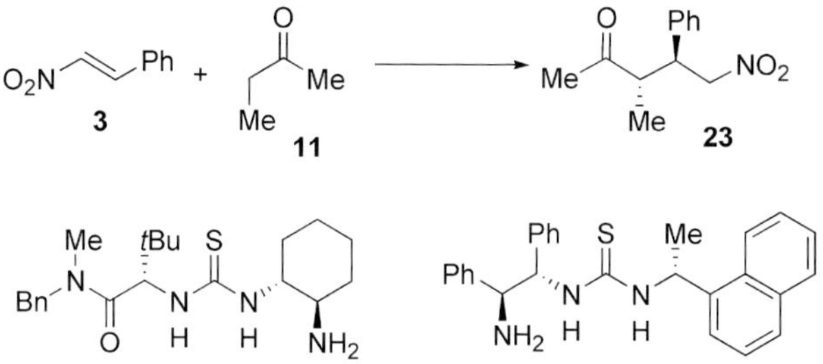

2.3 Direct Conjugate Additions via Enamine Activation | 83

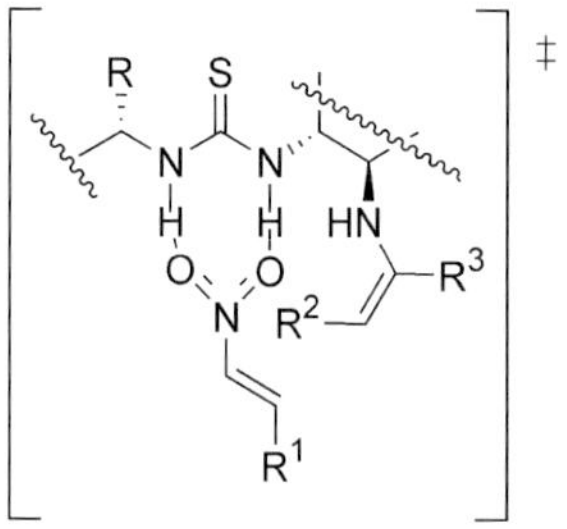

Scheme 2.42 The thiourea/amine catalyst-mediated addition of acyclic ketones to nitrostyrenes.

Brønsted acid (Scheme 2.42) [26–28]. (For experimental details see Chapter 14.9.4). These catalysts mediate the addition of ketones to nitroalkenes at room temperature in the presence of a weak acid co-catalyst, such as benzoic acid or n-butyric acid or acetic acid. The acid additive allows double alkylation to be avoided, and also increases the reaction kinetic. The Jacobsen catalyst **24** showed better enantio- and diastereoselectivity with higher n-alkyl-ethyl ketones or with branched substrates (ee = 86–99%; dr = 6/1 to 15/1), and forms preferentially the *anti* isomer (Scheme 2.42). The selectivity is the consequence of the preferred Z-enamine formation in the transition state; the catalyst also activates the acceptor, and orientates in the space. The regioselectively of the alkylation of nonsymmetric ketones is the consequence of this orientation. Whilst with small substrates the regioselectivity of the alkylation follows similar patterns (as described in the preceding section), leading to products of thermodynamic control, this selectivity can also be biased by steric factors.

Catalysts similar to **24** were also developed by the Tsogoeva group (Scheme 2.42) [28], with the best results being obtained with chiral naphthylamine-based thiourea catalysts having free NH_2 functions in the side chains (e.g., **25**). This led to *syn* products with cyclic ketones (Scheme 2.43), whereas acyclic ketones afforded the *anti* product (Scheme 2.42).

The Tang group combined the chiral pyrrolidine core with a thiourea function (Scheme 2.43) [29]. Optimal reaction conditions were obtained with **26** under solvent-free conditions and in the presence of n-butyric acid as additive (10 mol%) at 0 °C (Scheme 2.43). The high selectivity of the addition was attributed to the formation of a rigid three-dimensional H-bonded structure in the transition state, in which the enamine was positioned at the correct distance compared to the nitroolefin and allowed an addition from the *Re*-face of the enamine.

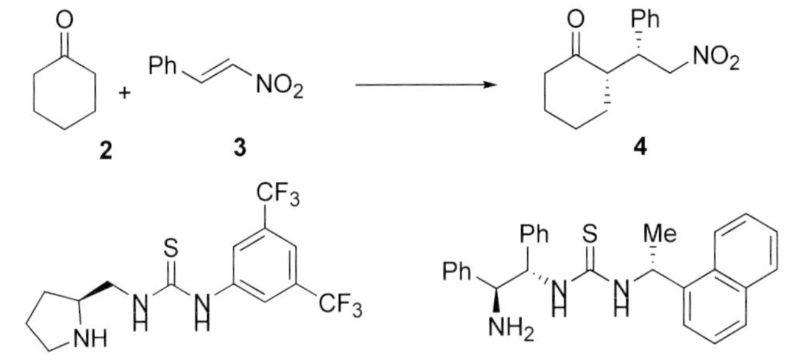

25 (15 mol%)
toluene, rt, (82%)
AcOH (15 mol%), H_2O (2 equiv)
ee = 96%; *syn*/*anti* = 80/20
Tsogoeva

26 (20 mol%)
n-butyric acid (10 mol%)
neat, 0° C, 38 h, (93%)
ee = 90%, *syn*/*anti* = 96/4
Tang

Scheme 2.43 The thiourea/amine catalyst-mediated addition of cyclic ketones to nitrostyrenes.

2.3.4
Addition of Aldehydes to Nitroolefins and Alkylidene Malonates

2.3.4.1 Aminopyrrolidine Catalysts

Aldehydes are more reactive than ketones, and may add to nitroolefins, alkylidene malonates, or to vinyl sulfones at temperatures below those at which ketones would react [30]. Most of the catalysts used for the addition of ketones and discussed in Section 2.3.2 were also seen to be efficient in mediating the addition of aldehydes to nitroolefins and alkylidene malonates. The addition of equimolar amounts of acid as co-catalyst (compared to the catalyst) is suggested in order to avoid base-catalyzed homoaldol coupling. As noted in Section 2.3.2, pyrrolidine catalysts promote *syn*-selective additions not only with ketones but also with aldehydes (Fig. 2.12). Interestingly, the same catalyst induces opposite selectivity with aldehydes, compared to ketones. This fact can be rationalized by the difference in the steric demand of the enamine forms in the transition state. As depicted in Figure 2.12, the equilibrating conformers gives rise to product via the less-congested transition state, thereby forming the thermodynamic product.

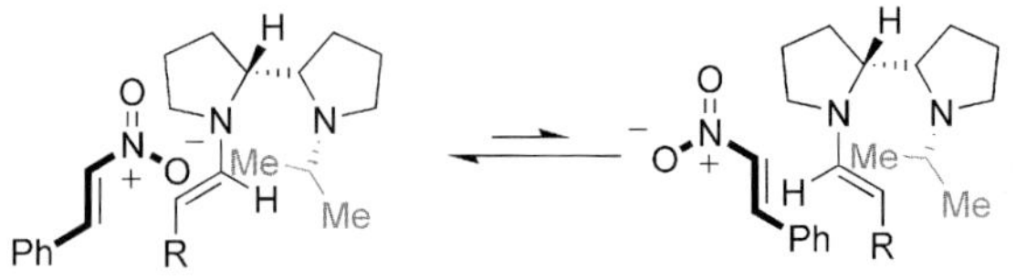

Re,Re favored

Si,Si

Fig. 2.12 The rationale of enantioselectivity of the pyrrolidine catalyst-mediated additions of aldehydes to nitroolefins.

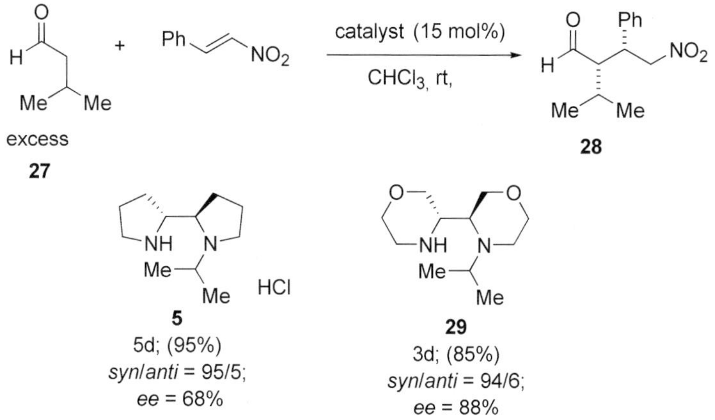

Scheme 2.44 Asymmetric addition of aldehydes to nitroolefins catalyzed by Alexakis's aminopyrrolinines **5** and **29**.

The replacement of the pyrrolidine by morpholine, such as **5** versus **29** results in increased efficiency of the catalyst (Scheme 2.44) [31, 32]. Hindered aldehydes, such as 3,3-dimethylbutyraldehyde or iso-butyraldehyde are unreactive, however, under these conditions.

Likewise, the TFA salts of diamines **6** and **19a** mediate Michael addition of α,α-disubstituted aldehydes to nitrostyrene allowing the preparation of quaternary centers [32, 33] (Scheme 2.45). The enantioselectivity of the transformation increases with the steric demand of the aldehyde, which results in diminished reactivity.

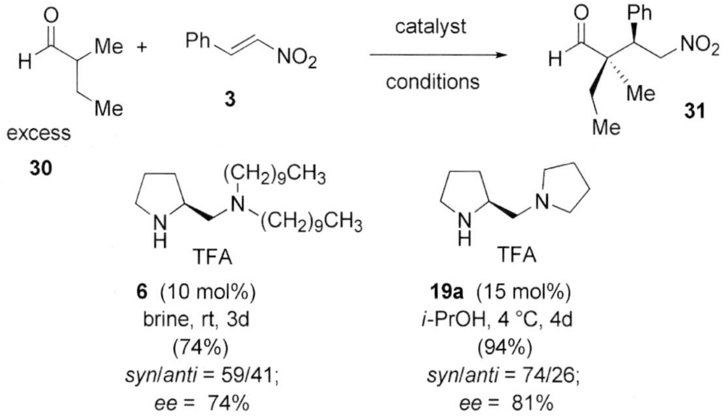

Scheme 2.45 Asymmetric addition of aldehydes to nitroolefins catalyzed by Barbas's aminopyrrolinines **6** and **19a**.

2 Enamine Catalysis

Due to the increased reactivity of the aldehyde, alkyl-substituted nitroolefins can also be used as substrates. Nevertheless, these reactions are usually low-yielding and afford moderate selectivity. Alexakis has shown, however, that the bispyrrolidine **5**-catalyzed additions may be used in multistep synthesis. The addition of propionaldehyde **34** to nitroolefin **33** resulted an approximate 2:3 mixture of *anti/syn* isomers in 92% yield and in high *ee* (93%), allowing the asymmetric synthesis of (−)-botryodiplodin (Scheme 2.46) [23b].

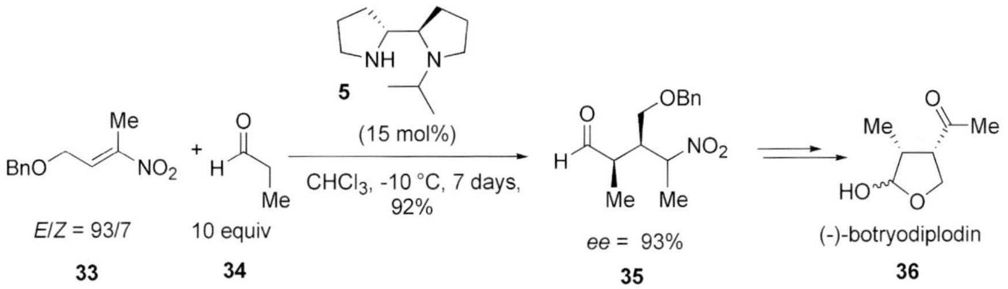

Scheme 2.46 Alexakis's synthesis of (−)-botryodiplodin.

The bifunctional pyrrolidine sulfonamide catalyst, **38**, having basic pyrrolidine nitrogen and Brønsted acidic sulfonamide function, mediates the addition of aldehydes to nitrostyrenes in high yield and selectivity (Scheme 2.47) [34]. Alkyl-substituted nitroolefins afford, however, product in low yield and in low selectivity.

The diphenylprolinol silyl ether **45a** catalyst was developed by the Hayashi's group for the addition of α-unbranched aldehydes to aryl and alkyl substituted nitroolefins [35]. This catalyst, as well as the perfluoroalkyl derivative **45b** [36],

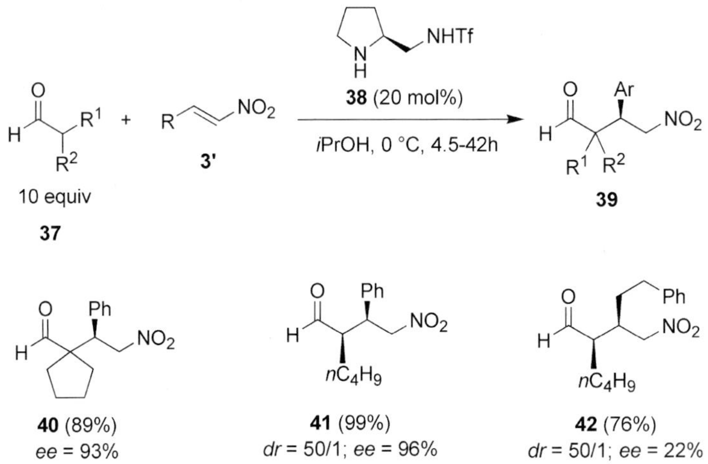

Scheme 2.47 Addition of aldehydes to nitroalkenes, catalyzed by **38**.

affords high yields, excellent diastereoselectivity, and near-perfect enantioselectivity with a broad range of aldehydes. Moreover, **45b** can also be used in aqueous media [37]. The bulky diphenylsilyloxymethyl group plays probably a double role by promoting selective formation of the *E*-enamine, as well as providing efficient shielding of the *Re*-face of the enamine (Scheme 2.48).

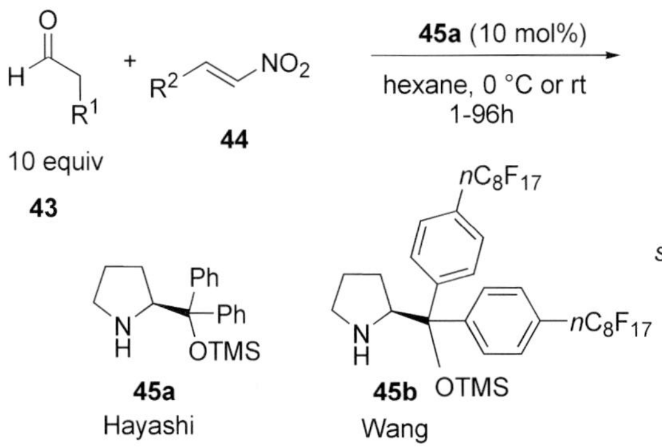

Scheme 2.48 The Michael addition of aldehydes on nitroolefins, catalyzed by **45a** and **45b**.

An elegant application of the **45a**-mediated conjugate addition was devised by Enders' group (Scheme 2.49) [38]. The three-component domino reaction between aldehyde **43**, nitroolefin **44** and α,β-unsaturated aldehyde **47**, used in near-stoichiometric amounts, afforded tetrasubstituted cyclohexene carbaldehydes in good yields, in high diastereoselectivity, and in almost complete enantioselectivity. The first step of the catalytic cycle involves the conjugate addition via enamine formation of aldehyde **43** and the nitroolefin **44**, followed by a Michael addition between the nitroalkane and the α,β-unsaturated aldehyde **47** via an iminium intermediate. The third step of the domino sequence is an intramolecular aldol reaction of the enamine and a subsequent dehydration, giving rise to the polyfunctional cyclohexene **48**. The overall high stereoselectivity of the transformation is the consequence of the high selectivity of the first Michael addition. The resulting product presumably dictates the stereochemistry of the reaction.

2.3.4.2 Addition of Aldehydes and Ketones to Enones

When the donor and the acceptor of the conjugate addition are carbonyl compounds, the amine catalyst may activate both reagents, forming the iminium and enamine intermediates, respectively. The major mechanistic path is dictated by the structure of the substrates and catalyst and, to a lesser extent, by the reaction conditions.

The addition of ketones to chalcone was reported with the pyrroline sulfonamide **38** (Scheme 2.50) [39]. The selectivity of the reaction was less sensitive to the chalcone substitution pattern than to the ketone. As in related transforma-

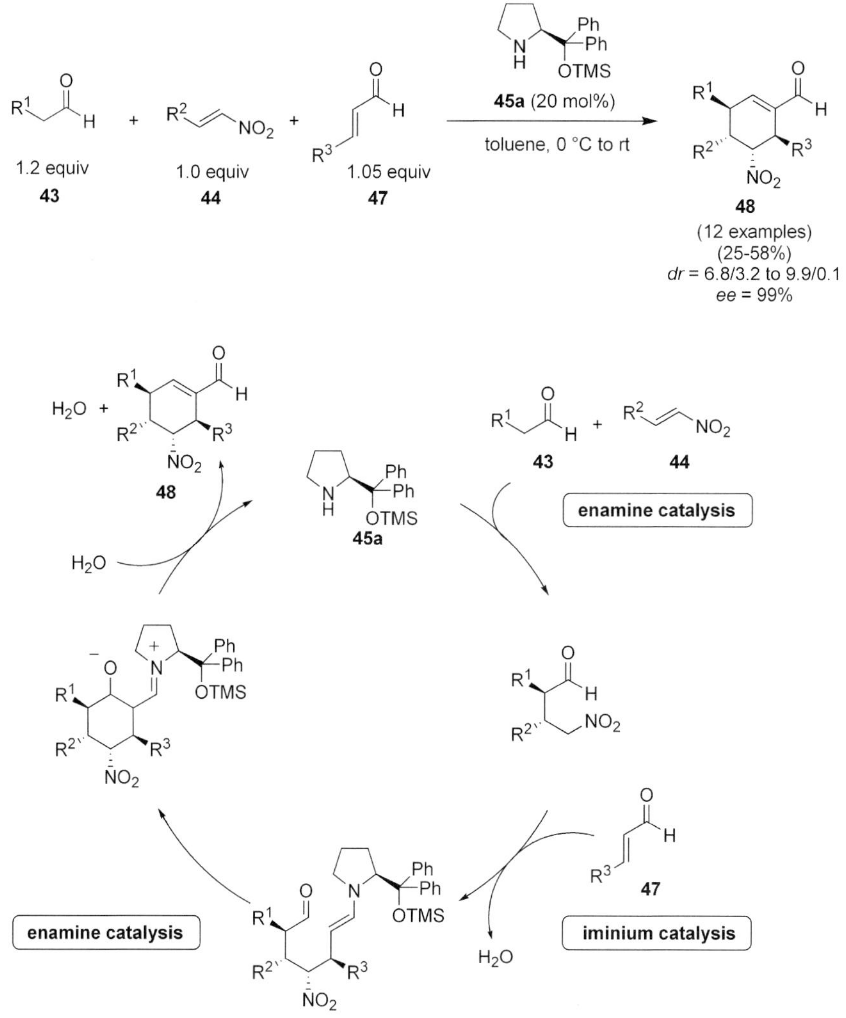

Scheme 2.49 Enders' three-component catalytic asymmetric domino reaction.

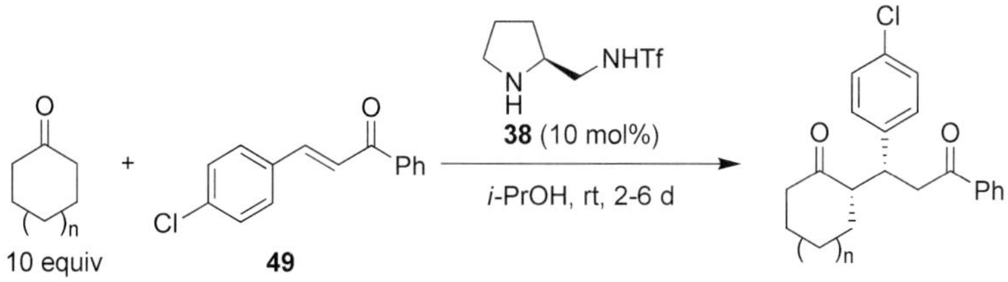

Scheme 2.50 Michael addition of cyclic ketones to the chalcone derivative **49**, catalyzed by **38**.

50 n= 1; (80%); syn/anti = 50/1; ee = 90%
51 n= 0; (87%); syn/anti = 3/1; ee = 73%

tions, the diastereoselectivity and enantioselectivity of the addition was optimal with cyclohexanone but less selective with cyclopentanone. Acyclic ketones such as acetone, acetophenone, and 3-pentanone were unreactive.

In the pioneering studies of Melchiorre and Jørgensen and colleagues, a variety of pyrrolidine-derived catalysts was tested (Scheme 2.51) [40]. Whilst modest selectivity was obtained with proline, and unhindered prolinol derivatives, (S)-2-(bis(phenyl)methyl)pyrrolidine derivative, **45c**, allowed good conversion and enantioselectivity in the addition of linear aldehydes to methyl-vinyl ketone (MVK) (*ee*-values up to 85%). Cyclic enones as well as β-substituted enones afforded no, or perhaps poor, results.

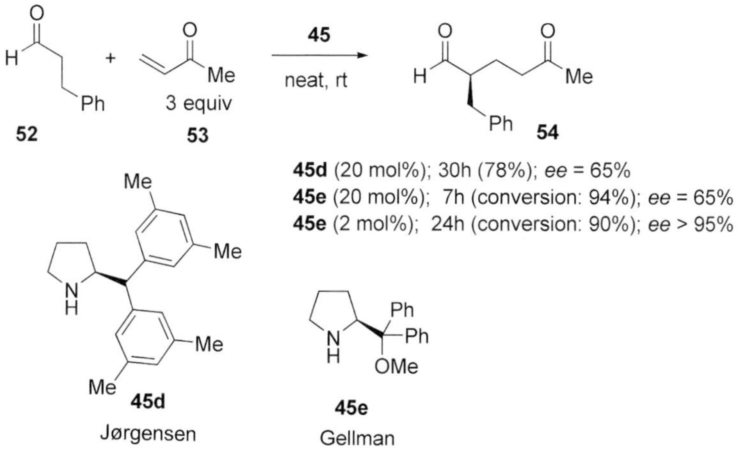

45d (20 mol%); 30h (78%); *ee* = 65%
45e (20 mol%); 7h (conversion: 94%); *ee* = 65%
45e (2 mol%); 24h (conversion: 90%); *ee* > 95%

Scheme 2.51 The **45d** and **45e**-catalyzed addition of dihydrocinnamaldehyde to methyl-vinyl ketone (MVK).

The observed small negative non-linear effect suggested the participation of more than one catalyst molecule in the enantiodifferentiating step [40]. This fact may be the consequence of a simultaneous activation of the nucleophile (enamine formation) and the electrophile partners (iminium formation). On the basis of a series of experimental and theoretical investigations, it was proposed that the major reaction path, however, was addition of the vinyl ketone to the *Si*-face of the enamine intermediate formed from reaction of the aldehyde with the chiral amine catalyst. Chi and Gellman [41] observed that a lower catalyst loading allowed a higher *ee*-value of the product – a fact explained by minimizing the competing epimerization of the Michael adduct in the presence of the catalyst (see Scheme 2.51) [41]. For the addition of ethyl-vinyl ketone, and the small increase in steric constraint compared to MVK, the addition of a 2H-donating ability co-catalyst such as 4-CO_2Et-catechol (**57**) improved either the conversion or/and the chemoselectivity of the process by minimizing formation of the self-aldol product.

MacMillan's organocatalyst, **56a**, which was used typically for electrophilic activation, was seen also to be efficient in promoting conjugate addition via enamine formation (Scheme 2.52) [42]. The proof of the enamine pathway was furnished by extended NMR studies. Gellman and colleagues noted an interesting dependence of selectivity on the catalyst structure: improved conversion and *ee*-value can occur with the spirocyclopentane derivative **56b**, and by the addition of a catechol derivative as acid additive (Scheme 2.52). The cyclohexane-derived catalyst **56c** was unreactive, however.

Similar to intermolecular reactions, the intramolecular addition of aldehydes to enones could be also promoted with catalyst **56a** (Scheme 2.53) [43]. The cycliza-

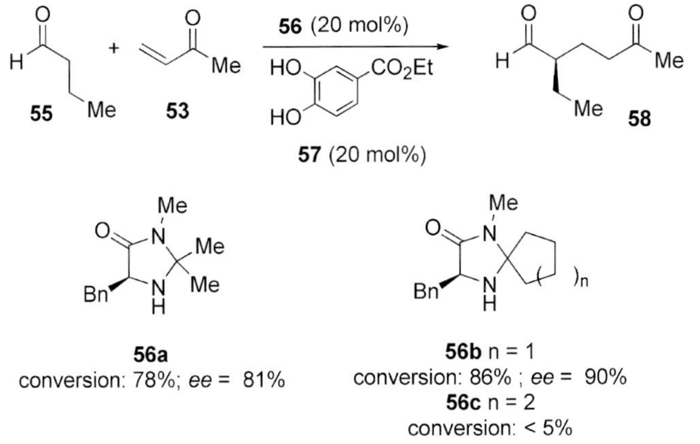

Scheme 2.52 The organocatalyzed asymmetric addition of butyraldehyde to MVK.

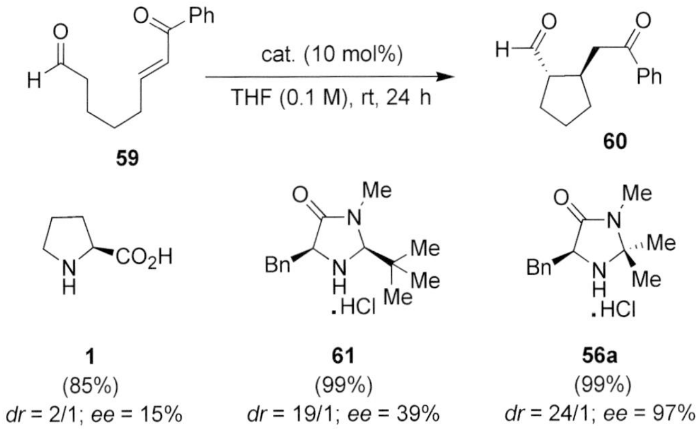

Scheme 2.53 Intramolecular addition of aldehydes to enones.

2.3 Direct Conjugate Additions via Enamine Activation

tion is highly *trans*-selective in forming 1,2-disubstituted cyclopentanes. Moreover, the reaction tolerates aryl and alkyl substituents on the enone, and can be carried out with enals, as well as in the presence of heteroatoms (Scheme 2.53).

MacMillan's catalysts **56a** and **61** allowed also the combination of the domino 1,4-hydride addition followed by intramolecular Michael addition [44]. The reaction is chemoselective, as the hydride addition takes place first on the iminium-activated enal. The enamine-product of the reaction is trapped in a rapid intramolecular reaction by the enone, as depicted in Scheme 2.54. The intramolecular trapping is efficient, as no formation of the saturated aldehyde can be observed. The best results were obtained with MacMillan's imidazolidinium salt **61** and Hantzsch ester **62** as hydride source. As was the case in the cyclization reaction, the reaction affords the thermodynamic *trans* product in high selectivity. This transformation sequence is particularly important in demonstrating that the same catalyst may trigger different reactions via different mechanistic pathways, in the same reaction mixture.

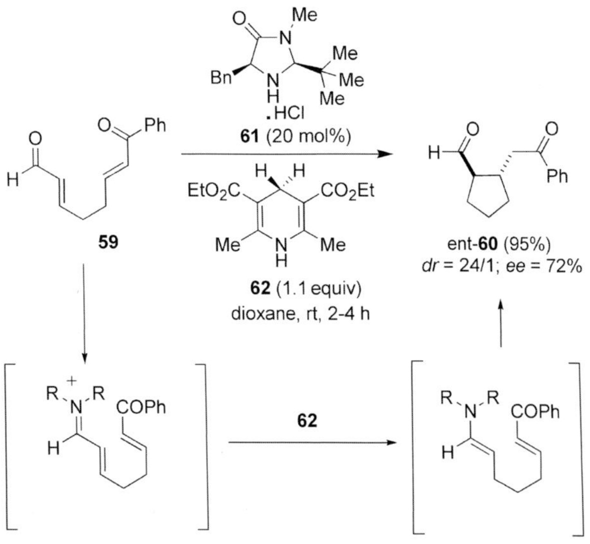

Scheme 2.54 The domino 1,4-hydride addition followed by intramolecular Michael addition.

Inverse (*cis*) diastereoselectivity was obtained in the intramolecular Michael addition with the cysteine-derived prolinamide analog, **63** (Scheme 2.55) [45].

The desymmetrization of 4-substituted-4-(3-formylpropyl)cyclohexa-2,5-dien-1-one **65** formed – in a single step – three contiguous stereocenters, including a quaternary stereocenter (Scheme 2.56).

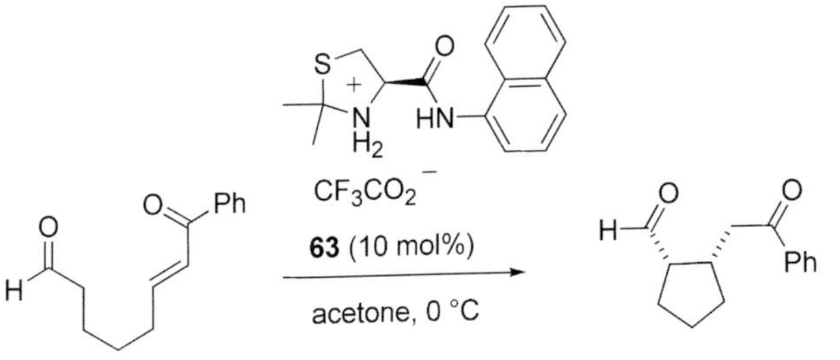

59

64 (quant)
dr = 89/11; *ee* = 99%

Scheme 2.55 The **63**-mediated intramolecular conjugate addition.

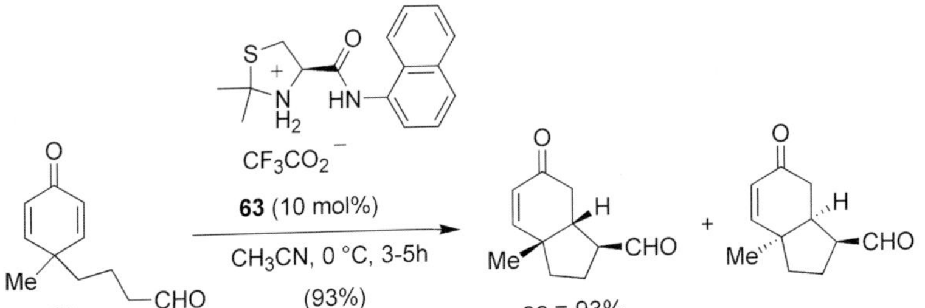

65

ee = 93%

66 *dr* = 96 / 4 **67**

Scheme 2.56 The desymmetrization of **65** by intramolecular Michael addition.

2.4 Conclusions

Enamine-activation is a recent addition to the arsenal of enantioselective organocatalytic reactions. A handful of chiral catalysts have been developed for the conjugate addition of ketones and aldehydes to electron-deficient alkenes. A mild acid co-catalyst was seen to be beneficial for the conversion and selectivity of the addition. The remarkably simple experimental conditions required undoubtedly represent a major asset compared to metal-mediated asymmetric transformations. Whilst high levels of diastereoselectivity and enantioselectivity can be obtained in selected cases – and in particular with cyclohexanone – the reactions are usually substrate-dependent. Further studies allowing a better understanding of the factors governing the selectivity of the reaction, as well as of the development of more general catalysts, would clearly help when using this addition for synthetic purposes.

Acknowledgment

Professor Alexandre Alexakis is kindly acknowledged for the critical reading of the manuscript.

References

1 For books and chapters, see: (a) P. Perlmutter, *Conjugate addition reactions in organic synthesis*, Pergamon Press, Oxford, UK, 1992; (b) M. Yamaguchi, in: E.N. Jacobsen, A. Pfaltz, H. Yamamoto (Eds.), *Comprehensive Asymmetric Catalysis I–III*. Springer-Verlag, Berlin, Germany, 1999, Chapter 31.2.

2 For reviews, see: (a) M.P. Sibi, S. Manyem, *Tetrahedron* 2000, 56, 8033–8061; (b) N. Krause, A. Hoffmann-Roder, *Synthesis* 2001, 2, 171–196; (c) H.-C. Guo, J.-A. Ma, *Angew. Chem. Int. Ed.* 2006, 45, 354–366.

3 For reviews, see: (a) O.M. Berner, L. Tedeschi, D. Enders, *Eur. J. Chem.* 2002, 1877–1894; (b) W. Notz, F. Tanaka, C.F. Barbas, III, *Acc. Chem. Res.* 2004, 37, 580–591; (c) B. List, *Acc. Chem. Res.* 2004, 37, 548–557.

4 (a) B. List, *Chem. Commun.* 2006, 819–824; (b) G. Lelais, D.W.C. MacMillan, *Aldrichimica Acta* 2006, 39, 79–87; (c) J. Seayad, B. List, *Org. Biomol. Chem.* 2005, 3, 719–724; (d) A.J.A. Cobb, D.M. Shaw, D.A. Longbottom, J.B. Gold, S.V. Ley, *Org. Biomol. Chem.* 2005, 3, 84–96; (e) X.F. Wei, *Chin. J. Org. Chem.* 2005, 25, 1619–1625; (f) S.B. Tsogoeva, *Lett. Org. Chem.* 2005, 2, 208–213; (g) C. Ó'Dálaigh, *Synlett* 2005, 875–876; (h) S.-K. Tian, Y. Chen, J. Hang, L. Tang, P. McDaid, L. Deng, *Acc. Chem. Res.* 2004, 37, 621–631; (i) B. List, *Acc. Chem. Res.* 2004, 37, 548–557; (j) S. France, D.J. Guerin, S.J. Miller, T. Lectka, *Chem. Rev.* 2003, 103, 2985–3012; (k) K. Maruoka, T. Ooi, *Chem. Rev.* 2003, 103, 3013–3028; (l) B. List, *Tetrahedron* 2002, 58, 5572–5590; (m) E.R. Jarvo, S.J. Miller, *Tetrahedron* 2002, 58, 2481–2495; (n) B. List, *Synlett* 2001, 1675–1686.

5 As defined by List in Ref. [4n]: "There are two aminocatalytic pathways. Iminium catalysis directly utilizes the higher reactivity of the iminium ion in comparison to the carbonyl species and facilitates Knoevenagel-type condensations, cyclo- and nucleophilic additions, and cleavage of σ-bonds adjacent to the α-carbon. Enamine catalysis on the other hand involves catalytically generated enamine intermediates that are formed *via* deprotonation of an iminium ion, and react with various electrophiles or undergo pericyclic reactions."

6 For preferred enamine geometry in related transformations, see: P.H.-Y. Cheong, H. Zhang, R. Thayumanavan, F. Tanaka, K.N. Houk, C.F. Barbas, III, *Org. Lett.* 2006, 8, 811–814.

7 D. Seebach, J. Golinski, *Helv. Chim. Acta* 1981, 64, 1413–1423.

8 (a) M. Yamaguchi, T. Shiraishi, M. Hirama, *J. Org. Chem.* 1996, 61, 3520; (b) M. Yamaguchi, Y. Igarashi, R.S. Reddy, T. Shiraishi, M. Hirama, *Tetrahedron* 1997, 53, 11223; (c) M. Yamaguchi, T. Shiraishi, Y. Igarashi, M. Hirama, *Tetrahedron Lett.* 1994, 35, 8233.

9 T. Bui, C.F. Barbas, III, *Tetrahedron Lett.* 2000, 41, 6951–6954.

10 J.M. Betancort, K. Sakthivel, R. Thayumanavan, C.F. Barbas, III, *Tetrahedron Lett.* 2001, 42, 4441–4444.

11 B. List, P. Pojarliev, H.J. Martin, *Org. Lett.* 2001, 3, 2423–2425.

12 D. Enders, A. Seki, *Synlett* **2002**, 26–28.
13 (a) A. Alexakis, O. Andrey, *Org. Lett.* **2002**, *4*, 3611–3614; (b) A. Alexakis, A. Tomassini, C. Chouillet, S. Roland, P. Mangeney, G. Bernardinelli, *Angew. Chem. Int. Ed.* **2000**, *39*, 4093–4095.
14 (a) J.M. Betancort, K. Sakthivel, R. Thayumanavan, F. Tanaka, C.F. Barbas, III, *Synthesis* **2004**, 1509–1521; (b) K. Sakthivel, W. Notz, T. Bui, C.F. Barbas, III, *J. Am. Chem. Soc.* **2001**, 5260–5267.
15 T. Ishii, S. Fujioka, Y. Sekiguchi, H. Kotsuki, *J. Am. Chem. Soc.* **2004**, *126*, 9558–9559.
16 S.V. Pansare, K. Pandya, *J. Am. Chem. Soc.* **2006**, *128*, 9624–9625.
17 S. Luo, X. Mi, L. Zhang, S. Liu, H. Xu, J.-P. Cheng, *Angew. Chem. Int. Ed.* **2006**, *45*, 3093–3097.
18 (a) A.J.A. Cobb, D.A. Longbottom, D.M. Shaw, S.V. Ley, *Chem. Commun.* **2004**, 1808–1809; (b) A.J.A. Cobb, D.M. Shaw, D.A. Longbottom, J.B. Gold, S.V. Ley, *Org. Biomol. Chem.* **2005**, *3*, 84–96.
19 C.E.T. Mitchell, A.J.A. Cobb, S.V. Ley, *Synlett* **2005**, 611–614.
20 Decreasing the reaction temperature results in an increase of the overall selectivity and a dramatic decrease in the conversion.
21 See for example in N. Mase, K. Watanabe, K. Yoda, F. Tanaka, C.F. Barbas, III, *J. Am. Chem. Soc.* **2006**, *128*, 4966–4967.
22 P. Kotrusz, S. Toma, H.-G. Schmalz, A. Adler, *Eur. J. Org. Chem.* **2004**, 1577–1583.
23 (a) O. Andrey, A. Alexakis, G. Bernardinelli, *Org. Lett.* **2003**, *5*, 2559–2561; (b) O. Andrey, A. Vidonne, A. Alexakis, *Tetrahedron Lett.* **2003**, *44*, 7901–7904.
24 Y. Xu, W. Zou, H. Sunden, I. Ibrahem, A. Cordova, *Adv. Synth. Catal.* **2006**, *348*, 418–424.
25 Y. Xu, A. Cordova, *Chem. Commun.* **2006**, 460–462.
26 H. Huang, E.N. Jacobsen, *J. Am. Chem. Soc.* **2006**, *128*, 7170–7171.
27 For related chiral (thio)urea catalyst-mediated electrophilic activation of enones in conjugate addition, see also Chapter 6.
28 S.B. Tsogoeva, S. Wei, *Chem. Commun.* **2006**, 1451–1453.
29 C.-L. Cao, M.-C. Ye, X.-L. Sun, Y. Tang, *Org. Lett.* **2006**, *8*, 2559–2562.
30 S. Mossé, A. Alexakis, *Org. Lett.* **2005**, *7*, 4361–4364.
31 (a) S. Mossé, M. Laars, K. Kriis, T. Kanger, A. Alexakis, *Org. Lett.* **2006**, *8*, 2559–2562; (b) T. Kanger, M. Laars, K. Kriis, T. Kailas, A.-M. Müürisepp, T. Pehk, M. Lopp, *Synthesis* **2006**, 1853–1857.
32 J.M. Betancort, C.F. Barbas, III, *Org. Lett.* **2001**, *3*, 3737–3740.
33 N. Mase, R. Thayumanavan, F. Tanaka, C.F. Barbas, III, *Org. Lett.* **2004**, *6*, 2527–2530.
34 W. Wang, J. Wang, H. Li, *Angew. Chem. Int. Ed.* **2005**, *44*, 1369–1371.
35 Y. Hayashi, H. Gotoh, T. Hayashi, M. Shoji, *Angew. Chem. Int. Ed.* **2005**, *44*, 4212–4215.
36 (a) L. Zu, H. Li, J. Wang, X. Yu, W. Wang, *Tetrahedron Lett.* **2006**, *47*, 5131–5134; (b) For related fluorous derivatives as catalysts in Michael-type additions, see: W. Wang, H. Li, J. Wang, L. Zu, *J. Am. Chem. Soc.* **2006**, *128*, 10354–10355.
37 L. Zu, J. Wang, H. Li, W. Wang, *Org. Lett.* **2006**, *8*, 3077–3079.
38 D. Enders, M.R.M. Hüttl, C. Grondal, G. Raabe, *Nature* **2006**, 861–863.
39 J. Wang, H. Li, L. Zu, W. Wang, *Adv. Synth. Catal.* **2006**, *348*, 425–428.
40 P. Melchiorre, K.A. Jørgensen, *J. Org. Chem.* **2003**, *68*, 4151–4157.
41 Y. Chi, S.H. Gellman, *Org. Lett.* **2005**, *7*, 4253–4256.
42 T.J. Peelen, Y. Chi, S.H. Gellman, *J. Am. Chem. Soc.* **2005**, *127*, 11598–11599.
43 M.T. Hechavarria Fonseca, B. List, *Angew. Chem. Int. Ed.* **2004**, *43*, 3958–3960.
44 J.W. Yang, M.T. Hechavarria Fonseca, B. List, *J. Am. Chem. Soc.* **2005**, *127*, 15036–15037.
45 Y. Hayashi, H. Gotoh, T. Tamura, H. Yamaguchi, R. Masui, M. Shoji, *J. Am. Chem. Soc.* **2005**, *127*, 16028–16029.

3
Iminium Catalysis

Gérald Lelais and David W. C. MacMillan

3.1
Introduction

The search for new catalytic approaches for the enantioselective preparation of chiral molecules has received increasing attention during recent years. In particular, the field of organocatalysis has grown at an incredible pace, from a small collection of chemically unique reactions to a thriving area of general concepts, atypical reactivity, and widely applicable reactions [1–5]. This chapter will discuss the discovery and development of one of the youngest subfields of organocatalysis, namely iminium activation [5]. The first section will introduce the concept of iminium catalysis, together with the rationale for the development of a general catalyst. The following sections will discuss the various transformations where this catalytic concept has been applied successfully, placing particular emphasis on the mechanisms, scopes, and limitations of the corresponding reactions. Applications towards the total synthesis of complex natural products will also be addressed.

3.2
The Catalysis Concept of Iminium Activation

During the early 2000s, MacMillan and co-workers [6] introduced a new strategy for asymmetric synthesis based upon design criteria borrowed from the area of Lewis acid catalysis. This catalytic concept – termed iminium activation – was founded on the mechanistic postulate that: (1) the electronic principles underpinning Lewis acid activation (LUMO-lowering activation); and (2) the kinetic lability towards ligand substitution enabling Lewis acid-catalyst turnover (Eq. 1) might also be available with an organic compound existing in a rapid equilibrium between an electron-deficient and a relatively electron-rich state (Eq. 2). Indeed, the MacMillan laboratory found that the reversible formation of iminium ions from α,β-unsaturated aldehydes and amines could emulate the equilibrium dynamics

Enantioselective Organocatalysis: Reactions and Experimental Procedures
Edited by Peter I. Dalko
Copyright © 2007 WILEY-VCH Verlag GmbH & Co. KGaA, Weinheim
ISBN: 978-3-527-31522-2

and π-orbital electronics that are inherent to Lewis acid catalysis. Importantly, this approach revealed the attractive prospect that chiral amines might function as enantioselective catalysts for a large range of transformations that traditionally employ metal salts.

$$\text{substrate} + \text{Lewis acid (LA)} \rightleftarrows \text{LUMO-activation} \quad (1)$$

$$\text{substrate} + R_2NH \rightleftarrows \text{iminium ion} \quad (2)$$

3.3
Development of the "First-Generation" Imidazolidinone Catalysts

Prior to MacMillan's studies, the iminium-catalysis strategy had never been documented. However, several established methodological investigations provided the experimental basis for this organocatalytic approach.

First, one of the most utilized transformations in organic synthesis is represented by reductive amination [7]. In this process, an aldehyde and an amine reversibly combine to generate an iminium ion in equilibrium quantities that rapidly undergoes hydride reduction to provide the corresponding alkyl amine. Of particular note in this equilibrium-based process is that only the iminium ion is of sufficient electronic deficiency to undergo hydride reduction, an important consideration with regard to LUMO-lowering activation. Second, the studies of Jung and co-workers [8] in their Diels–Alder investigations during the late 1980s revealed that α,β-unsaturated iminium ions are significantly more reactive as dienophiles than α,β-unsaturated aldehydes, acid chlorides, ketones, nitriles, or esters. This is an essential criterion for the amine-catalyzed strategy in that a significant rate acceleration of the enantioselective bond-forming step must accompany iminium ion formation. Finally, the seminal works of Grieco [9, 10] and Waldman [11] revealed that iminium ions generated under Mannich conditions will undergo aza-Diels–Alder reaction with electron-rich dienes. While these reactions incorporate the amine substrate into the Diels–Alder adducts, they do provide a strong precedent for the initial hypothesis that electron-rich substrates will undergo selective reaction with transiently generated iminium ions in the presence of aldehyde functionalities.

In order to test the iminium-activation strategy, MacMillan first examined the capacity of various amines to enantioselectively catalyze the Diels–Alder reaction between dienes and α,β-unsaturated aldehyde dienophiles [6]. Preliminary experimental findings and computational studies proved the importance of four objectives in the design of a broadly useful iminium-activation catalyst: (1) the chiral amine should undergo efficient and reversible iminium ion formation; (2) high

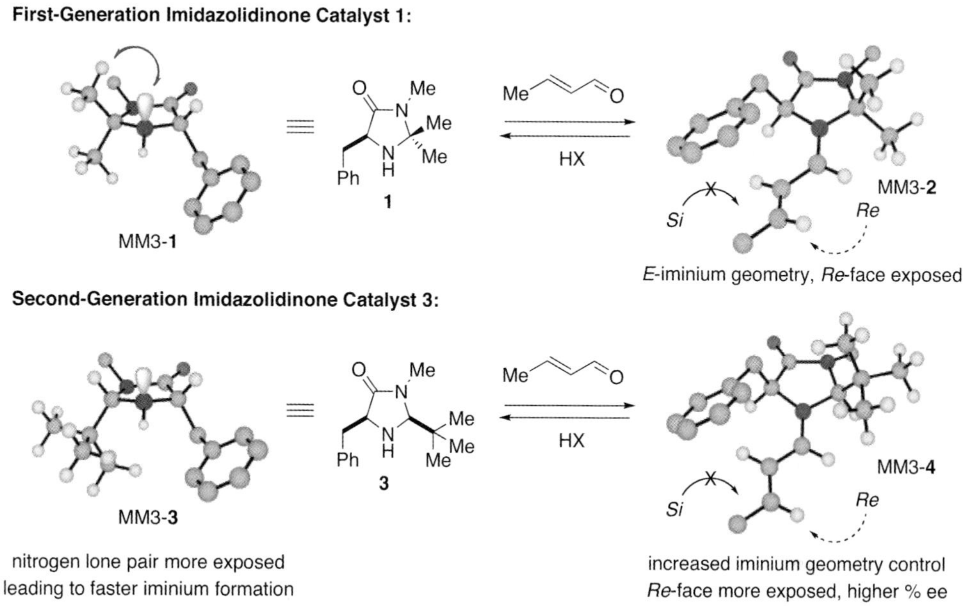

3.3 Development of the "First-Generation" Imidazolidinone Catalysts

First-Generation Imidazolidinone Catalyst 1:

E-iminium geometry, *Re*-face exposed

Second-Generation Imidazolidinone Catalyst 3:

nitrogen lone pair more exposed
leading to faster iminium formation

increased iminium geometry control
Re-face more exposed, higher % ee

Fig. 3.1 Computational studies of the first- and second-generation imidazolidinone catalysts (**1** and **3**) and of the corresponding iminium ions.

levels of iminium geometry control and (3) selective π-facial discrimination of the iminium ion should be achieved in order to control the enantioselectivity of the reaction, and (4) in addition, the ease of catalyst preparation and implementation would be essential for the widespread adoption of this catalytic technology. The first catalyst to fulfill these requirements was imidazolidione **1** (Fig. 3.1, top).

As indicated from computational studies, the catalyst-activated iminium ion MM3-2 was expected to form with only the (*E*)-conformation to avoid nonbonding interactions between the substrate double bond and the *gem*-dimethyl substituents on the catalyst framework. In addition, the benzyl group of the imidazolidinone moiety should effectively shield the iminium-ion *Si*-face, leaving the *Re*-face exposed for enantioselective bond formation. The efficiency of chiral amine **1** in iminium catalysis was demonstrated by its successful application in several transformations such as enantioselective Diels–Alder reactions [6], nitrone additions [12], and Friedel–Crafts alkylations of pyrrole nucleophiles [13]. However, diminished reactivity was observed when indole and furan heteroaromatics where used for similar conjugate additions, causing the MacMillan group to embark upon studies to identify a more reactive and versatile amine catalyst. This led ultimately to the discovery of the "second-generation" imidazolidinone catalyst **3** (Fig. 3.1, bottom) [14].

3.4
Development of the "Second-Generation" Imidazolidinone Catalysts

Preliminary kinetic studies with imidazolidinone catalyst **1** suggested that the overall rates of iminium-catalyzed processes were influenced by the efficiency of both the initial iminium ion and the C–C bond-forming steps. With the help of computational modeling, it was hypothesized that reduced reactivity of amine **1** in several iminium-catalyzed reactions was due to its diminished nucleophilicity towards carbonyl addition as the participating nitrogen lone pair is positioned adjacent to an eclipsing methyl group (Fig. 3.1, top). In order to overcome this unfavorable interaction, MacMillan postulated that the replacement of the methyl group with a hydrogen substituent would enable catalysts such as **3** to rapidly engage in iminium formation. At the same time, replacement of the *cis*-methyl group of **1** with a *tert*-butyl functionality as in **3** was proposed to increase iminium geometry control, providing better coverage of the blocked *Si*-enantioface (Fig. 3.1, bottom). Since its introduction in 2001, imidazolidinones of type **3** have been successfully applied to a broad range of transformations, which include cycloadditions [15, 16], conjugate additions [14, 17, 18], hydrogenations [19], epoxidations, and cascade reactions [20, 21].

3.5
Cyloaddition Reactions

3.5.1
Diels–Alder Reactions

The Diels–Alder reaction is a valuable transformation for the construction of complex carbocycles, and represents arguably one of the most powerful approaches in organic chemistry. In particular, catalytic enantioselective variants have received unprecedented attention [22], representing an appealing starting point for the development of MacMillan's concept of iminium catalysis.

In line with the mechanistic rationale of LUMO-lowering iminium activation, MacMillan hypothesized that intermediate **2**, generated from the secondary amine **1** and an α,β-unsaturated aldehyde, could be activated towards cycloaddition with an appropriate diene (Scheme 3.1). The Diels–Alder reaction would form iminium ion cycloadduct **5** that, in the presence of water, would hydrolyze to yield the enantioenriched product **6** and regenerate the chiral imidazolidinone catalyst **1**.

In 2000, the first highly enantioselective amine-catalyzed Diels–Alder reaction was disclosed [6], in which the addition of a range of α,β-unsaturated aldehydes (dienophiles) to a variety of dienes (symmetrical and unsymmetrical) in the presence of catalytic amounts of imidazolidinone **1** (5–20 mol%) afforded the corresponding cycloadducts in good yields (72–99%), and high regio- and enantioselectivities (Eq. 3). (For experimental details see Chapter 14.18.1).

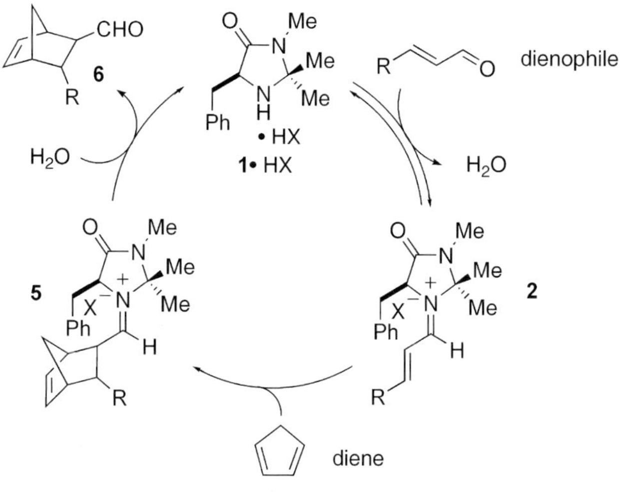

Scheme 3.1

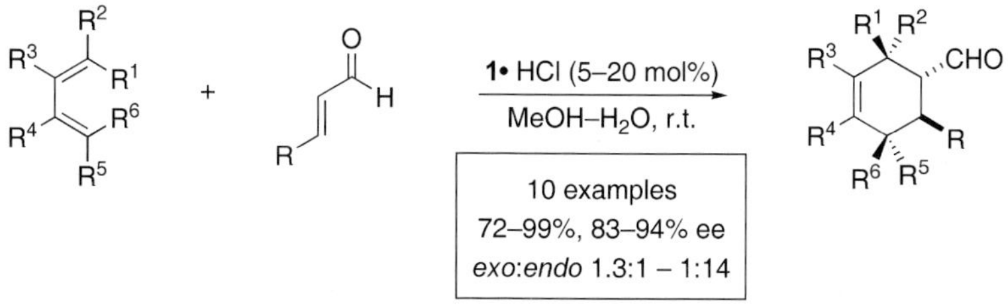

$$R^3 \underset{R^5}{\overset{R^2}{\underset{R^4}{\longrightarrow}}} R^1 + \underset{R}{\overset{O}{\longrightarrow}} H \quad \xrightarrow[\text{MeOH-H}_2\text{O, r.t.}]{\text{1• HCl (5–20 mol\%)}} \quad R^3 \underset{R^6 \ R^5}{\overset{R^1 \ R^2}{\underset{R^4}{\longrightarrow}}} \overset{\text{CHO}}{\underset{R}{\longrightarrow}} \quad (3)$$

10 examples
72–99%, 83–94% ee
exo:endo 1.3:1 – 1:14

Remarkably, the presence of water showed beneficial effects on both reaction rates and selectivities, while facilitating the iminium ion hydrolysis step in the catalytic cycle. Computational studies suggest an asynchronous mechanism for the reaction [23–25], where the attack of the diene to the β-carbon atom of the iminium ion is the rate-limiting step [23], and the π,π-interaction between the phenyl ring of the catalyst's benzyl group and the olefinic π-system of the iminium ion accounts for the selectivity of the reaction [6, 24].

Since the first publication, amine-catalyzed Diels–Alder reactions of α,β-unsaturated aldehydes have been investigated in much detail [15, 26–33]. Catalyst immobilization studies on solid support [26, 27], as well as in ionic liquids [29], have shown advantages for amine recycling, while partially maintaining good levels of asymmetric induction [34]. The use of this reaction in total synthesis has allowed the rapid preparation of (+)-hapalindole Q, a tricyclic alkaloid natural product containing four contiguous stereocenters (Scheme 3.2) [28].

A limitation of MacMillan's approach towards iminium-activated Diels–Alder reactions has been the use of α-substituted α,β-unsaturated aldehydes as dienophiles. Recently, Ishihara and Nakano [31] succeeded in partially overcoming this problem by identifying a novel primary amine organocatalyst for this type of

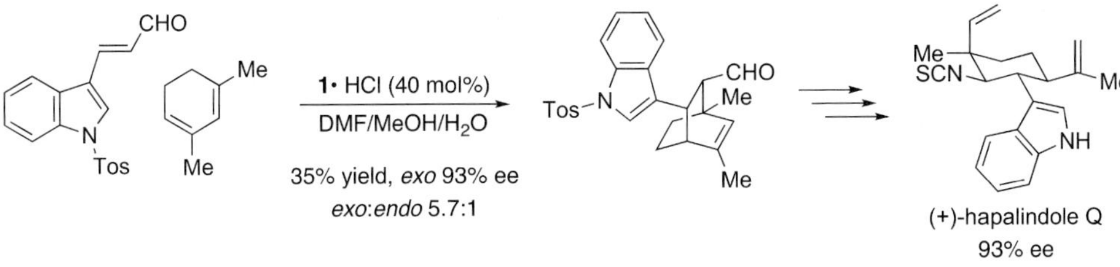

Scheme 3.2

transformation, where α-acyloxyacroleins underwent *exo*-selective cycloaddition with a variety of dienes in high yields and good enantioselectivity (79–92% *ee*; Scheme 3.3). (For experimental details see Chapter 14.18.2).

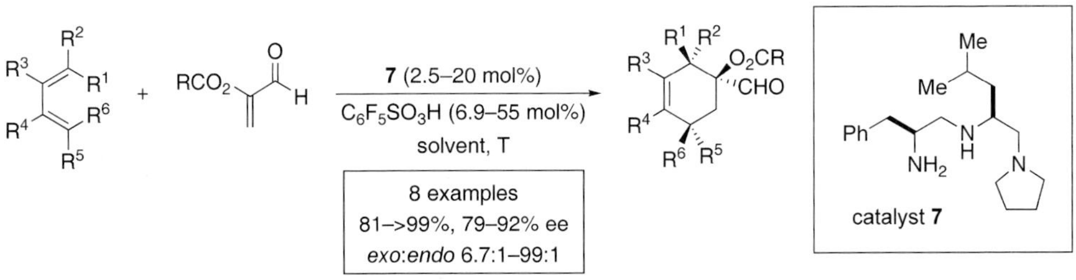

Scheme 3.3

Another important application of the iminium catalysis concept has been the development of enantioselective Type I [15, 30] and Type II [15] intramolecular Diels–Alder reactions (IMDA). (For experimental details see Chapter 14.18.3). For these transformations, both catalysts **1** and **3** proved to be highly efficient, as demonstrated by both the short and effective preparation of the marine methabolite solanapyrone D via Type I IMDA (Scheme 3.4, top) and the development of an early example of an enantioselective, catalytic Type II IMDA reaction (Scheme 3.4, bottom) [35]. Importantly, cycloadducts incorporating ether and quaternary carbon functionalities could be efficiently produced.

An important challenge in the asymmetric catalytic Diels–Alder reaction is the use of simple ketone dienophiles to obtain high enantioselectivity. Indeed, the success of chiral Lewis acid-mediated Diels–Alder reactions is founded upon the use of dienophiles such as aldehydes, esters, quinones [36–43], and bidentate chelating carbonyls [44–47], where high levels of lone pair discrimination are achieved in the metal association step, an organizational event that is essential for enantiocontrol. In contrast, Lewis acid coordination to ketone dienophiles is generally non-selective, since the participating lone pairs are positioned in similar steric and electronic environments (Eq. 4). The ability for diastereomeric activa-

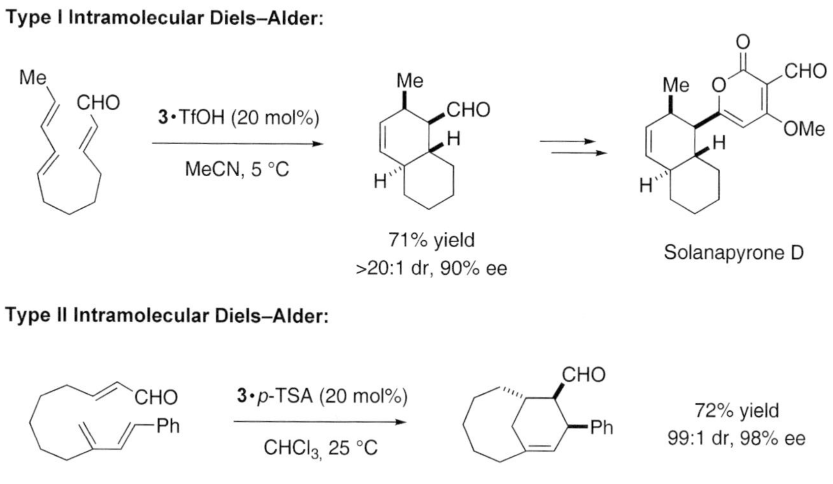

3.5 Cyloaddition Reactions

Type I Intramolecular Diels–Alder:

Scheme 3.4 shows the reaction of a Me/CHO-substituted triene with 3·TfOH (20 mol%) in MeCN at 5 °C, giving a bicyclic product in 71% yield, >20:1 dr, 90% ee, leading to Solanapyrone D.

Type II Intramolecular Diels–Alder:

The CHO/Ph-substituted macrocyclic substrate with 3·p-TSA (20 mol%) in CHCl$_3$ at 25 °C gives the bicyclic product in 72% yield, 99:1 dr, 98% ee.

Scheme 3.4

tion in this case often leads to poor levels of enantiocontrol and, ultimately, has prevented the use of simple ketone dienophiles in asymmetric Diels–Alder reactions [48]. The use of amines for the corresponding organocatalytic transformation was therefore a good opportunity to complement this deficiency [16]. In this case, the ability to perform substrate activation does not rely on specific lone pair activation, but alternatively on selective π-bond formation (iminium geometry control; see Eq. 5).

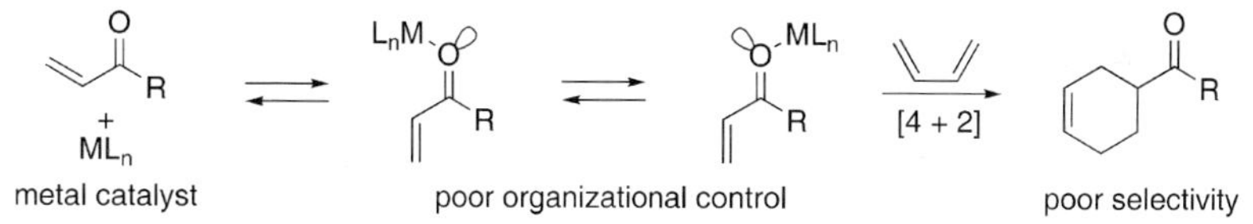

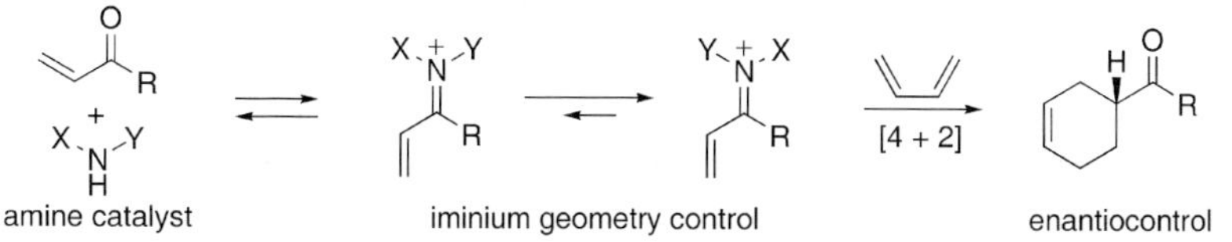

Equation (4): metal catalyst pathway — poor organizational control, [4 + 2], poor selectivity.

Equation (5): amine catalyst pathway — iminium geometry control, [4 + 2], enantiocontrol.

The fine-tuning of sterics and electronics governing the control of iminium geometry allowed the identification of imidazolidinone catalyst **8** to provide efficient catalytic Diels–Alder transformations using simple ketone dienophiles (Scheme 3.5) [16]. This reaction is quite general with respect to diene structure, allowing enantioselective access to a broad range of alkyl-, alkoxy-, amino-, and aryl-substituted cyclohexenyl ketones. In particular, single regio- and diastereomeric products were produced. Interestingly, whereas methyl ketones were usually poor substrates, higher-order derivatives (R = Et, Bu, i-amyl) afforded good levels

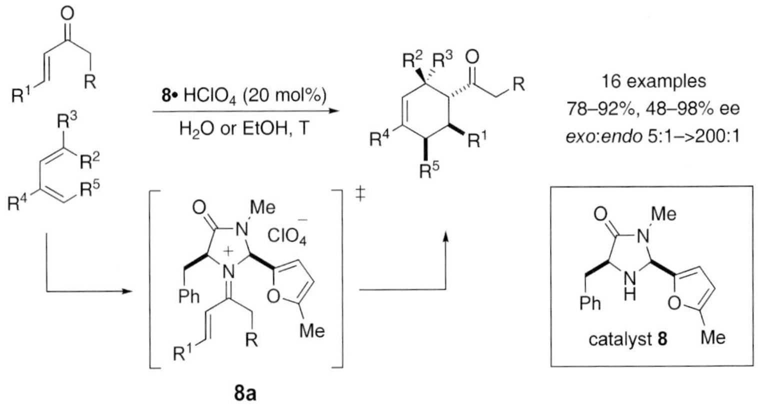

Scheme 3.5

of enantiocontrol and high *exo*-selectivity. The sense of induction observed in all cases is consistent with selective engagements of the diene substrate with the *Si*-face of the *cis*-iminium isomer **8a** [16, 25].

3.5.2
[3+2]-Cycloadditions

The 1,3-cycloaddition of nitrones to alkenes is a rapid and elegant method to prepare isoxazolidines, which are important building blocks for the construction of biologically active compounds [49]. Recently, asymmetric Lewis acid-catalyzed nitrone cycloadditions have successfully been used for their enantioselective preparation [50–52]. However, only a few examples used monodentate carbonyl compounds as substrate, most likely due to the competitive coordination of nitrones to Lewis acids [53–57]. As this problem does not occur in the realm of iminium catalysis, MacMillan and co-workers applied the LUMO-lowering activation strategy to the first organocatalytic [3+2]-cycloaddition (Scheme 3.6) [12]. Transformation of several α,β-unsaturated aldehydes with a variety of N-alkylated nitrones in the presence of catalytic amounts of imidazolidinone **1** afforded the corresponding isoxazolidines in high yields and good diastereo- and enantioselectivities. (For

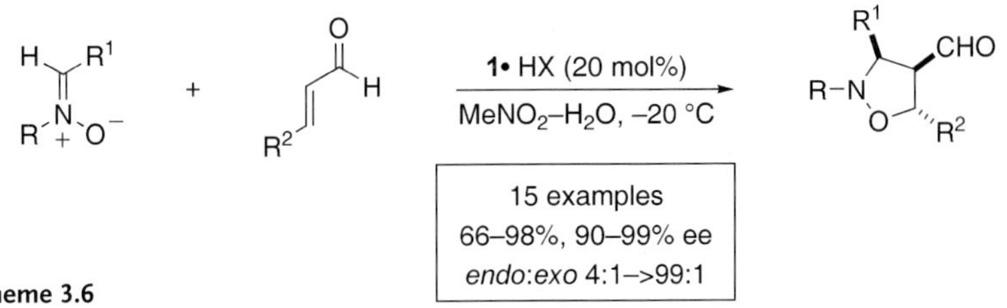

Scheme 3.6

experimental details see Chapter 14.19.1). The reaction appears quite general with respect to the nitrone structure (>66% yield, >92:8 *endo:exo*, >91% *ee*). Variation in the N-alkyl group (R = Me, Bn, Allyl) is possible without loss in enantioselectivity. The reaction is also tolerant to a range of aromatic substituents on the dipole (R^1 = Ph, *p*-Cl-Ph, *p*-OMe-Ph, *p*-Me-Ph, 2-naph, *c*-hex). Moreover, excellent levels of diastereo- and enantioselectivities can be achieved with an alkyl-substituted nitrone (R^1 = *c*-hex, 99:1 *endo:exo*, 99% *ee*). In contrast, only limited variation of the dipolarophile can be achieved; crotonaldehyde (R^2 = Me) and acrolein (R^2 = H) generate cycloadducts in good yields and selectivities, but other β-substituted enals are largely unsuccessful due to the sluggish nature of these reactions.

A polymer-supported version of catalyst **1** was also used in the nitrone cycloaddition, with promising results [58]. The enantioselectivity of the reaction was usually comparable to what has been observed in solution, although catalyst recycling was accompanied by substantial decrease in chemical yield.

The scope of the reaction was subsequently expanded to the 1,3-dipolar cycloaddition of nitrones to cyclic α,β-unsaturated aldehydes, allowing for the formation of fused bicyclic isoxazolidines (Scheme 3.7). Here, the use of catalyst **1** resulted in almost no reaction, whereas a proline-based diamine (**9**) afforded high levels of enantio-and diastereoselectivity [59, 60].

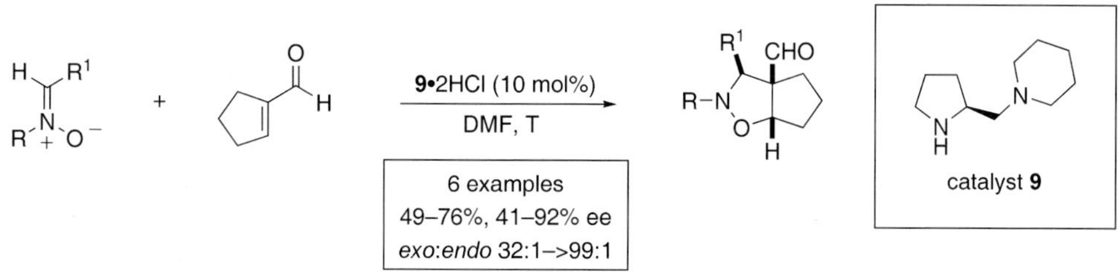

Scheme 3.7

3.5.3
Cyclopropanations

The biological activity of many natural isolates and therapeutic agents rely on the structural and reactivity properties of cyclopropane functionalities. Therefore, a vast array of asymmetric methods for their preparation has been developed [61]. Among all methodologies, few examples employ organocatalysts, but worthy of mention are the pioneering studies by Aggarwal and colleagues [62–64] and Gaunt and colleagues [65–67], who make use of catalyst-bound ylides to perform asymmetric cyclopropanations. Alternatively, Kunz and MacMillan have applied the iminium-catalysis concept to effectively activate olefin substrates towards enantioselective cyclopropanation [68]. In this context, a highly efficient protocol was found for the conversion of α,β-unsaturated aldehydes and stabilized ylides in

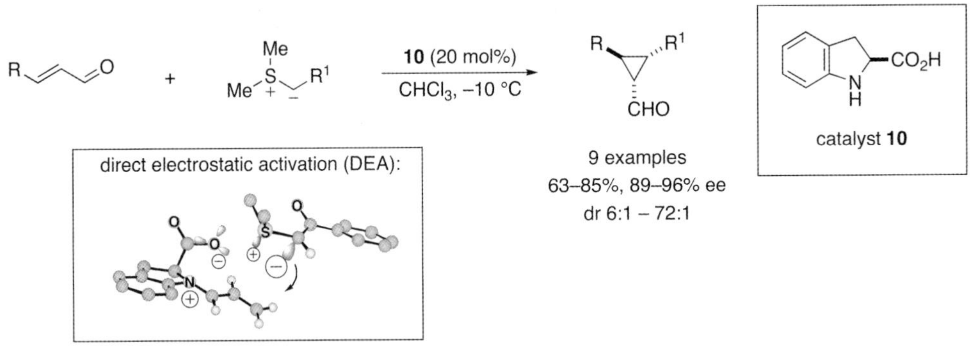

Scheme 3.8

the presence of 2-carboxylic acid dihydroindole catalyst (**10**) to enantioenriched highly substituted cyclopropanes (Scheme 3.8). (For experimental details see Chapter 14.11.1). Interestingly, the iminium ions derived from MacMillan imidazolidinones **1** or **3** and enals were completely inert to the sulfonium ylides used, and more importantly, proline (normally a poor catalyst for iminium catalysis) provided good levels of conversion and moderate enantioselectivity. Improvements in both chemical yield and stereoselectivity were obtained with the dihydroindole catalyst **10**.

In an effort to explain these atypical reactivity patterns, a mechanistic postulate based on direct electrostatic activation (DEA) was proposed (Scheme 3.8). Indeed, the zwitterionic iminium ions derived from catalyst **10** and α,β-unsaturated aldehydes enable both iminium geometry control and direct electrostatic activation of the approaching sulfonium ylides. The combination of geometric and electronic control seems to be essential for enantio- and diastereocontrol in the formation of the desired cyclopropyl compound.

3.5.4
Epoxidations

Chiral asymmetric epoxidations have been intensively investigated due to the fundamental importance of epoxides in organic chemistry [69, 70]. Nevertheless, catalytic asymmetric Lewis acid epoxidation of α,β-unsaturated aldehydes remains a challenge to chemists. Recently, Jørgensen and co-workers developed the first asymmetric approach to epoxides of enals, in which chiral pyrrolidine **11** was used as catalyst and H_2O_2 as oxidant, thus following the concept of iminium catalysis (Scheme 3.9) [71–73]. Importantly, reaction conditions are tolerant to a variety of functionalities and this chemical transformation proceeds in different solvents, with no loss of enantioselectivity. (For experimental details see Chapter 14.13.1).

From a mechanistic point of view, the first step is the formation of the iminium ion intermediate by reaction of the α,β-unsaturated aldehyde with the chiral

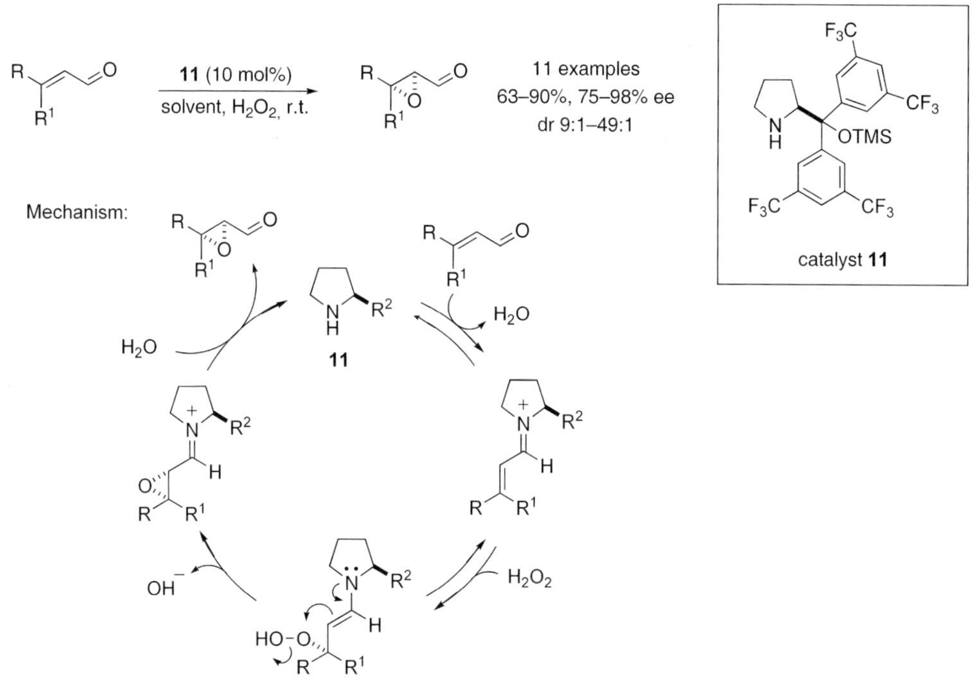

Scheme 3.9

amine. In the next step, the peroxide adds as a nucleophile to the electrophilic β-carbon atom producing a C–O-bond and leading to an enamine intermediate. The formation of the epoxide then takes place by attack of the nucleophilic enamine carbon atom to the electrophilic peroxygen atom, followed by hydrolysis of the iminium intermediate (Scheme 3.9).

3.5.5
[4+3]-Cycloadditions

Several laboratories are currently investigating the potential of iminium catalysis for the asymmetric catalytic synthesis of other cycloaddition products. For example, Harmata and co-workers recently disclosed an elegant approach for the preparation of enantioenriched seven-membered rings [74]. This approach involves an organocatalytic asymmetric [4+3]-cycloaddition of dienes with silyloxypentadienals in the presence of imidazolidinone catalyst **3**. Although showing promising prospects for the rapid synthesis of enantioenriched cycloheptanones, the scope of the reaction has yet to be determined. So far, only the reaction of 2,5-disubstituted furans with 4-trialkylsilylpentadienals has afforded the cycloadducts in modest yields and promising enantioselectivity (Eq. 6). It is notable that, among all asymmetric [4+3]-cycloaddition reactions, this represents the first organocatalytic version.

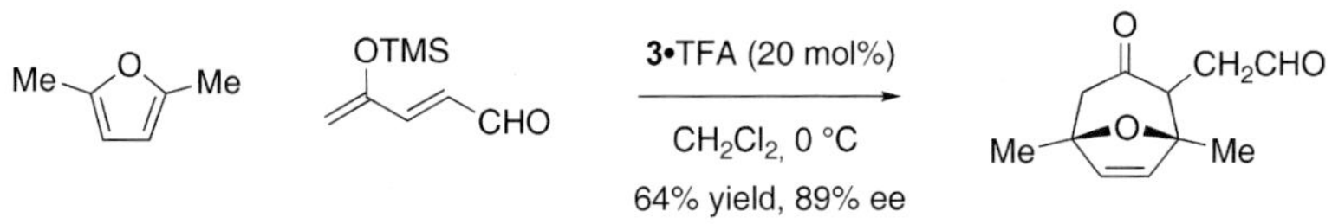

3.6
1,4-Addition Reactions

3.6.1
Friedel–Crafts Alkylations

The metal-catalyzed addition of aromatic substrates to σ- or π-systems, also known as Friedel–Crafts alkylation, belongs to one of the most powerful strategies for the formation of C–C bonds [75–77]. Nevertheless, relatively few enantioselective catalytic approaches have been reported that use this reaction manifold, despite the widespread availability of electron-rich aromatics and the chemical relevance of the resulting products.

Based on the mechanistic rationales discussed earlier, it is clear that π-facial selectivity and reaction rates of cycloaddition reactions result exclusively from the association of imidazolidinone catalysts to the electrophilic enal component. It is, therefore, conceivable that this platform should be amenable to a range of reactions of α,β-unsaturated carbonyl compounds, regardless of the nature of the HOMO-donor component.

To further demonstrate the value of iminium catalysis, MacMillan and coworkers undertook the development of asymmetric catalytic Friedel–Crafts alkylations that were previously unattainable using acid or metal catalysis. As such, amine-catalyzed 1,4-additions of aromatics and heteroaromatics to α,β-unsaturated aldehydes were investigated [13, 14, 17, 18]. Initial studies focused on the use of pyrroles as substrates to generate β-pyrrolyl carbonyls (Scheme 3.10) [13], useful synthons for the construction of a variety of biomedical agents [78–80]. As anticipated, N-protected and N-unprotected pyrroles underwent 1,4-addition to various α,β-unsaturated aldehydes in the presence of catalytic amounts of imidazolidinone **1** to provide the corresponding alkylated products in high yields and good enantioselectivity [81]. (For experimental details see Chapter 14.9.13).

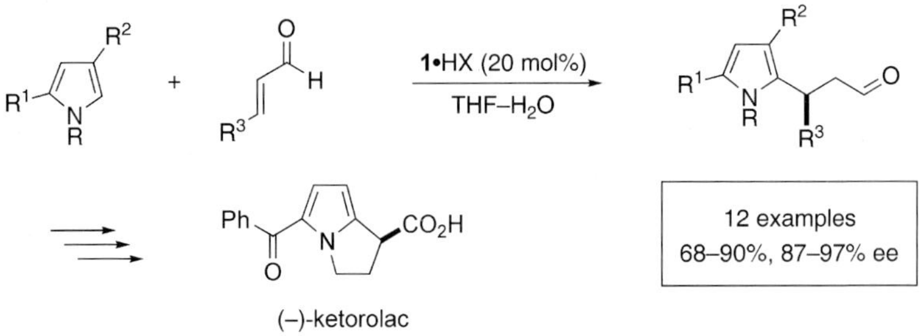

Scheme 3.10

It is important to note that products arising from 1,2-iminium additions were not observed in these reactions, in accord with the mechanistic postulate. The Friedel–Crafts alkylation is general with respect to the pyrrole architecture. Indeed, variation in the N-alkyl group (R = H, Me, Bn, Allyl) is possible without loss in yield or enantioselectivity. In addition, the incorporation of substituents at C(2)- or C(3)-pyrrole positions provides regioselective alkylations at C(5)- and C(2)-positions, respectively. Significant variation in the steric contribution of the olefin substituent (R^3 = Me, Pr, i-Pr) does not lead to any loss in yield or enantioselectivity. The reaction tolerates also β-aryl-substituted α,β-unsaturated aldehydes and accommodates electron-deficient aldehydes (R^3 = CO_2Me, CH_2OBn) that do not readily participate in iminium formation. Interestingly, the use of excess aldehydes in the reaction with N-methyl pyrrole allows for the formation of C_2-symmetric 2,5-disubstituted pyrroles in high diastereo- and enantioselectivity. The utility of this approach has been highlighted by the short and straightforward preparation of (−)-ketorolac [82].

The use of other nucleophilic species to perform Friedel–Crafts alkylations to α,β-unsaturated aldehydes has also been investigated, with success. For example, indoles [14, 83] and anilines [18, 84] have been added to α,β-unsaturated aldehydes, providing 1,4-addition products in good yields and excellent enantioselectivities. Again, the importance of such an approach is demonstrated by the suc-

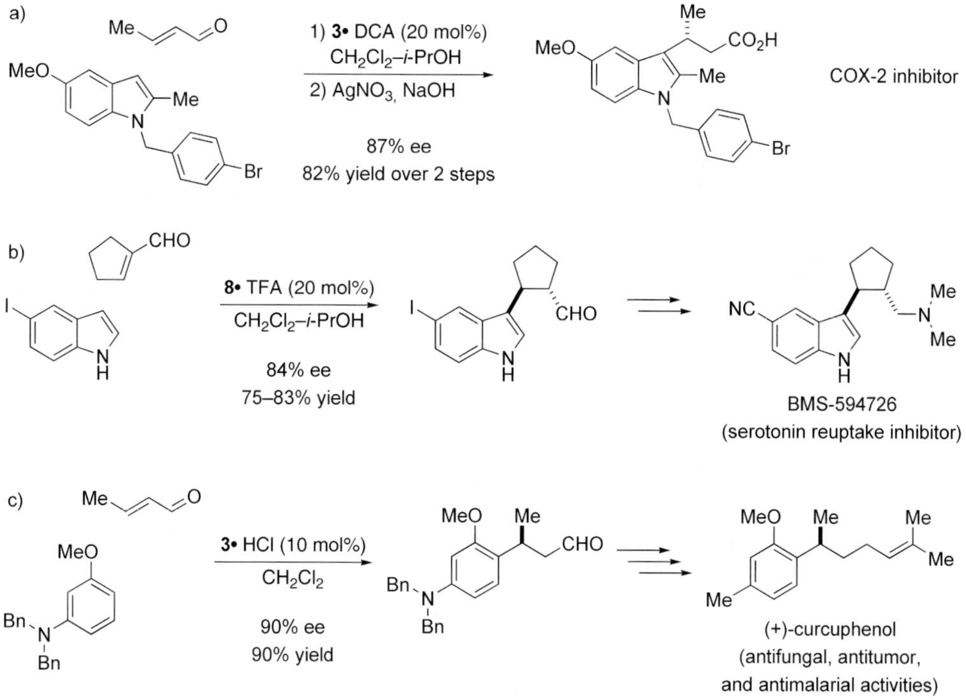

Scheme 3.11

cessful preparation of natural and biological relevant compounds (Scheme 3.11) [14, 83, 84].

3.6.2
Mukaiyama–Michael Reactions

Chiral imidazolidinone catalyst **3** can also catalyze the addition of silyloxy furans to α,β-unsaturated aldehydes to provide γ-butenolides [17]. Remarkably, whereas similar Lewis acid-catalyzed transformations give almost exclusively 1,2-addition products (Mukaiyama–aldol reaction), the organocatalyzed variant affords only 1,4-adducts (Mukaiyama–Michael reaction), thus providing a new strategy for the catalytic preparation of enantiomerically enriched butenolides [85]. Treatment of 2-silyloxyfurans with a variety of α,β-unsaturated aldehydes afforded the desired products in good enantiomeric purity with predominantly *syn*-selectivity. (For experimental details see Chapter 14.23.1). Significantly, the use of different reaction conditions – namely, changing co-catalyst and solvent – provided butenolide adducts of opposite sense of diastereoinduction, while retaining high levels of enantiocontrol. In a demonstration of the utility of this protocol, a four-step synthesis of spiculisporic acid, a *Penicillium spiculisporum* fermentation adduct that has found commercial application as bio-surfactant for metal decontamination processes and fine polymer production, and its 5-epi-diastereomer have been accomplished (Scheme 3.12) [17].

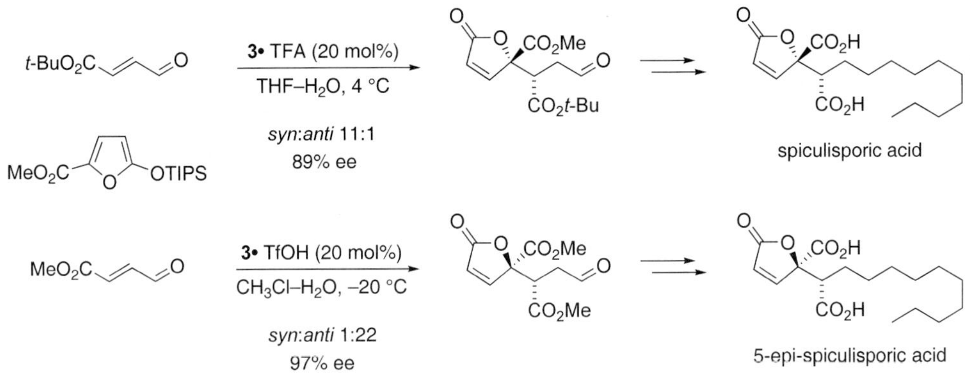

Scheme 3.12

3.6.3
Michael Reactions of α,β-Unsaturated Ketones

The Michael addition of carbogenic reagents to enones represents a challenging transformation for iminium catalysis. Given the inherent problem of forming tetrasubstituted iminium ions from ketones, along with the accordant issues associated with controlling the iminium ion geometry, it is noteworthy that significant

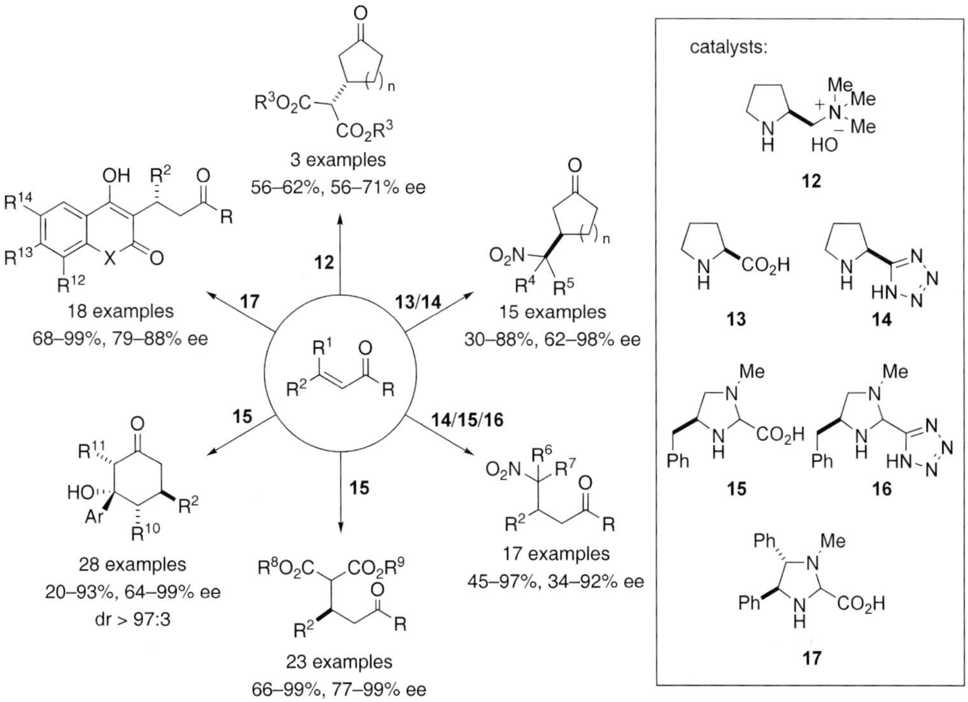

Scheme 3.13

progress has been made in the development of iminium catalysts for enone substrates during the past decade (Scheme 3.13). The asymmetric Michael addition to α,β-unsaturated carbonyl compounds was first catalyzed by metalloprolinates during the 1990s [86–90]. Several years later, Kawara and Taguchi reported the first organocatalyzed variant, in which a proline-derived catalyst (**12**) mediated the addition of malonates to cyclic and acyclic enones with moderate enantioselectivities (56–71% ee) [91]. Further improvements were reached by Hanessian and co-workers, who showed that a combination of L-proline (**13**) and trans-2,5-dimethylpiperazine could be used to facilitate the enantioselective addition of nitroalkanes to cyclic enones [92]. The use of tripeptides as catalysts was also investigated, and acceptable enantioselectivities were achieved with a trans-4-aminoproline-based tripeptide, in the presence of the same additive [93]. Although, no mechanistic interpretations have been proposed, it is likely that these reactions proceed via an iminium mechanism. Recently, Jørgensen and colleagues [94–99] and others [100–103] expanded the reaction scope by careful investigation of catalyst structure, developing organocatalyzed enantioselective conjugate addition reactions of a variety of carbogenic nucleophiles such as nitroalkanes [94, 99, 100, 103], malonates [95, 101], 1,3-dicarbonyl compounds [96–98, 102], β-ketosulphones [98] and aryl ketones [97, 98] to α,β-unsaturated enones (Scheme 3.13). (For experimental details see Chapter 14.9.1).

Remarkably, Jørgensen and co-workers demonstrated the effectiveness of this approach by synthesizing, in a very rapid fashion, enantiopure biologically active compounds such as warfarin and analogues [96]. Interestingly, some of these examples highlighted the use of highly enantio- and diastereoselective domino Michael–aldol reactions, furnishing optically active cyclohexanones with three or four contiguous chiral centers [97, 98].

3.7
Transfer Hydrogenation

The hydrogen atom is the most common substituent present on stereogenic centers. Not surprisingly, therefore, the field of asymmetric catalysis has focused great attention on the invention of hydrogenation methods during the past 50 years [104–106]. In demonstrating the importance of asymmetric hydrogenations, Knowles and Noyori were awarded the 2001 Nobel Prize in chemistry "... for their work on chirally catalyzed hydrogenation reactions" [107]. Interestingly, while most hydrogenation methodologies rely on the use of organometallic catalysts and hydrogen gas, living organisms typically use organic cofactors such as NADH or $FADH_2$ in combination with enzymes for similar transformations [108]. On this basis, the use of small organocatalysts in combination with dihydropyridine analogues to perform metal-free hydrogenations was a unique opportunity to challenge the LUMO-lowering activation concept in mimicking Nature. Studies from the MacMillan group [19] evidenced the possibility of selectively reducing α,β-unsaturated aldehydes of a broad steric and electronic composition using Hantzsch ester derivatives and imidazolidinone catalyst **18**, providing the products in good yields and excellent enantioselectivities (Scheme 3.14). (For experimental details see Chapter 14.22.1). List and co-workers reported a similar

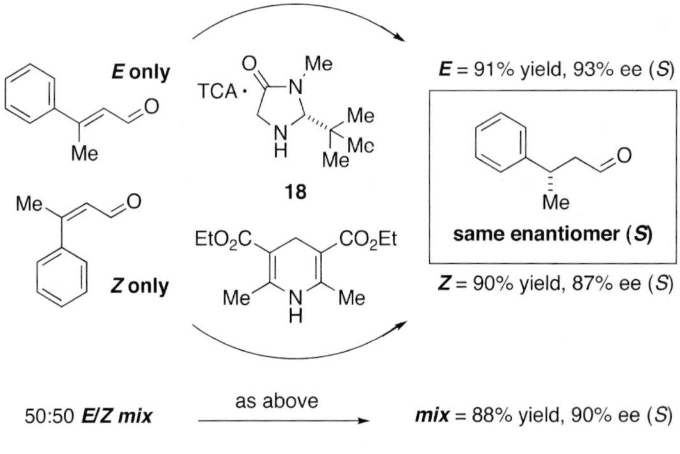

Scheme 3.14

variant of this transformation using MacMillan's imidazolidinone catalyst **3** [109, 110]. However, in this case, only β-aryl-β-methyl α,β-unsaturated aldehydes have been shown to undergo enantioselective reduction.

Importantly, mixtures of *E*- and *Z*-olefin substrates could be hydrogenated with comparable enantioselectivities, providing an enantioconvergent process; a highly desirable yet rare feature of a catalytic asymmetric reaction. In addition, this transformation effectively differentiates between β,β-olefin substituents of similar steric demand (e.g., Me/Et, Ar/*c*-hex), furnishing hydrogenated products with very high enantioselectivity.

3.8
Organocatalytic Cascade Reactions

3.8.1
Cascade Addition–Cyclization Reactions

Methodologies relying on cascade reactions have received increasing attention in modern chemical synthesis [111–114]. With this in mind, the MacMillan group investigated new approaches to expand the realm of iminium catalysis to include the activation of tandem bond-forming processes for the rapid construction of natural products [21]. In particular, based on mechanistic considerations, these authors sought to explore whether indole 1,4-addition reactions might be manipulated to allow the cascade formation of pyrroloindoline architectures in lieu of substituted indole production (Scheme 3.15).

It was envisioned that the addition of an indole derived from a tryptamine to the activated iminium ion, arising from imidazolidinone catalyst **3** and an α,β-unsaturated aldehyde, would generate a C(3)-quaternary carbon-substituted indolium ion. As a central feature this intermediate cannot undergo re-aromatization by means of proton loss, in contrast to the analogous 3-H indole addition pathway. As a result, 5-*exo*-heterocyclization of the pendant ethylamine would provide the corresponding pyrroloindoline compounds. In terms of molecular complexity, this cascade sequence should allow the rapid and enantioenriched formation of stereochemically defined pyrroloindoline architecture from tryptamines and simple α,β-unsaturated aldehydes.

Bringing these theoretical considerations into practice led to the development of an addition–cyclization cascade methodology, where tryptamines were added to α,β-unsaturated aldehydes in the presence of imidazolidinone catalysts, providing pyrroloindoline adducts in high yields and with excellent levels of enantioselectivity (Scheme 3.16) [21]. Interestingly, a large variation of enantioinduction was observed upon modification of the reaction solvent; a high-dielectric-constant media afforded one enantiomer, while a low-dielectric-constant solvent provided the corresponding enantiomer. (For experimental details see Chapter 14.9.14). The reaction appeared to be quite general with respect to the steric contribution of the carbamate substituent of the ethylenic amine functionality (Boc, CO_2Et,

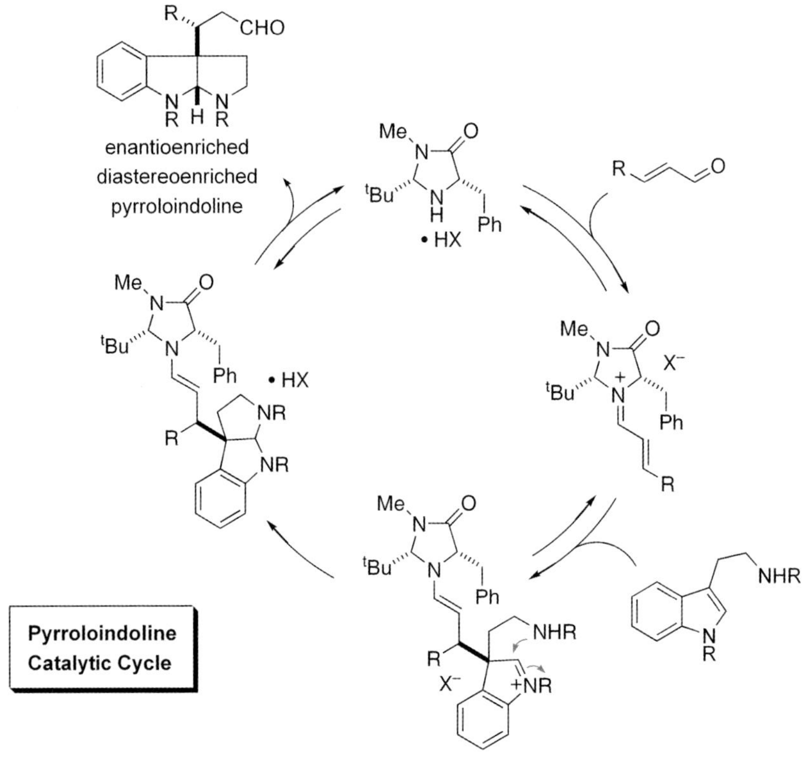

Scheme 3.15

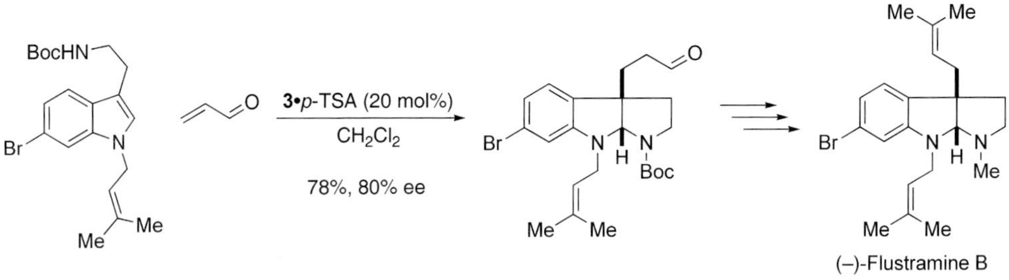

Scheme 3.16

CO₂allyl). In addition, the reaction could also accommodate a variety of electron-donating groups on the indole nitrogen (Bn, Allyl, Prenyl), as well as substitution on the aromatic ring. With regard to the α,β-unsaturated aldehyde substrate, β-substitution allowed the formation of highly diastereo- and enantioselective pyrroloindolines and the reaction was seen to be general for ester-, keto-, and CH₂OBz- groups. When acrolein was used as the enal, high levels of enantioselectivities were also observed. In these cases, enantioinduction was determined by the preference of tryptamines to react selectively from one of the indole faces.

Application of the pyrroloindoline-forming protocol in natural product synthesis was demonstrated by the first enantioselective synthesis of (−)-flustramine B (Scheme 3.16) [21]. Moreover, this amine-catalyzed transformation has also been extended to the enantioselective construction of furanoindoline frameworks, a widely represented substructure among natural isolates of biological relevance [21].

3.8.2
Cascade Catalysis: Merging Iminium and Enamine Activations

The synthesis of complex natural products has traditionally focused on a "stop-and-go" sequence of individual reactions. However, in biological systems, molecular complexity is achieved in a continuous process, where enzymatic transformations are combined in highly regulated catalytic cascade reactions [115]. With this in mind – and given the discovery from the MacMillan group that imidazolidinone catalysts can enforce orthogonal modes of substrate activation in the form of iminium (LUMO–lowering) [6, 12–19, 21] and enamine (HOMO-raising) [116–118] catalyses (Eqs. 7–9) – extensive research investigations have been conducted to translate the conceptual blueprints of biosynthesis into laboratory "cascade-catalysis" sequences. Specifically, several groups have succeeded in combining amine-catalyzed iminium- and enamine transformations to enable rapid access to structural complexity from simple starting materials, while achieving exquisite levels of enantiocontrol [20, 119, 120].

Imidazolidinones: Organocatalysts for LUMO or HOMO Activation Motives

Cascade Catalysis: Merging Iminium (Im) and Enamine (En) Activations

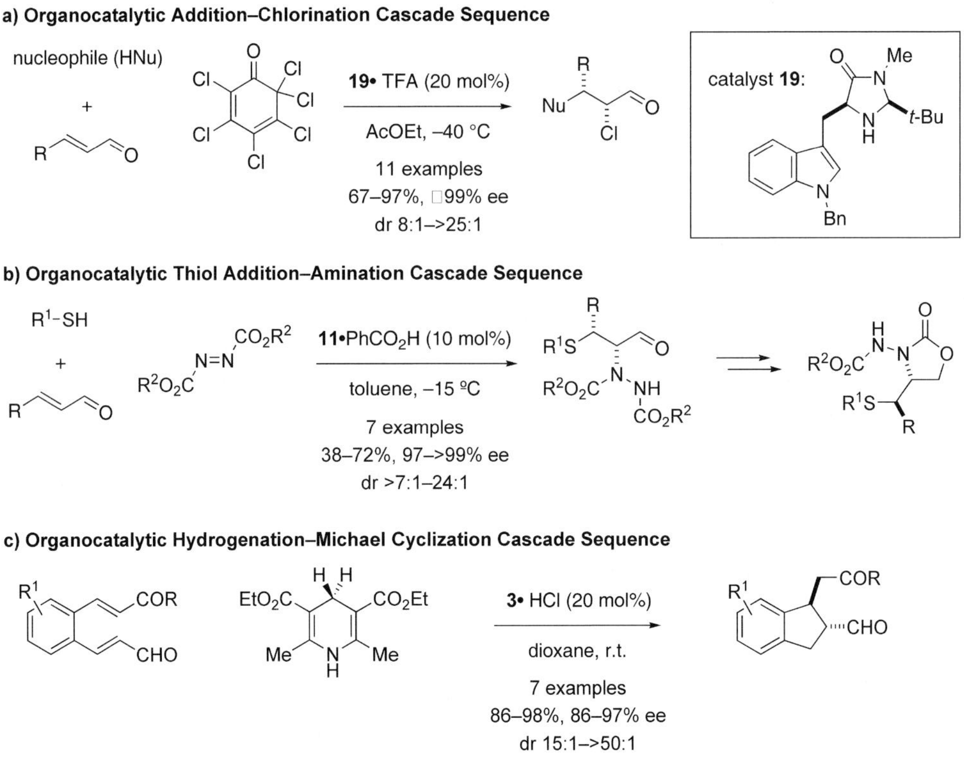

a) Organocatalytic Addition–Chlorination Cascade Sequence

b) Organocatalytic Thiol Addition–Amination Cascade Sequence

c) Organocatalytic Hydrogenation–Michael Cyclization Cascade Sequence

Scheme 3.17

As proof of concept, the MacMillan group [20] discovered that imidazolidinone **19** catalyzed the conjugate addition–chlorination cascade sequence of a diverse range of nucleophiles and α,β-unsaturated aldehydes to give the corresponding products with high levels of diastereo- and enantioselectivity (Scheme 3.17a). (For experimental details see Chapter 14.23.2). Further expansion of this new cascade approach allowed the invention of other enantioselective transformations, such as the asymmetric addition of HCl and HF across trisubstituted olefin systems [20]. Similarly, Jørgensen and co-workers developed an organocatalyzed conjugate addition–amination sequence of thiols to enals with the intermediacy of pyrrolidine **11** [119], affording 1,2-aminothiol derivatives with excellent enantioselectivities (>99% ee; Scheme 3.17b). (For experimental details see Chapter 14.23.3). In addition, List and co-workers reported a transfer hydrogenation–intramolecular Michael reaction cascade sequence, where MacMillans' imidazolidinone catalyst **3·HCl** gave the best selectivities (Scheme 3.17) [120].

Perhaps the most important point in these studies was the discovery that two discrete amine catalysts could be employed to enforce cycle-specific selectivities (Scheme 3.18) [20]. Conceptually, this achievement demonstrates that these cascade-catalysis pathways can be readily modulated to afford a required

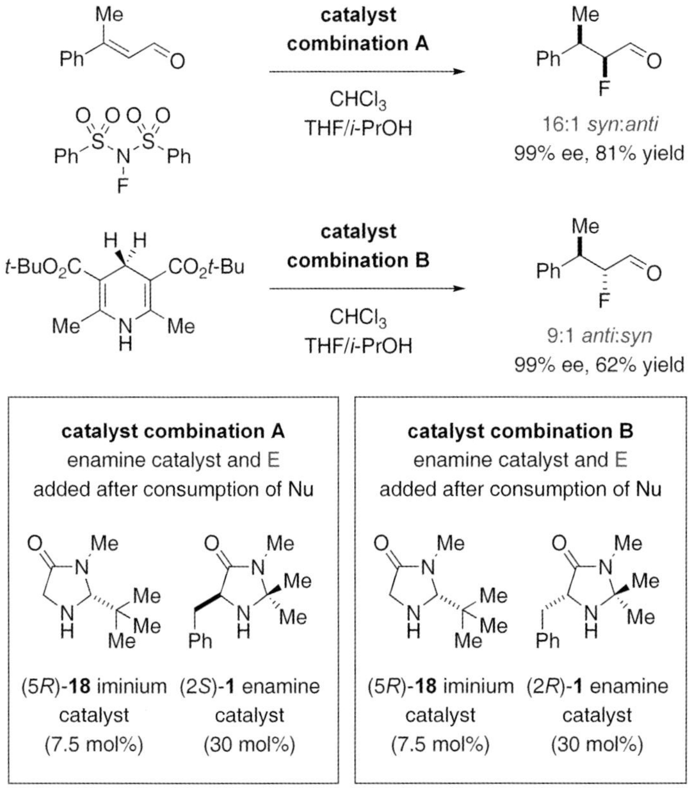

Scheme 3.18

catalyst combination A
enamine catalyst and E
added after consumption of Nu

(5R)-**18** iminium catalyst (7.5 mol%)
(2S)-**1** enamine catalyst (30 mol%)

catalyst combination B
enamine catalyst and E
added after consumption of Nu

(5R)-**18** iminium catalyst (7.5 mol%)
(2R)-**1** enamine catalyst (30 mol%)

diastereo- and enantioselective outcome via the judicious selection of simple amine catalysts.

3.9 Conclusions

Over the past six years, the field of asymmetric catalysis has bloomed extensively – and perhaps unexpectedly – with the introduction of a variety of metal-free-catalysis concepts that have collectively become known as "organocatalysis". Perhaps more impressively, the field of organocatalysis has quickly become a fundamental branch of catalysis, which can be utilized for the construction of enantiopure organic structures, thus providing a valuable complement to organometallic and enzymatic activations. Whilst substrate scope remains an important issue for many organocatalytic reactions, an increasingly large number of transformations are now meeting the requisite high standards of "useful" enantioselective processes. Most notably, the concept of iminium catalysis has grown almost hand in hand with the general field of organocatalysis. Since the introduction of the first highly enantioselective organocatalytic Diels–Alder reaction in 2000, there has been a large expansion in this amine-catalyzed subfield. Indeed, at the time of writing of this chapter, there exist currently over 40 discrete transformations that can be performed with useful levels of enantiocontrol ($\geq$90% *ee*). As such, the future for iminium catalysis and the field of organocatalysis appears to be bright, with perhaps application to industrial processes being the next major stage of development. One thing is certain, however – there are many new powerful enantioselective transformations waiting to be discovered using these novel modes of activation.

Acknowledgments

The authors acknowledge the financial support provided by the NIH National Institute of General Medical Sciences, and gifts from Amgen, Lilly, Bristol-Meyers Squibb, Johnson and Johnson, and Merck Research Laboratories. G.L. is grateful to the Swiss National Science Foundation (Stefano Franscini Fond), the Roche Foundation, and the Novartis Foundation for postdoctoral fellowship supports.

References

1 P.I. Dalko, L. Moisan, *Angew. Chem., Int. Ed.* **2001**, *40*, 3726.
2 P.I. Dalko, L. Moisan, *Angew. Chem., Int. Ed.* **2004**, *43*, 5138.
3 A. Berkessel, H. Gröger, *Asymmetric Organocatalysis: From Biomimetic Concepts to Applications in Asymmetric Synthesis*, Wiley-VCH, Weinheim, **2005**.
4 J. Seayad, B. List, *Org. Biomol. Chem.* **2005**, *3*, 719.
5 G. Lelais, D.W.C. MacMillan, *Aldrichim. Acta* **2006**, *39*, 79.
6 K.A. Ahrendt, C.J. Borths, D.W.C. MacMillan, *J. Am. Chem. Soc.* **2000**, *122*, 4243.
7 V.A. Tarasevich, N.G. Kozlov, *Russ. Chem. Rev.* **1999**, *68*, 55.
8 M.E. Jung, W.D. Vaccaro, K.R. Buszek, *Tetrahedron Lett.* **1989**, *30*, 1893.
9 P.A. Grieco, S.D. Larsen, *J. Org. Chem.* **1986**, *51*, 3553.
10 P.A. Grieco, S.D. Larsen, W.F. Fobare, *Tetrahedron Lett.* **1986**, *27*, 1975.
11 H. Waldman, *Angew. Chem., Int. Ed.* **1988**, *27*, 274.
12 W.S. Jen, J.J.M. Wiener, D.W.C. MacMillan, *J. Am. Chem. Soc.* **2000**, *122*, 9874.
13 N.A. Paras, D.W.C. MacMillan, *J. Am. Chem. Soc.* **2001**, *123*, 4370.
14 J.F. Austin, D.W.C. MacMillan, *J. Am. Chem. Soc.* **2002**, *124*, 1172.
15 R.M. Wilson, W.S. Jen, D.W.C. MacMillan, *J. Am. Chem. Soc.* **2005**, *127*, 11616.
16 A.B. Northrup, D.W.C. MacMillan, *J. Am. Chem. Soc.* **2002**, *124*, 2458.
17 S.P. Brown, N.C. Goodwin, D.W.C. MacMillan, *J. Am. Chem. Soc.* **2003**, *125*, 1192.
18 N.A. Paras, D.W.C. MacMillan, *J. Am. Chem. Soc.* **2002**, *124*, 7894.
19 S.G. Ouellet, J.B. Tuttle, D.W.C. MacMillan, *J. Am. Chem. Soc.* **2005**, *127*, 32.
20 Y. Huang, A.M. Walji, C.H. Larsen, D.W.C. MacMillan, *J. Am. Chem. Soc.* **2005**, *127*, 15051.

21 J.F. Austin, S.-G. Kim, C.J. Sinz, W.-J. Xiao, D.W.C. MacMillan, *Proc. Natl. Acad. Sci. USA* **2004**, *101*, 5482.
22 For recent reviews of enantioselective Diels–Alder reactions, see: (a) W. Oppolzer, in: B.M. Trost, I. Flemming (Eds.), *Comprehensive Organic Synthesis*, Vol. 5, Pergamon Press, New York, **1991**; (b) H.B. Kagan, O. Riant, *Chem. Rev.* **1992**, *92*, 1007; (c) L.C. Dias, *J. Braz. Chem. Soc.* **1997**, *8*, 289; (d) D.A. Evans, J.S. Johnson, in: E.N. Jacobsen, A. Pfaltz, H. Yamamoto (Eds.), *Comprehensive Asymmetric Catalysis*, Vol. 3, Springer, New York, **1999**; (e) E.J. Corey, *Angew. Chem., Int. Ed.* **2002**, *41*, 1650.
23 M. Zora, *J. Mol. Struct. (Theochem)* **2002**, *619*, 121.
24 M.C. Kozlowski, M. Panda, *J. Org. Chem.* **2003**, *68*, 2061.
25 R. Gordillo, K.N. Houk, *J. Am. Chem. Soc.* **2006**, *128*, 3543.
26 M. Benaglia, G. Celentano, M. Cinquini, A. Puglisi, F. Cozzi, *Adv. Synth. Catal.* **2002**, *344*, 149.
27 S.A. Selkälä, J. Tois, P.M. Pihko, A.M.P. Koskinen, *Adv. Synth. Catal.* **2002**, *344*, 941.
28 A.C. Kinsman, M.A. Kerr, *J. Am. Chem. Soc.* **2003**, *125*, 14120.
29 J.K. Park, P. Sreekanth, B.M. Kim, *Adv. Synth. Catal.* **2004**, *346*, 49.
30 S.A. Selkälä, A.M.P. Koskinen, *Eur. J. Org. Chem.* **2005**, 1620.
31 K. Ishihara, K. Nakano, *J. Am. Chem. Soc.* **2005**, *127*, 10504.
32 M. Lemay, W.W. Ogilvie, *Org. Lett.* **2005**, *7*, 4141.
33 J.L. Cavill, R.L. Elliott, I.L. Jones, J.A. Platts, A.M. Ruda, N.C.O. Tomkinson, *Tetrahedron* **2006**, *62*, 410.
34 For a recent review on polymer-supported organic catalysts, see: M. Benaglia, A. Puglisi, F. Cozzi, *Chem. Rev.* **2003**, *103*, 3401.
35 For an example of asymmetric Lewis acid-catalyzed Type II IMDA, see: C.P. Chow, K.J. Shea, *J. Am. Chem. Soc.* **2005**, *127*, 3678.
36 K. Mikami, M. Terada, Y. Motoyama, T. Nakai, *Tetrahedron: Asymmetry* **1991**, *2*, 643.
37 T.A. Engler, M.A. Letavic, F. Takusagawa, *Tetrahedron Lett.* **1992**, *33*, 6731.
38 T.A. Engler, M.A. Letavic, K.O. Lynch, Jr., F. Takusagawa, *J. Org. Chem.* **1994**, *59*, 1179.
39 K. Mikami, Y. Motoyama, M. Terada, *J. Am. Chem. Soc.* **1994**, *116*, 2812.
40 J.D. White, Y. Choi, *Org. Lett.* **2000**, *2*, 2373.
41 M. Breuning, E.J. Corey, *Org. Lett.* **2001**, *3*, 1559.
42 J.D. White, Y. Choi, *Helv. Chim. Acta* **2002**, *85*, 4306.
43 D.H. Ryu, G. Zhou, E.J. Corey, *J. Am. Chem. Soc.* **2004**, *126*, 4800.
44 Y. Honda, T. Date, H. Hiramatsu, M. Yamauchi, *Chem. Commun.* **1997**, 1411.
45 S. Otto, G. Boccaletti, J.B.F.N. Engberts, *J. Am. Chem. Soc.* **1998**, *120*, 4238.
46 S. Otto, J.B.F.N. Engberts, *J. Am. Chem. Soc.* **1999**, *121*, 6798.
47 D.A. Evans, J. Wu, *J. Am. Chem. Soc.* **2003**, *125*, 10162.
48 For recent examples of asymmetric Lewis acid-catalyzed Diels–Alder reactions of ketone dienophiles, see: (a) D.H. Ryu, T.W. Lee, E.J. Corey, *J. Am. Chem. Soc.* **2002**, *124*, 9992; (b) D.H. Ryu, E.J. Corey, *J. Am. Chem. Soc.* **2003**, *125*, 6388; (c) J.M. Hawkins, M. Nambu, S. Loren, *Org. Lett.* **2003**, *5*, 4293; (d) R.S. Singh, T. Harada, *Eur. J. Org. Chem.* **2005**, 3433.
49 M. Frederickson, *Tetrahedron* **1997**, *53*, 403.
50 J.-P.G. Seerden, A.W.A. Schotte op Reimer, H.W. Scheeren, *Tetrahedron Lett.* **1994**, *35*, 4419.
51 K.V. Gothelf, K.A. Jorgensen, *J. Org. Chem.* **1994**, *59*, 5687.
52 For recent reviews on catalytic asymmetric 1,3-cycloadditions: (a) K.V. Gothelf, K.A. Jørgensen, *Chem. Commun.* **2000**, 1449; (b) S. Kobayashi, K.A. Jørgensen, *Cycloaddition Reactions in Organic Synthesis*, Wiley-VCH, Weinheim, **2002**; (c) K. Rück-Braun, T.H.E. Freysold, F. Wierschem, *Chem. Soc. Rev.* **2005**, *34*, 507.

53 S. Kanemasa, N. Ueno, M. Shirahase, *Tetrahedron Lett.* **2002**, *43*, 657.
54 F. Viton, G. Bernardinelli, E.P. Kündig, *J. Am. Chem. Soc.* **2002**, *124*, 4968.
55 S. Kanemasa, in: S. Kobayashi, K.A. Jørgensen (Eds.), *Cycloaddition Reactions in Organic Synthesis*. Wiley-VCH, Weinheim, **2002**.
56 T. Mita, N. Ohtsuki, T. Ikeno, T. Yamada, *Org. Lett.* **2002**, *4*, 2457.
57 S. Kezuka, N. Ohtsuki, T. Mita, Y. Kogami, T. Ashizawa, T. Ikeno, T. Yamada, *Bull. Chem. Soc. Jpn.* **2003**, *76*, 2197.
58 A. Puglisi, M. Benaglia, M. Cinquini, F. Cozzi, G. Celentano, *Eur. J. Org. Chem.* **2004**, 567.
59 S. Karlsson, H.-E. Högberg, *Tetrahedron: Asymmetry* **2002**, *13*, 923.
60 S. Karlsson, H.-E. Högberg, *Eur. J. Org. Chem.* **2003**, 2782.
61 For a recent review on stereoselective cyclopropanation, see: H. Lebel, J.-F. Marcoux, C. Molinaro, A.B. Charette, *Chem. Rev.* **2003**, *103*, 977.
62 V.K. Aggarwal, H.W. Smith, R.V.H. Jones, R. Fieldhouse, *Chem. Commun.* **1997**, 1785.
63 V.K. Aggarwal, H.W. Smith, G. Hynd, R.V.H. Jones, R. Fieldhouse, S.E. Spey, *J. Chem. Soc., Perkin Trans. 1* **2000**, 3267.
64 V.K. Aggarwal, E. Alonso, G. Fang, M. Ferrara, G. Hynd, M. Porcelloni, *Angew. Chem., Int. Ed.* **2001**, *40*, 1433.
65 C.D. Papageorgiou, S.V. Ley, M.J. Gaunt, *Angew. Chem., Int. Ed.* **2003**, *42*, 828.
66 N. Bremeyer, S.C. Smith, S.V. Ley, M.J. Gaunt, *Angew. Chem., Int. Ed.* **2004**, *43*, 2681.
67 C.D. Papageorgiou, M.A. Cubillo de Dios, S.V. Ley, M.J. Gaunt, *Angew. Chem., Int. Ed.* **2004**, *43*, 4641.
68 R.K. Kunz, D.W.C. MacMillan, *J. Am. Chem. Soc.* **2005**, *127*, 3240.
69 For reviews of asymmetric catalytic epoxidations, see: (a) R.S. Johnson, K.B. Sharpless, in: B.M. Trost, I. Flemming (Eds.), *Comprehensive Organic Synthesis*, Vol. 7, Pergamon Press, New York, **1991**; (b) R. Noyori, *Asymmetric Catalysis in Organic Synthesis*, John Wiley & Sons, New York, **1994**; (c) A.-H. Li, L.-X. Dai, V.K. Aggarwal, *Chem. Rev.* **1997**, *97*, 2341; (d) E.N. Jacobsen, A. Pfaltz, H. Yamamoto (Eds.), *Comprehensive Asymmetric Catalysis*. Springer, New York, **1999**; (e) I. Ojima (Ed.), *Catalytic Asymmetric Synthesis*, 2nd edn. Wiley, New York, **2000**; (f) V.K. Aggarwal, J. Richardson, *Chem. Commun.* **2003**, 2644; (g) Z.-G. Zhang, X.-Y. Wang, C. Sun, H.-C. Shi, *Chin. J. Org. Chem.* **2004**, *24*, 7; (h) V.K. Aggarwal, C. Winn, *Acc. Chem. Res.* **2004**, *37*, 611; (i) Q.-H. Xia, H.-Q. Ge, C.-P. Ye, Z.-M. Liu, K.-X. Su, *Chem. Rev.* **2005**, *105*, 1603.
70 For reviews of asymmetric catalytic epoxidation with chiral ketones, see: (a) S.E. Denmark, Z. Wu, *Synlett* **1999**, 847; (b) M. Frohn, Y. Shi, *Synlett* **2000**, 1979; (c) Y. Shi, *J. Synth. Org. Chem. Jpn.* **2002**, *60*, 342; (d) Y. Shi, *Acc. Chem. Res.* **2004**, *37*, 488; (e) D. Yang, *Acc. Chem. Res.* **2004**, *37*, 497.
71 M. Marigo, J. Franzén, T.B. Poulsen, W. Zhuang, K.A. Jørgensen, *J. Am. Chem. Soc.* **2005**, *127*, 6964.
72 W. Zhuang, M. Marigo, K.A. Jørgensen, *Org. Biomol. Chem.* **2005**, 3883.
73 H. Sundén, I. Ibrahem, A. Córdova, *Tetrahedron Lett.* **2006**, *47*, 99.
74 M. Harmata, S.K. Ghosh, X. Hong, S. Wacharasindhu, P. Kirchhoefer, *J. Am. Chem. Soc.* **2003**, *125*, 2058.
75 G.A. Olah, *Friedel-Crafts and Related Reactions*, Vols. 1–4. Wiley-Interscience, New York, **1963–1965**.
76 G.A. Olah, *Friedel-Crafts Chemistry*. Wiley, New York, **1973**.
77 R.M. Roberts, A.A. Khalaf, *Friedel-Crafts Alkylation Chemistry: A Century of Discovery*, M. Dekker, New York, **1984**.
78 D. Della Bella, *Boll. Chim. Farm.* **1972**, *111*, 5.
79 A. Guzman, F. Yuste, R.A. Toscano, J.M. Young, A.R. Van Horn, J.M. Muchowski, *J. Med. Chem.* **1986**, *29*, 589.
80 A. Kleemann, J. Engel, B. Kutscher, D. Reichert, *Pharmaceutical*

Substances: Syntheses, Patents, Applications, 4th edn. Thieme, Stuttgart, **2001**.

81 For computational studies on the Friedel–Crafts alkylation of pyrroles and indoles catalyzed by chiral imidazolidinones, see: R. Gordillo, J. Carter, K.N. Houk, *Adv. Synth. Catal.* **2004**, *346*, 1175.

82 This methodology was further employed for the preparation of the medicinal agent (−)-ketorolac, making use of our second-generation catalyst **31** (*vide infra*): R.L. Pederson, I.M. Fellows, T.A. Ung, H. Ishihara, S.P. Hajela, *Adv. Synth. Catal.* **2002**, *344*, 728.

83 H.D. King, Z. Meng, D. Denhart, R. Mattson, R. Kimura, D. Wu, Q. Gao, J.E. Macor, *Org. Lett.* **2005**, *7*, 3437.

84 S.-G. Kim, J. Kim, H. Jung, *Tetrahedron Lett.* **2005**, *46*, 2437.

85 The use of simple silyl enol ethers for the asymmetric organocatalyzed Mukaiyama–Michael addition was recently reported. For reference, see: W. Wang, H. Li, J. Wang, *Org. Lett.* **2005**, *7*, 1637.

86 M. Yamaguchi, N. Yokota, T. Minami, *J. Chem. Soc., Chem. Commun.* **1991**, 1088.

87 M. Yamaguchi, T. Shiraishi, M. Hirama, *Angew. Chem., Int. Ed.* **1993**, *32*, 1176.

88 M. Yamaguchi, T. Shiraishi, Y. Igarashi, M. Hirama, *Tetrahedron Lett.* **1994**, *35*, 8233.

89 M. Yamaguchi, T. Shiraishi, M. Hirama, *J. Org. Chem.* **1996**, *61*, 3520.

90 M. Yamaguchi, Y. Igarashi, R.S. Reddy, T. Shiraishi, M. Hirama, *Tetrahedron* **1997**, *53*, 11223.

91 A. Kawara, T. Taguchi, *Tetrahedron Lett.* **1994**, *35*, 8805.

92 S. Hanessian, V. Pham, *Org. Lett.* **2000**, *2*, 2975.

93 S.B. Tsogoeva, S.B. Jagtap, Z.A. Ardemasova, V.N. Kalikhevich, *Eur. J. Org. Chem.* **2004**, 4014.

94 N. Halland, R.G. Hazell, K.A. Jørgensen, *J. Org. Chem.* **2002**, *67*, 8331.

95 N. Halland, P.S. Aburel, K.A. Jørgensen, *Angew. Chem., Int. Ed.* **2003**, *42*, 661.

96 N. Halland, T. Hansen, K.A. Jørgensen, *Angew. Chem., Int. Ed.* **2003**, *42*, 4955.

97 N. Halland, P.S. Aburel, K.A. Jorgensen, *Angew. Chem., Int. Ed.* **2004**, *43*, 1272.

98 J. Pulkkinen, P.S. Aburel, N. Halland, K.A. Jørgensen, *Adv. Synth. Catal.* **2004**, *346*, 1077.

99 A. Prieto, N. Halland, K.A. Jørgensen, *Org. Lett.* **2005**, *7*, 3897.

100 M. Benaglia, M. Cinquini, F. Cozzi, A. Puglisi, G. Celentano, *J. Mol. Cat. A: Chemical* **2003**, *204–205*, 157.

101 B. Dhevalapally, C.F. Barbas, III, *Chem. Eur. J.* **2004**, *10*, 5323.

102 D. Gryko, *Tetrahedron: Asymmetry* **2005**, *16*, 1377.

103 C.E.T. Mitchell, S.E. Brenner, S.V. Ley, *Chem. Commun.* **2005**, 5346.

104 S. Akabori, S. Sakurai, Y. Izumi, Y. Fujii, *Nature* **1956**, *178*, 323.

105 T. Ohkuma, M. Kitamura, R. Noyori, in: I. Ojima (Ed.), *Catalytic Asymmetric Synthesis*, 2nd edn. Wiley-VCH, New York, **2000**, p. 1.

106 R. Noyori, *Angew. Chem., Int. Ed.* **2002**, *41*, 2008.

107 http://nobelprize.org/chemistry/laureates/2001/press.html.

108 B. Alberts, A. Johnson, J. Lewis, M. Raff, K. Roberts, P. Walter, *Molecular Biology of the Cell*, 4th edn. Garland, New York, **2002**.

109 J.W. Yang, M.T. Hechavarria Fonseca, B. List, *Angew. Chem. Int. Ed.* **2004**, *43*, 6660.

110 J.W. Yang, M.T. Hechavarria Fonseca, N. Vignola, B. List, *Angew. Chem. Int. Ed.* **2005**, *44*, 108.

111 L.F. Tietze, *Chem. Rev.* **1996**, *96*, 115.

112 K.C. Nicolaou, T. Montagnon, S.A. Snyder, *Chem. Commun.* **2003**, 551.

113 D.J. Ramón, M. Yus, *Angew. Chem., Int. Ed.* **2005**, *44*, 1602.

114 J.-C. Wasilke, S.J. Obrey, R.T. Baker, G.C. Bazan, *Chem. Rev.* **2005**, *105*, 1001.

115 For selected reviews on this topic, see: (a) L. Katz, *Chem. Rev.* **1997**, *97*, 2557; (b) C. Khosla, *Chem. Rev.* **1997**, *97*, 2577; (c) C. Khosla, R.S. Gokhale, J.R. Jacobsen, D.E. Cane, *Annu. Rev. Biochem.* **1999**, *68*, 219; (d) J.

Staunton, K.J. Weissman, *Nat. Prod. Rep.* **2001**, *18*, 380.

116 M.P. Brochu, S.P. Brown, D.W.C. MacMillan, *J. Am. Chem. Soc.* **2004**, *126*, 4108.

117 I.K. Mangion, A.B. Northrup, D.W.C. MacMillan, *Angew. Chem., Int. Ed.* **2004**, *43*, 6722.

118 T.D. Beeson, D.W.C. MacMillan, *J. Am. Chem. Soc.* **2005**, *127*, 8826.

119 M. Marigo, T. Schulte, J. Franzén, K.A. Jørgensen, *J. Am. Chem. Soc.* **2005**, *127*, 15710.

120 J.W. Yang, M.T. Hechavarria Fonseca, B. List, *J. Am. Chem. Soc.* **2005**, *127*, 15036.

4
Ammonium Ions as Chiral Templates

Takashi Ooi and Keiji Maruoka

4.1
Introduction

Chiral, nonracemic quaternary ammonium salts have been emerging as a powerful metal-free catalyst for effecting various stereoselective bond-forming reactions under mild conditions. This organocatalysis functions in either homogeneous or heterogeneous system, and the reactivity and selectivity rely heavily on the three-dimensional (3-D) structure of chiral ammonium cations and the property of counteranions. The former homogeneous catalysis is based on the nucleophilic or basic character of ammonium fluorides and phenoxides, while the latter heterogeneous catalysis is well recognized as phase-transfer catalysis. In this chapter we first illustrate the utility of chiral quaternary ammonium fluorides and phenoxides as organocatalysts for homogeneous reactions [1]. The synthetic benefits of asymmetric phase-transfer catalysis are then described, with particular focus on the most significant application – the asymmetric synthesis of α-amino acid derivatives [2] – which is especially valuable in the utilization of chiral quaternary ammonium salts for routine experiments in both academic and industrial laboratories.

4.2
Homogeneous Catalysis with Chiral Quaternary Ammonium Fluorides

4.2.1
Aldol and Nitroaldol Reactions (Preparation of Chiral Quaternary Ammonium Fluorides)

Shioiri and co-workers systematically investigated the preparation of N-benzylcinchonium fluoride **2a** from the corresponding bromide **1a**, and established a standard procedure (Scheme 4.1) [3] which has been used repeatedly by research groups in this field (for examples, see below). The ^{1}H NMR analysis of the fluoride **2a** indicated no decomposition of the N-benzylcinchonium residue,

Enantioselective Organocatalysis: Reactions and Experimental Procedures
Edited by Peter I. Dalko
Copyright © 2007 WILEY-VCH Verlag GmbH & Co. KGaA, Weinheim
ISBN: 978-3-527-31522-2

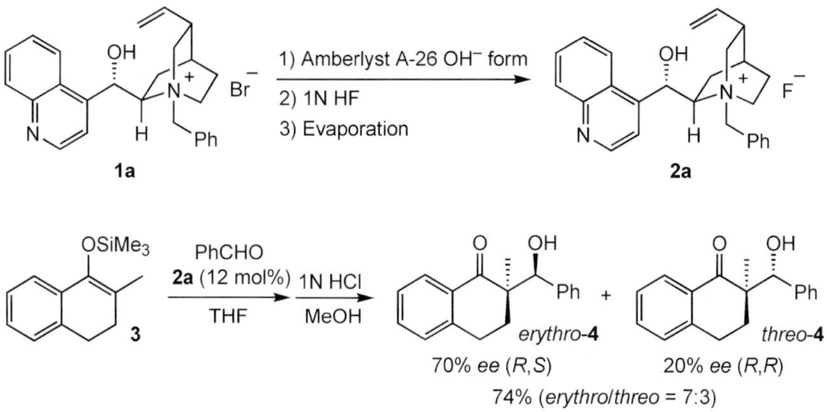

Scheme 4.1

and ^{19}F NMR measurements in CD$_2$Cl$_2$ showed a peak centered at ca. −124 ppm (using CFCl$_3$ as an internal standard) [4]. The catalytic activity and chiral efficiency of **2a** were evaluated in the asymmetric aldol reaction of enol silyl ether of 2-methyl-1-tetralone (**3**) with benzaldehyde as included in Scheme 4.1 [3].

Further examination on the fluoride ion-catalyzed asymmetric aldol reaction of the enol silyl ethers prepared from acetophenone (**5**) and pinacolone (**6**) with benzaldehyde using **2a** and its peudoenantiomer **7a** revealed the dependence of the stereochemistry of the reactions on the hydroxymethyl-quinuclidine fragment of the catalyst (Scheme 4.2) [3, 5].

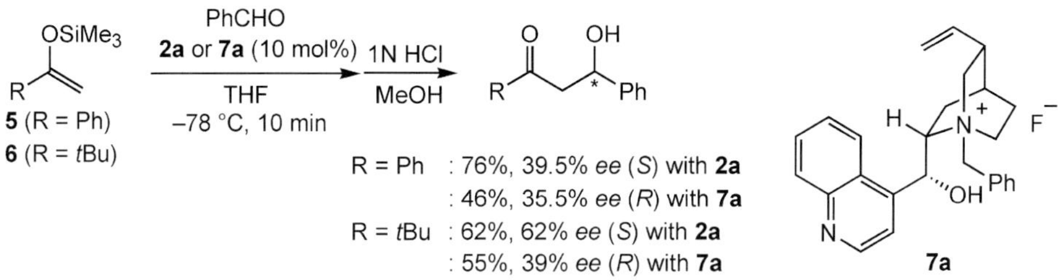

Scheme 4.2 (For experimental details see Chapter 14.1.4)

Campagne and Bluet reported the catalytic asymmetric vinylogous Mukaiyama (CAVM) reaction of aldehydes with dienol silyl ether **8** using chiral ammonium fluorides as an activator. For example, the CAVM reaction of isobutyraldehyde with **8** in the presence of **2a** (10 mol%) in THF at room temperature led to formation of the vinylogous aldol product **9** in 70% yield with 20% *ee*; moreover, the enantiomeric excess was improved to 30% by conducting the reaction at 0 °C (Scheme 4.3) [6].

Corey and Zhang utilized chiral quaternary ammonium fluoride **7b** possessing a 9-anthracenylmethyl group on nitrogen for the face-selective nitroaldol reaction

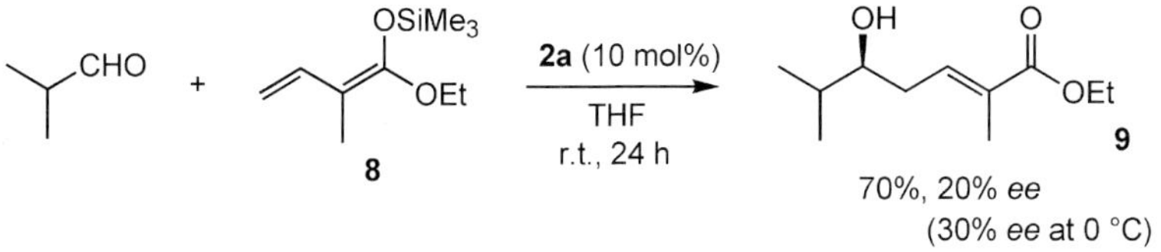

Scheme 4.3

of nitromethane with protected (S)-phenylalaninal **10**. This was directed toward the practical stereoselective synthesis of amprenavir, an important second-generation HIV protease inhibitor with a number of clinical advantages over first-generation agents. A THF solution of (S)-N,N-dibenzylphenylalaninal (**10**) was added to a mixture of **7b**, nitromethane, and finely ground potassium fluoride (KF) in THF at −10 °C. After stirring for 6 h, the desired nitro alcohol **11** was isolated in 86% yield with a diastereomeric ratio (dr) of 17:1, from which amprenavir was synthesized in a five-step sequence, as illustrated in Scheme 4.4 [7].

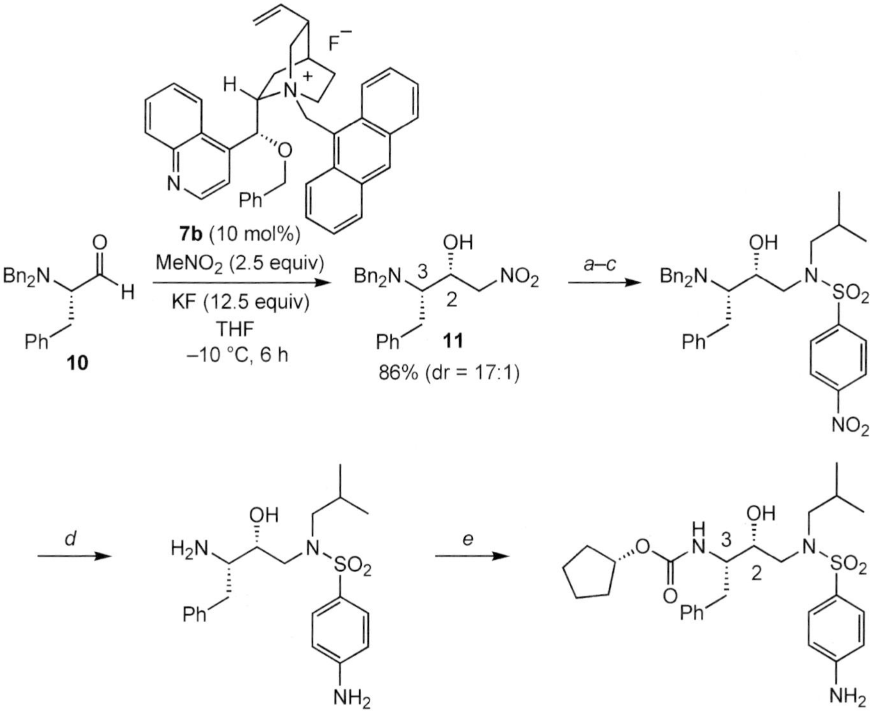

[a] NiCl$_2$, NaBH$_4$, MeOH, 0 °C (85%); [b] isobutyraldehyde, MgSO$_4$, then NaBH$_4$, EtOH, 0~23 °C (82%); [c] p-nitrobenzenesulfonyl chloride, Et$_3$N, CH$_2$Cl$_2$, 23 °C (94%); [d] H$_2$ (1 atm), Pd(OH)$_2$/C, MeOH, 23 °C; [e] (S)-3-tetrahydrofuranyl-N-oxysuccinimidyl carbonate, Et$_3$N, CH$_2$Cl$_2$, 23 °C (95% in 2 steps).

Scheme 4.4 (For experimental details see Chapter 14.4.1)

4.2.2
Trifluoromethylation

Iseki, Nagai and Kobayashi prepared cinchonine-derived ammonium fluoride **2b** from the corresponding bromide, and accomplished the asymmetric trifluoromethylation of aldehydes and ketones with trifluoromethyltrimethylsilane (Me$_3$SiCF$_3$) catalyzed by **2b** (Scheme 4.5) [8]. Although the enantioselectivities are not sufficiently high, this reaction system should offer a new access to various chiral trifluoromethylated molecules of analytical and medicinal interest through appropriate modifications.

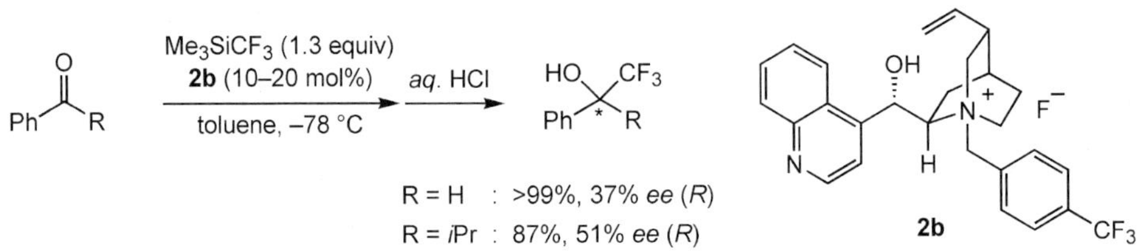

Scheme 4.5

4.2.3
Hydrosilylation

The hydrosilylation of carbonyl compounds with polymethylhydrosiloxane (PMHS) or other alkoxysilanes can be catalyzed by TBAF, at high efficiency [9]. The asymmetric version of this process has been developed by Lawrence and co-workers using chiral ammonium fluoride **7c** prepared via the method of Shioiri [10]. The reduction of acetophenone was performed with trimethoxysilane (1.5 equiv.) and **7c** (10 mol%) in THF at room temperature, yielding phenethyl alcohol quantitatively with 51% ee (R) (Scheme 4.6). A slightly higher enantioselectivity was observed in the reduction of propiophenone. When tris(trimethylsiloxy)silane was used as a hydride source, the enantioselectivity was increased, though a pro-

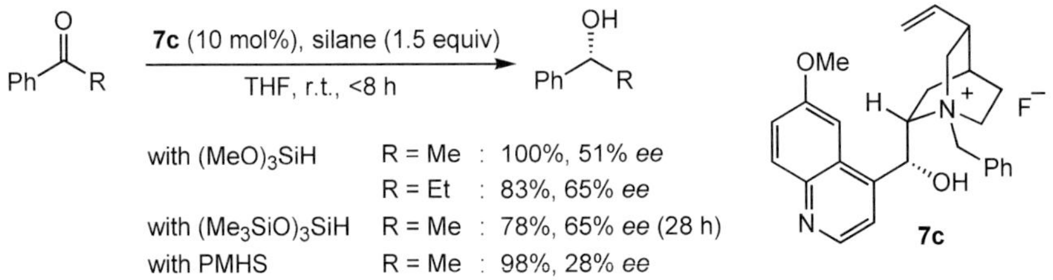

Scheme 4.6

longed reaction time was required. Although a significant rate acceleration was observed with PMHS, the stereoselectivity was, unfortunately, decreased.

4.3
Homogeneous Catalysis with Chiral Quaternary Ammonium Bifluorides

4.3.1
Aldol and Nitroaldol Reactions

Although tetraalkylammonium bifluoride, $R_4N^+HF_2^-$, is expected to be more stable and easy to handle compared to the corresponding fluoride [11], it is only recently that the reactivity and selectivity have been investigated in the field of asymmetric catalysis. Corey and co-workers prepared the cinchonidine-derived bifluoride **12** from the corresponding bromide by passage of a methanolic solution through a column of Amberlyst A-26 OH$^-$ form and subsequent neutralization with 2 equiv. of 1 N HF solution and evaporation (the modified Shioiri method). The catalytic activity and chiral efficiency of **12** (dried *in vacuo* over P_2O_5) have been demonstrated by the development of a Mukaiyama-type aldol reaction of ketene silyl acetal **13** with aldehydes under mild conditions, giving mostly *syn*-β-hydroxy-α-amino esters **14** as the major diastereomer, with good to excellent *ee*-values (Scheme 4.7 and Table 4.1) [12]. The highest *syn* selectivity was observed

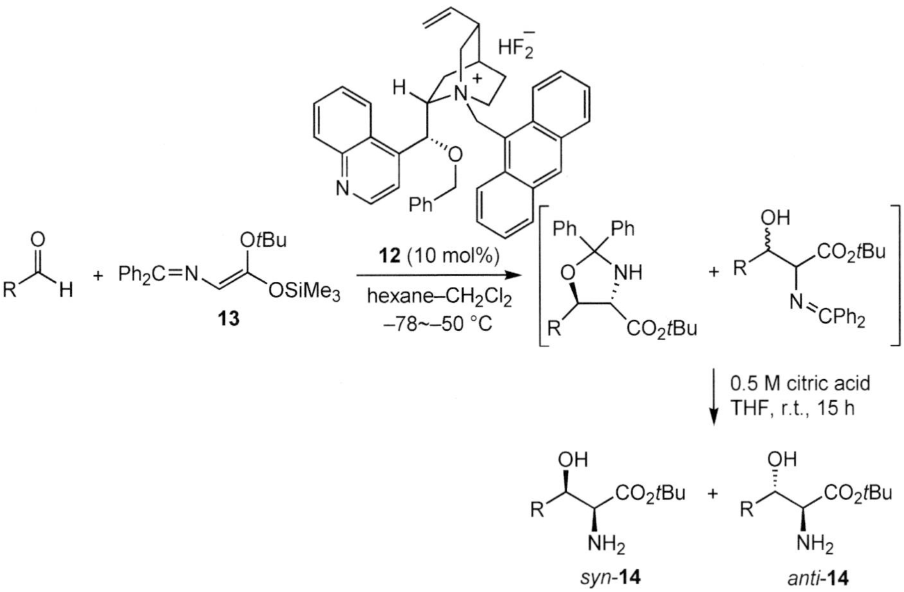

Scheme 4.7

Table 4.1 The chiral ammonium bifluoride **12**-catalyzed asymmetric Mukaiyama-type aldol reaction of ketene silyl acetal **13** with aldehydes. (For experimental details see Chapter 14.1.5)

Entry	R	Solvent ($Hex:CH_2Cl_2$)	Temperature [°C]	Time [h]	Yield [%]	dr (*syn*/*anti*)	*ee* [%] (*syn*:*anti*)
1	*i*Pr	3:1	−78	7	70 (**14a**)	6:1	95:83
2	*c*Hex	3:1	−50	1	81	13:1	88:46
3	Hex	3:1	−78	2	79	3:1	89:91
4	$Cl(CH_2)_3$	5:1	−78	2	48	1:1	82:86
5	$Ph(CH_2)_2$	3:1	−78	6	64	1:1	72:86
6	*i*Bu	5:1	−45	2	61	3:1	76:70

in the reaction with cyclohexanecarboxaldehyde (entry 2), while a lower diastereomeric ratio was generally associated with unbranched aldehydes (entries 3–6).

The nitroaldol reaction of silyl nitronates with aldehydes promoted by ammonium fluorides, which was originally introduced by Seebach and Colvin in 1978 [13], is a useful method for the preparation of 1,2-functionalized nitroalkanols. We have developed an asymmetric version of high efficiency and stereoselectivity by using a designer chiral quaternary ammonium bifluoride **15** as catalyst; this was readily prepared from the corresponding bromide using the modified Shioiri method (Scheme 4.8) [14]. For example, the treatment of trimethylsilyl nitronate **16a** with benzaldehyde in the presence of (*S*,*S*)-**15a** (2 mol%) in THF at −98 °C

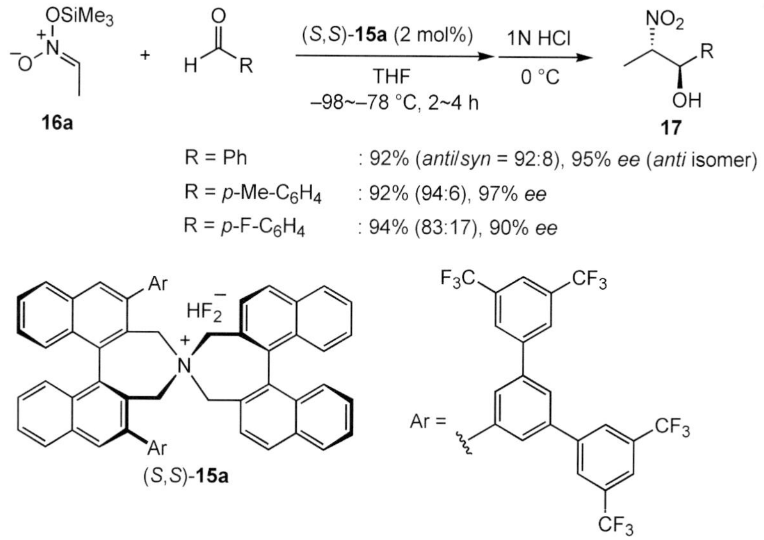
OSiMe3
16a
(S,S)-15a (2 mol%)
THF
−98~−78 °C, 2~4 h
1N HCl
0 °C
NO2
R
OH
17
R = Ph : 92% (anti/syn = 92:8), 95% ee (anti isomer)
R = p-Me-C6H4 : 92% (94:6), 97% ee
R = p-F-C6H4 : 94% (83:17), 90% ee
Ar
HF2
N
(S,S)-15a
Ar =
F3C
CF3

Scheme 4.8

for 1 h and at −78 °C for 1 h, with subsequent hydrolysis with 1 N HCl at 0 °C, resulted in clean formation of the corresponding nitroalkanol **17** (R = Ph) in 92% yield (*anti*:*syn* = 92:8) with 95% *ee* (*anti* isomer). This asymmetric nitroaldol protocol tolerates various aromatic aldehydes to afford *anti*-nitroaldols selectively, as included in Scheme 4.8.

4.3.2
Michael Reaction

The efficient homogeneous catalysis of chiral ammonium bifluorides of type **15** has been further utilized for achieving an asymmetric Michael addition of silyl nitronates to α,β-unsaturated aldehydes. Here, chiral ammonium bifluoride **15b** bearing a 3,5-di-*tert*-butylphenyl group was found to be the catalyst of choice, and the reaction of **16a** with *trans*-cinnamaldehyde under the influence of (*R*,*R*)-**15b** (2 mol%) in THF at −78 °C produced the 1,4-addition product **18** predominantly (**18/19** = 24:1) as a diastereomeric mixture (*syn*/*anti* = 78:22) with 85% *ee* of the major *syn* isomer (Scheme 4.9). Further, use of toluene as solvent led to almost exclusive formation of the 1,4-adduct (**18/19** = 32:1) with similar diastereoselectivity (*syn*/*anti* = 81:19), and critical enhancement of the enantioselectivity was attained (97% *ee*) [15].

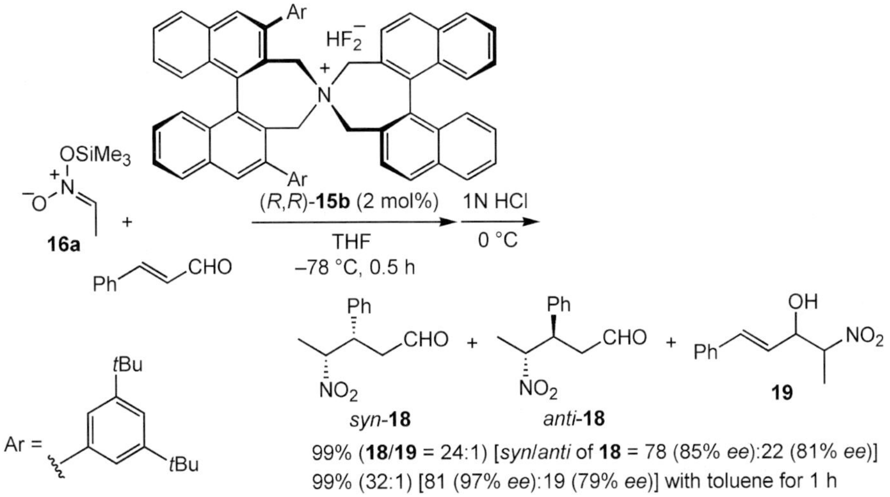

Scheme 4.9

The significant synthetic advantage of this approach is the isolation of regio- and stereo-defined enol silyl ethers of optically active γ-nitro aldehydes (Table 4.2). For example, after the reaction of **16a** with *trans*-cinnamaldehyde, the resulting mixture can be directly purified by silica gel column chromatography to produce the optically active enol silyl ether **20a** in 90% yield (Table 4.2, entry 1). High

4 Ammonium Ions as Chiral Templates

Table 4.2 Asymmetric Michael addition of silyl nitronate **16** to α,β-unsaturated aldehydes and cyclohexenone catalyzed by chiral quaternary ammonium bifluorides (R,R)-**15**. Isolation of optically active enol silyl ethers **20** and **21**.

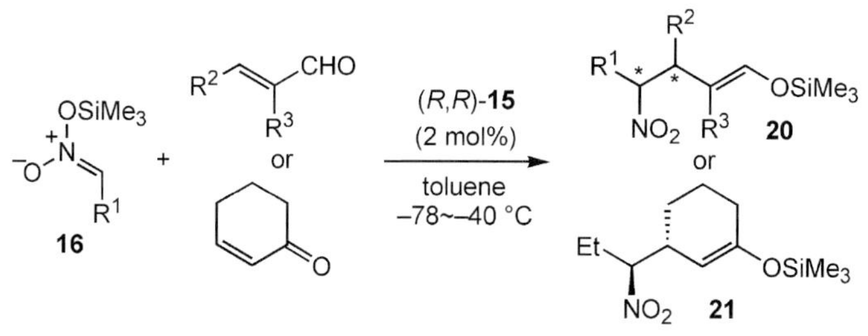

Entry	R^1 (**16**)	R^2, R^3 (aldehyde or cyclohexenone)	Yield [%]	dr (syn/anti)	ee of major isomer (**20** or **21**) [%]
1	Me (**16a**)	Ph, H **15b**	90	83:17 (**20a**)	97
2	Et (**16b**)	Ph, H	87	90:10	98
3	Me	Ph, Me	90	5:95	95
4	Et	cyclohexenone **15a**	91	1:99	96

levels of catalytic efficiency and stereoselectivity were also available in the Michael addition of silyl nitronate **16b** (entry 2). The introduction of an alkyl substituent at the α-carbon of enals can be well accommodated, as excellent diastereo- and enantiofacial differentiation have been achieved with α-methyl-*trans*-cinnamaldehyde (entry 3).

This unique Michael addition protocol has been successfully extended to cyclic α,β-unsaturated ketones such as cyclohexenone, where (R,R)-**15a** was suitable to

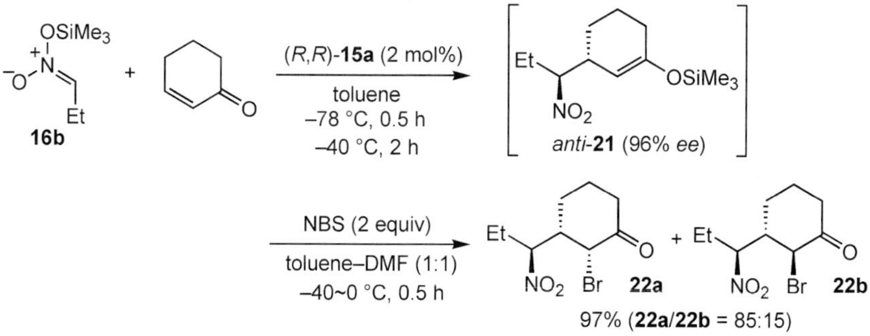

Scheme 4.10 (For experimental details see Chapter 14.9.9)

allow the isolation of stereochemically homogeneous enol silyl ether **21** (entry 4 in Table 4.2). The versatility of **21** as a chiral building block was highlighted by the stereoselective transformation to the corresponding α-bromo-γ-nitro ketone **22** (Scheme 4.10) [16].

4.4
Homogeneous Catalysis with Chiral Quaternary Ammonium Phenoxides

Recently, Mukaiyama and co-workers prepared cinchona alkaloid-derived chiral quaternary ammonium phenoxide-phenol complex **23** and used it as an efficient organocatalyst for the tandem Michael addition and lactonization between α,β-unsaturated ketones and a ketene silyl acetal **24** derived from phenyl isobutyrate. This approach permits the highly enantioselective synthesis of a series of 3,4-dihydropyran-2-ones (**25**), as shown in Scheme 4.11 [17].

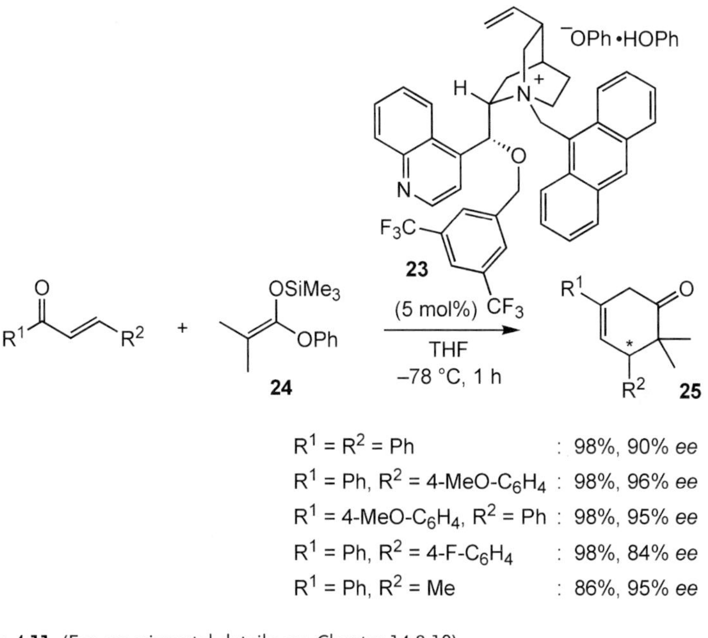

$R^1 = R^2 = Ph$: 98%, 90% ee
$R^1 = Ph, R^2 = 4\text{-MeO-}C_6H_4$: 98%, 96% ee
$R^1 = 4\text{-MeO-}C_6H_4, R^2 = Ph$: 98%, 95% ee
$R^1 = Ph, R^2 = 4\text{-F-}C_6H_4$: 98%, 84% ee
$R^1 = Ph, R^2 = Me$: 86%, 95% ee

Scheme 4.11 (For experimental details see Chapter 14.9.10)

4.5
Heterogeneous Catalysis: Chiral Phase-Transfer Catalysis

4.5.1
Pioneering Study

The development of asymmetric phase-transfer catalysis, which is based on the use of structurally well-defined chiral catalysts to create optically active organic molecules from prochiral substrates, was triggered by the pioneering studies of the Merck research group in 1984. By using the quaternary ammonium salt **1b** derived from cinchonine, the methylation of phenylindanone derivative **26** gave the corresponding alkylated product **27** in excellent yield, with high enantiomeric excess (Scheme 4.12) [18].

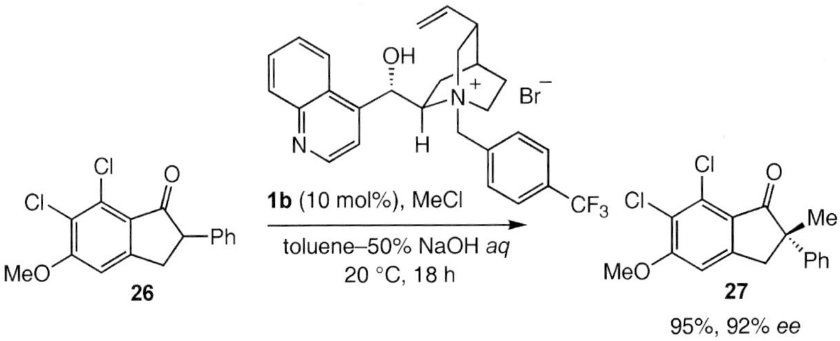

Scheme 4.12

4.5.2
Monoalkylation of Glycinate Schiff Base: Asymmetric Synthesis of α-Amino Acids

Following 5 years of epoch-creating investigations by the Merck group, this type of catalyst was used successfully for the asymmetric synthesis of α-amino acids by O'Donnell, using *tert*-butyl glycinate benzophenone Schiff base **28** as a key substrate [19]. Asymmetric alkylation of **28** proceeded smoothly under mild phase-transfer conditions with *N*-(benzyl)cinchoninium chloride [**1a** (Cl)] as a catalyst to give the alkylation product (*R*)-**29** in good yield, and moderate enantioselectivity (Scheme 4.13). By simply switching the catalyst from the cinchonine- to the cinchonidine-derived **30a**, the absolute configuration of the product was reversed (*S*), albeit with a similar degree of enantioselectivity. One important aspect of this reaction is the selective formation of the monoalkylated product **29**, without concomitant production of the undesired dialkylated product (this occurs as long as the benzophenone Schiff base is employed as a starting substrate [20]). This outcome is due to the much lower acidity of the remaining α-proton of **29** compared to that of **28**. This acidity-weakening effect is also crucial for securing the configurational stability of the newly created α-stereogenic center under the reaction conditions. Further optimization with hydroxy-protected catalyst **30b** (a second-generation catalyst) enhanced the enantioselectivity to 81% *ee* [21]. A single recrystallization and subsequent deprotection of **29** afforded essentially optically pure α-amino acids.

A significant advance in this field was made recently by two independent research groups through the development of a new class of cinchona alkaloid-

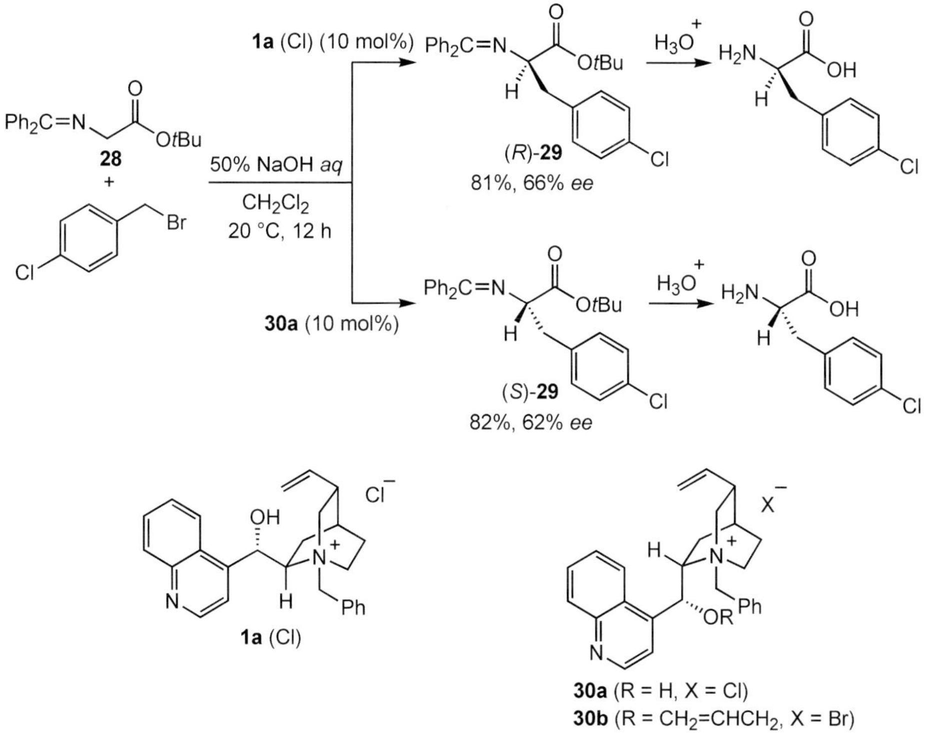

4.5 Heterogeneous Catalysis: Chiral Phase-Transfer Catalysis | 131

Scheme 4.13

derived catalysts bearing an N-anthracenylmethyl function (a third-generation catalyst). Lygo and colleagues designed N-anthracenylmethylammonium salts **1c** and **30c**, and applied them to the asymmetric phase-transfer alkylation of **28** to synthesize α-amino acids with much higher enantioselectivity (Scheme 4.14) [22].

In contrast, Corey and co-workers prepared O-allyl-N-anthracenylmethylcinchonidinium salt **30d** and achieved high asymmetric induction by the combined use of solid CsOH·H$_2$O at very low temperature, as also shown in Scheme 4.14 [23].

We designed and prepared the structurally rigid, chiral spiro ammonium salts **32** as a new C$_2$-symmetric chiral phase-transfer catalyst (Fig. 4.1), and successfully applied this to the highly efficient, catalytic enantioselective alkylation of **28** under mild conditions [24]. A key finding was the significant effect of an aromatic substituent at the 3,3′-position of one binaphthyl subunit of **32** (Ar) on the enantiofacial discrimination, and (S,S)-**32e** was revealed as the catalyst of choice for preparing a variety of essentially enantiopure α-amino acids by this transformation (Table 4.3).

Generally, 1 mol% of **32e** is sufficient for the smooth alkylation. In the reaction with simple alkyl halides such as ethyl iodide, the use of aqueous cesium hydrox-

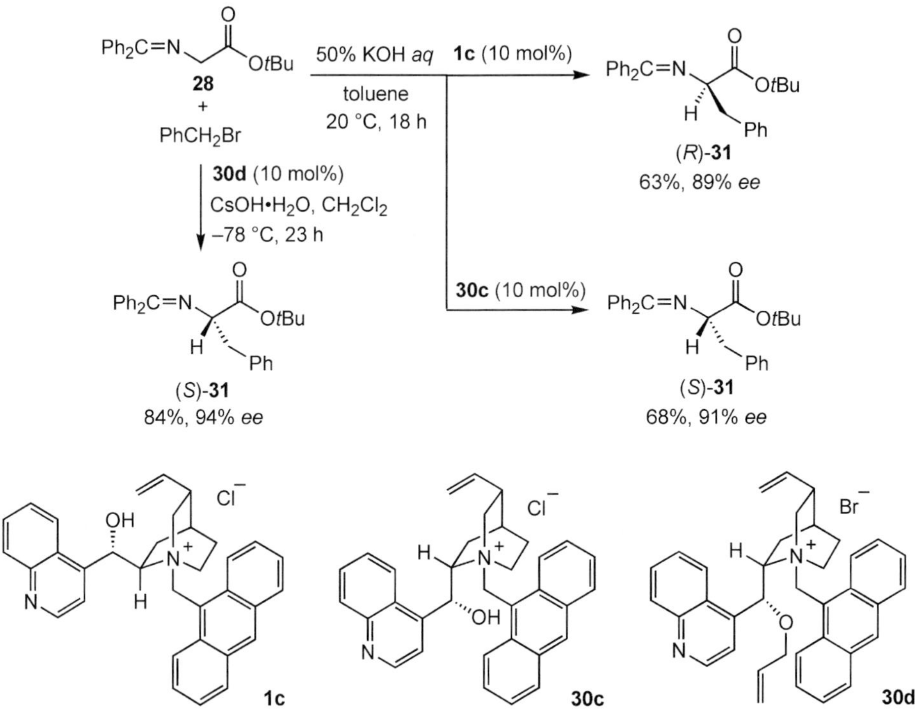

Scheme 4.14

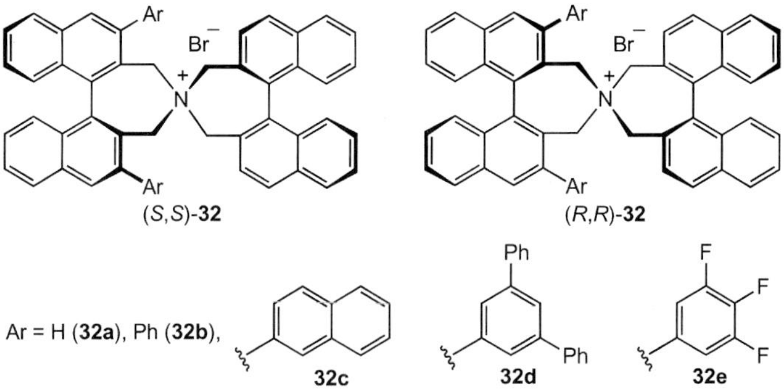

Fig. 4.1 The structurally rigid, chiral spiro ammonium salts.

ide (CsOH) as a basic phase at lower reaction temperature is recommended (entry 6).

These reports have accelerated research into improvements of the asymmetric alkylation of **28**, and have resulted in the emergence of a series of appropriately modified cinchona-alkaloid-based catalysts, as well as the elaboration of purely

Table 4.3 Effect of aromatic substituent (Ar) and applicability of 32e-catalyzed phase-transfer alkylation of **28**.

Ph₂C=N–CH(OtBu)(C=O) + RX →[(S,S)-**32** (1 mol%), toluene–50% KOH aq, 0 °C] Ph₂C=N–C(H)(R)(C=O)OtBu

Entry	Catalyst	RX	Yield [%]	ee [%] (config.)
1	32a	PhCH₂Br	73	79 (R)
2	32b	PhCH₂Br	81	89 (R)
3	32c	PhCH₂Br	95	96 (R)
4	32d	PhCH₂Br	91	98 (R)
5	32e	PhCH₂Br	90	99 (R)
6[a]	32e	EtI	89	98 (R)
7	32e	allyl-Br	80	99 (R)
8	32e	2,6-Me₂C₆H₃CH₂Br	98	99 (R)
9	32e	4-(PhC(O))C₆H₄CH₂Br	86	98 (R)

[a] With sat. CsOH at −15 °C.

synthetic chiral quaternary ammonium salts. The performance of the representative catalysts in the asymmetric benzylation of **28** are summarized in Table 4.4, in order to facilitate comparison for their preparative use.

While alkyl halides are typically employed as an electrophile for this transformation, Takemoto developed a palladium-catalyzed asymmetric allylic alkylation of **28** using allylic acetates and chiral phase-transfer catalyst **30h**, as shown in Scheme 4.15 [42]. The choice of triphenyl phosphite [(PhO)₃P] as an achiral palladium ligand was crucial to achieving high enantioselectivity.

Catalytic asymmetric alkylations of **28** have also been carried out with polymer-bound glycine substrates [43], or in the presence of polymer-supported cinchona alkaloid-derived ammonium salts as immobilized chiral phase-transfer catalysts [44], both of which feature their practical advantages especially for large-scale synthesis.

In addition, the potential synthetic utility of the asymmetric alkylation protocol discussed in this section has been fruitfully demonstrated by its application to the stereoselective synthesis of various biologically active natural products [45].

Table 4.4 Representative catalysts and their performance in the phase-transfer-catalyzed benzylation of **28**.

Entry	Catalyst	Conditions	Yield [%]	ee [%] (config.)	Ref.
1	**33**	5 mol%, 50% KOH aq., toluene:CH$_2$Cl$_2$ (7:3), 0 °C	93	94 (S)	25
2	**34**	3 mol%, 50% KOH aq., toluene:CH$_2$Cl$_2$ (7:3), −20 °C	94	94 (S)	26
3	**35**	1 mol%, 50% KOH aq., toluene:CH$_2$Cl$_2$ (7:3), 0 °C	95	97 (S)	27
4	**30e**	10 mol%, 50% KOH aq., toluene:CH$_2$Cl$_2$ (7:3), 0 °C	90	96 (S)	28

Table 4.4 (continued)

Entry	Catalyst	Conditions	Yield [%]	ee [%] (config.)	Ref.
5	30f	5 mol%, 50% KOH aq., toluene:CH$_2$Cl$_2$ (7:3), 0 °C	94	96 (S)	29
6	30g	10 mol%, 25% NaOH aq., toluene, 0 °C	88	76 (S)	30
7	(S)-36 (Ar^1 = 3,5-Ph$_2$-C$_6$H$_3$, Ar^2 = Ph)	1 mol%, 50% KOH aq., toluene, 0 °C	81	95 (R)	31
8	(S,S)-37 (Ar^1 = Ar^2 = 3,5-Ph$_2$-C$_6$H$_3$)	1 mol%, 50% KOH aq., toluene, 0 °C	88	96 (R)	32

Table 4.4 (continued)

Entry	Catalyst	Conditions	Yield [%]	ee [%] (config.)	Ref.
9	(S,S)-38 (Ar^1 = Ar^2 = 3,5-Ph$_2$-C$_6$H$_3$)	1 mol%, 50% KOH aq., toluene, 0 °C	87	97 (R)	33
10	(S)-39 (Ar = 3,4,5-F$_3$-C$_6$H$_2$)	0.05 mol%, 50% KOH aq., toluene, 0 °C	98	99 (R)	34
11	(S)-40 (Ar = 3,4,5-F$_3$-C$_6$H$_2$)	0.01 mol%, 50% KOH aq., toluene, 25 °C	95	96 (R)	35
12	41	2 mol%, NaOH, toluene, r.t.	>95	80 (R)	36
13	42	30 mol%, 1 M KOH aq., CH$_2$Cl$_2$, 0 °C	55	90 (R)	37

Table 4.4 (continued)

Entry	Catalyst	Conditions	Yield [%]	ee [%] (config.)	Ref.
14	43a	10 mol%, CsOH·H$_2$O, toluene:CH$_2$Cl$_2$ (7:3), −78 °C	87	93 (R)	38
15	44	20 mol%, 50% KOH aq., CH$_2$Cl$_2$, 0 °C	>95	95 (R)	39
16	45	1 mol%, 50% KOH aq., toluene, 0 °C	55	58 (S)	40
17	46	1 mol%, 15 M KOH aq., toluene, 0 °C	89	97 (R)	41

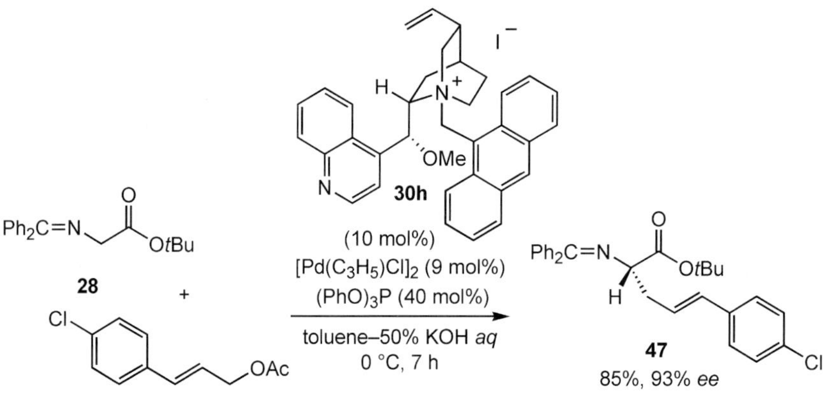

Scheme 4.15 (For experimental details see Chapter 14.16.1)

4.5.3
Dialkylation of Schiff Bases Derived from α-Alkyl-α-Amino Acids

Phase-transfer catalysis has made unique contributions in the development of a truly efficient method for the preparation of non-proteinogenic, chiral α,α-dialkyl-α-amino acids; these are often effective enzyme inhibitors and are also indispensable for the elucidation of enzymatic mechanisms.

In 1992, O'Donnell succeeded in obtaining optically active α-methyl-α-amino acid derivatives **49** in a catalytic manner through the phase-transfer alkylation of p-chlorobenzaldehyde imine of alanine tert-butyl ester **48** with cinchonine-derived **1a** as catalyst (see Scheme 4.16) [46]. Although the enantioselectivities are moderate, this study is the first example of preparing optically active α,α-dialkyl-α-amino acids by chiral phase-transfer catalysis.

RBr = PhCH$_2$Br : 80%, 44% ee
p-F-C$_6$H$_4$CH$_2$Br : 84%, 50% ee
CH$_2$=CHCH$_2$Br : 78%, 36% ee

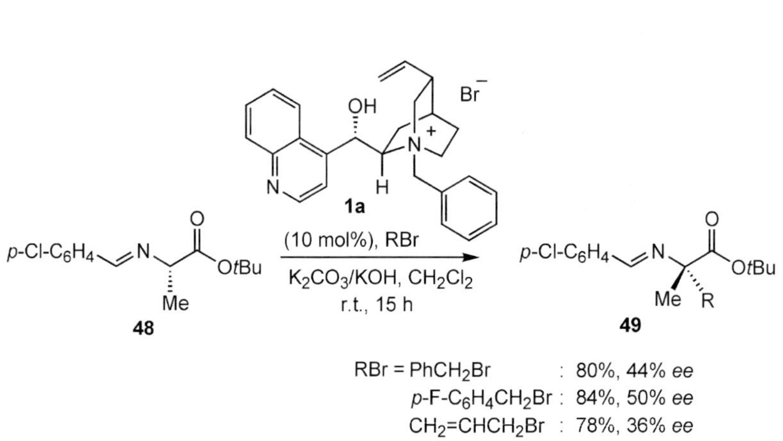

Scheme 4.16

Table 4.5 Phase-transfer-catalyzed enantioselective benzylation of aldimine Schiff bases derived from α-alkyl-α-amino acids.[a] (For experimental details see Chapter 14.15.1).

Entry	Substrate (Ar, R, Ak)	Catalyst	Conditions	Yield [%]	ee [%] (config.)	Ref.
1	p-Cl-C$_6$H$_4$, Me, tBu, (48)	30i	10 mol%, K$_2$CO$_3$/ KOH, toluene, r.t.	95	87 (S)	47
2[b]	Ph, Me, iPr (R,R)-TADDOL	(50)	3 mol%, 50% KOH aq., toluene, r.t.	94	94 (S)	48
3[b]	Ph, Me, iPr (R)-NOBIN	(51)	10 mol%, 50% KOH aq., toluene, r.t.	95	97 (S)	48
4	p-Cl-C$_6$H$_4$, Et, Me	41	2 mol%, NaOH, toluene, r.t.	91	82 (S)	36, 49

Table 4.5 (continued)

Entry	Substrate (Ar, R, Ak)	Catalyst	Conditions	Yield [%]	ee [%] (config.)	Ref.
5	p-Cl-C$_6$H$_4$, Me, tBu (S,S)-**32e** (Ar = 3,4,5-F$_3$-C$_6$H$_2$)		1 mol%, CsOH·H$_2$O, toluene, 0 °C	85	98 (R)	50
6	p-Cl-C$_6$H$_4$, iBu, tBu			64	92	50
7	2-Naph, Me, tBu	**30e**	10 mol%, RbOH, toluene, −35 °C	91	95 (S)	51
8	p-Cl-C$_6$H$_4$, Me, tBu **43a** (BF$_4$)		10 mol%, CsOH·H$_2$O, toluene:CH$_2$Cl$_2$ (7:3), −70 °C	83	89 (R)	38b

a) The alkylation products were usually isolated as α-amino esters after acidic hydrolysis under the following conditions: AcOH/THF-H$_2$O (entry 1), AcCl/MeOH (entry 4), 10% citric acid/THF (entries 5 and 6), HCl-H$_2$O/THF (entry 7), 0.2 M citric acid/THF (entry 8).
b) The product was obtained as α-amino acid hydrochloride after treatment with HCl-H$_2$O.

On the basis of this achievement, different types of catalysts have been designed and evaluated with aromatic aldimine Schiff bases of α-amino acid esters (mainly alaninate), and these are summarized in Table 4.5.

In addition to the high efficiency and broad generality, the characteristic feature of the **32e**-catalyzed asymmetric alkylation strategy is visualized by direct stereoselective introduction of two different side chains to glycine-derived aldimine

Schiff base **52** in one-pot under mild phase-transfer conditions. For example, the initial treatment of a toluene solution of **52** and (S,S)-**32e** (1 mol%) with allyl bromide (1 equiv.) and CsOH·H$_2$O at −10 °C, and the subsequent reaction with benzyl bromide (1.2 equiv.) at 0 °C, resulted in formation of the double alkylation product **53** in 80% yield with 98% ee after hydrolysis. Notably, in the double alkylation of **52** by the addition of the halides in reverse order, the absolute configuration of the product **53** was confirmed to be opposite, indicating intervention of the chiral ammonium enolate **54** at the second alkylation stage (Scheme 4.17) [50].

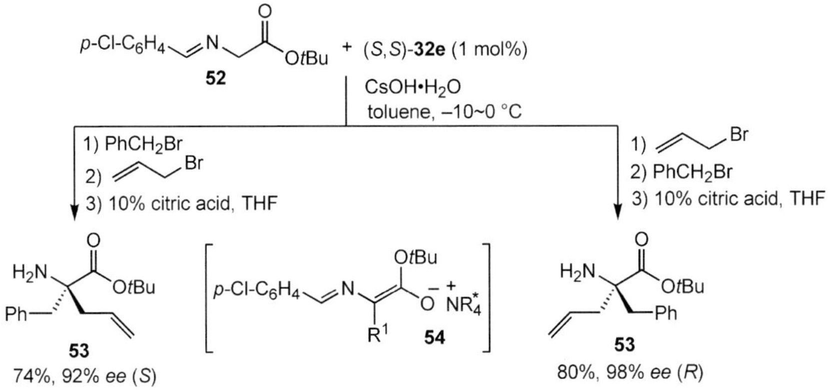

Scheme 4.17

Further, Jew and Park successfully applied the efficient phase-transfer catalysis of **32e** to the asymmetric synthesis of α-alkyl serines using phenyl oxazoline derivative **55a** as a requisite substrate. The reaction is general, and provides a practical access to a variety of optically active α-alkyl serines through acidic hydrolysis, as exemplified in Scheme 4.18 [52].

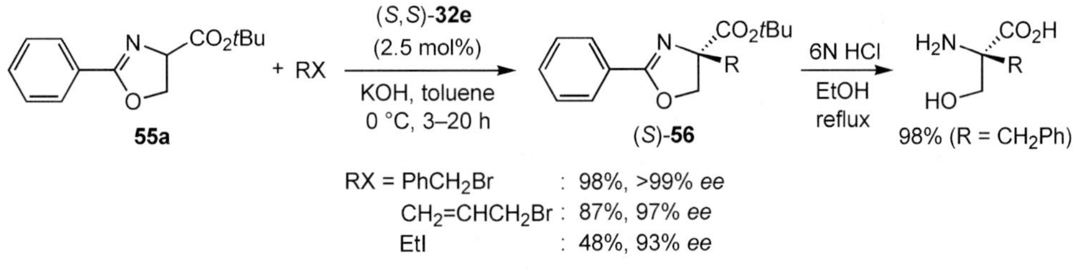

Scheme 4.18 (For experimental details see Chapter 14.15.2)

4.5.4
Michael Reaction of Glycinate Benzophenone Schiff Bases

Enantioselective Michael addition of glycine derivatives by means of chiral phase-transfer catalysis has been developed to synthesize various functionalized α-alkyl-amino acids. Corey and colleagues utilized **30d** as a catalyst for the asymmetric

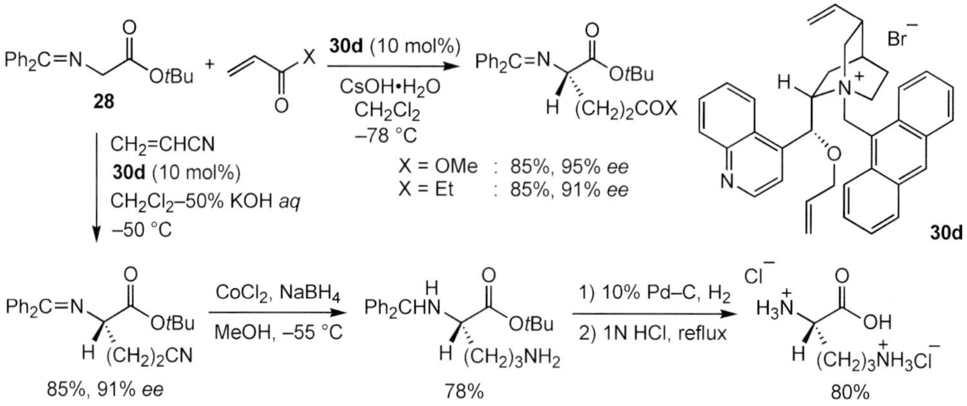

Scheme 4.19

Michael addition of glycinate Schiff base **28** to α,β-unsaturated carbonyl substrates with high enantioselectivity (Scheme 4.19) [53, 54]. With methyl acrylate as an acceptor, α-*tert*-butyl γ-methyl ester of (*S*)-glutamic acid can be formed; this functionalized glutamic acid derivative is very useful for synthetic applications because the two carboxyl groups are differentiated. Moreover, naturally occurring (*S*)-ornithine has been synthesized as its dihydrochloride in a concise manner by using acrylonitrile as an acceptor, as also included in Scheme 4.19 [54].

To date, this type of phase-transfer-catalyzed Michael reaction of **28** has been investigated with either acrylates or alkyl vinyl ketones as an acceptor, under the influence of different catalysts and bases. Typical results are listed in Table 4.6 in order to determine the characteristics of each system.

Belokon and co-workers designed a glycine-derived nickel complex **60** and examined its asymmetric addition to methyl acrylate under phase-transfer conditions. The screening of various NOBIN and *iso*-NOBIN derivatives in combination with NaH as a base revealed that N-pivaloyl-*iso*-NOBIN (**62b**) proved to be highly efficient catalyst, affording the product **61** in 80% yield with 96% *ee*, as illustrated in Scheme 4.20 [59].

Jew and Park achieved the highly enantioselective synthesis of (2*S*)-α-(hydroxymethyl)glutamic acid, a potent metabotropic receptor ligand, through the Michael addition of 2-naphthyl oxazoline derivative (**55b**) to ethyl acrylate under phase-transfer conditions. As shown in Scheme 4.21, the use of Schwesinger base BEMP at −60 °C with the catalysis of **32e** appeared to be essential for attaining an excellent selectivity [60].

4.5.5
Aldol and Mannich Reactions

The phase-transfer-catalyzed enantioselective direct aldol reactions of a glycine donor with aldehyde acceptors provide an ideal method for the simultaneous con-

Table 4.6 Catalytic enantioselective Michael addition of **28** to α,β-unsaturated carbonyl compounds under phase-transfer conditions. (For experimental details see Chapter 14.9.11).

Entry	Catalyst [mol%]	Acceptor	Conditions	Yield [%]	ee [%] (config.)	Ref.
1	**57** (10 mol%)	O*t*Bu	CsOH·H$_2$O (10 mol%), *t*BuOMe, −60 °C	73	77 (S)	55
2	**43b** (10 mol%)	OCH$_2$Ph	Cs$_2$CO$_3$ (50 mol%), Ph-Cl, −30 °C	84	81 (S)	38
3	**58** (20 mol%)	OEt	KO*t*Bu (20 mol%), CH$_2$Cl$_2$, −78 °C	65	96 (S)	56
4		Et		76	87 (S)	56
5	**59** (1 mol%)	Me	Cs$_2$CO$_3$ (200 mol%), Ph-Cl, −30 °C	100	75 (S)	57

Table 4.6 (continued)

Entry	Catalyst [mol%]	Acceptor	Conditions	Yield [%]	ee [%] (config.)	Ref.
6[a]	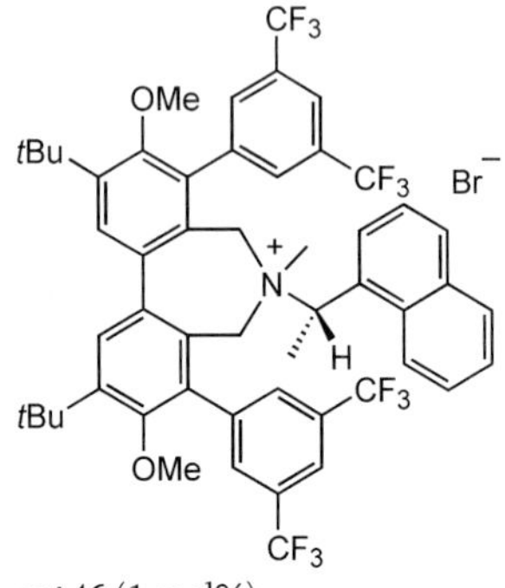 ent-46 (1 mol%)	Me	Cs$_2$CO$_3$ (50 mol%), iPr$_2$O, 0 °C	84	94 (S)	58
7[a]		nC$_5$H$_{12}$		60	94 (S)	58

[a] Benzhydryl glycinate benzophenone Schiff base as substrate.

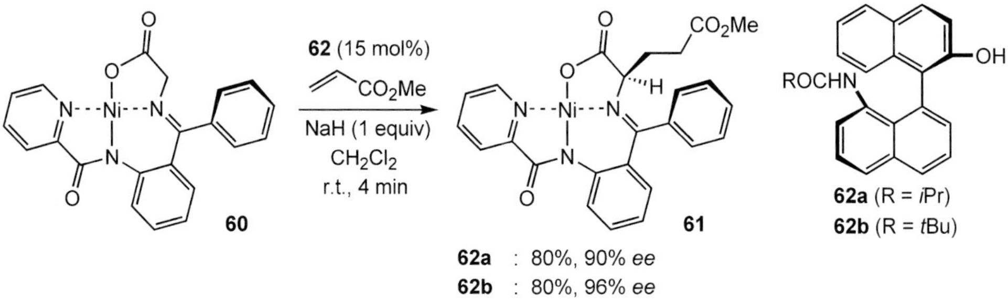

Scheme 4.20

struction of the primary structure and stereochemical integrity of β-hydroxy-α-amino acids, which are extremely important chiral units, especially from the pharmaceutical viewpoint. Prompted by the first report from the Miller's group on the possibility of this approach [61], we developed an efficient, highly diastereo- and enantioselective direct aldol reaction of **28** with a wide range of aliphatic aldehydes under mild phase-transfer conditions, employing **32f** as a key catalyst. Mechanistic investigations revealed the intervention of a highly stereoselective retro aldol reaction, which could be minimized by using a catalytic amount of 1% NaOH aqueous solution and ammonium chloride; this led to the establishment of a general and practical chemical process for the synthesis of optically active anti-β-hydroxy-α-amino esters **63** (Scheme 4.22) [62].

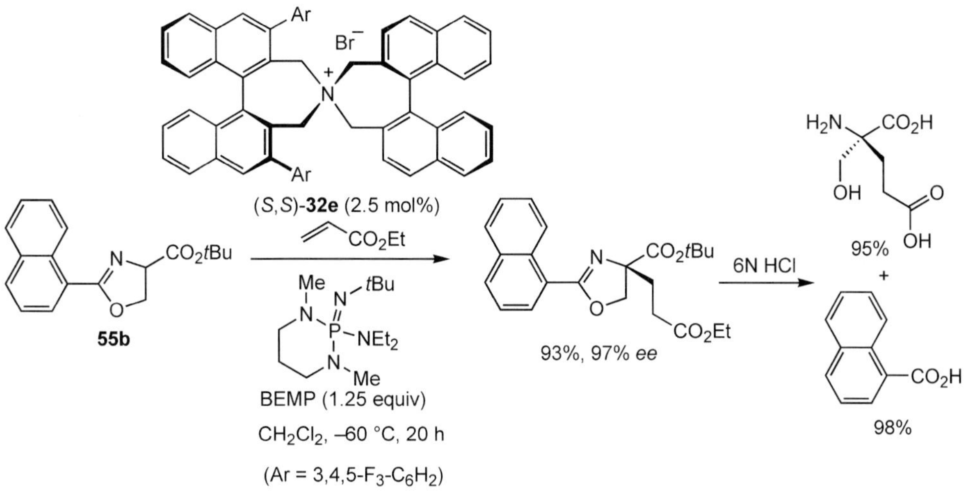

Scheme 4.21

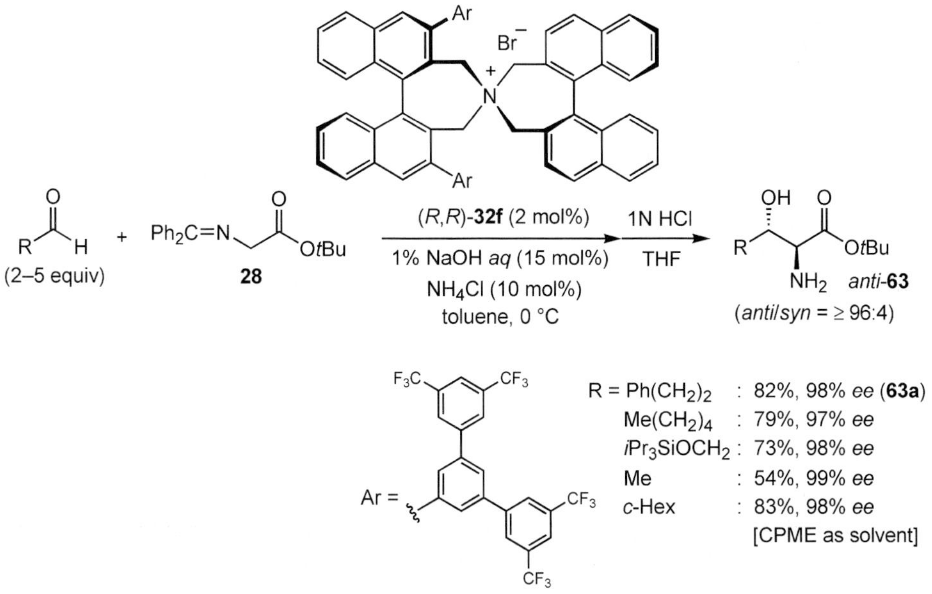

Scheme 4.22 (For experimental details see Chapter 14.1.6)

The phase-transfer-catalyzed direct Mannich reaction of **28** with α-imino ester **64** was achieved with high enantioselectivity by using **32e** as catalyst (Scheme 4.23) [63]. This method enables the catalytic asymmetric synthesis of differentially protected 3-aminoaspartate, a nitrogen analogue of dialkyl tartrate, the util-

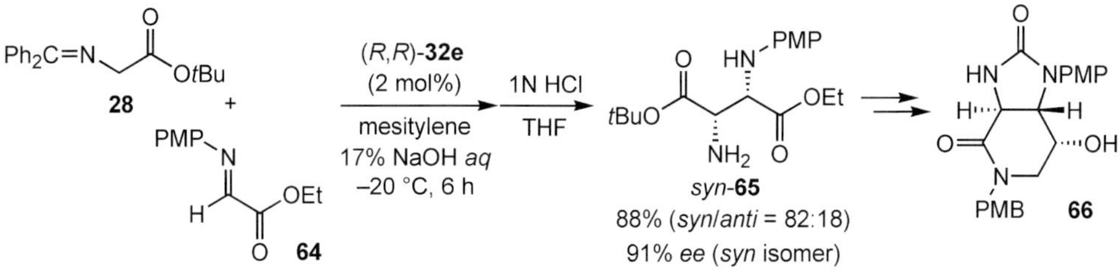

Scheme 4.23

ity of which was demonstrated by the product (*syn*-**65**) being converted into a precursor (**66**) of the streptolidine lactam.

A more general and highly diastereoselective Mannich-type reaction was developed by Ohshima and Shibasaki. The tartrate-derived diammonium salt **43c** possessing 4-fluorophenyl substituents was identified as an optimal catalyst for the reaction of **28** with various *N*-Boc imines under solid (Cs_2CO_3)-liquid (fluorobenzene) phase-transfer conditions, as exemplified in Scheme 4.24 [64]. The usefulness of the Mannich adduct **67b** was further demonstrated by the straightforward synthesis of the optically pure tripeptide **68**.

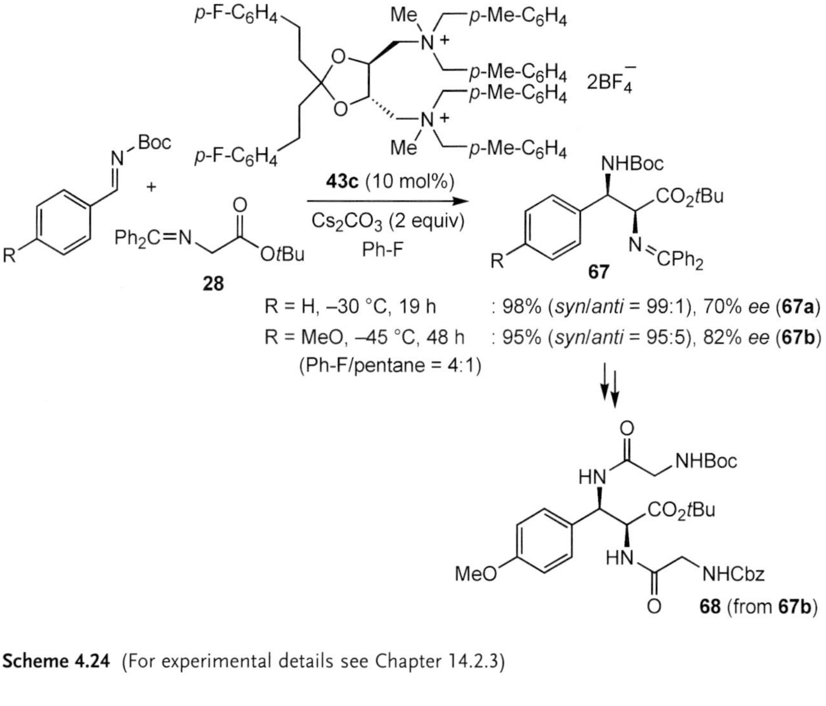

Scheme 4.24 (For experimental details see Chapter 14.2.3)

4.6
Conclusions

As described in this chapter, numerous recent efforts to elaborate chiral quaternary ammonium salts have greatly expanded their use as asymmetric catalysts in organic synthesis. However, the full potential of enantioselective organocatalysis is yet to be realized in terms of general applicability. In the case of the chiral ammonium cation, pairing with a requisite nucleophile (an anion) enhances its nucleophilicity, which in turn facilitates subsequent bond-forming reactions with electrophiles in a stereoselective manner. As the assembly of a chiral ion pair is based on simple electrostatic interactions, it is not a trivial task to control precisely the *direction* and *distance* of the ion pair, and this constitutes a major difficulty for their application to different types of transformation. Nevertheless, it should be possible to solve this intrinsic problem by the molecular design of chiral ammonium cations, so that their use as chiral templates becomes reliable and they acquire a general strategic role in practical asymmetric synthesis. These developments are eagerly anticipated.

References

1 For reviews, see: (a) T. Shioiri, A. Ando, M. Masui, T. Miura, T. Tatematsu, A. Bohsako, M. Higashiyama, C. Asakura. In: M.E. Halpern (Ed.), *Phase-Transfer Catalysis.* ACS Symposium Series 659; American Chemical Society: Washington, DC, **1997**; Chapter 11, p. 136; (b) T. Ooi, K. Maruoka, *Acc. Chem. Res.* **2004**, *37*, 526.

2 For recent reviews, see: (a) T. Shioiri. In: Y. Sasson, R. Neumann (Eds.), *Handbook of Phase-Transfer Catalysis.* Blackie Academic & Professional: London, **1997**, Chapter 14; (b) M.J. O'Donnell, *Phases – The Sachem Phase Transfer Catalysis Review* **1998**, Issue 4, p. 5; (c) M.J. O'Donnell, *Phases – The Sachem Phase Transfer Catalysis Review* **1999**, Issue 5, p. 5; (d) A. Nelson, *Angew. Chem.* **1999**, *111*, 1685; *Angew. Chem. Int. Ed.* **1999**, *38*, 1583; (e) T. Shioiri, S. Arai. In: F. Vogtle, J.F. Stoddart, M. Shibasaki (Eds.), *Stimulating Concepts in Chemistry.* Wiley-VCH: Weinheim, **2000**; p. 123; (f) M.J. O'Donnell. In: I. Ojima (Ed.), *Catalytic Asymmetric Syntheses,* 2nd edn. Wiley-VCH: New York, **2000**; Chapter 10; (g) M.J. O'Donnell, *Aldrichimica Acta* **2001**, *34*, 3; (h) K. Maruoka, T. Ooi, *Chem. Rev.* **2003**, *103*, 3013; (i) M.J. O'Donnell, *Acc. Chem. Res.* **2004**, *37*, 506; (j) B. Lygo, B.I. Andrews, *Acc. Chem. Res.* **2004**, *37*, 518.

3 A. Ando, T. Miura, T. Tatematsu, T. Shioiri, *Tetrahedron Lett.* **1993**, *34*, 1507.

4 D. Landini, H. Molinari, M. Penso, A.E. Rampoldi, *Synthesis* **1988**, 953.

5 T. Shioiri, A. Bohsako, A. Ando, *Heterocycles* **1996**, *42*, 93.

6 G. Bluet, J.-M. Campagne, *J. Org. Chem.* **2001**, *66*, 4293.

7 E.J. Corey, F.-Y. Zhang, *Angew. Chem.* **1999**, *111*, 2057; *Angew. Chem. Int. Ed.* **1999**, *38*, 1931.

8 K. Iseki, T. Nagai, Y. Kobayashi, *Tetrahedron Lett.* **1994**, *35*, 3137.

9 Y. Kobayashi, E. Takahisa, M. Nakano, K. Watatani, *Tetrahedron* **1997**, *53*, 1627.

10 M.D. Drew, N.J. Lawrence, W. Watson, S.A. Bowles, *Tetrahedron Lett.* **1997**, *38*, 5857.

11 Y. Sasson, N. Mushkin, E. Abu, S. Negussie, S. Dermeik, A. Zoran. In: M.E. Halpern (Ed.), *Phase-Transfer Catalysis*. ACS Symposium Series 659; American Chemical Society: Washington, DC, **1997**; Chapter 12, p. 148.

12 M. Horikawa, J. Busch-Petersen, E.J. Corey, *Tetrahedron Lett.* **1999**, *40*, 3843.

13 (a) E.W. Colvin, D. Seebach, *J. Chem. Soc. Chem. Commun.* **1978**, 689; (b) D. Seebach, A.K. Beck, T. Mukhopadhyay, E. Thomas, *Helv. Chim. Acta* **1982**, *65*, 1101.

14 T. Ooi, K. Doda, K. Maruoka, *J. Am. Chem. Soc.* **2003**, *125*, 2054.

15 T. Ooi, K. Doda, K. Maruoka, *J. Am. Chem. Soc.* **2003**, *125*, 9022.

16 T. Ooi, K. Doda, S. Takada, K. Maruoka, *Tetrahedron Lett.* **2006**, *47*, 145.

17 T. Tozawa, Y. Yamane, T. Mukaiyama, *Chem. Lett.* **2006**, *35*, 56.

18 (a) U.-H. Dolling, P. Davis, E.J.J. Grabowski, *J. Am. Chem. Soc.* **1984**, *106*, 446; (b) D.L. Hughes, U.-H. Dolling, K.M. Ryan, E.F. Schoenewaldt, E.J.J. Grabowski, *J. Org. Chem.* **1987**, *52*, 4745.

19 (a) M.J. O'Donnell, W.D. Bennett, S. Wu, *J. Am. Chem. Soc.* **1989**, *111*, 2353; (b) K.B. Lipkowitz, M.W. Cavanaugh, B. Baker, M.J. O'Donnell, *J. Org. Chem.* **1991**, *56*, 5181; (c) I.A. Esikova, T.S. Nahreini, M.J. O'Donnell. In: M.E. Halpern (Ed.), *Phase-Transfer Catalysis*. ACS Symposium Series 659, American Chemical Society: Washington, DC, **1997**; Chapter 7; (d) M.J. O'Donnell, I.A. Esikova, D.F. Shullenberger, S. Wu. In: M.E. Halpern (Ed.), *Phase-Transfer Catalysis*. ACS Symposium Series 659. American Chemical Society: Washington, DC, **1997**; Chapter 10.

20 M.J. O'Donnell, W.D. Bennett, W.A. Bruder, W.N. Jacobsen, K. Knuth, B. LeClef, R.L. Polt, F.G. Boldwell, S.R. Mrozack, T.A. Cripe, *J. Am. Chem. Soc.* **1988**, *110*, 8520.

21 M.J. O'Donnell, S. Wu, J.C. Huffman, *Tetrahedron* **1994**, *50*, 4507.

22 (a) B. Lygo, P.G. Wainwright, *Tetrahedron Lett.* **1997**, *38*, 8595; (b) B. Lygo, J. Crosby, T.R. Lowdon, P.G. Wainwright, *Tetrahedron* **2001**, *57*, 2391; (c) B. Lygo, J. Crosby, T.R. Lowdon, J.A. Peterson, P.G. Wainwright, *Tetrahedron* **2001**, *57*, 2403.

23 E.J. Corey, F. Xu, M.C. Noe, *J. Am. Chem. Soc.* **1997**, *119*, 12414.

24 (a) T. Ooi, M. Kameda, K. Maruoka, *J. Am. Chem. Soc.* **1999**, *121*, 6519; (b) T. Ooi, Y. Uematsu, K. Maruoka, *Adv. Synth. Catal.* **2002**, *344*, 288; (c) T. Ooi, M. Kameda, K. Maruoka, *J. Am. Chem. Soc.* **2003**, *125*, 5139.

25 (a) S.-s. Jew, B.-S. Jeong, M.-S. Yoo, H. Huh, H.-g. Park, *Chem. Commun.* **2001**, 1244; (b) H.-g. Park, B.-S. Jeong, M.-S. Yoo, J.-H. Lee, B.-s. Park, M.G. Kim, S.-s. Jew, *Tetrahedron Lett.* **2003**, *44*, 3497.

26 H.-g. Park, B.-S. Jeong, M.-S. Yoo, M.-k. Park, H. Huh, S.-s. Jew, *Tetrahedron Lett.* **2001**, *42*, 4645.

27 H.-g. Park, B.-S. Jeong, M.-S. Yoo, M.-s., J.-H. Lee, M.-k. Park, Y.-J. Lee, M.-J. Kim, S.-s. Jew, *Angew. Chem.* **2002**, *114*, 3162; *Angew. Chem., Int. Ed.* **2002**, *41*, 3036.

28 S.-s. Jew, M.-S. Yoo, B.-S. Jeong, I.-Y. Park, H.-g. Park, *Org. Lett.* **2002**, *4*, 4245.

29 M.-S. Yoo, B.-S. Jeong, J.H. Lee, H.-g. Park, S.-s. Jew, *Org. Lett.* **2005**, *7*, 1129.

30 (a) P. Mazón, R. Chinchilla, C. Nájera, G. Guillena, R. Kreiter, R.J.M. Klein Gebbink, G. van Koten, *Tetrahedron: Asymm.* **2002**, *13*, 2181; (b) G. Guillena, R. Kreiter, R. van de Coevering, R.J.M. Klein Gebbink, G. van Koten, P. Mazón, R. Chinchilla, C. Nájera, *Tetrahedron: Asymm.* **2003**, *14*, 3705.

31 T. Ooi, Y. Uematsu, M. Kameda, K. Maruoka, *Angew. Chem.* **2002**, *114*, 1621; *Angew. Chem. Int. Ed.* **2002**, *41*, 1551.

32 T. Hashimoto, K. Maruoka, *Tetrahedron Lett.* **2003**, *44*, 3313.

33 T. Hashimoto, Y. Tanaka, K. Maruoka, *Tetrahedron: Asymm.* **2003**, *14*, 1599.

34 M. Kitamura, S. Shirakawa, K. Maruoka, *Angew. Chem.* **2005**, *117*,

1573; *Angew. Chem., Int. Ed.* **2005**, *44*, 1549.
35 Z. Han, Y. Yamaguchi, M. Kitamura, K. Maruoka, *Tetrahedron Lett.* **2005**, *46*, 8555.
36 Y.N. Belokon, M. North, T.D. Churkina, N.S. Ikonnikov, V.I. Maleev, *Tetrahedron* **2001**, *57*, 2491.
37 T. Kita, A. Georgieva, Y. Hashimoto, T. Nakata, K. Nagasawa, *Angew. Chem.* **2002**, *114*, 2956; *Angew. Chem. Int. Ed.* **2002**, *41*, 2832.
38 (a) T. Shibuguchi, Y. Fukuta, Y. Akachi, A. Sekine, T. Ohshima, M. Shibasaki, *Tetrahedron Lett.* **2002**, *43*, 9539; (b) T. Ohshima, T. Shibuguchi, Y. Fukuta, M. Shibasaki, *Tetrahedron* **2004**, *60*, 7743.
39 H. Sasai, *Jpn. Kokai Tokkyo Koho* **2003**, JP2003335780.
40 N. Mase, T. Ohno, N. Hoshikawa, K. Ohishi, H. Morimoto, H. Yoda, K. Takabe, *Tetrahedron Lett.* **2003**, *44*, 4073.
41 B. Lygo, B. Allbutt, S.R. James, *Tetrahedron Lett.* **2003**, *44*, 5629.
42 (a) M. Nakoji, T. Kanayama, T. Okino, Y. Takemoto, *Org. Lett.* **2001**, *3*, 3329; (b) M. Nakoji, T. Kanayama, T. Okino, Y. Takemoto, *J. Org. Chem.* **2002**, *67*, 7418.
43 M.J. O'Donnell, F. Delgado, R.S. Pottorf, *Tetrahedron* **1999**, *55*, 6347.
44 For recent contributions, see: (a) R. Chinchills, P. Mazón, C. Nájera, *Tetrahedron: Asymm.* **2000**, *11*, 3277; (b) B. Thierry, J.-C. Plaquevent, D. Cahard, *Tetrahedron: Asymm.* **2001**, *12*, 983; (c) B. Thierry, T. Perrard, C. Audouard, J.-C. Plaquevent, D. Cahard, *Synthesis* **2001**, 1742; (d) T. Danelli, R. Annunziata, M. Benaglia, M. Cinquini, F. Cozzi, G. Tocco, *Tetrahedron: Asymm.* **2003**, *14*, 461; (e) B. Thierry, J.-C. Plaquevent, D. Cahard, *Tetrahedron: Asymm.* **2003**, *14*, 1671; (f) R. Chinchills, P. Mazón, C. Nájera, *Adv. Synth. Catal.* **2004**, *346*, 1186.
45 For representative examples, see: (a) K. Tohdo, Y. Hamada, T. Shioiri, *Synlett* **1994**, 247; (b) A.V.R. Rao, K.L. Reddy, A.S. Rao, T.V.S.K. Vittal, M.M. Reddy, P.L. Pathi, *Tetrahedron Lett.*

1996, *37*, 3023; (c) R.K. Boeckman, Jr., T.J. Clark, B.C. Shook, *Org. Lett.* **2002**, *4*, 2109; (d) S. Kim, J. Lee, T. Lee, H.-g. Park, D. Kim, *Org. Lett.* **2003**, *5*, 2703; (e) S.L. Castle, G.S.C. Srikanth, *Org. Lett.* **2003**, *5*, 3611; (f) T. Ohshima, V. Gnanadesikan, T. Shibuguchi, Y. Fukuta, T. Nemoto, M. Shibasaki, *J. Am. Chem. Soc.* **2003**, *125*, 11206; (g) S. Kumar, U. Ramachandran, *Tetrahedron Lett.* **2005**, *46*, 19.
46 M.J. O'Donnell, S. Wu, *Tetrahedron: Asymm.* **1992**, *3*, 591.
47 B. Lygo, J. Crosby, J.A. Peterson, *Tetrahedron Lett.* **1999**, *40*, 8671.
48 (a) Y.N. Belokon, K.A. Kochetkov, T.D. Churkina, N.S. Ikonnikov, A.A. Chesnokov, O.V. Larionov, V.S. Parmar, R. Kumar, H.B. Kagan, *Tetrahedron: Asymm.* **1998**, *9*, 851; (b) Y.N. Belokon, K.A. Kochetkov, T.D. Churkina, N.S. Ikonnikov, S. Vyskocil, H.B. Kagan, *Tetrahedron: Asymm.* **1999**, *10*, 1723; (c) Y.N. Belokon, K.A. Kochetkov, T.D. Churkina, N.S. Ikonnikov, A.A. Chesnokov, O.V. Larionov, I. Singh, V.S. Parmar, S. Vyskocil, H.B. Kagan, *J. Org. Chem.* **2000**, *65*, 7041.
49 (a) Y.N. Belokon, M. North, V.S. Kublitski, N.S. Ikonnikov, P.E. Krasik, V.I. Maleev, *Tetrahedron Lett.* **1999**, *40*, 6105; (b) Y.N. Belokon, R.G. Davies, M. North, *Tetrahedron Lett.* **2000**, *41*, 7245; (c) Y.N. Belokon, R.G. Davies, J.A. Fuentes, M. North, T. Parsons, *Tetrahedron Lett.* **2001**, *42*, 8093; (d) Y.N. Belokon, D. Bhave, D. D'Addario, E. Groaz, M. North, V. Tagliazucca, *Tetrahedron* **2004**, *60*, 1849; (e) Y.N. Belokon, J.A. Fuentes, M. North, J.W. Steed, *Tetrahedron* **2004**, *60*, 3191.
50 T. Ooi, M. Takeuchi, M. Kameda, K. Maruoka, *J. Am. Chem. Soc.* **2000**, *122*, 5228.
51 S.-s. Jew, B.-S. Jeong, J.-H. Lee, M.-S. Yoo, Y.-J. Lee, B.-s. Park, M.G. Kim, H.-g. Park, *J. Org. Chem.* **2003**, *68*, 4514.
52 S.-s. Jew, Y.-J. Lee, J. Lee, M.J. Kang, B.-S. Jeong, J.-H. Lee, M.-S. Yoo, M.-J. Kim, S.-h. Choi, J.-M. Ku, H.-g. Park,

Angew. Chem. **2004**, *116*, 2436; Angew. Chem. Int. Ed. **2004**, *43*, 2382.
53 E.J. Corey, M.C. Noe, F. Xu, Tetrahedron Lett. **1998**, *39*, 5347.
54 F.-Y. Zhang, E.J. Corey, Org. Lett. **2000**, *2*, 1097.
55 S. Arai, R. Tsuji, A. Nishida, Tetrahedron Lett. **2002**, *43*, 9535.
56 T. Akiyama, M. Hara, K. Fuchibe, S. Sakamoto, K. Yamaguchi, Chem. Commun. **2003**, 1734.
57 S. Arai, K. Tokumaru, T. Aoyama, Chem. Parm. Bull. **2004**, *52*, 646.
58 B. Lygo, B. Allbutt, E.H.M. Kirton, Tetrahedron Lett. **2005**, *46*, 4461.
59 Y.N. Belokon, N.B. Bespalova, T.D. Churkina, I. Cisarová, M.G. Ezernitskaya, S.R. Harutyunyan, R. Hrdina, H.B. Kagan, P. Kocovsky, K.A. Kochetkov, O.V. Larionov, K.A. Lyssenko, M. North, M. Polásek, A.S. Peregudov, V.V. Prisyazhnyuk, S. Vyskocil, J. Am. Chem. Soc. **2003**, *125*, 12860.
60 Y.-J. Lee, J. Lee, M.-J. Kim, B.-S. Jeong, J.-H. Lee, T.-S. Kim, J. Lee, J.-M. Ku, S.-s. Jew, H.-g. Park, Org. Lett. **2005**, *7*, 3207.
61 C.M. Gasparski, M.J. Miller, Tetrahedron **1991**, *47*, 5367.
62 (a) T. Ooi, M. Taniguchi, M. Kameda, K. Maruoka, Angew. Chem. **2002**, *114*, 4724; Angew. Chem. Int. Ed. **2002**, *41*, 4542; (b) T. Ooi, M. Kameda, M. Taniguchi, K. Maruoka, J. Am. Chem. Soc. **2004**, *126*, 9685.
63 T. Ooi, M. Kameda, J. Fujii, K. Maruoka, Org. Lett. **2004**, *6*, 2397.
64 A. Okada, T. Shibuguchi, T. Ohshima, H. Masu, K. Yamaguchi, M. Shibasaki, Angew. Chem. **2005**, *117*, 4640; Angew. Chem., Int. Ed. **2005**, *44*, 4564.

5
Organocatalytic Enantioselective Morita–Baylis–Hillman Reactions

Candice Menozzi and Peter I. Dalko

Nucleophilic amines or alkyl phosphines can mediate the addition of electron-deficient alkenes to reactive carbonyls such as aldehydes or ketones. This transformation, which affords functionalized allylic alcohols, is generally termed the Morita–Baylis–Hillman (MBH) reaction (Scheme 5.1) [1, 2].

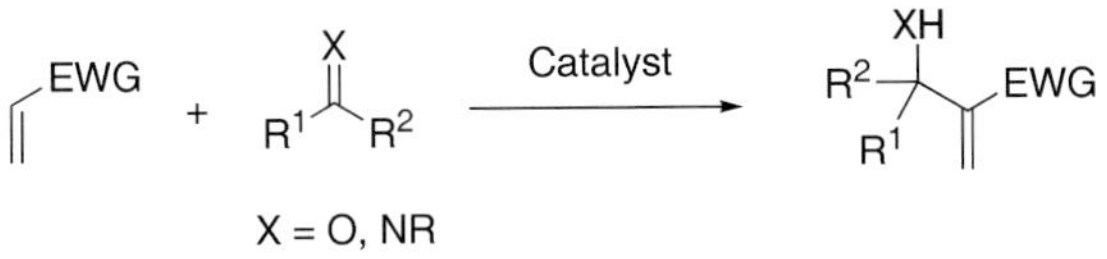

$$\text{EWG} + \underset{R^1 \quad R^2}{\overset{X}{\|}} \xrightarrow{\text{Catalyst}} R^2 \underset{R^1}{\overset{XH}{\diagdown}} \text{EWG}$$

$$X = O, NR$$

Scheme 5.1 The general reaction scheme of the Morita–Baylis–Hillman (MBH) reaction.

5.1
Addition of Ketones and Aldehydes to Activated Olefins

The reaction was first described by the Morita group, the discovery most likely being a case of serendipity, by virtue of the similarity of the original experimental procedure with that of Oda's olefin synthesis [3, 4]. Independently, some years later Baylis and Hillman discovered that the addition of acetaldehyde with ethyl acrylate and acrylonitrile could be realized in the presence of catalytic amounts of a strong Lewis nitrogen base such as 1,4-diazabicyclo[2,2,2]octane (DABCO) [5]. While these discoveries did not arouse particular attention during the following decade, the reaction began to receive increasing synthetic and mechanistic interest from the late 1880s onwards [1]. Both substrate compatibility problems and selectivity issues were considerably improved, though not completely solved, and the range of olefin reagents was extended to β-unsubstituted acrylic [6], and α,β-unsaturated thiol esters [7], acrylonitrile [8], vinyl ketones [8, 9], vinyl sulfones [10, 11], and sulfonates [12], vinyl phosphonates [13], acrylamides [14], and acrolein [15] (Scheme 5.2). More recently, allenes [16], acetylenes [17] and activated

Enantioselective Organocatalysis: Reactions and Experimental Procedures
Edited by Peter I. Dalko
Copyright © 2007 WILEY-VCH Verlag GmbH & Co. KGaA, Weinheim
ISBN: 978-3-527-31522-2

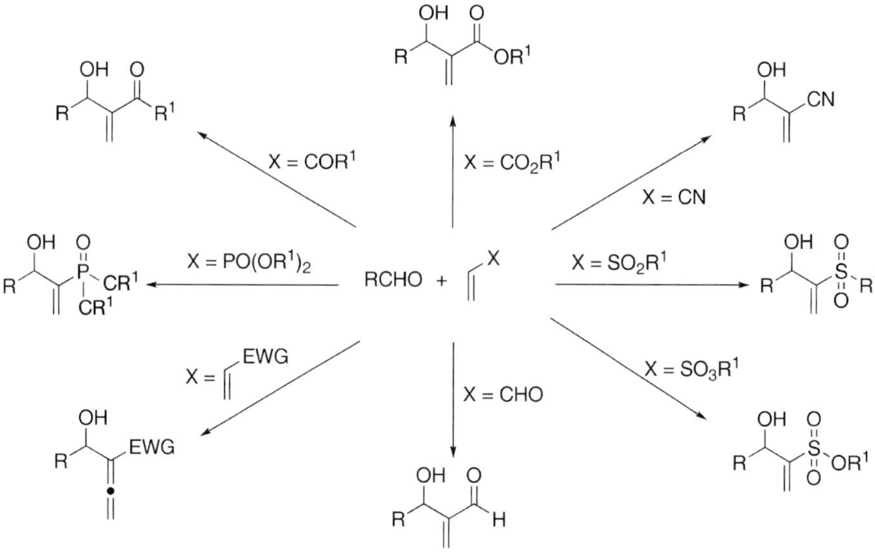

Scheme 5.2 Multifunctional products accessible by the MBH reaction.

dienes [18] were added as olefin equivalents, leading to products of more complex skeletons. Aromatic and aliphatic aldehydes, α-keto-esters [8b, 19], fluoroketones [20] or aldimines [21] were also seen in MBH reactions (Scheme 5.2). Although simple ketones and disubstituted olefins are unreactive under classical conditions, they can be used under higher pressure [22], or with the use of co-catalysts. Moreover, the organic nucleophilic catalysts can be replaced by metal complexes, which represent powerful alternatives in MBH reactions.

Historically, the most effective N-based organic catalysts were nucleophilic unhindered tertiary amines such as DABCO (diazabicyclo[2.2.2]octane, **1**) [23], quinuclidine (**2**), 3-hydroxy quinuclidine (3-HDQ, **3**), 3-quinuclidone (**4**) and indolizine (**5**) (Fig. 5.1) [24]. A direct correlation has been found between pK_a and the activity of the quinuclidine-based catalysts: the higher the pK_a, the faster the rate [25]. More recently, 1,8-diazabicyclo[5.4.0]undec-7-ene (DBU, **6**), considered as a hindered and non-nucleophilic base, was shown to be a better catalyst than DABCO, or 3-HDQ [26]. The reason for the increased reactivity for this catalyst was attributed to stabilization of the zwitterionic enolate by delocalization of the positive charge. Other N-based catalysts such as N,N-(dimethylamino)pyridine

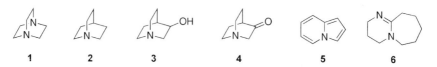

Fig. 5.1 Tertiary amines used as catalysts in the MBH reaction.

Table 5.1 The effect of Brønsted acid co-catalysts on the rate of the MBH reaction.

Entry	Co-catalyst	$10^{-2} k_{obs}/(h^{-1})$	k_{rel}
1	None	0.46	1
2	MeOH (40 mol%)	1.15	2.5
3	H$_2$O (40 mol%)	0.72	1.6

(DMAP) [27], imidazoles [28], guanidine [29], azole [30], and N-methylpiperidine [31] were also used successfully in non-asymmetric MBH reactions.

A basic amine catalyst may promote the self-aldol reaction of the aldehyde, having an enolizable carbonyl. This reaction can be particularly important in the case of slowly reacting hindered aldehydes. In order to avoid this secondary reaction, a number of trialkylphosphines were tested and, in non-asymmetric reactions, tributylphosphine was generally found to be the most effective [32, 33].

The low reaction rates usually associated with the MBH reaction can be increased either by pressure [15a, 22, 34], by the use of ultrasound [35] and microwave radiation [14a], or by the addition of co-catalysts. Various intra- or intermolecular Lewis acid co-catalysts have been tested [26, 36, 37]; in particular, mild Brønsted acids such as methanol [36, 57d], formamide [38], diarylureas and thioureas [39] and water [27a, 40] were examined and found to provide an additional acceleration of the MBH reaction rate (Table 5.1).

Typically, MBH reactions are conducted at or just below room temperature. The rate of product formation can be increased by warming the reaction mixture above room temperature, though the yields are usually not greater than those achieved when the reaction is run at room temperature [41]. At elevated temperatures, polymerization of the acrylate becomes a viable alternative; indeed, the formation of this byproduct makes purification of the desired MBH adducts difficult and should, if at all possible, be avoided.

The influence of solvents was extensively studied [38, 40b, 42], with reactions shown capable of being performed in neat, or, virtually in any polar medium. Whilst high dielectric constant oxygenated solvents such as tetrahydrofuran (THF), 1,4-dioxane, acetone (Et$_2$O), dimethyl sulfoxide (DMSO), and dimethylformamide (DMF) are used in non-asymmetric MBH reactions, dichloroethane (CH$_2$Cl$_2$) or acetonitrile are preferred for asymmetric transformations. MBH reactions between conjugated cyclic enones with a variety of aldehydes can also be managed in water in the presence of surfactants above their critical micelle concentrations (CMC) [43]. Contradictory effects were found, however, in ionic liquids [44].

5.1.1
Reaction Mechanism

Whilst the elementary steps of the reaction were postulated in the earliest publications [3], and remain (globally) even today as the core of the mechanistic discussion, the fine details of the reaction – and in particular those controlling the asymmetric induction – have been highlighted only recently. The first critical mechanism [15a, 45, 46], which is based on pressure-dependence data, established a reversible Michael addition of the nucleophilic base to the activated alkene (Scheme 5.3). In the following step, the formed zwitterionic enolate **11** adds to the electrophile and forms a second zwitterionic adduct **13**. This step was considered to be the rate-determining step (RDS) of the reaction. Subsequent proton transfer and release of the catalyst provides finally the desired product **14**.

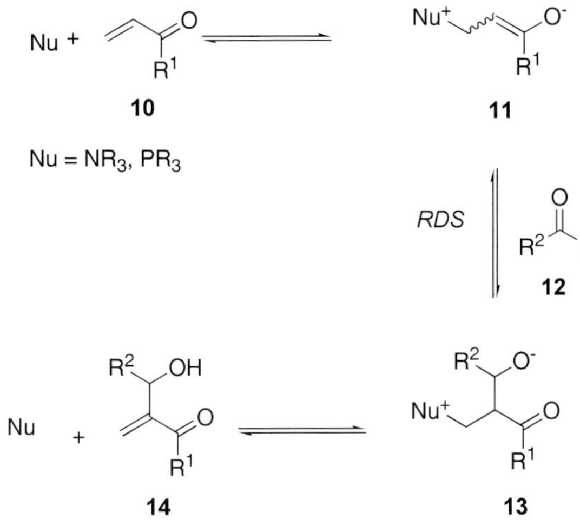

Scheme 5.3 The first critical mechanism of the MBH reaction.

While many of the observed events of the MBH reaction could be included in this scheme, the mechanism failed in some critical cases [47]. First, the mechanism did not provide any clue as to why stereocontrol is so difficult in MBH reactions. Privileged nucleophilic chiral catalysts [48], which in the past have usually allowed good results in related asymmetric transformations, afforded only modest asymmetric induction. This fact was surprising, and pointed to lack of understanding of the basic factors governing the selectivity of the reaction. Other observations, such as the rate acceleration by the build-up of product (i.e., the autocatalytic effect) [22], and also the formation of a considerable amount of "unusual" dioxanone byproduct, such as **23** (see Scheme 5.5) [49] in the addition of aryl aldehydes to acrylates warned of the limits of the discussed mechanism.

Detailed kinetic analysis showed that the RDS in the MBH reaction is not the 1,2-addition, as previously reported, but rather the proton abstraction followed by elimination of the catalyst from intermediate **15** (Scheme 5.4) [50]. This fact directs attention to the proton-donor ability of the catalyst, and has a major consequence in the catalyst design. If either the Brønsted acid or the Lewis base could be appropriately positioned on a chiral molecule, the Lewis base would react with the substrate (Michael addition), while the acid in an asymmetric environment would allow the chiral proton transfer. From all diastereoisomers generated in the reactions, only the structure allowing the fastest proton-transfer would evolve towards the product, while the other diastereoisomers generated in the reactions would revert to starting materials. The action of the Brønsted co-catalysts is not limited, however, to the stereo-determining protonation step. Rather, it also promotes conjugate addition by binding to the zwitterionic enolate, and stabilizing these intermediates. The Brønsted acid remains hydrogen-bonded to the resulting enolate in the enolate-addition step to the aldehyde [51], and finally ensures efficient proton transfer in the rate-determining proton abstraction step. This model also explains the autocatalytic effect of the product. With the increase in product concentration, the proton-transfer becomes increasingly efficient and the rate-limiting step shifts to step 2, as in the conventional model.

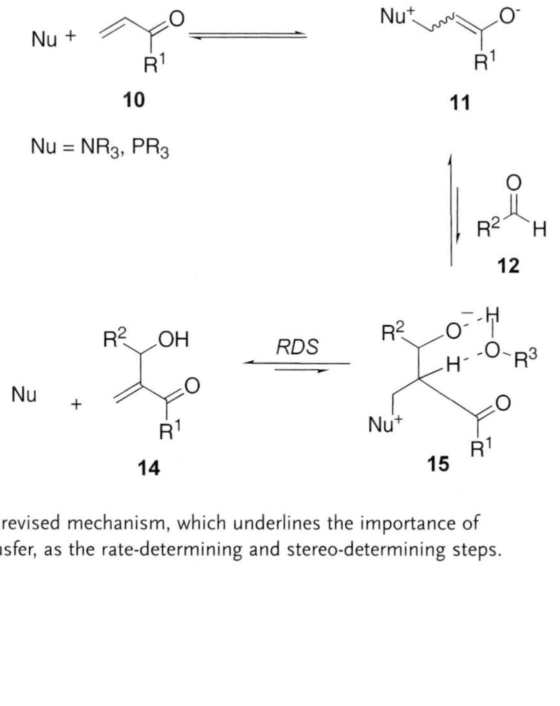

Scheme 5.4 A revised mechanism, which underlines the importance of the proton-transfer, as the rate-determining and stereo-determining steps.

Another support for this mechanism was found in the detailed kinetic analysis of the reaction of acrylates with aryl aldehyde, where the RDS is second order in aldehyde and first order in the catalyst and acrylate, a fact which is in accordance with the formation of a hemiacetal intermediate **19** (Scheme 5.5) [52]. This intermediate subsequently undergoes elimination and generates **21**, with liberation of the catalyst leading either to the "normal" adduct **22** or, alternatively, to the dioxanone product **23**. The formation of this byproduct is favored when O–R is a good leaving group and when the concentration in aldehyde is high.

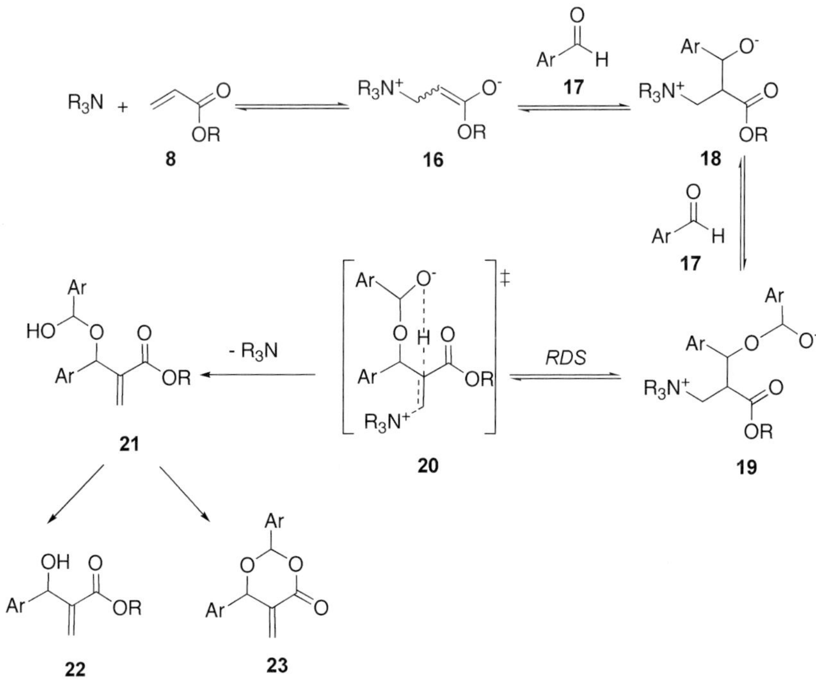

Scheme 5.5 An extended mechanistic model of the addition of acryl esters to aryl aldehydes.

5.1.2
Diastereoselectivity

The ease of racemization of chiral α-amino aldehydes under MBH conditions is undoubtedly a major difficulty in studying diastereoselective reactions [53]. Epimerization can be essentially avoided by conducting the reaction at low temperature [54, 67], or it can be minimized at room temperature when a conformationally restricted amino aldehyde, such or N-trityl-azetidine 2-(S)-carboxyaldehyde is used [54]. The use of ultrasound also increases the rate of the MBH reaction, avoiding racemization almost completely, even at room temperature [55]. When adding various α-amino acid-derived aldehydes to methyl acrylate using DABCO

Table 5.2 The diastereoselectivity of the MBH reaction in the addition of α-substituted aldehydes to acrylates under ultrasound irradiation.

$$R^1\underset{R^2}{}CHO \quad \xrightarrow{DABCO,))))}\quad R^1\underset{R^2}{\overset{OH}{}}\underset{}{\overset{O}{\|}}OMe \quad + \quad R^1\underset{R^2}{\overset{OH}{}}\underset{}{\overset{O}{\|}}OMe$$

24   8a   25   26
         1,2-anti   1,2-syn
         Felkin   anti-Felkin
         (major)   (minor)

Entry	Aldehyde	Time (h)	Yield (%) 25	anti:syn
1	N-Boc-L-phenylalaninal	40	75	7:1
2	N-Boc-L-phenylalaninal	32	73	4:1
3	N-Boc-L-phenylalaninal	24	72	7:1
4	N-Boc-L-phenylalaninal	40	82	6:1

as catalyst at room temperature with ultrasound irradiation, the *anti* product was obtained as the major isomer (Table 5.2) [55a].

Considerable effort has been devoted to the development of enantiocatalytic MBH reactions, either with purely organic catalysts, or with metal complexes. Paradoxically, metal complex-mediated reactions were usually found to be more efficient in terms of enantioselectivity, reaction rates and scope of the substrates, than their organocatalytic counterparts [36, 56]. However, this picture is actually changing, and during the past few years the considerable advances made in organocatalytic MBH reactions have allowed the use of viable alternatives to the metal complex-mediated reactions. Today, most of the organocatalysts developed are bifunctional catalysts in which the chiral N- and P-based Lewis base is tethered with a Brønsted acid, such as (thio)urea and phenol derivatives. Alternatively, these acid co-catalysts can be used as additives with the nucleophile base.

5.1.3
Chiral Amine Catalysts

The first chiral catalysts were developed from achiral molecules such as DABCO, quinuclidine, indolizine or pyrrolizine-derived catalysts by introducing asymmetric functions. Hirama and co-workers examined chiral C_2-symmetric 2,3-disubstituted 1,4-diazabicyclo[2.2.2]octanes such as 2,3-(dibenzoxymethyl)-DABCO (**29**) as catalysts for asymmetric MBH reaction between 4-nitrobenzaldehyde **27** and methyl vinyl ketone (MVK, **28**) (Scheme 5.6) [57]. The additive function of the catalyst compared to the achiral DABCO resulted, however, in diminished reactivity, and the reaction required high pressure in order to

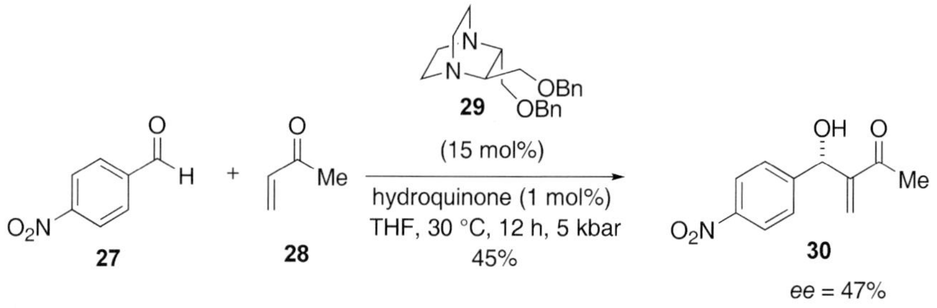

5 Organocatalytic Enantioselective Morita–Baylis–Hillman Reactions

Scheme 5.6 Asymmetric MBH reaction using chiral DABCO **29**.

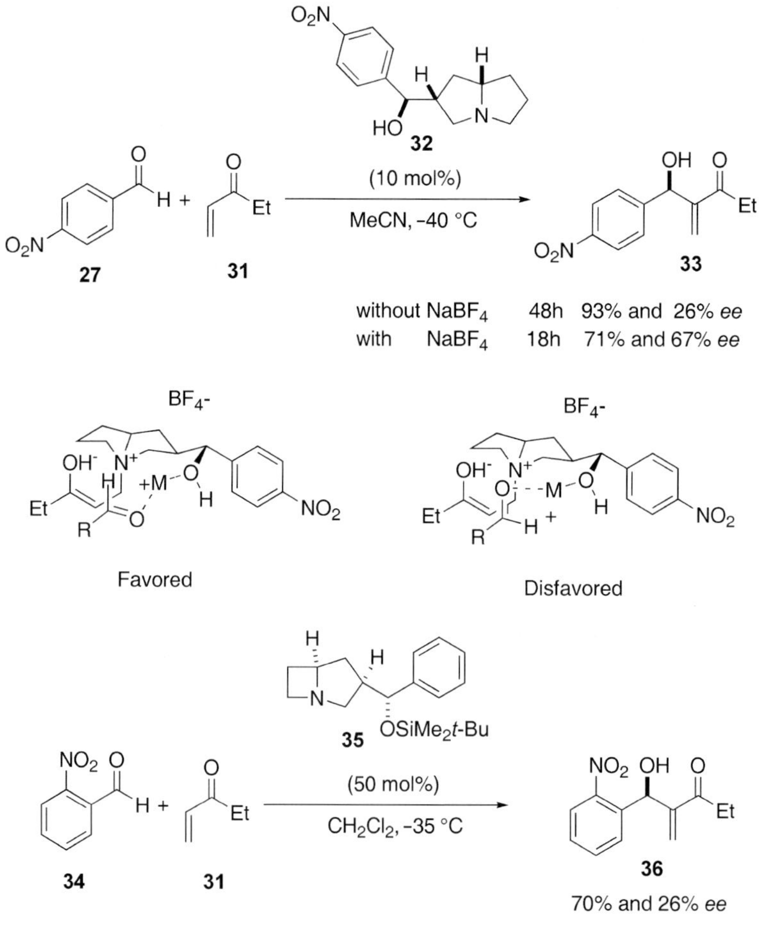

Scheme 5.7 Barrett's catalysts **32** and **35** in asymmetric MBH reactions.

ensure a reasonable reaction rate. The (S)-α-methylene-β-hydroxy-alkanone **30** was obtained in 47% ee.

The chiral pyrrolizine derivative **32** was developed by the Barrett group, and tested in the addition of methylvinylketone (MVK) or ethylvinylketone (EVK), and 4-nitrobenzaldehyde (Scheme 5.7) [58]. The reaction provided products in good yield, albeit in modest enantioselectivity (ee-value 21–47%). It was observed, that the addition of coordinating salts such as NaBF$_4$ increased the selectivity of the alkylation (ee-value up to 72%). The salt effect was rationalized by the formation of two possible chelate intermediates, among which the one leading to the (R)-product was preferred for steric reasons. Of note was the fact that the structural analogue bicyclic azetidine derivative **35**, having a silylated alcohol on the lateral chain, showed increased reactivity but low enantioselectivity (ee-value up to 26%) in similar transformation (Scheme 5.7) [59].

Natural products having chiral tertiary amine functions were tested among the first catalysts in asymmetric MBH reactions [24, 60]. The importance of the proton donor capacity of the catalyst in the rate and selectivity of the MBH reaction was recognized very quickly, and attention was turned to genuine α-amino alcohol structures, such as the compounds listed in Scheme 5.8 [61]. Results were modest, however. Apart from the earlier discussed (R)-3-HDQ, which catalyzed the MBH reaction at atmospheric pressure (though with no enantioselectivity),

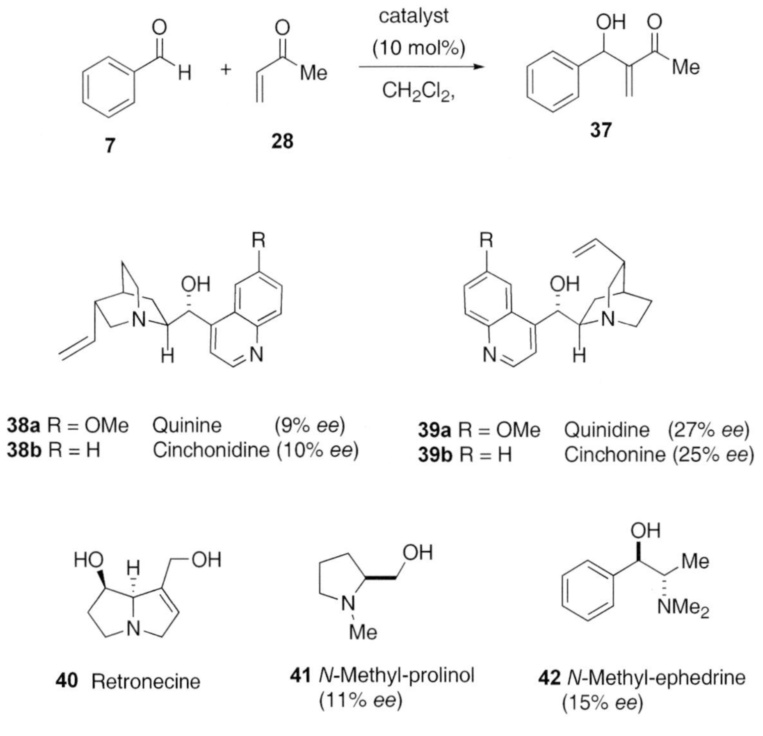

38a R = OMe Quinine   (9% ee)
38b R = H   Cinchonidine (10% ee)

39a R = OMe Quinidine (27% ee)
39b R = H   Cinchonine (25% ee)

40 Retronecine

41 N-Methyl-prolinol (11% ee)

42 N-Methyl-ephedrine (15% ee)

Scheme 5.8 Some natural product-derived catalysts in asymmetric MBH reactions.

all the other amino-alcohols required high-pressure conditions in the addition of MVK to the cyclohexyl carboxaldehyde in CH_2Cl_2 [62]. Among the various *p*-hydroxy-amines employed, the cinchona alkaloids displayed the highest level of enantioselection, followed closely by ephedrine and proline derivatives. As expected, quinine and cinchonidine gave lower but opposite enantioselectivities to quinidine and cinchonine. The crucial role played by the *p*-hydroxyl function was also noted, as the derivatization of quinidine into its *O*-acetyl analogue suppressed enantioselectivity.

Similar observations were made in the *N*-methylprolinol-mediated addition of MVK and benzaldehyde, which proceeded under atmospheric pressure in 1,4-dioxane:water (1:1, v/v) solvent mixture at 0 °C, and afforded product in up to 78% *ee* [63]. The catalytic activity was also suppressed in methylating the free hydroxyl function.

An important step in the development of asymmetric MBH reactions was the introduction of the quinidine-derived β-isocupreidine catalyst (β-ICD) **44**, by the Hatakeyama group [64, 65]. β-ICD mediated the addition of acrylate **43** to a variety of aromatic and aliphatic aldehydes even at −55 °C in DMF, and afforded (*R*)-adducts in good yields (40–58%) and excellent enantioselectivity (*ee* up to 99%) (Table 5.3). The contribution of the 1,1,1,3,3,3-hexafluoroisopropyl acryl ester (HFIPA) to the reactivity of the system deserves comment. This ester displayed

Table 5.3 Asymmetric MBH reaction of aldehydes and HFIPA (**43**) catalyzed by β-ICD (**44**).

Entry	Ar	Time (h)	Yield (%), configuraton and *ee* (%)	Yield (%), configuraton and *ee* (%)
1	C_6H_5	48	57, *R* (95)	-
2	*p*-$NO_2C_6H_5$	1	58, *R* (91)	11, *R* (4)
3	$(CH_3)_2CHCH_2$	4	51, *R* (99)	18, *S* (85)
4	*t*-Bu	72	-	-

5.1 Addition of Ketones and Aldehydes to Activated Olefins

an almost 200-fold rate acceleration compared to methyl acrylate, under identical reaction conditions. However, limitations were observed with bulky aldehydes such as butyraldehyde, where dimerization of the acrylate was observed. The enantioselectivity of the dioxanone byproduct **23** was found to be considerably lower, or even opposite compared to the "normal" adduct, **45** (Table 5.3). This fact can be rationalized by the different mechanism of the stereo-determining proton abstraction steps of these products, as outlined in Scheme 5.5.

The sense of stereo-induction leading to the (*R*)-enriched adduct can be rationalized by the formation of the betaine intermediates **A** and **B**, stabilized by intramolecular hydrogen bonding between the oxy anion and the phenolic hydroxyl group (Scheme 5.9) [66]. In the following transformations, intermediate **A** is preferred over **B**, while the intramolecular protonation of the formed alcoholate of **A** is configurationally more favorable: this intermediate gives rise to the (*R*)-product, while **B** will revert to the starting material. Of note in this arrangement is that the conformation is almost ideal for the subsequent elimination process that is necessary for regeneration of the catalyst.

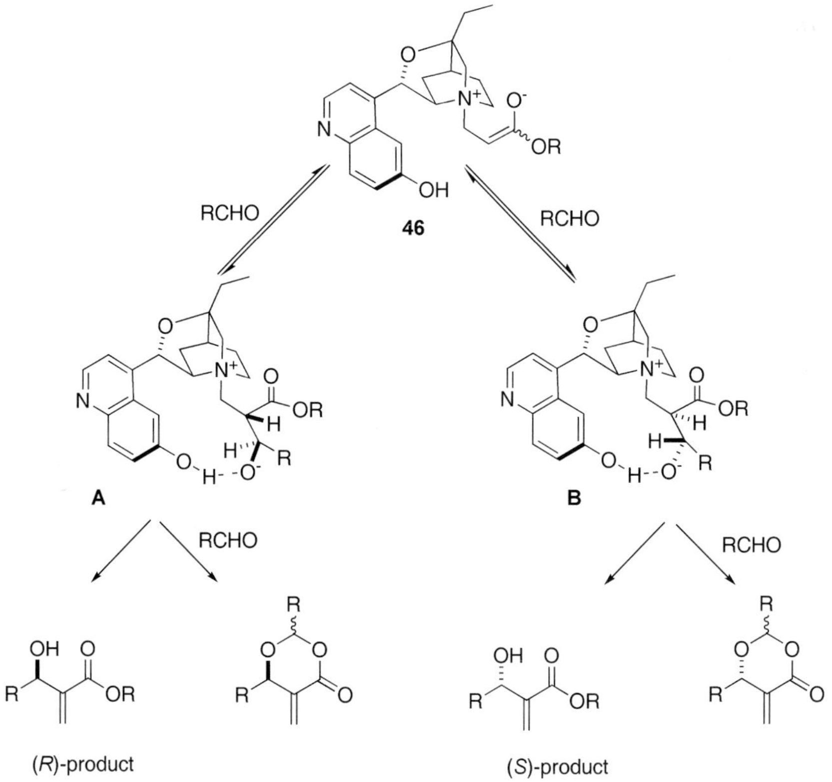

Scheme 5.9 The rationalization of stereo-induction in the β-ICD-mediated MBH reaction.

The β-ICD catalyst (**44**) was used in the synthesis of epopromycin B, a plant cell wall synthesis inhibitor [67], and also in the synthesis of mycestericin E, a potent immunosuppressor [68].

For the synthesis of epopromycin B, precursor **47** was prepared via the addition of (*S*)-*N*-Fmoc-leucinal, **46**, to HFIPA, **43**, at −55 °C (Scheme 5.10) [67]. The reaction afforded, after methanolysis, two diastereoisomers: the *syn* product **49** was obtained in 70% yield and in 99% *ee*, and the *anti* isomer **50** was obtained in 2% yield and in 99% *ee*. It should be noted that the hexafluoroisopropyl group was converted to the corresponding methyl ester in the alcoholysis step. The major *syn* isomer **49** was transformed to the epoxy end-product **51**.

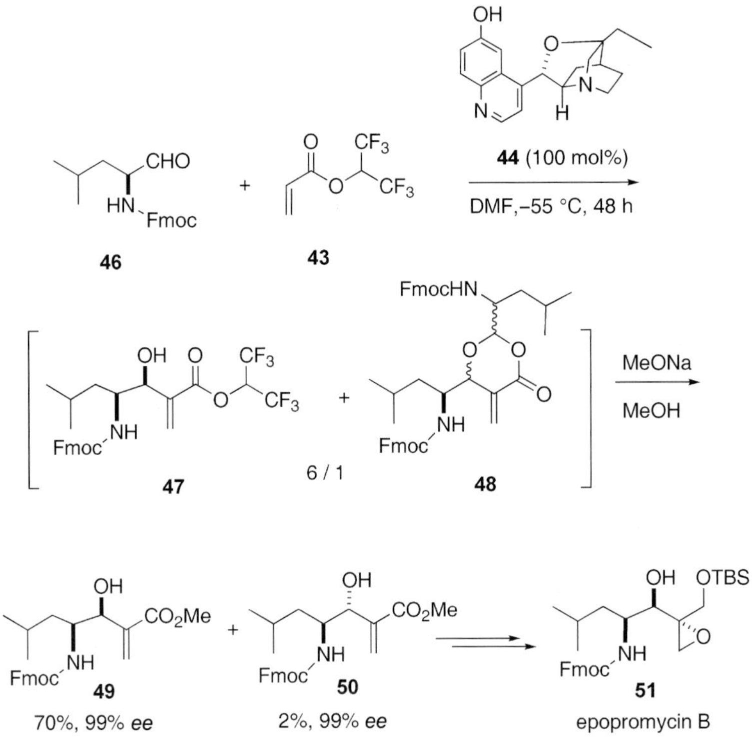

Scheme 5.10 The asymmetric MBH reaction in the synthesis of epopromycin B.

A similar strategy was used in the synthesis of mycestericin E [68]. The configuration of the chiral allylic alcohol **53** was established via a β-ICD-mediated MBH reaction between aldehyde **52** and the hexafluoroisopropyl acrylate **43** in a DMF/CH$_2$Cl$_2$ solvent mixture at low temperature (Scheme 5.11).

MVK derivatives are less reactive than the HFIPA. The β-ICD-mediated addition of activated aromatic aldehydes to MVK requires higher temperature and longer reaction times, resulting in a lower enantioselectivity (Table 5.4) [69].

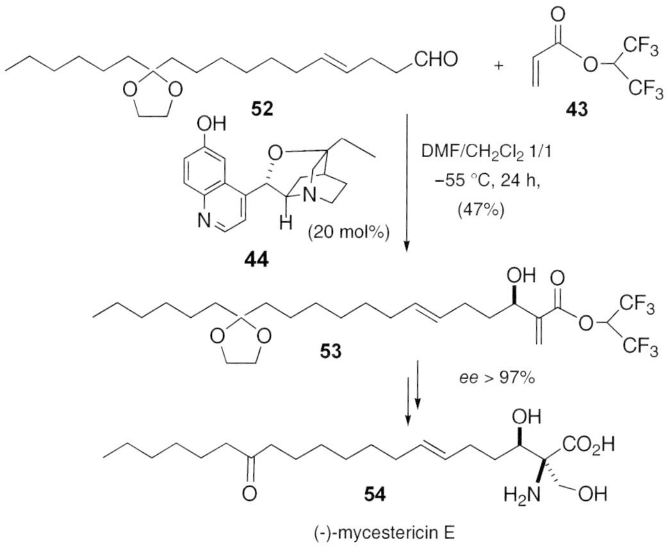

Scheme 5.11 The synthesis of mycestericin E.

Moreover, as highlighted in Table 5.4, the use of Brønsted or Lewis acid additives increases the reaction rate, yield and selectivity, but these results are normally substrate-dependent.

Although dimeric Sharpless ligands as catalysts showed impressive results in related organocatalytic transformations, they provided only limited success in asymmetric MBH reactions (Scheme 5.12) [70]. These compounds are bifunctional catalysts in the presence of acid additives: one of the two amine function of the dimers forms a salt and serves as an effective Brønsted acid, while another tertiary amine of the catalyst acts as a nucleophile. Whereas salts derived from (DHQD)$_2$PYR, or (DHQD)$_2$PHAL afforded trace amounts of products in the addition of methyl acrylate **8a** and electron-deficient aromatic aldehydes such as **27**, (DHQD)$_2$AQN, **56**, mediated the same transformation in *ee* up to 77%, albeit in low yield. It should be noted that, without acid, the reaction afforded the opposite enantiomer in a slow conversion.

Although cinchona-derived catalysts continue to play major role in asymmetric MBH reactions, peptide-based catalysts are beginning to emerge as alternatives [71, 72]. One exciting advance is the simultaneous use of a nucleophilic catalyst (i.e., cinchona or peptide) in combination with a conveniently chosen acid, such as proline, or proline-containing oligopeptides as co-catalyst in a MVK-based MBH reaction (Scheme 5.13) [39, 69, 73]. In these reactions the co-catalyst accelerates the reaction and acts synergistically, allowing higher *ee*-values to be obtained than in the absence of the additive. Contrary to the generally observed fact, the configuration of the proline co-catalyst has an important effect on the observed stereochemistry of the product [69, 71]. When pentapeptide **58** having a

5 Organocatalytic Enantioselective Morita–Baylis–Hillman Reactions

Table 5.4 MBH reaction of aryl aldehydes and MVK catalyzed by β-ICD.

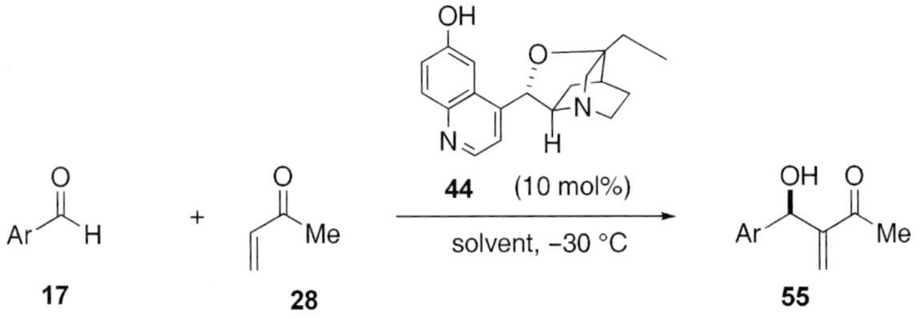

Entry	Ar	Additives (mol%)	Time (h)	Yield (%)	ee (%)
1	p-NO$_2$C$_6$H$_5$	None	48[a]	53	7 (R)
2	p-NO$_2$C$_6$H$_5$	D-proline (10)	42[a]	88	31 (R)
3	p-BrC$_6$H$_5$	None	96[a]	78	25 (R)
4	p-BrC$_6$H$_5$	D-proline (10)	60[a]	31	27 (S)
5	p-BrC$_6$H$_5$	LiClO$_4$ (20)	72[b]	43	49 (R)

[a] Reaction in DMF.
[b] Reaction in THF.

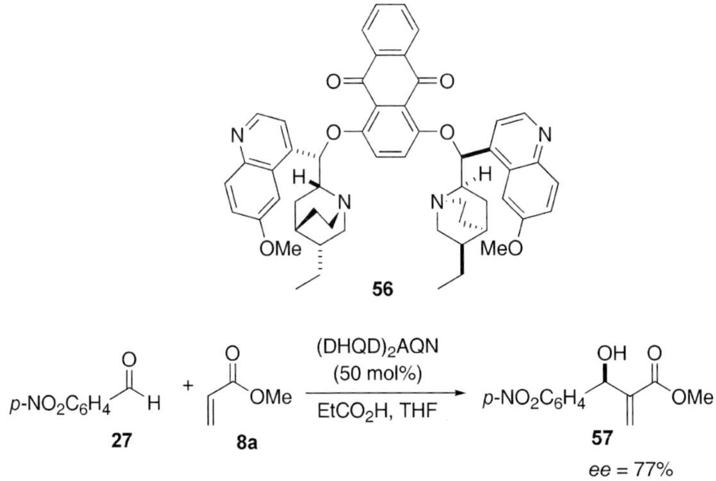

Scheme 5.12 (DHQD)$_2$AQN as catalyst in the MBH reaction.

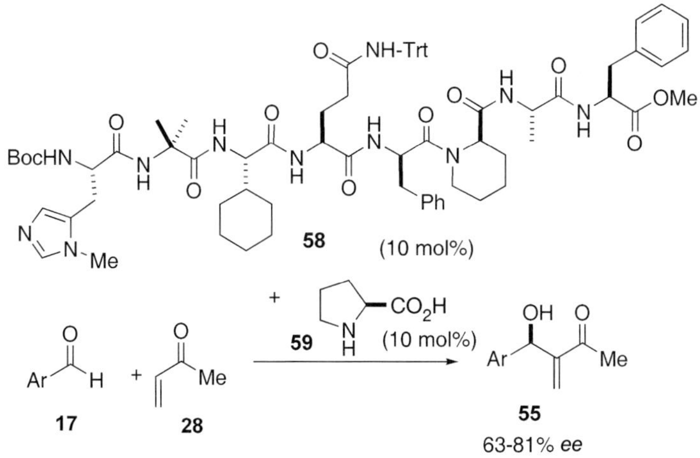

5.1 Addition of Ketones and Aldehydes to Activated Olefins | 165

Scheme 5.13 Dual stereocontrol in the MBH reaction.

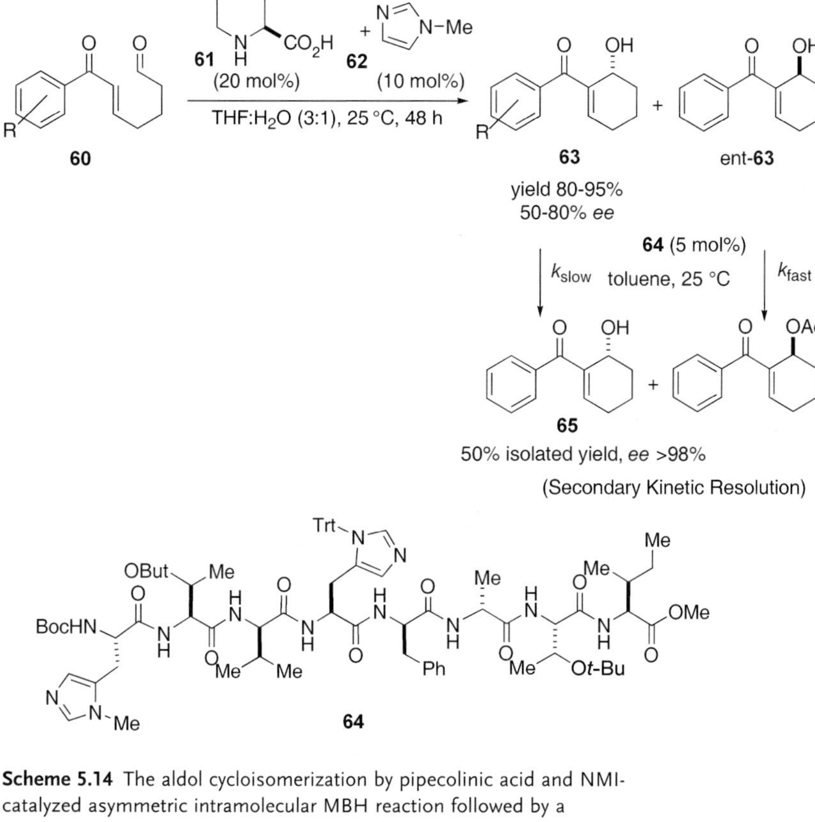

Scheme 5.14 The aldol cycloisomerization by pipecolinic acid and NMI-catalyzed asymmetric intramolecular MBH reaction followed by a "kinetic resolution quench".

modified histidine side-chain was used in the reaction of MVK (**28**) and benzaldehyde (**17**) in the presence of L-proline and D-proline co-catalysts, respectively, the reaction afforded the (*R*)-adduct in 78% *ee*, and the (*S*)-adduct in 31% *ee*, respectively, showing that the influence of the proline may outset that of the peptide in controlling the stereochemistry of the new center. In the control experience, proline alone provided no product during the same time window, and *N*-methylimidazole (NMI, 10 mol%) in the absence of proline catalyzed only a sluggish reaction between MVK and the aldehyde. It should be noted that, when NMI and L-proline were employed together, the reaction afforded a near-racemic product, indicating that the influence of the L-Pro chirality without the peptide was minimal.

A modification of this system was also used in intramolecular MBH reactions (also called as aldol cycloisomerization) [71, 74]. In this reaction, optically active pipecolinic acid **61** was found to be a better co-catalyst than proline, and allowed *ee*-values of up to 80% to be obtained, without a peptide catalyst. The intermolecular aldol dimerization, which is an important competing side-reaction of the basic amine-mediated intramolecular MBH reaction, was efficiently suppressed in a THF:H$_2$O (3:1) mixture at room temperature, allowing the formation of six-membered carbocycles (Scheme 5.14). The enantioselectivity of the reaction could be improved via a "kinetic resolution quench" by adding acetic anhydride as an acylating agent to the reaction mixture and a peptide-based asymmetric catalyst such as **64** that mediates a subsequent asymmetric acylation reaction. The non-acylated product **65** was recovered in 50% isolated yield with *ee* >98%.

An interesting inversion of enantioselectivity was observed in this reaction with proline catalyst, when the catalyst was used alone, or with imidazole co-catalyst (Table 5.5) [75]. When hep-2-enedial **66** was submitted to proline-mediated

Table 5.5 Proline-catalyzed intramolecular MBH reaction of hept-2-endial with and without imidazole additive.

Entry	Catalyst	Additive	T (°C)	Time (h)	Yield (%)	*ee* (%)
1	L-**59**	–	70	1	69	60 (*S*)
2	D-**59**	–	15	5	75	41 (*R*)
3	L-**59**	62	0	7	72	93 (*R*)
4	D-**59**	62	0	15	77	93 (*S*)

cyclization, the L-isomer afforded the (S)-adduct (entry 1, Table 5.5), while the D-isomer afforded the (R)-adduct (entry 2). When imidazole was added, an inversion of enantioselectivity was observed (entries 3 and 4). This phenomenon warrants divergent reaction mechanisms operating in these transformations, with or without imidazole.

5.1.3.1 Chiral (Thio)Urea/Amine Catalysts Systems

The simultaneous use of urea, or thiourea [76] and DABCO catalyst was introduced by the Connon group for the addition of methyl acrylate and benzaldehyde [39]. The study revealed that, although both ureas and thioureas accelerated the reaction relative to the uncatalyzed process, urea was superior to thiourea in terms of stability and efficiency. Chiral thiourea derivatives may offer, however, superior enantioselectivity. It was postulated, that the catalysts operate mainly via a Zimmerman–Traxler-type transition state **69** for addition of the resulting enolate anion to the aldehyde (Scheme 5.15).

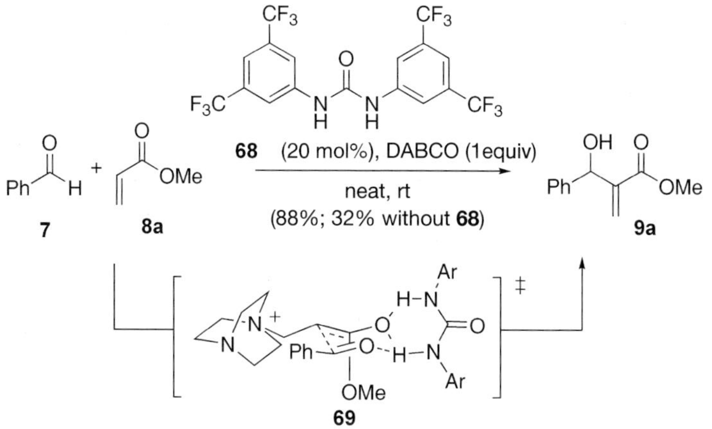

Scheme 5.15 DABCO and urea, **68**-promoted MBH reaction of benzaldehyde and methyl acrylate.

Cyclohexanediamine-derived amine thiourea **70**, which provided high enantioselectivities for the Michael addition [77] and aza-Henry reactions [78], showed poor activity in the MBH reaction. This fact is not surprising when one considers that a chiral urea catalyst functions by fundamentally different stereoinduction mechanisms in the MBH reaction, and in the activation of related imine substrates in Mannich or Strecker reactions [80]. In contrast, the binaphthylamine thiourea **71** mediated the addition of dihydrocinnamaldehyde **74** to cyclohexenone **75** in high yield (83%) and enantioselectivity (71% ee) (Table 5.6, entry 2) [79]. The more bulky diethyl analogue **72** displayed similar enantioselectivity (73% ee) while affording a lower yield (56%, entry 3). Catalyst **73** showed only low catalytic activity in the MBH reaction (18%, entry 4).

Table 5.6 The chiral tertiary amine/thiourea-catalyzed MBH reaction of dihydrocinnamaldehyde with cyclohexenone.

Entry	Catalyst	Yield (%)	ee (%)
1	70	21	39 (R)
2	71	83	71 (R)
3	72	56	73 (R)
4	73	18	-

Under optimized conditions, **71** mediated the addition of sterically demanding aldehydes with excellent enantioselectivity (90–94% *ee*) and in good yields (63–72%) (Table 5.7, entries 2 and 5). In turn, the catalyst was less efficient in the addition of aromatic aldehyde such as *o*-chlorobenzaldehyde, and produced both lower yield (55%) and *ee* (60%; Table 5.7, entry 3).

The C_2-symmetric chiral bis-thiourea catalyst **78** mediated the addition of cyclohexanone to a range of activated aldehydes in the presence of DMAP, or imida-

Table 5.7 Bifunctional tertiary amine/thiourea catalyst in MBH reaction of cyclohexenone and selected aldehydes.

R–CHO + cyclohexenone → 77 (with 71 (10 mol%), CH$_3$CN, 0 °C)

12   **75**   **77**

Entry	R	Time (h)	Yield (%)	ee (%)
1	PhCH$_2$CH$_2$	48	80	83
2	(CH$_3$)$_2$CH	72	63	94
3	o-ClC$_6$H$_4$	108	55	60
4	n-C$_4$H$_9$	48	84	81
5	Cyclohexyl	120	67	92

zole (40%) [81]. While aromatic aldehydes afforded generally mediocre results in terms of selectivity, better results were obtained with aliphatic aldehydes (up to 90% ee). A transition state **79**, in which both the aldehyde and the enone are coordinated to a thiourea group through hydrogen-bonding, was proposed (Table 5.8).

5.1.4
Chiral Phosphine-Catalyzed MBH Reactions [82]

In contrast to chiral amines, phosphorus-based chiral catalysts were less developed for asymmetric MBH transformations. As with amine-based reactions, the selectivity of the addition in phosphine-mediated reactions depends clearly on the nature of the complementary Brønsted acid co-catalysts used.

In the first asymmetric cycloisomerization reaction, the cyclopentenol derivative **82** was prepared from **80** in the presence of (−)-CAMP **81** (Scheme 5.16) [83]. The low asymmetric control (14% ee) was attributed to the reversibility of the cyclization. It should be noted that this reaction is not suitable for the preparation of six-membered rings.

The stereochemistry of the Michael acceptor plays an important role in the efficiency of the phosphine-mediated intramolecular MBH reactions. Enones having (Z)-alkenes (e.g., **83**) afforded a higher yield in the desired product than did the

Table 5.8 Chiral bis-thiourea and DMAP-mediated MBH reaction of cyclohexenone and aldehydes.

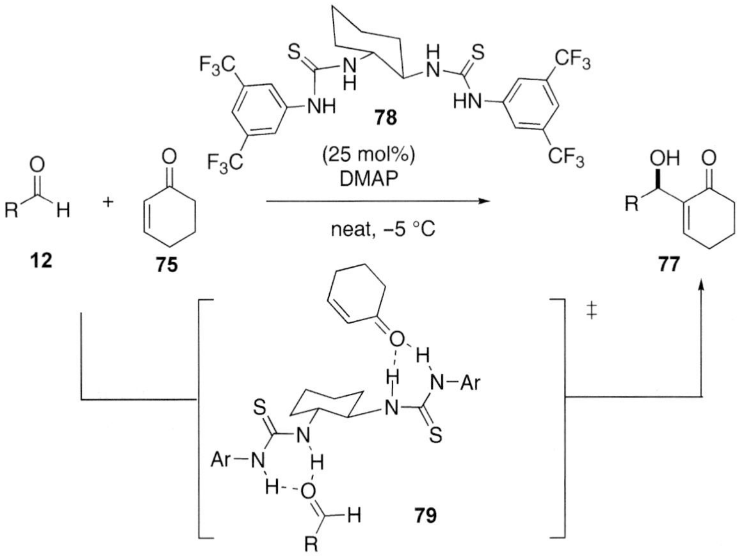

Entry	R	Time (h)	Yield (%)	ee (%)
1	PhCH$_2$CH$_2$	72	33	59
2	(CH$_3$)$_2$CH	72	67	60
3	C$_6$H$_5$	72	88	33
4	Cyclohexyl	72	72	90

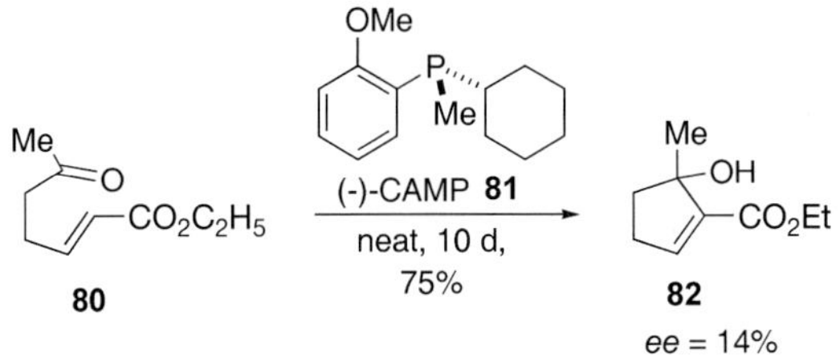

Scheme 5.16 The first asymmetric intramolecular MBH reaction.

5.1 Addition of Ketones and Aldehydes to Activated Olefins | 171

(*E*)-isomer **85** under identical conditions (Scheme 5.17) [84]. The reason for this difference in reactivity is most likely steric in nature, as substrates where the β-substituent is *cis* to the electron-withdrawing substituent are more accessible to the nucleophilic catalyst than are their *trans* counterparts.

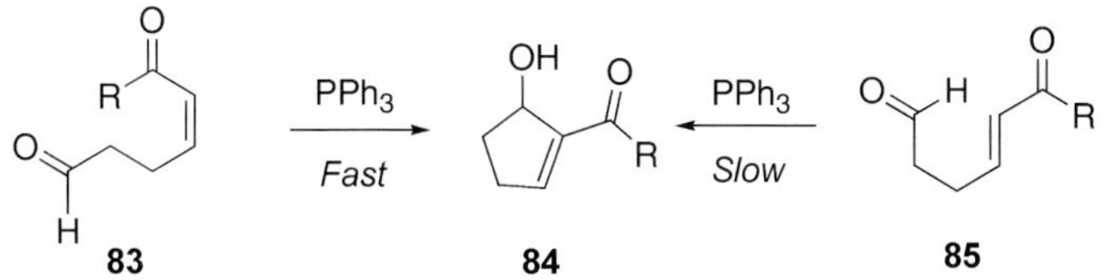

Scheme 5.17 The effect of enone stereochemistry on the rate of the phosphine-mediated MBH reaction.

Chiral phosphines were examined for intermolecular MBH reaction of pyrimidine 5-carboxaldehyde and methyl acrylate (Table 5.9) [85]. Most of the catalysts afforded no, or low, enantioselectivity. The best *ee*-value was obtained with BINAP catalyst (Table 5.9, entry 4), which afforded product in *ee*-values up to 44%.

Several D-mannitol-derived phosphines have also been examined in the asymmetric MBH reaction (Scheme 5.18) [86, 87]. Phospholane **89a** catalyzed this reaction with low conversion and low *ee* (19%). In the presence of the hydroxyl phospholane **89b** as catalyst, the reaction was accelerated significantly (83% in 9 h versus 29% in 70 h), while the enantioselectivity could not be ameliorated (Scheme 5.18).

Table 5.9 Chiral phosphine-catalyzed MBH reaction of methyl acrylate with pyrimidine 5-carboxaldehyde.

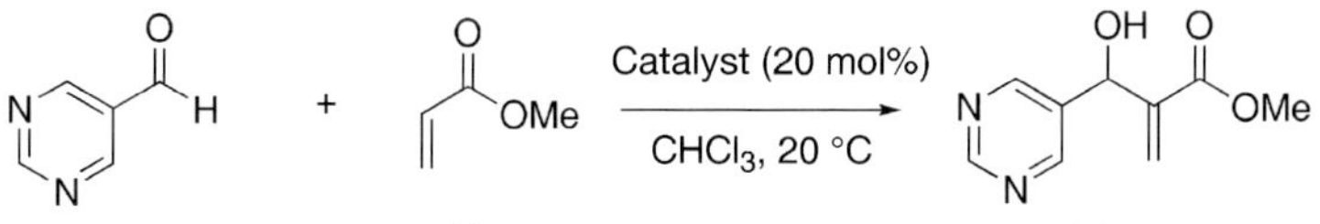

Entry	Catalyst	Time (h)	Yield (%)	ee (%)
1	(2R,3R)-DIOP	19	28	<1 (-)
2	(2R,3R)-NORPHOS	21	32	3 (-)
3	(R,S)-BPPFOH	20	46	2 (-)
4	(S)-BINAP	85	24	44 (-)

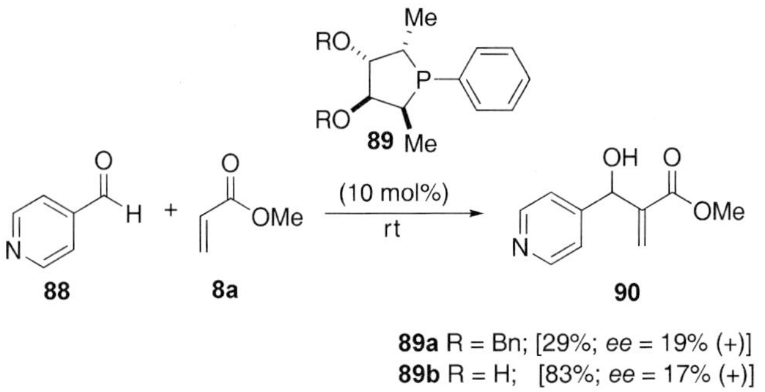

Me
RO
P
RO
89 Me
O
O
OH O
H +
OMe
(10 mol%)
rt
OMe
N
N
88
8a
90
89a R = Bn; [29%; ee = 19% (+)]
89b R = H; [83%; ee = 17% (+)]

Scheme 5.18 D-Mannitol-derived phospholane catalysts in asymmetric MBH reactions.

5.1.4.1 **Phosphine/Binaphthol-Derived Catalysts**

A remarkable acceleration of MBH reactions was observed in the presence of mild cooperative catalysts consisting of tributylphosphine and phenols or naphthols as Brønsted acids (Table 5.10) [88–90]. The non-asymmetric addition of

Table 5.10 The tributylphosphine and naphthol or phenol co-catalyzed MBH reaction of dihydrocinnamaldehyde **74** and cyclopentenone **91**.

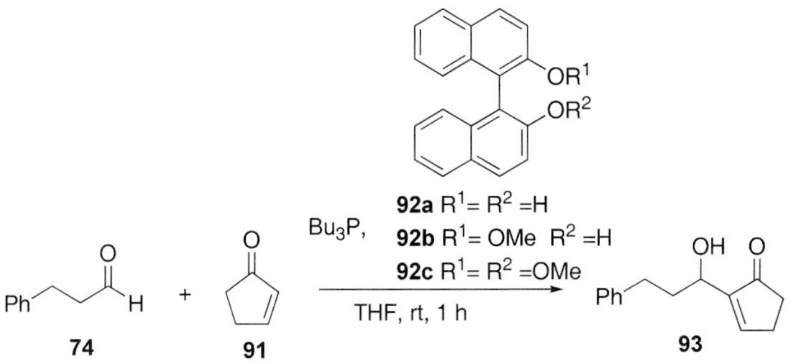

OR1
OR2
92a R1= R2 =H
Bu3P,
92b R1= OMe R2 =H
92c R1= R2 =OMe
THF, rt, 1 h
O
O
OH O
Ph
H
+
Ph
74
91
93

Entry	Co-catalyst	Yield (%)
1	-	23
2	MeOH (20 mol%)	23
3	**92a** (10 mol%)	quant
4	**92b** (10 mol%)	80
5	**92c** (10 mol%)	24
6	phenol (20 mol%)	quant

dihydrocinnamaldehyde **74** to cyclopentenone **91** was catalyzed with tributylphosphine (20 mol%), and afforded the corresponding MBH product **93**, albeit in 23% yield (Table 5.10, entry 1) [88]. When 10 mol% of racemic 1,1′-bi-2-naphthol (BINOL, **92a**), as a Brønsted acid co-catalyst was added, the reaction afforded the same product in quantitative yield during the same time window (entry 3). It should be noted that neither methanol, benzoic acid nor *p*-toluenesulfonic acid were efficient in this transformation (Table 5.10, entry 2); when the reaction was catalyzed by 2-hydroxy-2′-methoxy-1,1′-binaphthyl (**92b**), or 2,2′-dimethoxy-1,1′-binaphthyl (**92c**) the product **93** was formed in 80% and 24% yields, respectively (Table 5.10, entries 4 and 5), indicating that the synergetic action of both hydroxyls is beneficial to the reaction. DABCO or triphenylphosphine as Lewis bases were inefficient and afforded no products. Phenol and *p*-toluenesulfonamide, which has the same Brønsted acidity as phenol, were, in turn, effective (Table 5.10, entry 6). The attempted asymmetric reaction with (*R*)-BINOL and tributylphosphine afforded **93** in low *ee* (<10%) [88].

A similar system was studied a few years later by the Schaus group [89], who compared several binaphthol-derived chiral Brønsted acids such as **92a** and **94a–d** in the triethylphosphine-mediated MBH reaction between cyclohexenone and aldehydes. Optimized conditions were found with 2–20 mol% of chiral Brønsted acid and an excess of triethylphosphine (200 mol%) as the nucleophilic promoter at 0–10 °C in THF. Using PMe₃ or P(*n*-Bu)₃ in the reaction afforded **76** in yields similar to that of PEt₃, but in lower enantioselectivity (50% and 64% *ee*, respectively). The use of only (*R*)-BINOL in the MBH reaction of dihydrocinnamaldehyde **74** and cyclohexenone **75** resulted in the formation of **76** in 16% *ee*. Partially saturated BINOL derivatives such as **94a–d** offered high chemical yield and enantioselectivity (Scheme 5.19) [91]. Optimal results with the addition of aliphatic al-

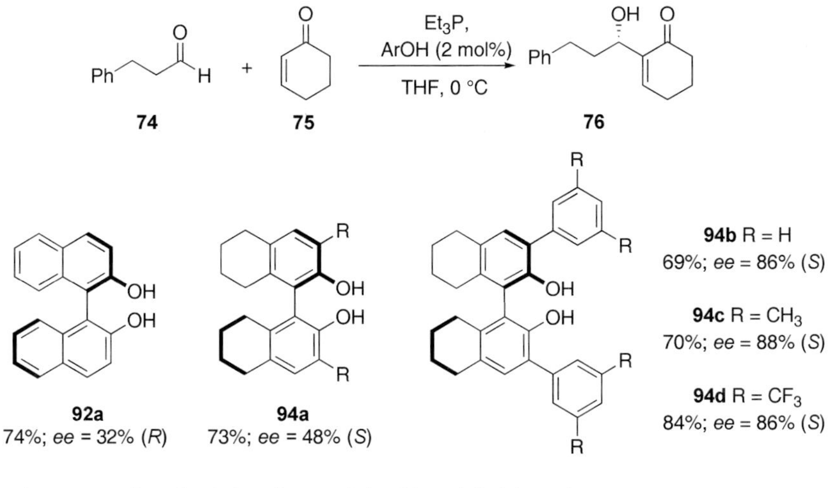

Scheme 5.19 The achiral phosphine and chiral binaphthol-derived Brønsted acid-catalyzed MBH reaction of cyclohexenone and 3-phenylpropionaldehyde.

dehydes was obtained with **94c**, while catalyst **94d** afforded the best results with more hindered aldehydes. Conjugated aldehydes such as benzaldehyde and cinnamaldehyde as acceptors resulted in low yields and enantioselectivity.

5.2
Asymmetric Aza-MBH Reactions

Aldehydes of MBH reactions can be replaced by activated aldimines such as N-tosyl, N-mesyl, N-nosyl, N-diphenylphosphinoyl, or N-SES aldimines. The addition of such aldimines to electron-deficient olefins can be mediated by nucleophilic tertiary N or P Lewis bases (Scheme 5.20).

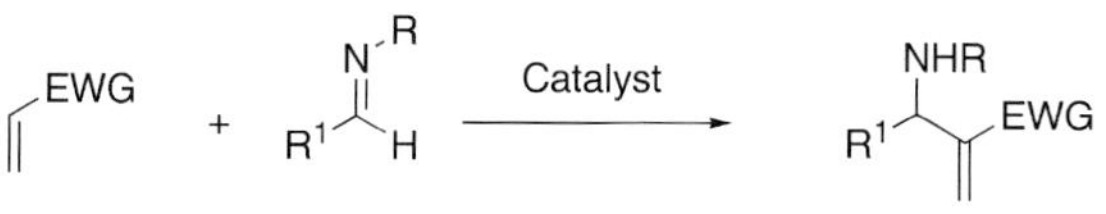

Scheme 5.20 The general scheme of the aza-MBH reaction.

While THF or CH$_2$Cl$_2$ are the most commonly used solvents, the solubility of the reagents or the catalyst may dictate the use of other solvents. Reactions are usually slow in DMF and in CH$_3$CN when DABCO, DBU or DMAP were used as catalysts. An often-observed byproduct of the aza-MBH reaction is a bridged compound of type **97**. This product is the result of a stepwise addition of **95** to **75** via the Mannich reaction, followed by an intramolecular conjugated addition (Michael addition) of the formed anion to the α,β-unsaturated ketone, and thus due to the elevated basicity of the catalyst (Scheme 5.21) [92].

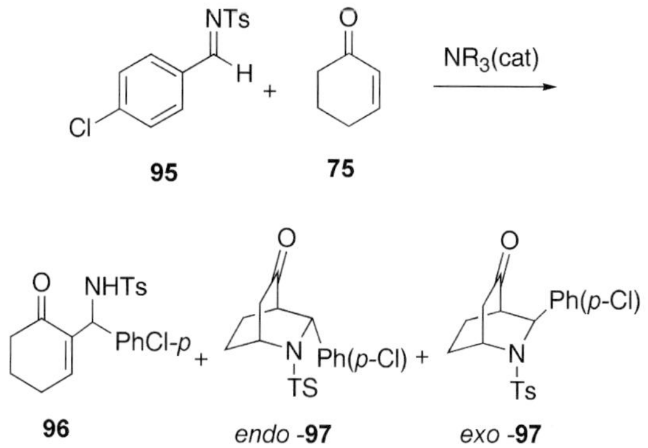

Scheme 5.21 Product and byproducts in the tertiary amine-catalyzed aza-MBH reaction of **95** and **75**.

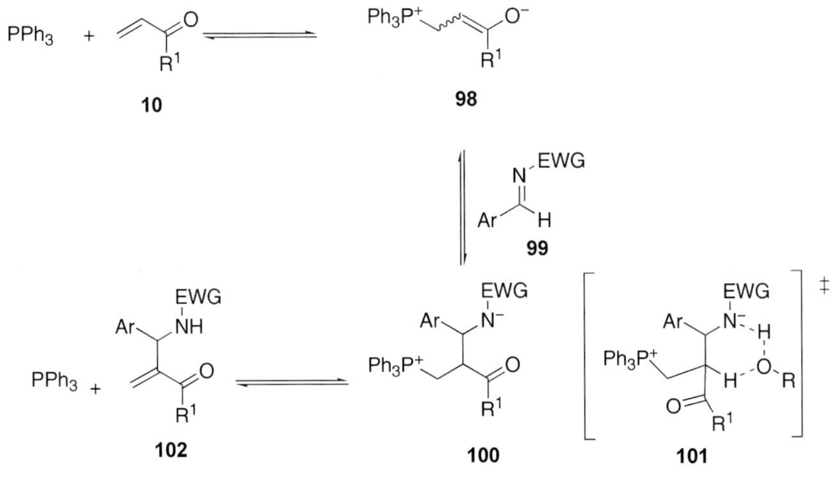

Scheme 5.22 The mechanistic model of the aza-MBH-reaction. EWG = electron-withdrawing group.

5.2.1
Reaction Mechanism

As discussed previously for the MBH reaction, the aza-MBH reaction involves rate-limiting proton transfer in the absence of added protic species (Scheme 5.22) [93]. In contrast to the MBH reaction, however, the aza-MBH exhibits no autocatalysis. Brønsted acidic additives lead to substantial rate enhancements through acceleration of the elimination step. It has been shown that phosphine catalysts – either alone or in combination with protic additives – may trigger epimerization of the aza-MBH product by proton exchange at the stereogenic center. This fact indicates that the spatial arrangement of a bifunctional chiral catalyst in this reaction is crucial not only for the stereodifferentiation within the catalytic cycle but also to prevent subsequent epimerization.

5.2.2
Chiral Amine Catalysts

5.2.2.1 Cinchona-Derived Catalysts

As in the MBH reactions, β-ICD (**44**) is also an efficient and remarkably general catalyst in aza-MBH reactions [94, 95]. This catalyst promotes the addition a variety of electron-deficient olefins such as acrylates, enones, and enals with activated aromatic aldimines. Of note, the absolute stereochemistry of the product is generally opposite compared to the analogous MBH reaction: with β-ICD catalyst imines gives rise to (S)-enriched adducts, in contrast to aldehydes which afford (R)-products [94]. Substitution patterns of the olefin may alter or even invert this trend, however.

Table 5.11 The β-ICD-mediated aza-MBH reaction of diphenyl-phosphinoyl imines and HFPA.

Entry	Ar	Time (h)	Yield (%)	ee (%)
1	phenyl	120	90	67
2	4-MeO-phenyl	96	42	73
3	4-O$_2$N-phenyl	2	97	54
4	naphthyl	48	79	72

The addition of acrylates to activated aryl imines (**103**) was studied by the groups of Hatakeyama and Shi [94, 95]. In Hatakeyama's study, various aryl diphenylphosphinoyl imines and HFIPA, **43**, afforded in DMF at low temperature the (S)-product (**104**) in yields up to 97% and high ee (Table 5.11) [94].

Similar to the addition to HFIPA, the β-ICD-mediated addition of methyl, phenyl, and naphthyl acrylates to N-sulfonyl imines afforded the (S)-adduct, which is opposite to that observed with aldehydes (Table 5.12) [95]. The best conditions were found using either dichloromethane or acetonitrile as solvent.

The elusive addition of N-tosyl arylaldimines to acrolein, and to acrylonitrile, was catalyzed by β-ICD, **44** (Scheme 5.23) [95]. The reaction afforded in both cases invariably the (S)-enriched adducts. While acrylonitrile is less reactive than

5.2 Asymmetric Aza-MBH Reactions | 177

Table 5.12 The β-ICD-mediated aza-MBH reaction of tosyl aldimines and acrylates.

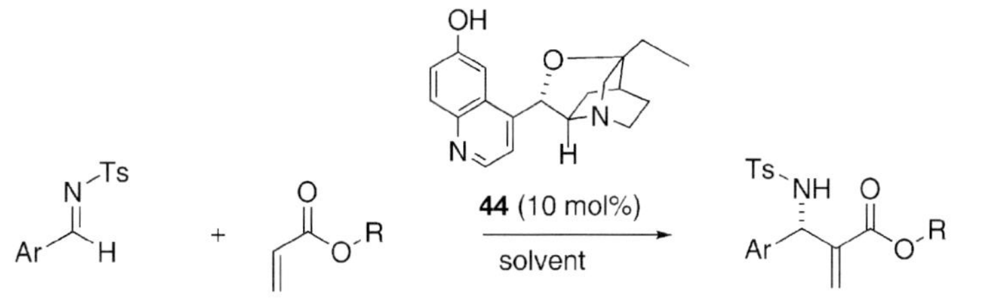

Entry	Ar	R	Temp. (°C)	Time (h)	Yield (%)	ee (%)
1	C₆H₅	Me	0	72	62	83
2	C₆H₅	Ph	−20	24	67	74
3	p-ClC₆H₅	Me	0	36	60	77
4	p-ClC₆H₅	α-(Nap)	0	12	67	78

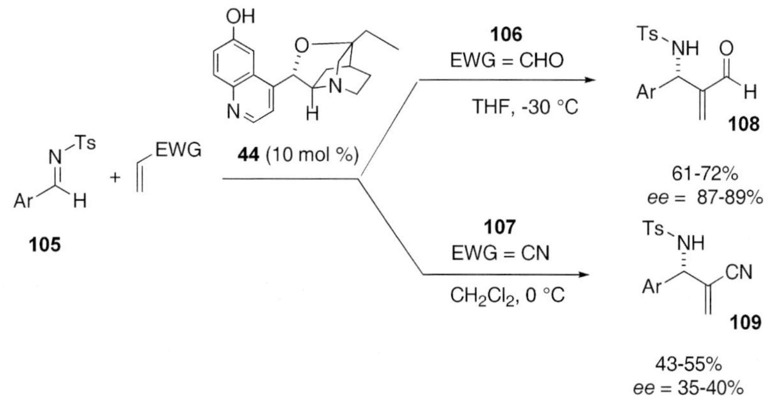

Scheme 5.23 The β-ICD-mediated aza-MBH reaction of aryl tosyl aldimines with acrolein and acrylonitrile. EWG = electron-withdrawing group.

Table 5.13 The β-ICD-mediated aza-MBH reaction of aryl tosyl aldimines with MVK and EVK.

Entry	Ar	R	Time (h)	Yield (%)	ee (%)
1	C_6H_5	Et	22	54	94
2	p-MeOC$_6$H$_5$	Me	36	67	99
3	p-MeC$_6$H$_5$	Me	24	80	96
4	p-MeC$_6$H$_5$	Et	41	49	87

acrolein, the reaction required a higher temperature, and afforded products in lower chemical yields and ee.

β-ICD mediated the addition of aryl methylbenzenesulfonamides, such as **105**, to ethyl or methyl vinyl ketones in MeCN:DMF (1:1) mixtures at a lower temperature (−30 °C) [95a]. The reaction afforded products with (R) absolute configuration (Table 5.13). Of note, this situation was opposite to that observed in the related aza-MBH reaction with acrylonitrile with acryl aldehyde or with acrylates. Related N-mesyl or N-SES-protected imine afforded similar results.

5.2.2.2 Non-Natural Tertiary Amine/Phenol Catalysts
A BINOL-dimethylaminopyridine hybrid was seen to be efficient in mediating the MBH reaction (Table 5.14) [96], with optimal reaction conditions being found as −15 °C with a mixed solvent system consisting of toluene and cyclopentyl methyl ether (CPME) in a 1:9 ratio. The reaction was sensitive to the structure of the catalyst **112**, the position of the Lewis base attached to BINOL, the substitution pattern of the amino group, and the length of the spacer. It should be noted that the bulky i-Pr substituent on the amino group showed the best selectivity and kinetic profile (Table 5.14, entry 5) [98]. (For experimental details see Chapter 14.10.4).

Table 5.14 The BINOL-dimethylaminopyridine hybrid catalyst-mediated aza-MBH reaction.

Entry	R	Time (h)	Yield (%)	ee (%)
1	H	240	62	87
2	Et	132	90	91
3	t-Bu	240	72	83
4	Bn	144	quant	93
5	i-Pr	72	93	93

5.2.2.3 Chiral Acid/Achiral Amine

In screening thiourea catalysts for the asymmetric aza-MBH reaction, the Jacobsen group identified catalysts of type **114**, and subsequently selected **114b** for further study due to its greater synthetic accessibility (Table 5.15) [97]. In the reaction of various N-nosyl imines and ethyl acrylate, high enantioselectivity was obtained by using chiral thiourea catalyst **114b** and DABCO as the nucleophilic additive (Table 5.15). As in the related aza-MBH reactions, pronounced solvent and concentration effects were observed. Under optimized reaction conditions, electron-donating, electron-withdrawing, and heteroaromatic substrates each underwent reaction with methyl acrylate in high enantioselectivity, albeit with modest chemical yield.

Table 5.15 Chiral thioureas **114** and DABCO-mediated aza-MBH reaction.

114a $R^1 = R^2 = Me$, $R^3 = OCOt$-Bu
114b $R^1 = Bn$, $R^2 = Me$, $R^3 = t$-Bu

Entry	Ar	Time (h)	Yield (%)	ee (%)
1	m-MeC$_6$H$_5$	24	40	93
2	m-MeOC$_6$H$_5$	24	42	96
3	1-Naphthyl	24	27	91
4	2-Thiophenyl	36	30	99
5	3-Furyl	36	25	98

5.2.3
Phosphine-Mediated Aza-MBH Reactions [98]

Bifunctional chiral phosphines such as **119** have been tested as catalysts in the addition of N-tosyl aryl imines to MVK (Scheme 5.24) [99, 100]; catalyst **119a**, in having free phenolic hydroxyl groups, showed superior results. The radially substituted analogues, such as **119f** and **119g**, which improved the catalyst efficiency in related organocatalytic reactions, did not significantly influence either the chemical yield or the ee-value.

Catalyst **119a** was also tested for the addition of various activated aromatic aldimines with MVK and EVK at low temperature in THF as solvent (Table 5.16).

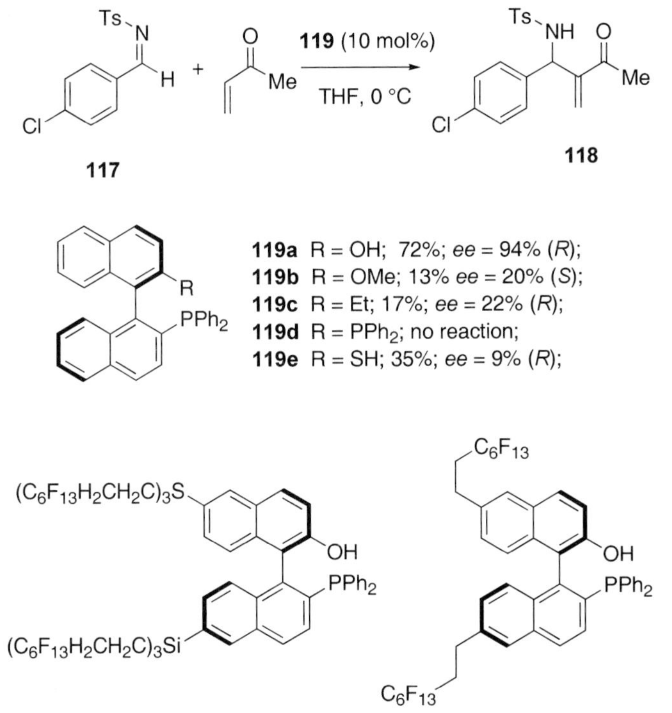

5.2 Asymmetric Aza-MBH Reactions

119a R = OH; 72%; ee = 94% (R);
119b R = OMe; 13% ee = 20% (S);
119c R = Et; 17%; ee = 22% (R);
119d R = PPh₂; no reaction;
119e R = SH; 35%; ee = 9% (R);

119f (15 °C) 78%; ee = 80% (S);   **119g** (15 °C) 76%; ee = 86% (S)

Scheme 5.24 Bifunctional chiral phosphines as catalysts in the aza-MBH reaction with imines and MVK.

The use of molecular sieves increased chemical yields. The N-mesyl and SES-activated aldimines afforded products in similar selectivity, while Ns derivatives decomposed under the reaction conditions. The replacement of MVK by EVK resulted in a diminished reactivity, with longer reaction times and lower chemical yields.

Cyclic enones such as cyclopentenone or cyclohexanone were inert under the above-mentioned conditions. The reaction of aryl aldimines to cyclic enones can be mediated, however, with the more nucleophilic phosphine **119f** [101], although the ee-value of the reactions is usually low (Table 5.17).

Aza-MBH reactions of alkyl or aryl acrylates, as well as acrolein, required higher reaction temperatures than were required with MVK or EVK. While phenyl- or naphthyl acrylates afforded products in CH₂Cl₂ with ee-values up to 69%, the best solvent in the addition of acrolein to activated aldimines proved to be THF (Scheme 5.25) [100].

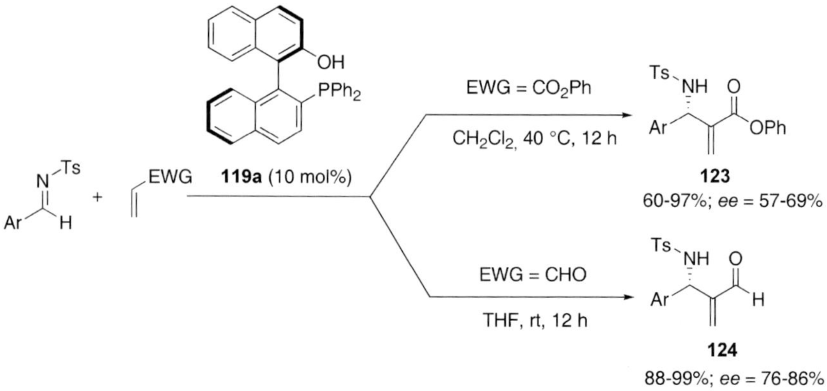
OH
PPh2
EWG = CO2Ph
CH2Cl2, 40 °C, 12 h
Ts
NH
O
Ar
OPh
123
60-97%; ee = 57-69%
N
Ts
Ar
H
+
EWG
119a (10 mol%)
EWG = CHO
THF, rt, 12 h
124
88-99%; ee = 76-86%

Scheme 5.25 The **119a**-mediated aza-MBH reaction of tosyl aryl aldimines with phenyl acrylate and acrolein. EWG = electron-withdrawing group.

Table 5.16 The **119a**-mediated aza-MBH reaction of aryl tosylated imines with MVK and EVK.

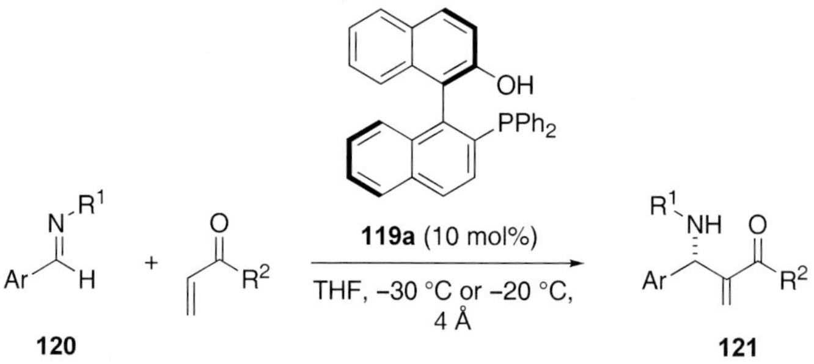
OH
PPh2
N
R1
Ar
H
+
O
R2
119a (10 mol%)
THF, −30 °C or −20 °C,
4 Å
R1
NH
O
Ar
R2
120
121

Entry	Ar	R^1	R^2	Time (h)	Yield (%)	*ee* (%)
1	*p*-ClC_6H_4	Ts	Et	48	77	86
2	C_6H_4-CH=CH	Ts	Me	24	94	95
3	*p*-ClC_6H_4	Ms	Me	24	94	82
4	*p*-ClC_6H_5	SES	Me	72	53	89
5	*p*-ClC_6H_5	Ns	Me	-	-	-

Table 5.17 The **119f**-mediated aza-MBH reaction with of aldimines with cyclohexenone and cyclopentenone.

Entry	Ar	n	Time (h)	Yield (%)	ee (%)
1	p-EtC$_6$H$_4$	1	12	89	32
2	p-EtC$_6$H$_4$	2	24	66	23
3	m-ClC$_6$H$_4$	1	12	80	44
4	p-ClC$_6$H$_4$	1	12	93	64

5.3 Conclusions

Asymmetric organocatalytic Morita–Baylis–Hillman reactions offer synthetically viable alternatives to metal-complex-mediated reactions. The reaction is best mediated with a combination of nucleophilic tertiary amine/phosphine catalysts, and mild Brønsted acid co-catalysts: usually, bifunctional chiral catalysts having both nucleophilic Lewis base and Brønsted acid site were seen to be the most efficient. Although many important factors governing the reactions were identified, our present understanding of the basic factors, and the control of reactivity and selectivity remains incomplete. Whilst substrate dependency is still considered to be an important issue, an increasing number of transformations are reaching the standards of current asymmetric reactions.

References and Notes

1 (a) For general reviews, see: L.J. Brzezinski, S. Rafel, J.W. Leahy, *J. Am. Chem. Soc.* **1997**, *119*, 4317–4318; (b) E. Ciganek. The Catalyzed α-Hydroxylation and α-Aminoalkylation of Activated Olefins (The Morita–Baylis–Hillman Reaction). In: L.A. Paquette (Ed.), *Organic Reactions*. Wiley, New York, **1997**, Vol. 51, p. 201; (c) D. Basavaiah, P.D. Rao, R.S. Hyma, *Tetrahedron* **1996**, *52*, 8001–8062; (d) P. Langer, *Angew. Chem. Int. Ed.* **2000**, *39*, 3049–3052; (e) A. Gilbert, T.W. Heritage, N.S. Isaacs, *Tetrahedron: Asymmetry* **1991**, *2*, 969–972; (f) D. Basavaiah, A.J. Rao, T. Satyanarayana, *Chem. Rev.* **2003**, *103*, 811–891; (g) K.Y. Lee, S. Gowrisankar, J.N. Kim, *Bull. Korean Chem. Soc.* **2005**, *26*, 1481–1490.

2 For related chalchogene-mediated additions, see: A. Kamimura, *J. Synth. Org. Chem. Jpn.* **2004**, *62*, 705–715.

3 (a) K. Morita, Z. Suzuki, H. Hirose, *Bull. Chem. Soc. Jpn.* **1968**, *41*, 2815–2815; (b) K. Morita, Japan Patent 6803364, 1968; *Chem. Abstr.* **1968**, *69*, 58828s.

4 R. Oda, T. Kawabata, S. Tanimoto, *Tetrahedron Lett.* **1964**, *5*, 1653–1657.

5 A.B. Baylis, M.E.D. Hillman; German Patent 2155113, **1972**; *Chem. Abstr.* **1972**, *77*, 34174q.

6 (a) S.E. Drewes, N.D. Emslie, *J. Chem. Soc. Perkins Trans. 1* **1982**, 2079–2085; (b) H.M.R. Hoffmann, J. Rabe, *Angew. Chem. Int. Ed. Engl.* **1983**, *22*, 795–796.

7 G.E. Keck, D.S. Welch, *Org. Lett.* **2002**, *4*, 3687–3690.

8 (a) H. Amri, J. Villieras, *Tetrahedron Lett.* **1986**, *27*, 4307–4308; (b) D. Basavaiah, T.K. Bharathi, V.V.L. Gowriswari, *Synth. Commun.* **1987**, *17*, 1893.

9 D. Basavaiah, V.V.L. Gowriswari, *Tetrahedron Lett.* **1986**, *27*, 2031–2032.

10 P. Auvray, P. Knochel, J.F. Normant, *Tetrahedron Lett.* **1986**, *27*, 5095–5098.

11 A. Weichert, H.M.R. Hoffmann, *J. Org. Chem.* **1991**, *56*, 4098–4112.

12 S.Z. Wang, K. Yamamoto, H. Yamada, T. Takahashi, *Tetrahedron* **1992**, *48*, 2333–2348.

13 H. Amri, M.M. El Gaied, J. Villieras, *Synth. Commun.* **1990**, *20*, 659.

14 (a) M.K. Kundu, S.B. Mukherjee, N. Balu, R. Padmakumar, S.V. Bhat, *Synlett* **1994**, 444–445; (b) C. Yu, L. Hu, *J. Org. Chem.* **2002**, *67*, 219–223; (c) C. Faltin, E.M. Fleming, S.J. Connon, *J. Org. Chem.* **2004**, *69*, 6496–6499.

15 (a) J.S. Hill, N.S. Isaacs, *Tetrahedron Lett.* **1986**, *27*, 5007–5010; (b) G.M. Strunz, R. Bethell, G. Sampson, P. White, *Can. J. Chem.* **1995**, *73*, 1666–1674.

16 (a) S. Tsuboi, S. Takatsuka, M. Utaka, *Chem Lett.* **1988**, 2003; (b) S. Tsuboi, H. Kuroda, S. Takatsuka, T. Fukawa, T. Sakai, M. Utaka, *J. Org. Chem.* **1993**, *58*, 5952–5957; (c) G.-L. Zhao, M. Shi, *J. Org. Chem.* **2005**, *70*, 9975–9984.

17 (a) Y. Matsuya, K. Hayashi, H. Nemoto, *J. Am. Chem. Soc.* **2003**, *125*, 646–647; (b) S. Xue, Q.-F. Zhou, L. Li, Q.-X. Guo, *Synlett* **2005**, 2990–2992.

18 T.G. Back, D.A. Rankic, J.M. Sorbetti, J.E. Wulff, *Org. Lett.* **2005**, *7*, 2377–2379.

19 (a) D. Basavaiah, V.V.L. Gowriswari, *Synth. Commun.* **1989**, *19*, 2461–2465; (b) C. Grundke, H.M.R. Hoffmann, *Chem. Ber.* **1987**, *120*, 1461–1461; (c) M. Shi, W. Zhang, *Tetrahedron* **2005**, *61*, 11887–11894.

20 A.S. Golubev, M.V. Galakhov, A.F. Kolomiets, A.V. Fokin, *Izv. Akad. Nauk, Ser. Khimi.* **1992**, *12*, 2763–2767.

21 (a) P. Perlmutter, C.C. Teo, *Tetrahedron Lett.* **1984**, *25*, 5951–5952; (b) K. Yamamoto, M. Takagi, M. Tsuji, *Bull. Chem. Soc. Jpn.* **1988**, *61*, 319; (c) M. Takagi, K. Yamamoto, *Tetrahedron* **1991**, *47*, 8869–8882.

22 J.S. Hill, N.S. Isaacs, *J. Chem. Res.* **1988**, 330–330.

23 W.D. Lee, K.S. Yang, K.M. Chen, *Chem. Commun.* **2001**, 1612–1613.

24 S.E. Drewes, G.H.P. Roos, *Tetrahedron* **1988**, *44*, 4653–4670.

25. V.K. Aggarwal, I. Emme, S.Y. Fulford, *J. Org. Chem.* **2003**, *68*, 692–700.
26. V.K. Aggarwal, A. Mereu, *Chem. Commun.* **1999**, 2311–2312.
27. (a) F. Rezgui, M.M. El Gaied, *Tetrahedron Lett.* **1998**, *39*, 5965–5966; (b) R. Octavio, M.A. de Souza, M.L. Vasconcellos, *Synth. Commun.* **2003**, *33*, 1383–1389.
28. (a) S. Luo, B. Zhang, J. He, A. Janczuk, P.G. Wang, J. Cheng, *Tetrahedron Lett.* **2002**, *43*, 7369–7371; (b) For the use of 1-methylimidazole 3-N-oxide, see: Y.-S. Lin, C.-W. Liu, T.Y.R. Tsai, *Tetrahedron Lett.* **2005**, *46*, 1859–1861.
29. (a) N.E. Leadbeater, C. Van der Pol, *J. Chem. Soc. Perkin Trans. 1* **2001**, 2831–2835; (b) R.S. Grainger, N.E. Leadbeater, A.M. Pamies, *Catal. Commun.* **2002**, *3*, 449–452.
30. S. Luo, X. Mi, G.W. Peng, J.P. Cheng, *Tetrahedron Lett.* **2004**, *45*, 5171–5174.
31. S.-H. Zhao, Z.-B. Chen, *Synth. Commun.* **2005**, *35*, 3045–3053.
32. T. Miyakoshi, H. Omichi, S. Saito, *Nippon Kagaku Kaishi* **1979**, 748–753; *Chem. Abstr.* **1979**, *91*, 123360d.
33. (a) S. Rafel, J.W. Leahy, *J. Org. Chem.* **1997**, *62*, 1521–1522; (b) M. Shi, Y.-H. Liu, *Org. Biomol. Chem.* **2006**, *4*, 1468; (c) Z. He, X. Tang, Y. Chen, Z. He, *Adv. Synth. Catal.* **2006**, *348*, 413–417.
34. (a) R.J.W. Schuurman, A.V. den Linden, R.P.F. Grimbergen, R.J.M. Nolte, H.W. Scheeren, *Tetrahedron* **1996**, *52*, 8307–8314; (b) N.S. Isaacs, *Tetrahedron* **1991**, *47*, 8463–8497; (c) Y. Hayashi, K. Okado, I. Ashimine, M. Shoji, *Tetrahedron Lett.* **2002**, *43*, 8683–8686; (d) See also: M. Shi, Y.-H. Liu, *Org. Biomol. Chem.* **2006**, *4*, 1468–1470.
35. (a) G.H.P. Roos, P. Rampersadh, *Synth. Commun.* **1993**, *23*, 1261–1266; (b) W.P. Almeida, F. Coehlo, *Tetrahedron Lett.* **1998**, *39*, 8609–8612; (c) F. Coehlo, W.P. Almeida, D. Veronese, C.R. Mateus, E.C. Silva Lopes, R.C. Rossi, G.P.C. Silveira, C.H. Pavam, *Tetrahedron* **2002**, *58*, 7437–7447.
36. V.K. Aggarwal, A. Mereu, G.J. Tarver, *J. Org. Chem.* **1998**, *63*, 7183–7189.
37. K.-S. Yang, W.-D. Lee, J.-F. Pan, K.-M. Chen, *J. Org. Chem.* **2003**, *68*, 915–919.
38. V.K. Aggarwal, D.K. Dean, A. Mereu, R. Williams, *J. Org. Chem.* **2002**, *67*, 510–514.
39. D.J. Maher, S.J. Connon, *Tetrahedron Lett.* **2004**, *45*, 1301–1305.
40. (a) H.S. Byurn, K.C. Reddy, R. Bitman, *Tetrahedron Lett.* **1994**, *35*, 1371–1374; (b) J. Auge, N. Lubin, A. Lubineau, *Tetrahedron Lett.* **1994**, *35*, 7947–7948; (c) J. Cai, Z. Zhou, G. Zhao, C. Tang, *Org. Lett.* **2002**, *4*, 4723–4725; (d) D. Basavaiah, M. Krishnamacharyulu, A. Rao, *Synth. Commun.* **2000**, *30*, 2061–2069; (e) C.Z. Yu, B. Liu, L.Q. Hu, *J. Org. Chem.* **2001**, *66*, 5413–5418.
41. Roos and Rampersadh have observed that "gentle warming" effects a marked rate increase in the reaction with acetaldehyde and a slight increase with other aldehydes, but they also observed polymerization at temperatures above 43 °C. G.H.P. Roos, P. Rampersadh, *Synth. Commun.* **1993**, *23*, 1261–1266.
42. P.R. Krishna, A. Manjuvani, V. Kannan, G.V.M. Sharma, *Tetrahedron Lett.* **2004**, *45*, 1183–1185.
43. A. Porzelle, C.M. Williams, B.D. Schwartz, I.R. Gentle, *Synlett* **2005**, 2923–2926.
44. (a) X. Mi, S. Luo, J.-P. Cheng, *J. Org. Chem.* **2005**, *70*, 2338–2341; (b) J.-C. Hsu, Y.-H. Yen, Y.-H. Chu, *Tetrahedron Lett.* **2004**, *45*, 4673–4676; (c) A. Kumar, S.S. Pawar, *J. Mol. Catal. A: Chemical* **2004**, *211*, 43–47; (d) E.J. Kim, S.Y. Ko, C.E. Song, *Helv. Chim. Acta* **2003**, *86*, 894–899; (e) V.K. Aggarwal, I. Emme, A. Mereu, *Chem. Commun.* **2002**, 1612–1613; (f) J.N. Rosa, C.A. Afonso, A.G. Santos, *Tetrahedron*, **2001**, *57*, 4189–4193.
45. (a) J.S. Hill, N.S. Isaacs, *J. Phys. Org. Chem.* **1990**, *3*, 285–288; For earlier discussions on the mechanism of the MBH reaction, see Ref. [6b].
46. (a) P.T. Kaye, M.L. Bode, *Tetrahedron Lett.* **1991**, *32*, 5611–5614; (b) Y. Fort, M.C. Berthe, P. Caubere, *Tetrahedron* **1992**, *48*, 6371–6384.

47 L.S. Santos, C.H. Pavam, W.P. Almeida, F. Coehlo, M.N. Eberlin, *Angew. Chem. Int. Ed.* **2004**, *43*, 4330–4333.

48 The term was coined in analogy of pharmaceutical compound classes that are active against a number of different biological targets. See: T.P. Yoon, E.N. Jacobsen, *Science* **2003**, *299*, 1691–1693.

49 (a) S.E. Drewes, N.D. Emslie, N. Karodia, A.A. Khan, *Chem. Ber.* **1990**, *123*, 1447–2456; (b) P. Perlmutter, E. Puniani, G. Westman, *Tetrahedron Lett.* **1996**, *37*, 1715–1718.

50 V.K. Aggarwal, S.Y. Fulford, G.C. Lloyd-Jones, *Angew. Chem. Int. Ed.* **2005**, *44*, 1706–1708.

51 Y. Takemoto, *Org. Biomol. Chem.* **2005**, *3*, 4299–4306.

52 (a) K.E. Price, S.J. Broadwater, H.M. Jung, D.T. McQuade, *Org. Lett.* **2005**, *7*, 147–150; (b) K.E. Price, S.J. Broadwater, B.J. Walker, D.T. McQuade, *J. Org. Chem.* **2005**, *70*, 3980–3987.

53 (a) G. Roos, T. Manickum, *Synth. Commun.* **1991**, *21*, 2269–2274; (b) S.E. Drewes, A. Khan, K. Rowland, *Synth. Commun.* **1993**, *23*, 183.

54 (a) B. Alcaide, P. Almendros, C. Aragoncillo, *J. Org. Chem.* **2001**, *66*, 1612–1620; (b) B. Alcaide, P. Almendros, C. Aragoncillo, R. Rodríguez-Acebes, *J. Org. Chem.* **2004**, *69*, 826–831.

55 (a) F. Coelho, G. Diaz, C.A.M. Abella, W.P. Almeida, *Synlett* **2006**, 435–439; (b) W.P. Almeida, F. Coelho, *Tetrahedron Lett.* **2003**, *44*, 937–937.

56 See for example: (a) M. Shi, J.-K. Jiang, Y.-S. Feng, *Org. Lett.* **2000**, *2*, 2397–2400; (b) M. Ono, K. Nishimura, Y. Nagaoka, K. Tomioka, *Tetrahedron Lett.* **1999**, *40*, 1509–1512; (c) T. Iwama, S.-I. Tsijiyama, H. Kinoshita, K. Kanamatsu, Y. Tsurukami, T. Iwamura, S.-I. Watanabe, T. Kataoka, *Chem. Pharm. Bull.* **1999**, *47*, 956–961; (d) M. Kawamura, S. Kobayashi, *Tetrahedron Lett.* **1999**, *40*, 1539–1542; (e) E.P. Kündig, L.H. Xu, P. Romanens, G. Bernardinelli, *Tetrahedron Lett.* **1993**, *34*, 7049–7052; (f) L.M. Walsh, C.L. Winn, J.M. Goodman, *Tetrahedron Lett.* **2002**, *43*, 8219–8222.

57 T. Oishi, H. Oguri, M. Hirama, *Tetrahedron: Asymm.* **1995**, *6*, 1241–1244.

58 A.G.M. Barrett, A.S. Cook, A. Kamimura, *Chem. Commun.* **1998**, 2533–2534.

59 A.G.M. Barrett, P. Dozzo, A.J.P. White, D.J. Williams, *Tetrahedron* **2002**, *58*, 7303–7313.

60 H.M.R. Hoffmann, J. Rabe, *J. Angew. Chem. Int. Ed. Engl.* **1985**, *24*, 94–109.

61 (a) S.E. Drewes, S.D. Freese, N.D. Emslie, G.H.P. Roos, *Synth. Commun.* **1988**, *18*, 1565–1572; (b) M. Bailey, I.E. Marko, W.D. Ollis, P.R. Rassmussen, *Tetrahedron Lett.* **1990**, *31*, 4509–4512; (c) The use of AcOH as co-catalyst was described earlier: H.M.R. Hoffmann, *J. Org. Chem.* **1988**, *53*, 3701–3710.

62 I.E. Marko, P.R. Giles, N.J. Hindley, *Tetrahedron* **1997**, *53*, 1015–1024.

63 P.R. Krishna, V. Kannan, P.V.N. Reddy, *Adv. Synth. Catal.* **2004**, *346*, 603–606.

64 (a) Y. Wabuchi, M. Nakatani, S. Yokoyama, S. Hatakeyama, *J. Am. Chem. Soc.* **1999**, *121*, 10219–10220; (b) A. Nakano, S. Kawahara, K. Morokuma, M. Nakatami, Y. Iwabuchi, K. Takahashi, J. Ishihara, S. Hatakeyama, *Tetrahedron* **2006**, *62*, 381–389.

65 β-ICD can be synthesized in approx. 60% yield in one step from (+)-quinidine by heating with 10 equiv. of KBr in 85% phosphoric acid at 100 °C for 5 days [64a].

66 F. Ameer, S.E. Drewes, S. Freese, P.T. Kaye, *Synth. Commun.* **1988**, *18*, 495–500.

67 Y. Iwabuchi, M. Sugihara, T. Esumi, S. Hatakeyama, *Tetrahedron Lett.* **2001**, *42*, 7867–7871.

68 Y. Iwabuchi, M. Furukawa, T. Esumi, S. Hatakeyama, *Chem. Commun.* **2001**, 2030–2031.

69 M. Shi, J.K. Jiang, *Tetrahedron: Asymm.* **2002**, *13*, 1941–1947.

70 C.M. Mocquet, S.L. Warriner, *Synlett* **2004**, 356–357.

71 (a) C.E. Aroyan, M.M. Vasbinder, S.J. Miller, *Org. Lett.* **2005**, *7*, 3849–3851; (b) J.E. Imbriglio, M.M. Vasbinder, S.J. Miller, *Org. Lett.* **2003**, *5*, 3741–3743.

72 For an excellent review, see: S.J. Miller, *Acc. Chem. Res.* **2004**, *37*, 601–610.

73 M. Shi, J.-K. Jiang, C.-Q. Li, *Tetrahedron Lett.* **2002**, *43*, 127–130.

74 For earlier intramolecular MBH reactions, see: (a) S.E. Drewes, O.L. Njamela, N.D. Emslie, N. Ramesar, J.S. Field, *Synth. Commun.* **1993**, *23*, 2807–2815; (b) G.P. Black, F. Dinon, S. Fratucello, P.J. Murphy, M. Nielson, H.L. Williams, N.D.A. Walshe, *Tetrahedron Lett.* **1997**, *38*, 8561–8564; (c) F. Dinon, E. Richards, P.J. Murphy, D.E. Hibbs, M.B. Hursthouse, K.M.A. Malik, *Tetrahedron Lett.* **1999**, *40*, 3279–3282; (d) E.L. Richards, P.J. Murphy, F. Dinon, S. Fratucello, P.M. Brown, T. Gelbrich, M.B. Hursthouse, *Tetrahedron* **2001**, *57*, 7771–7784.

75 S.-H. Chen, B.-C. Hong, C.-F. Su, S. Sarshar, *Tetrahedron Lett.* **2005**, *46*, 8899–8903.

76 Y. Takemoto, *Org. Biomol. Chem.* **2005**, *3*, 4299–4306.

77 (a) T. Okino, Y. Hoashi, Y. Takemoto, *J. Am. Chem. Soc.* **2003**, *125*, 12672–12673; (b) T. Okino, Y. Hoashi, T. Furukawa, X. Xu, Y. Takemoto, *J. Am. Chem. Soc.* **2005**, *127*, 119–125; (c) Y. Hoashi, T. Okino, Y. Takemoto, *Angew. Chem. Int. Ed.* **2005**, *44*, 4032–4035.

78 T. Okino, S. Nakamura, T. Furukawa, Y. Takemoto, *Org. Lett.* **2004**, *6*, 625–627.

79 J. Wang, H. Li, X. Yu, L. Zu, W. Wang, *Org. Lett.* **2005**, *7*, 4293–4296.

80 A.G. Wenzel, M.P. Lalonde, E.N. Jacobsen, *Synlett* **2003**, 1919–1922.

81 Y. Sohtome, A. Tanatani, Y. Hashimoto, K. Nagasawa, *Tetrahedron Lett.* **2004**, *45*, 5589–5592.

82 For revews, see: (a) T. Clark, C. Landis, *Tetrahedron: Asymm.* **2004**, *15*, 2123–2137; (b) M. Oestreich, *Nachr. Chem.* **2004**, *52*, 1257–1260; (c) J.L. Methot, W.R. Roush, *Adv. Synth. Catal.* **2004**, *346*, 1035–1050; See also: (d) X. Zhang (Ed.), 'Chiral Phosphorus Ligands'; Thematic issue; *Tetrahedron: Asymm.* **2004**, *15*, 2099–2311.

83 F. Roth, P. Gygax, G. Frater, *Tetrahedron Lett.* **1992**, *33*, 1045–1048.

84 W.-D. Teng, R. Huang, C.K.-W. Kwong, M. Shi, P.H. Toy, *J. Org. Chem.*, **2006**, *71*, 368–371.

85 T. Hayase, T. Shibata, K. Soai, Y. Wakatsuki, *Chem. Commun.* **1998**, 1271–1272.

86 W. Li, Z. Zhang, D. Xiao, X. Zhang, *J. Org. Chem.* **2000**, *65*, 3489–3496.

87 For a recent review on mannitol derivatives in enantiocatalytic reactions, see: S. Castillon, C. Claver, Y. Diaz, *Chem. Soc. Rev.* **2005**, *34*, 702–713.

88 Y.M.A. Yamada, S. Ikegami, *Tetrahedron Lett.* **2000**, *41*, 2165–2169.

89 N.T. McDougal, S.E. Schaus, *J. Am. Chem. Soc.* **2003**, *125*, 12094–12095.

90 N.T. McDougal, W.L. Trevellini, S.A. Rodgen, L.T. Kliman, S.E. Schaus, *Adv. Synth. Catal.* **2004**, *346*, 1231–1240.

91 Catalysts can be prepared by hydrogenation of naphthol derivatives over PtO_2 in acetic acid.

92 M. Shi, Y.M. Xu, *Chem. Commun.* **2001**, 1876–1877.

93 P. Buskens, J. Klankermayer, W. Leitner, *J. Am. Chem. Soc.* **2005**, *127*, 16762–16763.

94 S. Kawahara, A. Nakano, T. Esumi, Y. Iwabuchi, S. Hatakeyama, *Org. Lett.* **2003**, *5*, 3103–3105.

95 (a) M. Shi, Y.M. Xu, *Angew. Chem. Int. Ed. Engl.* **2002**, *41*, 4507–4510; (b) M. Shi, Y.-M. Xu, Y.-L. Shi, *Chem. Eur. J.* **2005**, *11*, 1794–1802.

96 K. Matsui, S. Takizawa, H. Sasai, *J. Am. Chem. Soc.* **2005**, *127*, 3680–3681.

97 I.T. Raheem, E.N. Jacobsen, *Adv. Synth. Catal.* **2005**, *347*, 1701–1708.

98 For a review, see: M. Shi, L.H. Chen, *Pure Appl. Chem.* **2005**, *77*, 2105–2110.

99 (a) M. Shi, L.-H. Chen, *Chem. Commun.* **2003**, 1310–1311; (b) M. Shi, L.-H. Chen, C.-Q. Li, *J. Am. Chem. Soc.* **2005**, *127*, 3790–3800.

100 M. Shi, L.H. Chen, W.D. Teng, *Adv. Synth. Catal.* **2005**, *347*, 1781–1789.

101 M. Shi, C.Q. Li, *Tetrahedron: Asymm.* **2005**, *16*, 1385–1391.

6
Asymmetric Proton Catalysis

Jeff D. McGilvra, Vijaya Bhasker Gondi, and Viresh H. Rawal

6.1
Introduction

The importance of hydrogen bonding in chemical and biological systems has been recognized in the scientific community for many years. This ubiquitous interaction is one of the central forces in Nature. Whereas individual hydrogen bonds are relatively weak compared to covalent bonds, collectively they can be of enormous importance. Apart from the crucial role that hydrogen bonds play in mediating the life-sustaining properties of water, they are found to be essential for maintaining the form and function of most biological systems [1]. Hydrogen bonds are important for the organization and base pairing of DNA and RNA, the secondary and tertiary structure of proteins [2], small molecule recognition [3], and the catalytic cycles of various enzymes [4, 5]. Despite the many vital functions that hydrogen bonds fulfill in biological systems, they had, until recently, been little utilized for the promotion of chemical reactions. This omission may have been due, in part, to the limited understanding of the reactivity-enhancing properties associated with hydrogen bonding and the hesitation to use what is recognizably a weak bonding interaction when other more established methods for catalysis were already available. Over the past six years, this situation has changed dramatically, and many enantioselective reactions have been developed in which a chiral hydrogen bond donor serves as the catalyst. This chapter provides a summary of the progress in this rapidly expanding field.

The historical recognition of hydrogen bonding is as nebulous as the phenomenon itself. Records dating back to the early 1900s allude to the observation of hydrogen bonding in various chemical systems. These interactions were referred to by using non-specific terms such as "inner complex building", "weak union", and "chelation" [6]. It was not until 1931, in Pauling's seminal paper on the nature of the chemical bond, that the term "hydrogen bond" was coined [7]. Even in 1936, research pioneers such as Huggins used the term "hydrogen bridge" [8] to describe the phenomenon of water's affinity to itself [9] and to carbohydrates,

Enantioselective Organocatalysis: Reactions and Experimental Procedures
Edited by Peter I. Dalko
Copyright © 2007 WILEY-VCH Verlag GmbH & Co. KGaA, Weinheim
ISBN: 978-3-527-31522-2

Table 6.1 Jeffrey's classification of strong, moderate and weak hydrogen bonds [6].

Bonding strength	Strong	Moderate	Weak
A–H···B interaction	Mostly covalent	Mostly electrostatic	Electrostatic
Bond lengths (Å)			
H ↔ B	~1.2–1.5	~1.5–2.2	~2.2–3.2
A ↔ B	~2.2–2.5	~2.5–3.2	~3.2–4.0
Bond angle AHB (°)	175–180	130–180	90–150
Bond energy (kcal mol^{-1})	14–40	4–15	<4

proteins, and other biological molecules [10, 11]. In *Nature of the Chemical Bond*, Pauling described the hydrogen bond between a donor (A–H) and an acceptor (B) based on the relative electronegativities of "A" and "B" [12]. The shared electrons in the covalently bonded "A–H" are drawn to the more electronegative "A", leaving the hydrogen atom partially deshielded of electrons. The hydrogen bond acceptor "B" typically possesses a lone pair of electrons or polarizable π electrons to interact with the deshielded hydrogen. As only one covalent bond to the 1s orbital of hydrogen is optimal, the additional bonding interaction between "H" and "B" is considerably ionic in character.

Unlike covalent bonds, which have a relatively narrow distribution of energies, lengths, and geometries within particular groupings of bonding partners, the nature of hydrogen bonds can vary dramatically. Countless variations can be found in bond strengths and geometric orientations with hydrogen bonds, even within the same bonding partners. To simplify the matter, hydrogen bonds have been grouped into three different categories: strong, moderate, and weak (Table 6.1) [6]. "Strong" hydrogen bonds are nearly covalent in nature, having linear bond angles, and bond distances that are shorter than the sum of the van der Waals radii of "A" and "B". More common are hydrogen bonds that fall under the "medium" category, and include interactions such as that between two water molecules. Further down the spectrum, "weak" hydrogen bonds can be as weak as van der Waals forces and can have near-perpendicular directionality with respect to the acceptor, with bonding distances far in excess of the van der Waals radii of the two atoms involved. The different forms of hydrogen bond interactions can be classified under the Lewis definition of acids and bases, where the hydrogen bond donor (A–H) is defined to be a Lewis acid, and the hydrogen bond acceptor (B), a Lewis base. As such, hydrogen bond-catalyzed reactions represent a type of Lewis acid-catalyzed reaction. The strength of a hydrogen bond donor is best correlated to its pK_a, although environmental factors, such as solvation, temperature, and dielectric constants, can also affect the nature of a hydrogen bond.

Given the infinite variations in bonding patterns, hydrogen bonding is best considered as a continuum of bonding interactions, from weak to strong. Moving

```
A-H·····:B          A-H··-:B          A—H--:B           A⁻    H-B⁺
   weak              moderate          strong           Brønsted-Lowry
```

Fig. 6.1 Hydrogen bonding continuum.

further along this spectrum, past even strong hydrogen bonds, there is the further possibility of full proton transfer between the hydrogen bond donor and acceptor, with the donor now being considered a Brønsted–Lowry acid (Fig. 6.1). Viewed in this capacity, the proton is unquestionably the most widely used catalyst for chemical reactions, utilized for the promotion of whole families of organic reactions. With catalyzed enantioselective reactions, however, the traditional Brønsted acid description may not be appropriate, particularly given the nonpolar solvents used for such reactions. In order to transfer chirality from the catalyst to the substrate, the chiral anion must be well ordered and, very likely, be tightly coordinated to the conjugate acid through a strong hydrogen bond. As such, the hydrogen bond donor can be thought of in the Lewis acid sense, capable of forming "very strong" hydrogen bonds. Concepts such as proton affinities and acidity/basicity associated with Brønsted acids may also be applied to systems that form hydrogen bonds [13]. The hydrogen bond itself can be considered a "frozen" stage of a proton transfer reaction, where each of the three bonding categories in Table 6.1 describes a different stage along the proton transfer spectrum.

Despite its prevalence in biological and chemical systems, hydrogen bonding has been little used as a design principle for asymmetric catalysts. These interactions can be weak and variable, such that catalyst–substrate interactions based on hydrogen bonding are expected to be disrupted easily at ambient temperatures. The electrostatic nature of these interactions also makes them susceptible to differences in solvent polarity. Unlike traditional metal-based Lewis acid catalysts, which have well-defined metal–ligand coordination geometries, hydrogen bonds exhibit a much broader distribution of donor–acceptor geometric orientations. Moreover, the distance between a hydrogen bond donor and acceptor is typically longer than that observed in many metal-based Lewis acid–Lewis base complexes, and this is expected to result in less effective chirality transfer. Similar issues of asymmetric induction arise in conceptualizing the design of a strongly acidic chiral catalyst, as the design must account for enantiofacial discrimination using what is essentially an achiral H^+ in the presence of a chiral anion. Solvent effects are particularly important for such "chiral proton" catalysts, since effective chirality transfer requires that the chiral counterion be held in close proximity, as a well-organized tight-ion pair. The wide variability observed for the strength and bonding geometries of hydrogen bonds not only provides unique challenges for catalyst design but also offers opportunities for the discovery and development of new concepts in catalysis.

The systematic use of hydrogen bonding for the promotion of racemic organic reactions began most noticeably during the 1980s, and these studies are summarized below. More recent developments on asymmetric catalysis of organic reactions using hydrogen bonds (which forms the focus of this chapter) are presented in Sections 6.2 to 6.16 [14–25]. In a seminal investigation, Hine and co-workers showed that hydrogen bonding with 1,8-biphenylene diols could effectively activate epoxides to nucleophilic attack [26]. Kinetic experiments showed that both hydroxyl groups, in their respective locations on the catalyst scaffold, were important for achieving maximum rates of epoxide opening. Kelly and co-workers found that achiral biphenylenediols, at 40–50 mol% catalyst loading, significantly accelerated the Diels–Alder reaction of cyclopentadiene with various acroleins [27]. The rate increase is ascribed to activation of the carbonyl group by the diol, via two hydrogen bonds. Through a series of elegant co-crystallization studies, the Etter group showed that achiral diaryl ureas form well-defined crystalline complexes with a number of carbonyl-containing compounds. The stability of the complex derives from a two-point hydrogen bond between the urea N–H bonds and the carbonyl oxygen [28–30]. The first report on the application of Etter's concept of using ureas as Lewis acids was by Curran and co-workers, who showed that the outcome of radical allylation reactions and Claisen rearrangements can be altered by the presence of ureas and thioureas [31–33]. At about the same time, Rawal and Eschbach examined the use of electron-deficient diaryl ureas for the promotion of a Diels–Alder reaction between cyclopentadiene and methyl vinyl ketone. Although the ureas clearly accelerated the cycloaddition reaction, the rate increase was modest (<2-fold with 10% catalyst), and consequently the study was not pursued further [34]. A thorough study on the use of Etter-type ureas for the promotion of Diels–Alder reactions has been described by Schreiner [35]. The use of Etter ureas has also been reported for the activation of nitrones [36]. Such ureas and numerous related chiral modifications have provided a treasure trove of useful catalysts for asymmetric synthesis (*vide infra*). Several other classes of hydrogen bond donors, such as guanidinium ions, alcohols, and phenols, have also been used as catalysts to promote organic reactions [37–39]. These early reports provide clear demonstrations of the ability of hydrogen bond donors to function as Lewis acids and thereby promote organic reactions. The formation of a hydrogen bond to the electrophile (typically a carbonyl or imine), through either a one-point or two-point interaction, increases its electrophilicity. In Frontier Molecular Orbital terms, hydrogen bond formation leads to lowering of the LUMO energy of the electrophile, thereby making it more reactive to the second reactant, such as a nucleophile or a diene.

Over the past half-dozen years, many laboratories have focused their efforts on the development of chiral hydrogen bond donors that function as catalysts for enantioselective organic reactions. One of the earliest successes in this area came from Jacobsen and co-workers, who reported the use of peptide-like chiral urea-based catalysts for the hydrocyanation of aldimines and ketoimines [40, 41]. Several other laboratories have also reported highly enantioselective transformations catalyzed by a chiral hydrogen bond donor. The following sections provide a summary of the many developments in hydrogen bond-catalyzed enantioselective reactions, along with a discussion of mechanisms and selectivity models.

6.2
Conjugate Addition Reactions

Hydrogen bond-catalyzed asymmetric conjugate addition reactions have received a great deal of attention dating back to 1981 and the initial investigations of Wynberg and co-workers on the mechanism of cinchona alkaloid-catalyzed thiol addition to cycloalkenones [42]. Asymmetric conjugate addition reactions are attractive targets for catalyst development due, in part, to the numerous substituted nucleophiles and electrophiles that are known to participate in these processes. These reactions give rise to densely functionalized products that are readily transformed into a variety of useful chiral intermediates. The conjugate additions reported in this section are divided into subclasses based on the electrophile employed in the reaction, namely α,β-unsaturated carbonyl substrates, nitroolefins and vinyl sulfones.

6.2.1
Conjugate Addition Reactions of α,β-Unsaturated Carbonyl Substrates

Hiemstra and Wynberg reported a comprehensive study on the mechanism of the cinchona alkaloid-catalyzed 1,4-addition of aromatic thiols to cyclic enones. In this landmark study, the authors discussed the different structural features of the catalyst and reaction parameters that are important for optimal asymmetric induction [42]. Catalytic quantities of cinchonidine (**1**), cinchonine (**2**) and other alkaloid derivatives were used to promote the addition of substituted aryl thiophenols to various substituted cyclic enones (Fig. 6.2). These catalyzed reactions afforded the corresponding thioethers in low to good enantioselectivities (5–75%). The authors proposed that these alkaloids served as bifunctional catalysts, such that they activated the nucleophile as well as the electrophile (Fig. 6.3). The basic amine of the quinuclidine ring activates the thiol group via deprotonation, and the resulting protonated ammonium, through a tight ion interaction, organizes the thiolate nucleophile for conjugate addition. The thiolate is further ordered through dispersion interactions between the sulfur anion and the quinoline π-system. The authors suggested that the catalyst also activates and organizes the enone for conjugate addition via a hydrogen bond between the hydroxyl group and the enone carbonyl. Support for this proposition comes from the observation that catalysts lacking the hydroxyl group gave significantly reduced product enan-

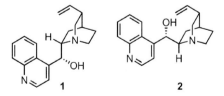

Fig. 6.2 Cinchona alkaloid catalysts for 1,4-additions of thiols to cyclic enones.

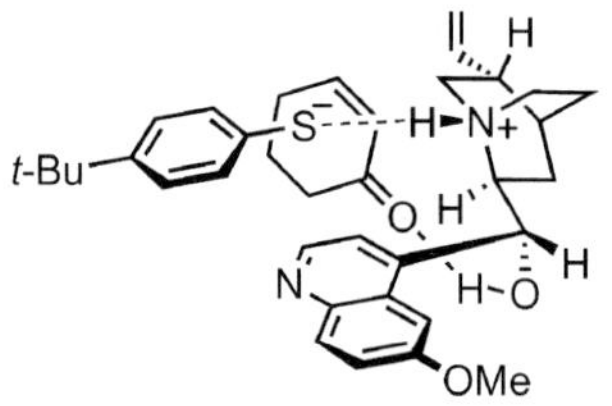

Fig. 6.3 Wynberg's model of bifunctional cinchona alkaloid catalysis.

tioselectivities. A combination of kinetics data, infra-red (IR) studies and reaction parameter variations provided the foundation for the above model.

Chen and co-workers later reported the successful asymmetric 1,4-addition of aryl thiols to α,β-unsaturated cyclic enones and imides using Takemoto's elegantly simple catalyst (3) [43]. This bifunctional amine-thiourea catalyst gives optimal reactivity and reproducibility when used at 10 mol% loading in the presence of freshly dried 4 Å molecular sieves (MS). This combination afforded the expected addition products in high yields (90–99%) and moderate to good enantioselectivities (55–85% ee) for a variety of cyclic and acyclic Michael acceptors (Table 6.2).

Table 6.2 Bifunctional thiourea-amine-catalyzed 1,4-additions to cyclic enones.

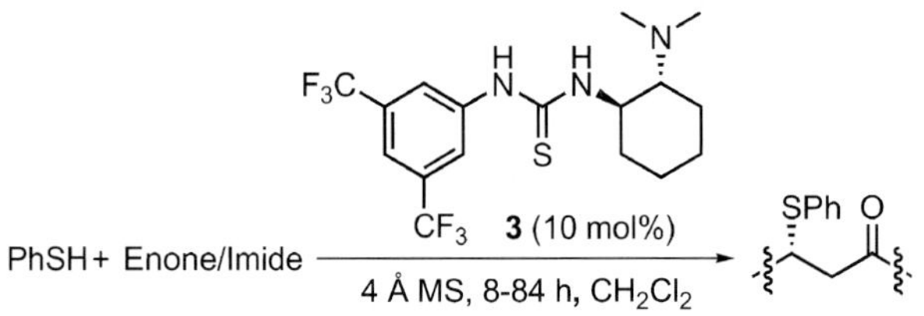

Entry	Enone/imide	Temp. [°C]	Yield [%]	ee [%]
1	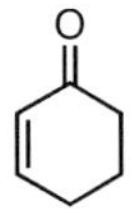 (cyclohexenone)	0	97	85
2	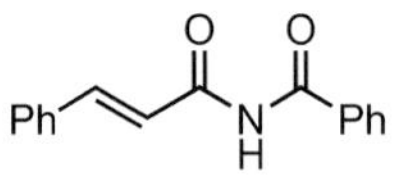 Ph-CH=CH-C(O)-N(H)-C(O)-Ph	−40	98	75
3	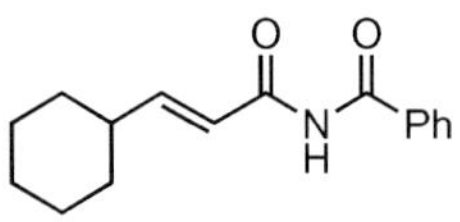 Cy-CH=CH-C(O)-N(H)-C(O)-Ph	−40	96	73
4	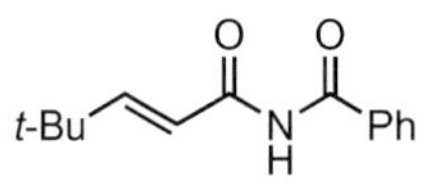 t-Bu-CH=CH-C(O)-N(H)-C(O)-Ph	−40	90	55

Catalyst **3** is proposed to function in a manner similar to the cinchona alkaloid catalysts (**1** and **2**), with the tertiary amine providing activation for the nucleophilic thiol, which is held in close proximity to the thiourea-bound carbonyl substrate.

The Chen group also demonstrated a successful conjugate addition/ asymmetric protonation of α-prochiral imide **4** using thiophenol in the presence of 10 mol% **3** (Scheme 6.1) [43]. It was hypothesized that the ammonium group of the catalyst serves as a chiral proton source for the catalyst-stabilized enone intermediate formed after initial 1,4-addition of the thiol (Fig. 6.4).

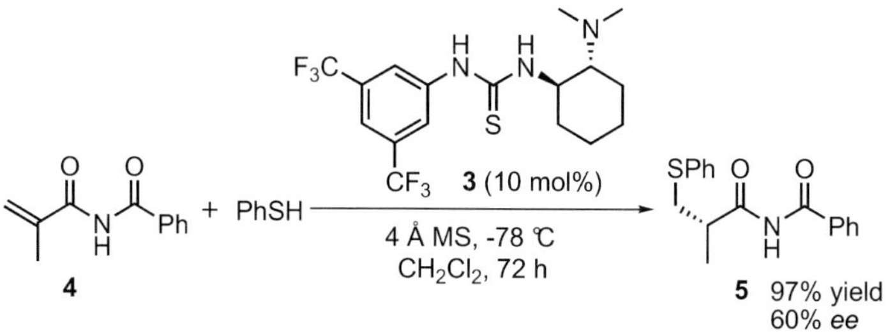

Scheme 6.1 Thiourea-catalyzed asymmetric protonation of a prochiral imide.

Soós and co-workers developed a bifunctional catalyst that combined the essential features of the catalysts used by Wynberg and Takemoto. The resultant cinchona alkaloid-derived thiourea catalyst **7** promoted the enantioselective conjugate addition of nitromethane to *trans*-chalcones (**6**) to give the desired nitroalkane adducts in high yields (80–94%) and high enantiopurities (89–96% *ee*) for the limited number of chalcone substrates studied (Scheme 6.2) [44]. Interestingly, only a 10% decrease in *ee*-value was observed when the reaction was run at 100 °C, though isolated yields were diminished (68% yield after 5 h at 100 °C versus 95% yield after 48 h at ambient temperature). The absolute configuration at C9 of the catalyst was found to be crucial for obtaining high selectivity, suggesting a synergistic interplay in the spatial orientation of the thiourea–enone complex and the ammonium nitronate nucleophile.

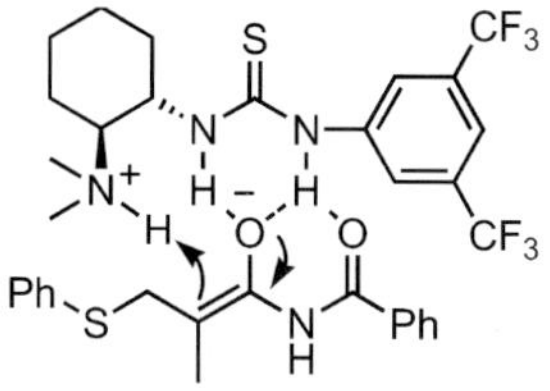

Fig. 6.4 Chen's selectivity model for protonation of an α-prochiral imide.

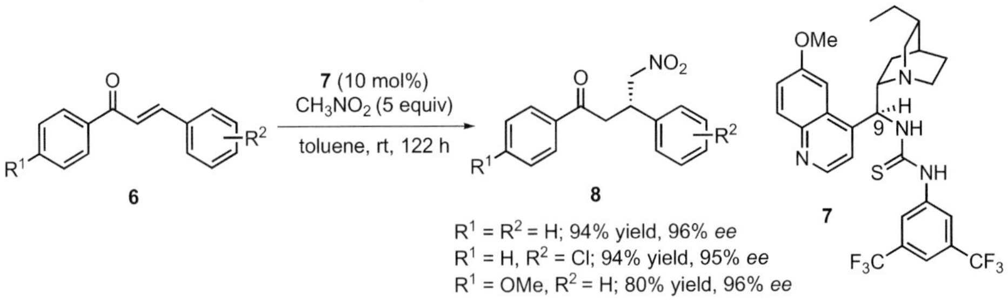

Scheme 6.2 Thiourea-catalyzed Michael additions of nitromethane to chalcones.

$R^1 = R^2 = H$; 94% yield, 96% ee
$R^1 = H, R^2 = Cl$; 94% yield, 95% ee
$R^1 = OMe, R^2 = H$; 80% yield, 96% ee

The Takemoto group reported the first organocatalytic enantioselective 1,4-additions of malononitrile to acyclic α,β-unsaturated imides (**9**) using their thiourea catalyst, **3** [45]. The authors found that, while additions of malononitrile to nitroolefins catalyzed by **3** proceeded with low enantioselectivities, the corresponding additions to the less reactive α,β-unsaturated imides (**9**) gave the desired adducts in high optical purities (84–92% ee) and good to high yields (77–99%; Table 6.3). Imides possessing aromatic, heteroaromatic and aliphatic β-substituents were employed successfully. Product enantiopurity was not highly dependent on the electronic properties of the imide β-substituent, though the rate of reaction was diminished for additions to electron-rich imides. A fivefold dilution of the reaction mixture gave a 4 to 9% increase in product ee, though the reaction time needed to be significantly increased in order to maintain high

Table 6.3 Thiourea-catalyzed additions of malononitrile to α,β-unsaturated imides.

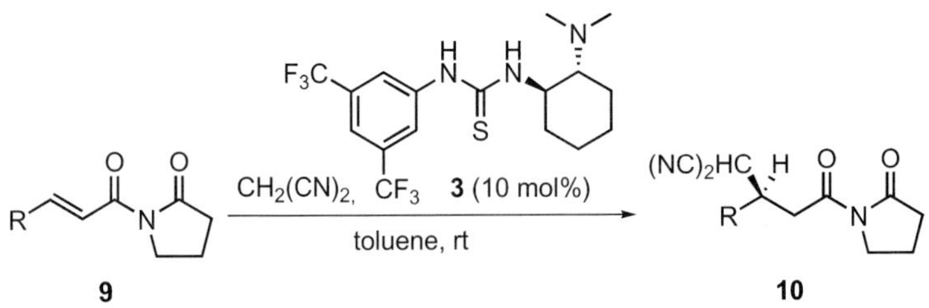

Entry	R	Time [h]	Yield [%]	ee [%]
1	p-Cl-C$_6$H$_4$	72	85	89
2	p-MeO-C$_6$H$_4$	216	77	85
3	2-furyl	192	79	85
4	t-Bu	168	78	92

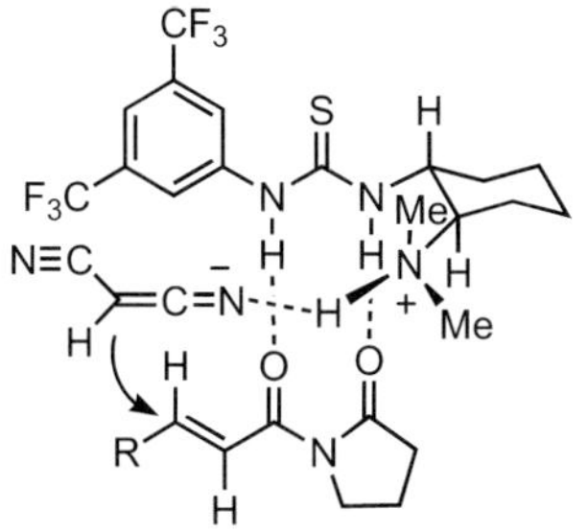

Fig. 6.5 Takemoto's rationale for selectivity of malononitrile enolate additions.

product yields. Malonates and β-ketoesters were found to be unreactive under the reported reaction conditions. (For experimental details see Chapter 14.9.5). The observed asymmetric induction was rationalized by suggesting a ternary complex wherein the imide carbonyls are coordinated to the thiourea through two hydrogen bonds, and the malononitrile enolate is associated with the ammonium group (Fig. 6.5). The authors provided ^{1}H NMR data in support of the proposed interaction between the thiourea and the imide.

6.2.2
Conjugate Addition Reactions of Nitroolefins

Takemoto and co-workers also reported the use of thiourea catalyst **3** to achieve highly enantioselective Michael additions of malonate esters to various β-substituted nitroolefins (**11**) [46]. Good to high yields and *ee*-values were reported for conjugate additions to nitroolefins possessing aromatic, heteroaromatic, and aliphatic β-substituents (Table 6.4). The method was effective even for 2-methylmalonate, which yielded a product having a quaternary carbon α to the newly formed stereocenter. Similar to the model proposed for the Michael additions to imides (see Fig. 6.5), the authors suggested a mechanism wherein the catalyst serves the dual function of activating both reaction partners: the amino group in the catalyst deprotonates the malonate and, through coordination, holds it in close proximity, while the thiourea moiety binds and activates the nitroolefin through hydrogen bonds (Fig. 6.6).

The Takemoto group reported additional studies that further expanded the scope of hydrogen bonding-catalyzed asymmetric conjugate addition reactions. Catalyst **3** was shown to be effective for the enantio- and diastereoselective Michael reactions of prochiral 1,3-dicarbonyl compounds (**13**) with nitroolefins [47]. Various cyclic ketoesters were reacted with β-substituted nitroolefins at ambient temperature in the presence of 10 mol% **3** to give the desired quaternary carbon-containing products in high yields (76–99%), moderate to high enantioselectivities (81–99% *ee* for major isomer), and moderate to high diastereoselectivities (57:43–99:1; Table 6.5). The authors reported that the acidity of the malonate

Table 6.4 Thiourea-catalyzed 1,4-additions of diethyl malonate to nitroolefins.

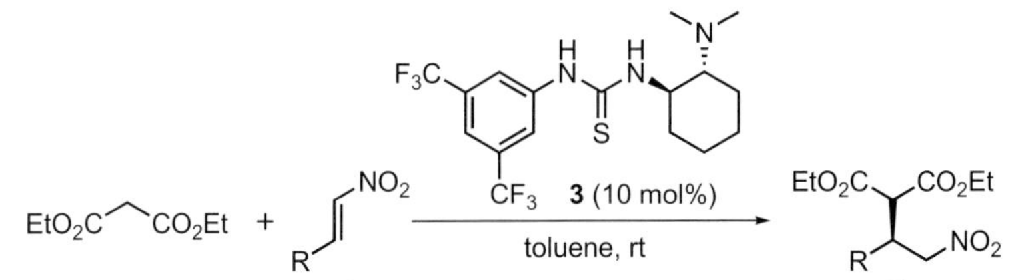

Entry	R	Time [h]	Yield [%]	ee [%]
1	Ph	24	86	93
2	4-F-C$_6$H$_4$	12	87	92
3	2-thienyl	48	74	90
4	i-Bu	48	88	81

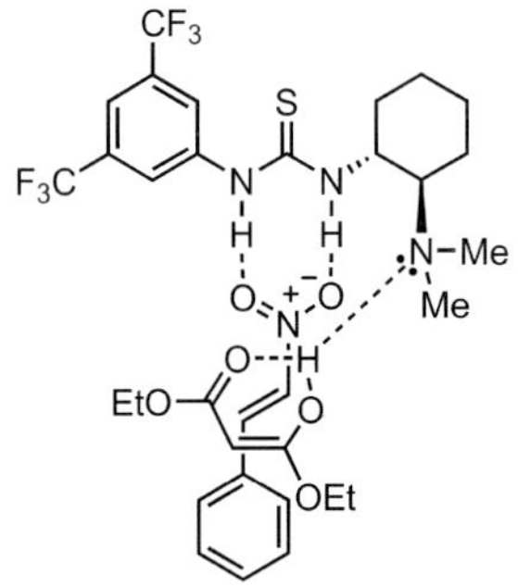

Fig. 6.6 Takemoto's model for bifunctional thiourea catalysis.

nucleophile, though it strongly influenced the overall rate of reaction, and had only a minimal effect on enantioselectivity. The usefulness of this methodology was demonstrated through the total synthesis of (R)-(−)-baclofen, a lipophilic analogue of γ-amino butyric acid, from 4-chlorobenzaldehyde, in six steps and 38% overall yield.

Takemoto and co-workers have also employed catalyst **3** in a sequence involving enantioselective double Michael reactions of γ,δ-unsaturated β-ketoesters (**15**) and nitroolefins to give chiral 4-nitrocyclohexanone derivatives containing three contiguous stereocenters [48]. Treatment of the initial acyclic Michael adducts, which were obtained in high enantiopurities (85–92% ee), with 1,1,3,3-tetramethylguanidine (TMG) at 0 °C furnished the desired cyclic adducts in good yields (63–87%)

Table 6.5 Thiourea-catalyzed diastereoselective 1,4-additions to nitroolefins.

$$R^2CO\underset{R^1}{\diagdown}CO_2R^3 \;+\; R^4\diagdown\!\!\!\diagup NO_2 \;\xrightarrow[\text{toluene}]{\textbf{3} \,(10\,\text{mol\%})}\; \underset{O}{R^2}\!\!-\!\!\overset{R^1}{\underset{R^4\,H}{C}}\!\!-\!\!CO_2R^3\diagdown NO_2$$

13   **11**   **14**

13	R⁴	Temp. [°C]	Time [h]	Yield [%]	dr (2R/2S)	ee (major) [%]
indanone-CO₂Me	Ph	−50	3	96	57/43	93
cyclopentanone-CO₂Me	Ph	−60	24	94	7/93	81
β-tetralone-CO₂Me	4-Br-C₆H₄	r.t.	–	99	97/3	92
β-tetralone-CO₂Me	2-thienyl	−20	–	98	99/1	90

and good to high diastereoselectivities (64–>99% de) for aliphatic and aromatic β-substituted nitroolefins (Table 6.6). The utility of such adducts was demonstrated through the total synthesis of (−)-epibatidine, wherein the enantioselective double Michael reaction played a key role in furnishing a synthetic intermediate in 75% ee.

Modified cinchona alkaloids **18** and **19**, derived from quinine and quinidine, respectively, were utilized by Deng and co-workers for the catalytic asymmetric Michael additions of malonates to nitroolefins [49]. These catalysts effectively promoted the conjugate additions of methylmalonate to a variety of aromatic (90–99% yield; 96–98% ee), heteroaromatic (97–99% yield; 96–98% ee) and aliphatic (71–86% yield; 94% ee) β-substituted nitroolefins (Table 6.7). As the two alkaloids

Table 6.6 Thiourea-catalyzed double Michael reactions of unsaturated β-ketoesters.

Entry	R	Temp. [°C]	Yield [%]	de [%] (16/17)	ee 16 [%]
1	Me	−20	87	>99	92
2	Ph	−40	79	90	89
3	OMe	r.t.	63	64	85

Table 6.7 Cinchona alkaloid-catalyzed Michael reactions of nitroolefins.

Entry	R	Catalyst	Time [h]	Yield [%]	ee [%]
1	Ph	18	36	97	96
2	Ph	19	36	99	93
3	2-thienyl	18	36	99	98
4	2-thienyl	19	44	96	95
5	i-Bu	18	72	86	94
6	i-Bu	19	72	84	92

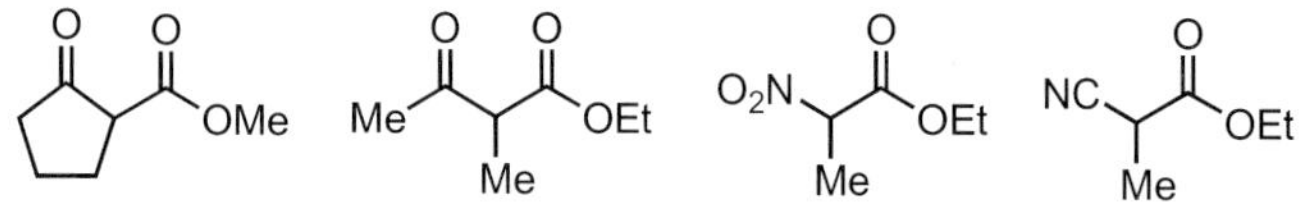

Fig. 6.7 Carbon nucleophiles for Deng's asymmetric conjugate addition reactions.

have a pseudoenantiomeric relationship, it is possible to access both enantiomers of the addition products. Slightly lower enantioselectivities were obtained with the quinidine-derived catalyst. Based on the observation that 6′-hydroxyquinoline-derived catalysts gave significantly higher reaction rates and enantioselectivities when compared to the corresponding 6′-methoxyquinoline derivatives (75–82% ee versus 6–24%), the authors suggested that the phenolic hydroxyl, functioning as a hydrogen bond donor, plays a key role in the organization of the transition state assembly. As with the other catalysts discussed so far, the cinchona-derived compounds are believed to behave as bifunctional catalysts.

In an extension of the above investigations, the Deng group reported the use of modified cinchona alkaloid catalysts **19a–c** for the stereocontrolled creation of adjacent quaternary and tertiary stereocenters by 1,4-additions of prochiral 1,3-dicarbonyl compounds to prochiral β-substituted nitroolefins [50]. Catalysts **19a–c** were excellent promoters of 1,4-additions of a wide range of trisubstituted cyclic and acyclic Michael donors (**21a–d**) to a range of nitroolefins bearing aryl, heteroaryl and aliphatic β-substituents (Fig. 6.7; Table 6.8, PHN = 9-phenanthrene). The conjugate addition products, possessing a high level of substitution, were formed in significant yields (73–94%), ee-values (92–>99%), and diastereomeric ratios (up to >98:2). A considerable improvement in diastereoselectivity was observed on lowering the reaction temperature. The authors rationalized the observed enantioselectivities by suggesting a transition-state complex in which the nitroolefin is activated by hydrogen bonding to the phenol, with the enolate held in close proximity by hydrogen bonds to both the ammonium and the phenol groups (Fig. 6.8).

Ricci and co-workers have reported an expansion of the nucleophile scope for the hydrogen bonding-catalyzed asymmetric 1,4-addition reaction using catalyst **25**, synthesized from commercially available 3,5-bis-trifluoromethyl-phenyl isothiocyanate and (1R,2S)-cis-1-amino-2-indanol, to catalyze the Friedel–Crafts alkylations of indoles [51]. Under optimized conditions, the reactions provided product in high yields (70–88%) and good enantioselectivities (71–89% ee) for a variety of aromatic, heteroaromatic, and aliphatic nitroolefins (Table 6.9). The reaction of an indole substrate containing electron-withdrawing functionality gave a similarly high ee-value, but with a significantly diminished rate of reaction (35% yield at 72 h). The Ricci group demonstrated the synthetic potential of these adducts through a two-step conversion of one adduct to the corresponding tryptamine derivative in high overall yield with no observed loss in ee-value. The authors reported that the combination of an unprotected indole nitrogen and a free

Table 6.8 Modified cinchona-alkaloid-catalyzed diastereoselective 1,4-additions.

Entry	21	R	Catalyst	Temp. [°C]	Time [h]	Yield [%]	dr	ee [%]
1	a	Ph	19b	−60	48	94	95:5	99
2	b	Ph	19c	−20	63	73	91:9	>99
3	c	n-pentyl	19a	−20	84	78	93:7	92
4	d	n-pentyl	19a	−20	84	75	93:7	98

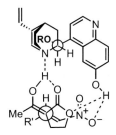

Fig. 6.8 Deng's selectivity model for 1,4-additions of 1,3-dicarbonyl substrates.

hydroxyl on the catalyst was important for inducing high enantiofacial selectivity for the alkylation reaction. The use of N-methylindole or catalysts lacking the indanone hydroxyl group gave lower conversion and enantioselectivity, indicating the presence of a hydrogen-bonding interaction between the catalyst and the nucleophile in addition to the hydrogen-bonding activation of the nitroolefin by thiourea (Fig. 6.9).

Jørgensen and co-workers employed chiral bis-sulfonamide catalyst **27**, a proven ligand for metal-based asymmetric catalysis, for the Friedel–Crafts alkylations of N-methylindoles (**24**) using β-substituted nitroolefins [52]. Using optimized conditions, 2 mol% **27** gave the desired indole alkylation products of substituted aryl and heteroaryl nitroolefins in moderate to high yields (20–91%) and moderate enantiopurities (13–63% *ee*; Scheme 6.3). Aliphatic β-substitution

Table 6.9 Thiourea-catalyzed Friedel–Craft alkylations of indoles.

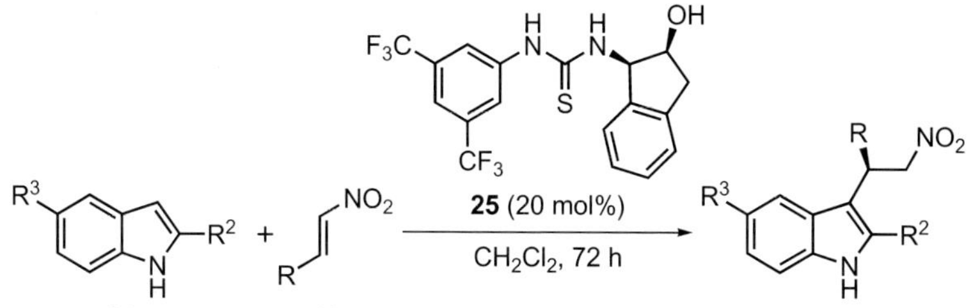

Entry	R^2	R^3	R	Temp. [°C]	Yield [%]	ee [%]
1	H	H	Ph	−24	78	85
2	Me	H	Ph	−45	82	74
3	H	OMe	Ph	−45	86	89
4	H	H	2-furyl	−24	88	73
5	H	H	n-pentyl	−24	76	83

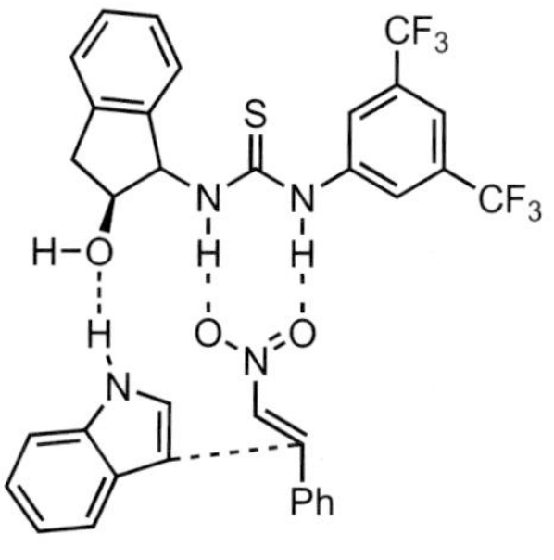

Fig. 6.9 Ricci's selectivity model for thiourea-catalyzed Friedel–Craft alkylations.

on the nitroolefin gave significantly lower optical purities (11–19%). In contrast to the results reported by Ricci and co-workers using **25** (see Table 6.9), catalyst **27** gave optimal results using N-methylated indoles. Other nitrogen-protecting groups did not garner any improved product yields or *ee*-values. The catalyst is proposed to function through a single hydrogen bonding interaction between the catalyst sulfonamide proton and the nitro group of the electrophile. The N–C–C–N dihedral angle of the catalyst appeared to have a large effect on observed enantiofacial selectivity.

Table 6.10 Bifunctional thiourea-catalyzed 1,4-additions to nitroolefins.

Entry	R	Time [h]	Yield [%]	ee [%]
1	4-Br-C_6H_4	40	94	93
2	2-thienyl	23	94	95
3	n-hexyl	69	88	86

values were comparable to those reported by Connon (see Table 6.10), though ee-values reported by Dixon for identical nitroolefin substrates were generally slightly lower. Michael addition to tert-butyl β-substituted nitroolefin was not observed in significant yield under the reported conditions. Rigorous purification to remove traces of water or oxygen from the reaction solvent was not necessary to obtain optimal results. Dixon reported observations similar to Connon regarding the necessity of the 9-epi catalyst stereochemistry to achieve optimal levels of enantioselectivity.

The Wang group reported the development of a new class of binaphthyl-derived hydrogen bonding thiourea catalyst (32) for the Michael reactions of 1,3-diones and nitroolefins [55]. The high catalytic activity of 32 allowed for the addition of less-reactive 2,4-pentanedione nucleophiles to a variety of aryl-substituted nitroolefins, with catalyst loadings as low as 1 mol%. The chiral adducts were synthesized in good yields (78–92%) and good to high ee-values (83–97%) for a variety of electron-withdrawing and electron-donating nitroolefin aryl substituents (Table 6.11). Product utility was demonstrated through the conversion of select 1,4-adducts to α-substituted-β-amino acids in four steps with retention of enantiopur-

Table 6.11 Thiourea-amine-catalyzed Michael additions of 2,4-pentanedione.

Entry	R	Time [h]	Yield [%]	ee [%]
1	Ph	26	87	95
2	4-MeO-C$_6$H$_4$	36	92	97
3	4-Cl-C$_6$H$_4$	24	91	97
4	2,4-(MeO)$_2$-C$_6$H$_3$	36	88	91

ities. This procedure represented the first highly enantioselective Michael reactions of a 1,3-diketone nucleophile with β-nitrostyrenes. (For experimental details see Chapter 14.9.6).

6.2.3
Conjugate Addition Reactions of Vinyl Sulfones

Deng and co-workers reported the first highly enantioselective catalytic additions of α-cyanoacetates to vinyl sulfones catalyzed by cinchona-derived catalysts **36** and **19c**, originally used for the Michael addition of malonate esters to nitroolefins [56]. Various α-aryl α-cyanoacetates (**34**) were shown to react with phenylvinyl sulfone in the presence of 20 mol% of catalysts **36** or **19c** to give 1,4-adducts bearing an all-carbon quaternary stereocenter in high yields (89–96%) and excellent enantioselectivities (93–97% ee; Table 6.12). A more electrophilic 3,5-bis-(trifluoromethyl)phenyl vinyl sulfone was used to obtain reasonable reaction rates for the additions of less-reactive α-alkyl α-cyanoacetates in high enantiopurities (92–94% ee), demonstrating the ability of the catalyst to tolerate electronically diverse vinyl sulfones. Reaction enantioselectivities were strongly dependent on catalyst structure, and highest enantioselectivities were obtained using a catalyst having a free C6'-OH and an aryl ether at C9. Synthetic utility of the products was demonstrated through conversion, via a seven-step sequence, to chiral α,α-disubstituted amino acids that previously were inaccessible via asymmetric catalysis. (For experimental details see Chapter 14.9.12).

Table 6.12 Cinchona-derivative-catalyzed conjugate additions to vinyl sulfones.

PHN = 9-phenanthrene

EtO$_2$C-CH(R)-CN (**34**) + CH$_2$=CH-SO$_2$R^1 (**35**) → **36** (20 mol%), toluene, rt → EtO$_2$C-C(R)(CN)-CH$_2$-CH$_2$-SO$_2$Ph (**37**)

Entry	R	R^1	Temp. [°C]	Time [h]	Yield [%]	ee [%]
1	Ph	Ph	−25	72	89	95
2	4-MeO-C$_6$H$_4$	Ph	0	70	92	94
3	4-F-C$_6$H$_4$	Ph	−25	72	90	94
4	2-thienyl	Ph	−25	48	95	97
5	Allyl	3,5-(CF$_3$)$_2$-C$_6$H$_3$	0	96	76	94

6.3 Hydrocyanation Reactions

The hydrocyanation reactions of electrophilic aldehydes, ketones and their corresponding imines gives direct access to synthetic derivatives of several important structures, including α-hydroxy carboxylic acids, β-amino alcohols and α-tertiary and α-quaternary-α-amino acids. The asymmetric hydrocyanation reaction provides access to chiral synthons, which have proven useful for the construction of many structurally complex and biologically active compounds. Catalysis of these reactions is especially attractive with respect to avoiding the cost and relative chemical inefficiency associated with the use of chiral auxiliaries.

6.3.1 Hydrocyanation Reactions of Aldehydes

The Inoue laboratory reported the first asymmetric hydrocyanation of an aldehyde using a synthetic peptide, cyclo[(S)-Phe-(S)-His] (**38**), to give the cyanohydrin of benzaldehyde in high optical purity (up to 90% ee at 40% conversion). The ee-value of the product was found to diminish with increased reaction time (Scheme 6.5) [57]. The catalytic activity of **38** is presumed to arise from the bifunctional character of the catalyst, wherein aldehyde activation occurs through hydrogen-

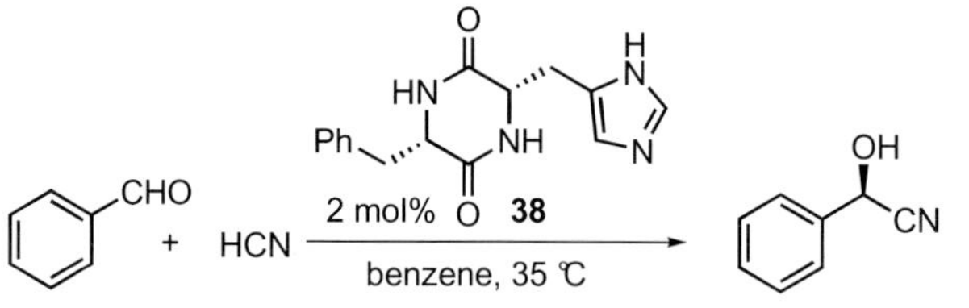

Scheme 6.5 Synthetic peptide-promoted hydrocyanation of benzaldehyde.

Product **39**: 90% ee at 40% conv; 21% ee at 90% conv.

bonding with the histidine residue, while the imidazolyl moiety functions to provide the cyanide anion by deprotonating HCN. Optimization studies indicated that high product optical purities could be obtained at higher conversions at −20 °C using toluene as the solvent. These optimized conditions provided cyanohydrin adducts for a variety of aromatic (57–97% yield; 78–97% ee), heteroaromatic (60–73% yield; 42–54% ee) and aliphatic (60–96% yield; 18–71% ee) substrates (Table 6.13) [58]. Aromatic aldehydes bearing electron-withdrawing functionality proved to be incompatible with catalyst **38**, providing products in low enantiopurities (quantitative yield; 4–53% ee). (For experimental details see Chapter 14.5.1).

Studies conducted at the Danda laboratory provided further insight into the activity of the Inoue catalyst system. Danda reported the importance of an amor-

Table 6.13 Synthetic peptide-promoted hydrocyanations of aldehydes.

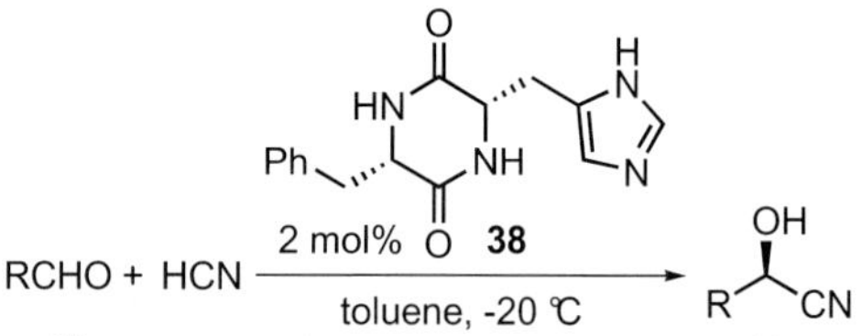

RCHO (**40**) + HCN → (2 mol% **38**, toluene, −20 °C) → R-CH(OH)-CN (**41**)

Entry	R	Time [h]	Yield [%]	ee [%]
1	Ph	8.0	97	97
2	p-MeO-C$_6$H$_4$	10.0	57	78
3	p-CN-C$_6$H$_4$	8.0	100	32
4	furfural	8.0	60	42
5	isobutyryl	5.0	79	71
6	isovaleryl	5.0	44	18

phous catalyst structure (versus highly crystalline **38**, obtained by recrystallization from aqueous methanol) and low reaction mixture viscosity for maintaining consistently high product yields and enantiopurities [59]. The Danda group later reported the observation of enantioselective autoinduction, where chiral cyanohydrin product was shown to be incorporated into the active catalyst species, promoting greater enantioselection as the reaction progressed to completion [60].

Discussion Catalyst **38** serves as a peptide mimic of enzymes used for hydrocyanations of aldehydes to provide access to a range of cyanohydrins in high yields and moderate to high enantioselectivities. Catalyst **38** was prepared in four steps from (benzyloxycarbonyl)-(S)-phenylalanine in good overall yield [58]. Hydrogen cyanide was produced according to literature precedent, via addition of an aqueous solution of sodium cyanide dropwise into dilute sulfuric acid and was stored in a freezer [61].

Caution: Hydrogen cyanide is extremely toxic and volatile. All manipulations of hydrogen cyanide and reactions thereof should be performed in a well-ventilated hood and care must be taken to avoid inhalation.

6.3.2
The Strecker Reaction

The Strecker reaction is defined as the addition of HCN to the condensation product of a carbonyl and amine component to give α-amino nitriles. Lipton and co-workers reported the first highly effective catalytic asymmetric Strecker reaction, using synthetic peptide **43**, a modification of Inoue's catalyst (**38**), which was determined to be inactive for the Strecker reactions of aldimines (see Scheme 6.5) [62]. Catalyst **43** provided chiral α-amino nitrile products for a number of N-benzhydryl imines (**42**) derived from substituted aromatic (71–97% yield; 64–>99% ee) and aliphatic (80–81% yield; <10–17% ee) aldehydes, presumably through a similar mode of activation to that for hydrocyanations of aldehydes (Table 6.14). Electron-deficient aromatic imines were not suitable substrates for this catalyst, giving products in low optical purities (<10–32% ee). The α-amino nitrile product of benzaldehyde was converted to the corresponding α-amino acid in high yield (92%) and ee (>99%) via a one-step acid hydrolysis.

While examining metal complexes of tridentate Schiff bases, Jacobsen and co-workers discovered that the free ligands served as catalysts for the asymmetric Strecker reactions of aldimines. A screen of ligand libraries, prepared through solid-supported parallel combinatorial methods, allowed the identification of highly effective catalysts (e.g., **45a**) for the enantioselective Strecker reaction. (Scheme 6.6) [40]. Using the Strecker reaction of N-allyl-benzaldimine as a model, it was determined that the thiourea moiety, the steric bulk of the amino acid derivative, and the 3-*tert*-butyl substitution pattern at the salicylaldehyde derivative played crucial roles in the enantioselection. Although **45a** was optimized for the reaction of N-allyl-benzaldimine, it proved to be an effective catalyst for

Table 6.14 Strecker reactions of aldimines using a synthetic peptide.

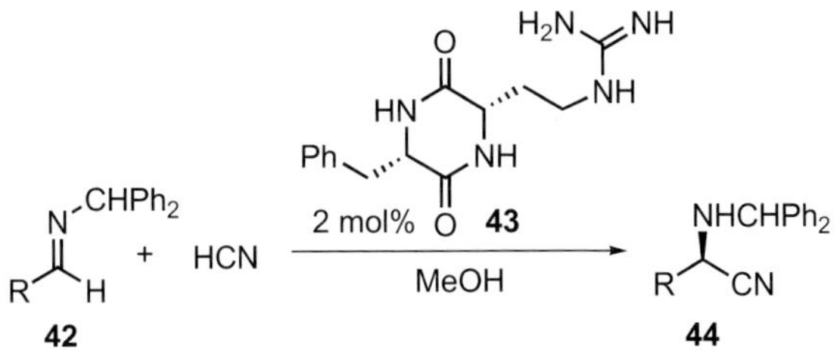

Entry	R	Temp. [°C]	Yield [%]	ee [%]
1	Ph	−25	97	>99
2	p-Cl-C$_6$H$_4$	−75	94	>99
3	p-MeO-C$_6$H$_4$	−75	90	96
4	m-NO$_2$-C$_6$H$_4$	−75	71	<10
5	t-Bu	−75	80	17

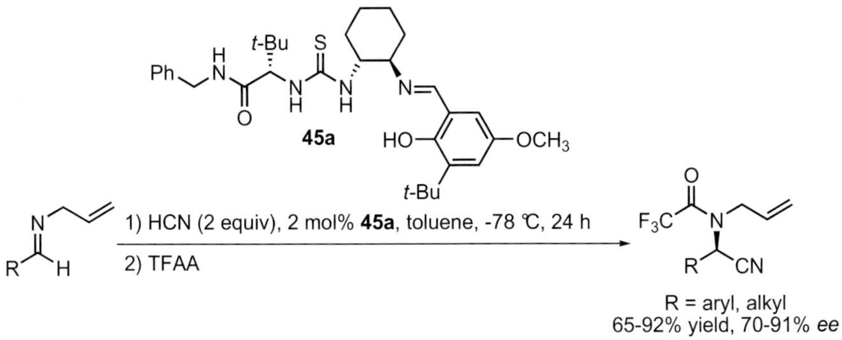

Scheme 6.6 Schiff base catalysts for the Strecker reactions of aldimines.

the hydrocyanations of a range of aromatic (65–92% yield; 70–91% ee) and aliphatic (70–77% yield; 83–85% ee) imines.

Further library optimization studies based on **45a**, incorporating seven new amino acids with large α-substituents and ten new salicylaldehyde derivatives, led to the discovery of urea catalyst **45b** [63]. Strecker reactions of N-allyl and N-benzyl aldimines (**46**) using catalyst **45b** at −70 °C showed significant improvements in yields, optical purities, and scope of the cyanohydrin products obtained (Table 6.15). In general, **45b** promoted hydrocyanations of aldimine substrates with slightly higher ee-values (+2–4%) than the corresponding resin-bound analogue **45c**. However, resin-bound catalyst **45c** offered an advantage in product iso-

Table 6.15 Urea-catalyzed Strecker reactions of aldimines.

$$46 \xrightarrow{\text{1) HCN (1.3 equiv), 2 mol\% 45b, toluene, -70 °C, 20 h}}_{\text{2) TFAA}} 47$$

Entry	R	R^1	Yield [%]	ee [%]
1	Ph	allyl	74	95
2	p-MeO-C$_6$H$_4$	allyl	98	95
3	p-Br-C$_6$H$_4$	allyl	89	89
4	t-butyl	allyl	75	95
5	cyclohexyl	allyl	88	86
6	CH$_3$(CH$_2$)$_4$	benzyl	69	78

lation, as pure product was isolated in essentially quantitative yield after simple filtration, to recover the catalyst, followed by solvent removal. Recovered catalyst was shown to retain reactivity and selectivity even after ten reaction–recovery cycles.

The first enantioselective catalytic Strecker reaction of ketoimines to give chiral quaternary cyanohydrins was demonstrated by the Jacobsen group using **45b** (Scheme 6.7) [41]. A series of substituted aryl and aliphatic N-benzyl methylketoimines (**48**) reacted with HCN in the presence of **45b** to provide essentially quantitative yields of the Strecker adducts in high optical purities (70–95% ee). The adducts were crystalline, and their recrystallization from hexanes increased their enantiopurities to >99.9% ee. The α-quaternary Strecker adducts (**49**) could be converted to α-quaternary α-amino acids through formamide protection of the secondary amine, followed by sequential hydrolysis of the nitrile and the formamide, followed by hydrogenolytic debenzylation.

In 2002, Jacobsen and co-workers reported their studies on the structural requirements of their urea and thiourea catalysts for high enantioselection in the asymmetric Strecker reaction [64]. Using a series of catalyst analogues, these authors confirmed that the two urea hydrogen are solely responsible for substrate activation. Energy minimization studies implicated an interesting bridging hydrogen-bonding structure between the imine nitrogen and both of the urea hy-

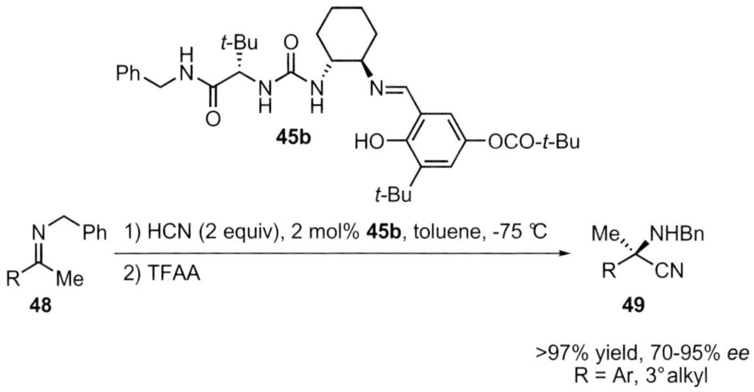

Scheme 6.7 Thiourea-catalyzed Strecker reactions of ketoimines.

drogen atoms. ^{1}H NMR titration studies using **45b** and N-p-methoxybenzyl acetophenone showed exclusive downfield shift of the Z-imine methyl resonance. This result, along with the experimental observation that cyclic Z-imines and acyclic, predominantly E-imines (capable of E–Z interconversion in solution) show an identical sense of stereoinduction, provided evidence for exclusive binding of the Z-imine in the catalyzed Strecker reaction. Numerous NOE experiments using **45b**–imine complexes, coupled with the observed product absolute stereochemistry, suggested that HCN approaches the reactive complex by passing over the diaminocyclohexane portion of the catalyst. New catalysts incorporating bulkier amino acids and amides were developed based on this hypothesis. Especially effective was thiourea catalyst **52** (Scheme 6.8), which gave good to excellent enantioselectivities for aliphatic imines (86–97%) that had proven to be problematic substrates for catalyst **45b** (70–80% ee). Substrates that had reacted reasonably well in the presence of **45b** were transformed with near-perfect enantiocontrol (99.3%) using catalyst **52**.

Corey and co-workers reported the use of a C_2-symmetric guanidine catalyst (**53**) for the Strecker reactions of N-benzhydryl aldimines [65]. The catalyst is be-

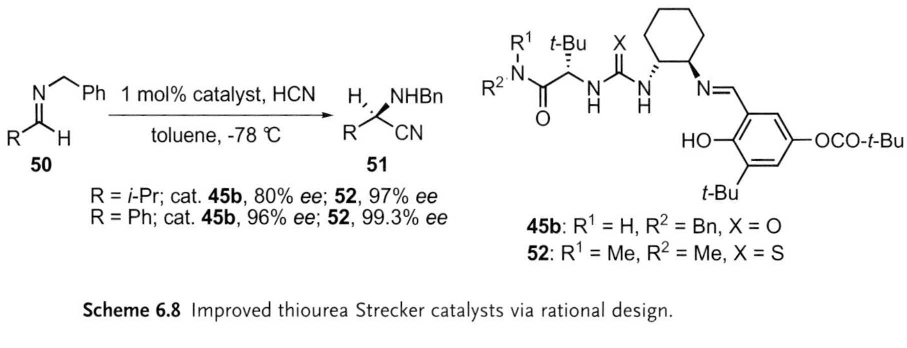

Scheme 6.8 Improved thiourea Strecker catalysts via rational design.

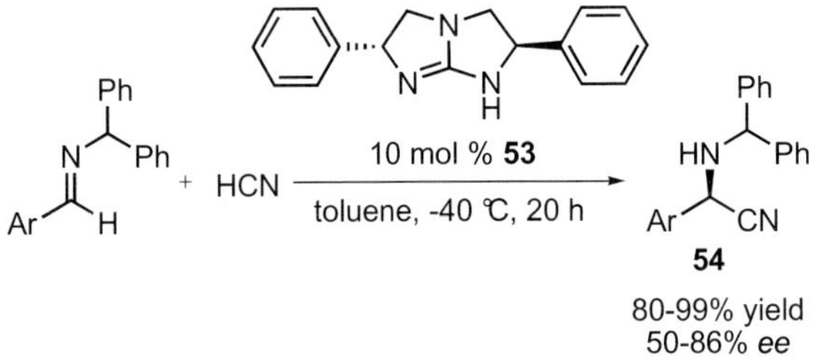

Scheme 6.9 Guanidine-catalyzed Strecker reactions of aldimines.

lieved to function via hydrogen-bonding activation of the aryl imine through a protonated guanidinium-cyanide complex to afford the (R) adduct (Scheme 6.9). The reaction was general for substituted aryl imines (80–99% yield; 50–88% ee). Interestingly, aliphatic N-benzhydrylaldimines were shown to give the (S)-adducts in high yields (~95%) and good enantiopurities (63–84% ee). As might be expected, the N-methyl derivative of **53** was not an active catalyst. The benzhydryl protecting group proved to be necessary for obtaining high enantioselectivities, presumably due to orientational effects arising from steric and π-stacking interactions in the transition state. Catalyst **53** was prepared from methyl-(R)-phenylglycinate in nine steps, with a 24% overall yield. The guanidine catalyst can be easily recovered from the crude reaction mixture by extraction with oxalic acid (80–90% yield).

Table 6.16 Chiral ammonium trifluoroacetate-catalyzed Strecker reactions.

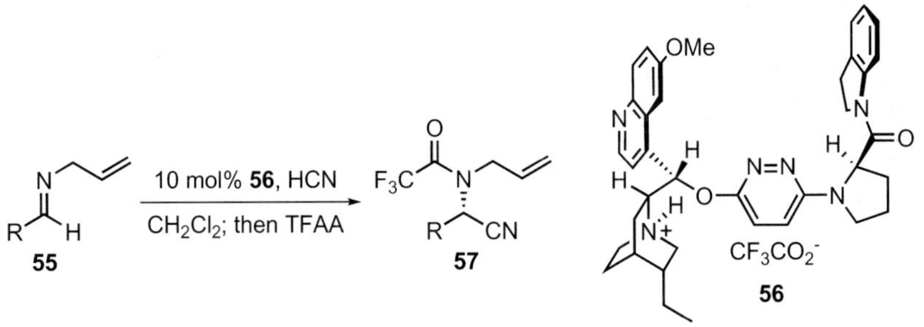

Entry	R	Time [h]	Yield [%]	ee [%]
1	Ph	36	95	92
2	p-MeO-C$_6$H$_4$	36	95	90
3	p-CN-C$_6$H$_4$	24	92	80
4	m-MeO-C$_6$H$_4$	40	96	>99
5	2-naphthyl	40	95	79

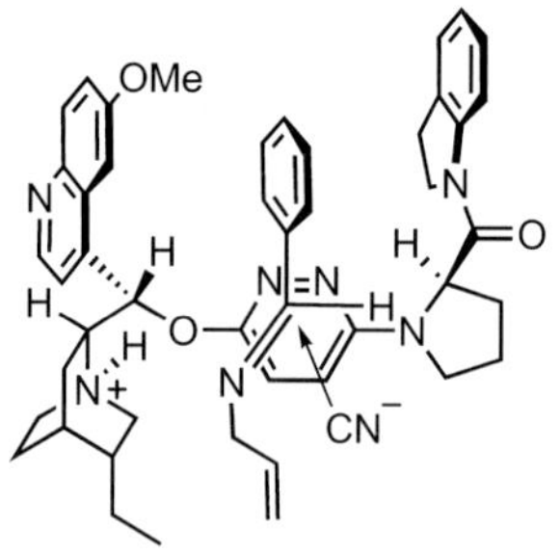

Fig. 6.10 Corey's selectivity model for the Strecker reactions of N-allyl aldimines.

Corey and co-workers also reported the successful Strecker reactions of N-allyl aldimines using a cinchona alkaloid-derived chiral ammonium salt (**56**) [66]. Catalyst **56**, a stable crystalline salt, was readily prepared from dihydroquinidine in three steps and in high yield (68% overall). Hydrocyanations of a variety of substituted aryl N-allyl aldimines were achieved using 10 mol% **56** and provided the corresponding (S)-α-amino nitriles in high yields and generally high ee-values (86–98% yield; 79–99% ee; Table 6.16). (For experimental details see Chapter 14.5.2). Formation of the (S)-adducts was rationalized by invoking polar ionic hydrogen-bonding activation of the imine by the protonated quinuclidine core, such that the aryl imine is held within the pocket formed by the dihydroindole and quinoline rings of the catalyst, which are held approximately 7.5 Å apart in near-parallel planes (Fig. 6.10). The model accounts for the observed facial selectivity since, under the above constraints, the imine si face is blocked by the pyridazine linker. The lower enantioselectivities observed for the Strecker reactions of aliphatic imines, especially those with bulky substituents, were consistent with the proposed binding model.

6.3.3
Hydrocyanation Reactions of Ketones

Catalytic asymmetric cyanations of ketones afford the corresponding cyanohydrins, which serves as precursors to many useful chiral building blocks, such as α-hydroxy acids, β-amino alcohols and others. Jacobsen and co-workers reported the synthesis of thiourea catalyst **59**, which proved effective for the promotion of highly enantioselective cyanosilylations of a range of ketones [67]. This catalyst was developed based on the observation that a derivative of **52** having a free amine in place of the salicylaldehyde moiety promoted the cyanosilylation of benzaldehyde (3 h, 100% conversion, 25% ee). The optimal catalyst was developed by fine-tuning the steric demands of the secondary amide as well as the tertiary amine. Alkyl-aryl ketones gave products in high optical purities, with only a slight dependence on alkyl group size. Likewise, α,β-unsaturated ketones were effective substrates and gave exclusively the 1,2-adducts (Table 6.17). (For experimental details see Chapter 14.5.3). High enantioselectivities were obtained only in

Table 6.17 Thiourea-catalyzed cyanosilylations of ketones.

Entry	R	Time [h]	Yield [%]	ee [%]
1	Ph	24	96	97
2	p-MeO-C$_6$H$_4$	48	93	95
3	p-Br-C$_6$H$_4$	12	94	93
4	2-naphthyl	12	98	97
5	2-thiophene	48	88	98
6	3-pentenyl	48	95	89

cyanosilylations of ketone substrates bearing one sp^2-hybridized substituent, implicating electronic rather than steric differentiation as the key element of asymmetric induction. The Brønsted basic amine was clearly of importance in the catalytic cycle, as analogues of **59** lacking the tertiary amine group were found to have no catalytic activity under the reported conditions.

6.4
Mannich Reactions

The asymmetric additions of enolate or ester enolate equivalents to imines are attractive and important routes to β-amino acid derivatives, as they represent the coupling of two components of similar complexity by a carbon–carbon bond-forming reaction. In continuation of their studies on nucleophilic additions to activated imines, the Jacobsen group explored the coupling of N-Boc aldimines with mono-silyl ketene acetals [68]. Initial studies using a urea catalyst showed promising results for the reaction of N-Boc benzaldimine and the trimethylsilyl-O,O-acetal of isopropyl acetate (90% yield, 54% ee), though the background reaction significantly diminished the observed product ee-value. The combined use of a thiourea catalyst and the *tert*-butyldimethylsilyl ketene acetal of isopropyl acetate served to increase the rate of the catalyzed reaction while decreasing the rate of background reaction. Further catalyst optimization provided catalyst **62**, which was found to promote the reactions of the ketene acetal with a range of substituted aryl and heteroaryl N-Boc aldimines (**61**) to give the corresponding N-protected β-amino esters in uniformly high yields (84–99%) and high optical purities (86–98% ee; Table 6.18). (For experimental details see Chapter 14.2.4).

Table 6.18 Thiourea-catalyzed Mukaiyama–Mannich reactions of aldimines.

Entry	R	Temp. [°C]	Yield [%]	ee [%]
1	Ph	−40	95	97
2	p-MeO-C$_6$H$_4$	4	91	86
3	p-F-C$_6$H$_4$	−30	88	93
4	2-thienyl	−30	95	92
5	3-pyridyl	−30	99	98

The Terada group reported an enantioselective route to β-amino ketones via the direct Mannich coupling reactions of acetyl acetone with N-Boc arylimines, catalyzed by the 3,3′-bis-aryl-substituted, BINOL-derived phosphoric acid catalyst **64** [69]. These chiral catalysts have not only strong Brønsted acidity but also a well-defined chiral environment around the proton. The aryl substituents at the ortho positions served the important purpose of extending the chirality of the BINOL backbone. The size and substitution pattern of the aryl groups were important for good enantioselectivities, with 4-(β-naphthyl)-C$_6$H$_4$ substitution providing both optimal yields and enantioselectivities for a range of substrates. Catalyst **64** was found to be effective at a relatively low loading for various ortho- and para-substituted aryl aldimines, providing β-amino ketones in short reaction times and in uniformly high yields (93–99%) and ee-values (90–98%; Table 6.19).

Mukaiyama–Mannich reactions of substituted silyl ketene acetals catalyzed by chiral Brønsted acids have also been reported by the Akiyama group [70]. Initial studies utilizing a phenyl aldimine and the trimethylsilylacetal of methyl 2,2-dimethylacetate in the presence of 30 mol% of the chiral phosphoric acid **68** gave the desired α-disubstituted-β-aminoester in essentially quantitative yield and good enantiopurity (89% ee, entry 1; Table 6.20). Monosubstituted silyl ketene acetals gave products in high yields (65–100%), high syn diastereoselectivities, and good to high enantiopurities (81–96% ee) for a range of substituted aryl, heteroaryl, and α,β-unsaturated aldimines. The observed selectivities imply that the protonated imine intermediate must be tightly coordinated to the chiral phos-

6.4 Mannich Reactions

Table 6.19 Chiral phosphoric acid-catalyzed direct Mannich reactions of aldimines.

61 (R-CH=N-Boc) + **Ac-CH(Ac)-Ac** → (2 mol% **64**, CH$_2$Cl$_2$, rt, 1 h) → **65** (R-CH(NHBoc)-CH(Ac)$_2$)

64: Ar = 4-(β-naphthyl)-C$_6$H$_4$

Entry	R	Yield [%]	ee [%]
1	Ph	99	95
2	p-MeO-C$_6$H$_4$	93	90
3	p-Br-C$_6$H$_4$	96	98
4	o-Me-C$_6$H$_4$	94	93

Table 6.20 Chiral Brønsted acid-catalyzed diastereoselective Mannich reactions.

66 + **67** → (10 mol% **68**, toluene, −78 °C, 24 h) → **69**

Entry	R^1	R^2	R^3	R^4	Yield [%]	syn/anti	ee [%]
1	Ph	Me	Me	Me	98	–	89
2	Ph	H	Me	Et	100	87:13	96
3	p-MeO-C$_6$H$_4$	H	Me	Et	100	92:8	88
4	2-thienyl	H	Me	Et	81	94:6	88
5	PhCH=CH	H	Me	Et	91	95:5	90
6	Ph	H	Ph$_3$SiO	Me	79	100:0	91

Table 6.21 Brønsted acid-catalyzed asymmetric Mannich reactions.

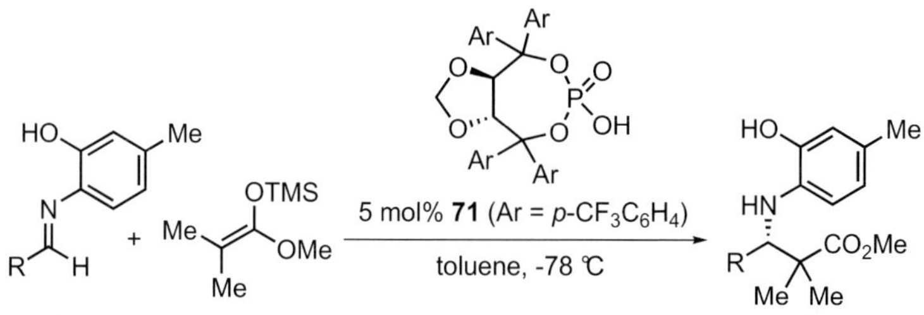

Entry	Ar	Time [h]	Yield [%]	ee [%]
1	Ph	30	100	89
2	p-F-C$_6$H$_4$	48	95	92
3	p-Me-C$_6$H$_4$	48	96	92

phate anion, so as to greatly differentiate the prochiral faces of the imine. The electron-withdrawing substituents on the 3,3′-aryl groups of the catalyst likely serve the dual purpose of creating a chiral environment about the phosphate while also increasing the acidity of the catalyst through an inductive effect. The o-hydroxy group of the aryl imine was critical for high enantioselection under the reported conditions (only 39% ee, R = Ph, N-benzylideneaniline).

A taddol-derived, cyclic chiral phosphoric acid catalyst (**71**) was reported by the Akiyama group for the Mukaiyama–Mannich reactions of aryl aldimines [71]. Examination of various catalysts and imine aryl substituents led to the discovery of conditions wherein 5 mol% of **71** was sufficient to promote the Mannich reactions of a silyl ketene acetal with several aryl aldimines (**70**) to give the corresponding β-amino esters in high yields (81–100%) and ee-values (85–92%; Table 6.21). The o-hydroxy aryl group on the imine nitrogen was once again critical for obtaining high product optical purities, suggesting a two-point interaction between the imine and the phosphoric acid catalyst (Fig. 6.11).

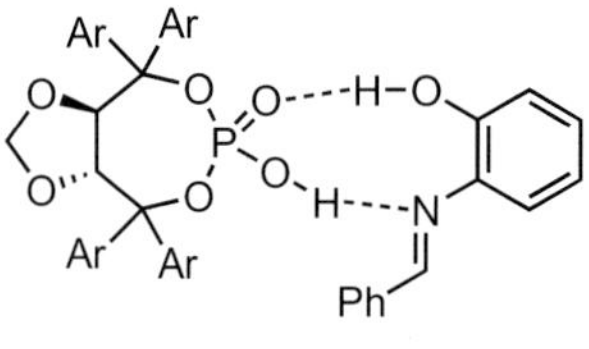

Fig. 6.11 Akiyama's two-point binding model for phosphoric acid catalysis.

Table 6.22 Thiourea-catalyzed acyl-Mannich reactions of substituted isoquinolines.

Entry	R	Yield [%]	ee [%]
1	H	80	86
2	3-Me	75	92
3	5-NO_2	71	71
4	6-OSO_2CF_3	67	83
5	7-OTBS	86	60

The Jacobsen group has reported the asymmetric Mannich reactions of acylated isoquinolines with ester silyl enolates promoted by thiourea catalyst **74**. This process provides access to enantiomerically enriched dihydroisoquinoline building blocks with potential utility in alkaloid synthesis [72]. In these reactions, N-acylation of the isoquinolines (**73**) by 2,2,2-trichloroethyl chloroformate (TrocCl) gave highly electrophilic iminium ions, which are further activated by hydrogen bonding with the thiourea catalyst. Nucleophilic attack by the silyl ketene acetal gave the Troc-protected addition products in good yields (67–86%) and enantioselectivities (60–92% ee; Table 6.22). Optical purity of the acyl-Mannich adducts depended significantly on the substituents on the pyrrole moiety in the catalyst. The precise mode of hydrogen bonding-activation of the electron-poor acyl-iminium ions remains unclear.

Discussion Catalyst **62** was synthesized in five steps in 86% overall yield from commercially available starting materials, with a single chromatographic purification [68]. The reaction as described by Jacobsen appears to be relatively insensitive to dilution, reagent stoichiometry, and rate of nucleophile addition. Different low-polarity solvents can be used with little effect on product *ee*-value. In highly polar aprotic solvents, however, the *ee*-values were significantly lower, whereas in protic solvents the imine rapidly decomposed. Aliphatic N-Boc imines were not examined due to a lack of useful methods for their synthesis. The inherent electrophilicity of the imine was shown to be important, with less-electrophilic N-allyl

6.5
Aza-Henry Reactions of Aldimines

The asymmetric aza-Henry (or nitro-Mannich) reaction is the addition of a nitroalkane anion to an imine. This reaction produces a carbon–carbon bond with generation of up to two contiguous nitrogen-containing chiral centers. The products of the reactions are precursors to useful nitrogen-containing building blocks, such as vicinal diamines and α-amino carbonyl compounds. The Takemoto group reported the first hydrogen bonding-promoted aza-Henry reaction, using the bifunctional, tertiary amine-containing thiourea catalyst, **78**. This catalyst promoted the direct reactions of nitroalkanes with N-phosphinoylimines (**76**), so as to produce the addition products in good to high yields (57–91%) and moderate enantioselectivities (64–76% ee) at ambient temperature (Table 6.23) [73]. Various substituted aryl and heteroaryl aldimines, bearing electron-donating or electron-withdrawing substituents on the imine aryl group, were effective as substrates, giving only minimal variations in yields and enantioselectivities. Based on their earlier studies, these authors expected the nitroalkane to coordinate to the thio-

Table 6.23 Thiourea-catalyzed aza-Henry reactions of N-phosphinoylimines.

Entry	R^1	R^2	Yield [%]	ee [%]	dr (syn:anti)
1	Ph	H	87	67	–
2	Ph	Me	76	67 (major)	73:27
3	p-Cl-C$_6$H$_4$	H	85	67	–
4	2-furyl	H	68	76	–
5	cinnamyl	H	83	65	–

dr = diastereomeric ratio.

urea through hydrogen bonds, such that the nearby tertiary amine could deprotonate the acidic proton. The resulting anion, ensconced in the chiral environment provided by the cyclohexyldiamine scaffold, was expected to participate in the asymmetric aza-Henry reaction. Binding and modulation of nitronate anion reactivity by thioureas had been reported previously [74].

Johnston and co-workers have reported the aza-Henry reactions of electron-poor N-Boc aldimines with nitroalkanes catalyzed by quinolinium salt catalyst **80**, capable of forming polar ionic hydrogen bonds [75]. A white, bench-stable crystalline solid, catalyst **80** is the triflic acid salt of a bis-amidine ligand. The free base was prepared as a single enantiomer in 86% yield via the palladium-catalyzed coupling of commercially available (+)-*trans*-cyclohexanediamine and 2-chloroquinoline. The aza-Henry reactions of activated imines were accomplished in the absence of base additives and gave adducts in moderate yields (51–69%) and moderate to high enantiopurities (59–95% *ee*; Scheme 6.10). The use of nitroethane as the nucleophile afforded addition products with two contiguous stereocenters in good to high diastereoselectivies (from 7:1 to 19:1, major isomer is *syn*). (For experimental details see Chapter 14.4.2). Although the nature of the catalyst–substrate complex remains to be defined, it goes without saying that the free amine must deprotonate the nitroalkane, and that the resulting anion must be tightly hydrogen-bonded to the chiral quinolinium catalyst. Hydrogen bonding-activation of the imine, whether by the catalyst that is coordinated to the nucleophile or by a second bis-amidine catalyst, is also likely to be involved. It is noteworthy that the unprotonated ligand does not enhance the rate of the reaction above nominal background levels.

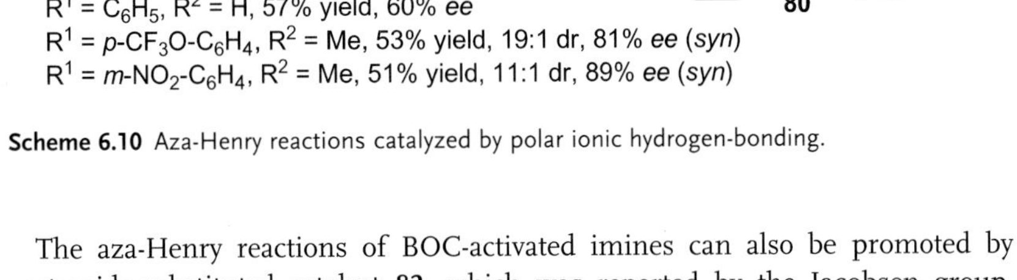

$R^1 = C_6H_5$, $R^2 = H$, 57% yield, 60% *ee*
$R^1 = p\text{-}CF_3O\text{-}C_6H_4$, $R^2 = Me$, 53% yield, 19:1 dr, 81% *ee* (*syn*)
$R^1 = m\text{-}NO_2\text{-}C_6H_4$, $R^2 = Me$, 51% yield, 11:1 dr, 89% *ee* (*syn*)

Scheme 6.10 Aza-Henry reactions catalyzed by polar ionic hydrogen-bonding.

The aza-Henry reactions of BOC-activated imines can also be promoted by acetamide-substituted catalyst **82**, which was reported by the Jacobsen group. The reactions were carried out at 4 °C using 10 mol% of the catalyst along with Hünig's base and powdered 4 Å MS [76]. High enantioselectivities (92–97%) were obtained with a number of heteroaryl- and substituted-aryl aldimine derivatives, bearing both electron-withdrawing and electron-donating substituents (Table

Table 6.24 Thiourea-catalyzed additions of nitroalkanes to aldimines.

Entry	R¹	R²	Yield [%]	dr (syn:anti)	syn ee [%]
1	Ph	Me	96	15:1	92
2	Ph	Et	99	7:1	95
3	Ph	TBSOCH$_2$	85	4:1	95
4	p-Cl-C$_6$H$_4$	Me	98	7:1	95
5	p-MeO-C$_6$H$_4$	Me	95	16:1	96
6	2-furyl	Me	95	6:1	97

6.24). Significant levels of *syn* diastereoselectivities (5:1 to 16:1) were observed for all substrates, with the exception of an *ortho*-chloro-substituted aryl imine, which provided only 2:1 *syn* selectivity. The catalyst was viable for a variety of nitroalkanes, and afforded adducts in uniformly high enantioselectivities (92–95% *ee*). The sense of enantiofacial selectivity in this reaction is identical to that reported for the thiourea-catalyzed Strecker (see Scheme 6.8) and Mannich (see Tables 6.18 and 6.22) reactions, suggesting a commonality in the mode of substrate activation. The asymmetric catalysis is likely to involve hydrogen bonding between the catalyst and the imine or the nitronate, or even dual activation of both substrates. The specific role of the 4 Å MS powder in providing more reproducible results remains unclear, as the use of either 3 Å or 5 Å MS powder was reported to have a detrimental effect on both enantioselectivities and rates of reaction.

6.6
Acyl Pictet–Spengler Reactions of Iminium Ions

The Pictet–Spengler reaction, the cyclization of an electron-rich aryl or heteroaryl group onto an imine electrophile, is the established method for the synthesis of tetrahydroisoquinoline and tetrahydro-β-carboline ring systems. Catalytic asymmetric approaches to these synthetically important structures are mostly restricted to asymmetric hydrogenations of cyclic imines [77, 78]. In a noteworthy

6.6 Acyl Pictet–Spengler Reactions of Iminium Ions

Table 6.25 Thiourea-catalyzed acyl-Pictet–Spengler reactions of imines.

Entry	R	R^1	Temp. [°C]	Yield [%]	ee [%]
1	H	$CH(CH_2CH_3)_2$	−30	65	93
2	H	$n\text{-}C_5H_{11}$	−60	65	95
3	H	$CH_2CH_2OTBDPS$	−60	77	90
4	5-OMe	$CH(CH_2CH_3)_2$	−40	81	93

contribution to the available methodology, the Jacobsen group reported a catalytic asymmetric route to tetrahydro-β-carbolines through hydrogen bond activation of substituted acylimine intermediates using the thiourea catalyst **84** [79]. Typically, racemic Pictet–Spengler reactions are carried out with imines using strong Brønsted acids, often at elevated temperatures. In initial studies, these authors found that catalyst **84** was incapable of promoting the cyclization of imines, most likely due to the low reactivity of simple imines. On the other hand, cyclization of the more electrophilic N-acyl-imine was effectively catalyzed by the pyrrole-derivatized thiourea **84**. Such asymmetric acyl-Pictet–Spengler reactions provided adducts in reasonable yields (65–81% over two steps) and good to high enantioselectivities (85–95% ee; Table 6.25). (For experimental details see Chapter 14.3.1). Catalyst **84** tolerated methoxy substitution at the indole 5- or 6-position, which is of interest given the prevalence of this substitution pattern in indole alkaloids. The ability of the thiourea catalyst to activate the weakly Lewis basic N-acyliminium ion toward cyclization raises interesting questions regarding the exact nature of the hydrogen-bonding interaction between the catalyst and substrate.

Discussion Jacobsen's incorporation of an unsymmetrical pyrrole subunit in catalyst **84** played an important role in achieving the reported levels of enantioselectivity. Pyrroles having symmetric substitution gave significantly lower product enantiopurities under the unoptimized conditions used during the catalyst structure–activity studies. Catalyst **84** is reported to be recoverable in essentially quantitative yield via column chromatography, with no loss in catalyst activity.

6.7
Aza-Friedel–Crafts Reactions of Aldimines

The Friedel–Crafts reaction, the electrophilic substitution of an alkenyl or aryl hydrogen with an electrophile, is one of the most general transformations in organic chemistry. With an imine as the electrophile, the reaction is called the aza-Friedel–Crafts, and provides useful, nitrogen-substituted chiral compounds. The first report of a hydrogen-bond promoted asymmetric aza-Friedel–Crafts reaction came from Terada and co-workers, who showed that 2 mol% of chiral phosphoric acid **86** catalyzes the reactions of commercially available 2-methoxyfuran with N-Boc aldimines (**61**) [80]. Under optimized conditions, the reaction gave consistently high yields (80–95%) and enantioselectivities (86–97% *ee*) for a range of aryl aldimine substrates, possessing either electron-withdrawing or electron-donating groups (Table 6.26). (For experimental details see Chapter 14.9.15). A slightly lower *ee*-value (86%) was observed for the product of the reaction with 2-furaldimine. The absolute configuration of these products was opposite to that

Table 6.26 1,2-Aza-Friedel–Craft reactions of 2-methoxyfuran and aldimines.

Entry	R	Yield [%]	ee [%]
1	Ph	95	97
2	p-MeO-C$_6$H$_4$	95	96
3	m-Br-C$_6$H$_4$	89	96
4	o-Me-C$_6$H$_4$	84	94
5	2-naphthyl	93	96

found in the Mannich reaction which the authors had earlier developed (see Table 6.19). As both processes are expected to involve substrate activation through hydrogen bond interaction with the chiral phosphoric acid catalyst, the difference in the outcome is intriguing and remains to be fully understood. The potential synthetic utility of the furan-2-ylamine products was demonstrated by the transformation of one of the adducts (entry 1, Table 6.26) to the corresponding γ-butenolide in two steps and 86% overall yield, with negligible loss of optical purity (96% ee).

6.8
Hydrophosphonylation Reactions of Aldimines

Nucleophilic additions of phosphites to imines provides direct access to α-amino phosphonic acid derivatives, some of which have been found to possess useful biological activities. Jacobsen and co-workers reported highly enantioselective syntheses of these compounds through the reactions of N-benzyl imines (**50**) with di-(2-nitrobenzyl) phosphite (**88**) in the presence of a chiral thiourea catalyst **52** [81]. High yields (52–93%) and enantioselectivities (typically >90% ee) were obtained using a broad range of aromatic, heteroaromatic, and aliphatic imines (Table 6.27). (For experimental details see Chapter 14.7.1). In general, the highest reac-

Table 6.27 Thiourea-catalyzed hydrophosphonylations of N-benzyl imines.

Entry	R	Temp. [°C]	Time [h]	Yield [%]	ee [%]
1	Ph	4	72	87	98
2	3-pentyl	4	24	90	96
3	p-MeO-C$_6$H$_4$	4	48	90	96
4	m-Cl-C$_6$H$_4$	23	48	83	98
5	3-pyridyl	23	48	77	96

tion rates were obtained with aliphatic imines, while electron-poor aromatic imines typically required longer reaction times or increased reaction temperatures. The hydrophosphonylation products were successfully subjected to global benzyl deprotection under mild hydrogenolytic conditions to give α-amino phosphonic acids in high yields (>87%), with little or no erosion of optical purities at the α-stereocenter.

Akiyama and co-workers reported similar transformations utilizing chiral phosphoric acid **68**, derived from (R)-BINOL. The catalyzed additions of diisopropyl phosphite to p-anisidine aldimines (**90**) took place at room temperature in m-xylene, and provided the corresponding α-amino phosphonates (**91**) in generally high yields (72–97%) with moderate to high enantioselectivities (52–90% ee) for a range of aryl- and cinnamyl-derived aldimines (Scheme 6.11) [82]. Based on experimental observations, the authors proposed a reaction mechanism wherein hydrogen-bonding of the imine to the phosphoric acid hydrogen, coupled with activation of the phosphite hydrogen by the phosphoryl oxygen of the catalyst, organizes the reactants in close proximity, allowing for the observed facial selectivities (Fig. 6.12).

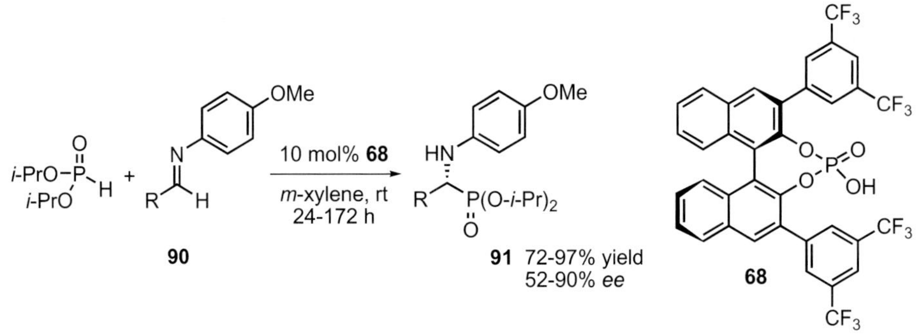

Scheme 6.11 Phosphoric acid-catalyzed hydrophosphonylations of imines.

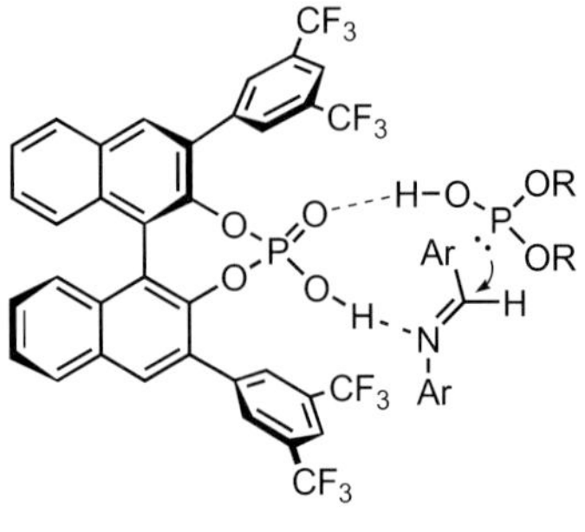

Fig. 6.12 Akiyama's hydrophosphonylation selectivity model.

Discussion Catalyst **52** is prepared from Boc-(L)-*tert*-leucine in five steps, with a 75% overall yield [41]. Details of imine and phosphite preparation are also provided by Jacobsen and co-workers [81]. The hydrophosphonylation reactions as reported by Jacobsen can be carried out without any special precautions, in unpurified commercial diethyl ether (Et_2O) and under an ambient atmosphere. A reduction in temperature was shown to have a beneficial effect on product enantiopurities, but with a decrease in reaction rates. Unbranched aliphatic aldehydes were incompatible with the reaction conditions as reported, due to their rapid decomposition prior to phosphonylation. Although phosphite ester groups that are more electron-withdrawing than o-nitrobenzyl significantly increase the overall reaction rates, products are obtained with diminished optical purities, possibly due to a retro-addition pathway.

6.9
Direct Alkylation Reactions of α-Diazoesters

Diazoacetates are commonly used for the formation of aziridines from imines under Lewis or Brønsted acidic conditions – a process known as the aza-Darzens reaction. A useful twist on the reaction is achieved if the 1,2-addition intermediate undergoes deprotonation of the α proton prior to intramolecular aziridine formation with N_2 extrusion. Such an interrupted aza-Darzens reaction accomplishes a

Table 6.28 Brønsted acid-catalyzed direct alkylations of α-diazoesters.

Entry	R	Yield [%]	ee [%]
1	Ph	81	97
2	*p*-F-C_6H_4	74	97
3	*p*-Ph-C_6H_4	71	97
4	*o*-MeO-C_6H_4	85	91
5	*m*-F-C_6H_4	84	93

net nucleophilic addition of the diazo compound to the imine, producing a β-amino diazoester with a new chiral center. Terada and co-workers found that 2 mol% of an achiral phosphoric acid catalyst promoted the clean addition of ethyl diazoacetate to the N-acylimine of benzaldehyde (70% yield, 1 h) [83]. Significantly, the use of chiral phosphoric acid **93**, also at 2 mol% loading, catalyzed the direct alkylations of *tert*-butyl diazoacetate to yield substituted β-amino-α-diazoester products (**94**) in good yields (62–89%) and high enantioselectivities (91–97% *ee*) for a range of substituted arylaldimines (Table 6.28). (For experimental details see Chapter 14.6.1). The bulky diazoacetate ester group and the electron-donating *para*-dimethylamine substituent on the acyl group of **92** were important in establishing high enantiofacial selectivity. While its precise role is unclear, it is likely that, through hydrogen bonding, the catalyst is able not only to activate the imine to nucleophilic addition but also to lower the nucleophilicity of the intermediate amide nitrogen that results from addition of the diazoester. The latter interaction may account for the complete selectivity for alkylation versus aziridine formation. Terada also demonstrated the utility of the β-amino-α-diazoester products by their conversion to chiral β-amino acids via a high-yielding three-step process.

6.10
Imine Amidation Reactions

The addition of amide nucleophiles to activated imines provides access to a class of unsymmetrical aminals. Such compounds have been incorporated into peptide chains for the synthesis of *retro-inverso* peptide mimics, which appear to have a range of biochemical applications [84]. These amide addition reactions typically require a doubly activated imine, with electron-withdrawing groups at both the carbon and nitrogen of the imine substrate. Antilla and co-workers reported an asymmetric catalytic version of this reaction, involving the addition of sulfonamides to N-Boc imines [85]. Preliminary studies showed that amide additions to N-Boc aldimines (**61**) using 2 mol% of an achiral phenyl phosphinic acid gave high yields of racemic aminals. Screening of chiral catalysts and amide substrates revealed that the amidation reactions using *p*-toluenesulfonamide in the presence of the highly hindered chiral VAPOL (vaulted biphenanthrol)-derived phosphoric acid catalyst **95** provided N,N-aminal products in high yields (88–95%) and *ee*-values (87–94%) for a variety of *para*-substituted arylaldimines as well as 2-thienylimine (Table 6.29). (For experimental details see Chapter 14.7.2). The nature of the amide nucleophile was critical, and the use of common amides, such as acrylamide or acetamide, gave the corresponding products in low enantiopurities (14 and 12% *ee* respectively, compared to 94% *ee* when using a sulfonamide). Catalyst **95** presumably promotes asymmetric amidation by activation of the imine substrate, and no addition product was observed under otherwise identical conditions lacking the catalyst. The use of rigorously dried catalyst, imine, and sulfonamide was critical for obtaining reproducible *ee*-values for the aminal products.

Table 6.29 Phosphoric acid-catalyzed imine amidation reactions.

Entry	R	95 [mol%]	Solvent	Time [h]	Yield [%]	ee [%]
1	Ph	5	Et$_2$O	1	95	94
2	p-CF$_3$-C$_6$H$_4$	10	toluene	20	99	99
3	p-MeO-C$_6$H$_4$	10	toluene	17	92	90
4	2-thienyl	10	toluene	17	94	87

6.11
Transfer Hydrogenation Reactions of Imines

The catalytic, asymmetric hydrogenations of alkenes, ketones and imines are important transformations for the synthesis of chiral substrates. Organic dihydropyridine cofactors such as dihydronicotinamide adenine dinucleotide (NADH) are responsible for the enzyme-mediated asymmetric reductions of imines in living systems [86]. A biomimetic alternative to NADH is the Hantzsch dihydropyridine, **97**. This simple compound has been an effective hydrogen source for the reductions of ketones and alkenes. A suitable catalyst is required to activate the substrate to hydride addition [87–89]. Recently, two groups have reported, independently, the use of **97** in the presence of a chiral phosphoric acid (**68** or **98**) catalyst for the asymmetric transfer hydrogenation of imines.

Rueping and co-workers reported the enantioselective reductions of ketoimines (**99**) using **97** and a catalytic amount of **68**, originally prepared by Akiyama and Terada [90]. In general, good ee-values (68–84%) and moderate to high yields (46–91%) were observed for the reductions of several N-aryl-ketoimines derived from methyl-aryl ketones (Scheme 6.12). In several instances, the ee-value of the products could be improved to >90% upon recrystallization from methanol. Lower reaction temperatures significantly impacted conversions. Activation of the substrate is proposed to occur via imine protonation, generating an iminium ion that is expected to be strongly bound to the chiral catalyst counterion.

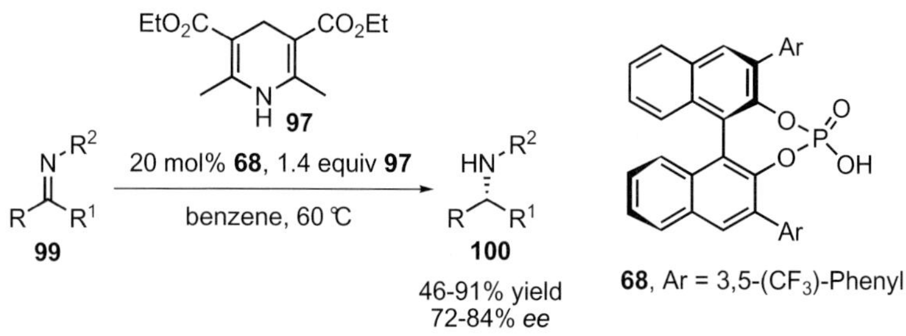

Scheme 6.12 Phosphoric acid-catalyzed transfer hydrogenations of imines.

List and co-workers reported a closely related asymmetric reduction, that of ketone-derived p-methoxyphenyl (PMP) imines (**101**), using **97** in the presence of chiral Brønsted acid catalyst **98** [91]. Aryl-methyl ketoimines, possessing various o-, m-, and p-substituents on the aryl portion, were reduced at 35 °C in high yields (84–96%) and good to high enantioselectivities (80–93% ee; Table 6.30). (For experimental details see Chapter 14.21.2). It is worth noting that the phosphoric acid catalyst, with its considerable steric demands (vs. **68**), is used at only 1 mol% loading. The asymmetric reduction of an aliphatic imine gave the corresponding amine product in 90% ee. The utility of the methodology was highlighted by the two-step conversion of acetophenone to the corresponding chiral amine in 81% overall yield (88% ee). The imine was formed *in situ* in the presence of 4 Å MS and subjected to asymmetric reduction. The PMP group was then removed under oxidative conditions to give the desired amine product.

Table 6.30 Phosphoric acid-catalyzed transfer hydrogenations of imines.

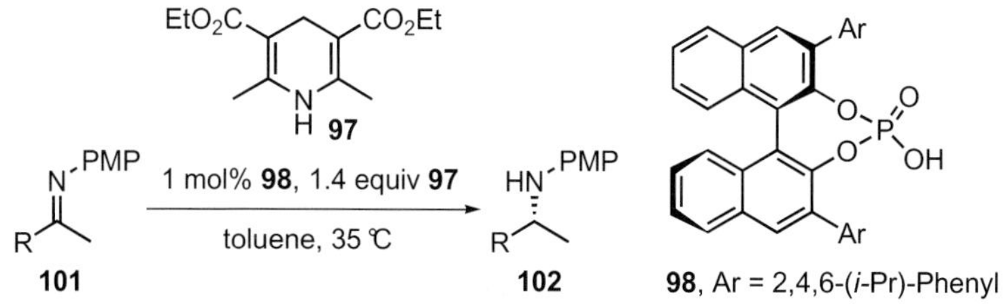

Entry	R	Time [h]	Yield [%]	ee [%]
1	Ph	45	96	88
2	p-(NO_2)-C_6H_4	42	96	80
3	2,6-$(Me)_2$-C_6H_3	71	88	92
4	i-propyl	60	80	90

6.12
Morita–Baylis–Hillman Reactions

The Morita–Baylis–Hillman (MBH) reaction – the reaction of an electron-deficient alkene with an aldehyde or an imine (aza-MBH) – provides a convenient route to highly functionalized allylic alcohols and amines. Although the reaction is catalyzed by simple amines or phosphines, its scope is somewhat limited due to slow reaction rates. The MBH reaction is discussed fully in Chapter 5, and consequently only those contributions that clearly involve hydrogen bond interactions will be described here.

In an early report, the Ikegami group showed that mild Brønsted acids such as phenol and BINOL have dramatic rate-increasing effects on the tri-*n*-butylphosphine-promoted MBH reactions of cyclic enones with aldehydes [92]. Over the past few years, several chiral hydrogen bond donor catalysts have been developed for the asymmetric MBH and aza-MBH reactions. One of the first practical catalytic asymmetric MBH reactions was reported by the Hatakeyama group, who utilized β-isocupreidine (**103**) for the additions of 1,1,1,3,3,3-hexafluoroisopropyl acrylate (HFIPA) to aldehydes [93]. A survey of cinchona alkaloid-derived catalysts identified **103**, which contains a rigid quinuclidine ether moiety and a C6′-hydroxy group, as the optimal catalyst for promoting the addition of the electron-deficient HFIPA to a range of aryl, aliphatic, and α,β-unsaturated aldehydes (**40**). The (α-methylene-β-hydroxy)ester products (**104**) were obtained in moderate yields (31–58%) and high *ee*-values (91–99%), but were accompanied by a dioxanone side product (**105**; Table 6.31). The high reac-

Table 6.31 Modified cinchona-catalyzed Morita–Baylis–Hillman reactions.

Entry	R	Time [h]	Yield [%] (% ee)	
			104	105
1	p-NO$_2$-C$_6$H$_4$	1	58 (91)	11 (4)
2	(E)-Ph-CH=CH	72	50 (92)	–
3	(CH$_3$)$_2$CHCH$_2$	4	51 (99)	18 (85)
4	cyclohexyl	72	31 (99)	23 (76)

tion rates observed using catalyst **103**, compared to those found with similar (but unconstrained) cinchona derivatives, presumably reflects the increased catalyst nucleophilicity achieved by restraining the steric congestion around the quinuclidine nitrogen. (For experimental details see Chapter 14.10.1).

Schaus and co-workers realized a novel strategy for the catalysis of MBH reactions. Whereas many previous workers had used chiral nucleophilic catalysts, in which chirality transfer takes place after formation of a covalent bond to the substrate, these authors used a chiral catalyst to activate and desymmetrize the substrate through noncovalent interactions. Building on the Ikegami precedent, the authors discovered that chiral BINOL catalyst **106** and an excess of triethylphosphine (PEt_3) promoted highly enantioselective MBH reactions of cyclohexanone and various substituted aldehydes [94]. The catalyst not only induces asymmetry, but also accelerates the reactions. For example, whereas the reaction of cyclohexanone and 3-phenylpropanal using 50 mol% PEt_3 yielded only 5% of the desired product, the use of 2 mol% (*R*)-BINOL afforded the desired product in 74% yield and 32% ee, over the same reaction time. Optimum results were obtained with H_8-BINOL-derived catalysts **106a** and **106b**, which afforded MBH products of cyclohexanone and aliphatic aldehydes in very good yields (70–88%) and enantioselectivities (82–96% ee; Table 6.32). (For experimental details see Chapter 14.10.3). Aryl and α,β-unsaturated aldehydes gave products in low conversions (39–40% yield) and only moderate enantioselectivities (67–80% ee). Partial saturation of the BINOL backbone, 3,3'-aryl-substitution, and the diol functionality were all critical for inducing high levels of enantioselectivity.

Table 6.32 BINOL-catalyzed Morita–Baylis–Hillman reactions of cyclic enones.

106a: Ar = 3,5-bis(CF_3)phenyl
b: Ar = 3,5-dimethylphenyl

Entry	R	Catalyst	Yield [%]	ee [%]
1	$PhCH_2CH_2$	106a	88	90
2	hexen-3-yl	106b	72	96
3	cyclohexyl	106b	71	96
4	*i*-Pr	106b	82	95
5	Ph	106b	40	67

6.12 Morita–Baylis–Hillman Reactions

Nagasawa and co-workers reported the use of a chiral bis-thiourea catalyst (**108**) for the asymmetric MBH reactions of cyclohexenone with aldehydes [95]. Since others had already shown that thioureas form hydrogen bonds with both aldehydes and enones, it was hypothesized that the inclusion of two thiourea moieties in close proximity on a chiral scaffold would organize the two partners of the MBH reaction and lead to enantiofacial selectivity. Initial studies showed that the achiral 3,5-bis-(trifluoromethyl)phenyl-substituted urea increased the rate of MBH reaction between benzaldehyde and cyclohexenone. These authors then showed that chiral 1,2-cyclohexyldiamine-linked bis-thiourea catalyst **108**, used at 40 mol% loading in the presence of 40 mol% DMAP, promoted the MBH reactions of cyclohexenone with various aliphatic and aromatic aldehydes (**40**) to produce allylic alcohols in moderate to high yields (33–99%) and variable enantioselectivities (19–90% *ee*; Table 6.33).

The Hatakeyama group later reported the use of catalyst **103** for the asymmetric aza-MBH reactions of HFIPA with activated aromatic imines [96]. The aza-MBH reactions of four different diphenylphosphinoyl aryl imines (**109**) with HFIPA were promoted using 10 mol% **103** and afforded the corresponding α-methylene-β-amino acid derivatives (**110**) in reasonable yields (42–97%) and moderate *ee*-values (54–73%; Scheme 6.13). Aliphatic imines were not suitable substrates for the reaction due to imine lability. (For experimental details see Chapter 14.10.3).

Table 6.33 Bis-thiourea-catalyzed Morita–Baylis–Hillman reactions of cyclic enones.

Entry	R	Yield [%]	*ee* [%]
1	Ph	88	33
2	p-CF$_3$-C$_6$H$_4$	99	33
3	CH$_3$(CH$_2$)$_5$	63	60
4	i-Pr	67	60
5	cyclohexyl	72	90

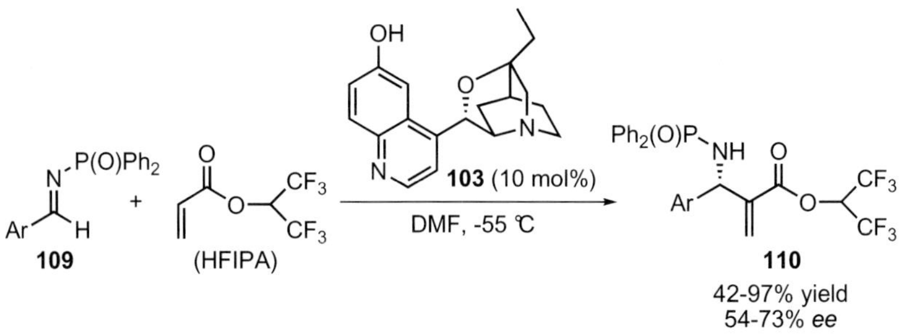

Scheme 6.13 β-Isocupreidine-catalyzed aza-MBH reactions of aldimines.

Sasai and co-workers reported the development of aza-MBH reactions catalyzed by pyridine-substituted BINOL **113**, which incorporated the chiral hydrogen bond donor and the nucleophile into a single catalyst [97]. This bifunctional catalyst positions the reaction partners in close proximity, so it was hoped that with proper scaffold design, high enantioselectivities may be achievable. Indeed, the aza-MBH reactions of vinyl ketones (**111**) and tosylimines (**112**), carried out using 10 mol% of **113** at −15 °C in a mixture of toluene and cyclopentyl methyl ether (CPME), provided the expected allylic amine products (**114**) in high yields (88–100%) and high enantioselectivities (87–95% *ee*; Table 6.34). (For experimental details see Chapter 14.10.4). In this reaction, it is likely that hydrogen bonding by the two phenol hydrogens activates the enone, which is then attacked at the β-position by the proximal aminopyridine. The resulting intermediate is held in a coordinated chiral environment, as required for enantioselectivity in the subsequent aldol addition.

6.13
Cycloaddition Reactions

Cycloaddition reactions, namely those occurring through a [4π+2π] concerted transition state, give rise to densely functionalized adducts, with multiple stereocenters set in a single step. Many Lewis acid metal catalysts have been developed over the past few decades, allowing for ready access to a broad range of highly functionalized Diels–Alder (DA) and hetero-Diels–Alder (HDA) adducts in high regio-, enantio-, and diastereoselectivities. Catalytic enantioselective cycloaddition reactions have proven to be exceedingly useful as key steps in the syntheses of a number of complex natural products [98, 99]. Over the past decade, there has been growing interest in the development of organocatalysts capable of promoting asymmetric cycloaddition reactions, and most of the successes have been realized using a secondary amine catalyst, which condenses with the dienophile carbonyl and thus serves as an *in-situ* chiral auxiliary [100]. More recently, several laboratories have reported hydrogen bonding-promoted asymmetric cycloadditions, and these results are summarized below, starting with HDA reactions.

Table 6.34 Bifunctional BINOL-promoted aza-MBH reactions of aldimines.

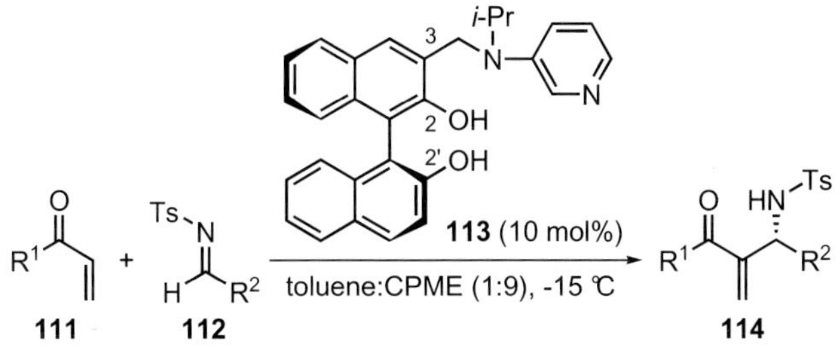

Entry	R¹	R²	Time [h]	Yield [%]	ee [%]
1	Me	Ph	168	93	87
2	Me	p-MeO-C$_6$H$_4$	132	93	94
3	Me	2-furyl	48	100	88
4	Et	p-NO$_2$-C$_6$H$_4$	96	88	88
5	H	p-NO$_2$-C$_6$H$_4$	36	95	94

6.13.1
Hetero Diels–Alder (HDA) Reactions

The HDA reaction allows for rapid access to chiral six-membered heterocyclic structures that serve as valuable intermediates in organic synthesis. The first highly enantioselective HDA reaction promoted by a chiral hydrogen bond donor was reported from the Rawal laboratory. While investigating the cycloaddition reactions of amino-siloxy diene **115**, it was observed that this diene was exceptionally reactive to heterodienophiles, and underwent HDA reactions with various aldehydes at room temperature, even in the absence of any added catalyst (Scheme 6.14). Subsequent treatment of the intermediate cycloadducts (**116**) with acetyl chloride afforded the corresponding dihydro-4-pyrones (**117**) in good overall yields [101]. Further studies of this reaction revealed a pronounced solvent effect,

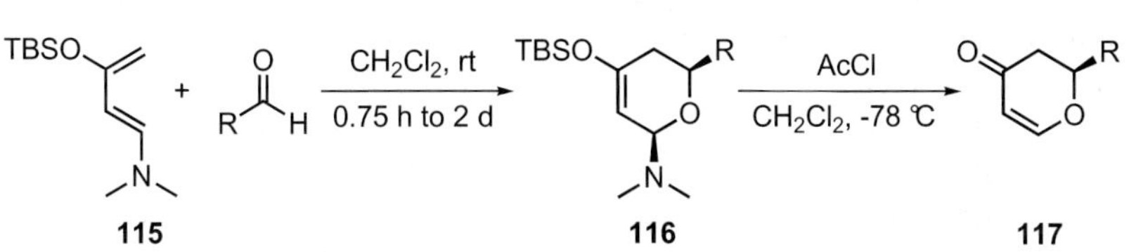

Scheme 6.14 Uncatalyzed HDA reactions of amino-siloxy diene with aldehydes.

such that the cycloaddition reactions were significantly faster in chloroform (~10–30-fold rate increase) than in other common aprotic organic solvents, and even polar solvents. The rate difference was greater still in alcoholic solvents [38]. The higher rates in chloroform and alcohols were attributed to hydrogen bond activation of the hetero-dienophile, which lowers its LUMO energy and, thereby, reduces the HOMO-LUMO gap for cycloaddition. Remarkably, alcohols were even found to promote the HDA reactions of simple ketones, known to be notoriously poor heterodienophiles (Scheme 6.15) [38].

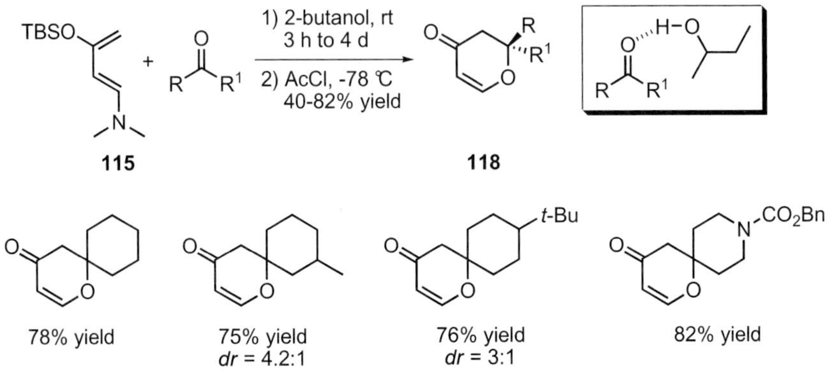

Scheme 6.15 Alcohol catalysis of HDA reactions of diene **115** with ketones.

An examination of several chiral alcohols as hydrogen bond donor catalysts led to the identification of taddols as highly effective organic catalysts for the enantioselective HDA reactions between electron-rich dienes and aldehydes. These tartrate-derived diols were popularized by Seebach and co-workers as ligands for metal-based catalysis [102]. Additionally, Toda and co-workers had shown that taddols, when used in stoichiometric quantities, were effective asymmetry-inducing agents for solid-state photochemical reactions and for resolutions [103]. The use of taddol as a substoichiometric catalyst was first noted by the Rawal group. These authors found that commercially available 1-naphthyl-taddol (**119**), when used at 20 mol% loading, was highly effective as a hydrogen bond donor catalyst for the HDA reactions between **115** and various aromatic aldehydes, affording the expected dihydropyranone products (**117**) on treatment with acetyl chloride in good overall yields (67–97%) and excellent enantioselectivities (92–>98% *ee*; Table 6.35) [104]. (For experimental details see Chapter 14.18.4). Aliphatic aldehydes also proved to be suitable dienophiles, with only a small decrease observed in enantioselectivities. A model that rationalizes the observed selectivities for taddol-catalyzed HDA reactions is presented at the end of this section.

Taddols can also catalyze the HDA reactions of other electron-rich dienes, as demonstrated by Ding and co-workers, who showed that taddol **119** catalyzed the cycloaddition of Brassard's diene with various aromatic aldehydes to give

Table 6.35 Taddol-catalyzed HDA reactions of **115** and aromatic aldehydes.

119 Ar = 1-naphthyl

Reaction conditions:
i) **119** (20 mol%), toluene, -78 °C, 2 d
ii) AcCl, CH$_2$Cl$_2$/toluene, -78 °C, 30 min

115 + H(C=O)R → **117**

Entry	R	Yield [%]	ee [%]
1	Ph	70	>98
2	4-MeO-C$_6$H$_4$	68	>94
3	1-naphthyl	69	99
4	cyclohexyl	64	86

δ-lactone products directly on reaction work-up [105]. The products were obtained in good yields (up to 85%) and good to high enantioselectivities (69–91% ee; Table 6.35).

Through a collaborative effort, the Rawal and Yamamoto groups developed chiral 1,1'-biaryl-2,2'-dimethanols (e.g., **121**, BAMOLs) as hydrogen bond donor catalysts. Structurally different from taddols, the axially chiral BAMOL scaffold allowed for hydrogen-bonding activation similar to taddols, but presented greater opportunities for tuning the steric and electronic properties of the catalyst. Several variations of these novel diols were prepared and examined as catalysts for the HDA reactions between 1-amino-3-siloxybutadienes and various aldehydes [106, 107]. Of the BAMOLs examined, compounds **121a** and **121b** were shown to be the most effective catalysts for the HDA reactions of both aromatic and aliphatic aldehydes, affording cycloadducts in moderate to high yields (67–99%) and generally excellent enantioselectivities (84–>99% ee; Table 6.37). Notably, high ee-values were obtained for acetaldehyde and 2-butynal (97% and 98% respectively), both of which contain small α-substituents. A crystal structure of 1,1'-biphenyl-BAMOL and benzaldehyde revealed the presence of an intramolecular hydrogen bond between the two hydroxyl groups of the catalyst, and also an intermolecular hydrogen bond to the carbonyl of benzaldehyde, similar to that observed in the crystal structure of a taddol-aldehyde complex (see Fig. 6.14).

Sigman and co-workers developed a novel oxazoline-based hydrogen-bonding catalyst (**122**) for HDA reactions. The modular nature of the catalyst design al-

Table 6.36 Taddol-catalyzed HDA reactions of Brassard's diene.

Entry	R	Yield [%]	ee [%]
1	Ph	67	83
2	m-Br-C$_6$H$_4$	67	89
3	m-MeO-C$_6$H$_4$	45	91
4	(CH$_2$)$_2$Ph	50	69

Table 6.37 Taddol-catalyzed HDA reactions of 1-amino-3-siloxydienes.

121a Ar = 4-F-3,5-Me$_2$C$_6$H$_2$
121b Ar = 4-F-3,5-Et$_2$C$_6$H$_2$

Entry	Catalyst	R	Yield [%]	ee [%]
1	121b	Me	75	97
2	121a	PhS(CH$_2$)$_2$	76	94
3	121b	2-(NO$_2$)-C$_6$H$_4$	93	98
4	121b	2-furyl	96	>99

Table 6.38 HDA reactions catalyzed by a modified oxazoline.

Entry	R	Temp. [°C]	Yield [%]	ee [%]
1	Ph	−55	62	90
2	4-MeO-C$_6$H$_4$	−30	80	91
3	1-naphthyl	−55	72	90

lowed for easy variation of the nature and relative configuration of the chiral centers, as well as for steric and electronic tuning of different parts of the scaffold [108]. Aromatic aldehydes were shown to be suitable substrates, providing the corresponding dihydropyranones in reasonable yields (42–80%) with enantioselectivities of up to 92% in the presence of 20 mol% **122** (Table 6.38). Electron-rich aryl aldehydes required an increase in reaction temperature to obtain good yields.

Mikami and Jørgensen independently reported the use of bis-triflamide catalysts (**27a** and **27b**) for the HDA reactions of Danishefsky's diene (**123**) with various reactive aldehydes and ketones. Jørgensen and co-workers found that, in the presence of bis-nonaflamide catalyst **27b**, *p*-substituted aromatic aldehydes and alkyl pyruvates gave dihydropyranone products with enantioselectivities of up to 73% (Table 6.39) [109]. The Mikami group found that the reactions of Danishefsky's diene with glyoxalates and aromatic glyoxals can be catalyzed with bis-triflylamide catalyst **27a** to give products with enantioselectivities up to 87% [110]. A ^{1}H NMR study of **27a** coordinated to a symmetrical ketone indicated a dual hydrogen-bonding interaction between the two amide hydrogens and the carbonyl oxygen.

Discussion The ready commercial availability of **119** and operational ease of the reactions make the taddol-catalyzed HDA reactions attractive processes. A rationale has been put forth by Rawal et al., for the high enantioselectivities observed in the taddol-catalyzed HDA reactions. From the many crystal structures in the literature, it was recognized that taddols are locked in well-defined conformations

Table 6.39 Bis-triflylamide-catalyzed HDA reactions of Danishefsky's diene.

R^2	R	Catalyst	Solvent	Time [h]	Temp. [°C]	Yield [%]	ee [%]
TMS	p-NO$_2$-C$_6$H$_4$	27b	CHCl$_3$	16	−24	76	50
TMS	p-CN-C$_6$H$_4$	27b	CHCl$_3$	16	−40	74	49
TIPS	COOn-Bu	27a	toluene	0.5	−78	87	86
TIPS	COPh	27a	toluene	0.5	−78	67	87

in the solid state, with one of the hydroxyls internally hydrogen bonded to the other (Fig. 6.13). This arrangement not only positions the opposing pairs of aryl groups in pseudo-axial and pseudo-equatorial orientations, but also renders the remaining hydroxyl more acidic and available for intermolecular hydrogen bonding. Going on the assumption that the same arrangement is likely in solution, the more-acidic proton is expected to form a hydrogen bond to the carbonyl group, and thereby lower its LUMO energy. The facial selectivity was rationalized by suggesting that the electron-deficient π-bond of the carbonyl group may be further stabilized by an electrostatic, donor–acceptor interaction with the electron-rich distal ring of the naphthalene unit. This stabilizing interaction serves not only to increase the population of the activated aldehyde, but also to block one face of the aldehyde to nucleophile attack. In support of some of the tenets of this hypothesis, the authors later reported the crystal structure of a complex between taddol **117** and p-anisaldehyde (Fig. 6.14) [112].

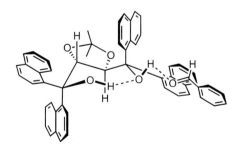

Fig. 6.13 Internally hydrogen-bonded conformation of 1-naphthyl-taddol.

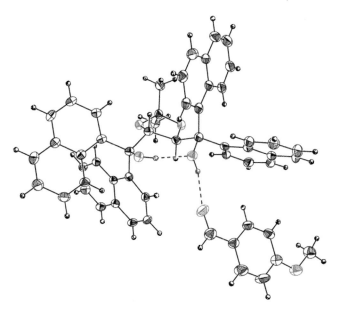

Fig. 6.14 X-ray crystal structure of a 1-naphthyl-taddol/anisaldehyde complex.

6.13.2
Diels–Alder (DA) Reactions

The enantioselective DA reaction is perhaps the most important reaction for complex molecule synthesis, providing access to chiral six-membered carbocyclic compounds containing up to four stereogenic centers in a single step. Early investigations into hydrogen bonding-promoted asymmetric DA reactions were reported by Göbel and co-workers, who used axially chiral amidinium ions **126a** and **126b** for the cycloadditions of diene **124** and diketones (**125**) to produce intermediates for the synthesis of (–)-norgestrel [113]. Under optimized conditions, these amidinium salts induced regioselectivities of up to 2.5:1 in favor of the desired regioisomer, with enantioselectivities ranging from 30 to 50% (Table 6.40). These catalysts are expected to activate the dienophile through hydrogen bonding between the amidinium hydrogen and the diketone carbonyl groups.

The Göbel group later reported the development of a new C_2-symmetric chiral bis-amidinium salt **126c**, which was applied to the same diene-dienophile system [114]. Compound **126c** (1 equiv.) increased the rate of the cycloaddition reactions by more than 3000-fold compared to the non-promoted reactions. Also of interest is the dramatic increase – and reversal – in the regioselectivities of the reactions (Table 6.40).

The first highly enantioselective (>90% *ee*) DA reaction catalyzed by a chiral hydrogen bond donor was reported by the Rawal group. Based on the success realized with the HDA reaction, these authors examined the capacity of taddols to

Table 6.40 Chiral amidinium-catalyzed DA reactions of diketones.

R	Catalyst	Temp. [°C]	Yield [%]	127:128	ee (127) [%]	ee (128) [%]
Me	126a	−27	73	2.8:1	26	33
Et	126b	−27	94	3.2:1	43	50
Me	126c	−80	80	1:22	15	47
Et	126c	−22	100	1:11	48	7

promote the DA reactions between the highly reactive 1-amino-3-siloxybutadiene and various α-substituted acroleins (**129**). An examination of several taddols showed that useful results were obtained simply by using the commercially available 1-naphthyl-taddol (**119**) [115]. This diol was shown to catalyze the DA reactions at low temperature to give the corresponding cycloadducts (**130**), which were isolated as the corresponding cyclohexenones (**131** or **132**, depending on work-up) in good yields (77–85%) and high enantioselectivities (86–92% ee; Table 6.41. (For experimental details see Chapter 14.18.5). The enantioselectivities observed for the DA reactions of acroleins were rationalized through a model which parallels that used for the HDA reactions discussed previously (see Fig. 6.13). Thus, hydrogen bonding-activation of the dienophile carbonyl group by the non-hydrogen-bonded hydroxyl proton is expected to lower the LUMO energy of the dienophile. The proposed donor–acceptor "π-stacking" interaction between the dienophile and the distal aromatic ring of the nearby catalyst naphthalene not only stabilizes the hydrogen-bonded complex, but also blocks one face of the

6.13 Cycloaddition Reactions

Table 6.41 Taddol-catalyzed DA reactions of 1-amino-3-siloxydienes.

Entry	R	Yield of 130 [%]	Yield of 132 [%]	ee of 132 [%]
1	Me	85	83	91
2	i-Pr	77	81	92
3	CH_2Ph	84	82	89

dienophile. This model correctly predicts the major enantiomer of the DA reactions examined by the Rawal group. Optical purities of the cycloadducts were found to increase steadily with decreasing reaction temperatures, consistent with the formation of more persistent hydrogen bond-based catalyst–aldehyde complexes.

Discussion Taddol-catalyzed DA reactions provide simple and direct routes to functionalized cyclohexenones in enantiomerically enriched form. As with the HDA reactions, the best results were obtained when pure diene and dienophile were used, and the reaction temperature was rigorously maintained. Traces of moisture or acid impact negatively upon diene stability and product enantiopurities. One important advantage of the method is the commercial availability of both taddol catalyst **119** and the diene. The best substrates for the taddol-catalyzed DA reactions were α-substituted acroleins. This reactivity profile is complementary to that found for the secondary amine-based organocatalysis developed by MacMillan and co-workers, in which β-substituted acroleins provided the best results [116].

6.14
Aldol and Related Reactions

Hydrogen bond-promoted asymmetric aldol reactions and related processes represent an emerging facet of asymmetric proton-catalyzed reactions, with the first examples appearing in 2005. Nonetheless, given their importance, these reactions have been the subject of investigation in several laboratories, and numerous advances have already been recorded. The substrate scope of such reactions already encompasses the use of enamines, silyl ketene acetals and vinylogous silyl ketene acetals as nucleophiles, and nitrosobenzene and aldehydes as electrophiles.

Yamamoto and co-workers reported highly regio- and enantioselective nitroso aldol reactions between achiral enamines and nitrosobenzene catalyzed by chiral hydrogen bond donors [117]. The regioselectivity (O versus N) of enamine addition to nitrosobenzene was controlled exclusively by the choice of both catalyst and enamine. Cyclic enamines derived from pyrrolidine (133a) and morpholine (133b) with varied ring substitutions gave exclusively O-nitroso adducts in moderate to high yields (64–91%) and good to high enantioselectivities (70–93% ee) when the addition was catalyzed by (S)-1-naphthyl glycolic acid (134; Table 6.42). In contrast, catalysis of the addition reactions of morpholino (133b) and thiomorpholino enamines by taddol 119 gave exclusive formation of N-nitroso adducts in moderate to high yields (63–91%) and good to high ee-values (65–91%;

Table 6.42 (S)-1-Naphthyl glycolic acid-catalyzed O-nitroso aldol reactions.

Entry	Enamine	R, R	Yield [%]	ee [%]
1	133a	H, H	69	70
2	133b	H, H	77	92
3	133b	Me, Me	91	90
4	133b	–(OCH$_2$CH$_2$O)–	83	93

Table 6.43 Taddol-catalyzed N-nitroso aldol reactions.

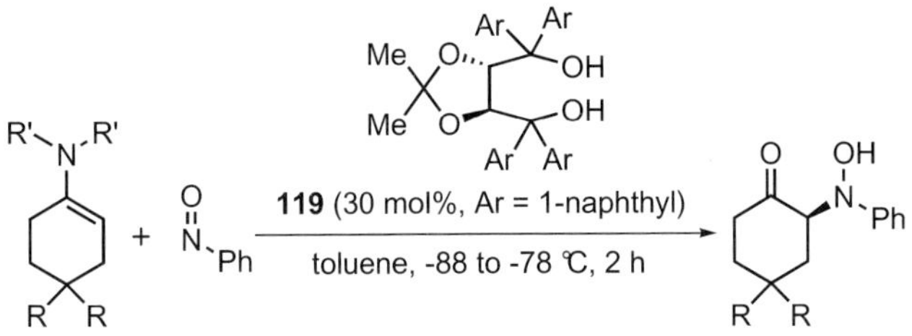

Entry	Enamine	R, R	Yield [%]	ee [%]
1	133b	H, H	81	83
2	133b	Me, Me	78	82
3	133b	–(OCH$_2$CH$_2$O)–	63	91

Table 6.43). These authors proposed activation of nitrosobenzene to aldol addition through N-coordination when using chiral carboxylic acid catalyst **134** and through O-coordination when using the less-acidic taddol catalyst **119** to explain the observed regioselectivities.

Jørgensen and co-workers reported the asymmetric additions of a silyl ketene acetal to aldehydes (**40**) using the chiral bis-sulfonamide catalyst **27** [109]. Among the limited number of aldehydes examined, adducts were obtained in moderate to high yields (41–90%) and modest levels of ee (30–56%; Table 6.44). The corresponding mono-sulfonamide catalyst was inactive under the reported conditions.

Rawal and co-workers reported the highly diastereo- and enantioselective aldol reactions of N,O-silyl ketene acetal **138** with various aldehydes catalyzed by taddol **139** [112]. The aldol reactions proceeded rapidly at low temperature in the presence of 10 mol% of **139**, and generated the corresponding silyl ether adducts. The β-hydroxy amide products were obtained upon treatment of the crude products with HF, which removed the silyl ethers. A wide range of aromatic and heteroaromatic aldehydes were found to be effective partners in the reaction. The aldol adducts were obtained with moderate to high levels of diastereoselectivity (from 2:1 to >25:1) and high enantioselectivities (86–98% ee; Table 6.45). (For experimental details see Chapter 14.1.7). The aldol reaction with n-butyraldehyde gave the corresponding adduct in high ee (91%) but modest yield. The authors also reported mild conditions, using Schwartz's reagent, for the direct reduction of the amide products (**141**) to the corresponding aldehydes (**142**), with minimal loss in product diastereomeric ratios (Scheme 6.16). The high enantioselectivities observed in the above reactions can be understood through the transition-state

Table 6.44 Bis-triflamide-catalyzed Mukaiyama aldol reactions.

Entry	R	Time [h]	Yield [%]	ee [%]
1	p-NO$_2$-C$_6$H$_4$	16	90	43
2	Ph	40	41	30
3	4-pyridine	16	85	56

model proposed by Rawal to explain selectivity in the HDA and DA reactions (*vide supra*) [112, 115]. Thus, the aldehyde carbonyl is activated by a single-point, intermolecular hydrogen bond to the internally hydrogen-bonded taddol, and is further organized through electrostatic interactions with a naphthalene ring of taddol (see Fig. 6.13).

Table 6.45 Taddol-catalyzed Mukaiyama aldol reactions of N,O-acetals.

Entry	R	Yield [%]	dr	ee syn [%]
1	Ph	94	15:1	98
2	p-Cl-C$_6$H$_4$	86	20:1	97
3	m-(MeO)-C$_6$H$_4$	81	13:1	97
4	2-thiophene	88	10:1	95
5	n-propyl	47	9:1	91

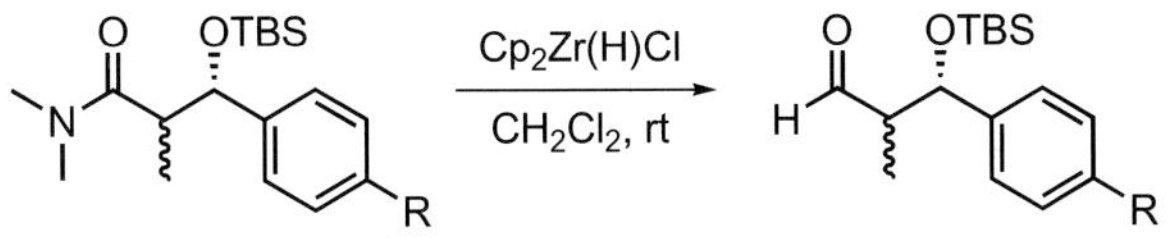

Scheme 6.16

141a: R = H, 35:1 syn:anti
b: R = Cl, 24:1 syn:anti
c: R = OMe, >50:1 syn:anti

142a: 88% yield, 30:1 syn:anti
b: 84% yield, 24:1 syn:anti
c: 85% yield, >50:1 syn:anti

Scheme 6.16 Mild reduction of amides to aldehydes using Schwartz's reagent.

The scope of hydrogen bonding-catalyzed aldol reactions was further extended to the vinylogous Mukaiyama aldol (VMA) reaction. Rawal and co-workers showed that the reactions of silyldienol ether **143** with a range of electron-deficient aldehydes (**40**) can be catalyzed with taddol **119** to give exclusively the δ-addition products in good yields (40–73%) and ee-values (67–90%; Table 6.46) [118]. Hunig's base was used as a scavenger of adventitious water or acid when using glyoxalate substrates in order to suppress a competing background reac-

Table 6.46 Taddol-catalyzed VMA reactions of reactive aldehydes.

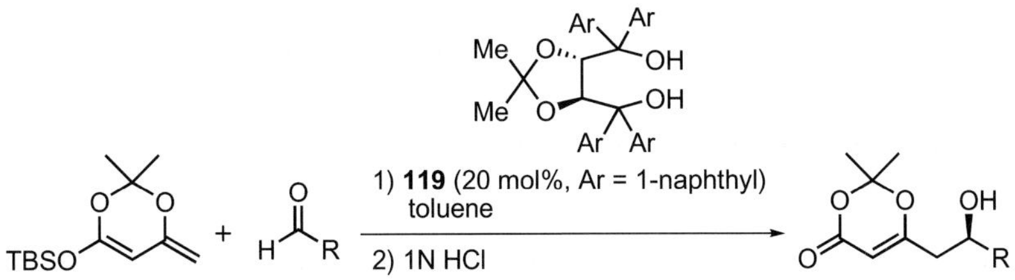

Entry	Aldehyde	Temp. [°C]	Time [h]	Yield [%]	ee [%]
1	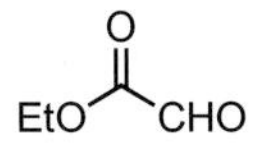 EtO-C(O)-CHO	−80	1	60	87
2	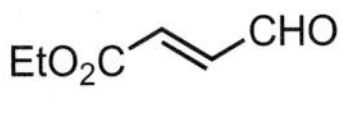 EtO₂C-CH=CH-CHO	−60	120	66	71
3	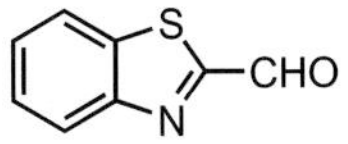 benzothiazole-CHO	−80	98	73	90
4	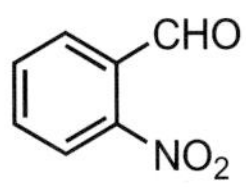 2-NO₂-C₆H₄-CHO	−60	130	58	75

tion. Typical aryl and aliphatic aldehydes were unsuitable substrates under the reported conditions due to prohibitively slow rates of reaction.

6.15
Conclusion and Prospects

Enantioselective reactions have, for decades, been the domain of metal-based Lewis acid catalysts. Hundreds of catalysts, all with a metal or metalloid atom at their center, have been developed to promote a plethora of organic reactions. Within the past ten years, however, there has been a paradigm shift, and metal-free catalysts have been developed to accomplish a variety of enantioselective transformations. Prominent among these organic catalysts are chiral hydrogen bond donors, many of which have been discussed in this chapter. These small organic compounds activate substrates in much the same way as metal-based Lewis acids, the difference being that an electron-deficient hydrogen atom (or atoms) serves as the Lewis acid in place of the metal atom.

As discussed above, this new form of asymmetric catalysis has proven to be highly effective and powerful, and today many enantioselective reactions can be catalyzed using simple, chiral organic compounds capable of donating a hydrogen bond. Whilst in many regards this area represents a new frontier in asymmetric catalysis, hydrogen bond catalysis also rests on well-established principles of catalysis and reactivity. For example, whereas the strength of a metal-based catalyst depends on the Lewis acidity of the metal used, with a hydrogen bond catalyst it likely reflects the Lewis acid strength – that is, the pK_a – of the promoter hydrogen. The broad capability enjoyed by metal-based Lewis acids is the fruit of exhaustive searches, by thousands of scientists worldwide, for the optimal pairing of chiral ligands and metals. It is fair to say that hydrogen bond-based catalysts, with just hydrogen as the tunable Lewis acidic atom, are unlikely to equal the enormous impact of traditional metal-based catalysts. The former, however, offers capabilities that may prove valuable in many situations. For example, the absence of a metal – besides the obvious advantage that this offers in the synthesis of pharmaceuticals – makes hydrogen bond-based catalysts well suited for attachment to a solid support. These catalysts should prove to be more robust and recyclable, as the normally encountered issue of catalyst deactivation, which results from metal leaching from the ligand, is a non-issue.

It is expected that the many advances in asymmetric catalysis, which have led to the identification of numerous "privileged" scaffolds, will greatly facilitate the discovery of useful hydrogen bond-based catalysts. Indeed, as a starting point for a catalyst search, there is no better place to explore than the free ligands of many privileged metal-based Lewis acid catalysts. Such ligands are expected to be effective as hydrogen bond-based promoters of asymmetric reactions, as the hydrogen atoms in the free ligand and the metal in the metal complex occupy approximately the same position. This analysis, when taken in conjunction with the many accomplishments summarized above, implies that the prospects are bright for new discoveries in this burgeoning field.

6.16
Addendum

During the six months since the initial writing of this chapter, numerous additional reports have appeared in the literature detailing the use of chiral hydrogen bond donors as catalysts for asymmetric reactions. Within the field of hydrogen bonding catalysis, thiourea- and urea-based catalyst scaffolds continue to show promise for a variety of reactions, most notably for the addition of activated nucleophiles to nitroolefins. Jacobsen and co-workers have described the use of a primary amine-containing bifunctional thiourea catalyst for the enantioselective conjugate addition of ketones to nitroolefins [119]. A similar strategy was applied, albeit with slightly lower selectivity, by Tsogoeva and Wei for the conjugate additions of ketones to β-aryl nitroolefins [120, 121]. Tang and co-workers reported the use of a chiral pyrrolidine-substituted thiourea to induce high levels of enantio- and diastereoselectivity in the Michael addition reactions of cyclohexanone to nitroolefins [122]. Additional nucleophiles utilized for the thiourea-catalyzed conjugate addition to nitroolefins include in-situ-generated enols [123], indoles [124], aryl methyl ketone-derived enamines [125], and thioacetic acid [126]. Nagasawa and co-workers have developed a bifunctional guanidine-bis-thiourea catalyst for the asymmetric Henry reactions of aliphatic aldehydes and nitroalkanes [127, 128], while Chen and colleagues reported the use of Takemoto's thiourea catalyst for the construction of quaternary carbon stereocenters with good to high ee-values via the additions of α-substituted cyanoacetates to vinyl sulfones [129]. In addition, Takemoto and co-workers have reported the use of a bifunctional urea catalyst for the enantioselective hydrazinations of β-keto esters [130].

The scope of cinchona alkaloid-catalyzed asymmetric reactions has also continued to grow, due in large part to the ability of the incorporated quinuclidine and quinoline functionalities to activate both the electrophilic and nucleophilic components of the reaction. Deng and co-workers have reported the use of an unusual pyrimidine-substituted dihydroquinidine catalyst for the conjugate additions of α-substituted carbonyl donors to α,β-unsaturated aldehydes to give chiral aldehydes with synthetically useful levels of stereoselection [131]. The Jørgensen laboratory has utilized a hydrosilylated cinchonine catalyst to provide deconjugative Michael addition reactions of activated alkylidenes to acroleins in moderate yields and selectivities [132]. Petterson and Fini used a quinine catalyst for the hydrophosphonylations of imines to yield α-amino phosphonates in good yields and enantioselectivities [133]. Additional reports highlight the use of cinchona derivatives for the Michael additions of triazoles to nitroolefins [134] and the conjugate additions of 1,3-dicarbonyl compounds to maleimides [135].

The use of bifunctional thiourea-substituted cinchona alkaloid derivatives has continued to garner interest, with the Deng laboratory reporting the use of a 6′-thiourea-substituted cinchona derivative for both the Mannich reactions of malonates with imines [136] and the Friedel–Crafts reactions of imines with indoles [137]. In both reports, a catalyst loading of 10–20 mol% provided the desired products in almost uniformly high yields and high enantioselectivities. Thiourea-substituted cinchona derivatives have also been used for the enantioselective aza-Henry reactions of aldimines [138] and the enantioselective Henry reactions of nitromethane with aromatic aldehydes [139].

Multiple new reports have described the use of BINOL-derived phosphoric acids as chiral Lewis acid catalysts. The Terada group reported using substrate:catalyst ratios of up to 2000:1 in the highly enantioselective Aza-Ene-type reactions of N-benzoylimines and enamides/enecarbamates to give β-aminoimine adducts in high yields [140]. Akiyama and co-workers found that aza-Diels–Alder reactions of Brassard's diene with imines were promoted by the pyridinium salt of a chiral phosphoric acid to give chiral piperidinone derivatives in high yields and very high enantioselectivities [141]. The reactions catalyzed by the pyridinium salt of the catalyst showed improved product yields with respect to reactions catalyzed by the acid alone. Akiyama also reported the beneficial effects of added acetic acid, which improved both the yields and enantioselectivities of the asymmetric aza-Diels–Alder reactions of Danishefsky's diene and aldimines [142]. Additional publications have described the use of chiral BINOL-derived phosphoric acid catalysts for the reductive amination of ketones [143], the Pictet–Spengler reactions of substituted tryptamines [144], and the Strecker reactions of aryl aldimines [145].

The Yamamoto laboratory has reported the synthesis of two novel chiral Lewis acid catalysts for use in the Diels–Alder and Mannich-type reactions. A chiral BINOL-derived N-triflyl phosphoramide was used as a strong acid catalyst to promote the Diels–Alder reactions of ethyl vinyl ketone and substituted siloxydienes to give cycloadducts in generally high yields and good enantioselectivities [146]. A conceptually different Brønsted acid-assisted chiral Brønsted acid catalyst was prepared by substituting one hydroxy group of an optically active BINOL with a more acidic bis-(trifluoromethanesulfonyl)methyl group [147]. This catalyst was used for the enantioselective Mannich-type reactions of silyl ketene acetals and aldimines in the presence of a stoichiometric amount of an achiral proton source to give (S)-β-amino esters in high yields and moderate to good enantioselectivities.

References

1 G.A. Jeffrey, W. Saenger, *Hydrogen Bonding in Biological Structures*. Springer-Verlag, New York, **1991**.
2 G.J. Quigley, *Trans. Am. Cryst. Assoc.* **1986**, *22*, 121–130.
3 L.J. Prins, D.N. Reinhoudt, P. Timmerman, *Angew. Chem., Int. Ed.* **2001**, *40*, 2383–2426.
4 S.O. Shan, D. Herschlag, *Proc. Natl. Acad. Sci. USA* **1996**, *93*, 14474–14479.
5 F.E. Romesberg, B. Spiller, P.G. Schultz, R.C. Stevens, *Science* **1998**, *279*, 1929–1933.
6 G.A. Jeffrey, *An Introduction to Hydrogen Bonding*. Oxford University Press, New York, **1997**.
7 L. Pauling, *J. Am. Chem. Soc.* **1931**, *53*, 1367–1400.
8 M.L. Huggins, *J. Phys. Chem.* **1922**, *26*, 601–625.
9 M.L. Huggins, *J. Phys. Chem.* **1936**, *40*, 723–731.
10 M.L. Huggins, *J. Org. Chem.* **1936**, *1*, 407–456.
11 M.L. Huggins, *Chem. Rev.* **1943**, *32*, 195–218.
12 L. Pauling, *The Nature of the Chemical Bond and the Structure of Molecules and Crystals: An Introduction to Modern Structural Chemistry*. Cornell University Press, Ithica, New York, **1940**.
13 T. Steiner, *Angew. Chem., Int. Ed.* **2002**, *41*, 48–76.
14 S.J. Connon, *Chem. Eur. J.* **2006**, *12*, 5418–5427.
15 S.J. Connon, *Angew. Chem., Int. Ed.* **2006**, *45*, 3909–3912.
16 T. Akiyama, J. Itoh, K. Fuchibe, *Adv. Syn. Catal.* **2006**, *348*, 999–1010.
17 M.S. Taylor, E.N. Jacobsen, *Angew. Chem., Int. Ed.* **2006**, *45*, 1520–1543.
18 H. Yamamoto, K. Futatsugi, *Angew. Chem., Int. Ed.* **2005**, *44*, 1924–1942.
19 C. Bolm, T. Rantanen, I. Schiffers, L. Zani, *Angew. Chem., Int. Ed.* **2005**, *44*, 1758–1763.
20 J. Seayad, B. List, *Org. Biomol. Chem.* **2005**, *3*, 719–724.
21 P.M. Pihko, *Lett. Org. Chem.* **2005**, *2*, 398–403.
22 P.I. Dalko, L. Moisan, *Angew. Chem., Int. Ed.* **2004**, *43*, 5138–5175.
23 P.M. Pihko, *Angew. Chem., Int. Ed.* **2004**, *43*, 2062–2064.
24 P.R. Schreiner, *Chem. Soc. Rev.* **2003**, *32*, 289–296.
25 P.I. Dalko, L. Moisan, *Angew. Chem., Int. Ed.* **2001**, *40*, 3726–3748.
26 J. Hine, K. Ahn, J.C. Gallucci, S.M. Linden, *J. Am. Chem. Soc.* **1984**, *106*, 7980–7981.
27 T.R. Kelly, P. Meghani, V.S. Ekkundi, *Tetrahedron Lett.* **1990**, *31*, 3381–3384.
28 M.C. Etter, Z. Urbanczyklipkowska, M. Ziaebrahimi, T.W. Panunto, *J. Am. Chem. Soc.* **1990**, *112*, 8415–8426.
29 M.C. Etter, S.M. Reutzel, *J. Am. Chem. Soc.* **1991**, *113*, 2586–2598.
30 M.C. Etter, *Acc. Chem. Res.* **1990**, *23*, 120–126.
31 D.P. Curran, L.H. Kuo, *J. Org. Chem.* **1994**, *59*, 3259–3261.
32 D.P. Curran, L.H. Kuo, *Tetrahedron Lett.* **1995**, *36*, 6647–6650.
33 C.S. Wilcox, E. Kim, D. Romano, L.H. Kuo, A.L. Burt, D.P. Curran, *Tetrahedron* **1995**, *51*, 621–634.
34 A. Eschbach, V.H. Rawal, The Ohio State University **1994**, unpublished results.
35 P.R. Schreiner, A. Wittkopp, *Org. Lett.* **2002**, *4*, 217–220.
36 T. Okino, Y. Hoashi, Y. Takemoto, *Tetrahedron Lett.* **2003**, *44*, 2817–2821.
37 T. Schuster, M. Kurz, M.W. Gobel, *J. Org. Chem.* **2000**, *65*, 1697–1701.
38 Y. Huang, V.H. Rawal, *J. Am. Chem. Soc.* **2002**, *124*, 9662–9663.
39 D.C. Braddock, I.D. MacGilp, B.G. Perry, *Synlett* **2003**, 1121–1124.
40 M.S. Sigman, E.N. Jacobsen, *J. Am. Chem. Soc.* **1998**, *120*, 4901–4902.
41 P. Vachal, E.N. Jacobsen, *Org. Lett.* **2000**, *2*, 867–870.
42 H. Hiemstra, H. Wynberg, *J. Am. Chem. Soc.* **1981**, *103*, 417–430.
43 B.J. Li, L. Jiang, M. Liu, Y.C. Chen, L.S. Ding, Y. Wu, *Synlett* **2005**, 603–606.
44 B. Vakulya, S. Varga, A. Csampai, T. Soos, *Org. Lett.* **2005**, *7*, 1967–1969.
45 Y. Hoashi, T. Okino, Y. Takemoto, *Angew. Chem., Int. Ed.* **2005**, *44*, 4032–4035.

46 T. Okino, Y. Hoashi, Y. Takemoto, *J. Am. Chem. Soc.* **2003**, *125*, 12672–12673.
47 T. Okino, Y. Hoashi, T. Furukawa, N. Xu, Y. Takemoto, *J. Am. Chem. Soc.* **2005**, *127*, 119–125.
48 Y. Hoashi, T. Yabuta, Y. Takemoto, *Tetrahedron Lett.* **2004**, *45*, 9185–9188.
49 H.M. Li, Y. Wang, L. Tang, L. Deng, *J. Am. Chem. Soc.* **2004**, *126*, 9906–9907.
50 H.M. Li, Y. Wang, L. Tang, F.H. Wu, F. Liu, C.Y. Guo, B.M. Foxman, L. Deng, *Angew. Chem., Int. Ed.* **2005**, *44*, 105–108.
51 R.P. Herrera, V. Sgarzani, L. Bernardi, A. Ricci, *Angew. Chem., Int. Ed.* **2005**, *44*, 6576–6579.
52 W. Zhuang, R.G. Hazell, K.A. Jorgensen, *Org. Biomol. Chem.* **2005**, *3*, 2566–2571.
53 S.H. McCooey, S.J. Connon, *Angew. Chem., Int. Ed.* **2005**, *44*, 6367–6370.
54 J. Ye, D.J. Dixon, P.S. Hynes, *Chem. Commun.* **2005**, 4481–4483.
55 J. Wang, H. Li, W.H. Duan, L.S. Zu, W. Wang, *Org. Lett.* **2005**, *7*, 4713–4716.
56 H.M. Li, J. Song, F. Liu, L. Deng, *J. Am. Chem. Soc.* **2005**, *127*, 8948–8949.
57 J.I. Oku, S. Inoue, *J. Chem. Soc., Chem. Commun.* **1981**, 229–230.
58 K. Tanaka, A. Mori, S. Inoue, *J. Org. Chem.* **1990**, *55*, 181–185.
59 H. Danda, *Synlett* **1991**, 263–264.
60 H. Danda, H. Nishikawa, K. Otaka, *J. Org. Chem.* **1991**, *56*, 6740–6741.
61 G. Brauer (Ed.), *Handbook of Preparative Inorganic Chemistry, Vol. 2.* 2nd edn., **1965**.
62 M.S. Iyer, K.M. Gigstad, N.D. Namdev, M. Lipton, *J. Am. Chem. Soc.* **1996**, *118*, 4910–4911.
63 M.S. Sigman, P. Vachal, E.N. Jacobsen, *Angew. Chem., Int. Ed.* **2000**, *39*, 1279–1281.
64 P. Vachal, E.N. Jacobsen, *J. Am. Chem. Soc.* **2002**, *124*, 10012–10014.
65 E.J. Corey, M.J. Grogan, *Org. Lett.* **1999**, *1*, 157–160.
66 J.K. Huang, E.J. Corey, *Org. Lett.* **2004**, *6*, 5027–5029.
67 D.E. Fuerst, E.N. Jacobsen, *J. Am. Chem. Soc.* **2005**, *127*, 8964–8965.
68 A.G. Wenzel, E.N. Jacobsen, *J. Am. Chem. Soc.* **2002**, *124*, 12964–12965.
69 D. Uraguchi, M. Terada, *J. Am. Chem. Soc.* **2004**, *126*, 5356–5357.
70 T. Akiyama, J. Itoh, K. Yokota, K. Fuchibe, *Angew. Chem., Int. Ed.* **2004**, *43*, 1566–1568.
71 T. Akiyama, Y. Saitoh, H. Morita, K. Fuchibe, *Adv. Syn. Catal.* **2005**, *347*, 1523–1526.
72 M.S. Taylor, N. Tokunaga, E.N. Jacobsen, *Angew. Chem., Int. Ed.* **2005**, *44*, 6700–6704.
73 T. Okino, S. Nakamura, T. Furukawa, Y. Takemoto, *Org. Lett.* **2004**, *6*, 625–627.
74 B.R. Linton, M.S. Goodman, A.D. Hamilton, *Chem. Eur. J.* **2000**, *6*, 2449–2455.
75 B.M. Nugent, R.A. Yoder, J.N. Johnston, *J. Am. Chem. Soc.* **2004**, *126*, 3418–3419.
76 T.P. Yoon, E.N. Jacobsen, *Angew. Chem., Int. Ed.* **2005**, *44*, 466–468.
77 R. Noyori, M. Ohta, Y. Hsiao, M. Kitamura, T. Ohta, H. Takaya, *J. Am. Chem. Soc.* **1986**, *108*, 7117–7119.
78 L.S. Santos, R.A. Pilli, V.H. Rawal, *J. Org. Chem.* **2004**, *69*, 1283–1289.
79 M.S. Taylor, E.N. Jacobsen, *J. Am. Chem. Soc.* **2004**, *126*, 10558–10559.
80 D. Uraguchi, K. Sorimachi, M. Terada, *J. Am. Chem. Soc.* **2004**, *126*, 11804–11805.
81 G.D. Joly, E.N. Jacobsen, *J. Am. Chem. Soc.* **2004**, *126*, 4102–4103.
82 T. Akiyama, H. Morita, J. Itoh, K. Fuchibe, *Org. Lett.* **2005**, *7*, 2583–2585.
83 D. Uraguchi, K. Sorimachi, M. Terada, *J. Am. Chem. Soc.* **2005**, *127*, 9360–9361.
84 M. Chorev, M. Goodman, *Acc. Chem. Res.* **1993**, *26*, 266–273.
85 G.B. Rowland, H.L. Zhang, E.B. Rowland, S. Chennamadhavuni, Y. Wang, J.C. Antilla, *J. Am. Chem. Soc.* **2005**, *127*, 15696–15697.
86 B. Albert, et al., *Molecular Biology of the Cell*, 4th edn., **2003**.

87 D.M. Stout, A.I. Meyers, *Chem. Rev.* **1982**, *82*, 223–243.
88 J.W. Yang, M.T.H. Fonseca, B. List, *Angew. Chem., Int. Ed.* **2004**, *43*, 6660–6662.
89 S.G. Ouellet, J.B. Tuttle, D.W.C. MacMillan, *J. Am. Chem. Soc.* **2005**, *127*, 32–33.
90 M. Rueping, E. Sugiono, C. Azap, T. Theissmann, M. Bolte, *Org. Lett.* **2005**, *7*, 3781–3783.
91 S. Hoffmann, A.M. Seayad, B. List, *Angew. Chem., Int. Ed.* **2005**, *44*, 7424–7427.
92 Y.M.A. Yamada, S. Ikegami, *Tetrahedron Lett.* **2000**, *41*, 2165–2169.
93 Y. Iwabuchi, M. Nakatani, N. Yokoyama, S. Hatakeyama, *J. Am. Chem. Soc.* **1999**, *121*, 10219–10220.
94 N.T. McDougal, S.E. Schaus, *J. Am. Chem. Soc.* **2003**, *125*, 12094–12095.
95 Y. Sohtome, A. Tanatani, Y. Hashimoto, K. Nagasawa, *Tetrahedron Lett.* **2004**, *45*, 5589–5592.
96 S. Kawahara, A. Nakano, T. Esumi, Y. Iwabuchi, S. Hatakeyama, *Org. Lett.* **2003**, *5*, 3103–3105.
97 K. Matsui, S. Takizawa, H. Sasai, *J. Am. Chem. Soc.* **2005**, *127*, 3680–3681.
98 S.A. Kozmin, T. Iwama, Y. Huang, V.H. Rawal, *J. Am. Chem. Soc.* **2002**, *124*, 4628–4641.
99 K.C. Nicolaou, S.A. Snyder, T. Montagnon, G. Vassilikogiannakis, *Angew. Chem., Int. Ed.* **2002**, *41*, 1668–1698.
100 G. Lelais, D.W.C. MacMillan, *Aldrichimica Acta* **2006**, *39*, 79–87.
101 Y. Huang, V.H. Rawal, *Org. Lett.* **2000**, *2*, 3321–3323.
102 D. Seebach, A.K. Beck, A. Heckel, *Angew. Chem., Int. Ed.* **2001**, *40*, 92–138.
103 K. Tanaka, F. Toda, *Chem. Rev.* **2000**, *100*, 1025–1074.
104 Y. Huang, A.K. Unni, A.N. Thadani, V.H. Rawal, *Nature (London)* **2003**, *424*, 146–146.
105 H.F. Du, D.B. Zhao, K.L. Ding, *Chem. Eur. J.* **2004**, *10*, 5964–5970.
106 A.K. Unni, N. Takenaka, H. Yamamoto, V.H. Rawal, *J. Am. Chem. Soc.* **2005**, *127*, 1336–1337.
107 A.K. Unni, *Asymmetric Catalysis of Hetero-Diels-Alder Reactions Through Hydrogen Bonding*. PhD Thesis, The University of Chicago, Chicago, IL, June **2005**.
108 S. Rajaram, M.S. Sigman, *Org. Lett.* **2005**, *7*, 5473–5475.
109 W. Zhuang, T.B. Poulsen, K.A. Jorgensen, *Org. Biomol. Chem.* **2005**, *3*, 3284–3289.
110 T. Tonoi, K. Mikami, *Tetrahedron Lett.* **2005**, *46*, 6355–6358.
111 Y. Huang, *Exploring the New Horizon of Diels-Alder Reactions: Asymmetric Catalysis*. PhD Thesis, The University of Chicago, Chicago, IL, August **2002**.
112 J.D. McGilvra, A.K. Unni, K. Modi, V.H. Rawal, *Angew. Chem., Int. Ed.* **2006**, *45*, 6130–6133.
113 T. Schuster, M. Bauch, G. Durner, M.W. Gobel, *Org. Lett.* **2000**, *2*, 179–181.
114 S.B. Tsogoeva, G. Durner, M. Bolte, M.W. Gobel, *Eur. J. Org. Chem.* **2003**, 1661–1664.
115 A.N. Thadani, A.R. Stankovic, V.H. Rawal, *Proc. Natl. Acad. Sci. USA* **2004**, *101*, 5846–5850.
116 K.A. Ahrendt, C.J. Borths, D.W.C. MacMillan, *J. Am. Chem. Soc.* **2000**, *122*, 4243–4244.
117 N. Momiyama, H. Yamamoto, *J. Am. Chem. Soc.* **2005**, *127*, 1080–1081.
118 V.B. Gondi, M. Gravel, V.H. Rawal, *Org. Lett.* **2005**, *7*, 5657–5660.
119 H.B. Huang, E.N. Jacobsen, *J. Am. Chem. Soc.* **2006**, *128*, 7170–7171.
120 S.B. Tsogoeva, S.W. Wei, *Chem. Commun.* **2006**, 1451–1453.
121 D.A. Yalalov, S.B. Tsogoeva, S. Schmatz, *Adv. Syn. Catal.* **2006**, *348*, 826–832.
122 C.L. Cao, M.C. Ye, L. Sun, Y. Tang, *Org. Lett.* **2006**, *8*, 2901–2904.
123 S.B. Tsogoeva, D.A. Yalalov, M.J. Hateley, C. Weckbecker, K. Huthmacher, *Eur. J. Org. Chem.* **2005**, 4995–5000.
124 E.M. Fleming, T. McCabe, S.J. Connon, *Tetrahedron Lett.* **2006**, *47*, 7037–7042.
125 D.J. Dixon, R.D. Richardson, *Synlett* **2006**, 81–85.

126 H. Li, J. Wang, L. Zu, W. Wang, *Tetrahedron Lett.* **2006**, *47*, 2585–2589.

127 Y. Sohtome, Y. Hashimoto, K. Nagasawa, *Adv. Syn. Catal.* **2005**, *347*, 1643–1648.

128 Y. Sohtome, Y. Hashimoto, K. Nagasawa, *Eur. J. Org. Chem.* **2006**, 2894–2897.

129 T.Y. Liu, J. Long, B.J. Li, L. Jiang, R. Li, Y. Wu, L.S. Ding, Y.C. Chen, *Org. Biomol. Chem.* **2006**, *4*, 2097–2099.

130 N. Xu, T. Yabuta, P. Yuan, Y. Takemoto, *Synlett* **2006**, 137–140.

131 F.H. Wu, R. Hong, J.H. Khan, F. Liu, L. Deng, *Angew. Chem., Int. Ed.* **2006**, *45*, 4301–4305.

132 M. Bell, K. Frisch, K.A. Jorgensen, *J. Org. Chem.* **2006**, *71*, 5407–5410.

133 D. Pettersen, M. Marcolini, L. Bernardi, F. Fini, R.P. Herrera, V. Sgarzani, A. Ricci, *J. Org. Chem.* **2006**, *71*, 6269–6272.

134 J. Wang, H. Li, L.S. Zu, W. Wang, *Org. Lett.* **2006**, *8*, 1391–1394.

135 G. Bartoli, M. Bosco, A. Carlone, A. Cavalli, M. Locatelli, A. Mazzanti, P. Ricci, L. Sambri, P. Melchiorre, *Angew. Chem., Int. Ed.* **2006**, *45*, 4966–4970.

136 J. Song, Y. Wang, L. Deng, *J. Am. Chem. Soc.* **2006**, *128*, 6048–6049.

137 Y.Q. Wang, J. Song, R. Hong, H.M. Li, L. Deng, *J. Am. Chem. Soc.* **2006**, *128*, 8156–8157.

138 L. Bernardi, F. Fini, R.P. Herrera, A. Ricci, V. Sgarzani, *Tetrahedron* **2005**, *62*, 375–380.

139 T. Marcelli, R.N.S. van der Haas, J.H. van Maarseveen, H. Hiemstra, *Angew. Chem., Int. Ed.* **2006**, *45*, 929–931.

140 M. Terada, K. Machioka, K. Sorimachi, *Angew. Chem., Int. Ed.* **2006**, *45*, 2254–2257.

141 J. Itoh, K. Fuchibe, T. Akiyama, *Angew. Chem., Int. Ed.* **2006**, *45*, 4796–4798.

142 T. Akiyama, Y. Tamura, J. Itoh, H. Morita, K. Fuchibe, *Synlett* **2006**, 141–143.

143 R.I. Storer, D.E. Carrera, Y. Ni, D.W.C. MacMillan, *J. Am. Chem. Soc.* **2006**, *128*, 84–86.

144 J. Seayad, A.M. Seayad, B. List, *J. Am. Chem. Soc.* **2006**, *128*, 1086–1087.

145 M. Rueping, E. Sugiono, C. Azap, *Angew. Chem., Int. Ed.* **2006**, *45*, 2617–2619.

146 D. Nakashima, H. Yamamoto, *J. Am. Chem. Soc.* **2006**, *128*, 9626–9627.

147 A. Hasegawa, Y. Naganawa, M. Fushimi, K. Ishihara, H. Yamamoto, *Org. Lett.* **2006**, *8*, 3175–3178.

7
Chiral Lewis Bases as Catalysts

Pavel Kočovský and Andrei V. Malkov

7.1
Introduction

Nucleophilic addition to carbonyl compounds is the cornerstone of organic synthesis, highlighted by the classical examples of Grignard reactions, aldol condensation, and LiAlH$_4$ reductions. However, when less-reactive nucleophiles are employed, such as allylsilane [1], allylstannane [1, 2] or Cl$_3$SiH [3], activation is required. Here, we have a choice of activating either the electrophile or the nucleophile. Activation of the electrophile – that is, the carbonyl reactant – can be achieved by coordination of a Lewis acid to the oxygen [4], which increases the electrophilicity of the carbonyl carbon and, consequently, the reactivity of the C=O group. Alternatively, coordination of a Lewis base to the nucleophile would increase its nucleophilicity, which should also allow the addition to occur [5]. As only the coordinated species will react, the activator can be used in just a catalytic amount, provided that it can be regenerated after completion of the reaction and returned to the next catalytic cycle. Naturally, if the catalyst is chiral, preferential formation of one enantiomer of the product can be expected.

Of the numerous examples of asymmetric reactions catalyzed by Lewis bases, this chapter focuses mainly on the activation of silicon reagents and related processes. Various other types of Lewis basic (nucleophilic) activation, namely the Morita–Baylis–Hillman (MBH) reaction, acyl transfer, nucleophilic carbenes, and carbonyl reduction, are described in the other chapters of this book.

Although this chapter describes details of studies published up until the end of 2005, selected examples of the early 2006 literature are also included. The schemes and tables were chosen to illustrate mainly the catalytic asymmetric reactions. Stoichiometric or non-asymmetric examples, relevant to the main focus, are briefly mentioned in the text, especially when further development is expected but, in general, they are not presented in a graphical form. The key catalytic reactions exhibiting high enantioselectivity and the most successful catalysts are highlighted by being displayed in frames (e.g., Fig. 7.1). We believe that this layout

Enantioselective Organocatalysis: Reactions and Experimental Procedures
Edited by Peter I. Dalko
Copyright © 2007 WILEY-VCH Verlag GmbH & Co. KGaA, Weinheim
ISBN: 978-3-527-31522-2

will benefit the reader who wishes to obtain initial information quickly by browsing through the chapter's pages.

7.2
Allylation Reactions

7.2.1
Catalytic Allylation of Aldehydes

Lewis acids are known to catalyze the allylation of aldehydes **1** with $AllylSiMe_3$ and its congeners by coordination to the carbonyl group [1, 6]. By contrast, Lewis bases are typically inert to $AllylSiMe_3$ and require $AllylSiCl_3$ (**2a**) [7], the silicon atom of which is more Lewis-acidic (Scheme 7.1) [6–13].

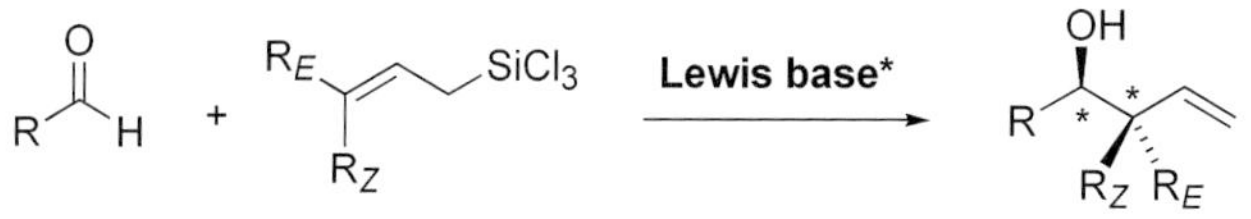

1

2a, R_E = H, R_Z = H
2b, R_E = Me, R_Z = H
2c, R_E = H, R_Z = Me
2d, R_E = Me, R_Z = Me

3a, R_E = H, R_Z = H
3b, R_E = Me, R_Z = H
3c, R_E = H, R_Z = Me
3d, R_E = Me, R_Z = Me

Scheme 7.1 Asymmetric allylation of aldehydes.

If the latter reaction proceeds through a closed transition state (e.g., **5** in Scheme 7.2), good diastereocontrol can be expected in the case of *trans*- and *cis*-$CrotylSiCl_3$ (**2b**/**2c**) [14, 15]. Here, the *anti*-diastereoisomer **3b** should be obtained from *trans*-crotyl derivative **2b**, whereas the *syn*-isomer **3c** should result from the reaction of the *cis*-isomer **2c** (Scheme 7.2). Furthermore, this mechanism creates an opportunity for transferring the chiral information if the Lewis base employed is chiral. Provided that the Lewis base dissociates from the silicon in the intermediate **6** at a sufficient rate, it can act as a catalyst (rather than as a stoichiometric reagent). Typical Lewis bases that promote the allylation reaction are the common dipolar aprotic solvents, such as dimethylformamide (DMF) [8, 12], dimethyl sulfoxide (DMSO) [8, 9], and hexamethylphosphoramide (HMPA) [9, 16], in addition to other substances that possess a strongly Lewis basic oxygen, such as various formamides [17] (in a solution or on a solid support [7, 8, 18]), urea derivatives [19], and catecholates [10] (and their chiral modifications [5c], [20]). It should be noted that, upon coordination to a Lewis base, the silicon atom becomes more Lewis acidic (*vide infra*), which facilitates its coordination to the carbonyl in the cyclic transition state **5**.

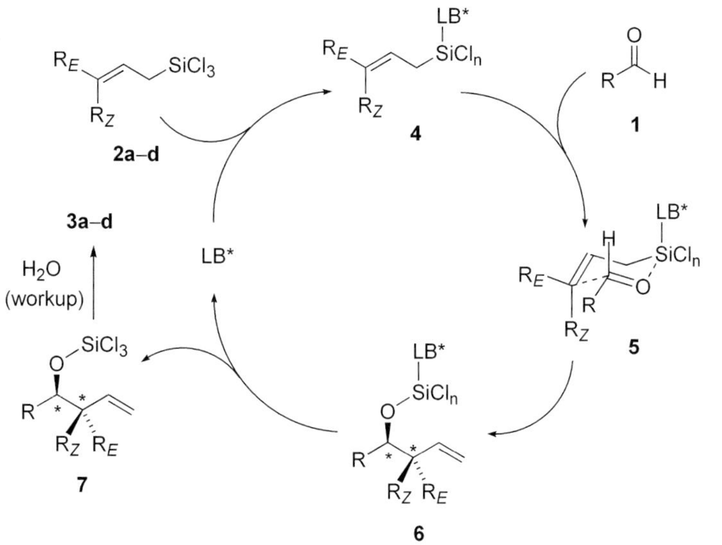

Scheme 7.2 Lewis base-catalyzed allylation of aldehydes **1** with allyl trichlorosilanes **2a–d**.

Fluoride ion (from CsF, CdF$_2$, or AgF) can also catalyze the allylation using AllylSiF$_3$ and AllylSi(OMe)$_3$, respectively [10, 11]; the asymmetric version (with ≤56% ee) requires a combination of Bu$_4$N$^+$[SiPh$_3$F$_2$]$^-$ with the Lewis acidic complex of CuCl and BINAP [21].

Chiral Lewis-basic catalysts (Figs. 7.1 and 7.2), in particular phosphoramides **8–12** [9, 14c, 15c, 22–24], formamide **13** [17], pyridine N,N'-bisoxides **17** and **18** [25–27], N-monoxides (**19–26**) [27–32], and N,N'N''-trisoxides (**27**) [33] exhibit good to high enantioselectivities for the allylation of aromatic, heteroaromatic, and cinnamyl-type aldehydes (**1**) with allyl, trans- and cis-crotyl, and prenyl trichlorosilanes (**2a–d**). Chiral formamides (with the exception of **13**, as discussed below) [17], pyridine-oxazolines [34], urea derivatives [19] and sulfoxides [35] are effective only in stoichiometric quantities (or in excess) and, as a rule, exhibit lower enantioselectivities.

The early phosphoramides **8** and **9** (Fig. 7.1), developed by Denmark [22], exhibited modest enantioselectivity in the allylation reaction (Scheme 7.1 and Table 7.1), but played an important role in the mechanistic elucidation and development of the second generation of catalysts (vide infra). The proline-derived phosphoramide **10** proved more selective (Table 7.1, entry 2) [14c, 15c], but its preparation is hampered by the fact that, unlike with **8** and **9**, its phosphorus atom is stereogenic, so that two diastereoisomers are formed and need to be separated. Phosphoramide **10** has been reported to give higher asymmetric induction than its diastereoisomer (entry 2) [22f]. However, the results are further affected by the variation of the N-substituents; generally, the sense of asymmetric induc-

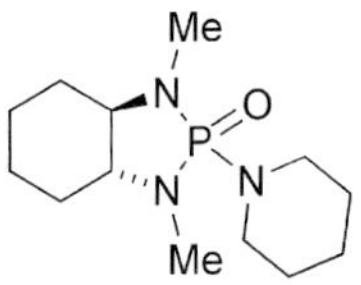

8 (≤ 43% ee, R)[9] 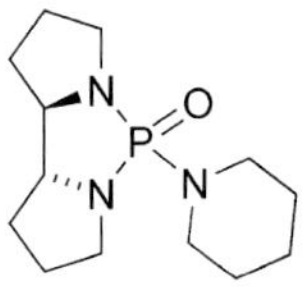   **9** (≤ 56% ee, S)[22] 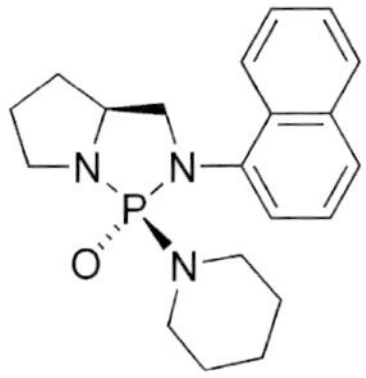   **10** (≤ 72% ee, R)[14b,15c,22,23]

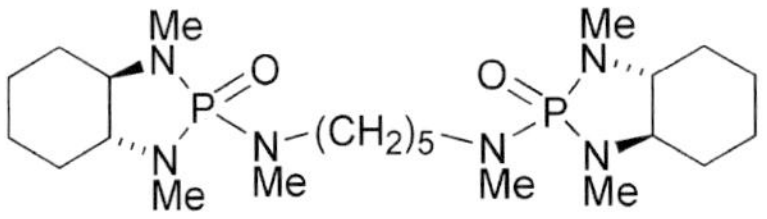

11 (≤ 72% ee, S)[22] 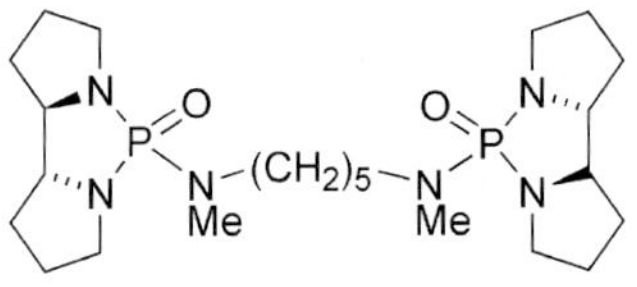   **12** (≤ 87% ee, S)[22]

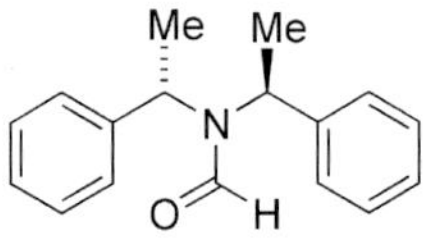

13 (≤ 98% ee, R)[17]

Fig. 7.1 Selected Lewis basic catalysts for allylation of aldehydes with **2a**. The enantioselectivities attained are shown in parentheses, and the absolute configuration refers to the product obtained from PhCHO.

tion appears to be independent of the configuration at the phosphorus, which only affects the level of enantiocontrol [22f].

Kinetic measurements and the observation of a non-linear relationship between the enantiopurity of **8/9** and the product **3** led Denmark to the conclusion that

Table 7.1 Allylation of benzaldehyde (**1**) with **2a–d**, catalyzed by phosphoramides at −78 °C (Scheme 7.1 and Fig. 7.1) [9, 22].

Entry	Catalyst	Silane	Yield (%)	syn:anti	ee (%)
1	**8**[a]	**2a**	40	–	52
2	**10**[b]	**2a**	59	–	72
3	**11**[c]	**2a**	58	–	72
4	**12**[c]	**2a**	85	–	87
5	**12**[c]	**2b**	82	1:99	86
6	**12**[c]	**2c**	89	99:1	94

[a] 10 mol%.
[b] 25 mol%.
[c] 5 mol%.

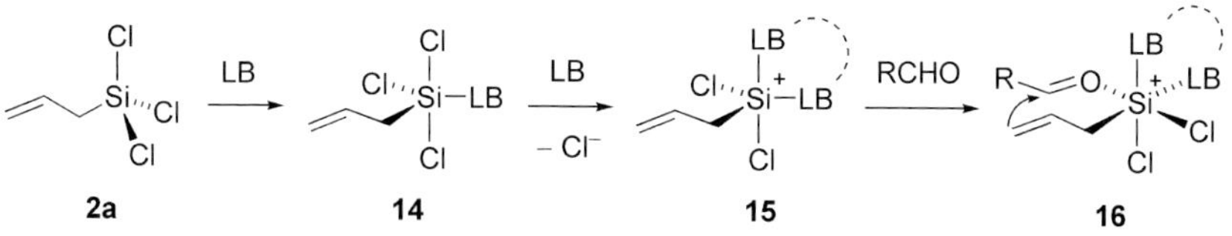

Scheme 7.3 Coordination of a Lewis base (LB) to 2a.

two molecules of the catalyst are coordinated to the silicon center (**15**; Scheme 7.3) [22]. However, when the concentration of the catalyst is low, a second mechanism may compete, namely that with only one molecule of the catalyst coordinated (**14**), which apparently attenuates the enantioselectivity [22]. Therefore, Denmark designed bidentate catalysts, such as **11** and **12**, which proved not only to exhibit higher enantioselectivities (Table 7.1; compare entries 1, 3, and 4) but also to react faster, so that their loading could be reduced from 10–20 mol% to 5 mol% [22]. Excellent diastereoselectivity was observed for crotylation (entries 5 and 6), the sense of which is consistent with the transition state **5**, as discussed earlier (Scheme 7.2) [22f].

In addition to his synthetic studies, Denmark carried out a very detailed investigation, using complexes of bisphosphoramides **11** and **12** and their analogues, all with varying length of the tether connecting the two ligating units. Instead of Si, Denmark employed Sn, which exhibits a similar coordination pattern but with stronger bonding, so that the complexes of Sn are more stable and can be studied. This switch to $SnCl_4$, in conjunction with ^{119}Sn and ^{31}P NMR and crystallographic studies, allowed Denmark to identify the bidentate, the *cis*-coordination of the ligand, the octahedral geometry of the Sn, and the details of the shape of the chiral cavity, as dictated by the nature of the chiral scaffold and the length of the tether (with the optimum of five methylene units, as shown in **11** and **12**). Extrapolation of these findings to Si resulted in the formulation of a working mechanistic model for the allylation, and for the sense of asymmetric induction, presented in this chapter [22e]. Thus, ideally, in the transition state, the nucleophilicity of the allyl is increased by the *trans*-donation from the coordinated Lewis base, whereas the carbonyl of the aldehyde should be coordinated *trans* to the electron-withdrawing chloride – the site of maximal Lewis acidity, as shown in **16** [22e].

Chiral formamide **13** (Fig. 7.1) typically requires ≥20 mol% loading and HMPA (1 equiv.) as a co-catalyst; in order to attain 80% yield and 98% *ee*, the reaction was left at −78 °C for 2 weeks, which is less than practical. Interestingly, **13** is the only catalyst reported to date that exhibits very high enantioselectivities (≤98% *ee*) with aliphatic aldehydes, while with benzaldehydes it gives an almost racemic product (8% *ee*) [17].

Pyridine-type *N*-oxides (Fig. 7.2) represent another, no less-successful class of catalysts for the allylation reaction. Thus, Nakajima first demonstrated that the axially chiral biquinoline *N,N'*-bisoxide **17** can indeed catalyze the allylation

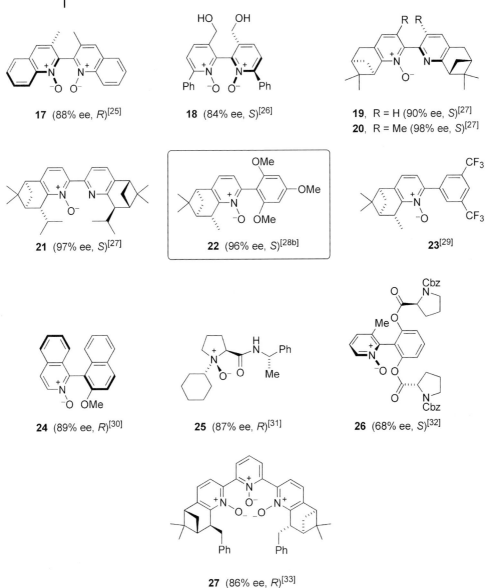

Fig. 7.2 Selected Lewis basic catalysts for allylation reactions and the enantioselectivities attained for the reaction of PhCHO with **2a**. The absolute configuration of the product **3** is shown in parentheses.

(Scheme 7.1) with high yield and enantioselectivity (71–92% *ee* at −78 °C) [25]. Nakajima's efforts were later followed by Hayashi, who reported similar levels of asymmetric induction attained with the bipyridine analogue **18** and its congeners (with 56–98% *ee*) [26]. Catalyst **18** is remarkably reactive, so that its loading can be reduced to the 0.1 mol% level, and moderate activity is retained even at 0.01 mol% loading [26], which renders **18** the most reactive organocatalyst reported to date. A chelation model, where both oxygens of the catalyst coordinate Si of the reagent, has been proposed to account for the reactivity, mirroring the chelation model **16** suggested by Denmark for **11** and **12** [22].

Prior to Hayashi's report [26], Malkov and Kočovský [27] showed the terpene-derived bipyridine *N*-monoxides **19–21** (PINDOX, Me$_2$PINDOX, and *iso*-PINDOX) to be even more enantioselective than the bisoxide **17**, although the reaction rate dropped, especially with the severely hindered *iso*-PINDOX (**21**). The most successful derivative of this series, Me$_2$PINDOX (**20**), combines the effects of both central and axial chirality, since the rotation about the bond connecting the two pyridine moieties is restricted by the two methyls and the N–O group. However, the barrier to rotation is rather low, and **20** isomerizes within several days (in solution) to a 1:2 mixture of **20** and its atropoisomer, which attenuates the asymmetric induction [27b]. PINDOX (**19**) and *iso*-PINDOX (**21**) lack the restriction to the rotation, so that a suitable configuration is apparently established on coordination to the silicon atom of the allylating reagent [27]. In analogy to the chelation model, proposed for **17** and **18**, chelation of Si in AllylSiCl$_3$ by O and N was considered for **19–21** [27]. However, METHOX (**22**), lacking the pyridine ring, was found to be much more reactive than **19–21** with the same level of enantioselectivity attained (≤96% *ee* at 1–5 mol% loading at −40 °C in MeCN) [28, 29], suggesting that coordination to nitrogen in **19–21** may not play an important role. Instead, arene–arene interactions between the catalyst and the substrate have been suggested to account for the high reactivity and selectivity of the electron-rich **22**. This hypothesis is supported by the lack of reactivity of the electron-poor analogue **23** [29]. Furthermore, the case of METHOX (**22**) shows clearly that the axial chirality, whether predetermined (as in **17** and **18**), or induced during the reaction (**19** and **21**), is not an absolute prerequisite for attaining high enantioselectivity in the allylation reaction [28].

Another three pyridine-type *N*-monoxides, namely **24** (QUINOX) [30], **25** [31] and **26** [32], have also been reported as successful organocatalysts, and the series is complemented by the *N*,*N*′*N*″-trisoxide **27** and its congeners [33].

The effect of the electronic properties of the substituted benzaldehydes (**1a–c**) on the allylation reaction is another interesting issue. Whilst the majority of catalysts shown in Figures 7.1 and 7.2 generally exhibit rather minor variation of the *ee*-value (typically with less than 20% difference between the electron-rich and electron-poor aldehydes), METHOX (**22**) appears to be a particularly tolerant catalyst, exhibiting practically the same enantioselectivity (93–96% *ee*; Table 7.2, entries 1–3) and reaction rate across the range of substrates [28b, 29]. By contrast, QUINOX (**24**) stands on the opposite side of the spectrum, showing the most dramatic differences between the electron-poor and electron-rich substrate aldehyde (12–96% *ee*; entries 4–6) [30].

Crotylation catalyzed by the *N*-oxides **17–22** [25–30] (and also by phosphoramides **8–12**) [22] is highly diastereoselective, with the *trans*-isomer **2b** affording almost exclusively the *anti*-product **3b**, while the *cis*-isomer **2c** gives mostly the

Table 7.2 Electronic effects in the allylation of *p*-substituted benzaldehydes **1a–c** with **2a**, catalyzed by pyridine *N*-oxides **22** and **24** at −40 °C (Scheme 7.1 and Fig. 7.2) [28–30].

Entry	Catalyst	Aldehyde	R	Solvent	Yield (%)	ee (%)
1	22	1a	C_6H_5	MeCN	≥95	96
2	22	1b	4-$MeOC_6H_4$	MeCN	≥95	96
3	22	1c	4-$CF_3C_6H_4$	MeCN	86	93
4	24	1a	C_6H_5	CH_2Cl_2	60	87
5	24	1b	4-$MeOC_6H_4$	CH_2Cl_2	70	12
6	24	1c	4-$CF_3C_6H_4$	CH_2Cl_2	85	96

syn-product **3c** (although in a slightly slower reaction). (For experimental details see Chapter 14.16.2). This behavior is consistent with the cyclic transition state **5** (Scheme 7.2) that guarantees a high degree of stereocontrol. METHOX (**22**) represents an extreme case, as it reacts well only with **2b**, affording **3b** with excellent diastereocontrol (>99:1) and enantioselectivity (95% *ee*), while the reaction with **2c** is very slow and less diastereoselective (6:1) [28b]. By contrast, QINOX (**24**) has been shown to exhibit rather low diastereocotrol [30], which can be attributed to the participation of either a cyclic boat-like or open transition state [36].

Of the catalysts listed in Figures 7.1 and 7.2, **12**, **18**, and **22** are the most efficient in terms of diastereo- and enantioselectivity, catalyst loading, and reaction rate. While **18** is synthesized in seven steps (including resolution or the use of a chiral auxiliary) [26], the other two are accessible more efficiently. Thus, phosphoramide **12** was prepared in four steps (including resolution) from pyrrolidine (**28**) via **29** (Scheme 7.4) [22f]. The key step here is the photochemical dimerization of pyrrolidine (**28** → **29**), originally reported by Crabtree [37]. METHOX (**22**) was synthesized in three steps from the commercially available pinocarvone (**33**) using the Kröhnke annulation (**33** + **34** → **36**). Pinocarvone itself can be readily obtained from the much cheaper α-pinene (**31**) via an ene-reaction with singlet oxygen; as both enantiomers of α-pinene are commercially available, this procedure can be used to prepare both enantiomers of METHOX (**22**) [28b]. The Kröhnke reagent **34** [38] is readily obtained in one pot via iodination of the corresponding acetophenone **35** in pyridine [28b].

While aromatic, heteroaromatic, and cinnamyl-type aldehydes react readily (to give 1,2-addition products in the latter case) [8, 12, 27–30], aliphatic aldehydes have been known to react sluggishly and to require stoichiometric conditions (*vide infra*). This problem has now been addressed by Denmark (Scheme 7.5), who showed that instead of allylation, aliphatic aldehydes preferentially undergo a reaction with Cl^- as the nucleophile, while the oxygen is silylated (**15** → **38** → **39**). The unstable chlorosilyloxy derivative **39**, generated from aliphatic aldehydes, is then hydrolyzed on work-up to regenerate the starting alde-

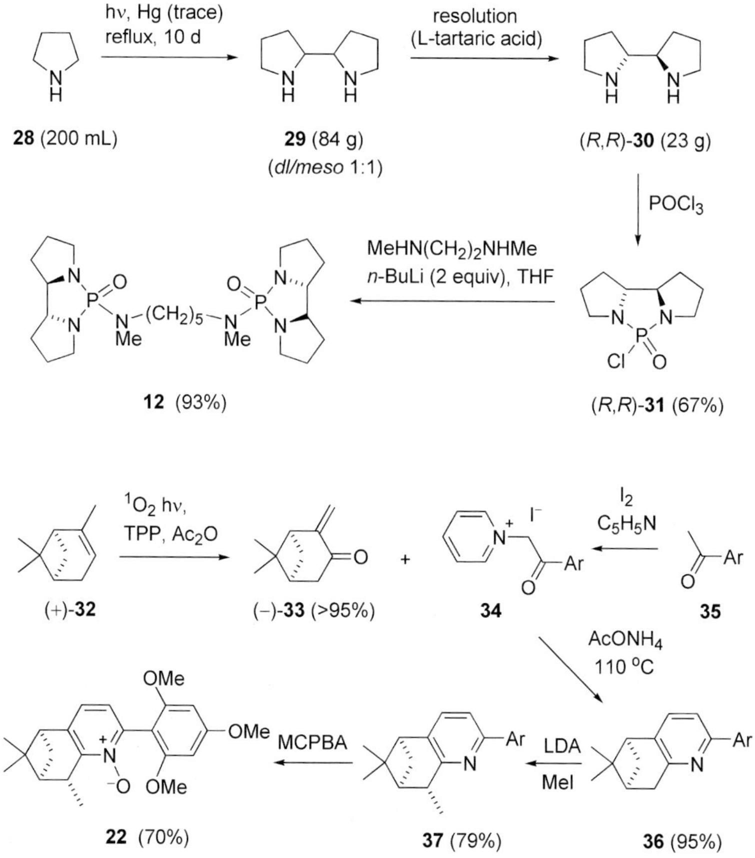

Scheme 7.4 Synthesis of catalysts **12** and **22** (TPP = tetraphenylporphine).

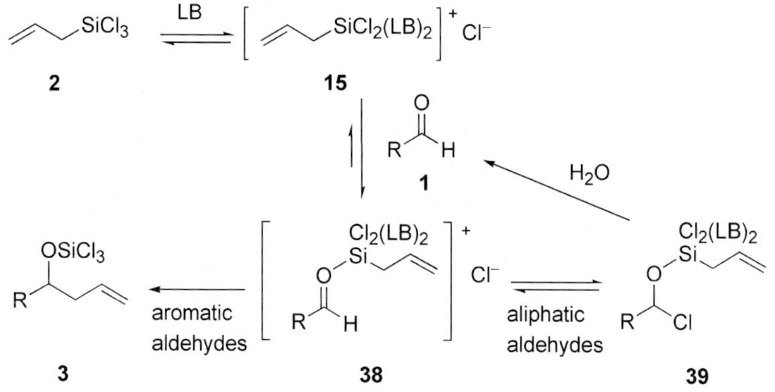

Scheme 7.5 Competing processes in allylation (LB = Lewis base).

hyde **1** – a result that had been previously interpreted as if no reaction had occurred at all. In attempting to block the latter pathway, Denmark used various metal cations, in particular Hg^{2+}, which sequestered Cl^- by formation of the corresponding ate complex [39]. Adding $HgCl_2$ (10 mol%) to the mixture of $PhCH_2CH_2CH=O$, **2a**, and catalyst **12** (5 mol%) in CH_2Cl_2 proved sufficient to reach 56% conversion at room temperature in 2 h. At $-25\,°C$, the yield fell to 32% and the enantioselectivity observed was 11% ee. Similarly, $HgCl_2$ had an accelerating effect on the allylation of benzaldehyde catalyzed by **12**, but the product was found to be racemic [39]. Hence, although promising, this protocol might require extensive further development before it can be considered practical.

7.2.2
Stoichiometric Allylation of Aldehydes and Ketones

As discussed above, catalytic allylation with $AllSiCl_3$ is generally confined to conjugated aldehydes (aromatic, heteroaromatic, and cinnamyl). However, if a stoichiometric activator is employed (e.g., DMF or HMPA as an additive or solvent), the substrate range broadens to aliphatic aldehydes and α-keto acids $RCOCO_2H$ (R = aryl, alkyl) [16]. In the case of chiral aldehydes, such as N-protected alaninal, diastereoselectivity attained in DMF is modest (~3:1) [40]. $AllylSiF_3$ can transfer the allyl group to α-hydroxy ketones and β-diketones on heating with Et_3N in THF [41].

7.2.3
Allylation of Imines

N-Benzoyl hydrazones **40**, derived from aromatic and aliphatic aldehydes, undergo allylation with $AllylSiCl_3$ in DMF at 20 °C [42] or in CH_2Cl_2 with excess of DMSO or Ph_3PO as activator at $-78\,°C$ (Scheme 7.6) [43]. Three equivalents (!) of

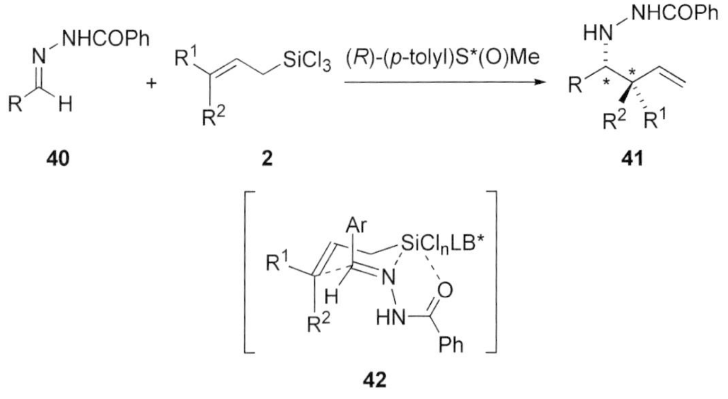

Scheme 7.6 Allylation of acylhydrazones.

(R)-(p-tolyl)S*(O)Me were required to induce the reaction of **40** (R = PhCH$_2$CH$_2$) with **2a**, which gave **41** in up to 93% *ee*. Slightly higher reactivity has been reported for (S)-BINAPO (P,P'-bisoxide of BINAP), of which only 1 equiv. was required to effect the allylation of benzoylhydrazone of ethyl glyoxylate, affording the corresponding (R)-product in up to 98% *ee* [44]. The allylation of hydrazones with (E/Z)-CrotylSiCl$_3$ exhibits the opposite diastereoselectivity to that observed for aldehydes, apparently owing to the different arrangement in the transition state (compare **42** with **5**) [42, 43, 45].

The allylation of hydrazones derived from arylalkyl ketones [42] and of *N*-aryl aldimines in DMF at 0 °C [46] have also been reported.

Aldimines react with AllylSiMe$_3$ in the presence of iodine in MeCN to afford the products of allylation, but an asymmetric version of this method has not yet been developed [47]. Nevertheless, with a chiral auxiliary attached to the imine nitrogen, dual catalysis was reported. Here, In(III) was employed as a Lewis acid to coordinate the imine nitrogen, while Bu$_4$N$^+$F$^-$ was used as a Lewis basic activator of Si, resulting in excellent diastereocontrol [48, 49].

7.3
Propargylation, Allenylation, and Addition of Acetylenes

7.3.1
Addition to Aldehydes

Reactions of propargyl trichlorosilane (**43**) with aromatic, heteroaromatic, and cinnamyl-type aldehydes (**1**), mediated by DMF (acting as a Lewis-base), have been shown to occur in analogy with the allylation, producing allenyl alcohols **44** with >99:1 selectivity (Scheme 7.7). Similarly, allenyl trichlorosilane (**45**) gives homopropargyl alcohols (**46**) [50, 51]. The reagents **43** and **45** were synthesized via a metal-catalyzed reaction of propargyl chloride with Cl$_3$SiH; CuCl catalysis led to **43** (>99:1), whereas the reaction catalyzed by (acac)$_2$Ni produced **45** (>99:1) [50]. The diastereoselectivity of these reactions has not yet been elucidated, but its asymmetric version has been reported. Thus, using formamide **13** (20 mol%) and HMPA (1 equiv.), Kobayashi was able to obtain **44** in up to 95% *ee* with pivalaldehyde as substrate (R = *t*-Bu), while less-hindered aldehydes exhibited lower enantioselectivity (43–79% *ee*) [51c]. Interestingly, benzaldehyde gave ($\pm$)-**44**, even if the catalyst loading was increased to 40 mol%.

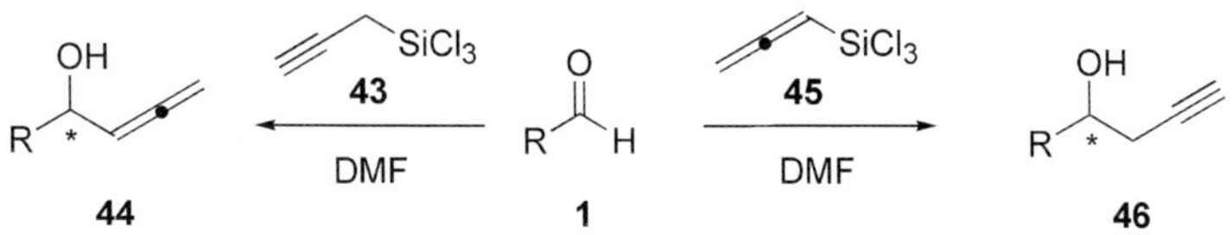

Scheme 7.7 Propargylation and allenylation of aldehydes.

It is noteworthy that, unlike with other propargylic and allenic organometallics, no metallotropic interconversion of **43** and **45** has been observed, which points to an interesting potential for this methodology [50, 51].

An analogous propargylation can be attained with allenyl tributyl stannane; however, this reaction is catalyzed by $SiCl_4$ that is activated by a bisphosphoramide catalyst. Note that in this case, the role of the Lewis basic phosphoramide is to increase the Lewis acidity of $SiCl_4$ rather than to increase the nucleophilicity of the stannane [51b]. A discussion of these effects is provided in Section 7.4.

7.3.2
Addition to Imines

Allenylation of N-tosylimines **47** has been shown to effect the Kwon annulation with the allene derivative **48** on catalysis with the axially chiral phosphine **50** (Scheme 7.8), affording the piperidine derivative **49** with good diastereocontrol (9:1) and high enantioselectivity (93% ee) [52, 53].

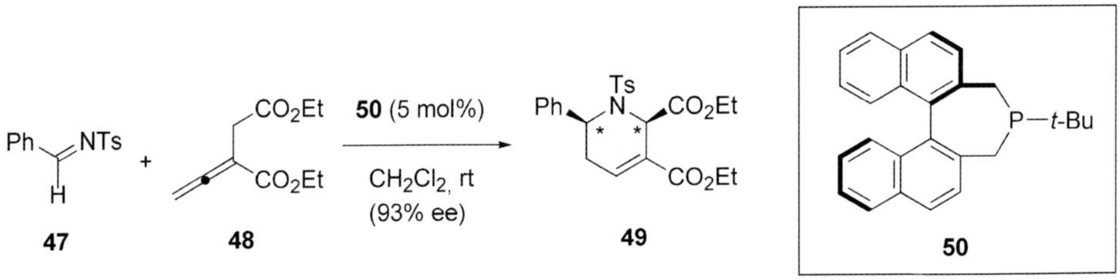

Scheme 7.8 Kwon annulation.

Triethoxysilyl acetylene (**51**) allows a new organocatalytic approach toward the introduction of the alkyne moiety via a nucleophilic addition to aromatic aldehydes, ketones, and aldimines, with EtOK as catalyst (10 mol%). Although a catalytic asymmetric version has not yet been developed, the application of a chiral auxiliary, in the case of imines **53** (Scheme 7.9), led to an impressively high diastereoselectivity (20:1) [54], unparalleled by other acetylenic organometallics.

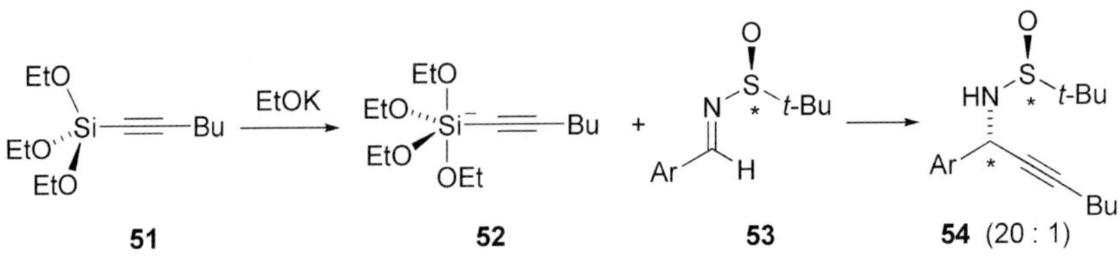

Scheme 7.9 Alkynylation of aldimines.

7.4
Aldol-Type Reactions

According to Mayr's nucleophilicity scale (N), silyl enol ethers derived from aldehydes ($N \geq 3.5$) and ketones ($N \geq 5$) and, in particular, silyl ketene acetals ($N \geq 8$) [55] represent powerful nucleophilic reagents. Indeed, aldol-type addition of trichlorosilyl enol ethers **55a–d** to aldehydes **1** proceeds readily at room temperature, without a catalyst (Scheme 7.10); this contrasts with the lack of reactivity of allyl silanes in the absence of a catalyst. As a result, the reaction exhibits a simple first-order kinetics in each component [56, 57]. Nevertheless, the reaction is substantially accelerated by Lewis bases, which provide a solid ground for the development of an asymmetric variant. The required trichlorosilyl enol ethers **55** can be generated in various ways, for example from the corresponding trimethylsilyl enol ethers on reaction with $SiCl_4$, catalyzed by $(AcO)_2Hg$; from carbonyl compounds or trimethylsilyl enol ethers on treatment with Cl_3SiOTf; from α-chloroketones on reaction with Cl_3SiH and Et_3N; or from the corresponding tributylstannyl enol ethers, etc. [58]. (For experimental details see Chapter 14.1.8).

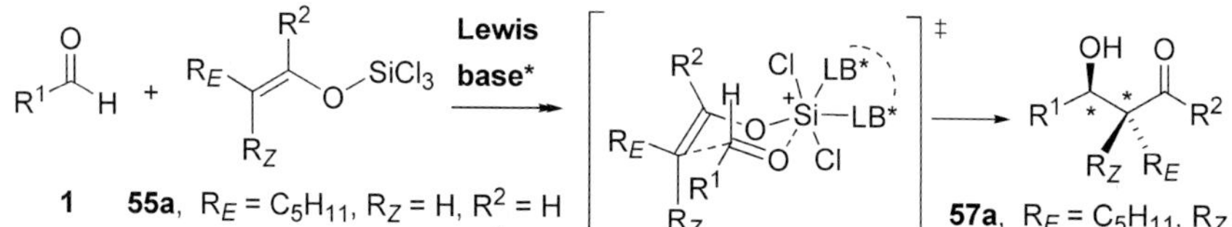

1

55a, $R_E = C_5H_{11}$, $R_Z = H$, $R^2 = H$
55b, $R_E = H$, $R_Z = C_5H_{11}$, $R^2 = H$
55c, $R_E = H$, $R_Z = Me$, $R^2 = Ph$
55d, R_E, $R^2 = (CH_2)_4$, $R_Z = H$
55e, $R_E = H$, $R_Z = H$, $R^2 = Me$

56

57a, $R_E = C_5H_{11}$, $R_Z = H$, $R^2 = H$
57b, $R_E = H$, $R_Z = C_5H_{11}$, $R^2 = H$
57c, $R_E = H$, $R_Z = Me$, $R^2 = Ar$
57d, R_E, $R^2 = (CH_2)_4$, $R_Z = H$
57e, $R_E = H$, $R_Z = H$, $R^2 = Me$

Scheme 7.10 Asymmetric aldol addition of trichlorosilyl enol ethers to aldehydes.

Denmark introduced an array of efficient chiral phosphoramides as nucleophilic activators (Fig. 7.3) for the enantioselective C–C bond formation, and also carried out a detailed mechanistic investigation [56, 59, 60]. Bidentate (e.g., **60**) and smaller monodentate catalysts (**58**) have been shown by Denmark to react via a cationic chair-like transition state **56** with octahedral extracoordinate silicon (Scheme 7.10). Following this manifold, (Z)-enol ethers **55b** and **55c** produced *syn*-adducts **57b** and **57c**, whereas (E)-derivatives **55a** and **55d** furnished *anti*-diastereoisomers **57a** and **57d**. By contrast, with a bulky monodentate activator (e.g., **59**), where coordination of the second catalyst molecule is precluded by steric factors, the reaction exhibits opposite diastereoselectivity, presumably due to the cationic boat-like transition state, where silicon is pentacoordinate. Along this manifold, cyclohexanone-derived enol ether **55d** with the fixed (E)-

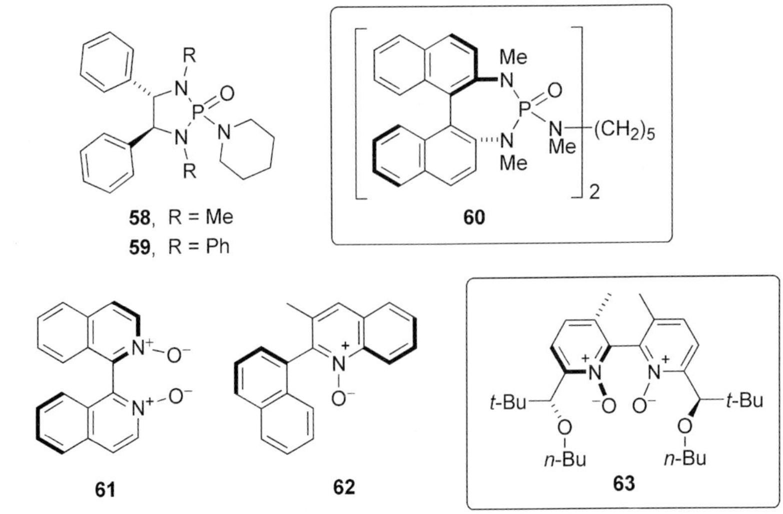

Fig. 7.3 Catalysts for aldol addition.

configuration of the double bond gave rise to the corresponding *syn*-product [56, 60]. Some of these stereoselectivities are highlighted in Table 7.3.

The effect of a remote chiral center in this reaction has also been investigated. In most cases, the reaction was found to be controlled mainly by the chiral catalyst (e.g., **58**), whereas the resident center generally exercised a weak to modest influence [61].

Chiral N-oxides have also been employed as catalysts to promote aldol addition [62], but their true potential remains to be realized. Catalysis by N-oxides follows the same general trends that were established for the phosphoramide activators, though with reduced enantioselectivity. Thus, Nakajima [62] has demonstrated that the reaction of aldehydes **1** with silyl enol ethers **55**, catalyzed by bidentate

Table 7.3 Aldol reaction (Scheme 7.10 and Fig. 7.3) [5a, 60d].

Entry	Catalyst	R^1	R^2	R_E	R_Z	Yield (%)	syn:anti	ee (%)
1	58	Ph	Me	H	H	98	–	87
2	58	PhCH=CH	Bu	H	H	94	–	84
3	58	c-C_6H_{11}	–$(CH_2)_4$–		H	95	1:49	93
4	58	Ph	Ph	H	Me	95	18:1	95
5	58	Ph	Me	H	H	98	–	87
6	60	Ph	H	C_5H_{11}	H	91	3:97	82
7	60	Ph	H	H	C_5H_{11}	92	99:1	90

bis-N-oxides **17** and **61** (catalyst 3 mol%, 1 equiv. (i-Pr$_2$)NEt, CH$_2$Cl$_2$, −78 °C), proceeds via transition state **56**, resulting in high diastereoselectivity and moderate-to-good enantioselectivity (≤82% ee). On the other hand, aldol reaction of cyclic silyl enol ethers of type **55d** with aromatic aldehydes, catalyzed by bulky mono-oxides **24** and **62**, displayed *syn*-selectivity, consistent with participation of the boat-like transition state. The enantioselectivity remained modest in the latter case (up to 72% ee) [62].

The range of substrates in the aldol reaction (and in allylation, as shown above) employing trichlorosilyl reagents is generally restricted to aldehydes, while less-reactive ketones remain essentially inert. However, the exceptionally high nucleophilicity of silyl ketene acetals [57] provides an opportunity to use ketones as substrates (Scheme 7.11). In the absence of an activator, the addition of trichlorosilyl ketene acetal **65** to acetophenone (**64**, R^1 = Ph, R^2 = Me; Scheme 7.11) takes place slowly at 0 °C, paving the way for the development of a catalytic asymmetric variant. Among the large number of Lewis basic promoters investigated, bis-N-oxides emerged as the most promising class in terms of reactivity and enantioselectivity (catalyst 10 mol%, −20 °C, CH$_2$Cl$_2$) [62–64]. Thus, bis-N-oxide (S_{ax},R,R)-**63**, with a matched combination of axial and central chirality, delivered the best results (up to 86% ee), while its mismatched diastereoisomer furnished the opposite enantiomers of **66** with substantially reduced selectivities. Catalysts **17** and **61** also proved inferior [62].

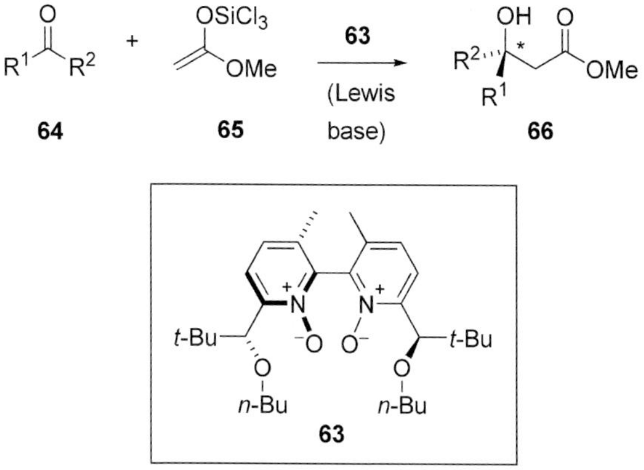

Scheme 7.11 Aldol addition of trichlorosilyl ketene acetals to ketones.

As a further reflection of their high nucleophilicity, silyl ketene acetals, such as **65**, proved to be reactive not only toward aromatic, heteroaromatic, and cinnamyl aldehydes but also, very importantly, even to aliphatic aldehydes. (For experimental details see Chapter 14.1.9). Furthermore, the catalyst loading can be reduced to 1 mol%, without erosion of enantioselectivity [64].

In stark contrast to trichlorosilyl enol ethers (**55** and **65**, etc.), their more readily obtainable trimethylsilyl counterparts, such as **67**, are insensitive to the presence of Lewis bases, owing to the lower Lewis acidity of the silicon atom. Therefore, a different approach is required if these nucleophiles are to be used in aldol chemistry. Here, Denmark turned to the well-established electrophilic activation of the aldehyde, but in an interestingly innovative manner. He employed SiCl$_4$, which itself is too weak a Lewis acid to promote the reaction. However, its coordination to a suitable Lewis base generates a new species that is sufficiently Lewis-acidic (*vide infra*) to facilitate the aldol reaction. Since only the activated species is reactive, the Lewis base can be used in a catalytic amount. Again, as the Lewis base, Denmark used a portfolio of phosphoramides, of which **60** turned out to be optimal (Scheme 7.12). With the latter catalyst (15 mol%) and with a supra-stoichiometric amount of SiCl$_4$ (2 equiv.), the reaction of aromatic aldehydes (**1**, R = aryl, cinnamyl) with the acetaldehyde-derived silyl enol ether **67** proceeded readily to afford the corresponding aldol adducts **68** in good yields and with high *ee*-value (≤94%) [65, 66]. With bulkier trialkylsilyl group (Ph$_2$MeSi or *t*-BuMe$_2$Si) or with aliphatic aldehydes (R = Bu), the reaction does not proceed [65, 66].

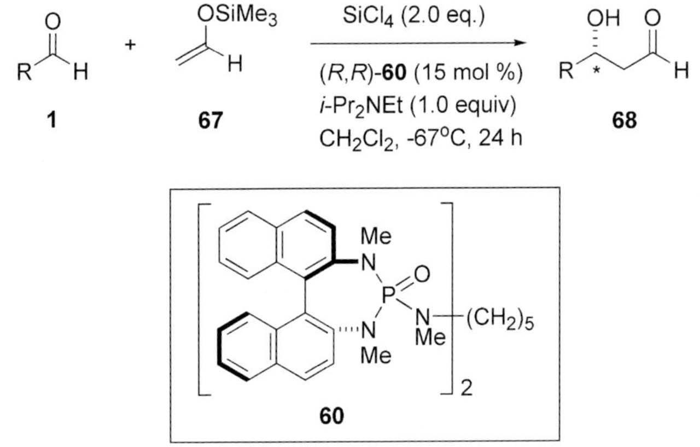

Scheme 7.12 Aldol reaction of TMS silyl enol ether.

Trimethylsilyl ketene acetals **69** react in a similar way (note that a bulky trialkylsilyl group works well); ketene acetals derived from *t*-Bu esters proved to exhibit slightly higher enantioselectivities than their Me, Et, or Ph counterparts (Scheme 7.13 and Table 7.4) [67]. Owing to the higher reactivity of ketene acetals, the catalyst loading could be reduced down to 1–5 mol%. However, in contrast to the trichlorosilyl enolates, this reaction is believed to proceed via an open transition state, since both (*E*) and (*Z*) isomers **69b,c** produce *anti* aldol adducts **70** with high diastereo- and enantioselectivity (Scheme 7.13 and Table 7.5) [67]. Vinylogous aldol-type reaction [67, 68] of aromatic and heteroaromatic aldehydes with silyl dienol ethers **71**, derived from α,β-unsaturated esters, also proceeded

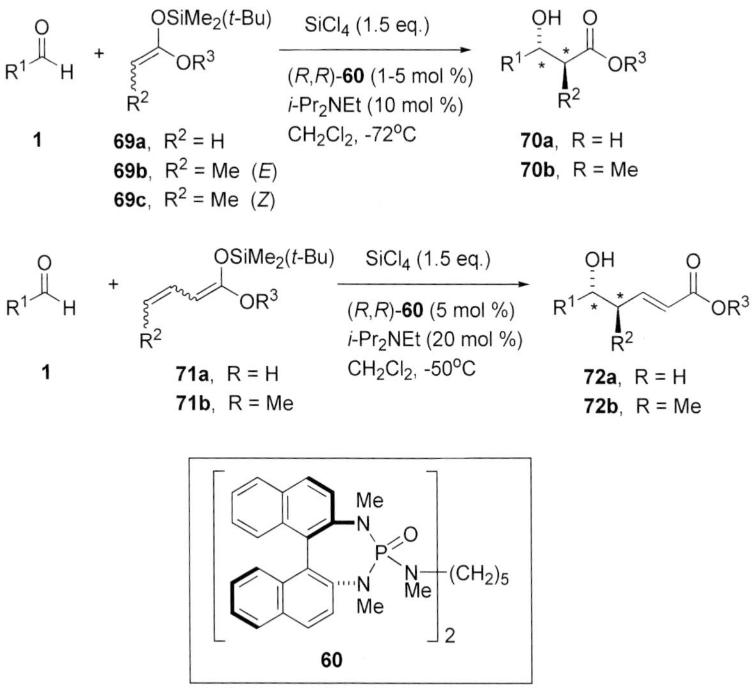

Scheme 7.13 Aldol-type condensation of aldehydes with TBS ketene acetals.

Table 7.4 Aldol-type condensation of aldehydes **1** with silyl ketenes **69a** (R^2 = H, R^3 = Me), catalyzed by **60** (Scheme 7.13) [67].

Entry	R^1	Yield (%)	ee (%)
1	Ph	97	86
2	(E) PhCH=CH	95	94
3	(E) PhCH=CH(Me)	98	45
4	2-furyl	94	87
5	c-C_6H_{11}	86	88
6	$PhCH_2CH_2$	72	81

smoothly, furnishing exclusively the γ-addition products **72** (≥99:1) with high enantioselectivity (Scheme 7.13 and Table 7.6) [67b]. This reaction has also been reported for vinylogous amide analogues of **71** [67c].

The latter activation of Si with a Lewis base requires several comments. Computational analysis [69] revealed that silanes behave like weak Lewis acids towards

Table 7.5 Aldol-type condensation of aldehydes **1** with silyl ketenes **69b/69c** (R^2 = Me, R^3 = t-Bu) catalyzed by **60** (Scheme 7.13) [67].

Entry	1, R^1	69b/c (E:Z)	Yield (%)	anti:syn	ee (%)
1	Ph	69b (95:5)	93	≥99:1	≥99
2	Ph	69c (12:88)	73	≥99:1	99
3	(E) PhCH=CH	69b (95:5)	98	≥99:1	≥99
4	(E) PhCH=CH(Me)	69b (95:5)	90	≥99:1	92
5	c-C_6H_{11}	69b (95:5)	49	89:11	36
6	$PhCH_2CH_2$	69b (95:5)	55	93:7	89

Table 7.6 Aldol-type condensation of aldehydes **1** with vinylogous silyl ketenes **71a/71b** (R^3 = t-Bu), catalyzed by **60** (Scheme 7.13) [67].

Entry	R^1	R^2	Yield (%)	anti:syn	ee (%)
1	Ph	H, 71a	89	–	98
2	(E) PhCH=CH	H, 71a	84	–	97
3	$PhCH_2CH_2$	H, 71a	68	–	90
4	Ph	Me, 71b	92	≥99:1	89
5	(E) PhCH=CH	Me, 71b	71	≥99:1	82
6	$PhCH_2CH_2$	Me, 71b	–	ND	ND

ND = not determined.

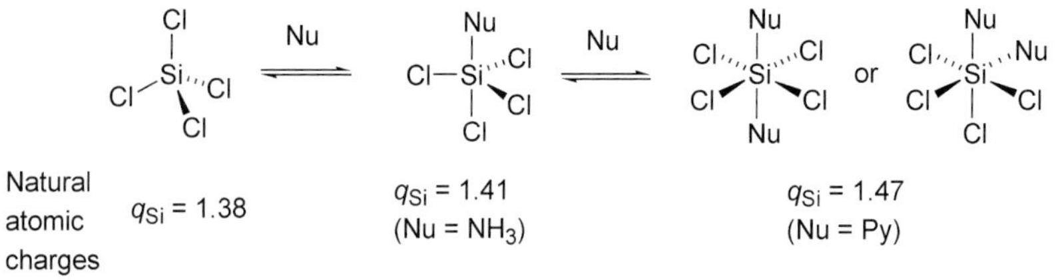

Natural atomic charges: q_{Si} = 1.38 ; q_{Si} = 1.41 (Nu = NH_3) ; q_{Si} = 1.47 (Nu = Py)

Scheme 7.14 The charge on silicon as a function of extra-coordination.

neutral N- and O-donor Lewis bases. The equilibrium (Scheme 7.14) is greatly shifted to the left (i.e., towards tetracoordinate silanes), but stronger Lewis bases or stabilizing intermolecular forces in the solution can enhance the formation of extracoordinate adducts [3e]. When a Lewis base is employed in catalytic quantity, the instability of the latter adduct should be viewed as a positive factor, because this effect can facilitate the catalyst turnover.

Another important feature of the extracoordinate silicon compounds (Scheme 7.14) is the increase in natural atomic charge at the central atom compared to the tetracoordinate precursors [69]. The counter-intuitive increase in the positive charge on silicon, which becomes even more substantial in the case of anionic nucleophiles, such as F$^-$, is compensated by a more negative character of the surrounding groups (X), and this results in an enhanced ionic nature of the Si–X bond. This polarization then favors intermolecular charge–dipole interaction, which results in an increased Lewis acidity of the hypercoordinate silicon [70].

7.5
Hydrocyanation and Isonitrile Addition

7.5.1
Cyanation of Aldehydes

Activation of Me$_3$SiCN by coordination of the Si to lithium BINOL-ate as catalyst has been shown to result in the enantioselective formation of cyanohydrins **73** from aromatic and heteroaromatic aldehydes with 82–98% *ee* (Scheme 7.15) [71]. (For experimental details see Chapter 14.5.4). Several other groups have used dual activation with a chiral Lewis acid and a non-chiral Lewis base [72]. Asymmetric cyanosilylation of PhCOMe and its congeners has also been reported to occur in the presence of sodium phenyl glycinate as catalyst, with up to 94% *ee* [73].

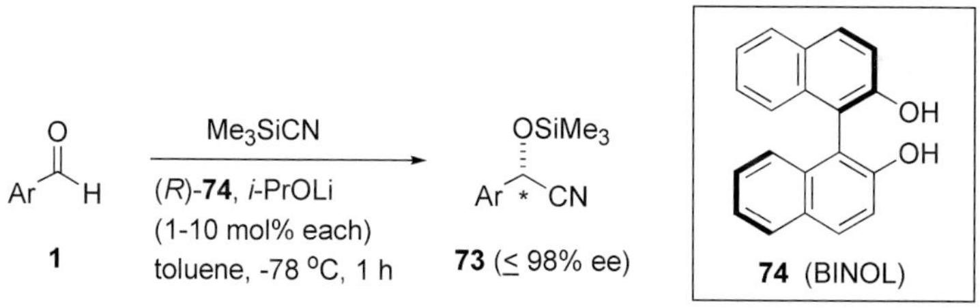

Scheme 7.15 Cyanohydrin formation.

A related reaction is the addition of isonitriles **75** to aldehydes **1** (the Passerini reaction). Denmark has demonstrated that SiCl$_4$, upon activation by a chiral Lewis base, which increased the Lewis acidity of the silicon (*vide supra*; Scheme 7.14), can mediate this reaction to produce α-hydroxy amides **77** after aqueous work-up (Scheme 7.16). Phosphoramide **60** was employed as the chiral Lewis-basic catalyst [74]. Modification of the procedure for hydrolysis of **76** gives rise to the corresponding methyl ester (rather than the amide **77**) [74]. (For experimental details see Chapter 14.5.5).

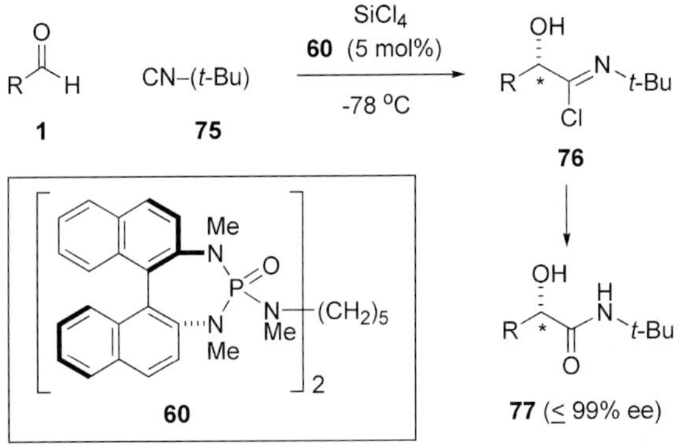

Scheme 7.16 The Passerini reaction. R = aromatic, heteroaromatic, c-Hex.

7.5.2
Cyanation of Imines (Strecker Reaction)

Axially chiral bis-isoquinoline N,N'-bisoxide (S)-**17** has been reported to promote the addition of Me$_3$SiCN (1.5–2.0 equiv.) to imines **78**, derived from aromatic aldehydes (Scheme 7.17); here, CH$_2$Cl$_2$ was identified as an optimal solvent. The reaction is stoichiometric in **17**, and exhibits partial dependence on the imine electronics (62–78% ee), but much less than that observed for the allylation of PhCHO catalyzed by QINOX (**24**) (*vide supra*). The o-substitution had a positive effect in the case of Cl (95% ee), but a very negative effect in the instance of MeO (12% ee). Chelation of the silicon by the N-oxide groups was suggested to account for the stereochemical outcome. Analogues of **17** were much less successful [75].

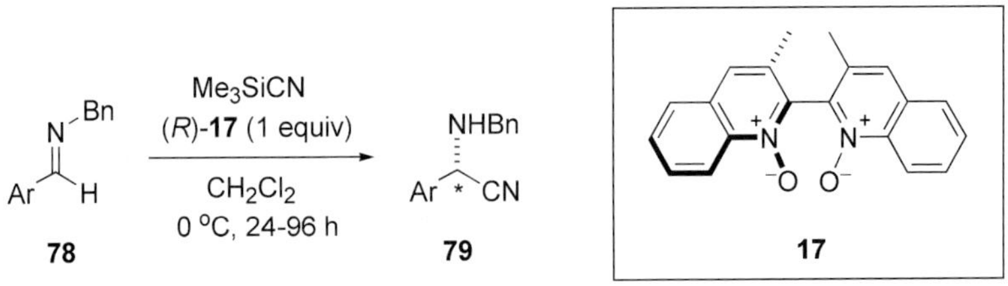

Scheme 7.17 The Strecker reaction.

The combination of Me$_2$AlCN and BINOL (1.2 equiv.) also induced the cyanation of imines derived from benzaldehyde, affording the Strecker products in up to 70% ee [76]. Finally, N-allylbenzaldimines have been reported to react with HCN in the presence of chiral ammonium salt catalysts [77, 78].

7.6
Reduction of Imines

Silanes are widely recognized as efficient reagents for the reduction of carbonyl and heterocarbonyl functionality (Scheme 7.18). In the case of alkyl- and arylsilanes, the reaction requires catalysis by transition-metal complexes [1]; however, with more Lewis acidic trichloro- or trialkoxysilanes, an alternative metal-free activation can be accomplished. It has been demonstrated that extracoordinate silicon hydrides, formed by the coordination of silanes to Lewis bases, such as tertiary amines [3a] or DMF [3b], can serve as mild reagents for the reduction of imines to amines. In the case of trichlorosilane (an inexpensive and relatively easy-to-handle reducing reagent) and DMF acting as a Lewis base, the intermediacy of the hexacoordinate species has been confirmed using ^{29}Si NMR spectroscopy [3b].

Scheme 7.18 Asymmetric reduction of ketimines; see Table 7.5 for R^1-R^3, and Figure 7.4 for catalysts.

Not surprisingly, chiral formamides emerged as prime candidates for the development of an asymmetric variant of this reaction. A selection of the most efficient amide catalysts based on amino acids is shown in Figure 7.4; representative examples of enantioselective hydrosilylation are collected in Tables 7.7 and 7.8. Proline-derived anilide **82a** and its naphthyl analogue **82b**, introduced by Matsumura [3c], produced moderate enantioselectivity in the reduction of aromatic ketimines with trichlorosilane at 10 mol% catalyst loading (Table 7.7, entries 1 and 2). Formamide functionality proved to be crucial for the activation of the silane, as the corresponding acetamides failed to initiate the reaction.

Replacing the rigid proline core with a more flexible pipecolinic acid, as in **83**, led to an increased selectivity (Table 7.7, entry 3) [79]. However, a real improvement was attained with catalysts **84a,b**, where 1,2-diphenylamino ethanol derivatives were used instead of aniline to form the amide (entries 4 and 5). Configuration at the α-carbon of the amino alcohol moiety proved to be critical in forming the matched pair (compare entries 4 and 6), while configuration at the β-carbon bore no influence on the enantioselectivity (compare entries 4 and 5).

An alternative approach to improve the catalyst design was taken by Malkov and Kočovský [80], who employed acyclic N-methyl amino acids in the core of the catalyst structure, as illustrated by catalysts **85** and **86** (Table 7.8). (For experimental details see Chapter 14.21.3). Intrinsic rotational flexibility of the bond linking the stereogenic center with the formamide nitrogen allowed the molecule to adopt the most favorable conformation in the transition state. As a result,

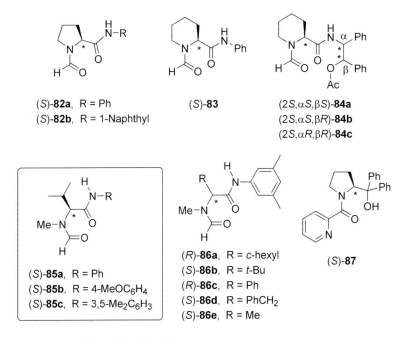

Fig. 7.4 Amino acid-derived chiral amides.

valine-derived catalysts **85a–c**, having a single chiral center, delivered the level of enantioselectivity similar to more complex systems **84a,b**. Furthermore, despite the same configuration of the parent amino acids in **82–85** [81], the valine-based catalysts **85** induced the formation of the opposite enantiomer of the product (compare Table 7.7, entries 1 and 3, to Table 7.8, entries 1–3). Variation of the side chain of the amino acid scaffold revealed that the cyclohexyl (**86a**) and *t*-butyl (**86b**) analogues showed only a marginally reduced efficiency compared to their valine counterpart (Table 7.8, entries 4 and 5), whereas modest to low enantioselectivities were observed for the phenylglycine-, phenylalanine- and alanine-derived catalysts **86c–e** [80b].

In general, catalysis by **84a** and **85c** resulted in good to excellent enantioselectivities in the reduction of ketimines derived from methyl aryl ketones and aromatic amines (**80**, R^1, R^3 = aryl, R^2 = Me), where the electronic effects of substituents in both aromatic groups did not show any significant influence [79, 80]. On the other hand, imines obtained from aliphatic amines (**80**, R^3 = alkyl) gave virtually racemic products with **85a** [80b]. In the reduction of non-aromatic imines, such as **80c**, only catalyst **84a** maintained high enantioselectivity (Table 7.7, entry 9), whereas the valine-derived congener **85c** proved less efficient (Table 7.8, entry 6). Good enantioselectivities were recorded when catalysts **84a** and **85c** were employed in the hydrosilylation of α,β-unsaturated imine **80d** (Table 7.7, entry 10; and Table 7.8, entry 7) [79, 80], while the synthesis of *N*-phenyl phenylglycine was accomplished in moderate **ee** (Table 7.8, entry 8).

Table 7.7 Asymmetric reduction of imines **80** with trichlorosilanes catalyzed by chiral amides (10 mol%) derived from cyclic amino acids in CH_2Cl_2 (Scheme 7.18 and Fig. 7.4) [3c, 79, 81, 82].

Entry	Imine R^1, R^2, R^3	Catalyst	Temp (°C)	Yield (%)	ee (%) (config.)
1	**80a**, Ph, Me, Ph	(S)-**82a**	ambient	91	55 (R)
2	**80a**	(S)-**82b**	ambient	52	66 (R)
3	**80a**	(S)-**83**	0	94	73 (R)
4	**80a**	(S,S,S)-**84a**	0	97	94 (R)
5	**80a**	(S,S,R)-**84b**	0	97	93 (R)
6	**80a**	(S,R,R)-**84c**	0	71	47 (R)
7	**80a**	(S)-**87**	ambient	86	73 (S)
8	**80b**, Ph, Me, PMP	(S)-**87**	ambient	96	75 (S)
9	**80c**, c-C_6H_{11}, Me, Ph	(S,S,S)-**84a**	0	81	95 (?)
10	**80d**, PhCH=CH, Me, Ph	(S,S,S)-**84a**	0	81	87 (?)

PMP = p-methoxyphenyl.

Table 7.8 Asymmetric reduction of imines **80** with trichlorosilanes catalyzed by chiral amides (10 mol%) derived from non-cyclic amino acids (Scheme 7.18 and Fig. 7.4) [80].

Entry	Imine R^1, R^2, R^3	Catalyst	Solvent	Temp (°C)	Yield (%)	ee (%) (config.)
1	**80a**	(S)-**85a**	$CHCl_3$	ambient	79	86 (S)
2	**80a**	(S)-**85b**	$CHCl_3$	ambient	62	85 (S)
3	**80a**	(S)-**85c**	Toluene	ambient	81	92 (S)
4	**80b**, Ph, Me, PMP	(R)-**86a**	Toluene	ambient	95	82 (R)
5	**80b**	(S)-**86b**	Toluene	ambient	95	83 (S)
6	**80c**	(S)-**85a**	$CHCl_3$	20	53	59 (S)
7	**80d**	(S)-**85c**	Toluene	20	64	81 (S)
8	**80e**, Ph, CO_2Me, Ph	(S)-**85c**	Toluene	20	69	59 (S)

PMP = p-methoxyphenyl.

The proline motif was recently rekindled by Matsumura, who modified the original structure of **82a** by replacing the formamide group with picolinoylamide and the anilide moiety with a diphenylcarbinol segment [82]. The new catalyst **87** thus obtained exhibited increased enantioselectivities and, compared with **82a**, induced the opposite configuration of the products (Table 7.7, entries 7 and 8).

Malkov and Kočovský [83] recently introduced a different type of Lewis-basic activator, built around a (pyridyl)oxazoline framework, which featured in the

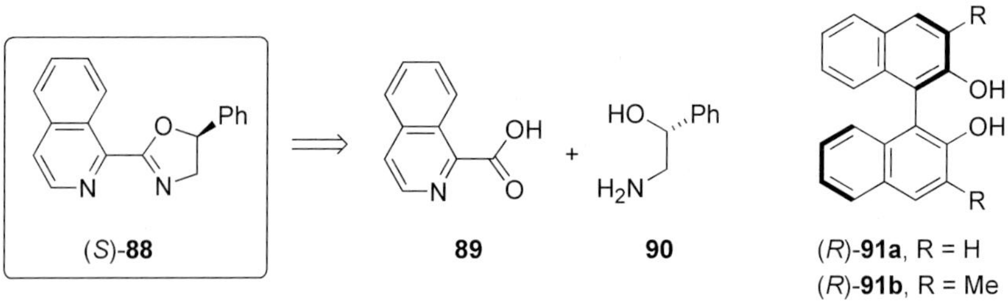

Fig. 7.5 Catalyst **88** combines isoquinoline and chiral oxazoline derived from mandelic acid.

chiral ligand design in the related transition metal-mediated hydrosilylation of ketones [1]. Catalyst **88** (20 mol%), which combines isoquinoline and chiral oxazoline derived from mandelic acid (Fig. 7.5), exhibited high enantioselectivity (85–87% *ee*) in the reduction of a range of aromatic imines (**80**, R^1, R^3 = aryl, R^2 = Me). The remote position of the stereogenic center was found to be vital for catalyst performance as the positional isomer of **88**, derived from phenylglycinol, where the phenyl substituent is adjacent to the carbon next to the nitrogen atom, was significantly less efficient in terms of reactivity and enantioselectivity. It is important to note that catalyst **88** was equally effective in the asymmetric reduction of aromatic ketones, which makes it one of the rare systems (including metal catalysts) [84] applicable to both imines and ketones. (Note that ketone reduction is treated in Chapter 11.)

Trialkoxysilanes are less sensitive to moisture and thus represent an attractive alternative to trichlorosilane as reducing reagents. However, the activation of trialkoxysilanes requires stronger bases, such as alkoxide-anions, and this area remains in its early stages of development. Hosomi [85] investigated the catalytic reduction of tosylimine **80** (R^1 = Ph, R^2 = Me, R^3 = Ts) with $(MeO)_3SiH$, screening a number of chiral alcoholates. The dilithium salt of BINOL **91a** (20 mol%) emerged on top, producing the corresponding tosylamide **81** in 65% *ee*. Attempts to introduce an additional steric bias to the catalyst structure, as in **91b**, proved fruitless. However, enantioselectivity improved to 72% when *p*-nitrobenzenesulfonyl group was used in place of tosyl protection.

An alternative method for the organocatalytic reduction of imines employs Brønsted acids as catalysts and Hantzsch dihydropyridine as a reducing reagent. This topic is described in Chapter 11 (on Brønsted acids) [86].

7.7
Epoxide Opening

Epoxides are versatile intermediates in organic synthesis owing to the possibility of controlled stereoselective ring opening. The ability of various chlorosilanes to serve as a source of chloride ion for the opening of epoxides was first recognized

50 years ago [87]. Later, it was shown that the reaction could be dramatically accelerated by nucleophilic catalysts, such as phosphines, imidazole [88] and, most recently, by phosphorous heterocycles [89] and HMPA [90].

Although the mechanism of the nucleophile-assisted opening of epoxides has not been established in detail, the following picture can be suggested (Scheme 7.19). Initially, tetrachlorosilane and a monodentate Lewis base (LB) form a pentacoordinate donor–acceptor complex **92**, while in the case of a bidentate Lewis base, the formation of hexacoordinate complex **93** can be envisaged. The silicon center in the extracoordinate derivatives, such as **92**, becomes more Lewis acidic due to the increased positive charge (*vide supra*; Scheme 7.14), thus promoting the interaction with epoxide **94**. The displacement of chloride to form salt **95** can follow either the associative or dissociative pathway, the latter proceeding via tetracoordinate cationic species [90]. It is pertinent to note that only dissociative mechanisms may operate in the case of complex **93**; this is due to saturation of the valence shell of the silicon. Opening of the activated epoxide by chloride ion then proceeds in an S_N2 fashion to yield intermediate **96**. Due to the reversible nature of the donor–acceptor interactions between the Lewis base and silicon, complex **92** is then regenerated, releasing the resulting silylated chlorohydrin **97**, which completes the catalytic cycle.

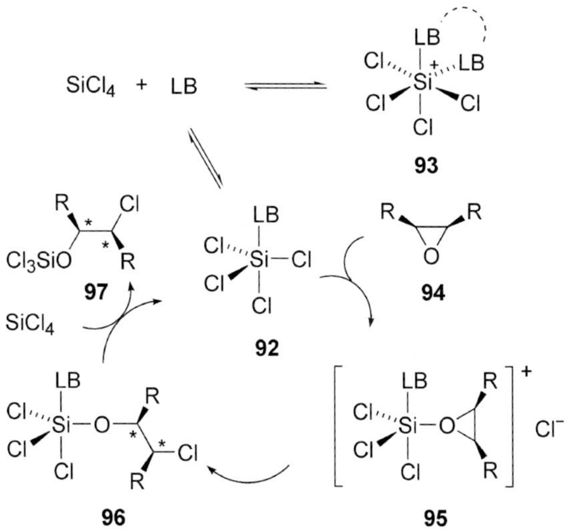

Scheme 7.19 Epoxide ring-opening by extra-coordinate $SiCl_4$.

According to this mechanism (Scheme 7.19), the process is amenable to asymmetric modifications. Among a variety of silicon reagents that were examined, only tetrachlorosilane proved to be suitable for the asymmetric process, while the application of other chlorosilanes resulted in the formation of racemates. A

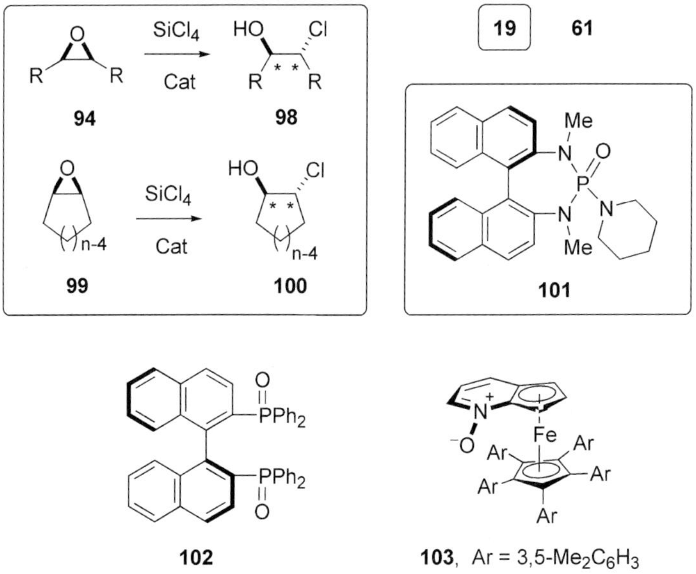

Scheme 7.20 Ring opening in *meso*-epoxides.

number of chiral Lewis bases were investigated in the opening of *meso*-epoxides **94** and **99** (Scheme 7.20) to produce the corresponding chlorohydrins **98** and **100**. The most successful catalysts **19**, **61**, and **101–103** are shown in Scheme 7.20, with a selection of representative examples highlighted in Table 7.9.

Table 7.9 Enantioselective ring opening of *meso*-epoxides catalyzed by chiral Lewis bases [51a, 90–93].

Entry	Epoxide	Catalyst (mol%)	Time (h)	Temp (°C)	Yield (%)	ee (%)
1	94a, R = Ph	101 (10)	3	−78	94	87
2	94a	103 (5)	24	−85	88	94
3	94a	61 (10)	6	−78	95	90
4	94a	102 (10)	4	−78	94	90
5	94b, R = CH$_2$OBn	101 (10)	4	−78	95	71
6	94b	103 (5)	24	−85	91	50
7	94b	61 (10)	6	−78	98	74
8	94b	102 (10)	12	−78	97	31
9	99a, n = 6	101 (10)	0.3	−78	90	51
10	99a	102 (10)	1	−78	81	71
11	99b, n = 8	102 (10)	72	−78	81	50
12	99b	19 (10)	48	−90	80	85

In general, the efficiency of the catalysts was found to be substrate-dependent; none of the catalysts was able to provide a good level of enantioselectivity with a wide range of acyclic **94** and cyclic **99** epoxides (Scheme 7.20). Axially chiral phosphoramide **101** [90], bis-isoquinoline N,N'-dioxide **61** [91], BINAPO **102** [51a], and planar chiral pyridine N-oxide **103** [92] exhibited high enantioselectivity in the opening of derivatives of cis-stilbene oxide **94a** (R = Ar) (Table 7.9, entries 1–4). In the case of the less-sterically demanding aliphatic **94b** (R = CH_2OBn) and cyclic epoxides **99**, the selectivity with these catalysts fell to a moderate level, and below (entries 5–10). Cyclooctene oxide **99b** ($n = 8$) proved to be a particularly difficult substrate, showing low reactivity and selectivity with most of the Lewis bases except for the pyridine mono-N-oxide **19**, which furnished the corresponding chlorohydrin in 85% ee (see footnote 25 in Ref. [27]); however, catalyst **19** was ineffective with other substrate types [93]. (For experimental details see Chapter 14.14.1).

7.8
Conclusions and Outlook

Organocatalysis, in general, is experiencing a fascinating boom, and Lewis-basic catalysts are no exception. The chemistry described in this chapter is dominated by silicon, an inexpensive and environmentally friendly element. Hence, in addition to the fundamental issues being addressed by this research, significant practical applications can be expected in the near future. Further developments of the catalytic, enantioselective processes discussed in this chapter will undoubtedly continue, aiming at more efficient catalysts, lower catalyst loadings, and the simplification of their syntheses, all supported by detailed mechanistic studies. Clearly, more valuable methodology and new concepts involving Lewis bases as catalysts are on the horizon.

References

1 Jacobsen, E.N., Pfaltz, A., Yamamoto, H., *Comprehensive Asymmetric Catalysis*, Springer: Heidelberg, Vols. I–III, **1999**.

2 (a) Marshal, J.A., *Chem. Rev.* **1996**, *96*, 31; (b) Marshal, J.A. In: Schlosser, M. (Ed.), *Organometallics in Synthesis*. J. Wiley & Sons, Chichester, **2002**, p. 399.

3 Activation of Cl_3SiH: (a) Benkeser, R.A., Snyder, D., *J. Organomet. Chem.* **1982**, *225*, 107; (b) Kobayashi, S., Yasuda, M., Hachiya, I., *Chem. Lett.* **1996**, 407; (c) Iwasaki, F., Onomura, O., Mishima, K., Maki, T., Matsumura, Y., *Tetrahedron Lett.* **1999**, *40*, 7507; (c) Iwasaki, F., Onomura, O., Mishima, K., Kanematsu, T., Maki, T., Matsumura, Y., *Tetrahedron Lett.* **2001**, *42*, 2525; (e) Brook, M.A., *Silicon in Organic, Organometallic and Polymeric Chemistry*. Wiley, New York, **2000**, pp. 133–136.

4 For the coordination of the C=O group to Lewis acids (E vs. Z; η^1 vs η^2 fashion), see, e.g.: (a) Shambayati, S., Crowe, W.E., Schreiber, S.L., *Angew. Chem., Int. Ed. Engl.* **1990**, *29*, 256; (b) Lenges, C.P., Brookhart, M., White, P.S., *Angew. Chem., Int. Ed.* **1999**, *38*, 552.

5 For leading reviews, see: (a) Denmark, S.E., Stavenger, R.A., *Acc. Chem. Res.* **2000**, *33*, 432; (b) Denmark, S.E., Fu, J., *Chem. Commun.* **2003**, 167; (c) Denmark, S.E., Fu, J., *Chem. Rev.* **2003**, *103*, 2763; (d) Kennedy, J.W.J., Hall, D.G., *Angew. Chem., Int. Ed.* **2003**, *42*, 4732. For a detailed discussion of changing the electronics of the silicon upon extra coordination to Lewis bases, see: (e) Rendler, S., Oestreich, M., *Synthesis* **2005**, 1727.

6 For the use of stoichiometric, achiral Lewis acids, see, e.g. (a) Kira, M., Kobayashi, M., Sakurai, H., *Tetrahedron Lett.* **1987**, *28*, 4081; (b) Kira, M., Sato, K., Sakurai, H., *J. Am. Chem. Soc.* **1990**, *112*, 257. For an overview of chiral Lewis acids, see Ref. [1].

7 Allyl trichlorosilane is conveniently prepared from allyl chloride, Cl_3SiH, CuCl, and $(i\text{-}Pr)_2EtN$ (or Et_3N) in ether at 20 °C: (a) Kobayashi, S., Nishio, K., *Chem. Lett.* **1994**, 1773; (b) Nakajima, M., Saito, M., Hashimoto, S., *Chem. Pharm. Bull.* **2000**, *48*, 306. For an overview of the various other methods, see: (c) Kočovský, P. In: Paquette, L.A. (Ed.), *Encyclopedia of Reagents for Organic Synthesis (eEROS)*. J. Wiley & Sons, New York, **2004**, RN00568.

8 (a) Kobayashi, S., Nishio, K., *Tetrahedron Lett.* **1993**, *34*, 3453; (b) Kobayashi, S., Nishio, K., *J. Org. Chem.* **1994**, *59*, 6620; (c) Kobayashi, S., Nishio, K., *Synthesis* **1994**, 457.

9 Denmark, S.E., Coe, D.M., Pratt, N.E., Griedel, B.D., *J. Org. Chem.* **1994**, *59*, 6161.

10 (a) Kira, M., Sato, K., Sakurai, H., *J. Am. Chem. Soc.* **1988**, *110*, 4599; (b) Kira, M., Zhang, L.C., Kabuto, C., Sakurai, H., *Organometallics* **1998**, *17*, 887. For an overview, see: (c) Sakurai, H., *Synlett* **1989**, 1.

11 (a) Chemler, S.R., Roush, W.R., *J. Org. Chem.* **2004**, *68*, 1319; (b) Aoyama, N., Hamada, T., Manabe, K., Kobayashi, S., *Chem. Commun.* **2003**, 676; (c) Aoyama, N., Hamada, T., Manabe, K., Kobayashi, S., *J. Org. Chem.* **2003**, *68*, 7329.

12 Short, J.D., Attenoux, S., Berrisford, D.J., *Tetrahedron Lett.* **1997**, *38*, 2351.

13 For the reaction catalyzed by Lewis acids, see the following: Fleming, I. Allylsilanes, Allylstannanes and Related Systems. In: Trost, B.M., Fleming, I. (Eds.), *Comprehensive Organic Synthesis*. Pergamon, Oxford, **1991**, Vol. 2, p. 563.

14 *trans*-Crotyl trichlorosilane can be synthesized from *trans*-crotyl chloride, Cl_3SiH, CuCl, and $(i\text{-}Pr)_2EtN$ in ether at 20 °C; see Ref. [7] and the following: (a) Aoki, S., Mikami, K., Terada, M., Nakai, T., *Tetrahedron* **1993**, *49*, 1783; (b) D'Aniello, F., Falorni, M., Mann, A., Taddei, M., *Tetrahedron: Asymm.* **1966**, *7*, 1217; (c) Iseki, K., Kuroki, Y., Takahashi, M., Kishimoto, S., Kobayashi, Y., *Tetrahedron* **1997**, *53*, 3513; (d) Shibato, A., Itagaki, Y., Tayama, E., Hokke, Y., Asao, N., Maruoka, K., *Tetrahedron* **2000**, *56*, 5373. For an alternative route, utilizing *trans*-crotyl chloride, $SiCl_4$, and Cp_2Ni in HMPA at 90 °C, see: (e) Calas, R., Dunogues, J., Deleris, G., Duffaut, N., *J. Organomet. Chem.* **1982**, *225*, 117; (f) Lefort, M., Simmonet, C., Birot, M., Deleris, G., Dunogues, J., Calas, R., *Tetrahedron Lett.* **1980**, *21*, 1857. Yet another alternative, using *trans*-crotyl chloride with Cl_3SiH and Bu_4PCl at 150 °C has also been reported: (g) Cho, Y.-S., Kang, S.-H., Han, J.S., Yoo, B.R., Jung, N., *J. Am. Chem. Soc.* **2001**, *123*, 5584.

15 *cis*-Crotyl trichlorosilane can be prepared via the 1,4-addition of Cl_3SiH to butadiene, catalyzed by $(PhCN)_2PdCl_2$ [10] or $(Ph_4P)_4Pd$ (at 20 or −78 °C); for the latter catalyst, see Ref. [14c] and the following: (a) Tsuji, J., Hara, M., Ohno, K., *Tetrahedron* **1974**, *30*, 2143; (b) Wadamoto, M., Ozasa, N., Yanagisawa, A., Yamamoto, H., *J. Org. Chem.* **2003**, *63*, 5593; (c) Iseki, K., Kuroki, Y., Takahashi, M., Kobayashi, Y., *Tetrahedron Lett.* **1996**, *37*, 5149. This hydrosilylation can also be catalyzed by Ni: (d) Čapka, M., Hetflejš, J., *Collect. Czech. Chem.*

Commun. **1975**, *40*, 2073 and 3020. An uncatalyzed reaction occurs at 500–600 °C: (e) Heinicke, J., Kirst, I., Gehrhus, B., Tzschach, A., *Z. Chem.* **1988**, *28*, 261.

16 Wang, Z., Xu, G., Wang, D., Pierce, M.E., Confalone, P.N., *Tetrahedron Lett.* **2000**, *41*, 4523.

17 (a) Iseki, K., Mizuno, S., Kuroki, Y., Kobayashi, Y., *Tetrahedron Lett.* **1998**, *39*, 2767; (b) Iseki, K., Mizuno, S., Kuroki, Y., Kobayashi, Y., *Tetrahedron* **1999**, *55*, 977.

18 Ogawa, C., Sugiura, M., Kobayashi, S., *Chem. Commun.* **2003**, 192.

19 Chataigner, I., Piarulli, U., Gennari, C., *Tetrahedron Lett.* **1999**, *40*, 3633.

20 For reviews, see, e.g.: (a) Malkov, A.V., Kočovský, P., *Curr. Org. Chem.* **2003**, *7*, 1737; (b) Chelucci, G., Murineddu, G., Pinna, G.A., *Tetrahedron: Asymm.* **2004**, *15*, 1373. (c) Malkov, A.V., Kočovský, P., *Eur. J. Org. Chem.* **2007**, 29.

21 Yamasaki, S., Fuji, K., Wada, R., Kanai, M., Shibasaki, M., *J. Am. Chem. Soc.* **2002**, *124*, 6536.

22 (a) Denmark, S.E., Coe, D.M., Pratt, N.E., Griedel, B.D., *J. Org. Chem.* **1994**, *59*, 6161; (b) Denmark, S.E., Fu, J., *J. Am. Chem. Soc.* **2000**, *122*, 12021; (c) Denmark, S.E., Fu, J., *J. Am. Chem. Soc.* **2001**, *123*, 9488; (d) Denmark, S.E., Fu, J., *Org. Lett.* **2002**, *4*, 1951; (e) Denmark, S.E., Fu, J., *J. Am. Chem. Soc.* **2003**, *125*, 2208; (f) Denmark, S.E., Fu, J., Coe, D.M., Su, X., Pratt, N.E., Griedel, B.D., *J. Org. Chem.* **2006**, *71*, 1513.

23 Hellwig, J., Belser, T., Müller, J.F.K., *Tetrahedron Lett.* **2001**, *42*, 5417.

24 Chiral phosphoramidates attached to polystyrene have also been reported to exhibit asymmetric induction in the allylation of PhCHO with AllylSiCl$_3$ ($\leq$ 63% ee): Oyama, T., Yoshioka, H., Tomoi, M., *Chem. Commun.* **2005**, 1857.

25 (a) Nakajima, M., Saito, M., Shiro, M., Hashimoto, S., *J. Am. Chem. Soc.* **1998**, *120*, 6419; (b) Nakajima, M., Saito, M., Hashimoto, S., *Chem. Pharm. Bull.* **2000**, *48*, 306.

26 (a) Shimada, T., Kina, A., Ikeda, S., Hayashi, T., *Org. Lett.* **2002**, *4*, 2799; (b) Shimada, T., Kina, A., Hayashi, T., *J. Org. Chem.* **2003**, *68*, 6329; (c) Kina, A., Shimada, T., Hayashi, T., *Adv. Synth. Catal.* **2004**, *346*, 1169.

27 (a) Malkov, A.V., Orsini, M., Pernazza, D., Muir, K.W., Langer, V., Meghani, P., Kočovský, P., *Org. Lett.* **2002**, *4*, 1047; (b) Malkov, A.V., Bell, M., Orsini, M., Pernazza, D., Massa, A., Herrmann, P., Meghani, P., Kočovský, P., *J. Org. Chem.* **2003**, *68*, 9659.

28 (a) Malkov, A.V., Bell, M., Vassieu, M., Bugatti, V., Kočovský, P., *J. Mol. Catal. A* **2003**, *196*, 179; (b) Malkov, A.V., Bell, M., Castelluzzo, F., Kočovský, P., *Org. Lett.* **2005**, *7*, 3219.

29 Malkov, A.V., Kočovský, P., unpublished results.

30 Malkov, A.V., Dufková, L., Farrugia, L., Kočovský, P., *Angew. Chem., Int. Ed.* **2003**, *42*, 3674.

31 Traverse, J.F., Zhao, Y., Hoveyda, A.H., Snapper, M.L., *Org. Lett.* **2005**, *7*, 3151.

32 Pignataro, L., Benaglia, M., Annunziata, R., Cinquini, M., Cozzi, F., *J. Org. Chem.* **2006**, *71*, 1458.

33 Wong, W.-L., Lee, C.-S., Leung, H.K., Kwong, H.-L., *Org. Biomol. Chem.* **2004**, 1967.

34 Angell, R.M., Barrett, A.G.M., Braddock, D.C., Swallow, S., Vickery, B.D., *Chem. Commun.* **1997**, 919.

35 (a) Kobayashi, S., Ogawa, C., Konishi, H., Sugiura, M., *J. Am. Chem. Soc.* **2003**, *125*, 6610; (b) Massa, A., Malkov, A.V., Kočovský, P., Scettri, A., *Tetrahedron Lett.* **2003**, *44*, 7179; (c) Rowlands, G.J., Barnes, W.K., *Chem. Commun.* **2003**, 2712.

36 It is pertinent to note that analogues of QUINOX (**24**) with a substituent next to the N–O group exhibit low reactivity and drastically decreased enantioselectivity, apparently owing to interference of the substituent in the transition state: (a) Hrdina, R., Stará, I.G., Dufková, L., Mitchell, S., Císařová, I., Kotora, M., *Tetrahedron* **2006**, *62*, 968; (b) Malkov, A.V., Gutnov, A., Kočovský, P., unpublished results.

37 (a) Ferguson, R.R., Boojmra, C.G., Brown, S.H., Crabtree, R.H.,

Heterocycles **1989**, *28*, 121; (b) Krajnik, P., Ferguson, R.R., Crabtree, R.H., *New J. Chem.* **1993**, *17*, 559.

38 For the Kröhnke annulation, see: (a) Kröhnke, F., *Chem. Ber.* **1937**, *70*, 864. For a review, see: (b) Kröhnke, F., *Synthesis* **1976**, 1.

39 Denmark, S.E., Fu, J., Lawler, M.J., *J. Org. Chem.* **2006**, *71*, 1523.

40 (a) Gryko, D., Urbanczyk-Lipkowska, Z., Jurczak, J., *Tetrahedron* **1997**, *53*, 13373; (b) Gryko, D., Jurczak, J., *Helv. Chim. Acta* **2000**, *83*, 2705.

41 (a) Sato, K., Kira, M., Sakurai, H., *J. Am. Chem. Soc.* **1989**, *111*, 6429; (b) Kira, M., Sato, K., Sekimoto, K., Gewald, R., Sakurai, H., *Chem. Lett.* **1995**, 281; (c) Gewald, R., Kira, M., Sakurai, H., *Synthesis* **1996**, 111; (d) Chemler, S.R., Roush, W.R., *Tetrahedron Lett.* **1999**, *40*, 4643; (e) Chemler, S.R., Roush, W.R., *J. Org. Chem.* **2003**, *68*, 1319.

42 (a) Kobayashi, S., Hirabayashi, R., *J. Am. Chem. Soc.* **1999**, *121*, 6942; (b) Hirabayashi, R., Ogawa, C., Sugiura, M., Kobayashi, S., *J. Am. Chem. Soc.* **2001**, *123*, 9493.

43 (a) Kobayashi, S., Ogawa, C., Konishi, H., Sugiura, M., *J. Am. Chem. Soc.* **2003**, *125*, 6610; (b) Ogawa, C., Konishi, H., Sugiura, M., Kobayashi, S., *Chem. Commun.* **2004**, 446.

44 Ogawa, C., Sugiura, M., Kobayashi, S., *Angew. Chem., Int. Ed.* **2004**, *43*, 6491. For an overview on the allylation of acylhydrazones, using both Lewis acids and bases and chiral auxiliaries, see: (b) Sugiura, M., Kobayashi, S., *Angew. Chem., Int. Ed.* **2005**, *44*, 5176.

45 Ogawa, C., Sugiura, M., Kobayashi, S., *J. Org. Chem.* **2002**, *67*, 5359.

46 Sugiura, M., Robvieux, F., Kobayashi, S., *Synlett* **2003**, 1749.

47 Phukan, P., *J. Org. Chem.* **2004**, *69*, 4005.

48 (a) Friestad, G.K., Ding, H., *Angew. Chem., Int. Ed.* **2001**, *40*, 4491; (b) Friestad, G.K., Korapala, C.S., Ding, H., *J. Org. Chem.* **2006**, *71*, 281.

49 For a general review on enantioselective additions to imines, see: Kobayashi, S., *Chem. Rev.* **1999**, *99*, 1069.

50 (a) Kobayashi, S., Nishio, K., *J. Am. Chem. Soc.* **1995**, *117*, 6392; (b) Schneider, U., Sugiura, M., Kobayashi, S., *Tetrahedron* **2006**, *62*, 496.

51 (a) Nakajima, M., Saito, M., Hashimoto, S., *Tetrahedron: Asymm.* **2002**, *13*, 2449; (b) Denmark, S.E., Wynn, T., *J. Am. Chem. Soc.* **2001**, *123*, 6199; (c) Iseki, K., Kuroki, Y., Kobayashi, Y., *Tetrahedron: Asymm.* **1998**, *9*, 2889.

52 Wurz, R.P., Fu, G., *J. Am. Chem. Soc.* **2005**, *127*, 12234.

53 The same catalyst (**50**) has been employed to facilitate [3+2] cycloadditions: Wilson, J.E., Fu, G., *Angew. Chem., Int. Ed.* **2006**, *45*, 1426.

54 Lettan, R.B., Scheidt, K.A., *Org. Lett.* **2005**, *7*, 3227.

55 (a) Mayr, H., Bug, T., Gotta, M.F., Hering, N., Irrang, B., Janker, B., Kempf, B., Loos, R., Ofial, A.R., Remennikov, G., Schimmel, H., *J. Am. Chem. Soc.* **2001**, *123*, 9500; (b) Mayr, H., Kempf, B., Ofial, A.R., *Acc. Chem. Res.* **2003**, *36*, 66.

56 (a) Denmark, S.E., Bui, T., *Proc. Natl. Acad. Sci. USA* **2004**, *101*, 5439; (b) Denmark, S.E., Bui, T., *J. Org. Chem.* **2005**, *70*, 10393; (c) Denmark, S.E., Pham, S.M., Stavenger, R.A., Su, X., Wong, K.-T., Nishigaichi, Y., *J. Org. Chem.* **2006**, *71*, 3904.

57 Mayr, H., Kempf, B., Ofial, A.R., *Acc. Chem. Res.* **2003**, *36*, 66.

58 (a) Denmark, S.E., Winter, S.B.D., Su, X., Wong, K.-T., *J. Am. Chem. Soc.* **1996**, *118*, 7404; (b) Denmark, S.E., Stavenger, R.A., Wong, K.-T., *J. Org. Chem.* **1998**, *63*, 918; (c) Denmark, S.E., Stavenger, R.A., Winter, S.B.D., Wong, K.-T., Barsanti, P.A., *J. Org. Chem.* **1998**, *63*, 9517; (d) Denmark, S.E., Stavenger, R.A., *J. Org. Chem.* **1998**, *63*, 9524. (e) For other approaches, see also Ref. [61b].

59 For an overview, see: Denmark, S.E., Stavenger, R.A., *Acc. Chem. Res.* **2000**, *33*, 432.

60 (a) Denmark, S.E., Su, X., Nishigaichi, Y., *J. Am. Chem. Soc.*

1998, *120*, 12990; (b) Denmark, S.E., Su, X., Nishigaichi, Y., Coe, D.N., Wong, K.-T., Winter, S.B.D., Choi, J.Y., *J. Org. Chem.* **1999**, *64*, 1958; (c) Denmark, S.E., Stavenger, R.A., *J. Am. Chem. Soc.* **2000**, *122*, 8837. (d) Denmark, S.E., Pham, S.M., *J. Org. Chem.* **2003**, *68*, 5045; (d) Denmark, S.E., Gosh, S.K., *Angew. Chem., Int. Ed.* **2001**, *40*, 4759.
61 (a) Denmark, S.E., Fujimori, S., *Org. Lett.* **2002**, *4*, 3477; (b) Denmark, S.E., Fujimori, S., Pham, S.N., *J. Org. Chem.* **2005**, *70*, 10823.
62 Nakajima, M., Yokota, T., Saito, M., Hashimoto, S., *Tetrahedron Lett.* **2004**, *45*, 61.
63 (a) Denmark, S.E., Fan, Y., *J. Am. Chem. Soc.* **2002**, *124*, 4233; (b) Denmark, S.E., Fan, Y., Eastgate, M.D., *J. Org. Chem.* **2005**, *70*, 5235.
64 Denmark, S.E., Fan, Y., *Tetrahedron: Asymm.* **2006**, *17*, 687.
65 Denmark, S.E., Bui, T., *J. Org. Chem.* **2005**, *70*, 10190.
66 Denmark, S.E., Heemstra, J.R., *Org. Lett.* **2003**, *5*, 2303.
67 (a) Denmark, S.E., Wynn, T., Beutner, G.L., *J. Am. Chem. Soc.* **2002**, *124*, 13405; (b) Denmark, S.E., Beutner, G.L., Wynn, T., Eastgate, M.D., *J. Am. Chem. Soc.* **2005**, *127*, 3774; (c) Denmark, S.E., Heemstra, J.R., *J. Am. Chem. Soc.* **2006**, *128*, 1038.
68 (a) Denmark, S.E., Heemstra, J.R., *Synlett* **2004**, 2411; (b) For a review on vinylogous aldol reaction, see: Denmark, S.E., Heemstra, J.R., Beutner, G.L., *Angew. Chem., Int. Ed.* **2005**, *44*, 4682.
69 Fleischer, H., *Eur. J. Inorg. Chem.* **2001**, 393.
70 For earlier calculations, see: (a) Gordon, M.S., Carroll, M.T., Davis, L.P., Burggraf, L.W., *J. Phys Chem.* **1990**, *94*, 8125; (b) Kira, M., Sato, K., Sakurai, H., Hada, M., Izawa, M., Ushio, J., *Chem. Lett.* **1991**, 387.
71 Hatano, M., Ikeno, T., Miyamoto, T., Ishihara, K., *J. Am. Chem. Soc.* **2005**, *127*, 10776.
72 (a) Casas, J., Nájera, C., Sansano, J.M., Saá, J.M., *Org. Lett.* **2002**, *4*, 2589; (b) Che, F.-X., Zhou, H., Liu,
X., Qin, B., Feng, X., Zhang, G., Jiang, Y., *Chem. Eur. J.* **2004**, *10*, 4790; (c) Ryu, D.H., Corey, E.J., *J. Am. Chem. Soc.* **2005**, *127*, 5384; (d) Lundgren, S., Wingstrand, E., Penhoat, M., Moberg, C., *J. Am. Chem. Soc.* **2005**, *127*, 11592. For a recent review on cyanohydrin formation, see: (e) Brunel, J.-M., Holmes, I.P., *Angew. Chem., Int. Ed.* **2004**, *43*, 2752.
73 Liu, X., Qin, B., Zhou, X., He, B., Feng, X., *J. Am. Chem. Soc.* **2005**, 12224.
74 (a) Denmark, S.E., Fan, Y., *J. Am. Chem. Soc.* **2003**, *125*, 7825; (b) Denmark, S.E., Fan, Y., *J. Org. Chem.* **2005**, *70*, 9667.
75 Jiao, Z., Feng, X., Liu, B., Chen, F., Zhang, G., Jiang, Y., *Eur. J. Org. Chem.* **2003**, 3818.
76 Nakamura, S., Sato, N., Sugimoto, M., Toru, T., *Tetrahedron: Asymm.* **2004**, *15*, 1513.
77 Huang, J., Corey, E.J., *Org. Lett.* **2004**, *6*, 5027.
78 For a recent overview on asymmetric cyanation of ketimines, using various Lewis and Brønsted acids, see: Spino, C., *Angew. Chem., Int. Ed.* **2004**, *43*, 1764.
79 (a) Wang, Z., Ye, X., Wei, S., Wu, P., Zhang, A., Sun, J., *Org. Lett.* **2006**, *8*, 999. (b) Wang, Z., Cheng, M., Wu, P., Wei, S., Sun, J., *Org. Lett.* **2006**, *8*, 3045.
80 (a) Malkov, A.V., Mariani, A., MacDougall, K.N., Kočovský, P., *Org. Lett.* **2004**, *6*, 2253; (b) Malkov, A.V., Stončius, S., MacDougall, K.N., Mariani, A., McGeoch, G.D., Kočovský, P., *Tetrahedron* **2006**, *62*, 264.
81 The absolute configuration of the products have not been revealed in Sun's paper (Ref. [79]), but could be deduced, in some instances, from the chiral HPLC data given in the Supporting Information and their comparison with our own experiments (Ref. [80]).
82 (a) Onomura, O., Kouchi, Y., Iwasaki, F., Matsumura, Y., *Tetrahedron Lett.* **2006**, *47*, 3751. See also: (b) Iwasaki,

F., Omonura, O., Mishima, K., Kanematsu, T., Maki, T., Matsumura, Y., *Tetrahedron Lett.* **2001**, *42*, 2525.

83 Malkov, A.V., Stewart-Liddon, A.J.P., Ramíréz-López, P., Bendová, L., Haigh, D., Kočovský, P., *Angew. Chem., Int. Ed.* **2006**, *45*, 1432.

84 (a) Lipshutz, B.H., Noson, K., Chrisman, W., Lower, A., *J. Am. Chem. Soc.* **2003**, *125*, 8779; (b) Lipshutz, B.H., Shimizu, H., *Angew. Chem., Int. Ed.* **2004**, *43*, 2228.

85 Nishikori, H., Yoshihara, R., Hosomi, A., *Synlett* **2003**, 561.

86 (a) Singh, S., Batra, U.K., *Indian J. Chem.* **1989**, 1; (b) Rueping, M., Sugiono, E., Azap, C., Theissmann, T., Bolte, M., *Org. Lett.* **2005**, *7*, 3781; (c) Hoffmann, S., Seayad, A.M., List, B., *Angew. Chem., Int. Ed.* **2005**, *44*, 7424, (d) Storer, R.I., Carrera, D.E., Ni, Y., MacMillan, D.W.C., *J. Am. Chem. Soc.* **2006**, *128*, 84.

87 Malinovskii, M.S., Romantsevich, M.K., *Zh. Obshch. Khim.* **1957**, *27*, 1873.

88 Andrews, G.C., Crawford, T.C., Contillo, L.G., *Tetrahedron Lett.* **1981**, *22*, 3803.

89 Carrett, C.E., Fu, G.C., *J. Org. Chem.* **1997**, *62*, 4534.

90 (a) Denmark, S.E., Barsanti, P.A., Wong, K.-T., Stavenger, R.A., *J. Org. Chem.* **1998**, *63*, 2428; (b) Denmark, S.E., Chung, W.-J., *J. Org. Chem.* **2006**, *71*, 4002.

91 Nakajima, M., Saito, M., Uemura, M., Hashimoto, S., *Tetrahedron Lett.* **2002**, *43*, 8827.

92 Tao, B., Lo, M.M.-C., Fu, G.C., *J. Am. Chem. Soc.* **2001**, *123*, 353.

93 Malkov, A.V., Kočovský, P., unpublished results.

8
Asymmetric Acyl Transfer Reactions

Alan C. Spivey and Paul McDaid

8.1
Introduction

The preparation of stereochemically enriched compounds by asymmetric acyl transfer in a *stoichiometric* sense can be divided into two broad classes [1]: (1) those in which a racemic or achiral/*meso* nucleophile reacts diastereoselectively with an enantiomerically highly enriched acyl donor (Type I, see below); and (2) those in which an enantiomerically highly enriched nucleophile reacts diastereoselectively with a racemic or achiral/*meso* acyl donor (Type II, see below). When a racemic component is involved, the process constitutes a kinetic resolution (KR) and the maximum theoretical yield of diastereomerically pure product – given perfect diastereoselectivity – is 50%. When an achiral/*meso* component is involved, the process can constitute a site-selective asymmetric desymmetrization (ASD) or, in the case of π-nucleophiles and reactions involving ketenes, a face-selective addition process and the maximum theoretical yield of diastereomerically pure product – given perfect diastereoselectivity – is 100% (Scheme 8.1).

An analogous classification can be applied when the enantiomerically highly enriched component is replaced by an achiral component and an enantiomerically highly enriched *catalyst* is employed, ideally in a substoichiometric amount. Now the same yield limitations apply but, again given perfect stereoselectivity, the products are enantiomerically rather than diastereomerically pure and, advantageously, only a substoichiometric quantity of potentially expensive enantiomerically highly enriched material (the catalyst) has been deployed (Scheme 8.2).

General theoretical [2] and practical [3, 4] aspects of catalyzed KR and of ASD [5–7] processes mediated by both enzymes [8] and also by non-enzyme-based [9, 10] catalysts have been widely reviewed. Moreover, two well-documented variants on the standard KR reaction: (1) parallel KR (PKR) [11, 12], in which the selectivity of a KR process can be significantly improved by running enantiocomplementary KRs in parallel in one-pot processes; and (2) dynamic KR (DKR) [13, 14], in which the 50% yield limit can be circumvented by achieving *in-situ* racemization of the substrate during the KR process, have also been reviewed recently. Conse-

Enantioselective Organocatalysis: Reactions and Experimental Procedures
Edited by Peter I. Dalko
Copyright © 2007 WILEY-VCH Verlag GmbH & Co. KGaA, Weinheim
ISBN: 978-3-527-31522-2

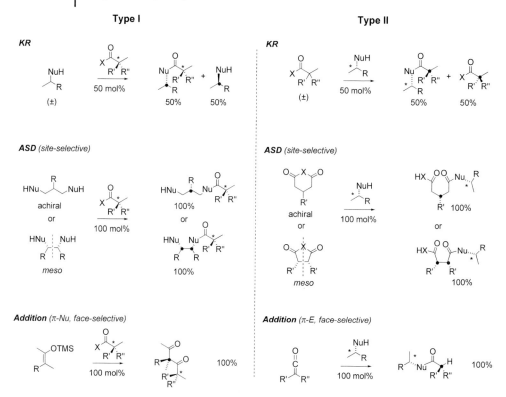

Scheme 8.1 Classification of stoichiometric asymmetric acyl transfer processes. ASD = asymmetric desymmetrization.

quently, a detailed discussion is not provided here, save to note that there is significant variation in the literature as to how best to define/report the levels of enantioselectivity achieved in KR processes. This is because the levels of enantioenrichment in both the product and the recovered starting material vary as a function of conversion [2]. The selectivity value (s), which is defined as the ratio of the relative rate constants for the two reacting enantiomers, is accepted to be the key parameter, but this suffers from the fact that it is not directly measurable [2]. Several methods for the determination of s-values are available, all of which are based on measurements of the extent of conversion (C), the enantiomeric composition of the substrate and product (enantiomeric excess, ee or enantiomeric ratio, er) and the time elapsed (t) [15]. These are all subject to errors arising from the assumptions made about the molecularity of the processes and from the accuracy of the measured parameters [15], in particular the enantiomeric purity of the catalyst [16–19]. Indeed, the most commonly employed method for obtaining s values used by synthetic chemists involves performing a single reaction run

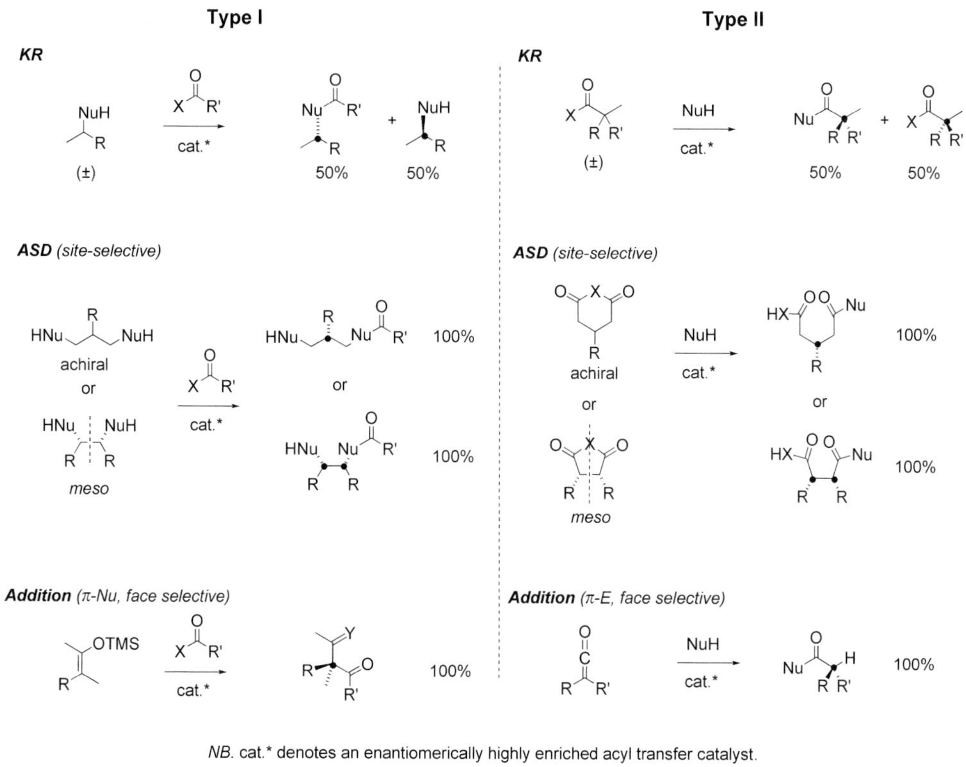

Scheme 8.2 Classification of catalyzed asymmetric acyl transfer processes.

NB. cat.* denotes an enantiomerically highly enriched acyl transfer catalyst.

(or sometimes triplicate runs) and measuring the *ee* of both the substrate and the product [typically by chiral-high-performance liquid chromatography (HPLC) or chiral-gas chromatography (GC)] and then using some equations first derived by Kagan [2] to determine the corresponding *s*-value. Not only is this method prone to the above-described compound errors, which vary as a function of conversion, but this method will also fail to identify non-linear effects that may be operating [20–22]. Despite these limitations, and also a growing realization in the community that *er* probably constitutes a more convenient descriptor of enantiomeric composition than *ee* as a starting point for these calculations [23], we have in this chapter tried to record *s*- and C-values as well as *ee*-values where available.

Until the last decade or so, the only synthetically useful catalytic asymmetric acyl transfer processes were biotransformations using hydrolase enzymes; particularly lipases and esterases [24]. Various lipases and esterases provide high levels of stereoselectivity (*s*) for the acylative KR and ASD of a wide variety of *sec*-alcohols and some amines, although the latter transformations have been less thoroughly explored [25–28]. However, the preparative use of enzymes is associated with a number of well-documented limitations, including their generally high cost, stringent operating parameters, batch-to-batch irreproducibility, and availability in just one enantiomeric form. Consequently, over the past decade or so there has been significant interest in developing small-molecule chiral catalysts capable of mediating these important asymmetric transformations in order to circumvent these limitations. The use of chiral Lewis acids for this purpose has met with limited success until recently [29, 30], and this is in any case beyond the remit of this survey, which is focused on the use of chiral organocatalysts to promote asymmetric acyl transfer via chiral nucleophile, chiral general base and chiral Brønsted acid catalytic manifolds. In particular, notable successes have been achieved using chiral catalysts based on amines, phosphines, carbenes and alcohols and their protonated congeners. Indeed, for a number of important classes of substrate, organocatalytic acylative KR, ASD and addition processes have been developed to a stage where they constitute the most practical and attractive preparative methods available. The early development of this field has been reviewed [29, 31], as has the use of nucleophilic chiral amines as catalysts in asymmetric synthesis [32, 33]. Mechanistic aspects have also been overviewed [34], particularly relating to catalysis by chiral derivatives of 4-(dimethylamino)pyridine (DMAP) [35–37], so the intention here is to focus on and provide an evaluation of the most practical and well-developed catalytic systems as a function of reaction class (Scheme 8.2), whilst also providing lead references to emerging and/or less established studies in the area.

8.2
Type I Acyl Transfer Processes

All chiral organocatalysts developed to date that mediate Type I acyl transfer processes are believed to impart their acceleration and stereoinduction primarily via a nucleophilic catalytic cycle (Fig. 8.1) [35].

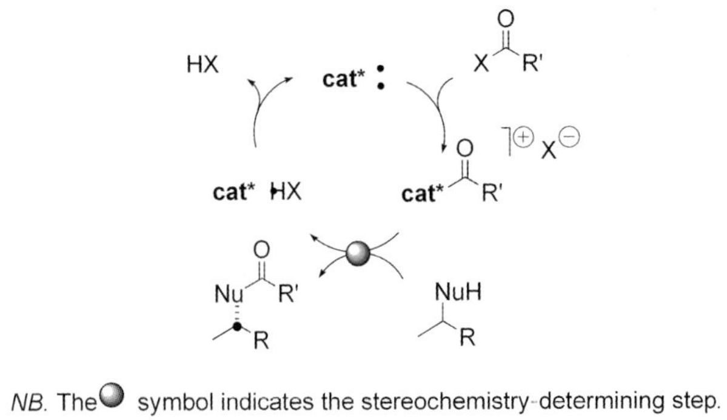

NB. The ● symbol indicates the stereochemistry-determining step.

Fig. 8.1 Nucleophilic catalysis cycle for a "Type I" acyl transfer process [35].

This means that the chiral catalyst participates in nucleophilic attack on an achiral acyl donor to afford a reactive chiral acyl salt. Nucleophilic attack on this salt by an appropriate nucleophile (an alcohol, amine or π-nucleophile) then provides the acylated product and regenerates the catalyst. This latter step determines the stereochemistry, but knowledge of the precise mechanism by which stereochemical information is transferred in most of these processes is still rather limited.

As most of the chiral nucleophilic catalysts that have been described to date were initially developed or tested for Type I acylative KR of sec-alcohols, these are the first class of reactions considered here.

8.2.1
Type I Acylative KR of Racemic Alcohols and Amines

8.2.1.1 Aryl Alkyl sec-Alcohols

Enantiopure aryl alkyl sec-alcohols are valuable building blocks for the synthesis of natural products, chiral ligands, auxiliaries, catalysts and biologically active compounds, and although there are numerous methods for their preparation, catalytic acylative KR of racemates is attractive for many applications [38]. Aryl alkyl sec-alcohols are also by far the most widely studied class of substrate for nucleophilic organocatalytic acylative KR.

The first practical method, using C_2-symmetric chiral phosphine **1** as catalyst, was disclosed in 1996 by Vedejs [39]. Employing (3-Cl-C_6H_4CO)$_2$O (2.5 equiv.) as the acylating agent in the presence of phosphine **1** (16 mol%), various aryl alkyl sec-alcohols were surveyed and s-values of 12 to 15 were obtained for the optimal substrate: 2,2-dimethyl-1-phenyl-1-propanol (Scheme 8.3).

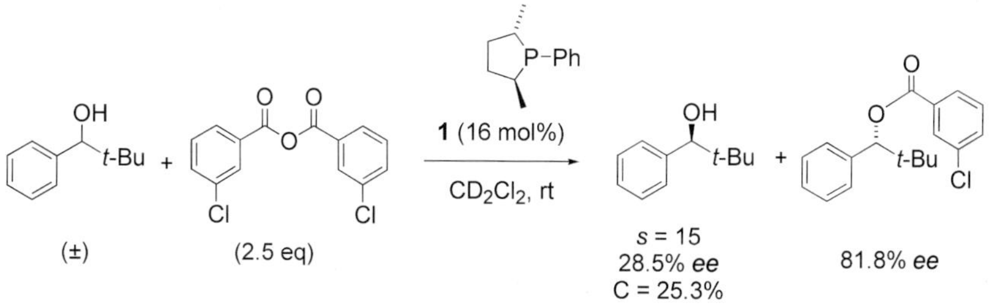

Scheme 8.3 Vedejs' first-generation phosphine-catalyzed KR of a sec-alcohol [39].

Subsequently, Vedejs developed a series of P-aryl-2-phosphabicyclo[3.3.0]octane (PBO) chiral phosphines (**2a–c**) as much more efficient catalysts for this type of transformation (Fig. 8.2) [40].

In particular, 3,5-di-tert-butylphenyl derivative **2c** (2.5–12.1 mol%) with (i-PrCO)$_2$O (2.5 equiv.) in heptane at −20 °C to −40 °C was shown to provide s-values of 42 to 369 for a wide range of aryl alkyl sec-alcohols (Table 8.1) [17, 41, 42].

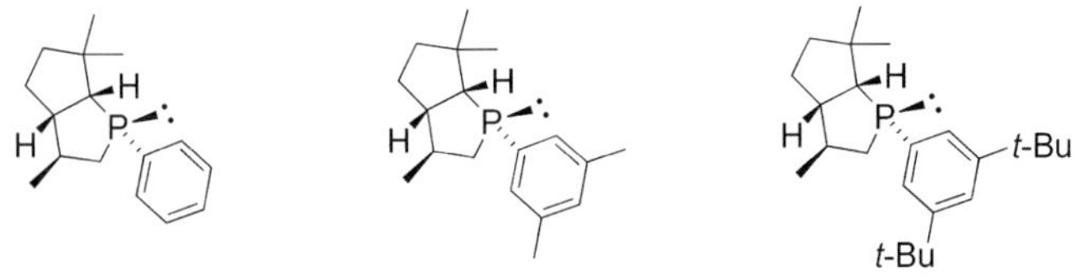

Fig. 8.2 Vedejs' PBO catalysts [40].

Table 8.1 Vedejs' PBO-catalyzed KR sec-alcohols [17, 42].

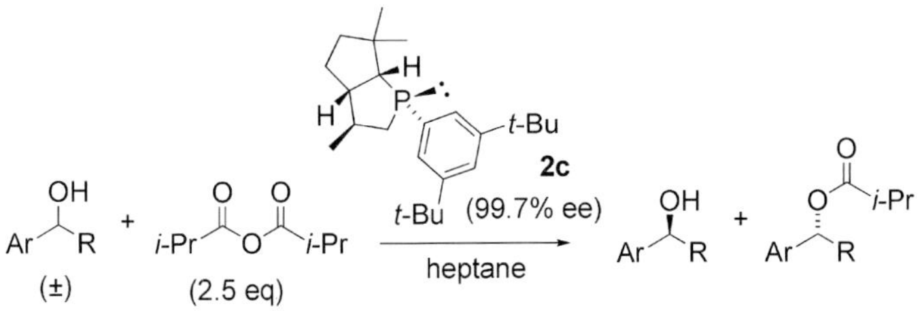

Entry	Ar	R	T (°C)	mol% cat.	C (%)	ee$_A$ (%)	ee$_E$ (%)	s
1	Ph	Me	-20	2.5	29.2	38.4	93.3	42
2	Ph	Bu	-40	3.9	51.3	93.3	88.6	57
3	Ph[a,b]	t-Bu	-40	4.9	45.8	78.7	93.1	67
4	2-Tol	Me	-40	3.5	48.5	90.2	95.7	142
5	Mesityl[b,c]	Me	-40	12.1	44.4	78.8	98.7	369
6	1-Nap	Me	-40	3.9	29.8	41.2	97.0	99

a) Bz$_2$O used in place of (i-PrCO)$_2$O.
b) Toluene used as solvent.
c) Catalyst **2c** of >99.9% ee used.

The levels of selectivity achieved in these reactions are amongst the highest reported for non-enzymatic acylative KR, and the scope of the method has been reviewed by Vedejs [40], as has its application in PKR [43]. The PBO catalysts **2a–c** are prepared by a multi-step enantioselective synthesis from lactate esters [42, 44] and are air-sensitive; hence, the reactions are generally run in de-oxygenated solvents. However, the air-stable tetrafluoroboric acid salts of these catalysts can also be employed with *in-situ* deprotonation by Et$_3$N; these conditions give results comparable with those obtained using the original protocol [45]. (For experimental details see Chapter 14.17.1).

The first class of amine-based nucleophilic catalysts to give acceptable levels of selectivity in the KR of aryl alkyl sec-alcohols were a series of planar chiral pyrrole derivatives (e.g., **3a** and **3b**) initially disclosed by Fu and colleagues in 1996 [46, 47]. In these seminal studies diketene (1.2 equiv.) was employed as the acyl donor, and by using 10 mol% of catalyst **3a** selectivities up to $s = 6.5$ could be obtained for KR of 1-naphthylethanol (Scheme 8.4) [46].

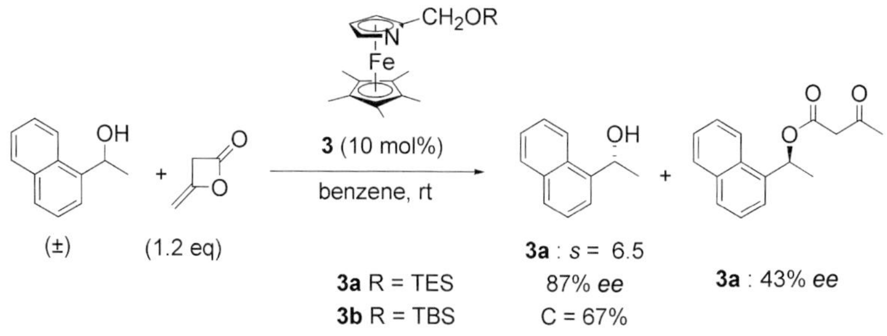

Scheme 8.4 Fu's first-generation planar chiral pyrrole-catalyzed KR of a *sec*-alcohol [46].

Subsequently, Fu developed a series of planar chiral ferrocenyl DMAP and 4-(pyrrolidino)pyridine (PPY) derivatives (**4a–d**) that have proved to be highly versatile and efficient catalysts for many acyl transfer processes (Fig. 8.3) [46–68].

Although these catalysts are not strictly organocatalysts, they are included here because the iron atom does not play a *direct* role in their mode of action. However, the metal does render the pyridine nitrogen significantly more nucleophilic than in the non-complexed heterocycle, in addition to fulfilling a scaffold role.

Fu found that a range of aryl alkyl *sec*-alcohols could be resolved highly efficiently using pentaphenylcyclopentadienyl DMAP catalyst **4b** (1–2 mol%) in conjunction with Ac$_2$O (0.75 equiv.) as the acyl donor and Et$_3$N (0.75 equiv.) as an auxiliary base [48, 49]. Fu noted that both the rate and selectivity of these reactions were strongly solvent-dependent: using Et$_2$O as solvent and 2 mol% catalyst **4b** provided *s*-values of 12 to 52 at room temperature (rt), whereas using *t*-amyl alcohol and 1 mol% catalyst **4b** provided *s*-values of 32 to 95 at 0 °C (Table 8.2).

The most serious limitation associated with the use of Fu's planar chiral ferrocenyl catalysts (**4a–d**) is that their preparation is via a multi-step synthesis followed by chiral-HPLC resolution; however, catalysts **4b** and **4c** are commercially

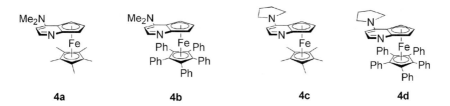

Fig. 8.3 Fu's planar chiral ferrocenyl DMAP and PPY catalysts [40].

Table 8.2 Fu's planar chiral ferrocenyl DMAP-catalyzed KR of sec-alcohols [48, 49].

Entry	Ar	R	2 mol% 46, Et$_2$O, rt				1 mol% 46, t-amyl alcohol, 0 °C			
			C (%)	ee$_A$ (%)	ee$_E$ (%)[a]	s[b]	C (%)	ee$_A$ (%)	ee$_E$ (%)[a]	s[b]
1	Ph	Me	61.9	95.2	58.7	14	55.5	98.9	79.2	43
2	Ph	t-Bu	51.8	92.2	88.0	52	51.0	96.1	92.2	95
3	4-F-C$_6$H$_4$	Me	64.4	99.2	55.9	18	54.9	99.9	82.0	68
4	Ph	CH$_2$Cl	68.4	98.9	44.5	12	56.2	97.5	76.1	32
5	2-Tol	Me	60.3	98.7	64.9	22	53.2	98.6	86.6	71
6	1-Nap	Me	63.1	99.7	57.7	22	51.6	95.1	89.3	65

[a] Determined following reduction to the alcohol using LiAlH$_4$.
[b] Average of 2-3 runs.

available from Strem Chemicals Ltd. [69]. (For experimental details see Chapter 14.17.2).

In addition to the planar chiral ferrocenyl catalysts 4a–d developed by Fu, a number of other chiral derivatives of DMAP and PPY [35–37] have been explored by other groups as organocatalysts for KR of sec-alcohols (and other transformations). Contributions have been made by the groups of Vedejs [71–75], Fuji and Kawabata [76–79], Morken [80], Spivey [81–91], Kotsuki [92, 93], Inanaga [94], Campbell [95–98], Jeong [99], Yamada [100], Connon [101], Johannsen [102], Díez [103], Levacher [104], and Richards [105] (Fig. 8.4).

None of these derivatives has been shown to provide higher levels of selectivity for the KR of aryl alkyl sec-alcohols than Fu's catalyst 4b, although several are more readily prepared. For example, the axially chiral biaryl DMAP 12a developed by Spivey [81–90] is relatively readily prepared but provides only modest levels of selectivity for the KR of aryl alkyl sec-alcohols: $s \leq 30$ at -78 °C over 8–12 h or $s \leq 15$ at rt in ~20 min (Table 8.3) [91].

Comparable levels of selectivity ($s = 7.6$–24) can also be achieved for the KR of a similar range of aryl alkyl sec-alcohols using Yamada's reasonably readily accessed chiral thiazolidine-2-thione-based DMAP 19 (Fig. 8.4) at between 0 °C and rt over 3 to 12 h [100]. The N-(4'-pyridinyl)-α-methylproline derived catalysts 16, and the solid-supported variant 17 (Fig. 8.4) developed by Campbell [95–98]

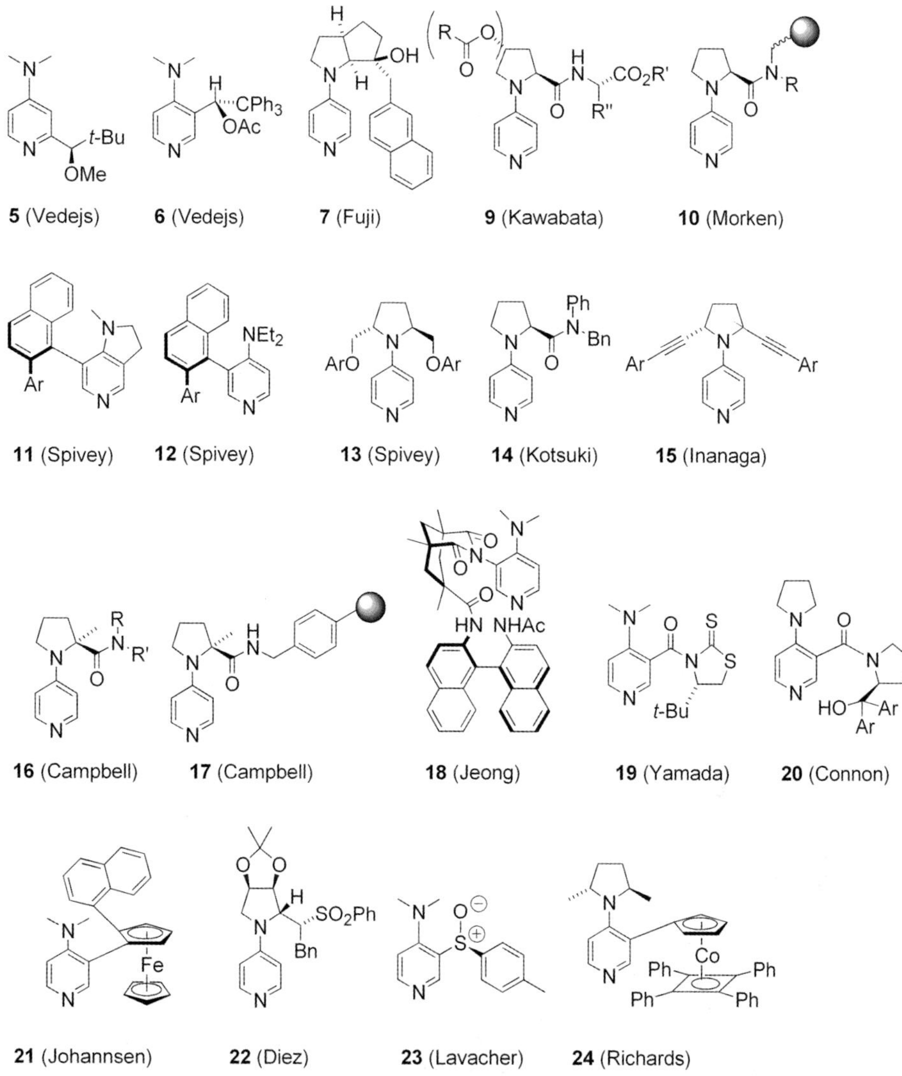

8.2 Type I Acyl Transfer Processes | 295

5 (Vedejs) **6** (Vedejs) **7** (Fuji) **9** (Kawabata) **10** (Morken)

11 (Spivey) **12** (Spivey) **13** (Spivey) **14** (Kotsuki) **15** (Inanaga)

16 (Campbell) **17** (Campbell) **18** (Jeong) **19** (Yamada) **20** (Connon)

21 (Johannsen) **22** (Diez) **23** (Lavacher) **24** (Richards)

Fig. 8.4 Chiral derivatives of DMAP and PPY.

are also particularly notable for their ease of preparation using well-developed peptide-coupling techniques, but are not currently able to resolve aryl alkyl sec-alcohols efficiently. These catalysts do provide promising levels of selectivity ($s \leq 19$) with certain 1,2-amino alcohol derivatives which allow for a H-bonding interaction with the catalyst (see Section 8.2.1.4).

In 2004, Birman reported a new class of chiral catalysts based on the (2S)-phenyl-2,3-dihydroimidazo[1,2a]-pyridine (PIP) core which were highly effective for the acylative KR of aryl alkyl sec-alcohols [106]. An initial structure–activity

Table 8.3 Spivey's axially chiral DMAP-catalyzed KR of sec-alcohols [83].

Entry	R	Ar	(i-PrCO)$_2$O	C (%)	ee$_A$ (%)	ee$_E$ (%)	s
1	Me	1-Nap	2 eq	17.2	18.6	89.3	21
2	Me	1-Nap	1 eq	22.3	26.3	91.4	29
3	Me	Ph	2 eq	39.0	49.9	78.1	13
4	Me	2-Tol	2 eq	41.4	60.7	86.0	25
5	t-Bu	Ph	2 eq	17.5	18.8	88.8	20

relationship study identified the easily accessible 6-trifluoromethyl derivative (CF$_3$-PIP) 25 as the optimal catalyst. A combination of CF$_3$-PIP 25 (2 mol%) with (n-PrCO)$_2$O in CHCl$_3$, with i-PrNEt$_2$ as an auxiliary base resolved a variety of aryl alkyl sec-alcohols with good to excellent selectivities (s = 26–87) (Table 8.4).

Birman proposed that a π-π stacking interaction between the catalyst and the substrate in the transition state was crucial to the observed enantioselectivity. In

Table 8.4 Birman's CF$_3$-PIP-catalyzed KR of sec-alcohols [106].

Entry	Ar	R	t (h)	C (%)	ee$_A$ (%)	ee$_A$ (%)	s
1	Ph	Me	6	29	36.7	89.9	27
2	Ph	Et	6	42	64.7	90.2	38
3	Ph	i-Pr	30	55	97.6	80.9	41
4	Ph	t-Bu	52	48	88.2	93.5	87
5	1-Nap	Me	10	55	98.8	82.5	52
6	3-MeO-C$_6$H$_4$	Me	8	41	63.2	89.4	34

order to maximize this interaction, a second-generation catalyst with an extended π-system was designed based on a (2S)-phenyl-1,2-dihydroimidazo[1,2a]quinoline (PIQ) core. The 7-chloro derivative (ClPIQ) **26** was found to provide even better selectivity and reactivity than CF$_3$-PIP **25** for aryl alkyl *sec*-alcohols and, moreover, was effective for certain cinnamyl-based allylic *sec*-alcohol substrates (Scheme 8.5) [107, 108].

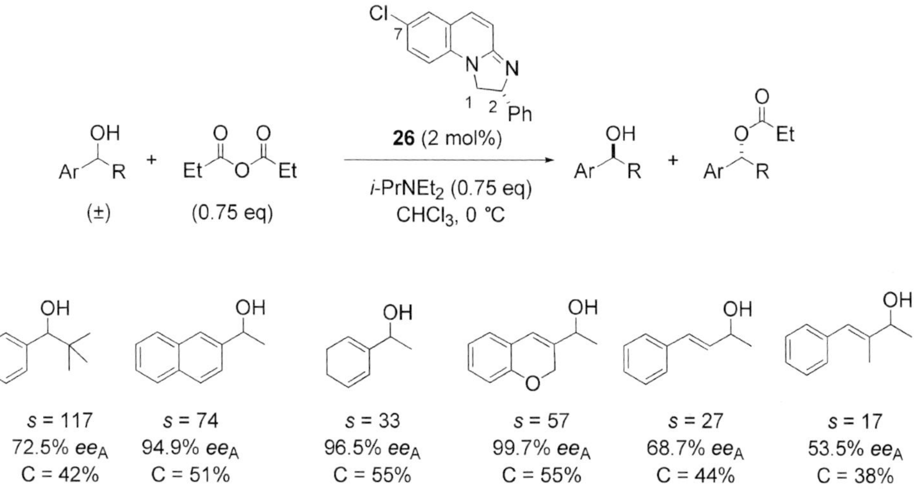

Scheme 8.5 Birman's Cl-PIQ-catalyzed KR of *sec*-alcohols [107, 108]. (For experimental details see Chapter 14.17.3).

Given that the Birman catalysts are readily prepared in just two steps from commercially available, enantiomerically pure phenylalaninol, these catalysts constitute attractive alternatives to Fu's planar chiral ferrocenyl catalysts **4a–d** (Fig. 8.3).

Miller has developed various short peptides containing 3-(1-imidazolyl)-(S)-alanine as the catalytic core; these "biomimetic" enantioselective acyl transfer catalysts allow the formation of an acyl imidazolium intermediate in a chiral environment formed by folding of the peptide [109–123]. The first catalyst of this type to be reported was tripeptide **27** in 1998 [109]. This peptide incorporates a C-terminal (R)-methylbenzylamide to encourage order-inducing stacking interactions in the acylimidazolium intermediate (Scheme 8.6).

This tripeptide provided moderate levels of selectivity ($s \leq 12.6$) for the KR of certain amide-containing *sec*-alcohols (see Section 8.2.1.4), but displayed essentially no selectivity for KR of aryl alkyl *sec*-alcohols. However, Miller subsequently devised an elegant fluorescent chemosensor-based screening protocol to assay for acylation [111, 115, 117, 119], and this facilitated the identification of octapeptide **28** as a much more efficient catalyst for KR of *sec*-alcohols via acylation [115]. The substrate scope includes not only aryl alkyl *sec*-alcohols but also alkyl alkyl *sec*-alcohols for which lipases and other organocatalysts invariably perform poorly (Scheme 8.7).

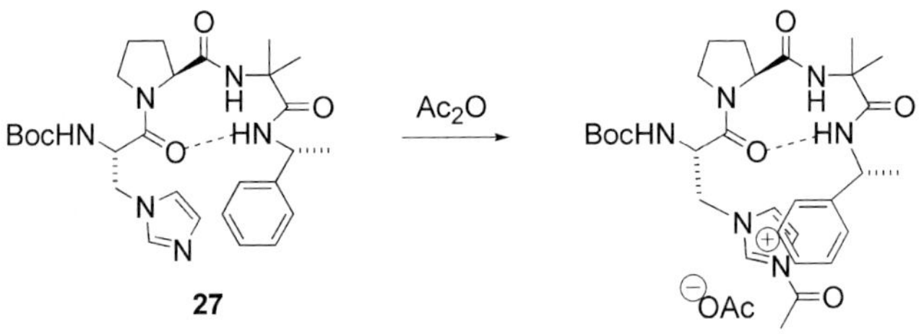

Scheme 8.6 Miller's first-generation 3-(1-imidazolyl)-(S)-alanine-containing peptide [109].

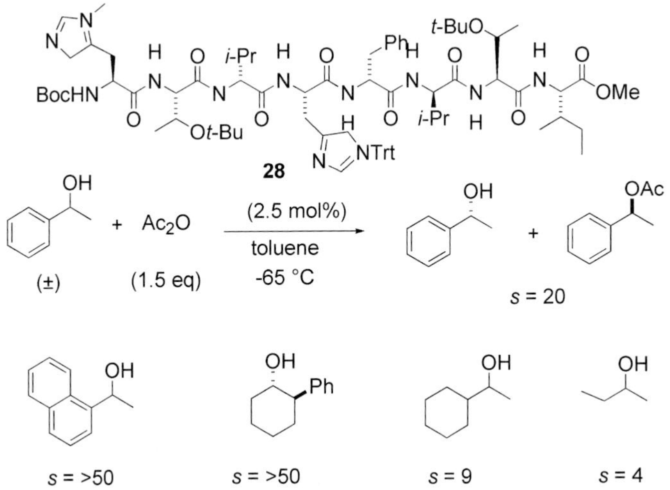

Scheme 8.7 Miller's octapeptide-catalyzed KR of sec-alcohols [115].

The combination of rapid automated synthesis of libraries of peptides and fluorescent screening for acylation efficiency distinguishes these organocatalysts from the small-molecule alternatives described above, and uniquely allows for the identification of a "designer" nucleophilic peptide for a given application. This has been nicely demonstrated by the identification of specific peptides for the KR of an intermediate en route to an aziridomitosane [116, 121], for the KR of certain tert-alcohols [117] (Section 8.2.1.4), and for the regioselective acylation of carbohydrates [120]. The scope of peptides as organocatalysts has been reviewed [118], and evidence that bifunctional catalysis may be responsible for the enantioselectivity when employing the Miller peptides has been presented [122, 124].

The KR of aryl alkyl sec-alcohols using various chiral nucleophilic N-heterocyclic carbenes (NHCs) has also recently been achieved by the groups of Suzuki [125, 126] and Maruoka [127]. These studies build on an emerging body of information showing that achiral NHCs are extremely efficient nucleophilic organocatalysts for transesterification [128]. The levels of selectivity achieved by Maruoka using the C_2-symmetric NHC **29b** for the KR of aryl alkyl sec-alcohols (and two allylic alcohols) lie in the range 16 to 80 (Scheme 8.8).

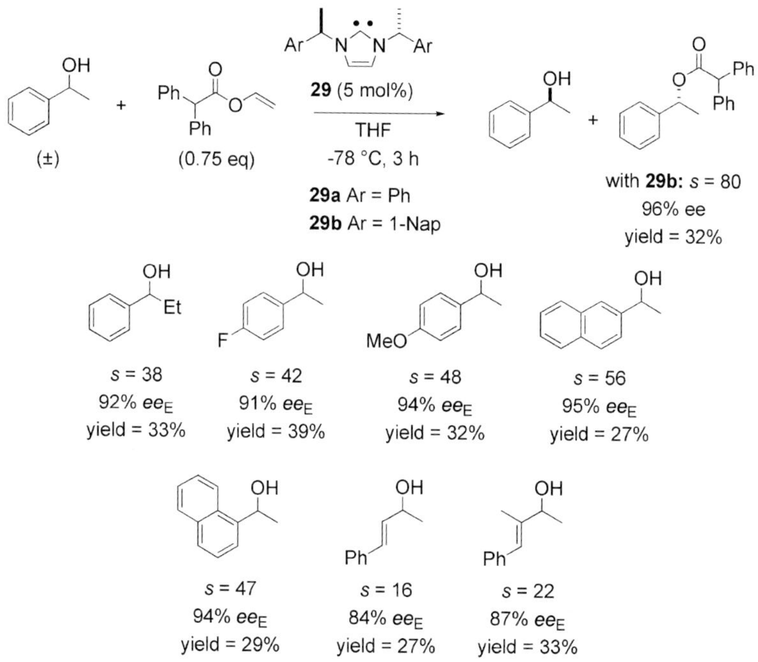

Scheme 8.8 Maruoka's chiral NHC-catalyzed KR of sec-alcohols [127].

The NHC-catalyzed processes are particularly notable for their operation at low temperature (down to −78 °C), despite the use of a relatively unactivated vinyl ester as the acyl donor. (For experimental details see Chapter 14.17.4).

8.2.1.2 Allylic sec-Alcohols

Enantiopure allylic alcohols are employed widely as building blocks for asymmetric synthesis, and particularly as substrates for various diastereoselective alkene functionalization reactions such as cyclopropanation and epoxidation directed by the hydroxyl group [129].

Vedejs has explored the KR of allylic alcohols via enantioselective acylation mediated by his 2,5-di-*tert*-butylphenyl PBO catalyst **2b** (Scheme 8.9) [40, 42, 45, 130].

Moderate selectivities ($s \leq 21$) were obtained with β-aryl substituted allylic alcohols, whereas carbocyclic allylic alcohols and allylic alcohols with highly substituted alkenes gave higher selectivities ($s = 25$–82). (For experimental details see Chapter 14.17.5).

Fu has used planar chiral ferrocenyl DMAP **4b** for the KR of a variety of allylic alcohols with good selectivities [54]. Allylic alcohols possessing a substituent geminal to the hydroxy-bearing group and *trans*-cinnamyl type substrates were resolved with good selectivities ($s = 4.7$–64). Allylic alcohols with a substituent *syn* to the hydroxy-bearing group and tetrasubstituted allylic alcohols were also reasonable substrates ($s = 5.3$–18) (Scheme 8.10).

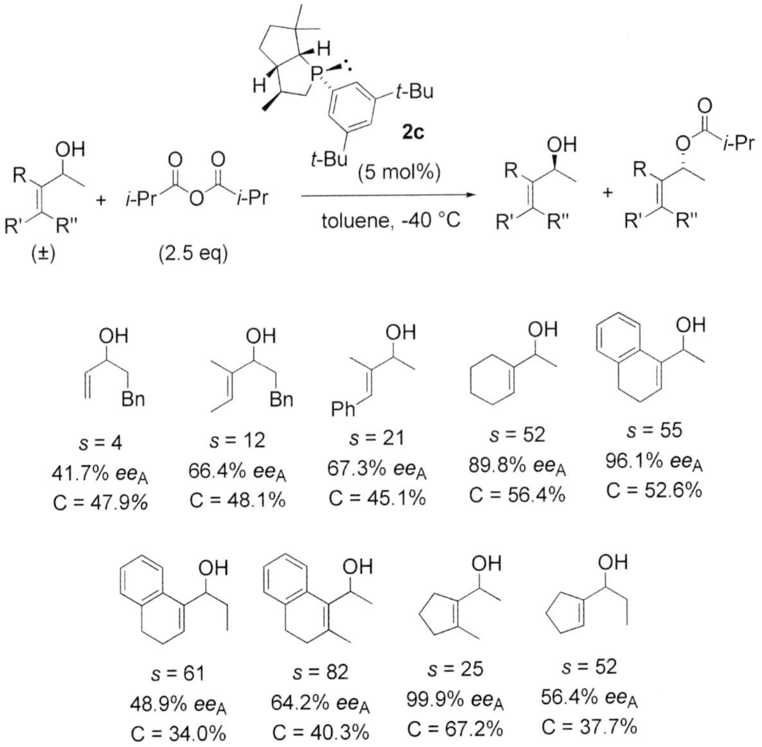

Scheme 8.9 Vedejs' PBO-catalyzed KR of sec-allylic alcohols [130].

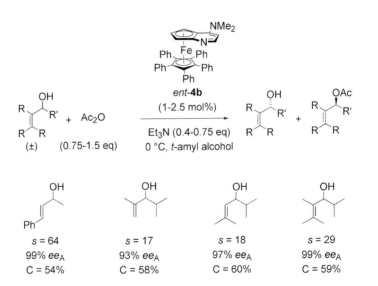

Scheme 8.10 Fu's chiral planar ferrocenyl DMAP-catalyzed KR of sec-allylic alcohols [54].

8.2 Type I Acyl Transfer Processes | 301

Fu demonstrated the synthetic utility of this method by preparing an intermediate in Brenna's synthesis of (−)-baclofen by a KR protocol which gave the desired compound in 40% yield and with 99.4% *ee* ($s = 37$) on a 2-g scale (Scheme 8.11).

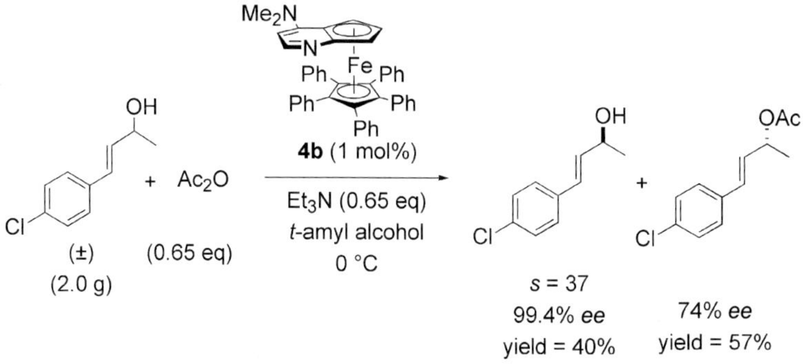

Scheme 8.11 Preparation of a (−)-baclofen intermediate using Fu's planar chiral DMAP **4b** [54].

Similarly, a racemic aldol intermediate **30** in the Sinha–Lerner synthesis of epothilone A was resolved on a 1.2-g scale to provide the natural dextrorotatory enantiomer in 47% yield and with 98% *ee* ($s = 107$) (Scheme 8.12) [54]. (For experimental details see Chapter 14.17.6).

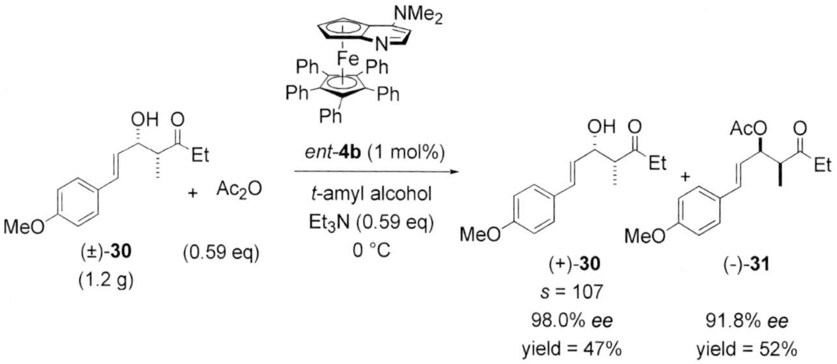

Scheme 8.12 Preparation of an epothilone A intermediate **30** using Fu's planar chiral DMAP **4b** [54].

Birman has also shown that *trans*-cinnamyl type allylic *sec*-alcohols can be efficiently resolved using his CF$_3$-PIP catalyst **25** ($s = 6$–26) or Cl-PIQ catalyst **26** ($s = 17$–57; see Scheme 8.5) [107]. Additionally, as noted earlier, Maruoka's NHC catalyst **29b** was shown to resolve two *trans*-cinnamyl-type allylic *sec*-alcohols efficiently ($s = 16$ and 22; see Scheme 8.8).

8.2.1.3 Propargylic sec-Alcohols

Enantiopure propargylic sec-alcohols are valuable synthetic intermediates that can be prepared by a number of efficient asymmetric transformations [131]. KR of the racemic propargylic sec-alcohols is a valuable approach, particularly as subsequent exhaustive hydrogenation of the alkyne provides an indirect approach to enantio-enriched alkyl alkyl sec-alcohols which themselves are challenging substrates for both enzymatic and non-enzymatic KR.

Fu's planar chiral ferrocenyl DMAP derivative **4b** is the only organocatalyst that has been shown to be generally effective for the KR of propargylic sec-alcohols [53]. Optimal conditions employ 1 mol% of catalyst and Ac$_2$O as acylating agent in t-amyl alcohol at 0 °C in the absence of base (Et$_3$N was found to catalyze a nonselective background reaction). Moderate selectivities were achieved ($s = 3.8$–20) with the selectivity value being sensitive to substrate structure: an increase in the size of the alkyl group (R = Me, Et, i-Pr, t-Bu) leads to a deterioration in the selectivity value, and best results are obtained with unsaturated groups at the remote position of the alkyne (Table 8.5).

8.2.1.4 1,2-Amino Alcohols and Mono-Protected 1,2-Diols

Diastereo- and enantiopure 1,2-amino alcohols and differentially protected 1,2-diols are motifs that occur widely in bioactive natural products and pharmaceutical substances [132]. Their preparation by KR of the corresponding racemates has been explored widely [133], particularly by asymmetric acyl transfer because these substrates provide a convenient scaffold for probing the influence of H-bonding and π-π-stacking effects on the efficiency of chirality transfer by various catalyst systems.

Table 8.5 Fu's planar chiral DMAP-catalyzed KR of sec-propargylic alcohols [53]. (For experimental details see Chapter 14.17.7).

Entry	R	R'	s	ee$_A$ (%)	ee$_E$ (%)	C (%)
1	Me	Ph	20	96	69	58
2	Et	Ph	18	94	67	58
3	i-Pr	Ph	11	93	55	63
4	t-Bu	Ph	3.8	95	18	86
5	Me	n-Bu	3.9	–	–	–

Table 8.6 Fuji and Kawabata's chiral PPY-catalyzed KR of cyclic cis-amino alcohol derivatives [77].

P = 4-Me$_2$N-C$_6$H$_4$CO

Entry	n	(i-PrCO)$_2$O	C (%)	ee$_A$ (%)	ee$_E$ (%)	s
1	2	0.6 eq	58	93	68	17
2	1	0.7 eq	69	>99	44	>12
3	3	0.7 eq	69	97	46	10

One of the first effective chiral PPY derivatives to be developed for asymmetric acyl transfer was catalyst **7**, which was shown by Fuji and Kawabata in 1997 to be effective for the acylative KR of various racemic mono-benzoylated cis-diol derivatives [76]. Subsequently, it was also successfully applied in the KR of N-protected cyclic cis-amino alcohols [77]. Using 5 mol% of PPY **7** in the presence of a stoichiometric amount of collidine in CHCl$_3$ at rt, a variety of cyclic cis-amino alcohol derivatives were resolved with moderate selectivities ($s = 10$–21) (Table 8.6).

Fuji and Kawabata proposed that catalyst **7** was selective, despite the distance between the chiral centers and the acyl pyridinium carbonyl "active site", as a result of remote chirality transfer by face to face π-π-stacking interactions between the naphthalene substituent and the pyridinium ring. Thus, analysis of ^{1}H NMR chemical shifts and nuclear Overhauser (nOe) measurements revealed that catalyst **7** interconverted between the two conformations (open and closed), depending on whether it was in the "free" or acyl pyridinium state (Fig. 8.5).

In the closed conformation the carbonyl group is orientated towards the naphthalene system, and consequently its Si-face is blocked such that only the Re-face is available to react with an incoming alcohol nucleophile. The relative orientation of the nucleophile was also suggested to be ordered by π-π-stacking interactions because sec-alcohol nucleophiles incorporating an electron-rich aryl amide gave the highest selectivities.

These seminal insights into the possibilities offered by harnessing π-π-ordering interactions to aid chirality transfer inspired many subsequent researchers in this

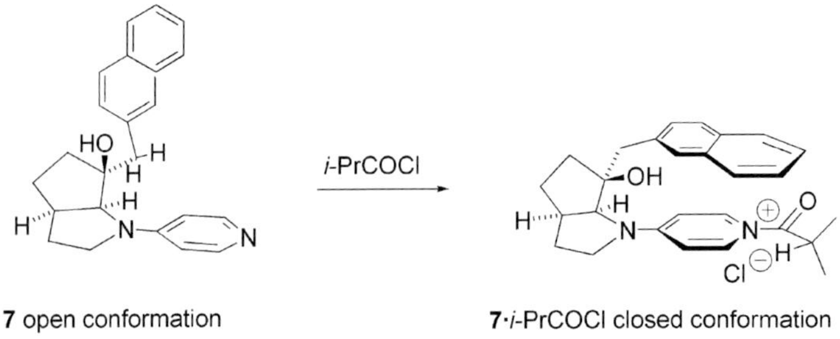

Fig. 8.5 Proposed conformations of Fuji and Kawabata's free and acylated PPY catalyst **7** [76].

area to design systems that could benefit from π-π, cation-α and related ordering interactions [134] to achieve/enhance chirality transfer [84, 89, 100, 101, 107–109].

The situation vis-à-vis H-bonding interactions is not dissimilar. The importance of H-bonding interactions was first demonstrated in Miller's studies on acylative KR using imidazole-containing peptides beginning in 1998, and appears to be a key feature of these catalysts [109–123]. As described earlier, tripeptide **27** provides essentially no selectivity in attempted acylative KRs of aryl alkyl sec-alcohols, but catalyzes KR of trans-2-(N-acetylamino)-cyclohexan-1-ol with good selectivity ($s = 12.6$) due at least in part to the H-bond donor properties of the acetamide NH group (Scheme 8.13) [109].

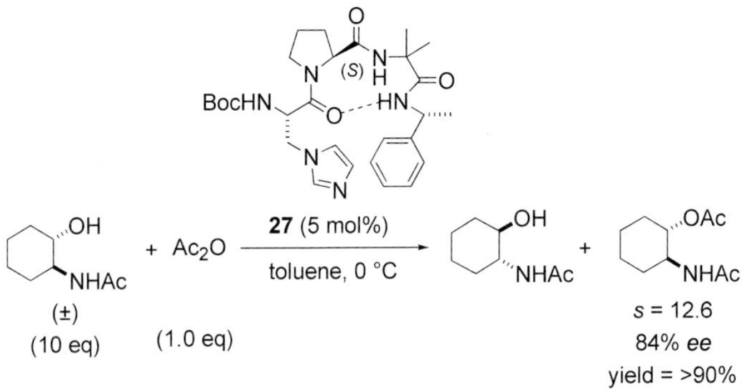

Scheme 8.13 Miller's tripeptide-catalyzed KR of a cyclic cis-amino alcohol derivative [109].

Miller showed that changing the configuration of the proline from (S) to (R) in related tetrapeptide **32** effects a dramatic reversal of the sense of induced enantioselectivity and an increase in the level of selectivity ($s = 3 \rightarrow 28$) for this same substrate. He also suggested that both changes were probably the result of changes in key H-bonding interactions within the peptide and between the peptide and the substrate (Scheme 8.14) [110].

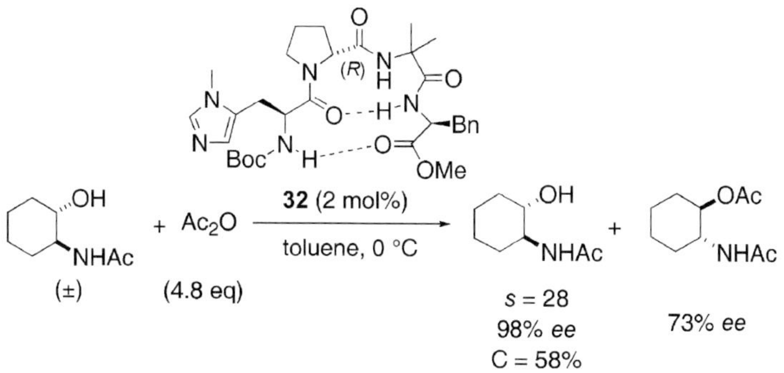

Scheme 8.14 Miller's tetrapeptide-catalyzed KR of a 1,2-amino alcohol derivative [110].

As for aryl alkyl alcohol KR, Miller demonstrated that longer peptides can give higher levels of selectivity. In this series, an octapeptide (containing the β-hairpin structure of tetrapeptide **32**) gave a selectivity of $s = 51$ with trans-2-(N-acetylamino)-cyclohexan-1-ol (and $s = 15$ and 27 with the analogous five- and seven-membered ring substrates) under otherwise identical conditions [112]. As mentioned earlier, fluorescent chemosensor-based screening was also employed to identify a tetrapeptide (**33**) that was competent for the KR of N-acylated tert-amino alcohols with s values from 19 to >50. Despite the long reaction times required (3 days), this represents a significant achievement given that tert-alcohols are invariably very poor substrates for lipases and other organocatalysts (Scheme 8.15) [117].

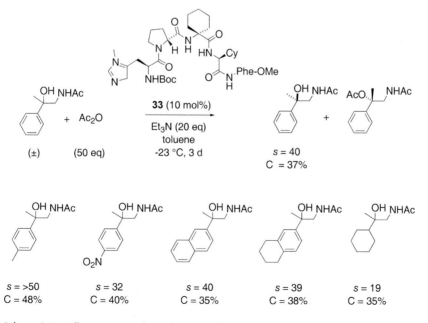

Scheme 8.15 Miller's tetrapeptide-catalyzed KR of tert-alcohols [117].

H-bond interactions between the catalyst and substrate were suggested by Campbell to explain why his N-(4′-pyridinyl)-α-methylproline-derived catalysts **16** and **17** (Fig. 8.4) are more enantioselective for N-acylated 1,2-amino alcohols ($s = 9$–18.8) than for other classes of substrate [95–98]. Recently, Ishihara has designed the histidine derivative **34** as a "minimal artificial acylase" for the KR of mono-protected cis-1,2-diols and N-acylated 1,2-amino alcohols [135]. This catalyst incorporates a sulfonamide linkage specifically to allow the NH group to engage as a H-bond donor with the substrates, and gives impressive levels of selectivity with a range of appropriate substrates (Scheme 8.16).

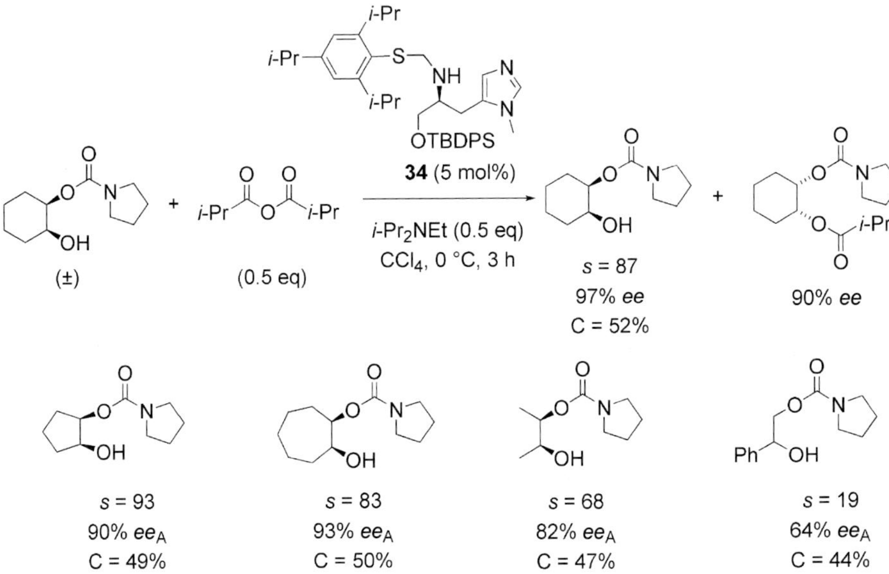

Scheme 8.16 Ishihara's histidine derivative-catalyzed KR of mono-protected cis-diols [135]. (For experimental details see Chapter 14.17.8).

8.2.1.5 α-Chiral Primary Amines

Amines bearing a chiral center α to the nitrogen are widespread constituents of bioactive alkaloids and pharmaceuticals. Face-selective addition reactions to imine derivatives [136, 137] and hydroamination of alkenes are among the most straightforward strategies for their asymmetric preparation [138] but, given the availability of racemic amines, acylative KR is an attractive approach. Indeed BASF use a proprietory immobilized lipase from *Burkholderia plantarii* to resolve a range of primary amines on an industrial scale. Notwithstanding this, the literature relating to enzymatic KR of amines is considerably less voluminous than that for alcohols, but is expanding rapidly [28]. In contrast, the literature relating to KR of amines by non-enzymatic acylation is very limited.

Although a number of methods for the asymmetric acylation of α-chiral amines using stoichiometric amounts of chiral acylating agents have been developed (notably by Shibuya [139], Atkinson [140–145], Murakami [146], Krasnov [147], Arseniyadis [148, 149], and Toniolo [150]), the development of a catalytic process represents a far greater challenge. In particular, the development of organo-nucleophile-catalyzed processes is hampered by the high nucleophilicity of amines (cf. alcohols) and the consequent relatively high rate of the uncatalyzed, non-enantioselective background reaction. The only effective catalytic system reported to date is that developed by Fu in which O-carbonyloxyazlactone **35** is employed as the stoichiometric acyl donor in combination with 10 mol% of planar chiral ferrocenyl PPY **4c** as the catalyst. The key component in this system appears to be the unusual acyl donor **35**, as these conditions evolved from a previously reported protocol in which an acylpyridinium salt had to be preformed and employed stoichiometrically [57]. After optimization studies, a variety of racemic primary amines were successfully resolved with moderate to good s-values (11–27) (Scheme 8.17) [58].

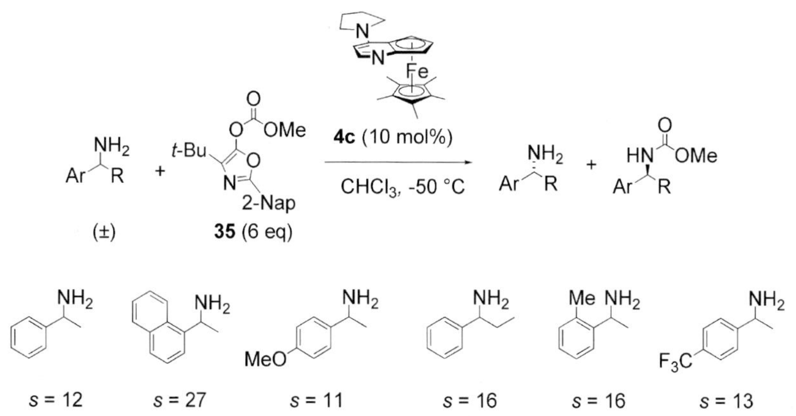

Scheme 8.17 Fu's planar chiral ferrocenyl PPY-catalyzed amine KR [58].

Advantageously, in the context of subsequent synthetic manipulation, the acylated products in these processes are carbamates (rather than amides). Fu proposed a mechanistic pathway that involves rapid initial reaction of the catalyst with the O-carbonyloxyazlactone to form an ion-pair, followed by slow transfer of the methoxycarbonyl group from this ion-pair to the amine in the enantioselectivity-determining step (Fig. 8.6).

8.2.2
Type I Acylative ASD of Achiral/*meso*-Diols

ASD differs from KR in that it offers the opportunity for 100% conversion of an achiral or, advantageously from the standpoint of number of stereocenters created, a *meso* substrate into an enantiomerically pure product, provided that perfect stereoselection can be achieved. Consequently, many groups have explored the ASD of *meso*-diols using organocatalytic acylation.

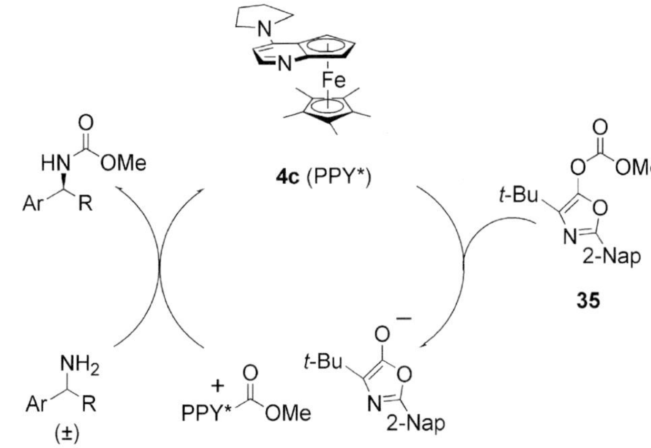

Fig. 8.6 Fu's proposed mechanism of planar chiral ferrocenyl PPY-catalyzed KR of amines [58]. (For experimental details see Chapter 14.17.9).

Oriyama was the first to provide a practical protocol for the ASD of this class of substrate [151–154]. Thus, employing just 0.5 mol% of (S)-proline-derived chiral diamine **36** in conjunction with benzoyl chloride as the stoichiometric acyl donor in the presence of Et$_3$N, asymmetric benzoylation of a variety of *meso*-diols could be achieved with good to excellent enantioselectivities (66–96% *ee*) and ≥80% yields (Scheme 8.18).

Oriyama subsequently showed that this catalyst system was also effective for the KR of various classes of *sec*-alcohols, notably β-halohydrins [155, 156] and also certain α-chiral primary alcohols such as glycerol derivatives [157]. Miller has also explored the ASD of glycerol derivatives using a pentapeptide incorporating an N-terminal nucleophilic 3-(1-imidazolyl)-(S)-alanine residue [123]. Janda

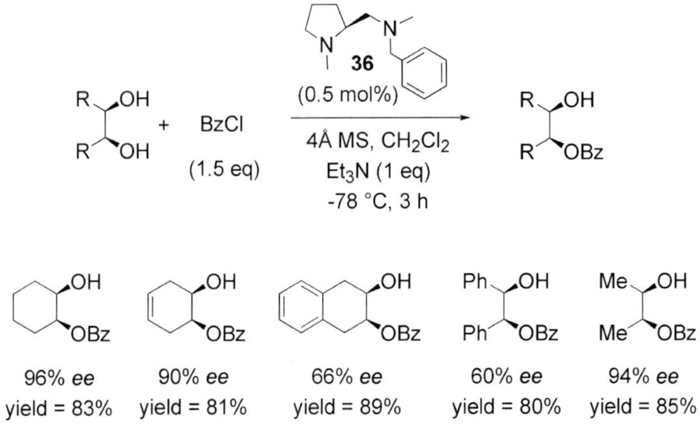

Scheme 8.18 Oriyama's proline diamine-catalyzed ASD of *meso* diols [152]. (For experimental details see Chapter 14.17.10).

has developed a solid-supported version of Oriyama's catalyst which delivers comparable selectivity values [158–160], and Kundig has employed Oriyama's catalyst and related diamines to desymmetrize a *meso*-diol chromium tricarbonyl-complex [161, 162]. Fujimoto has also described an asymmetric benzoylation system that is effective for ASD of cyclic *meso*-1,3- and 1,4-diols and which employs a phosphinite derivative of quinidine as the catalyst [163, 164].

Fu's planar chiral ferrocenyl DMAP catalyst **4b** and also Vedejs' PBO catalysts, particularly *p*-phenyl-PBO catalyst **2a**, are also effective for this type of ASD reaction, as illustrated for the cases of unusual *meso*-diol **37** [49] and *meso*-hydrobenzoin [165], respectively (Scheme 8.19).

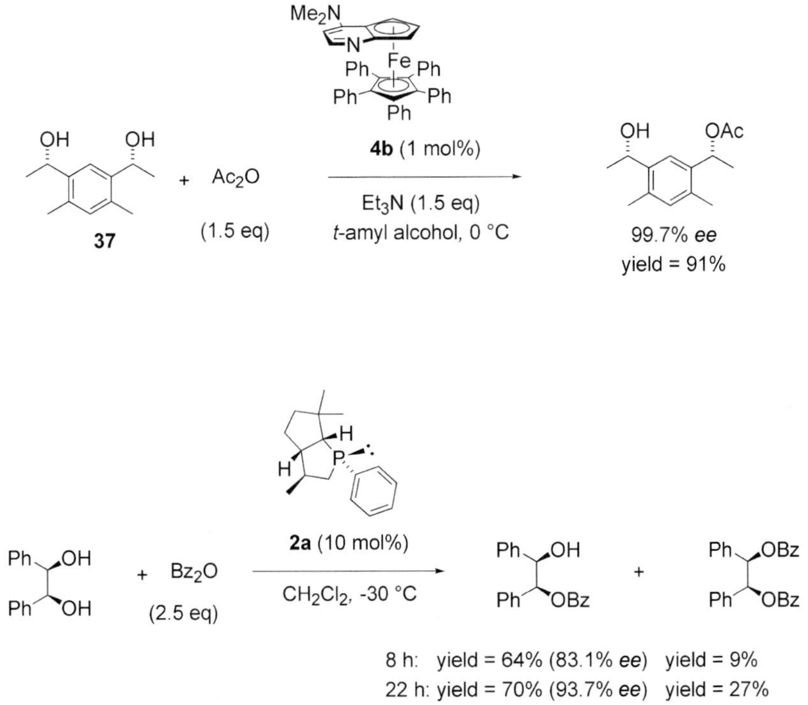

Scheme 8.19 Examples of acylative ASD reactions catalyzed by Fu's planar chiral ferrocenyl PPY catalyst **4b** and Vedejs' PBO catalyst **2a** [49, 165].

8.2.3
Type I Asymmetric Acyl Addition to π-Nucleophiles: Steglich and Related Rearrangements and Additions to Silyl Ketene Acetals/Imines

In 1970, Steglich reported that DMAP catalyzed the rearrangement of *O*-acylated azlactones to their *C*-acylated isomers (the Steglich rearrangement) [166, 167]. This process effects C–C bond formation and concomitant construction of a qua-

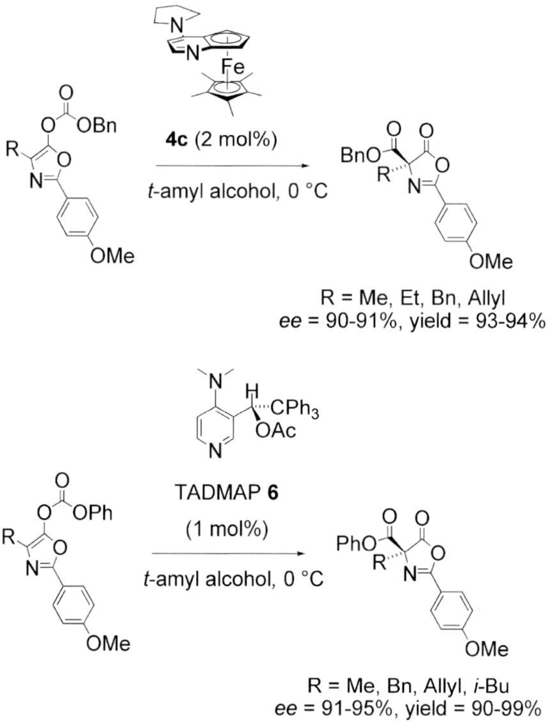

Scheme 8.20 Fu's and Vedejs' chiral PPY/DMAP-catalyzed rearrangements of O-acyl azlactones [50, 74].

ternary stereocenter. Building upon this foundation, first Fu [50] and later Vedejs [74, 75], Johannsen [102] and Richards [105] have explored the utility of chiral DMAP/PPY derivatives to effect this type of rearrangement. Fu's planar chiral ferrocenyl DMAP catalyst **4c** and Vedejs' TADMAP catalyst **6** are very effective, giving products generally with *ee*-values >90% and in almost quantitative yields (Scheme 8.20) [50, 74].

Vedejs' PBO catalyst **2c** is also effective for this type of transformation, but requires higher loading levels (10 mol% versus 1 mol% for TADMAP **6**). Johannsen's [102] ferrocenyl DMAP **21** (Fig. 8.4) and Richards' [105] cobalt metallocenyl PPY **24** (Fig. 8.4) give significantly lower levels of selectivity (25% and 45–67% *ee*, respectively), but have been less thoroughly investigated. Analogous rearrangements have also been performed by both Fu [61] and Vedejs [75] on O-acyl benzofuranones and O-acyl oxindoles to provide synthetic intermediates potentially suitable for elaboration to diazonamide A and various oxindole-based alkaloids such as gelsemine, respectively.

Fu has also explored intermolecular C-acetylation of silyl ketene acetals by Ac$_2$O for the formation of quaternary stereocenters catalyzed by planar chiral fer-

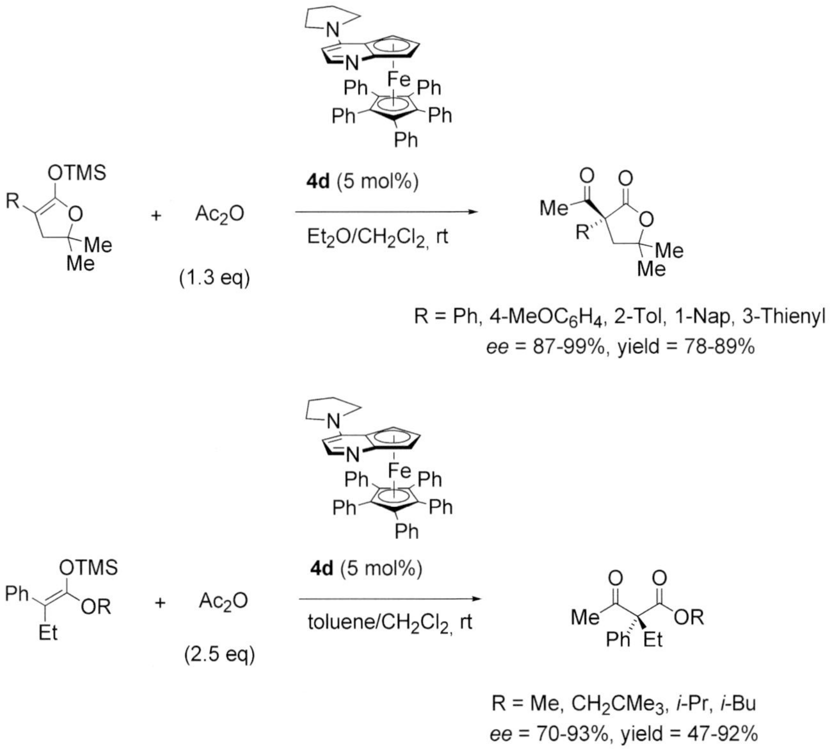

Scheme 8.21 Fu's planar chiral ferrocenyl PPY-catalyzed C-acetylation of silyl ketene acetals [60, 64].

rocenyl PPY **4d** (5 mol%) [60, 64]. Cyclic and acyclic silyl ketene acetals react to give β-ketoesters with good levels of stereocontrol (Scheme 8.21).

Fu has demonstrated that acetate anion attack on the silicon center of the silyl ketene acetal, as well as formation of an acyl pyridinium salt, contribute towards the promotion of these reactions [62]. Additionally, silyl ketene imines have also been shown to participate in analogous asymmetric C-acylation reactions to yield chiral quaternary nitriles, and this method was employed as a key step in the synthesis of verapamil [65].

8.3
Type II Acyl Transfer Processes

In contrast to the situation for the Type I processes discussed thus far, where all the catalysts are believed to act primarily as nucleophilic catalysts, the mechanisms by which acceleration and stereoinduction are primarily achieved in Type

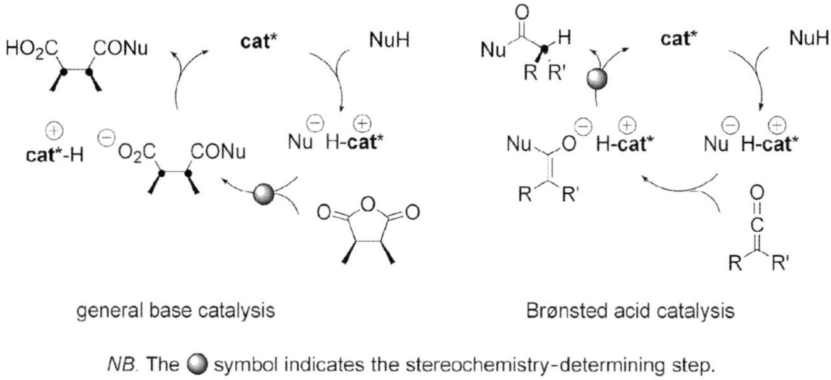

general base catalysis **Brønsted acid catalysis**

NB. The ● symbol indicates the stereochemistry-determining step.

Fig. 8.7 Lewis base and Brønsted acid catalysis cycles for "Type II" acyl transfer processes [30, 62].

II processes extend to encompass general base and Brønsted acid catalytic cycles (Fig. 8.7) [30, 62].

These catalytic scenarios differ from the nucleophilic catalysis discussed previously in that in both cases the chiral organocatalyst acts as a base rather than a nucleophile. In the case of anhydride ring-opening reactions, the stereoinducing step is believed to be the subsequent site-selective attack by the deprotonated nucleophile on one carbonyl group (general base catalysis; Fig. 8.7) [30]. In the case of ketene addition reactions there is evidence to suggest that the stereoinducing step is the face-selective protonation of the enolate, generated by attack of the deprotonated nucleophile on the ketene, by the protonated chiral catalyst (Brønsted acid catalysis; Fig. 8.7).

Type II catalytic asymmetric acyl transfer processes have been most extensively developed for the case of ASD of *meso*-anhydrides by nucleophilic ring-opening with alcohols, and so these processes will be the first type II processes considered here.

8.3.1
Type II Alcoholative ASD of Achiral/*meso*-Cyclic Anhydrides

ASD of achiral and *meso*-anhydrides by ring opening with alcohols can be catalyzed by either chiral Lewis acids or bases, and this area has been reviewed [30, 31, 33]. Pioneering studies using organocatalysts were conducted by the groups of Oda [168, 169] and Aitken [170, 171] with cinchona alkaloids during the 1980s. These studies provided the foundation for the development of a significantly more enantioselective system for the ASD of cyclic *meso* anhydrides by Bolm, who employed a stoichiometric quantity of the cinchona alkaloid quinidine (or its pseudoenantiomer, quinine) as the catalyst [172]. The reactions of bicyclic and tricyclic *meso*-anhydrides **38a–h** with methanol in the presence of 110 mol% quinidine in a 1:1 toluene:CCl$_4$ solvent system at −55 °C provided the corre-

Table 8.7 Bolm's quinidine/quinine-promoted ASD of *meso*-anhydrides [172].

		Quinidine		Quinine[a]	
Entry	Anhydride	ee (%)	Yield (%)	ee (%)	Yield (%)
1	38a	93	98	87	91
2	38b	99	98	99	92
3	38c	96	96	93	94
4	38d	85	96	85	95
5	38e	95	97	93	99
6	38f	94	99	87	93
7	38g	95	93	93	99
8	38h	94	84	94	86

[a] Quinine-catalyzed reactions give enantiomeric products.

sponding hemiesters with ≥93% *ee* and ≥84% yields. The use of quinine instead of quinidine generally provided entiomeric products with similar levels of selectivity (Table 8.7).

The practicality of the method was enhanced by the fact that simple extraction provided access to both the hemiester product and the alkaloid without chromatography, and the recovered cinchona alkaloid could be reused with no deterioration in the *ee* or yield of the reactions. This method has found use by other groups in the synthesis of β-amino alcohols and in natural product synthesis [173, 174], and has recently been reported as an accepted method in *Organic Syntheses* [175].

Table 8.8 Bolm's quinidine-catalyzed ASD of *meso*-anhydrides [176].

Entry	Anhydride	ee (%)	Yield (%)
1	38b	90	98
2	38c	91	94
3	38e	89	96
4	38f	81	97
5	38i	74	98

A variant of this procedure that employs a substoichiometric quantity of cinchona alkaloid has also been developed by Bolm [176]. In this method, 10 mol% quinidine was used in conjunction with a stoichiometric amount of 1,2,2,6,6-pentamethylpiperidine (pempidine) to prevent sequestration of the cinchona alkaloid by the acidic hemiester product. The chiral hemiester products derived from various *meso*-anhydrides were obtained with ≥74% *ee* and ≥94% yields (Table 8.8).

Although both quinidine and pempidine can be quantitatively recovered and reused following these processes, it is noteworthy that pempidine is more expensive than quinidine. Moreover, the erosion in the levels of enantioselectivity observed in the hemiester products using the catalytic protocol relative to the stoichiometric protocol, as well as the extended reaction time, limit the utility of this modification.

Bolm has demonstrated the utility of the quinidine-mediated ASD of cyclic *meso*-anhydrides by developing protocols for the conversion of the hemiester products into enantiomerically enriched unnatural β-amino alcohols by means of Curtius degradation [177]. A particularly practical variant of this procedure utilizes benzyl alcohol rather than methanol as the nucleophile in the quinidine-mediated ASD reaction, thereby allowing – following Curtius degradation – for hydrogenolytic deprotection of both the benzyl ester and a *N*-Cbz group to afford free β-amino alcohols in a single step [178].

In 2000, Deng reported that commercially available "Sharpless ligands" also catalyze the highly enantioselective alcoholysis of *meso*-cyclic anhydrides [179]. This was the first example of the use of these modified cinchona alkaloids as organocatalysts. The reaction of a variety of monocyclic, bicyclic and tricyclic succinic anhydrides with methanol in ether in the presence of a catalytic amount of the biscinchona alkaloid $(DHQD)_2AQN$ provided the corresponding hemiesters with excellent enantioselectivities (91–98%) and good to excellent yields (72–99%). The reactions were carried out at 20–30 °C, employed catalyst loadings of 5 to 30 mol%, and required one to three days to reach completion. The antipodal products could be accessed, albeit with slightly reduced enantioselectivity, by employing the pseudo-enantiomeric $(DHQ)_2AQN$ as catalyst (Table 8.9).

Table 8.9 Deng's $(DHQD)_2AQN$-catalyzed ASD of achiral/*meso*-anhydrides [179].

Entry	Anhydride	mol% cat[a]	T (°C)[a]	Yield (%)[a]	ee (%)[a]
1	38a	5 (5)	-20 (-20)	97 (95)	97 (93)
2	38b	10 (20)	-30 (-20)	82 (82)	95 (90)
3	38e	8 (8)	-30 (-30)	99 (90)	95 (93)
4	38g	7 (7)	-20 (-20)	95 (92)	98 (96)
5	38j	5 (5)	-20 (-20)	93 (88)	98 (98)
6	38k	30 (30)	-40 (-35)	70 (56)	91 (82)
7	38l	30 (30)	-40 (-35)	72 (62)	90 (83)

[a] Values in parentheses are for reactions using $(DHQ)_2AQN$ that give enantiomeric products.

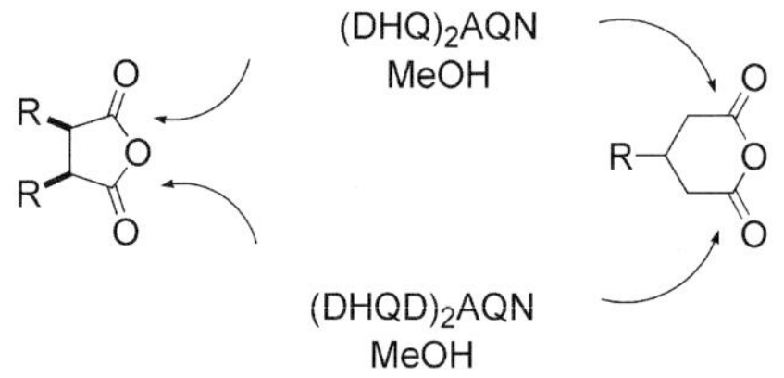

Fig. 8.8 Mnemonic for prediction of stereoselection in Deng's ASD of achiral/*meso*-anhydrides [179].

As for the Bolm procedure, simple extraction enables essentially quantitative product purification and catalyst recovery without chromatography, and the recovered catalyst can be reused without any deterioration in its efficiency. The sense of enantioselectivity in these processes can be predicted with the aid of a mnemonic (Fig. 8.8).

A noteworthy feature of the method is that turnover of the catalyst is achieved in the absence of a stoichiometric achiral base, despite the generation of an acidic product (*cf*. the use of pempidine in the Bolm catalytic protocol) [180–182]. The synthetic utility of this approach has been demonstrated in a formal total synthesis of (+)-biotin [183]. Recently a silica-supported heterogeneous analogue of (DHQD)$_2$AQN has also been developed by Han [184, 185]. (For experimental details see Chapter 14.17.11).

Surprisingly few studies have been directed towards the development of non-cinchona alkaloid-based catalysts for the alcoholative ASD of *meso*-anhydrides, or indeed any of the enantioselective alcoholysis processes. Uozumi has reported a series of (2*S*, 4*R*)-4-hydroxyproline-derived 2-aryl-6-hydroxyhexahydro-1*h*-pyrrolo[1,2-*c*] imidazolones which mediate the methanolytic ASD of *cis*-hexahydrophthalic anhydride in up to 89% *ee* when employed at the 10 mol% level for 20 h at −25 °C in toluene [186]. Additionally, Nagao has described the use of a bifunctional chiral sulfonamide for the *thiolytic* ASD of *meso*-cyclic anhydrides in up to 98% *ee* when employed at the 5 mol% level for 20 h at rt in ether [187].

8.3.2
Type II Alcoholative KR of Racemic Anhydrides, Azlactones, *N*-Carboxyanhydrides, Dioxolanediones and *N*-Acyloxazolodinethiones

It might be anticipated that, if a racemic unsymmetrically substituted cyclic anhydride were to be used as a substrate for asymmetric alcoholysis, a KR would ensue. In fact, Deng has shown that for monosubstituted succinic anhydrides, because both carbonyl groups have comparable reactivity, what actually occurs on subjection to his (DHQD)$_2$AQN-catalyzed asymmetric alcoholysis conditions, is a PKR [188]. Thus, the reaction of 2-methyl succinic anhydride (**39a**) with 2,2,2-trifluoroethanol (10 equiv.) in ether at −24 °C in the presence of (DHQD)$_2$AQN (15 mol%) provided a mixture of two regioisomeric hemiesters

Table 8.10 Deng's (DHQD)$_2$AQN-catalyzed ASD of *meso*-anhydrides [188].

(±)-39a–g + (DHQD)$_2$AQN (15 mol%), CF$_3$CH$_2$OH (10.0 eq), Et$_2$O, −24 °C → 40a–g + 41a–g

Entry	R	40/41	ee (%)		Yield (%)	
			40	41	40	41
1[a]	a Me	44/55	93	80	36	41
2	b Et	40/60	91	70	38	50
3	c n-C$_8$H$_{17}$	42/56	98	66	38	41
4	d Allyl	46/53	96	82	40	49
5[b]	e Ph	N/A	95	87	44	32
6[b]	f 3-MeO-C$_6$H$_4$	N/A	96	83	45	30
7[b]	g 4-Cl-C$_6$H$_4$	N/A	96	76	44	29

[a] 20 mol% catalyst was used.
[b] Yields are overall values for the derived 2- and 3-aryllactones.

40a and **41a** in a ∼1:1 ratio, with 93% and 80% *ee*, respectively. A variety of other 2-alkyl and 2-aryl succinic anhydrides (**39b–g**) could be resolved via this process with good to excellent enantioselectivities (66–98% *ee*) (Table 8.10) [188].

The synthetic utility of this PKR process was exemplified in a formal total synthesis of the γ-aminobutyric acid (GABA) receptor agonist (*R*)-baclofen [188]. (For experimental details see Chapter 14.17.12).

The first type of reaction for which organocatalytic type II KR was explored was the enantioselective synthesis of protected α-amino acids from racemic azlactones by ring opening with alcohols. This reaction is closely related to the Dakin–West reaction, which Steglich had shown to be efficiently catalyzed by DMAP during the 1970s [167]. Building on these findings, and also some enantioselective DKR variants using enzymes [189], Fu found that planar chiral ferrocenyl DMAP **4a** (5 mol%) gave moderate levels of enantioselectivity (44–61% *ee*) for the DKR of a series of 5-phenyl azlactones in toluene at rt in the presence of benzoic acid (10 mol%) (Scheme 8.22) [51].

This process proceeds as a DKR [13, 190] because the DMAP catalyst promotes not only the asymmetric alcoholysis of the azlactone but also its racemization under the reaction conditions; the *N*-benzoyl α-amino acid ester product does not racemize under these conditions. Johannsen has also screened chiral DMAP **21** (Fig. 8.4) for this transformation, but obtained poorer yields and selectivities [102].

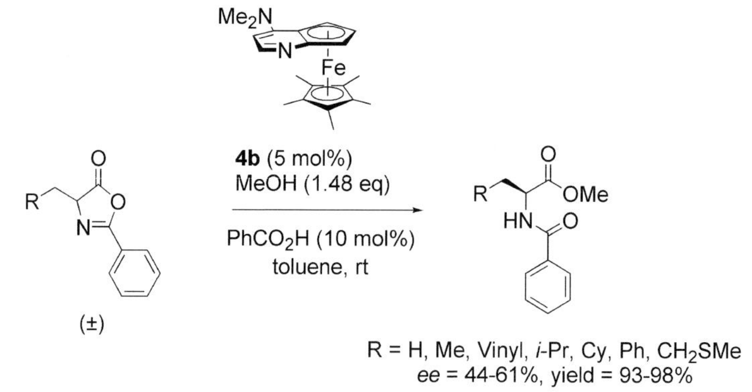

Scheme 8.22 Fu's planar chiral ferrocenyl DMAP-catalyzed DKR of 5-phenyl azlactones [51].

Deng has found that the urethane-protected α-amino acid N-carboxy anhydrides (UNCAs) offer a similar entry to enantiomerically enriched α-amino acid derivatives using (DHQD)$_2$AQN-catalyzed asymmetric alcoholysis [191]. UNCAs are readily prepared from the corresponding racemic α-amino acids and are structurally analogous to mono-substituted succinic anhydrides. They differ in that the two carbonyl groups in UNCAs are chemically distinct, and so the nucleophile attacks only the carbonyl adjacent to the chiral center (as for azlactones). The reactions of a variety of alkyl and aryl UNCAs containing various carbamate-protecting groups (**42a–h**) with methanol in ether at −60 °C provided carbamate-protected amino esters **43a–h** and unreacted starting material via a regular KR process with impressive selectivity values ($s = 23$ to 170). Achiral hydrolysis of the unreacted anhydrides provided a reaction mixture containing an acidic component (the carbamate-protected amino acids **44a–h**), a neutral component (the esters), and a basic component (the catalyst), thereby allowing all three compounds to be isolated in pure form via an extractive procedure (Table 8.11).

Moreover, when these reactions were performed at higher temperatures (~rt), DKR could be achieved. This was first demonstrated for α-aryl UNCAs [192], but subsequently also for α-alkyl UNCAs [193]. Allyl alcohol was found to be the optimal nucleophile, allowing a variety of UNCAs to be resolved with high stereoselectivities (90–92% ee) and good yields (93–98%) [192]. The resulting allyl esters could be converted to the optically active α-amino acids via Pd-catalyzed deallylation (Table 8.12).

The ready availability of the starting materials, the lack of special precautions to exclude air and moisture from the reaction mixture, and the ease of recovery of products make these DKR protocols attractive for the preparation of enantiomerically highly enriched N-protected-α-amino acids. (For experimental details see Chapter 14.17.13).

Substituted 1,3-dioxolane-2,4-diones correspond to UNCAs in which the protected nitrogen atom has been replaced with an oxygen atom, and thus represent potential precursors to enantiomerically enriched α-hydroxy acid derivatives.

Table 8.11 Deng's (DHQD)$_2$AQN-catalyzed KR of UNCAs [191].

Entry		R	P	T (°C)	t (h)	C (%)	ee/% (Yield/%)		s
							(S)-44	(R)-43	
1	a	Bn	Cbz	-60	17	51	98 (48)	93 (48)	114
2	b	4-F-C$_6$H$_4$CH$_2$	Cbz	-78	31	50	93 (42)	92 (48)	79
6	c	CH$_3$(CH$_2$)$_5$	Cbz	-60	37	51	94 (42)	91 (49)	78
7	d	BnOCH$_2$	Cbz	-78	72	52	96 (44)	89 (49)	69
9	e	Ph[a]	Cbz	-78	16	46	84 (46)	97 (45)	170
11	f	Bn	Fmoc	-78	46	51	96 (47)	92 (50)	93
12	g	Bn	Boc	-40	15	59	98 (41)	67 (56)	19
13	h	Bn	Alloc	-60	15	50	91 (45)	91 (45)	67

[a] EtOH was used as the nucleophile.

Table 8.12 Deng's (DHQD)$_2$AQN-catalyzed DKR of UNCAs [192].

Entry		R	T (°C)[a]	t (h)[a]	(R)-45		(R)-44	
					ee (%)[a]	Yield (%)[a]	ee (%)	Yield (%)
1	e	Ph	23 (34)	1(1)	91(83)	97(96)	90	91
2	i	4-F-C$_6$H$_4$	23	1	90	96	90	93
3	j	4-Cl-C$_6$H$_4$	23	1	92	97	92	92
4	k	4-CF$_3$-C$_6$H$_4$	23	1	90	95	90	88
5	l	2-Thienyl	-30	2	92	93	92	93
6	m	2-Furyl	23(-30)	0.5(1)	91(92)	98(91)	89	86

[a] Values in parentheses are for reactions using (DHQ)$_2$AQN that give enantiomeric products.

Table 8.13 Deng's (DHQD)$_2$AQN-catalyzed KR of 1,3-dioxolane-2,4-diones [194].

Entry	R	R'	t(h)	ee(%) (Yield/%)		s
				(R)-47	(S)-48	
1	a Bn	Et	12	96(47)	95(39)	133
2	b PhCH$_2$CH$_2$	Et	24	93(46)	85(40)	67
3	c n-Bu	Et	36	92(46)	95(36)	57
4	d i-Pr	Allyl	6	90(48)	93(32)	49

Deng found that alcoholative KR of α-alkyl-1,3-dioxolane-2,4-diones **46a–d** using (DHQD)$_2$AQN as the catalyst provides chiral α-hydroxy esters **47a–d** and enantiomerically enriched unreacted starting materials with excellent selectivities ($s = 49$–133) (Table 8.13) [194].

Again, achiral hydrolysis of the unreacted starting material and separation of the three components of the reaction mixture via an extractive procedure provides access to both the α-hydroxy acid and the α-hydroxy ester, and allows quantitative recovery of the catalyst.

As for the UNCAs, Deng found that under appropriate conditions 1,3-dioxolane-2,4-diones could also be induced to undergo DKR. As these compounds are even more readily racemized than UNCAs, DKR of many α-aryl-1,3-dioxolane-2,4-diones occurs even at $-78\,°C$, although temperatures up to $-20\,°C$ proved optimal for certain substrates. Thus, for a range of α-aryl-1,3-dioxolane-2,4-diones (**46a–g**), (DHQD)$_2$AQN (10 mol%) catalyzed DKR to the corresponding esters **47a–g** with excellent stereoselectivities (91–96% ee) and good yields (65–85%) (Table 8.14) [194].

Sammakia has developed a rather unique chiral O-nucleophilic acyl transfer catalyst **49**, and shown that it is also effective for the KR of certain α-hydroxy acid [195] and α-amino acid [196] derivatives. It was found that by employing this catalyst at the 10 mol% level in toluene at -26 to $0\,°C$, it was possible to resolve α-acetoxy-N-acyloxazolidinethiones with s-values in the range 17 to 32 [195], and α-(N-trifluoroacetyl)-N-acyloxazolidinethiones with s-values in the range 20 to 86 [196] (Scheme 8.23). This KR method is particularly notable for its success with cyclic amino acid derivatives, which makes it complementary to the Deng protocol. Moreover, Sammakia has shown that the recovered oxazolidinethiones can be used directly in peptide-coupling reactions using (i-Pr)$_2$EtN and HOBt. Previous studies with an achiral variant of the catalyst had shown that this class

Table 8.14 Deng's (DHQD)$_2$AQN-catalyzed DKR of 1,3-dioxolane-2,4-diones [194].

Entry	R	T (°C)	t (h)	Yield (%)	ee (%)
1	**a** Ph	-78	24	71	95
2	**b** 4-Cl-C$_6$H$_4$	-78	24	70	96
3	**c** 4-CF$_3$-C$_6$H$_4$	-78	24	85	93
4	**d** 4-*i*-Pr-C$_6$H$_4$	-20	8	68	91
5	**e** 1-Nap[a]	-40	14	74	91
6	**f** 2-Cl-C$_6$H$_4$	-60	10	66	62
7	**g** 2-Me-C$_6$H$_4$	-20	4	61	60

[a] THF used as solvent and *n*-PrOH in place of EtOH.

of catalyst act as *O*-nucleophiles, and that the trifluoromethyl group is essential for optimal activity [197].

8.3.3
Type II Asymmetric Alcohol, Phenol, Enol, and Amine Addition to Ketenes

Chiral aryl acetic acids constitute a "privileged" class of target structures due to their prevalence in bioactive natural products and pharmaceuticals and so, unsurprisingly, they constitute attractive targets for asymmetric synthesis [198]. The face-selective addition of a nucleophile to an aryl alkyl ketene provides a very direct entry for the preparation of such compounds. Although this can be achieved by the use of a chiral nucleophile or acid (*cf.* Scheme 8.1) [199], catalysis of the addition of an achiral nucleophile is clearly attractive from the standpoint of efficiency.

The pioneering studies in this area were conducted by Pracejus during the 1960s using alkaloids to promote the addition of methanol to phenylmethylketene (maximum *ee*-value 76%) [200–206]. This approach was significantly extended by Simpkins using silylketenes (maximum *ee*-value 94%) [207]. However, more recently Fu has considerably extended the levels of selectivity attainable and the range of ketenes and nucleophiles that can be employed [56, 62].

Fu's initial investigations used the planar chiral pyrrole catalyst **3b** (see Scheme 8.4) at the 10 mol% level, with 2,6-di-*tert*-butylpyridinium triflate (12 mol%) as cocatalyst, to effect the asymmetric addition of methanol to a range of arylmethylketenes to give arylpropionic esters in excellent yields (80–97%) and with good selectivity (68–80% *ee*) (Scheme 8.24) [47]. Subsequently, drawing on insight gained from mechanistic investigations which suggested that a Brønsted acid catalytic

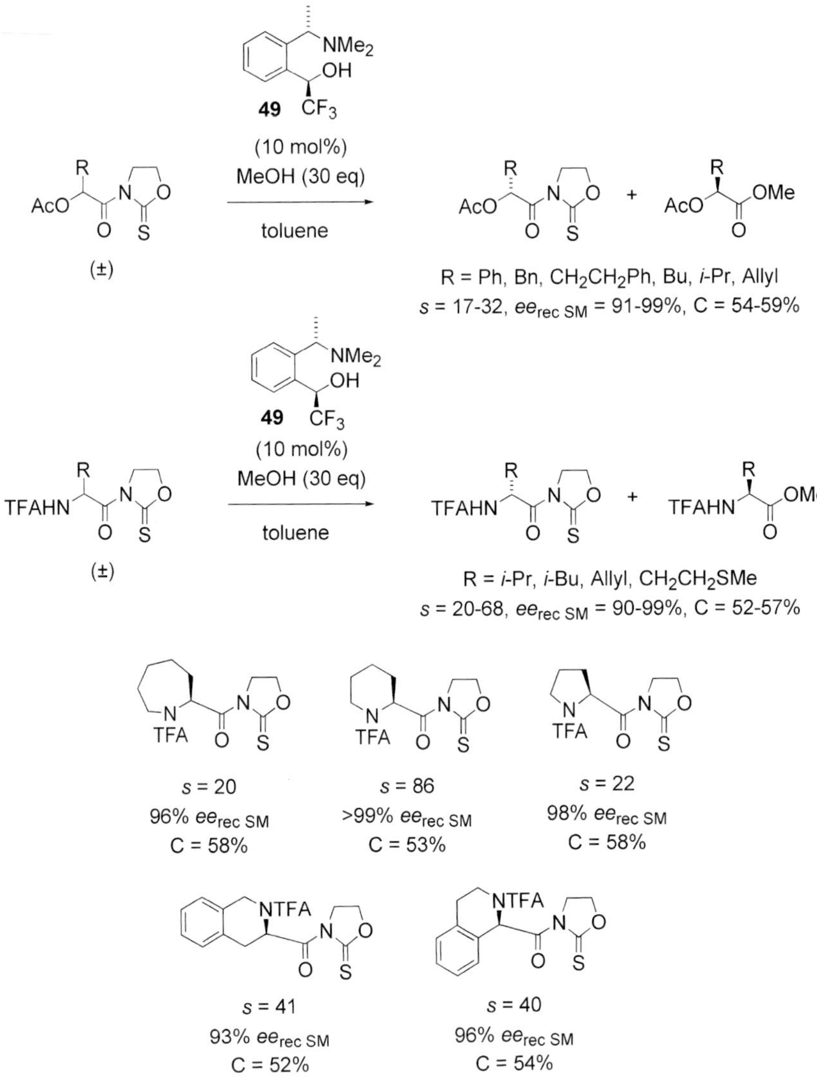

Scheme 8.23 Sammakia's chiral alcohol-catalyzed KR of α-acetoxy- and α-(N-trifluoroacetyl) amino acid-N-acyloxazolidinethiones [195, 196].

cycle was operational (Fig. 8.7), two much-improved systems were developed using planar chiral ferrocenyl PPY **4c** as catalyst. The first system used 2-tert-butylphenol as the nucleophile, employed just 3 mol% of PPY **4c** in toluene at rt, and provided the corresponding aryl esters in excellent selectivities (79–94% ee) and yields (66–97%) [68]. The second system used the enol of diphenylacetaldehyde as the nucleophile and employed 10 mol% of **4c** in chloroform at 0 °C to provide the corresponding vinyl ester products with comparable selectivities (77–98% ee) and even better yields (74–99%) (Scheme 8.24) [66].

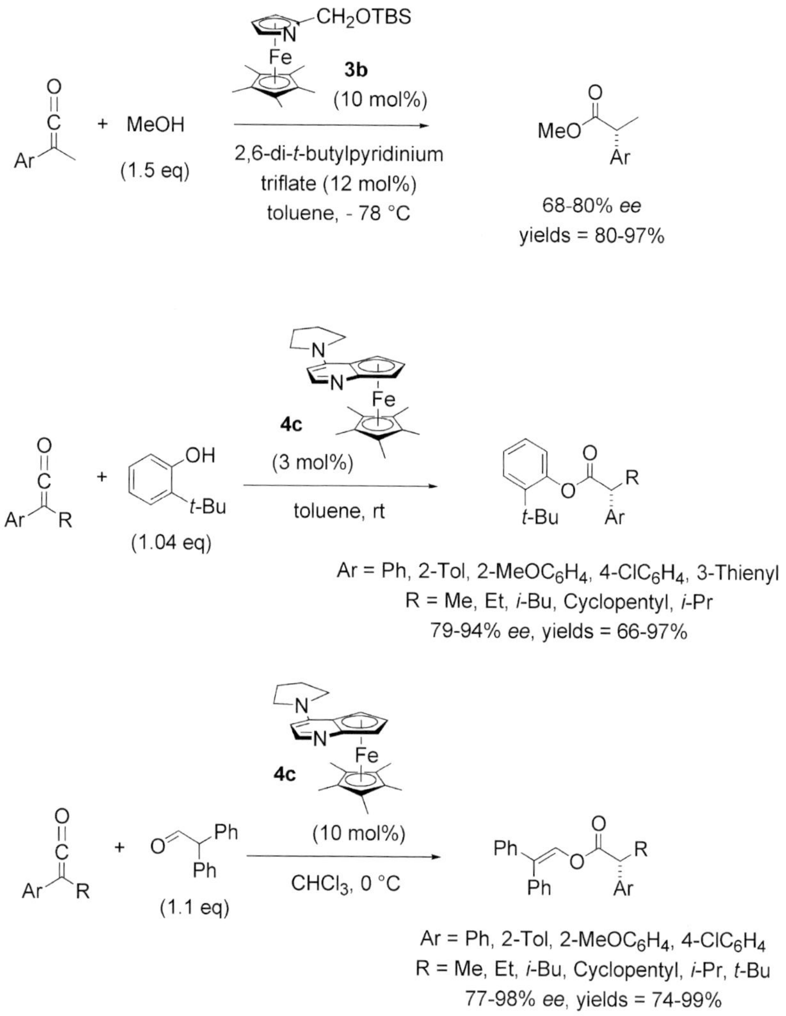

Scheme 8.24 Fu's planar chiral pyrrole and PPY-catalyzed addition of alcohols to arylalkylketenes [47, 66, 68].

For both the aryl and vinyl ester products, Fu was able to demonstrate that hydrolysis or reduction could be performed under mild conditions to yield the corresponding chiral acids and primary alcohols, without racemization.

Fu has also achieved the addition of 2-cyanopyrrole to arylalkylketenes, again using PPY **4c** (2 mol%) as catalyst in toluene at rt, and with similarly high selectivity and yield [59]. Additionally, related processes have been developed for the asymmetric preparation of β-lactams [67] and β-lactones [63] by the formal [2+2]-cycloaddition of N-tosylaldimines and aryl aldehydes, respectively, to ketenes. However, these and other catalyzed asymmetric cycloadditions fall somewhat outside the remit of this review and have been reviewed elsewhere [32].

8.4
Concluding Remarks

From the foregoing survey it is clear that the past decade has witnessed great strides in the area of organocatalyzed asymmetric acyl transfer processes. Many versatile and important chiral compounds including *sec*-alcohols, 1,2-diols, 1,2-amino alcohols, α-chiral amines, α-hydroxy acids, α-amino acids, α-chiral arylacetic acids, α-chiral oxindoles, etc. can be efficiently prepared in highly enantiomerically enriched form using asymmetric acyl transfer reactions promoted by small-molecule organocatalysts. However, many challenges remain, particularly with regard to the development of easily accessible chiral catalysts and the achievement of levels of reactivity and selectivity that will allow even wider substrate scope. An integral part of the achievement of this goal will undoubtedly be an increasing understanding of the detailed catalytic cycles involved and the precise factors that regulate chirality transfer in organocatalytic systems. Given the current worldwide interest in all aspects of asymmetric organocatalysis, we can confidently anticipate rapid progress in this area, and we hope that this chapter will prove helpful to researchers planning to make contributions towards that end.

References

1 The designation of "Type I" and "Type II" processes here is essentially arbitrary; the opposite definition has been employed in one biotransformation textbook (see Ref. [26]).
2 H.B. Kagan, J.C. Fiaud, *Top. Stereochem.* **1988**, *18*, 249–330.
3 J.M. Keith, J.F. Larrow, E.N. Jacobsen, *Adv. Synth. Catal.* **2001**, *343*, 5–26.
4 E.N. Jacobsen, N.S. Finney, *Chem. Biol.* **1994**, *1*, 85–90.
5 R.S. Ward, *Chem. Soc. Rev.* **1990**, *19*, 1–19.
6 M.C. Willis, *J. Chem. Soc., Perkin Trans. 1* **1999**, 1765–1784.
7 A.C. Spivey, B.I. Andrews, A. Brown, *Recent Res. Develop. Org. Chem.* **2002**, *6*, 147–167.
8 C.J. Sih, S.H. Wu, *Top. Stereochem.* **1989**, *19*, 63–125.
9 D.E.J.E. Robinson, S.D. Bull, *Tetrahedron: Asymm.* **2003**, *14*, 1407–1446.
10 E. Vedejs, M. Jure, *Angew. Chem. Int. Ed.* **2005**, *44*, 3974–4001.
11 J. Eames, *Angew. Chem. Int. Ed.* **2000**, *39*, 885–888.
12 J.R. Dehli, V. Gotor, *Chem. Soc. Rev.* **2002**, *31*, 365–370.
13 S. Caddick, K. Jenkins, *Chem. Soc. Rev.* **1996**, *25*, 447–456.
14 F.F. Huerta, A.B.E. Minidis, J.-E. Backvall, *Chem. Soc. Rev.* **2001**, *30*, 321–331.
15 A.J.J. Straathof, J.A. Jongejan, *Enzyme Microb. Technol.* **1997**, *21*, 559–571.
16 R.F. Ismagliov, *J. Org. Chem.* **1998**, *63*, 3772–3774.
17 E. Vedejs, O. Daugulis, *J. Am. Chem. Soc.* **1999**, *121*, 5813–5814.
18 T.O. Luukas, C. Girard, D.R. Fenwick, H.B. Kagan, *J. Am. Chem. Soc.* **1999**, *121*, 9299–9306.

19 D.G. Blackmond, *J. Am. Chem. Soc.* **2001**, *123*, 545–553.
20 C. Girard, H.B. Kagan, *Angew. Chem. Int. Ed.* **1998**, *37*, 2923–2959.
21 D.W. Johnson, D.A. Singleton, *J. Am. Chem. Soc.* **1999**, *121*, 9307–9312.
22 H.B. Kagan, *Synlett* **2001**, 888–899.
23 R.E. Gawley, *J. Org. Chem.* **2006**, *71*, 2411–2416.
24 U.T. Bornscheuer, R.J. Kazlauskas, *Hydrolases in Organic Synthesis. Regio- and Stereoselective Biotransformations.* Wiley-VCH, Weinheim, **1999**.
25 C.-H. Wong, G.M. Whitesides, *Enzymes in Synthetic Organic Chemistry.* Pergamon, Oxford, **1994**.
26 K. Faber, *Bio-transformations in Organic Chemistry – A Text Book.* Springer-Verlag, Berlin, **1995**.
27 H. Waldmann, K. Drauz, *Enzyme Catalysis in Organic Synthesis – A Comprehensive Handbook.* Wiley-VCH, Weinheim, **2002**.
28 F. van Rantwijk, R.A. Sheldon, *Tetrahedron* **2004**, *60*, 501–519.
29 A.C. Spivey, A. Maddaford, A. Redgrave, *Org. Prep. Proc. Int.* **2000**, *32*, 331–365. For state-of-the-art Lewis acid acylation see: C. Mazet, S. Roseblade, V. Kohler, A. Pfaltz, *Org. Lett.* **2006**, *8*, 1879–1882 and K. Matsumoto, M. Mitsuda, N. Ushijima, Y. Demizu, O. Onomura, Y. Matsumara *Tetrahedron Lett.* **2006**, *47*, 8453–8456.
30 A.C. Spivey, B.I. Andrews, *Angew. Chem. Int. Ed.* **2001**, *40*, 3131–3134.
31 Y. Chen, P. McDaid, L. Deng, *Chem. Rev.* **2003**, *103*, 2965–2983.
32 S. France, D.J. Guerin, S.J. Miller, T. Lectka, *Chem. Rev.* **2003**, *103*, 2985–3012.
33 S.-K. Tian, Y. Chen, J. Hang, L. Tang, P. McDaid, L. Deng, *Acc. Chem. Res.* **2004**, *37*, 621–631.
34 J. Seayad, B. List, *Org. Biomol. Chem.* **2005**, *3*, 719–724.
35 A.C. Spivey, S. Arseniyadis, *Angew. Chem. Int. Ed.* **2004**, *43*, 5436–5441.
36 S. Xu, I. Held, B. Kempf, H. Mayr, W. Steglich, H. Zipse, *Chem. Eur. J.* **2005**, *11*, 4751–4757.
37 I. Held, A. Villinger, H. Zipse, *Synthesis* **2005**, 1425–1430.
38 G. Helmchen, R.W. Hoffmann, J. Mulzer, E. Schaumann, *Methods in Organic Chemistry – Houben-Weyl 6 Volume Set – E21a–f: Stereoselective Synthesis.* Georg Thieme Verlag, Stuttgart, **1996**.
39 E. Vedejs, O. Daugulis, S.T. Diver, *J. Org. Chem.* **1996**, *61*, 430–431.
40 E. Vedejs, O. Daugulis, J.A. MacKay, E. Rozners, *Synlett* **2001**, 1499–1505.
41 E. Vedejs, O. Daugulis, *Latv. Kim. Z.* **1999**, *1*, 31–38.
42 E. Vedejs, O. Daugulis, *J. Am. Chem. Soc.* **2003**, *125*, 4166–4173.
43 E. Vedejs, E. Rozners, *J. Am. Chem. Soc.* **2001**, *123*, 2428–2429.
44 J.A. MacKay, E. Vedejs, *J. Org. Chem.* **2006**, *71*, 498–503.
45 J.A. MacKay, E. Vedejs, *J. Org. Chem.* **2004**, *69*, 6934–6937.
46 J.C. Ruble, G.C. Fu, *J. Org. Chem.* **1996**, *61*, 7230–7231.
47 B.L. Hodous, J.C. Ruble, G.C. Fu, *J. Am. Chem. Soc.* **1999**, *121*, 2637–2638.
48 J.C. Ruble, H.A. Latham, G.C. Fu, *J. Am. Chem. Soc.* **1997**, *119*, 1492–1493.
49 J.C. Ruble, J. Tweddell, G.C. Fu, *J. Org. Chem.* **1998**, *63*, 2794–2795.
50 J.C. Ruble, G.C. Fu, *J. Am. Chem. Soc.* **1998**, *120*, 11532–11533.
51 J. Liang, J.C. Ruble, G.C. Fu, *J. Org. Chem.* **1998**, *63*, 3154–3155.
52 C.E. Garrett, G.C. Fu, *J. Am. Chem. Soc.* **1998**, *120*, 7479–7483 and erratum p. 10276.
53 B. Tao, J.C. Ruble, D.A. Hoic, G.C. Fu, *J. Am. Chem. Soc.* **1999**, *121*, 5091–5092 and erratum p. 10452.
54 S. Bellemin-Laponnaz, J. Tweddell, J.C. Ruble, F.M. Breitling, G.C. Fu, *Chem. Commun.* **2000**, 1009–1010.
55 M. Suginome, G.C. Fu, *Chirality* **2000**, *12*, 318–324.
56 G.C. Fu, *Acc. Chem. Res.* **2000**, *33*, 412–420.
57 Y. Ie, G.C. Fu, *Chem. Commun.* **2000**, 119–120.
58 S. Arai, S. Bellemin-Laponnaz, G.C. Fu, *Angew. Chem. Int. Ed.* **2001**, *40*, 234–236.
59 B.L. Hodous, G.C. Fu, *J. Am. Chem. Soc.* **2002**, *124*, 10006–10007.

60 A.H. Mermerian, G.C. Fu, *J. Am. Chem. Soc.* **2003**, *125*, 4050–4051.
61 I.D. Hills, G.C. Fu, *Angew. Chem. Int. Ed.* **2003**, *42*, 3921–3924.
62 G.C. Fu, *Acc. Chem. Res.* **2004**, *37*, 542–547.
63 J.E. Wilson, G.C. Fu, *Angew. Chem. Int. Ed.* **2004**, *43*, 6358–6360.
64 A.H. Mermerian, G.C. Fu, *J. Am. Chem. Soc.* **2005**, *127*, 5604–5607.
65 A.H. Mermerian, G.C. Fu, *Angew. Chem. Int. Ed.* **2005**, *44*, 949–952.
66 C. Schaefer, G.C. Fu, *Angew. Chem. Int. Ed.* **2005**, *44*, 4606–4608.
67 E.C. Lee, B.L. Hodous, E. Bergin, C. Shih, G.C. Fu, *J. Am. Chem. Soc.* **2005**, *127*, 11586–11587.
68 S.L. Wiskur, G.C. Fu, *J. Am. Chem. Soc.* **2005**, *127*, 6176–6177.
69 Catalogue numbers: *R*-(+)-4b: 26–1410; *S*-(−)-1414b: 1426–1411; *R*-(+)-1414c: 1426–3700; *R*-(−)-1414d: 1426–3701.
70 G.C. Fu, J.C. Ruble, J. Tweddell, *J. Org. Chem.* **1998**, *63*, 2794–2795.
71 E. Vedejs, X. Chen, *J. Am. Chem. Soc.* **1996**, *118*, 1809–1810.
72 E. Vedejs, X. Chen, *United States Patent No. 1997/5646287* **1997**.
73 E. Vedejs, X. Chen, *J. Am. Chem. Soc.* **1997**, *119*, 2584–2585.
74 S.A. Shaw, P. Aleman, E. Vedejs, *J. Am. Chem. Soc.* **2003**, *125*, 13368–13369.
75 S.A. Shaw, P. Aleman, J. Christy, J.W. Kampf, P. Va, E. Vedejs, *J. Am. Chem. Soc.* **2006**, *128*, 925–934.
76 T. Kawabata, M. Nagato, K. Takasu, K. Fuji, *J. Am. Chem. Soc.* **1997**, *119*, 3169–3170.
77 T. Kawabata, K. Yamamoto, Y. Momose, H. Yoshida, Y. Nagaoka, K. Fuji, *Chem. Commun.* **2001**, 2700–2701.
78 T. Kawabata, R. Stragies, T. Fukaya, Y. Nagaoka, H. Schedel, K. Fuji, *Tetrahedron Lett.* **2003**, *44*, 1545–1548.
79 T. Kawabata, R. Stragies, T. Fukaya, K. Fuji, *Chirality* **2003**, *15*, 71–76.
80 S.J. Taylor, J.P. Morken, *Science* **1998**, *280*, 267–270.
81 A.C. Spivey, T. Fekner, H. Adams, *Tetrahedron Lett.* **1998**, *39*, 8919–8922.
82 A.C. Spivey, T. Fekner, S.E. Spey, H. Adams, *J. Org. Chem.* **1999**, *64*, 9430–9443.
83 A.C. Spivey, T. Fekner, S.E. Spey, *J. Org. Chem.* **2000**, *65*, 3154–3159.
84 A.C. Spivey, A. Maddaford, T. Fekner, A. Redgrave, C.S. Frampton, *J. Chem. Soc., Perkin Trans. 1* **2000**, 3460–3468.
85 A.C. Spivey, T. Fekner, *PCT Int. Patent No. WO 2001039884*, **2001**.
86 A.C. Spivey, A. Maddaford, A.J. Redgrave, *J. Chem. Soc., Perkin Trans. 1* **2001**, 1785–1794.
87 A.C. Spivey, P. Charbonnier, T. Fekner, D. Hochmuth, A. Maddaford, C. Malardier-Jugroot, J.-P. Plog, A. Redgrave, M. Whitehead, *J. Org. Chem.* **2001**, *66*, 7394–7401.
88 C. Malardier-Jugroot, A.C. Spivey, M.A. Whitehead, *J. Mol. Struct. (Theochem)* **2003**, *623*, 263–276.
89 A.C. Spivey, F. Zhu, M.B. Mitchell, S.G. Davey, R.L. Jarvest, *J. Org. Chem.* **2003**, *68*, 7379–7385.
90 A.C. Spivey, D.P. Leese, F. Zhu, S.G. Davey, R.L. Jarvest, *Tetrahedron* **2004**, *60*, 4513–4525.
91 A.C. Spivey, S. Arseniyadis, T. Fekner, A. Maddaford, D.P. Leese, *Tetrahedron* **2006**, *62*, 295–301.
92 H. Kotsuki, H. Sakai, T. Shinohara, *Synlett* **2000**, 116–118.
93 T. Ishii, S. Fujioka, Y. Sekiguchi, H. Kotsuki, *J. Am. Chem. Soc.* **2004**, *126*, 9558–9559.
94 G. Naraku, N. Shimomoto, T. Hanamoto, J. Inanaga, *Enantiomer* **2000**, *5*, 135–138.
95 G. Priem, M.S. Anson, S.J.F. Macdonald, B. Pelotier, I.B. Campbell, *Tetrahedron Lett.* **2002**, *43*, 6001–6003.
96 G. Priem, B. Pelotier, S.J.F. Macdonald, M.S. Anson, I.B. Campbell, *J. Org. Chem.* **2003**, *68*, 3844–3848.
97 B. Pelotier, G. Priem, I.B. Campbell, S.J.F. Macdonald, M.S. Anson, *Synlett* **2003**, 679–683.
98 B. Pelotier, G. Priem, S.J.F. Macdonald, M.S. Anson, R.J. Upton, I.B. Campbell, *Tetrahedron Lett.* **2005**, *46*, 9005–9007.
99 K.-S. Jeong, S.-H. Kim, H.-J. Park, K.-J. Chang, K.S. Kim, *Chem. Lett.* **2002**, 1114–1115.
100 S. Yamada, T. Misono, Y. Iwai, *Tetrahedron Lett.* **2005**, *46*, 2239–2242.

101 C.O. Dalaigh, S.J. Hynes, D.J. Maher, S.J. Connon, *Org. Biomol. Chem.* **2005**, *3*, 981–984.
102 J.G. Seitzberg, C. Dissing, I. Sotofte, P.-O. Norrby, M. Johannsen, *J. Org. Chem.* **2005**, *70*, 8332–8337.
103 D. Diez, M.J. Gil, R.F. Moro, N.M. Garrido, I.S. Marcos, P. Basabe, F. Sanz, H.B. Broughton, J.G. Urones, *Tetrahedron: Asymm.* **2005**, *16*, 2980–2985.
104 T. Poisson, M. Penhoat, C. Papamicaël, G. Dupas, V. Dalla, F. Marsais, V. Levacher, *Synlett* **2005**, 2285–2288.
105 H.V. Nguyen, D.C.D. Butler, C.J. Richards, *Org. Lett.* **2006**, *8*, 769–772.
106 V.B. Birman, E.W. Uffman, H. Jiang, X. Li, C.J. Kilbane, *J. Am. Chem. Soc.* **2004**, *126*, 12226–12227.
107 V.B. Birman, H. Jiang, *Org. Lett.* **2005**, *7*, 3445–3447.
108 V.B. Birman, X. Li, H. Jiang, E.W. Uffman, *Tetrahedron* **2006**, *62*, 285–294.
109 S.J. Miller, G.T. Copeland, N. Papaioannou, T.E. Horstmann, E.M. Ruel, *J. Am. Chem. Soc.* **1998**, *120*, 1629–1630.
110 G.T. Copeland, E.R. Jarvo, S.J. Miller, *J. Org. Chem.* **1998**, *63*, 6784–6785.
111 G.T. Copeland, S.J. Miller, *J. Am. Chem. Soc.* **1999**, *121*, 4306–4307.
112 E.R. Jarvo, G.T. Copeland, N. Papaioannou, P.J. Bonitatebus, S.C. Miller, *J. Am. Chem. Soc.* **1999**, *121*, 11638–11643.
113 E.R. Jarvo, M.M. Vasbinder, S.J. Miller, *Tetrahedron* **2000**, *56*, 9773–9779.
114 M.M. Vasbinder, E.R. Jarvo, S.J. Miller, *Angew. Chem. Int. Ed.* **2001**, *40*, 2824–2827.
115 G.T. Copeland, S.J. Miller, *J. Am. Chem. Soc.* **2001**, *123*, 6496–6502.
116 N. Papaioannou, C.A. Evans, J.T. Blank, S.J. Miller, *Org. Lett.* **2001**, *3*, 2879–2882.
117 E.R. Jarvo, C.A. Evans, G.T. Copeland, S.J. Miller, *J. Org. Chem.* **2001**, *66*, 5522–5527.
118 E.R. Jarvo, S.J. Miller, *Tetrahedron* **2002**, *58*, 2481–2495.
119 C.A. Evans, S.J. Miller, *Curr. Opin. Chem. Biol.* **2002**, *6*, 333–338.
120 K.S. Griswold, S.J. Miller, *Tetrahedron* **2003**, *59*, 8869–8875.
121 N. Papaioannou, J.T. Blank, S.J. Miller, *J. Org. Chem.* **2003**, *68*, 2728–2734.
122 M.B. Fierman, D.J. O'Leary, W.E. Steinmetz, S.J. Miller, *J. Am. Chem. Soc.* **2004**, *126*, 6967–6971.
123 C.A. Lewis, B.R. Sculimbrene, Y. Xu, S.J. Miller, *Org. Lett.* **2005**, *7*, 3021–3023.
124 F. Formaggio, A. Barazza, A. Bertocco, C. Toniolo, Q.B. Broxterman, B. Kaptein, E. Brasola, P. Pengo, L. Pasquato, P. Scrimin, *J. Org. Chem.* **2004**, *69*, 3849–3856.
125 Y. Suzuki, K. Yamauchi, K. Muramatsu, M. Sato, *Chem. Commun.* **2004**, 2770–2771.
126 Y. Suzuki, K. Muramatsu, K. Yamauchi, Y. Morie, M. Sato, *Tetrahedron* **2006**, *62*, 302–310.
127 T. Kano, K. Sasaki, K. Maruoka, *Org. Lett.* **2005**, *7*, 1347–1349.
128 G.A. Grasa, R. Singh, S.P. Nolan, *Synthesis* **2004**, 971–985.
129 A.H. Hoveyda, D.A. Evans, G.C. Fu, *Chem. Rev.* **1993**, *93*, 1307–1370.
130 E. Vedejs, J.A. MacKay, *Org. Lett.* **2001**, *3*, 535–536.
131 K. Matsumura, S. Hashiguchi, T. Ikariya, R. Noyori, *J. Am. Chem. Soc.* **1997**, *119*, 8738–8739 and references therein.
132 S. Masamune, W. Choy, *Aldrichimica Acta* **1982**, *15*, 47–63.
133 M. Periasamy, *Aldrichimica Acta* **2002**, *35*, 89–101.
134 C.A. Hunter, K.R. Lawson, J. Perkins, C.J. Urch, *J. Chem. Soc., Perkin Trans. 2* **2001**, 651.
135 K. Ishihara, Y. Kosugi, M. Akakura, *J. Am. Chem. Soc.* **2004**, *126*, 12212–12213.
136 R. Bloch, *Chem. Rev.* **1998**, *98*, 1407–1438.
137 D. Enders, U. Reinhold, *Tetrahedron: Asymm.* **1997**, *8*, 1895–1946.
138 A. Johansson, *Contemp. Org. Synth.* **1995**, *2*, 393–407.
139 T. Yokomatsu, A. Arakawa, S. Shibuya, *J. Org. Chem.* **1994**, *59*, 3506–3508.
140 R.S. Atkinson, E. Barker, C.J. Price, D.R. Russell, *J. Chem. Soc., Chem. Commun.* **1994**, 1159–1160.

141 R.S. Atkinson, E. Barker, P.J. Edwards, G.A. Thomson, *J. Chem. Soc., Perkin Trans. 1* **1996**, 1047–1055.
142 A.G. Al-Sehemi, R.S. Atkinson, J. Fawcett, D.R. Rusell, *Tetrahedron Lett.* **2000**, *41*, 2239–2242.
143 A.G. Al-Sehemi, R.S. Atkinson, J. Fawcett, D.R. Rusell, *Tetrahedron Lett.* **2000**, *41*, 2243–2246.
144 A.G. Al-Sehemi, R.S. Atkinson, D.R. Rusell, *Chem. Commun.* **2000**, 43–44.
145 A.G. Al-Sehemi, R.S. Atkinson, C.K. Meades, *Chem. Commun.* **2001**, 2684–2685.
146 K. Kondo, T. Kurosaki, Y. Murakami, *Synlett* **1998**, 725–726.
147 V.P. Krasnov, G.L. Levit, M.I. Kodess, V.N. Charushin, O.N. Chupakhin, *Tetrahedron: Asymm.* **2004**, *15*, 859–862 and references therein.
148 S. Arseniyadis, A. Valleix, A. Wagner, C. Mioskowski, *Angew. Chem. Int. Ed.* **2004**, *43*, 3314–3317.
149 S. Arseniyadis, P.V. Subhash, A. Valleix, S.P. Mathew, D.G. Blackmond, A. Wagner, C. Mioskowski, *J. Am. Chem. Soc.* **2005**.
150 A. Moretto, C. Peggion, F. Formaggio, M. Crisma, B. Kaptein, Q.B. Broxterman, C. Toniolo, *Chirality* **2005**, *17*, 481–487.
151 T. Oriyama, K. Imai, T. Hosoya, T. Sano, *Tetrahedron Lett.* **1998**, *39*, 397–400.
152 T. Oriyama, K. Imai, T. Sano, T. Hosoya, *Tetrahedron Lett.* **1998**, *39*, 3529–3532.
153 T. Oriyama, T. Hosoya, T. Sano, *Heterocycles* **2000**, *52*, 1065–1069.
154 T. Oriyama, H. Taguchi, D. Terakado, T. Sano, *Chem. Lett.* **2002**, 26–27.
155 T. Sano, K. Imai, K. Ohashi, T. Oriyama, *Chem. Lett.* **1999**, 265–266.
156 T. Sano, H. Miyata, T. Oriyama, *Enantiomer* **2000**, *5*, 119–123.
157 D. Terakado, H. Koutaka, T. Oriyama, *Tetrahedron: Asymm.* **2005**, *16*, 1157–1165.
158 A. Cordova, K.D. Janda, *J. Org. Chem.* **2001**, *66*, 1906–1909.
159 A. Córdova, M.R. Tremblay, B. Clapham, K.D. Janda, *J. Org. Chem.* **2001**, *66*, 5645–5648.
160 B. Clapham, C.-W. Cho, K.D. Janda, *J. Org. Chem.* **2001**, *66*, 868–873.
161 E.P. Kundig, T. Lomberget, R. Bragg, C. Poulard, G. Bernardinelli, *Chem. Commun.* **2004**, 1548–1549.
162 E.P. Kundig, A.E. Garcia, T. Lomberget, G. Bernardinelli, *Angew. Chem. Int. Ed.* **2006**, *45*, 98–101.
163 S. Mizuta, M. Sadamori, T. Fujimoto, I. Yamamoto, *Angew. Chem. Int. Ed.* **2003**, *42*, 3383–3385.
164 S. Mizuta, T. Tsuzuki, T. Fujimoto, I. Yamamoto, *Org. Lett.* **2005**, *7*, 3633–3635.
165 E. Vedejs, O. Daugulis, N. Tuttle, *J. Org. Chem.* **2004**, *69*, 1389–1392.
166 W. Steglich, G. Hofle, *Tetrahedron Lett.* **1970**, 4727–4730.
167 W. Steglich, H. Vorbruggen, G. Höfle, *Angew. Chem. Int. Ed. Engl.* **1978**, *17*, 569–583.
168 J. Hiratake, Y. Yamamoto, J.-I. Oda, *J. Chem. Soc., Chem. Commun.* **1985**, 1717–1719.
169 J. Hiratake, M. Inagaki, Y. Yamamoto, J.-I. Oda, *J. Chem. Soc., Perkin Trans. I* **1987**, 1053–1058.
170 R.A. Aitken, J. Gopal, J.A. Hirst, *J. Chem. Soc., Chem. Commun.* **1988**, 632–634.
171 R.A. Aitken, J. Gopal, *Tetrahedron: Asymm.* **1990**, *1*, 517–520.
172 C. Bolm, A. Gerlach, C.L. Dinter, *Synlett* **1999**, 195–196.
173 J.T. Starr, G. Koch, E.M. Carreira, *J. Am. Chem. Soc.* **2000**, *122*, 8793–8794.
174 A. Bernardi, D. Arosio, D. Dellavecchia, F. Micheli, *Tetrahedron: Asymm.* **1999**, *10*, 3403–3407.
175 C. Bolm, I. Atodiresei, I. Schiffers, *Org. Synth.* **2005**, *82*, 120–124.
176 C. Bolm, I. Schiffers, C.L. Dinter, A. Gerlach, *J. Org. Chem.* **2000**, *65*, 6984–6991.
177 C. Bolm, I. Schiffers, C.L. Dinter, L. Defrere, A. Gerlach, G. Raabe, *Synthesis* **2001**, 1719–1730.
178 C. Bolm, I. Schiffers, I. Atodiresei, C.P.R. Hackenberger, *Tetrahedron: Asymm.* **2003**, *14*, 3455–3467.
179 Y.G. Chen, S.-K. Tian, L. Deng, *J. Am. Chem. Soc.* **2000**, *122*, 9542–9543.
180 L. Deng, X. Liu, Y. Chen, S.-K. Tian, *PCT Int. Patent No. WO 2004110609* **2004**.

181 L. Deng, Y. Chen, S.-K. Tian, *United States Patent No. 2004/0082809* **2004**.
182 L. Deng, J. Hang, L. Tang, *United States Patent No. 2002/0151744* **2002**.
183 C. Choi, S.-K. Tian, L. Deng, *Synthesis* **2001**, 1737–1741.
184 Y.-M. Song, J.S. Choi, J.W. Yang, H. Han, *Tetrahedron Lett.* **2004**, *45*, 3301–3304.
185 H.S. Kim, Y.-M. Song, J.S. Choi, J.W. Yang, H. Han, *Tetrahedron* **2004**, *60*, 12051–12057.
186 Y. Uozumi, K. Yasoshima, T. Miyachi, S.-i. Nagai, *Tetrahedron Lett.* **2001**, *42*, 411–414.
187 T. Honjo, S. Sano, M. Shiro, Y. Nagao, *Angew. Chem. Int. Ed.* **2005**, *44*, 5838–5841.
188 Y.G. Chen, L. Deng, *J. Am. Chem. Soc.* **2001**, *123*, 11302–11303.
189 S.A. Brown, M.-C. Parker, N.J. Turner, *Tetrahedron: Asymm.* **2000**, *11*, 1687–1690.
190 H. Pellissier, *Tetrahedron* **2003**, *59*, 8291–8327.
191 J. Hang, S.-K. Tian, L. Tang, L. Deng, *J. Am. Chem. Soc.* **2001**, *123*, 12696–12697.
192 J. Hang, H. Li, L. Deng, *Org. Lett.* **2002**, *4*, 3321–3324.
193 J. Hang, L. Deng, *Synlett* **2003**, 1927–1930.
194 L. Tang, L. Deng, *J. Am. Chem. Soc.* **2002**, *124*, 2870–2871.
195 G.T. Notte, T. Sammakia, P.J. Steel, *J. Am. Chem. Soc.* **2005**, *127*, 13502–13503.
196 G.T. Notte, T. Sammakia, *J. Am. Chem. Soc.* **2006**, *128*, 4230–4231.
197 K.A. Wayman, T. Sammakia, *Org. Lett.* **2003**, *5*, 4105–4108.
198 H.R. Sonawane, N.S. Bellur, J.R. Ahuja, D.G. Kulkarni, *Tetrahedron: Asymm.* **1992**, *3*, 162.
199 C. Fehr, *Angew. Chem., Int. Ed.* **1996**, *35*, 2567–2587.
200 G. Pracejus, *Liebigs Ann. Chem.* **1959**, *622*, 10–22.
201 H. Pracejus, *Liebigs Ann. Chem.* **1960**, *634*, 9–22.
202 H. Pracejus, *Liebigs Ann. Chem.* **1960**, *634*, 23–29.
203 H. Pracejus, A. Tille, *Chem. Ber.* **1963**, *96*, 854–865.
204 H. Pracejus, H. Matje, *J. Prakt. Chem.* **1964**, *24*, 195–205.
205 H. Pracejus, G. Kohl, *Liebigs Ann. Chem.* **1969**, *722*, 1–11.
206 T. Yamashita, H. Yasueda, N. Nakamura, *Bull. Chem. Soc. Jpn.* **1979**, *52*, 2165–2166.
207 Blake, A.J., Friend, C.L., Outram, R.J., Simpkins, N.J., Whitehead, A.J., *Tetrahedron Lett.* **2001**, *42*, 2877–2881.

9
Nucleophilic N-Heterocyclic Carbenes in Asymmetric Organocatalysis

Dieter Enders, Tim Balensiefer, Oliver Niemeier, and Mathias Christmann

9.1
The Benzoin Condensation

When, in 1832, Wöhler and Liebig first discovered the cyanide-catalyzed coupling of benzaldehyde that became known as the "benzoin condensation", they laid the foundations for a wide field of growing organic chemistry [1]. In 1903, Lapworth proposed a mechanistical model with an intermediate carbanion formed in a hydrogen cyanide addition to the benzaldehyde substrate and subsequent deprotonation [2]. In the intermediate "active aldehyde", the former carbonyl carbon atom exhibits an inverted, nucleophilic reactivity, which exemplifies the "Umpolung" concept of Seebach [3]. In 1943, Ukai et al. reported that thiazolium salts also surprisingly catalyze the benzoin condensation [4], an observation which attracted even more attention when Mizuhara et al. found, in 1954, that the thiazolium unit of the coenzyme thiamine (vitamin B_1) (**1**, Fig. 9.1) is essential for its activity in enzyme biocatalysis [5]. Subsequently, the biochemistry of thiamine-dependent enzymes has been extensively studied, and this has resulted in widespread applications of the enzymes as synthetic tools [6].

Thus, the question raised was whether an explanation of the thiazolium-catalyzed benzoin condensation also lead to an elucidation of the biochemistry

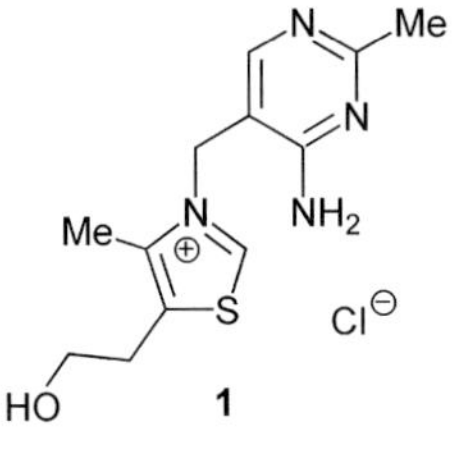

Fig. 9.1 Thiamine (vitamin B_1) (**1**), a coenzyme.

Enantioselective Organocatalysis: Reactions and Experimental Procedures
Edited by Peter I. Dalko
Copyright © 2007 WILEY-VCH Verlag GmbH & Co. KGaA, Weinheim
ISBN: 978-3-527-31522-2

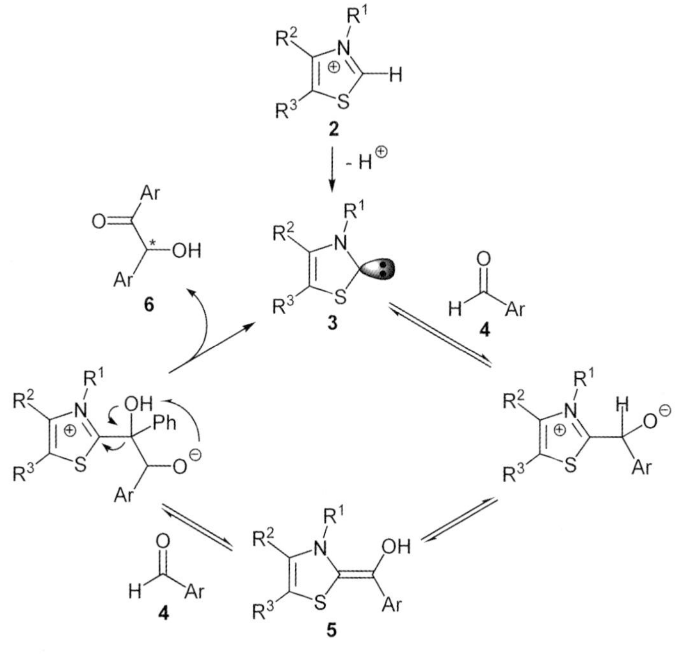

Scheme 9.1 The catalytic cycle of the benzoin condensation, as proposed by Breslow.

of thiamine? In 1958, Ronald Breslow proposed the mechanistic model shown in Scheme 9.1 [7], and assumed the catalytically active species to be thiazolin-2-ylidene (**3**), a carbene that would be formed *in situ* by deprotonation of the thiazolium salt **2**. The resulting heterocyclic carbene couples with an aromatic aldehyde molecule **4** to generate the "active aldehyde", the hydroxy-enamine "Breslow intermediate" **5**. This nucleophilic acylation reagent (which is equivalent to a d^1-synthon in the terminology of Seebach et al.) reacts with an electrophilic substrate such as a second aldehyde molecule to form the product benzoin **6**, at the same time regenerating the carbene catalyst. An alternative mechanistic model proposed by López Calahorra et al. was based on the observed facile formation of carbene dimers [8].

In the benzoin condensation, a new stereogenic center is formed, as the product is an α-hydroxy ketone. Consequently, many chemists aspired to develop heterazolium-catalyzed asymmetric benzoin condensations and, later, other nucleophilic acylation reactions [9]. For example, Sheehan et al. presented the first asymmetric benzoin condensation in 1966, with the chiral thiazolium salt **7** (Fig. 9.2) as catalyst precursor [10].

As a typical pioneering result, the enantiomeric excess (*ee*) of the synthesized benzoin was only 2%. With modified thiazolium salts such as **8**, Sheehan et al. obtained *ee*-values of up to 52%, but the yield was unfortunately reduced to 6% (48% in an improved protocol by Rawal et al. published many years later) [11,

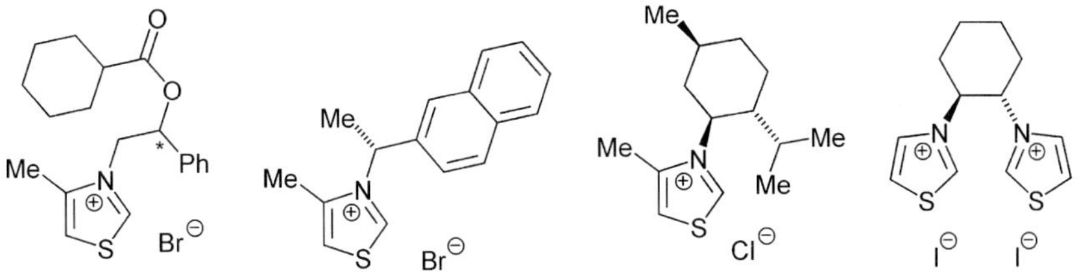

Fig. 9.2 Chiral thiazolium salts for enantioselective benzoin condensation.

12]. Tagaki et al. subsequently employed chiral menthyl-substituted thiazolium salts such as compound **9** in a micellar two-phase reaction system, reaching an *ee* of 35% and an improved yield of 20% [13]. Zhao et al. obtained moderate *ee*-values of 47 to 57% and yields of 20 to 30% when combining the Sheehan catalysts with the Tagaki reaction conditions [14]. Based on their mechanistic model, López Calahorra et al. developed bisthiazolium salt catalysts such as compound **10**, yielding 21% of benzoin in 27% *ee* [15].

In parallel with research studies on the carbene-catalyzed benzoin reaction, much effort has been devoted to isolate the actual nucleophilic carbenes. The studies of Breslow were extended by Wanzlick et al., who investigated the chemistry of *N*-heterocyclic nucleophilic carbenes during the 1960s [16]. Until then, the divalent carbenes had only been regarded as highly reactive intermediates, and indeed, Wanzlick was unable to isolate the carbenes because of their inevitable dimerization. It was in fact some 20 years later that Bertrand et al. [17] and Arduengo et al. [18] marked an essential progress when presenting carbene compounds that were stable at room temperature. However, Bertrand's phosphinocarbene was suspected not to have a predominant carbene character, but rather to react as a phosphaacetylene [19], although more recent studies have seemed to support again the carbene character [20]. Thus, the imidazolin-2-ylidene **11** (R = adamantyl) synthesized by Arduengo et al. in 1991 is now referred to as the first *N*-heterocyclic carbene to be isolated and characterized (Fig. 9.3). At this point it should be noted that the carbenes **11** have the same structure as some of

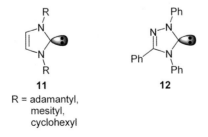

11
R = adamantyl,
mesityl,
cyclohexyl

12

Fig. 9.3 Stable carbenes synthesized by Arduengo et al. (**11**) and Enders, Teles et al. (**12**).

the carbenes that Wanzlick et al. attempted to isolate during the 1960s. Subsequently, many stable N-heterocyclic carbenes of various types have been synthesized, including many variations on the imidazolin-2-ylidene **11** [21].

In cooperation with Teles and colleagues, our research group has studied the triazole heterocycle as an alternative core structure of nucleophilic carbenes. First, the triazol-5-ylidene **12** (Fig. 9.3; see also Scheme 9.2) was synthesized and shown to be stable at temperatures up to 150 °C in the absence of air and moisture [22]. Compound **12** exhibited the typical behavior of a nucleophilic N-heterocyclic carbene, and was found to be sufficiently stable to become the first commercially available carbene [23]. As shown in Scheme 9.2, the crystalline carbene was obtained from the corresponding triazolium salt precursor **13** by the addition of methanolate and subsequent thermal decomposition of the adduct **14** *in vacuo* via α-elimination of methanol [24].

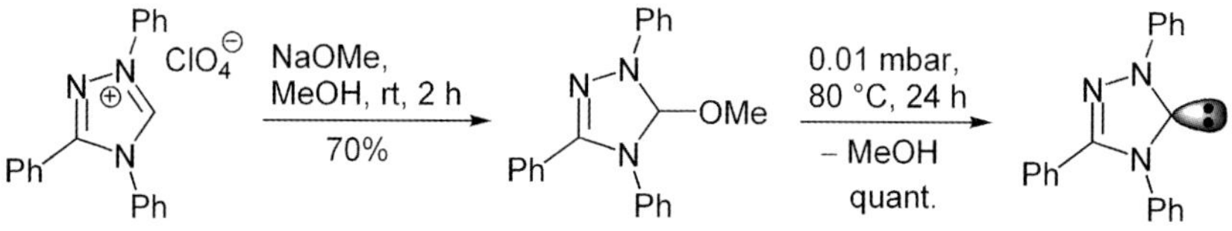

Scheme 9.2 Synthesis of the stable carbene **12**, as developed by Enders, Teles, and colleagues.

The triazol-5-ylidene **12** was found to be a powerful catalyst for the conversion of formaldehyde to glycolaldehyde in a "formoin reaction" [25.] The concept of triazolium salt catalysis appeared to show promise, and consequently our research group undertook the synthesis of a variety of chiral triazolium salts for the asymmetric benzoin reaction [26]. However, the *ee*-values and catalytic activities shifted widely with slight structural changes in the substitution pattern of the triazolium system. The most active catalyst **15** (Fig. 9.4) afforded benzoin (**6**, Ar = Ph) in its (*R*)-configuration with 75% *ee* and a satisfactory yield of 66%.

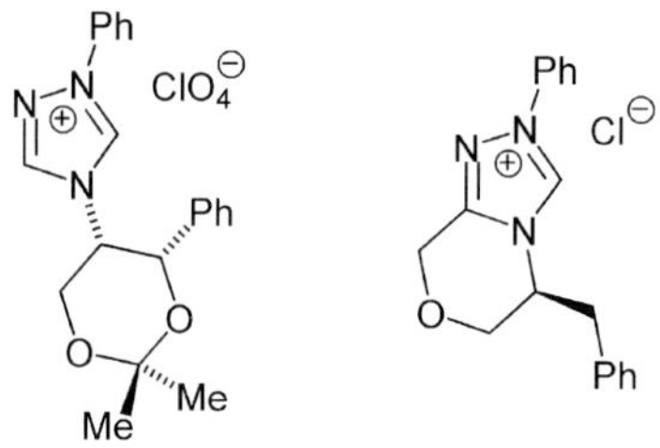

Fig. 9.4 Triazolium salts produced by Enders et al. (**15**) and Leeper et al. (**16**).

9.1 The Benzoin Condensation

The low catalyst loading of only 1.25 mol% indicated an activity that had increased by almost two orders of magnitude compared to the chiral thiazolium salts used previously.

An important contribution to the research investigations into the asymmetric benzoin condensation was made by Leeper and colleagues, who introduced the concept of bicyclic heterazolium salts. In 1997, this group reported the use of chiral, bicyclic thiazolium catalysts for the synthesis of aromatic benzoins and aliphatic butyroins [27], and subsequently (in 1998) presented details of improved chiral, bicyclic triazolium salts (e.g., **16**; Fig. 9.4) that produced various aromatic acyloins with modest to good enantioselectivities (20–83% *ee*) [28]. Based on the protocol of Leeper et al., another chiral, bicyclic triazolium salt derived from (*S*)-*tert*-leucine was developed by our research group, and in 2002 the results were published regarding the asymmetric benzoin condensation (shown in Scheme 9.3) using the novel triazolium salt **17** as catalyst precursor [29].

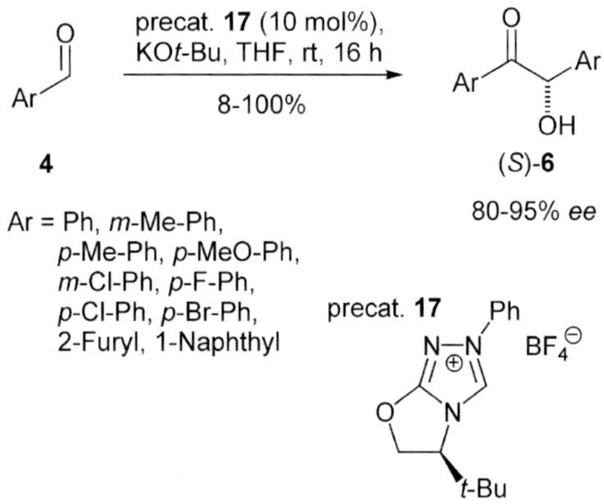

Scheme 9.3 Asymmetric condensation of aromatic aldehydes, as proposed by Enders and colleagues.

The precatalyst **17** produced (*S*)-benzoin (**6**, Ar = Ph) in very good yield (83%) and enantioselectivity (90% *ee*). The condensation of numerous other aromatic aldehydes **4** yielded the corresponding α-hydroxy ketones **6** with excellent *ee*-values of up to 95%. (For experimental details see Chapter 14.20.2). Electron-rich aromatic aldehydes gave consistently higher asymmetric inductions than electron-deficient (i.e., activated) aromatic aldehydes, with lower reaction temperatures or lower amounts of catalyst leading to slightly higher enantioselectivities coupled with lower yields.

The observed stereoselectivity of the reaction might be explained by transition state **18**, as shown in Figure 9.5. The *Si*-face of the assumed Breslow-type intermediate would be shielded by the *tert*-butyl group of the bicyclic catalyst, thus limiting any attack to the *Re*-face of the hydroxy enamine. Furthermore, the substituents of the enol moiety might cause a pre-orientation of the approaching

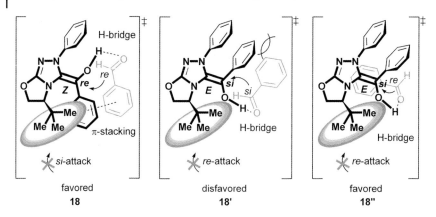

Fig. 9.5 Possible transition states for the asymmetric benzoin condensation, as proposed by Enders et al. and Houk et al.

second aldehyde (via π-stacking and H-bridge activation of the aldehyde carbonyl group). Thus, the *Re*-face of the approaching aldehyde molecule would be preferred, and this in consequence would lead to (*S*)-benzoin, as observed in the experiments.

However, the *E/Z*-geometry of the Breslow intermediate has not yet been determined. The olefine geometry is crucial for the pre-orientation of the second aldehyde molecule, and the *E*-isomer would, if one was to assume H-bridge interactions, probably favor a *Si-Si*-attack as shown in transition state **18′**, leading to (*R*)-benzoin. Presumably, due to an unfavorable steric interaction in transition state **18′** between the twisted phenyl substituent of the enamine moiety and the phenyl substituent of the attacking aldehyde, this is not observed. In computational calculations, Houk et al. found **18″** to be the most favorable transition state [30]. In this intermediate state, π-stacking does not occur, but the substituent of the approaching aldehyde resides in an open pocket of the catalyst with minimum steric repulsion, again favoring the formation of (*S*)-benzoin.

As an obvious extension of the benzoin reaction, the cross-coupling of aldehydes or of aldehydes and ketones was first achieved with the thiamine-dependent enzyme benzoylformate decarboxylase. This linked a variety of mostly aromatic aldehydes to acetaldehyde to form the corresponding α-hydroxy ketones, both chemo- and stereoselectively [31]. Synthetic thiazolium salts, developed by Stetter and co-workers and similar to thiamine itself [32], have been successfully used by Suzuki et al. for a diastereoselective intramolecular crossed aldehyde-ketone benzoin reaction during the course of an elegant natural product synthesis [33]. Stereocontrol was exerted by pre-existing stereocenters in the specific substrates, the catalysts being achiral.

In order to develop a general method, the present authors began investigations on the use of simple aldehyde ketones as substrates for the carbene-catalyzed crossed intramolecular benzoin condensation. Synchronously to the studies of

Suzuki et al., we presented our observation that indeed various five- and six-membered cyclic acyloins could be obtained as their racemates employing commercially available thiazolium salts as precatalysts [34].

Unfortunately, the chiral bicyclic triazolium salt that had been found to be an excellent catalyst for the enantioselective intermolecular benzoin condensation proved to be ineffective in the intramolecular reaction. In searching for alternative catalysts, we synthesized the novel triazolium salts **19** and **20**, starting from easily accessible enantiopure polycyclic γ-lactams (Schemes 9.4 and 9.5) that finally delivered good results in the enantioselective intramolecular cross-benzoin condensation [35].

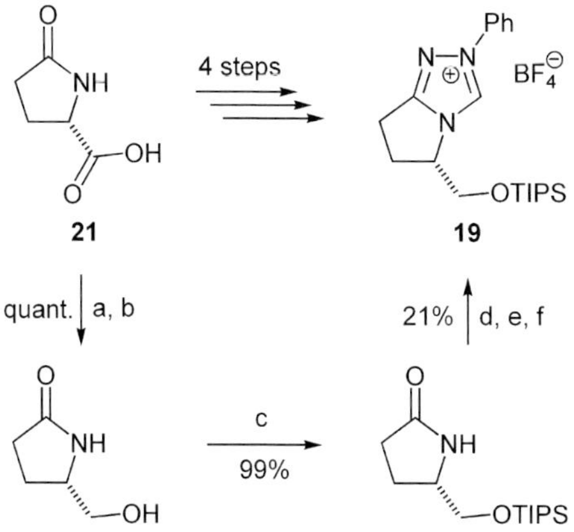

Scheme 9.4 Preparation of the L-pyroglutamic acid-derived triazolium salt **19**. (a) $SOCl_2$, MeOH, −15 °C, 2 h. (b) $NaBH_4$, EtOH, 0 °C, 12 h. (c) TIPSCl, imidazole, DMF, rt, 20 h. (d) Me_3OBF_4, DCM, rt, 3 h. (e) Phenylhydrazine, DCM, rt, 3 h. (f) $HC(OMe)_3$, MeOH, 80 °C, 12 h.

Precatalyst **19** was built up from cheap L-pyroglutamic acid (**21**) in a short and easy synthesis sequence involving a one-pot variant developed by Rovis et al. to form the triazolium core (Scheme 9.4) [36]. The synthesis of **20** started from α-tetralone (**22**), which was stereoselectively converted to the polycyclic γ-lactam **23** with inexpensive (R)-phenylglycinol as chiral source. After cleavage of the auxiliary, the one-pot-procedure that had already been used before yielded the triazolium salt **20** (Scheme 9.5).

The precatalyst **20** led to excellent results in the enantioselective intramolecular crossed benzoin condensation of the aldehyde ketones **24**, as shown in Scheme 9.6. The quaternary stereocenter of the acyloins **25** was created with good to very good yields and excellent *ee*-values. (For experimental details see Chapter 14.20.1). The precatalyst **19** proved to be even more active, and the yields were consistently excellent, albeit accompanied by lower *ee*-values (63–84%).

The substrates of the enantioselective intramolecular crossed benzoin condensation were varied to widen the scope of the reaction. Promising results were

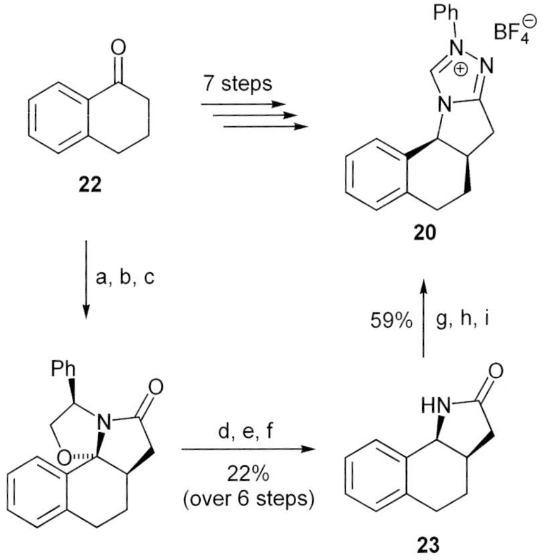

Scheme 9.5 Preparation of the tetracyclic triazolium salt **20**. (a) LDA, BrCH$_2$CO$_2$Et, THF, DMPU, −78 °C to rt, 18 h. (b) LiOH·H$_2$O, THF/H$_2$O (2/1), rt, 20 h. (c) (R)-Phenylglycinol, toluene, 4 Å molecular sieves, reflux, 20 h. (d) Et$_3$SiH, TiCl$_4$, DCM, −78 °C to rt, 3 h. (e) LiOH·H$_2$O, DMSO, 140 °C, 72 h. (f) 1 M HCl, THF, reflux, 8 h. (g) Me$_3$OBF$_4$, DCM, rt, 3 h. (h) Phenylhydrazine, DCM, rt, 3 h. (i) HC(OMe)$_3$, MeOH, 80 °C, 12 h.

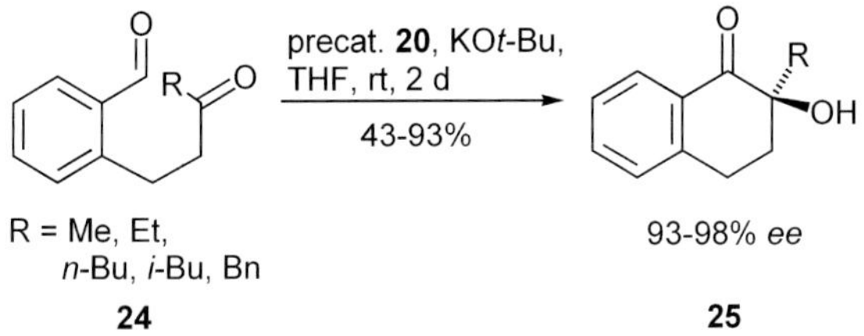

Scheme 9.6 Asymmetric intramolecular crossed benzoin reaction, according to Enders et al.

achieved with substrates where the aldehyde and ketone functions were interchanged, and with substrates where a five-membered ring was formed.

9.2
The Stetter Reaction

During the early 1970s, Hermann Stetter investigated the reaction of the acylanion equivalent **26** with Michael acceptors such as **27** (Scheme 9.7) [37]. Since that

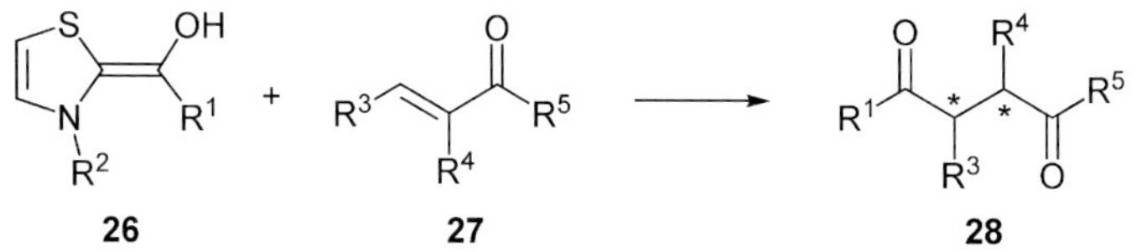

Scheme 9.7 Addition of the acyl anion equivalent **26** to a conjugate acceptor **27**.

time, the 1,4-addition of aldehydes to α,β-unsaturated carbonyl compounds carries Stetter's name [38]. In this reaction, chiral 1,4-dicarbonyl compounds such as **28** are obtained, which represent important intermediates in the synthesis of biologically active molecules.

Inspired by Sheehan and Hunneman's seminal studies on the asymmetric benzoin condensation [10], many research groups have designed novel chiral heterazolium salt precatalysts for benzoin and Stetter reactions. In order to compare the catalyst efficiency, the cyclization of the salicylaldehyde-derived substrate **29** to the corresponding chromanone **30** has become a benchmark reaction (Scheme 9.8). Following the pioneering studies of the Enders group, who reported the first asymmetric Stetter reaction in 1990 [39], very little attention has been paid to this important transformation. During the past few years, however, Rovis et al. have achieved significant progress [40]. By using 20 mol% of the aminoindanol-derived triazolium salt **31**, the asymmetric intramolecular Stetter reaction of **29** afforded the desired product **30** in 94% yield and 94% *ee*. With regard to the electronic nature of the phenyl ring on the triazole nitrogen, *p*-methoxy substitution gave the best results.

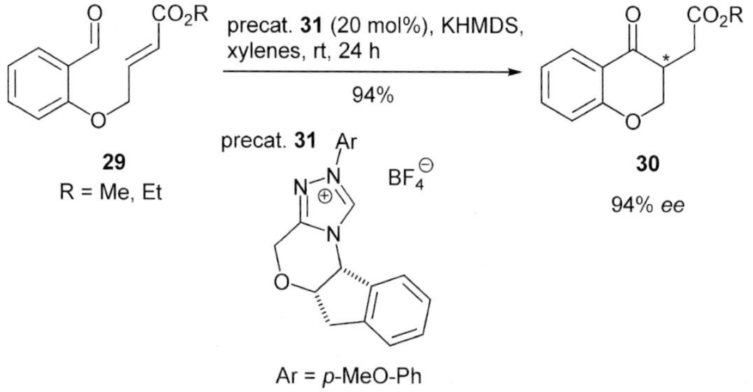

Scheme 9.8 Intramolecular Stetter reaction, as described by Rovis et al.

While this catalyst provided good yields and *ee*-values with substrates bearing a heteroatom in the γ-position of the acceptor, in some cases – for example with the aliphatic substrate **32** – the catalyst **33** was superior in the formation of the cyclopentanones **34** (Scheme 9.9).

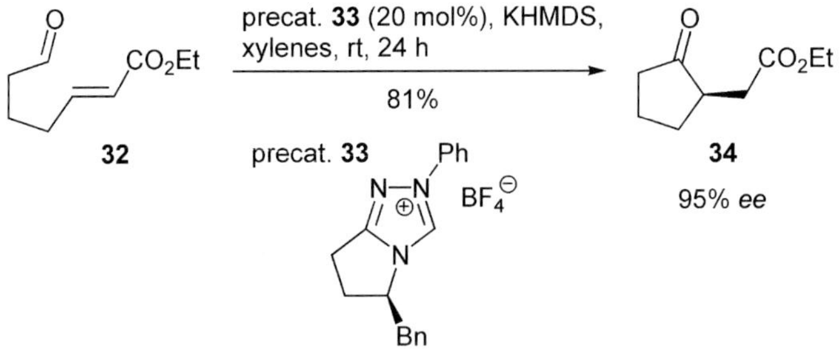

Scheme 9.9 Intramolecular Stetter reaction of aliphatic substrates.

Subsequently, Rovis et al. – and later also Hamada et al. – accepted the challenge of generating quaternary stereocenters [41]. As in catalyst **31**, the substitution pattern of the phenyl ring was found to be the decisive factor. In this case, as shown in Scheme 9.10, the electron-deficient pentafluorophenyl catalyst **35** proved to be the most effective for β,β-disubstituted substrates **36** (up to 99% ee of the products **37**), suggesting subtle mechanistic differences.

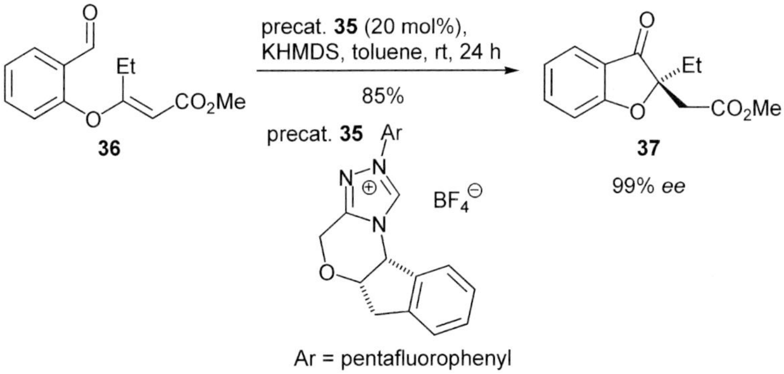

Scheme 9.10 The generation of quaternary stereocenters.

It is possible to achieve much lower catalyst loadings at the price of slightly diminished selectivity and yield. For example, the reaction of **36** in the presence of 1 mol% of the triazolium salt **35** in isopropanol affords the desired product **37** in 78% yield with 83% ee. This holds promise for catalysts with improved efficiency.

Very recently, Rovis examined the intramolecular Stetter reaction with α,β-disubstituted acceptors **38** (Scheme 9.11) [42]. The key challenge is to secure a diastereoselective proton transfer onto the enolate intermediate. HMDS (from the KHMDS base) was shown to cause a deterioration in diastereoselectivity, but this problem was overcome by using the free carbene catalyst – that is, HMDS was removed in high vacuum prior to the reaction. Using the free carbene catalyst **39**, the desired chromanone **40** could be obtained in excellent yield (94%),

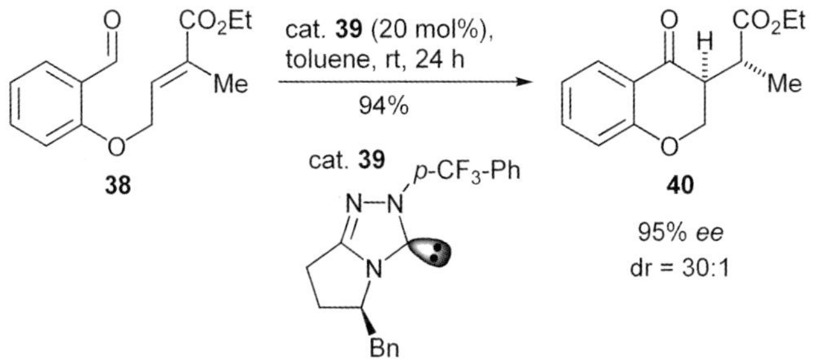

Scheme 9.11 The generation of adjacent stereocenters.

with 95% ee and a diastereomeric ratio (d.r.) of 30:1. The high selectivity was proposed to originate from an intramolecular proton transfer, which is supported by the fact, that double-bond isomers afford the complementary diastereoselectivity.

In related investigations, the group of Rovis examined the effect of pre-existing stereocenters in the intramolecular asymmetric Stetter reaction [43]. The same group also exploited the concept of desymmetrization for the asymmetric synthesis of hydrobenzofuranones in an intramolecular Stetter reaction [44].

Bach and co-workers developed the axially chiral N-arylthiazolium catalyst **41** bearing a menthol-derived backbone (Fig. 9.6) [45]. Using 20 mol% of this catalyst, they were able to isolate the Stetter product **30** (R = Me) in 75% yield with 50% ee. The low stereoselectivity was ascribed to an atropo-isomerization of the catalyst during the course of the reaction.

By applying the concept of small peptide catalysts to the intramolecular asymmetric Stetter reaction, the Miller group [46] replaced a histidine residue of amino acid derivatives with thiazolylalanine (Taz) derivatives. An initial screening with **29** (R = t-Bu) as the substrate and 20 mol% of the catalyst **42** afforded the corresponding chromanone **30** in 40% yield and 80% ee. It was speculated that both yield and selectivity might be improved by imbedding Taz into a peptidic framework. Using catalyst **43**, the product **30** could be obtained in promising 67% yield and 73% ee.

In the intermolecular Stetter reaction, self-condensation of the donor aldehyde usually dominates over the conjugate addition, leading to benzoins as the major

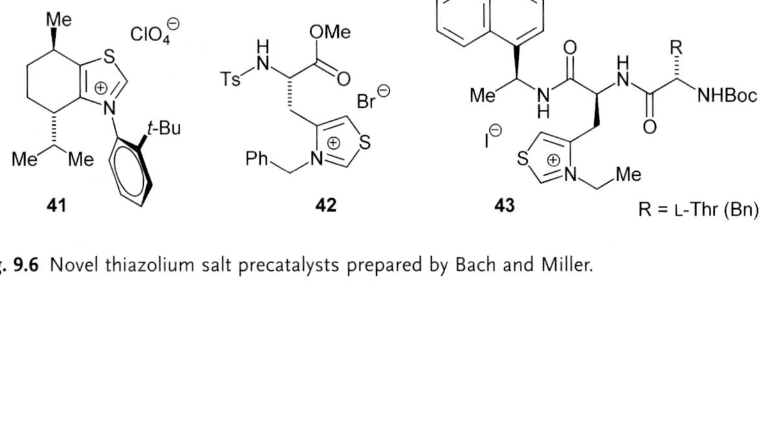

Fig. 9.6 Novel thiazolium salt precatalysts prepared by Bach and Miller.

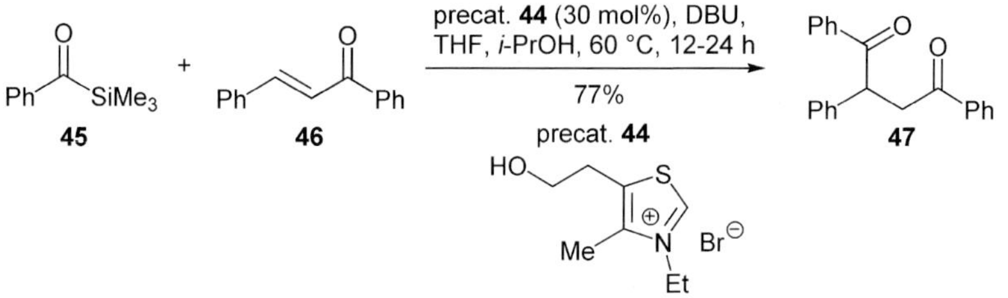

Scheme 9.12 The Sila–Stetter reaction, as reported by Scheidt et al.

products. In order to circumvent this problem, Scheidt et al. [47] devised a strategy that employs acylsilanes as acyl anion precursors (Scheme 9.12) [48]. In fact, the thiazolium-catalyzed (**44**) reaction of the acyl silane **45** and chalcone **46** as the conjugate acceptor proceeded smoothly to give the 1,4-dicarbonyl product **47** in 77% yield.

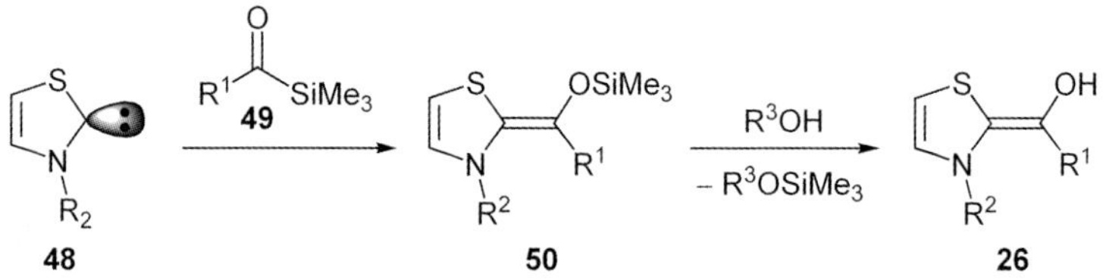

Scheme 9.13 Mechanistic proposal for the Sila–Stetter reaction.

In the rationale of Scheidt et al., a carbene catalyst (**48**) adds to the acylsilane **49** to produce – after 1,2-silyl migration – the intermediate **50** (Scheme 9.13). The alcohol is believed to induce a desilylation, leading to the Breslow-intermediate **26**. Due to the reduced electrophilicity of the acylsilane (in comparison with an aldehyde), the conjugate addition predominates.

Scheidt and co-workers were also able to carry out Stetter reactions under neutral aqueous conditions [49]. In a biomimetic fashion, a thiazolium catalyst adds to the keto group of a pyruvate ($\rightarrow$ **51**), which leads to decarboxylation ($\rightarrow$ **52**) (Scheme 9.14). This intermediate can be trapped by substituted α,β-unsaturated 2-acyl imidazoles **53**, leading to the usual Stetter products **54**.

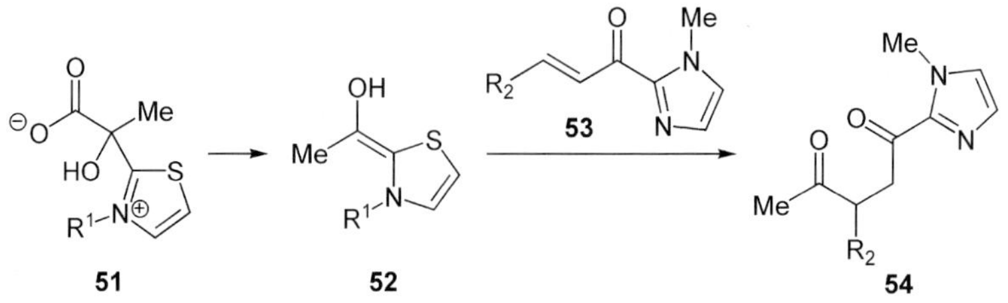

Scheme 9.14 Stetter reactions in water, according to Scheidt.

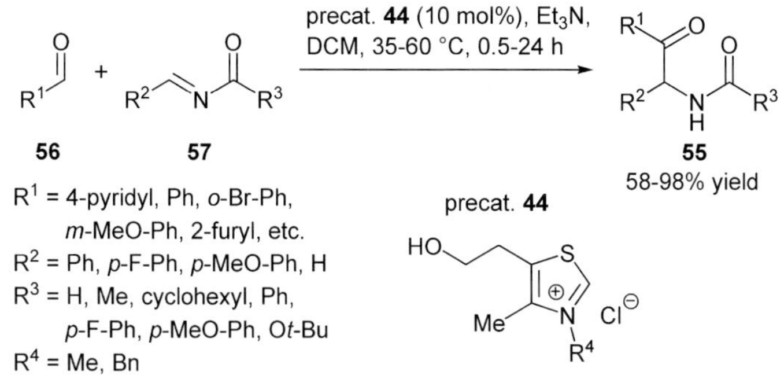

Scheme 9.15 Imino-Stetter reaction, as reported by Murry, Frantz and colleagues.

In order to address the synthesis of biologically relevant α-amido ketones **55**, Murry, Frantz and colleagues have devised a thiazolium-catalyzed (**44**) conjugate addition of aldehydes **56** to acyclimines **57** (Scheme 9.15) [50] In their synthetic design, arylsulfonylamides serve as a precursor for the acylimines, which are also referred to as imino esters.

A successful asymmetric variant of the imino-Stetter reaction has recently been presented by Miller et al., who employed chiral peptidic thiazolium salts similar to compound **43** (see Fig. 9.6) [51].

The 1,4-dicarbonyl compounds resulting from Stetter reactions have been used by Müller and colleagues [52] and by Bharadwaj and Scheidt [53] in efficient one-pot Stetter–Paal–Knorr protocols for the synthesis of highly substituted pyrroles. In an analogous fashion, Frantz et al. converted their α-ketoamides into the corresponding imidazoles by treatment with a primary amine [54].

Despite the unrivaled easy access to 1,4-dicarbonyl compounds, only a few examples of the application of Stetter reactions in the synthesis of natural products have been reported to date [55]. Tius et al. have employed a diastereoselective intermolecular Stetter reaction and a ring-closing metathesis reaction as the key steps in their elegant synthesis of roseophilin (**58**; Scheme 9.16) [56]. The 1,4-

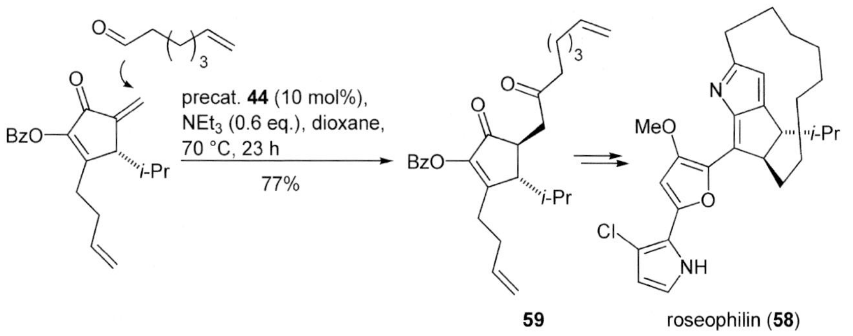

Scheme 9.16 The synthesis of roseophilin according to Tius et al.

dicarbonyl functionality in **59** served as a precursor for the central pyrrole unit of the natural product.

In order to carry out Stetter (and other) reactions in a highly parallel fashion and with minimal purification, it is necessary to immobilize the catalysts. This strategy should also allow for catalyst recycling, and was first addressed by Enders et al. [57]. Starting from the thiazole **60**, Barrett and co-workers [58] have synthesized a ROMP gel-supported thiazolium iodide **61** which was successfully applied to Stetter reactions and afforded the products in high yield and purity (Scheme 9.17) [59].

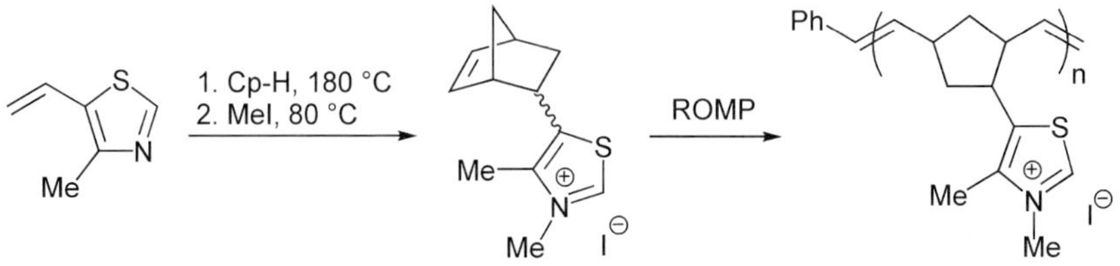

Scheme 9.17 Synthesis of ROMP gel-supported thiazolium iodide **61**.

This chapter exemplifies how the development of highly active catalysts and creative reaction design have revitalized the interest in Stetter reactions as a valuable tool for efficient C–C-bond-forming reactions. The next challenges to be met in this field are the reduction of the catalyst loading and the catalytic asymmetric intermolecular Stetter reactions.

9.3
Further Applications

9.3.1
a^3 to d^3 Umpolung

The reactions discussed above have focused on an a^1 to d^1 umpolung, as exemplified by the reaction of the resulting d^1 nucleophile with aldehydes (benzoin condensation [9]) or with Michael acceptors (Stetter reaction [38]) [3]. In contrast, α,β-unsaturated aldehydes can react as d^3 nucleophiles by utilizing a conjugated umpolung that involves a homoenolate reactivity.

One application of this catalytic generation of homoenolate type intermediates is in the stereoselective formation of γ-butyrolactones **64** from α,β-unsaturated aldehydes **62** and their reaction with aldehydes or ketones **63** [60]. (For experimental details see Chapter 14.19.2). Glorius [60a] and Bode [60b] almost simultaneously published their results utilizing a N-heterocyclic carbene generated from a bisarylimidazolium salt **65** (IMes). The corresponding disubstituted γ-butyrolactones

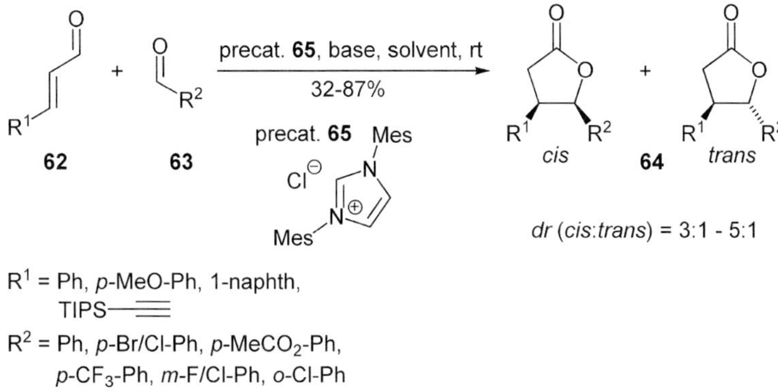

Scheme 9.18 Generation of γ-butyrolactones by Glorius and colleagues [60a] and by Bode and colleagues [60b]. Reaction conditions: (a) for Glorius et al.: 5 mol% cat. **65**, 10 mol% KOt-Bu, THF, 16 h; (b) for Bode et al.: 8 mol% cat. **65**, 7 mol% DBU, THF/t-BuOH (10:1), 3–15 h.

64 are generated in moderate to good yields, and with preference for the *cis*-diastereomer (Scheme 9.18). The acceptor is limited to either aromatic or α,β-unsaturated aldehydes.

The postulated catalytic cycle is shown in Scheme 9.19. The *N*-heterocyclic carbene catalyst **11** (1,3-dimesityl-imidazolin-2-ylidene, IMes) attacks the α,β-unsaturated aldehyde **62**, generating the zwitterionic substance **66**. After

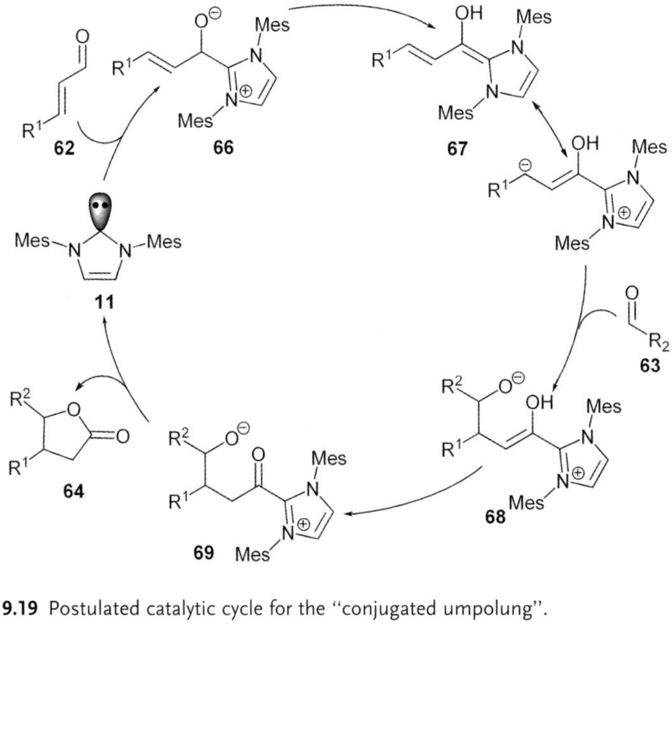

Scheme 9.19 Postulated catalytic cycle for the "conjugated umpolung".

protonation/deprotonation, the conjugated dienamine **67** is formed which attacks the aromatic aldehyde **63** to yield the alcoholate **68**. Tautomerization to the intermediate **69** enables the intramolecular nucleophilic attack of the alcoholate to the carbonyl group, thereby generating the lactone **64** and the catalyst for further use.

The formation of a quaternary stereocenter could be achieved with α,α,α-trifluoroacetone as electrophile. An enantioselective procedure for this novel carbon–carbon bond-forming process has not been reported to date, the first results with a chiral imidazolium salt as precatalyst developed by Glorius et al. resulted in only low enantiomeric excesses (12–25% ee).

Employing a similar strategy, γ-lactams could be synthesized by addition of the homoenolate equivalent to an appropriate imine (Scheme 9.20) [61]. A variety of functionalized α,β-unsaturated aldehydes **62** and N-4-methoxybenzenesulfonyl imines **70** produced disubstituted γ-lactams **71** in good yields and with a preference for the *cis* diastereomer. One crucial point is the reversibility of the addition of the catalyst to the imine to enable a reaction with the aldehyde. N-Aryl, N-alkyl, N-tosyl and N-phosphinoyl imines where either unreactive or inhibited any catalytic reaction due to the formation of a stable adduct with the catalyst.

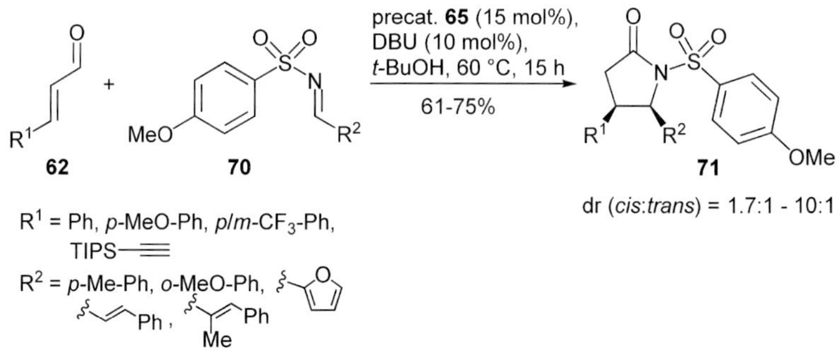

Scheme 9.20 Catalytic synthesis of γ-lactams, according to Bode et al.

Scheidt et al. [62] reported the formation of saturated esters utilizing benzimidazolium salts when protonating the intermediate **67** (see Scheme 9.19) and trapping the resulting activated carbonyl unit with an alcohol nucleophile (Scheme 9.21). These authors were able to show that the electrophile (phenol) and the nucleophile (primary or secondary alcohols) can be decoupled, enabling a broad substrate scope, though they had to employ an excess of phenol (2 eq.) and the nucleophile (5 eq.), accompanied by rather harsh conditions (100 °C).

A one-step construction of γ,γ-difunctionalized γ-butyrolactones from benzoins or benzaldehydes via a tandem reaction promoted by 1,3-dimethyl-imidazol-2-ylidene in the presence of methyl acrylate was reported by Zhai et al. [63]. So far, the exact mechanism of the reaction, as well as the role of the catalyst, has not been clarified.

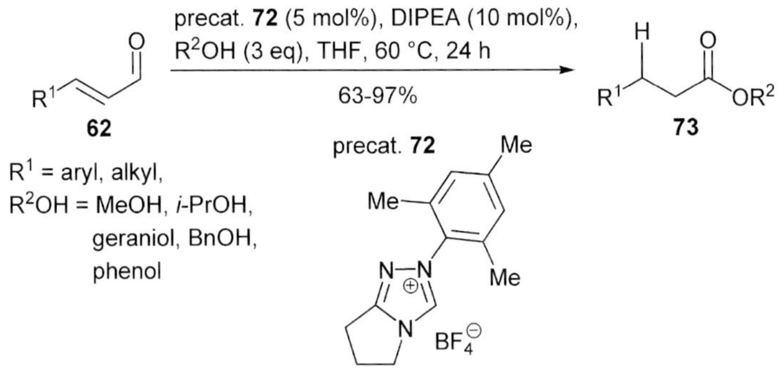

Scheme 9.21 Catalytic esterification of α,β-unsaturated aldehydes, according to Bode et al.

Recently, Bode et al. were able to demonstrate that the products formed after generation of the homoenolate equivalents **67** are determined by the catalytic base [64]. Strong bases such as KO*t*-Bu led to carbon–carbon bond-formation (γ-butyrolactones), while weaker bases such as diisopropylethylamine (DIPEA) allowed for protonation of the homoenolate and the subsequent generation of activated carboxylates. The combination of triazolium catalyst **72** and DIPEA in THF as solvent required no additional additives and enabled milder reaction conditions (60 °C), accompanied by still high conversions in the formation of saturated esters out of unsaturated aldehydes (Scheme 9.21). Aliphatic and aromatic enals **62**, as well as primary alcohols, secondary alcohols and phenols, are suitable substrates. α-Substituted unsaturated aldehydes did not yield the desired products **73**.

An alternative to generate esters **73** was reported by Rovis et al., who converted α-haloaldehydes into the corresponding dehalogenated esters (Scheme 9.22) [65]. In the proposed mechanism, the Breslow intermediate **74** (X^1 = Br, X^2 = H) contains a β-leaving group which allows the formation of the enol **75** that gives rise to the ester **73**.

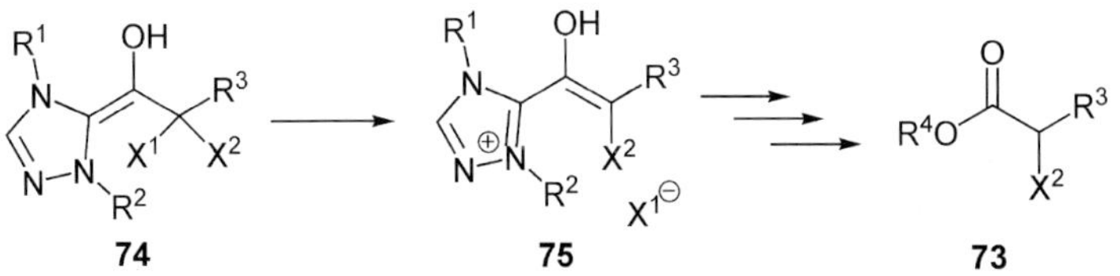

Scheme 9.22 Key steps in the generation of esters from α-haloaldehydes, according to Rovis et al.

An enantioselective version was recently reported by the same group (Scheme 9.23) [66]. 2,2-Dichloroaldehydes **76** react with phenols **77** in the presence of the chiral triazolin-5-ylidene carbene **35** to form α-chloroesters **78** in good yield and enantioselectivity. The key step in the proposed catalytic cycle is an enantioselec-

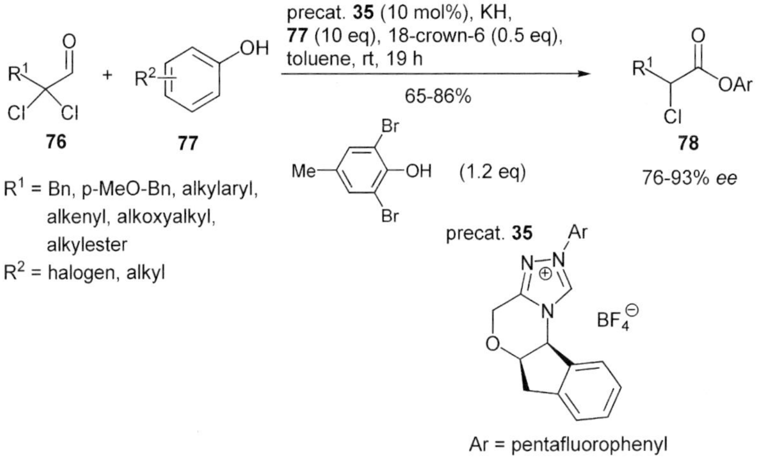

Scheme 9.23 Asymmetric synthesis of α-chloro esters, according to Rovis et al.

tive protonation of the chiral enolate **75** ($X^1, X^2 = Cl$). The addition of 18-crown-6 was necessary to provide a homogeneous reaction medium, as well as 2,6-dibromo-4-methylphenol in order to suppress epimerization. The reaction is limited to aldehydes lacking β-branching.

The broad scope of the catalytic generation of activated carboxylates was demonstrated by Bode et al. in the diastereoselective synthesis of β-hydroxy esters **79** from α,β-epoxy aldehydes **80** employing achiral thiazolium salts **81** as precatalysts (Scheme 9.24) [67]. The incorporation of a reducible functionality into the aldehyde substrate is the premise for a catalyst-induced intramolecular redox reaction generating the activated carboxylate **82**.

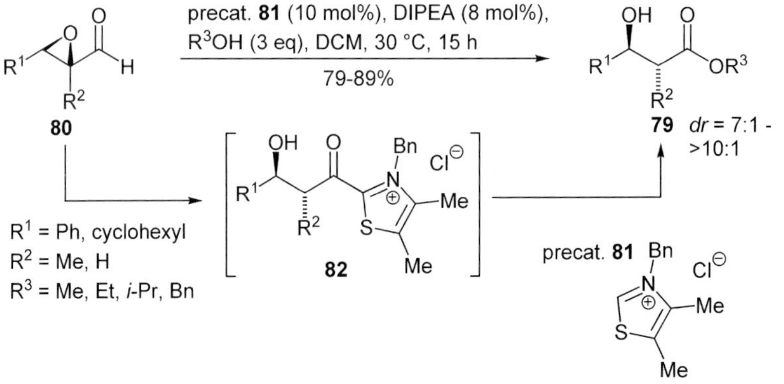

Scheme 9.24 Synthesis of β-hydroxyesters from epoxyaldehydes, according to Bode et al.

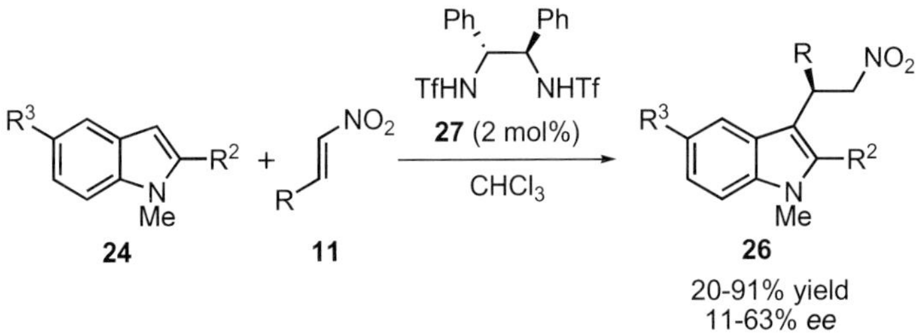

Scheme 6.3 Bis-sulfonamide-catalyzed Friedel–Craft reactions of indoles.

Connon and co-workers reported the use of cinchona alkaloid-derived thiourea catalysts **28** and **29** (compare to Scheme 6.2) for the 1,4-additions of dimethyl malonate to various nitroolefins [53]. Using 2 mol% of catalysts **28** or **29** in the presence of dimethyl malonate allowed for the conversion of both electronically activated and deactivated β-nitrostyrenes to give chiral 1,4-adducts in uniformly high yields (91–95%) and high enantiopurities (87–99% *ee*), though more electron-rich nitrostyrenes generally gave higher product enantiopurities (Table 6.10). Aliphatic nitroolefins gave slightly lower observed *ee*-values (75–86%). An interesting characteristic of these cinchona-derived thiourea catalysts is the observation of an apparent synergistic co-catalytic effect of the thiourea moiety and the quinuclidine nitrogen, moderated by the stereochemistry of the catalyst at C9. While thiourea catalysts derived from natural cinchona alkaloids gave measurable rate increases, reactions catalyzed by 9-*epi*-thiourea catalysts derived from dihydroquinine and dihydroquinidine exhibited significantly higher enantioselectivities and rates of reaction.

Dixon and co-workers independently reported the asymmetric hydrogen-bonding catalyzed 1,4-additions of dimethyl malonate to nitroolefins using a cinchonine-derived thiourea catalyst **30** [54]. Catalyst **30** gave good to high yields (81–99%) and good to high enantioselectivities (82–97% *ee*) for a variety of aromatic, heteroaromatic and aliphatic nitroolefins (Scheme 6.4). Optical purity

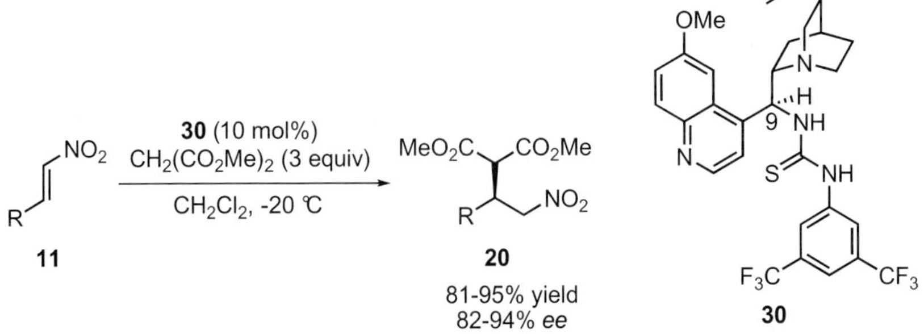

Scheme 6.4 Thiourea-catalyzed 1,4-additions of dimethyl malonate to nitroolefins.

Alternatively, a chiral imidazolium salt generates a chiral-activated ester which allows for the kinetic resolution of chiral secondary alcohols. Chan and Scheidt described this reaction of racemic 1-phenylethanol and cinnamaldehyde [62]. The enantiodiscrimination was explained by the chiral intermediate formed between the chiral imidazolium salt and cinnamaldehyde, which has sufficient facial selectivity to react preferentially with the (R)-stereoisomer of 1-phenylethanol.

Fu et al. recently presented their results on carbene-catalyzed intramolecular Heck-type cyclizations [68]. The N-heterocyclic carbene generated *in situ* from the triazolium salt **83** catalyzed cyclization of the substrates **84** to form the cycloalkanes **85** (Scheme 9.25). The lowest yields were obtained when the reaction products were cyclobutanes. These authors assumed that the carbene initially reacted with the Michael acceptor site of the substrate to the intermediate **86**, which subsequently cyclized via **86'** to the final products **85**.

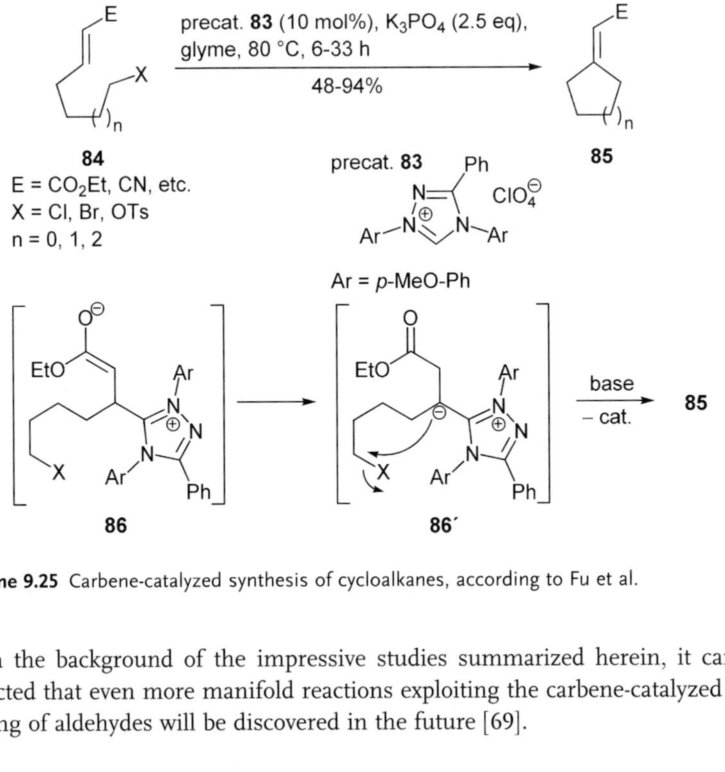

Scheme 9.25 Carbene-catalyzed synthesis of cycloalkanes, according to Fu et al.

On the background of the impressive studies summarized herein, it can be expected that even more manifold reactions exploiting the carbene-catalyzed umpolung of aldehydes will be discovered in the future [69].

9.3.2
Transesterifications and Polymerizations

In 1994, Smith et al. reported the first transfer of alkoxycarbonyl groups from 2-alkoxycarbonylimidazolium salts to benzyl alcohols to yield the corresponding benzyl alkyl carbonates [70]. Unfortunately, stoichiometric amounts of the substituted imidazolium salts had to be used, and the imidazole-alkoxycarbonyl adduct had to be prepared separately in advance. Nolan et al. later reported the use of imidazol-2-ylidenes as efficient catalysts in the transesterification of esters with primary alcohols [71]. Only low catalyst loadings (0.5–5 mol%) of aryl-substituted (IMes) or alkyl-substituted imidazolin-2-ylidenes **11** were necessary to synthesize various esters in high yields, under mild reaction conditions, and in short reaction times. Using methylesters as substrates, the addition of 4 Å molecular sieves led to quantitative conversions, and a selective protection of primary alcohols in the presence of secondary alcohols was possible. Benzyl alcohol was almost exclusively acylated by vinyl acetate in the presence of 2-butanol.

The transesterification with secondary alcohols was also reported by the same group [72]. By utilizing 1,3-dicyclohexyl-imidazolin-2-ylidene (ICy, **11**) as catalyst, various alcohols **87** (including aliphatic cyclic and aromatic alcohols) reacted with diverse esters **88** to yield the desired esters **89** (Scheme 9.26). The steric bulk at the α position of the alcohol reduces the reaction rates, and longer reaction times are required; hence, the reaction proceeds slowly (5 days) with tertiary alcohols (1-adamantanol) and requires higher catalyst loadings (20 mol%) to yield moderate yields (54%).

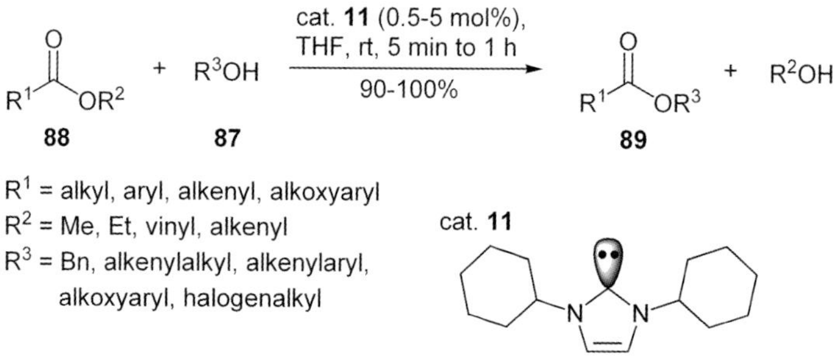

Scheme 9.26 Nolan's transesterifications.

An interesting version of the transesterification reactions was reported by Movassaghi et al., with the amidation of unactivated esters with amino alcohols (Scheme 9.27) [73]. The amidation was explained by carbene–alcohol interactions. A nucleophilic activation of the hydroxyl group of the aminoalcohol **90** by the catalyst **11** is followed by transesterification to the ester **91** which is *in-situ*-converted to the amide **92** through a N → O acyl transfer. Various aliphatic and aromatic esters with different functionalities, as well as chiral aminoalcohols, are suitable for this reaction.

Suzuki et al. [74] and Maruoka and colleagues [75] made further progress in the enantioselective acylation of secondary alcohols **93**. Suzuki et al. reported moderate enantiomeric excesses (up to 51% *ee*) when employing C_2 symmetric chiral imidazolium salts **94** as precatalysts and vinyl acetate as acylation agent. How-

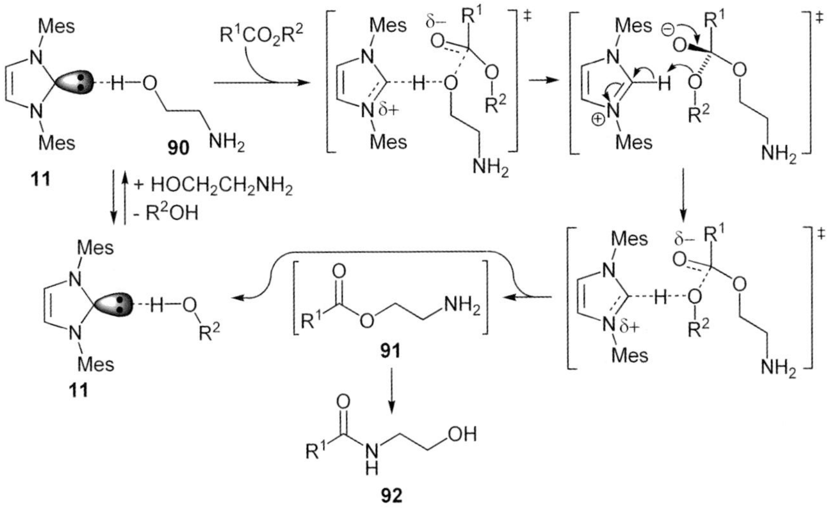

Scheme 9.27 Proposed mechanism of the amidation of unactivated esters, according to Movassaghi et al.

ever, the induction could not be significantly enhanced by employing bulkier N-substituents at the catalyst. Maruoka and colleagues subsequently showed that more sterically hindered acylation agents improve the enantioselectivities. Indeed, selectivities of up to 96% ee accompanied by moderate to good yields could be achieved for the de-racemized esters **95** by utilizing vinyl diphenylacetate **96** as the acylation agent (Scheme 9.28).

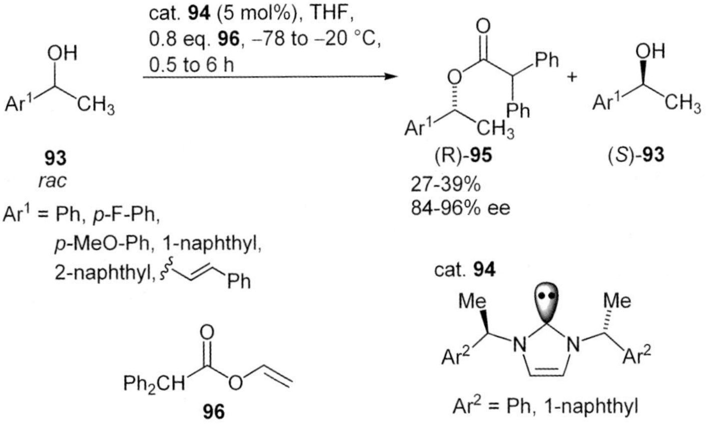

Scheme 9.28 Enantioselective acylation of secondary alcohols, as reported by Maruoka and colleagues.

The concept of transesterifications was used for polymerization reactions by Hedrick and colleagues [76]. Various biodegradable polyesters were synthesized with the 1,3-dimethylimidazol-2-ylidene carbene in THF at 25 °C. Polymers such as poly(ε-caprolactone) were obtained with no need of organometallic catalysts, as in classical methods. Poly(ethylene terephthalate) (PET) **97** was synthesized in the ionic liquid **98**, which functions as the reaction medium and, at the same time, as a precatalyst that is activated (**99**) with KO*t*-Bu. Dimethyl terephthalate (DMT) **100** was condensed with an excess of ethylene glycol **101** to generate **102**. The melt condensation of **102** was performed under vacuum using a heating ramp to 280 °C.

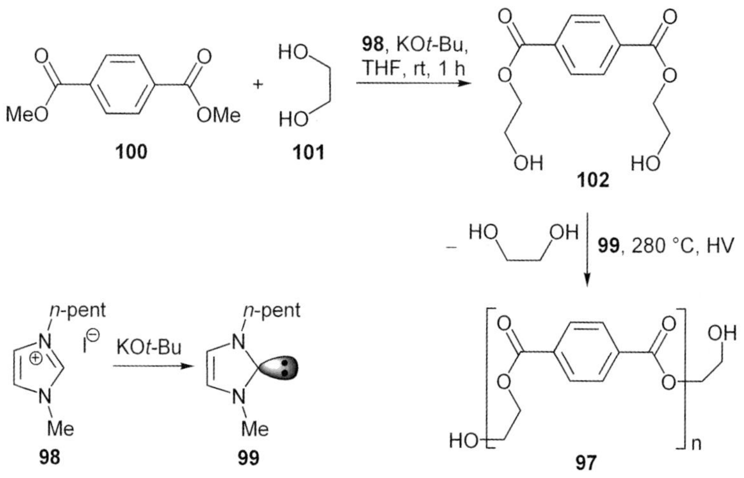

Scheme 9.29 Polymerizing esterifications, according to Hedrick et al.

A further example of this strategy was presented by the same group for the on-demand living polymerization of lactide [77]. Alkoxytriazolines were shown to reversibly dissociate at 90 °C to generate an initiating/propagating alcohol. The ring-opening polymerization can be activated/deactivated with temperature on demand, making it ideally suited for the preparation of block copolymers and complex macromolecular architectures.

The research areas in organic chemistry where nucleophilic carbenes are applied as catalysts are continuously growing, and these now include nucleophilic acylation reactions (benzoin reactions, Stetter reactions, a^3/d^3 umpolung, generation of activated carboxylates) as well as manifold transesterifications. Nonetheless, many more interesting contributions are to be expected in the future. Publication titles such as "Extending mechanistic routes in heterazolium catalysis – promising concepts for versatile synthetic methods" announce fruitful further developments in organocatalysis with nucleophilic carbenes [69].

References

1. F. Wöhler, J. Liebig, *Ann. Pharm.* **1832**, *3*, 249.
2. A. Lapworth, *J. Chem. Soc.* **1903**, *83*, 995.
3. D. Seebach, *Angew. Chem.* **1979**, *91*, 259–278; *Angew. Chem. Int. Ed. Engl.* **1979**, *18*, 239.
4. T. Ukai, R. Tanaka, T. Dokawa, *J. Pharm. Soc. Jpn.* **1943**, *63*, 296 (*Chem. Abstr.* **1951**, *45*, 5148).
5. S. Mizuhara, P. Handler, *J. Am. Chem. Soc.* **1954**, *76*, 571.
6. (a) F. Jordan, *Nat. Prod. Rep.* **2003**, *20*, 184; (b) U. Schoerken, G. A. Sprenger, *Biochem. Biophys. Acta* **1998**, *1385*, 229; (c) G. A. Sprenger, M. Pohl, *J. Mol. Catalysis B: Enzymatic* **1999**, *6*, 145.
7. R. Breslow, *J. Am. Chem. Soc.* **1958**, *80*, 3719.
8. (a) D.M. Lemal, R.A. Lovald, K.I. Kawano, *J. Am. Chem. Soc.* **1964**, *86*, 2518; (b) J. Castells, F. López Calahorra, F. Geijo, R. Pérez-Dolz, M. Bassedas, *J. Heterocyclic Chem.* **1986**, *23*, 715; (c) J. Castells, F. López Calahorra, L. Domingo, *J. Org. Chem.* **1988**, *53*, 4433; (d) J. Castells, L. Domingo, F. López Calahorra, J. Martí, *Tetrahedron Lett.* **1993**, *34*, 517; (e) R. Breslow, R. Kim, *Tetrahedron Lett.* **1994**, *35*, 699; (f) Y.-T. Chen, G.L. Barletta, K. Haghjoo, J.T. Cheng, F. Jordan, *J. Org. Chem.* **1994**, *59*, 7714; (g) F. López Calahorra, R. Rubires, *Tetrahedron* **1995**, *51*, 9713; (h) F. López Calahorra, E. Castro, A. Ochoa, J. Martí, *Tetrahedron Lett.* **1996**, *37*, 5019; (i) F. López Calahorra, J. Castells, L. Domingo, J. Martí, J.M. Bofill, *Heterocycles* **1994**, *37*, 1579; (j) J. Martí, F. López Calahorra, J.M. Bofill, *J. Mol. Struct. (Theochem)* **1995**, *339*, 179; (k) R. Breslow, C. Schmuck, *Tetrahedron Lett.* **1996**, *37*, 8241.
9. Review: D. Enders, T. Balensiefer, *Acc. Chem. Res.* **2004**, *37*, 534.
10. J. Sheehan, D.H. Hunneman, *J. Am. Chem. Soc.* **1966**, *88*, 3666.
11. J. Sheehan, T. Hara, *J. Org. Chem.* **1974**, *39*, 1196.
12. C.A. Dvorak, V.H. Rawal, *Tetrahedron Lett.* **1998**, *39*, 2925.
13. W. Tagaki, Y. Tamura, Y. Yano, *Bull. Chem. Soc. Jpn.* **1980**, *53*, 478.
14. C. Zhao, S. Chen, P. Wu, Z. Wen, *Huaxue Xuebao* **1988**, *46*, 784.
15. J. Martí, J. Castells, F. López Calahorra, *Tetrahedron Lett.* **1993**, *34*, 521.
16. (a) H.-W. Wanzlick, E. Schikora, *Angew. Chem.* **1960**, *72*, 494; (b) H.-W. Wanzlick, H.-J. Kleiner, *Angew. Chem.* **1963**, *75*, 1204.
17. (a) A. Igau, H. Gruetzmacher, A. Baceiredo, G. Bertrand, *J. Am. Chem. Soc.* **1988**, *110*, 6463; (b) A. Igau, A. Baceiredo, G. Trinquier, G. Bertrand, *Angew. Chem.* **1989**, *101*, 617; *Angew. Chem. Int. Ed. Engl.* **1989**, *28*, 621.
18. (a) A.J. Arduengo, III, R.L. Harlow, M. Kline, *J. Am. Chem. Soc.* **1991**, *113*, 361; (b) A.J. Arduengo, III, H.V.R. Dias, R.L. Harlow, M. Kline, *J. Am. Chem. Soc.* **1992**, *114*, 5530.
19. (a) D.A. Dixon, K.D. Dobbs, A.J. Arduengo, III, G. Bertrand, *J. Am. Chem. Soc.* **1991**, *113*, 8782; (b) M. Soleihavoup, A. Baceiredo, O. Treutler, R. Ahlrichs, M. Nieger, G. Bertrand, *J. Am. Chem. Soc.* **1992**, *114*, 10959; (c) D. Bourissou, G. Bertrand, *Adv. Organomet. Chem.* **1999**, *44*, 175.
20. E. Despagnet, H. Gornitzka, A.B. Rozhenko, W.W. Schoeller, D. Bourissou, G. Bertrand, *Angew. Chem.* **2002**, *114*, 2959; *Angew. Chem. Int. Ed. Engl.* **2002**, *41*, 2835.
21. Reviews: (a) W.A. Herrmann, C. Koecher, *Angew. Chem.* **1997**, *109*, 2256; *Angew. Chem. Int. Ed. Engl.* **1997**, *36*, 2162; (b) A.J. Arduengo, III, R. Krafczyk, *Chem. unserer Zeit* **1998**, *32*, 6; (c) A.J. Arduengo, III, *Acc. Chem. Res.* **1999**, *32*, 913; (d) D. Bourissou, O. Guerret, F.P. Gabbaï, G. Bertrand, *Chem. Rev.* **2000**, *100*, 39; (e) W.A. Herrmann, *Angew. Chem.* **2002**, *114*, 1342; *Angew. Chem. Int. Ed. Engl.* **2002**, *41*, 1290; (f) M.C. Perry, K. Burgess, *Tetrahedron:*

22 D. Enders, K. Breuer, G. Raabe, J. Runsink, J.H. Teles, J.-P. Melder, K. Ebel, S. Brode, *Angew. Chem.* **1995**, *107*, 1119; *Angew. Chem. Int. Ed. Engl.* **1995**, *34*, 2021.

23 (a) G. Raabe, K. Breuer, D. Enders, J.H. Teles, *Z. Naturforsch.* **1996**, *51a*, 95; (b) D. Enders, K. Breuer, J.H. Teles, J. Runsink, *Liebigs Ann. Chem.* **1996**, 2019; (c) D. Enders, K. Breuer, J.H. Teles, K. Ebel, *J. Prakt. Chem.* **1997**, *339*, 397; (d) D. Enders, K. Breuer, G. Raabe, J. Simonet, A. Ghanimi, H.B. Stegmann, J.H. Teles, *Tetrahedron Lett.* **1997**, *38*, 2833.

24 D. Enders, K. Breuer, U. Kallfass, T. Balensiefer, *Synthesis* **2003**, 1292.

25 J.H. Teles, J.-P. Melder, K. Ebel, R. Schneider, E. Gehrer, W. Harder, S. Brode, D. Enders, K. Breuer, G. Raabe, *Helv. Chim. Acta* **1996**, *79*, 61.

26 (a) D. Enders, K. Breuer, J.H. Teles, *Helv. Chim. Acta* **1996**, *79*, 1217; (b) D. Enders, K. Breuer, in: *Comprehensive Asymmetric Catalysis*. Springer Verlag, Heidelberg **1999**, Vol. 3, 1093; (c) J.H. Teles, K. Breuer, D. Enders, H. Gielen, *Synth. Commun.* **1999**, *29*, 1.

27 (a) R.L. Knight, F.J. Leeper, *Tetrahedron Lett.* **1997**, *38*, 3611; (b) A.U. Gerhards, F.J. Leeper, *Tetrahedron Lett.* **1997**, *38*, 3615.

28 R.L. Knight, F.J. Leeper, *J. Chem. Soc. Perkin Trans. 1* **1998**, 1891.

29 D. Enders, U. Kallfass, *Angew. Chem.* **2002**, *114*, 1822; *Angew. Chem. Int. Ed. Engl.* **2002**, *41*, 1743.

30 T. Dudding, K.N. Houk, *Proc. Natl. Acad. Sci. USA* **2004**, *101*, 5770.

31 H. Iding, T. Duennwald, L. Greiner, A. Liese, M. Mueller, P. Siegert, J. Groetzinger, A.S. Demir, M. Pohl, *Chem. Eur. J.* **2000**, *6*, 1483.

32 H. Stetter, *Angew. Chem.* **1976**, *88*, 695; *Angew. Chem. Int. Ed.* **1976**, *15*, 639.

33 Y. Hachisu, J.W. Bode, K. Suzuki, *J. Am. Chem. Soc.* **2003**, *125*, 8432.

34 (a) D. Enders, O. Niemeier, *Synlett* **2004**, 2111; (b) Y. Hachisu, J.W. Bode, K. Suzuki, *Adv. Synth. Catal.* **2004**, *346*, 1097.

35 D. Enders, O. Niemeier, T. Balensiefer, *Angew. Chem.* **2006**, *118*, 1491; *Angew. Chem. Int. Ed.* **2006**, *45*, 1463.

36 M.S. Kerr, J.R. de Alaniz, T. Rovis, *J. Org. Chem.* **2005**, *70*, 5725.

37 (a) H. Stetter, M. Schreckenberg, *Angew. Chem.* **1973**, *85*, 89; *Angew. Chem. Int. Ed.* **1973**, *12*, 81; (b) H. Stetter, *Angew. Chem.* **1976**, *88*, 695; *Angew. Chem. Int. Ed.* **1976**, *15*, 639.

38 For a short review, see: M. Christmann, *Angew. Chem.* **2005**, *117*, 2688; *Angew. Chem. Int. Ed.* **2005**, *44*, 2632.

39 (a) J. Tiebes, Diploma Thesis, RWTH Aachen, **1990**; (b) D. Enders, in: *Stereoselective Synthesis*. Springer-Verlag, Heidelberg **1993**, p. 63; (c) D. Enders, B. Bockstiegel, H. Dyker, U. Jegelka, H. Kipphardt, D. Kownatka, H. Kuhlmann, D. Mannes, J. Tiebes, K. Papadopoulos, in: *Dechema-Monographies*. VCH, Weinheim **1993**, Vol. 129, p. 209; (d) D. Enders, K. Breuer, J. Runsink, J.H. Teles, *Helv. Chim. Acta.* **1996**, *79*, 1899.

40 M.S. Kerr, J. Read de Alaniz, T. Rovis, *J. Am Chem. Soc.* **2002**, *124*, 10298.

41 (a) M.S. Kerr, T. Rovis, *J. Am. Chem. Soc.* **2004**, *126*, 8876; (b) T. Nakamura, O. Hara, T. Tamura, K. Makino, Y. Hamada, *Synlett* **2005**, 155.

42 J. Read de Alaniz, T. Rovis, *J. Am. Chem. Soc.* **2005**, *127*, 6284.

43 N.T. Reynolds, T. Rovis, *Tetrahedron* **2005**, *61*, 6368.

44 Q. Liu, T. Rovis, *J. Am. Chem. Soc.* **2006**, *128*, 2552.

45 J. Pesch, K. Harms, T. Bach, *Eur. J. Org. Chem.* **2004**, 2025.

46 S.M. Mennen, J.T. Blank, M.B. Tran-Dubé, J.E. Imbriglio, S.J. Miller, *Chem. Commun.* **2005**, 195.

47 A.E. Mattson, A.R. Bharadwaj, K.A. Scheidt, *J. Am. Chem. Soc.* **2004**, *126*, 2315.

48 X. Linghu, J.S. Johnson, *Angew. Chem.* **2003**, *115*, 2638; *Angew. Chem. Int. Ed.* **2003**, *43*, 2534.

49 M.C. Myers, A.R. Bharadwaj, B.C. Milgram, K.A. Scheidt, *J. Am. Chem. Soc.* **2005**, *127*, 14675.
50 J.A. Murry, D.E. Frantz, A. Soheili, R. Tillyer, E.J.J. Grabowski, P.J. Reider, *J. Am. Chem. Soc.* **2001**, *123*, 9696.
51 M.S. Mennen, J.D. Gipson, Y.R. Kim, S.J. Miller, *J. Am. Chem. Soc.* **2005**, *127*, 1654.
52 R.U. Braun, K. Zeitler, T.J.J. Müller, *Org. Lett.* **2001**, *3*, 3297.
53 A.R. Bharadwaj, K.A. Scheidt, *Org. Lett.* **2004**, *6*, 2465.
54 D.E. Frantz, L. Morency, A. Soheili, J.A. Murry, E.J.J. Grabowski, R.D. Tillyer, *Org. Lett.* **2004**, *6*, 843.
55 B.M. Trost, C.D. Shuey, F. DiNinno, S.S. McElvain, *J. Am. Chem. Soc.* **1979**, *101*, 1284.
56 P.E. Harrington, M.A. Tius, *J. Am. Chem. Soc.* **2001**, *123*, 8509.
57 D. Enders, H. Gielen, K. Breuer, *Molecules Online* **1998**, *2*, 105.
58 A.G.M. Barrett, A.C. Love, L. Tedeschi, *Org. Lett.* **2004**, *6*, 3377.
59 For an example with immobilized substrates, see: S. Raghavan, K. Anuradha, *Tetrahedron Lett.* **2002**, *43*, 5181.
60 (a) C. Burstein, F. Glorius, *Angew. Chem.* **2004**, *116*, 6331; *Angew. Chem. Int. Ed.* **2004**, *43*, 6205; (b) S.S. Sohn, E.V. Rosen, J.W. Bode, *J. Am. Chem. Soc.* **2004**, *126*, 14370.
61 M. He, J.W. Bode, *Org. Lett.* **2005**, *7*, 3131.
62 A. Chan, K.A. Scheidt, *Org. Lett.* **2005**, *7*, 905.
63 W. Ye, G. Cai, Z. Zhuang, X.H. Zhai, *Org. Lett.* **2005**, *7*, 3769.
64 S.S. Sohn, J.W. Bode, *Org. Lett.* **2005**, *7*, 3873.
65 N.T. Reynolds, J.R. de Alaniz, T. Rovis, *J. Am. Chem. Soc.* **2004**, *126*, 9518.
66 N.T. Reynolds, T. Rovis, *J. Am. Chem. Soc.* **2005**, *127*, 16406.
67 K. Yu-Kin, J.W. Bode, *J. Am. Chem. Soc.* **2004**, *126*, 8126.
68 C. Fischer, S.W. Smith, D.A. Powell, G.C. Fu, *J. Am. Chem. Soc.* **2006**, *128*, 1472.
69 K. Zeitler, *Angew. Chem.* **2005**, *117*, 7674; *Angew. Chem. Int. Ed.* **2005**, *44*, 7506.
70 C. Bakhtiar, E.H. Smith, *J. Chem. Soc. Perkin Trans. 1* **1994**, 239.
71 (a) G.A. Grasa, R.M. Kissling, S.P. Nolan, *Org. Lett.* **2002**, *4*, 3583; (b) G.A. Grasa, T. Güveli, R. Singh, S.P. Nolan, *J. Org. Chem.* **2003**, *68*, 2812.
72 R. Singh, R.M. Kissling, M.-A. Letellier, S.P. Nolan, *J. Org. Chem.* **2004**, *69*, 209.
73 M. Movassaghi, M.A. Schmidt, *Org. Lett.* **2005**, *7*, 2453.
74 Y. Suzuki, K. Yamauchi, K. Muramatsu, M. Sato, *Chem. Commun.* **2004**, 2770.
75 T. Kano, K. Sasaki, K. Maruoka, *Org. Lett.* **2005**, *7*, 1347.
76 (a) E.F. Connor, G.W. Nyce, M. Myers, A. Möck, J.L. Hedrick, *J. Am. Chem. Soc.* **2002**, *124*, 914; (b) G.W. Nyce, J.A. Lamboy, E.F. Connor, R.M. Waymouth, J.L. Hedrick, *Org. Lett.* **2002**, *4*, 3587; (c) G.W. Nyce, T. Glauser, E.F. Connor, A. Möck, R.M. Waymouth, J.L. Hedrick, *J. Am. Chem. Soc.* **2003**, *123*, 3046.
77 (a) O. Coulembier, A.P. Dove, R.C. Pratt, A.C. Sentman, D.A. Culkin, L. Mespouille, P. Dubois, R.M. Waymouth, J.L. Hedrick, *Angew. Chem.* **2005**, *117*, 5044; *Angew. Chem. Int. Ed.* **2005**, *44*, 4964; (b) S. Csihony, D.A. Culkin, A.C. Sentman, A.P. Dove, R.M. Waymouth, J.L. Hedrick, *J. Am. Chem. Soc.* **2005**, *127*, 9079.

10
Ylide-Based Reactions

Eoghan M. McGarrigle and Varinder K. Aggarwal

10.1
Introduction

In this chapter, we will review the use of ylides as enantioselective organocatalysts. Three main types of asymmetric reaction have been achieved using ylides as catalysts, namely epoxidation, aziridination, and cyclopropanation. Each of these will be dealt with in turn. The use of an ylide to achieve these transformations involves the construction of a C–C bond, a three-membered ring, and two new adjacent stereocenters with control of absolute and relative stereochemistry in one step. These are potentially very efficient transformations in the synthetic chemist's arsenal, but they are also challenging ones to control, as we shall see. Sulfur ylides dominate in these types of transformations because they show the best combination of ylide stability [1] with leaving group ability [2] of the onium ion in the intermediate betaine. In addition, the use of nitrogen, selenium and tellurium ylides as catalysts will also be described.

10.2
Epoxidation

Whilst epoxides are important functional groups in organic synthesis as targets in their own right, they are perhaps more important as intermediates that yield bifunctional compounds after stereoselective ring-opening by a nucleophile. Starting from an enantiopure epoxide, the absolute stereochemistry at two adjacent chiral centers can be controlled, and consequently epoxides not only have a long history as targets for asymmetric catalysis, but a variety of enantioselective catalytic methods for their synthesis also exists [3–12]. Most of these methods concern the oxidation of an alkene, which is often synthesized from a carbonyl compound. Conceptually, however, it is also possible to epoxidize a carbonyl compound by enantioselective alkylidenation using an ylide, a carbene, or a Darzens' reaction.

Enantioselective Organocatalysis: Reactions and Experimental Procedures
Edited by Peter I. Dalko
Copyright © 2007 WILEY-VCH Verlag GmbH & Co. KGaA, Weinheim
ISBN: 978-3-527-31522-2

In this section, we review the development of organocatalytic ylide-based epoxidation methods which allow a one-step route from carbonyl compounds, and therefore compete with the more traditional two-step approach of olefination followed by epoxidation of the resulting alkene (see Scheme 10.1) [13]. Indeed, ylide-based methodologies side-step the construction of a C=C double bond and achieve the whole transformation in one step; thus, they are potentially more atom-efficient (see Scheme 10.1). However, there are greater challenges as both the absolute and relative stereochemistries must be controlled in one step [14–22].

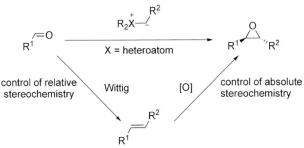

Scheme 10.1 Comparison of routes to epoxides from carbonyl compounds. Top: Asymmetric carbonyl epoxidation. Bottom: Wittig olefination followed by asymmetric alkene epoxidation.

10.2.1
Sulfur Ylide-Catalyzed Epoxidation

Stoichiometric sulfur ylide epoxidation was first reported by A.W. Johnson [23] in 1958, and subsequently the method of Corey and Chaykovsky has found widespread use [24–26]. The first enantioselective epoxidations using stoichiometric amounts of ylide were reported in 1968 [27, 28]. In another early example, Hiyama et al. used a chiral phase-transfer catalyst (20 mol%) and stoichiometric amounts of Corey's ylide to effect asymmetric epoxidation of benzaldehyde in moderate to good enantiomeric excess (*ee*) of 67 to 89% [29]. Here, we will focus on epoxidations using catalytic amounts of ylide [30–32]. A general mechanism for sulfur ylide epoxidation is shown in Scheme 10.2, whereby an attack by the ylide on a carbonyl group yields a betaine intermediate which collapses to yield

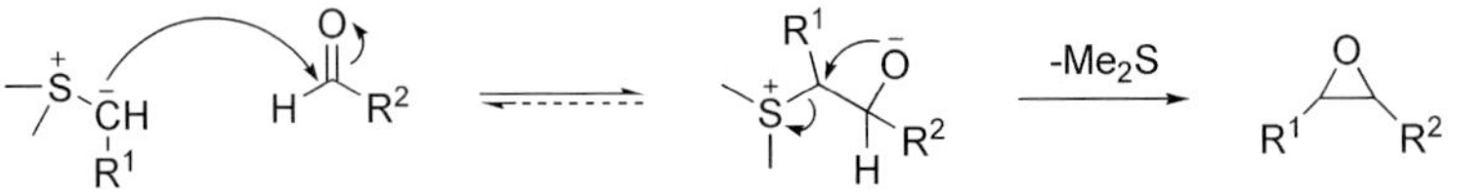

Scheme 10.2 General mechanism of sulfur ylide epoxidation.

epoxide and sulfide. In order to develop a catalytic method, the substoichiometric amount of sulfur ylide must be regenerated from the sulfide under reaction conditions. This has been achieved in two different ways: (1) by alkylation of the sulfide followed by deprotonation of the newly formed sulfonium salt; or (2) by direct reaction of the sulfide with a carbene source. Both of these approaches are discussed below.

10.2.1.1 Catalysis via Sulfide Alkylation/Deprotonation
The first enantioselective catalytic sulfur ylide epoxidation was reported by Furukawa et al. (a general catalytic cycle is illustrated in Scheme 10.3) [33]. Benzyl bromide was used to alkylate chiral sulfide **1**, and deprotonation of the resulting sulfonium salt by potassium hydroxide gave a sulfur ylide which epoxidized an aldehyde and regenerated the sulfide. Furukawa et al. were able to demonstrate some catalyst turnover at sulfide loadings of 10–50 mol%. A number of research groups have reported catalytic epoxidations based on this cycle, using a variety of sulfide structures [34–36]. Figure 10.1 shows examples, along with further details of the catalytic results, for the coupling of benzyl bromide and benzaldehyde to form *trans*-stilbene oxide.

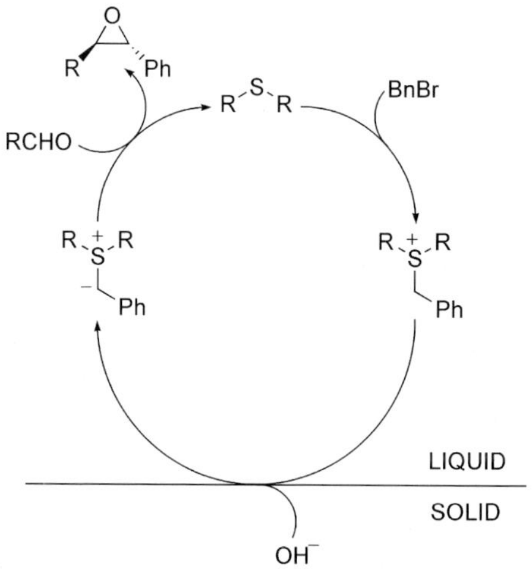

Scheme 10.3 Catalytic cycle for epoxidation via sulfide alkylation and subsequent deprotonation.

10.2.1.2 Conditions and Results for Alkylation/Deprotonation Cycle
Typically, reactions are carried out open to air, at room temperature (rt), using a mixture of tBuOH/water (9:1) or MeCN/water (9:1) as solvent and sodium or potassium hydroxide as base. The presence of water suppresses undesired side reactions (e.g., Cannizzaro reaction, Williamson alkylation, solvent reactivity) [37]. Benzyl bromides are most commonly reported as the halide component.

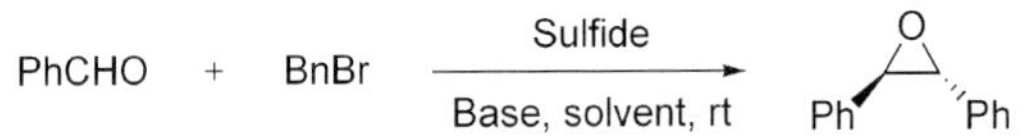

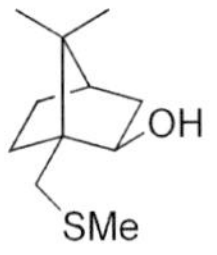

PhCHO + BnBr →(Sulfide / Base, solvent, rt) Ph-epoxide-Ph

1[33]
50%
dr = 100:0
ee = 47% (R,R)
1.5 d, 50 mol%

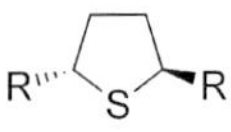

a R = Me
b R = Et

2a/b[35]
82/90%
dr = 92:8
ee = 85/92% (S,S)
4/6 d, 10 mol%

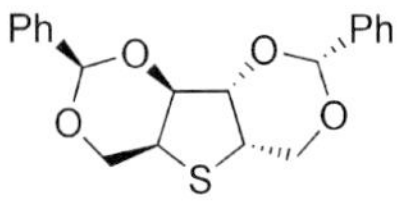

3[36]
41%
dr = 90:10
ee = 97% (R,R)
4 d, 10 mol%

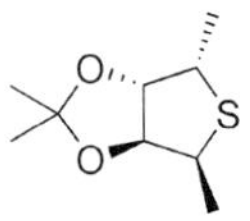

2c[34]
88%
dr = 83:17
ee = 96% (R,R)
1 d, 20 mol%

Fig. 10.1 Selected chiral sulfides and results obtained using alkylation/deprotonation catalytic methodology for the asymmetric synthesis of *trans*-stilbene oxide. dr = *trans*:*cis*; solvents and additives vary.

Sulfide loadings of 10–50 mol% have been used, although low yields or long reaction times are reported at low loadings for some sulfides. Reported reaction times vary from 1 day to 1 month, depending on the sulfide and reaction conditions chosen. The rates can be improved by increasing the reaction concentration, but side reactions (e.g., Cannizzaro reactions) may compete and so a balance must be found. The alkylation step is slow and reversible, whereas the deprotonation step is rapid but reversible, and in protic solvents the equilibrium lies towards the starting materials [38]. Additives such as Bu_4NI and sodium iodide have been used to speed up the alkylation step through halogen exchange with the alkyl bromide [35]. Other additives such as catechol (0.5 mol%) and Bu_4NHSO_4 can also have positive effects. It has been proposed that the ammonium cations may act as phase-transfer catalysts [34]. (For experimental details see Chapter 14.13.2). Increasing the bulk of the sulfide can increase the selectivity, but generally slows the alkylation step. Factors which affect the selectivity in the reaction between ylide and aldehyde are discussed in Sections 10.2.1.9 and 10.2.1.10. The epoxidation of aromatic aldehydes with benzyl ylides can be achieved in good yield, with high diastereo- and enantioselectivity, although the scope is limited. The basic reaction conditions restrict the process to non-enolizable aldehydes. Cinnamaldehyde and heteroaromatic aldehydes are the only other aldehydes to produce reasonable results under catalytic conditions [34, 35, 39, 40]. Overall, this is an operationally simple method for the production of *trans*-epoxides in high enantio- and diastereoselectivity, although its usefulness is restricted by its narrow scope.

10.2.1.3 Catalysis via Ylide Formation from a Carbene Source

Aggarwal et al. [41] first reported the use of an alternative catalytic cycle in 1994 (see Scheme 10.4), wherein a diazo compound reacts with a sulfide in the pres-

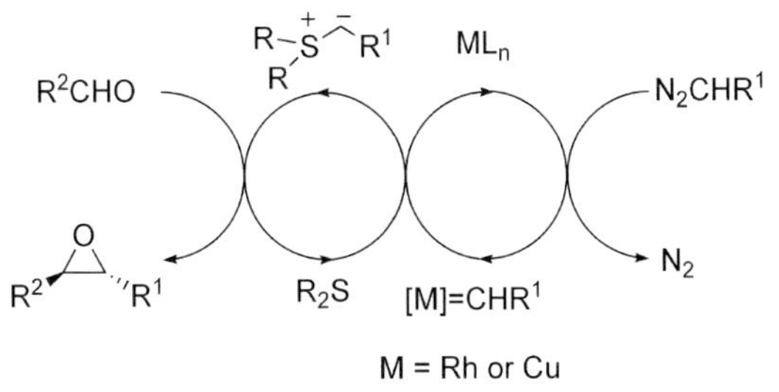

Scheme 10.4 Catalytic cycle for ylide epoxidation via the carbene route.

ence of a metal catalyst to form an ylide; the latter then reacts with the carbonyl compound to give the desired epoxide and regenerate the sulfide. The reaction is carried out under neutral conditions, and therefore this method has the advantage that base-sensitive substrates can be used [42, 43]. In addition, because the metal carbene intermediate is more reactive than alkyl halides, ylide formation with less-reactive sulfides is possible.

Increasing the sulfide concentration (reducing the solvent volume) and adding the diazo compound slowly are crucial to the successful catalytic process. This is to ensure rapid reaction of the metal carbenoid with the sulfide; otherwise, side reactions involving the metallocarbene dominate [43, 44]. This reduction in side reactions (especially stilbene formation) does not compromise the enantioselectivity. Both $Cu(acac)_2$ and $Rh_2(OAc)_4$ were seen to effect the formation and transfer of carbene to sulfide, although the reaction was quite sensitive to the quality of the copper reagent [21]. Notably, $Cu(acac)_2$ gave better results, especially with hindered sulfides, compared to $Rh_2(OAc)_4$ [45]. The main problem with this method is the need to handle stoichiometric amounts of diazo compounds, which are potentially hazardous and not desirable in large-scale reactions. Thus, a modified catalytic cycle was introduced with the *in situ* generation of the diazo compound.

10.2.1.4 Carbenes from *In Situ* Diazo Compound Generation

In this modified catalytic cycle, tosylhydrazones were used as a source of diazo compounds. $Rh_2(OAc)_4$ gave better results than $Cu(acac)_2$ as the metal source for the generation of the metallocarbene from the diazo compound (see Scheme 10.5) [46, 47]. Sodium tosylhydrazone salts decomposed to give diazo compounds in acetonitrile at 30–40 °C in the presence of a phase-transfer catalyst (5–20 mol%) (PTC). Sulfide loadings as low as 5 mol% were achieved with reasonable reaction times (<48 h). The tosylhydrazone could be generated from an aldehyde *in situ*, thus providing a one-pot procedure for coupling two aldehydes to give an epoxide. The process is also amenable to scale-up (50-mmol scale demonstrated), and further reductions in rhodium and PTC loadings were achieved on scale-up. Optimized procedures have been described in detail [48]. (For experimental details see Chapter 14.13.3). This method provides access to ylides that cannot be formed from the direct reaction of a diazo compound and sulfide due to instability of the diazo compound (e.g., *p*-methoxybenzaldehyde-derived hydrazone salt

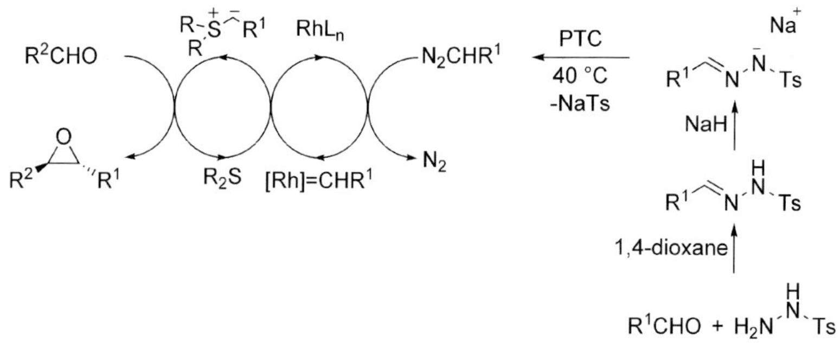

Scheme 10.5 *In situ* generation of diazo compound and resulting catalytic epoxidation cycle.

works well, but the corresponding diazo compound decomposes at −80 °C and can detonate when isolated) [29]. In addition to operational benefits, yields and diastereoselectivities were higher using this process compared to using a preformed diazo compound.

10.2.1.5 Scope of Carbene-Based Methodologies

Aromatic, certain heteroaromatic, aliphatic and α,β-unsaturated aldehydes could all be epoxidized in high enantio- and diastereoselectivity with moderate to good yields in reactions using sulfide 4 and the sodium salt of benzaldehyde tosylhydrazone [29].

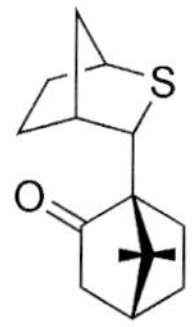

4

Figure 10.2 illustrates selected examples of these epoxide products. Aromatic and heteroaromatic aldehydes proved to be excellent substrates, regardless of steric or electronic effects, with the exception of pyridine carboxaldehydes. Yields of aliphatic and α,β-unsaturated aldehydes were more varied, though the enantioselectivities were always excellent. The scope of tosylhydrazone salts that could be reacted with benzaldehyde was also tested (Fig. 10.3) [29]. Electron-rich aromatic tosylhydrazones gave epoxides in excellent selectivity and good yield, except for the mesitaldehyde-derived hydrazone. Heteroaromatic, electron-poor aromatic and α,β-unsaturated-derived hydrazones gave more varied results, and some substrates were not compatible with the catalytic conditions described. The use of stoichiometric amounts of preformed sulfonium salt derived from 4 has been shown to be suitable for a wider range of substrates, including those that are incompatible with the catalytic cycle, and the sulfide can be recovered quantitatively afterwards [31]. Overall, the demonstrated scope of this *in situ* protocol is wider than that of the alkylation/deprotonation protocol, and the extensive substrate

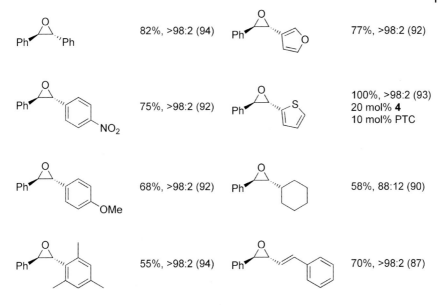

Fig. 10.2 Selected epoxides formed from the reaction of PhCHNNTsNa. Yield, *trans:cis* (% ee). Reaction conditions: PhCHNNTsNa, RCHO, sulfide **4** (5 mol%), $Rh_2(OAc)_4$ (1 mol%), PTC (5 mol%), CH_3CN, 40 °C, 48 h, unless stated otherwise [47].

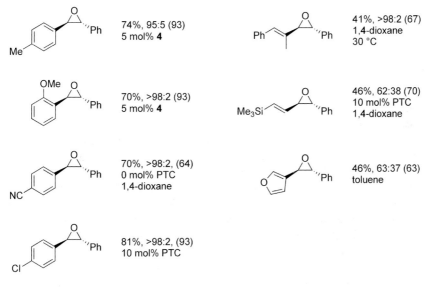

Fig. 10.3 Selected epoxides formed from the reaction of PhCHO and RCHNNTsNa. Yield, *trans:cis* (% ee). Reaction conditions: RCHNNTsNa, PhCHO, sulfide **4** (20 mol%), $Rh_2(OAc)_4$ (1 mol%), PTC (5 mol%), 40 °C, unless stated otherwise [47].

screening that has been conducted allows the best disconnection for a given epoxidation to be chosen [29].

10.2.1.6 Catalytic Cycle Using the Simmons–Smith Reagent

The use of the Simmons–Smith reagent as an alternative source of carbene has been reported [49–51]. Thus, the combination of Et_2Zn, $ClCH_2I$ (2 equiv.) and sulfide gave an ylide which epoxidized aldehydes to give terminal epoxides in good yield. Base-sensitive aldehydes were tolerated, and those bearing stereocenters gave mixtures of diastereomers, though the pre-existing stereocenters were not racemized. Aggarwal et al. reported that sulfides tethered to zinc with chiral ligands led to terminal epoxides from aliphatic and aromatic aldehydes in up to 54% ee [50]. Goodman reported ee-values of up to 76% for terminal epoxides using sulfide **3** and a chloromethyl zinc complex [51], while Aggarwal and co-workers showed that the zinc ions were intimately involved in the epoxidation step, which is in sharp contrast to other catalytic methods where only the sulfur ylide is involved in the epoxidation step [50, 52].

10.2.1.7 Synthesis of Glycidic Amides

Diazoacetamides have also been used as diazo compounds in sulfur ylide-mediated epoxidations yielding glycidic amides **6** [53]. Seki and co-workers have reported the use of diazoacetamides and $Cu(acac)_2$ to generate the sulfur ylides *in situ* [54]. Chiral sulfides **5a/b** (20 mol%) were used to catalyze the reaction of aromatic aldehydes and diazoacetamides to give glycidic amides in low to moderate yield (18–71%) and moderate *ee* (up to 64%) (see Scheme 10.6). Currently, only stoichiometric ylide epoxidations give access to glycidic amides in high enantioselectivity and yield [55].

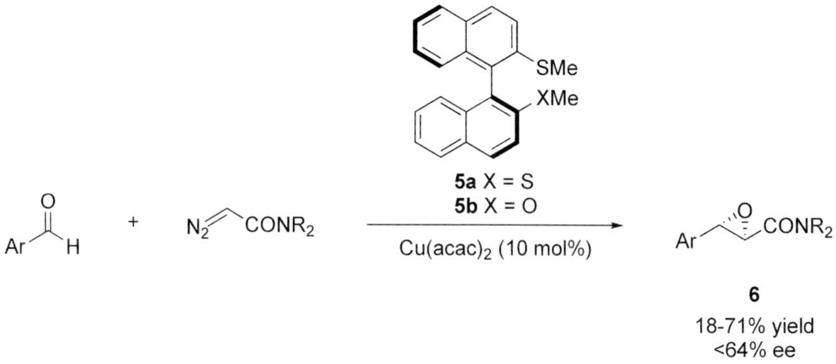

Scheme 10.6 Sulfur ylide-catalyzed asymmetric synthesis of glycidic amides.

10.2.1.8 Optimum Sulfide in Carbene-Based Cycle

Sulfide **4** is the best sulfide reported so far for use in carbene-based catalytic cycles. This gives excellent enantio- and diastereoselectivities in many cases, and also displays sufficient reactivity and stability. It can be used at loadings of 5 mol% in many cases, and it can be recovered from the reaction and recycled. It can be synthesized on a 20-g scale in four steps from camphorsulfonyl chloride, which is available in both enantiomeric forms [46, 56]. The origin of the selectivities observed in epoxidation reactions are discussed below. Many of these factors are also relevant in sulfide-catalyzed aziridination and cyclopropanation (see Sections 10.3 and 10.4.1).

10.2.1.9 Diastereoselectivity in S-Ylide Epoxidations

The diastereoselectivity in sulfur ylide-mediated epoxidation is controlled by the reversibility of the initial betaine formation step [57]. Scheme 10.7 shows the individual steps in the epoxidation. Computational studies have shown that the aldehyde approaches the ylide on a "cisoid" or "head-on" trajectory (due to favorable Coloumbic interactions) and initially forms **8a** and **7a** [58]. Rotation about the newly formed C–C bond leads to conformers **8b** and **7b**. The subsequent ring-closing of the anti-periplanar rotamers is rapid, and leads to formation of the *cis*- and *trans*-epoxides **10** and **9**, respectively. Surprisingly, calculations suggest that the rate-determining process can be the C–C bond rotation for *cis*-epoxide formation. If betaine formation is non-reversible, then the diastereoselectivity is controlled by the relative energies of the transition states for formation of the *syn*- and *anti*-betaines, **8a** and **7a**, the difference of which (from calculation) is small. Alternatively, if betaine formation is reversible then the rates of conversion of the *syn*- and *anti*-betaines to *cis*- and *trans*-epoxides will affect the diastereoselectivity observed. In reactions of benzaldehyde with dimethylsulfonium benzylide, it has been shown that the *anti*-betaine **7a** (R = Ph) forms non-reversibly and ring-closes to form *trans*-epoxide **9** [57]. The *syn*-betaine **8a** forms reversibly with only slow conversion to the anti-periplanar rotamer, which can form *cis*-epoxide **10**. Thus, a high degree of *trans*-selectivity is observed in these cases.

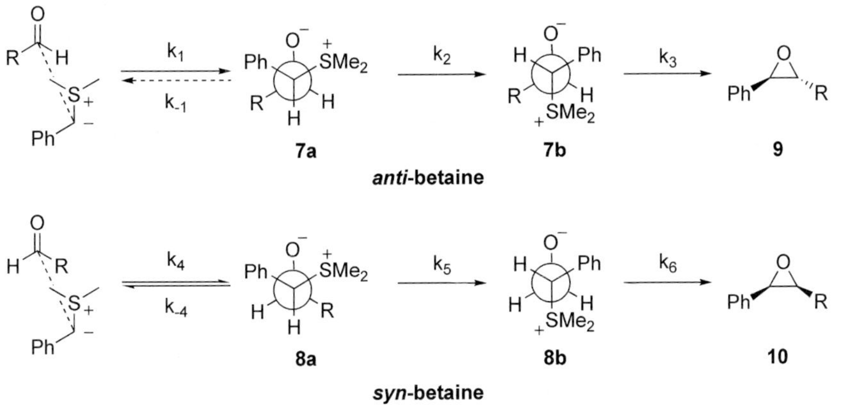

Scheme 10.7 Individual steps involved in epoxide formation.

The structure of the aldehyde and ylide, as well as the solvent and metal ions, can influence the relative stabilities of the betaines and the relative rates of the individual steps [43, 47, 57–60]. The effects of these factors on *syn*-betaine reversibility outweigh any effect on *anti*-betaine reversibility, and so increasing reversibility corresponds to improved diastereoselectivity. Ylides or aldehydes bearing bulky groups increase the barrier to C–C bond rotation in the betaines and tend to favor reversion. Polar solvents and metal ions can stabilize the alkoxide, thus allowing a more facile separation of charge and thereby lowering the barrier to rotation in the betaines, leading to lower diastereoselectivity. Stabilized aldehydes and ylides favor reversible betaine formation, and hence good diastereoselectivities but less-stable aldehydes and ylides tend to give poorer diastereocontrol.

10.2.1.10 Enantioselectivity in S-Ylide Epoxidations

There are four main factors that control the enantioselectivity in the sulfur ylide-mediated epoxidation: (1) selectivity for one sulfide lone pair; (2) control of ylide conformation; (3) control of facial selectivity; and (4) the reversibility of *anti*-betaine formation.

Enantioselectivity can be controlled by presenting only one face of one ylide isomer to the approaching aldehyde and ensuring that this step is non-reversible and therefore enantiodetermining [45–47, 61]. Sulfide **4** was designed so that one of the lone pairs on sulfur is much more hindered than the other, and so only one pair reacts to form an ylide (**11**). There are two potentially important conformers of the ylide **11A** and **11B**, as an orthogonal arrangement of the lone pairs on sulfur and carbon is favored [62]. Steric interactions disfavor **11B**, and so **11A** is the major conformer available to react (Scheme 10.8). Aldehyde approach to **11** is only possible to one face due to steric blocking of the other face [47]. It should be noted that improving selectivity with bulky sulfides can lead to decreasing reactivity [43], and so a compromise must be found in sulfide design. In the case of 1,3-oxathianes such as **12**, electronic effects were shown to improve the face selectivity of the sulfide in addition to steric effects [61]. In the cases of C_2-symmetric sulfides such as **2a** and **3**, the issue of lone pair selectivity is avoided. Calculations by Goodman and co-workers have suggested that the resulting ylide conformations derived from **2a** are not well controlled, but in fact in these cases the reactivities of the ylide conformers are very different so that high enantioselectivities can still be obtained as the equilibrium between conformers is

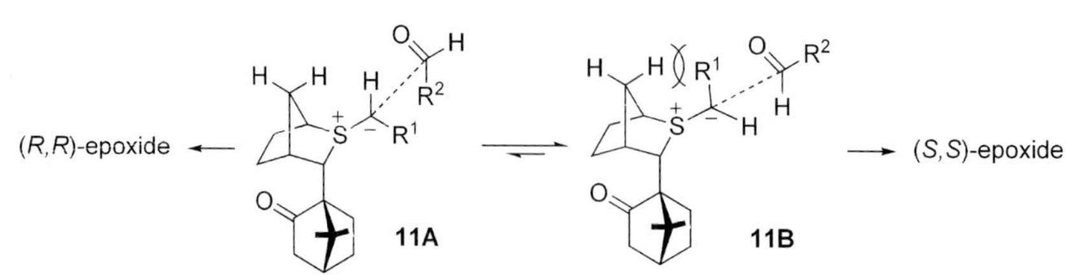

Scheme 10.8 Equilibration of ylide conformers and trajectory of aldehyde approach to **4**.

rapid [63]. Enantioselectivity should be controlled through non-bonding interactions in the transition state of the C–C-bond-forming step [34].

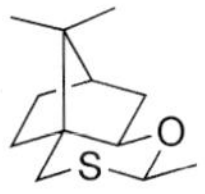

12

Control of the above-described factors leads to the selective formation of one betaine intermediate. If betaine formation is non-reversible, this will lead to high enantioselectivity in epoxide formation, which is observed for a wide range of substrates using **4** as catalyst. It is believed that in many cases the minor enantiomer of epoxide observed is formed from aldehyde reacting with the minor ylide conformer [47, 60]. If the *anti*-betaine is formed reversibly, the bond-rotation and/or ring-closure step can become enantiodifferentiating, and lower enantioselectivities can result because of differences in the reversibility of betaine formation and subsequent steps for the pathways starting from **11A** and **11B** [47, 60, 64]. This explains the slightly lower *ee*-values observed in reactions of benzaldehyde with ylides derived from electron-poor aromatic tosylhydrazone salts. Recently, it was reported that the stereochemistry of chiral aldehydes can also influence the reversibility of betaine formation and thus the selectivity observed [65]. A simple test has been developed to determine whether reduced enantioselectivity is caused by reversibility of betaine formation or imperfect control of ylide conformation. Reactions are carried out in MeCN and compared to the results in MeCN/H_2O. An improvement in *ee*-values in MeCN/H_2O mixtures, where the reversibility of betaine formation is reduced, indicates that reversibility is a problem [47]. Although the use of MeCN/H_2O mixtures can improve the *ee*-values, this does not provide a practical solution as substantially lower yields are usually obtained.

10.2.1.11 Applications of S-Ylide Epoxidation in Synthesis

Most studies on sulfur ylide-mediated asymmetric epoxidation have concentrated on the development of the methodology. The usefulness of this approach has been demonstrated in the synthesis of a number of biologically interesting compounds. Furaldehyde-derived epoxides can be oxidized to produce glycidic esters that are versatile intermediates in several syntheses (see Scheme 10.9) [46]. The

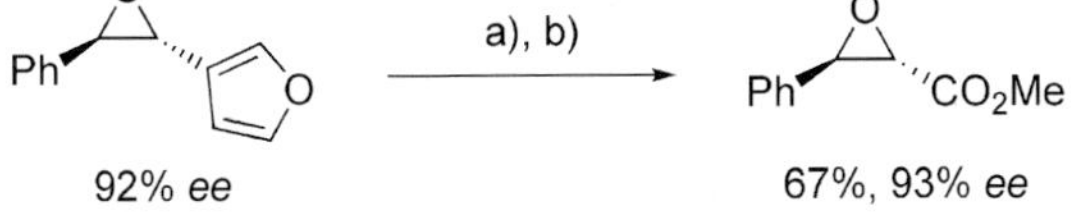

92% ee   67%, 93% ee

Scheme 10.9 Synthesis of glycidic esters. Reagents and conditions:
(a) $RuCl_3$ (2 mol%), $NaIO_4$ (4 equiv.), MeCN/CCl_4/H_2O, rt, 3 h;
(b) CH_2N_2, Et_2O, rt.

use of stoichiometric amounts of pre-formed sulfonium salt has been applied to the synthesis of the anti-inflammatory agent CDP-840 (Scheme 10.10) [31]. The suitability of this methodology for sensitive substrates was demonstrated in the first synthesis of a ferrocenyl epoxide (see Scheme 10.11) [66]. One equivalent of 4 mediated the synthesis of **13** from ferrocene carbaldehyde and benzaldehyde tosylhydrazone sodium salt in excellent enantioselectivity, but low yield. The epoxide was ring-opened with sodium azide and then purified due to the fragile nature of the ferrocenyl epoxide. Catalytic asymmetric epoxidation using sulfide 4 was used as the key step in the synthesis of β-hydroxy-γ-lactone **14**, an early metabolite in the biosynthesis of polyketide antibiotics (see Scheme 10.12) [67]. Additional synthetic applications of this useful methodology will, no doubt, be reported during the coming years.

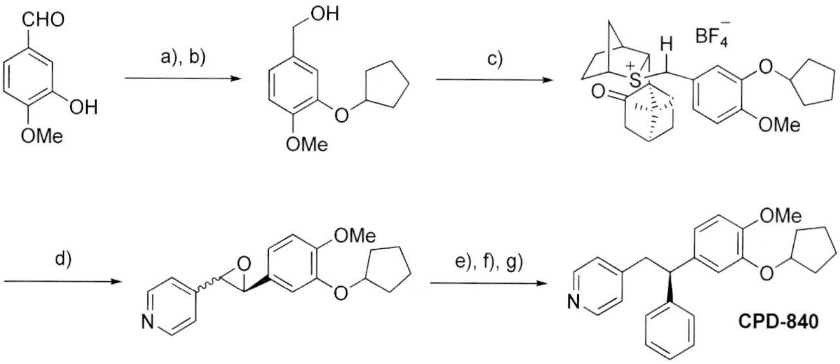

Scheme 10.10 Synthesis of CDP-840 involving asymmetric epoxidation as a key step. Reagents and conditions: (a) K_2CO_3, cyclopentyl bromide, DMF, 97%; (b) $NaBH_4$, MeOH, 99%; (c) ent-**4**, HBF_4, Et_2O, rt, 4 h, 80%; (d) EtP_2-base, CH_2Cl_2, −78 °C, 15 min, then 4-pyridine carboxaldehyde, −78 °C, 1 h, 89%, (trans:cis = 7:3), >98% ee; (e) PhMgBr, CuI, THF, 85%; (f) Et_3N, MsCl, CH_2Cl_2; (g) Zn, AcOH, 80% (two steps).

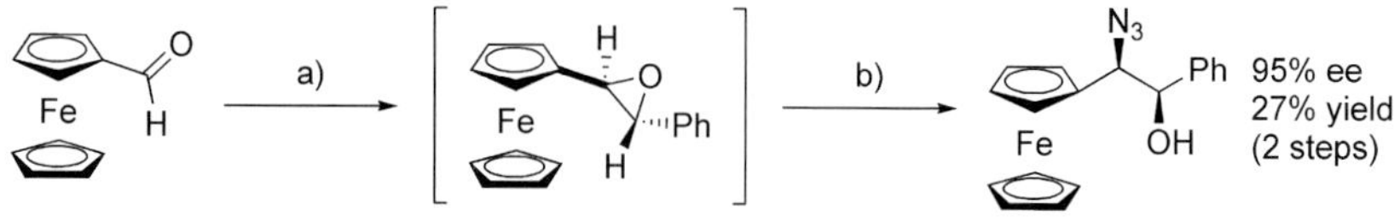

Scheme 10.11 Synthesis of ferrocenyl epoxide. Reaction conditions: (a) PhCHNNTsNa (1.5 equiv.), $Rh_2(OAc)_4$ (1 mol%), $BnEt_3NCl$ (20 mol%), **4** (1 equiv.), 1,4-dioxane, 40 °C, 47 h; (b) NaN_3 (3.0 equiv.), NH_4Cl (3.0 equiv.), EtOH, reflux, 3.5 h.

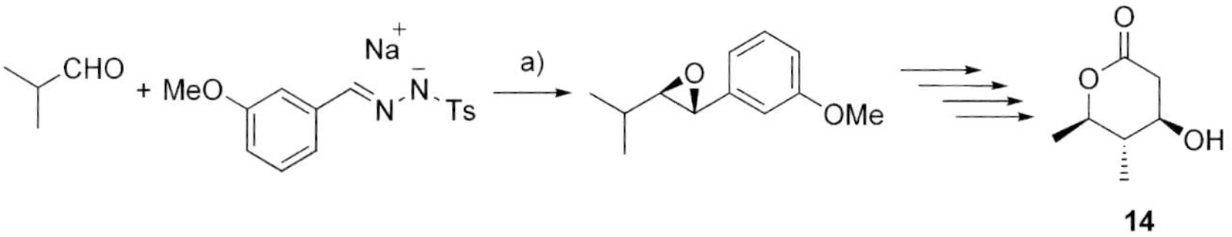

Scheme 10.12 Catalytic asymmetric ylide epoxidation as the key step in a six-step synthesis of Prelactone B (**14**), 10% yield, 93% ee. Reaction conditions: (a) Sulfide **4** (25 mol%), $Rh_2(OAc)_4$ (1 mol%), $BnEt_3NCl$ (10 mol%), MeCN, 40 °C, 36 h.

10.2.2
Te-Ylide-Catalyzed Epoxidation

The use of ylides derived from tellurium analogues of sulfides **2a** and **2b** for asymmetric epoxidation has been described [68, 69]. Preliminary experiments using **15** as catalyst gave high enantio- and diastereoselectivities, but poor yield (see Scheme 10.13).

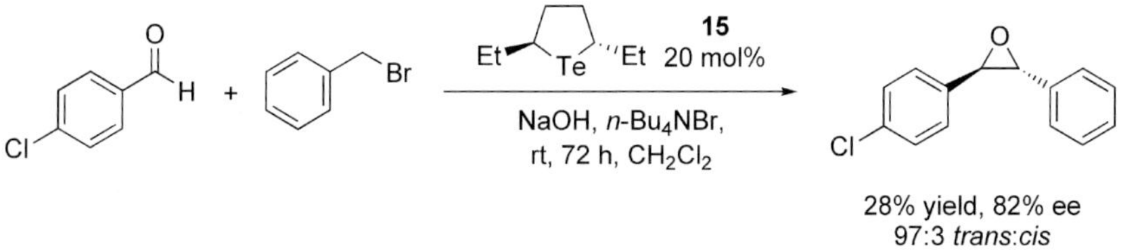

Scheme 10.13 Te-ylide-catalyzed asymmetric epoxidation.

10.2.3
Se-Ylide-Catalyzed Epoxidation

Catalytic asymmetric epoxidation using Se-ylides has been reported [68, 70]. Cinnamaldehyde, aromatic and heteroaromatic aldehydes gave high yields and enantioselectivities, but no diastereoselectivity was observed (see Scheme 10.14).

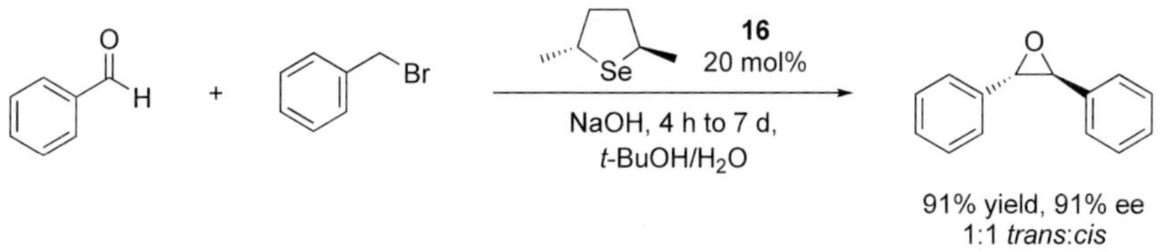

Scheme 10.14 Se-ylide-catalyzed asymmetric epoxidation using **16**.

10.2.4
Summary of Ylide-Mediated Epoxidation

Catalytic asymmetric sulfur ylide epoxidation provides access to *trans*-epoxides in high yield, together with enantio- and diastereoselectivity from carbonyl compounds. A range of complementary methods for the production of *cis*-, *tri*-, *tetra*-substituted and terminal epoxides exist [3–6, 9–12]. Unfunctionalized *trans*-epoxides can also be accessed from alkenes using dioxirane-based methods [7, 8]. Ylide-catalyzed carbonyl epoxidation provides an alternative to alkene epoxidation. Ylide epoxidations using sulfide **4**, and ylide generation from a diazo precursor generated *in situ*, are particularly effective for a broad substrate range, including base-sensitive substrates. For certain products (e.g., some α,β-unsaturated epoxides) the catalytic reaction performs poorly, but the stoichiometric ylide method works well and the sulfide can be recovered in good yield. The combined catalytic and stoichiometric methodologies allow the application of the sulfur ylide disconnection with confidence in the design of syntheses.

10.3
Asymmetric Aziridination

Like epoxides, chiral non-racemic aziridines are useful synthetic intermediates [3, 71], and a number of methodologies have been developed for their asymmetric synthesis [3, 6, 14, 72, 73]. Although several groups have developed stoichiometric methods using chiral ylides [16, 20, 22, 74, 75], catalytic asymmetric ylide aziridinations remain relatively rare. In fact, the first catalytic aziridination with an ylide was only reported ten years ago [76]. Progress in this area is reviewed in the following section.

10.3.1
Sulfur Ylide-Catalyzed Aziridination

10.3.1.1 Ylide Regeneration via Carbenes

The first asymmetric aziridination process to use catalytic amounts of ylide was reported by Aggarwal et al. in 1996 [77]. The proposed catalytic cycle (see Scheme 10.15) is analogous to that used in their asymmetric epoxidation process (see Sec-

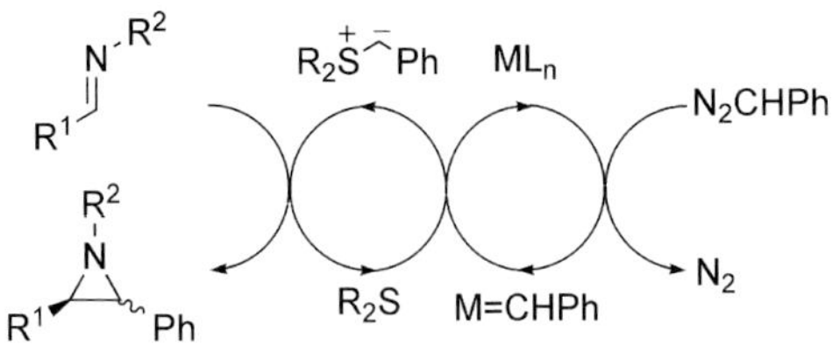

Scheme 10.15 Catalytic cycle for asymmetric aziridination via sulfur ylides.

tion 10.2.1.3). A diazocompound is slowly added to a dichloromethane solution of sulfide (20 mol%), imine and transition-metal catalyst (1 mol% $Rh_2(OAc)_4$ or 5 mol% $Cu(acac)_2$) at room temperature. The diazocompound reacts with a transition-metal catalyst to form a metal carbenoid which can, in turn, react with sulfide to generate an ylide. The sulfur ylide then reacts with imines to give the desired aziridines.

A complication in aziridination is that metal carbenoids can react directly with some imines to yield aziridines. Fortunately, imines bearing electron-withdrawing groups are less reactive to carbenoids and more reactive to ylides than electron-rich imines. Thus, *N*-tosyl, *N*-diphenylphosphinyl and *N*-[-(tri-methylsilyl)ethansulfonyl] imines (-(trimethylsilyl)ethansulfonyl = SES) were all suitable substrates using dimethylsulfide and $Rh_2(OAc)_4$, with no background reaction detected [77, 78]. Phenyldiazomethane, *N*,*N*-diethyl diazoacetamide and ethyl diazoacetate could be used as the diazo component, although the latter two required temperatures of 60 °C to decompose.

10.3.1.2 Scope and Limitations of the Carbene-Based Cycle

Having demonstrated the feasibility of their catalytic process, Aggarwal and colleagues explored the use of a chiral sulfide, **12** [77, 78]. It was found that *N*-SES benzaldimines yielded aziridines **18a–e** in excellent enantioselectivity (up to 97% *ee*), albeit in moderate yield and diastereoselectivity (*trans*:*cis* 3:1–5:1). At catalytic loadings of **12** (20 mol%), $Rh_2(OAc)_4$ gave slightly better yields and *ee*-values than $Cu(acac)_2$ in most cases. The use of alternative diazo compounds such as *N*,*N*-diethyl diazoacetamide and ethyl diazoacetate provided moderate enantioselectivities at best, and only stoichiometric sulfide loadings were reported.

As in the epoxidation reactions, the diazo compound could be generated *in situ* from tosylhydrazone salts (see Scheme 10.16) [79, 80]. This addressed the major

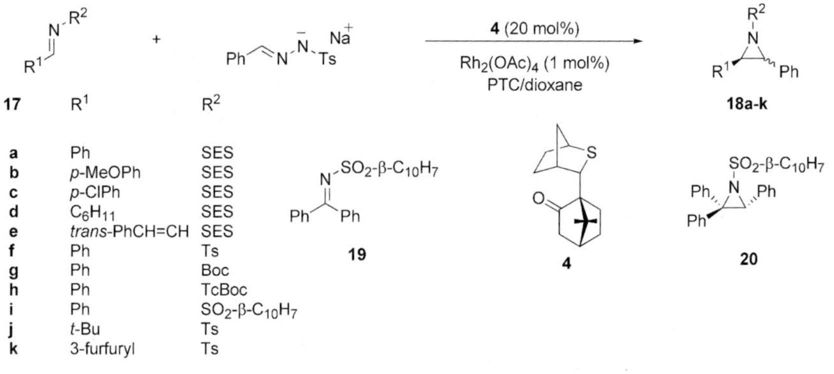

	R^1	R^2
a	Ph	SES
b	*p*-MeOPh	SES
c	*p*-ClPh	SES
d	C_6H_{11}	SES
e	*trans*-PhCH=CH	SES
f	Ph	Ts
g	Ph	Boc
h	Ph	TcBoc
i	Ph	SO_2-β-$C_{10}H_7$
j	*t*-Bu	Ts
k	3-furfuryl	Ts

Scheme 10.16 Asymmetric ylide-catalyzed aziridination [79, 82].

Table 10.1 Results obtained for the asymmetric sulfur ylide-catalyzed aziridination[a] (according to Scheme 10.16 [79, 82]).

Entry	Substrate	Product	Yield (%)	dr[b]	ee (%)[c]
1	17a	18a	75	2.5:1	94
2[d]			66	2.5:1	95
3	17b	18b	60	2.5:1	92 (78)
4	17c	18c	82	2:1	98 (81)
5	17d	18d	50	2.5:1	98 (89)
6	17e	18e	59	8:1	94
7	17f	18f	68	2.5:1	98
8	17g	18g	33[e]	8:1	89
9	17h	18h	71	6:1	90
10	17i	18i	70	3:1	97
11	17j	18j	53	2:1	73 (95)
12	17k	18k	72	8:1	95
13	19	20	50	–	84

[a] Reaction conditions: Imine (1 equiv.), **4** (20 mol%), $Rh_2(OAc)_4$ (1 mol%), PhCHNNTsNa (1.5 equiv.), $BnEt_3NCl$ (10 mol%), 1,4-dioxane, 40 °C.
[b] trans:cis ratio.
[c] ee of trans (cis).
[d] Using 5 mol% sulfide.
[e] 17g was unstable under the reaction conditions.

drawback in the system, as the use of stoichiometric amounts of phenyldiazomethane is potentially hazardous. Under the new protocol, the reaction was carried out at 40 °C in dioxane with a phase-transfer catalyst in addition to $Rh_2(OAc)_4$ (1 mol%) and 20 mol% of chiral sulfide for one to two days. Sulfide **4** gave excellent ee-values and good yields in the aziridination of a range of imines (see Table 10.1). (For experimental details see Chapter 14.11.2).

In many cases the ee-values and yields obtained using this *in situ* process were better than those reported using **12** and $PhCHN_2$. In one example, the sulfide loading was reduced to 5 mol% and no significant effect on either ee-value or yield was observed. The range of aziridines was expanded and included imines derived from ketones and aliphatic aldehydes (see Table 10.1), as well as a variety of substituents on nitrogen (cf. entries 1 and 7–10). The diastereoselectivities observed were moderate and highly dependent on the group on nitrogen. The tosylhydrazone salt was also varied; for example, unsaturated aziridine **22** was synthesized from the unsaturated tosylhydrazone salt **21** (see Scheme 10.17). Unsaturated aziridines are particularly useful as they can undergo ring-expansions to give either β- or δ-lactams (see Scheme 10.17 for an example).

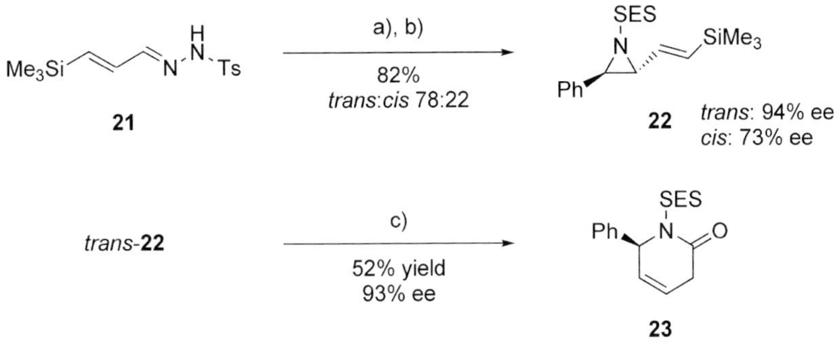

Scheme 10.17 Synthesis of a δ-lactam **23**. Reagents and conditions:
(a) LiHMDS (1.5 equiv.), THF, −78 °C, 1 h; (b) **17a** (1 equiv.), Rh$_2$(OAc)$_4$
(1 mol%), phase-transfer catalyst (10 mol%), sulfide **4** (20 mol%),
1,4-dioxane, 40 °C, 12 h; (c) [Pd$_2$(dba)$_3$·CHCl$_3$] (15 mol%), PPh$_3$
(1.3 equiv.), PhH, CO (1 atm), 50 °C, 12 h. HMDS = 1,1,1,3,3,3-
hexamethyldisilazane; SES = 2-(trimethylsilyl)ethanesulfonyl;
dba = trans,trans-dibenzylideneacetone.

10.3.1.3 Catalysis via Sulfonium Salt Formation and Deprotonation

As in catalytic ylide epoxidation (see Section 10.2.1.1), an alternative catalytic cycle can be based on generation of the ylide *in situ* by reaction of a sulfide with an alkyl halide to form a salt, which can then be deprotonated [76]. In 2001, Saito et al. reported the asymmetric version of this cycle using a 3:1 ratio of alkyl halide to sulfonyl imine (see Scheme 10.18) [81]. Good yields and *ee*-values were reported for aryl- and styryl-substituted aziridines using stoichiometric amounts of sulfide **24**, and the diastereoselectivities ranged from 1:1 to 4:1. Unfortunately, when loadings were reduced the reaction times became longer and lower yields were reported (see Table 10.2).

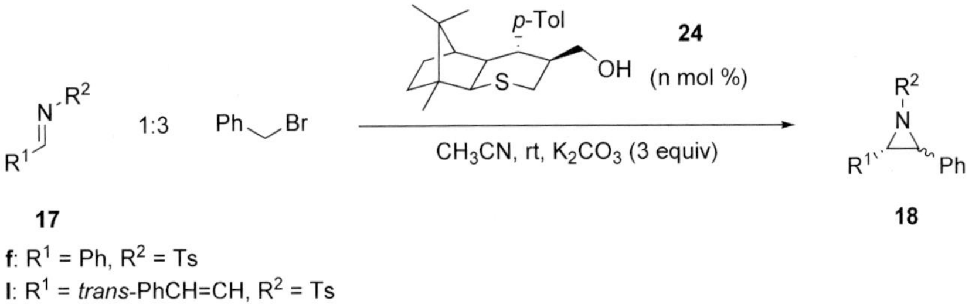

f: R^1 = Ph, R^2 = Ts
l: R^1 = *trans*-PhCH=CH, R^2 = Ts

Scheme 10.18 Catalytic asymmetric aziridination via alkylation of sulfide **24** and deprotonation.

Table 10.2 Results obtained for asymmetric aziridination using **24** (according to Scheme 10.18).

Substrate	Product	Sulfide loading (mol%)	Time (days)	Yield (%)	dr[a]	ee (%)[b]
17f	18f	100	1–2	>99	3:1	92
		50	1–2	91	3:1	92
		20	4	61	3:1	90
		10	7	<10	3:1	91
17l	18l	100	1–2	99	3:1	94 (84)
		50	1–2	73	4:1	93 (88)
		10	7	52	3:1	93 (87)

[a] *trans:cis* ratio.
[b] *ee* of *trans* (*cis*).

10.3.1.4 Selectivity in S-Ylide Aziridinations

For the reaction of phenyl-stabilized ylides, both the *syn*- and the *anti*-betaines were shown to form non-reversibly using crossover experiments (the betaines were generated from the corresponding sulfonium salts in the presence of a more reactive imine, but the reactive imine was not incorporated in the product) [82]. Thus, the diastereoselectivity is controlled by the relative rates of betaine formation [78, 82, 83]. This is not the case in ylide epoxidation, where *syn*-betaine formation is partially reversible and so higher diastereoselectivities are observed (see Section 10.2.1.9). Steric and Coulombic interactions can affect the relative energies of the calculated transition states for the addition steps. In addition, for *N*-sulfonyl imines hydrogen-bonding (from a methyl hydrogen on the ylide to one of the sulfonyl oxygens) can help stabilize the "head-on" or "*cisoid*" transition states. The *anti*-betaine is predicted to form through a "head-on" or "*cisoid*" approach (Fig. 10.4A), as in epoxidation [83]. For the *syn*-betaine the steric clashes that would occur in a "head-on" approach (Fig. 10.4C) dominate over the loss of

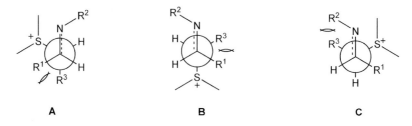

Fig. 10.4 Approaches in betaine formation. (A) favored "head-on" for *anti*-betaine; (B) favored "head-to-tail" for *syn*-betaine formation; (C) disfavored "head-on" for *syn*-betaine.

the favorable Coulombic interactions, and so a "head-to-tail" approach results (Fig. 10.4B). The results in Table 10.1 remain difficult to rationalize fully, but clearly the energies of the transition states leading to *syn-* and *anti*-betaines are finely balanced and so only low to moderate diastereoselectivities have been obtained at present.

For the reaction of more-stabilized ylides (e.g., ester- and amide-stabilized ylides), in the initial addition steps the same approaches are favored (i.e., "cisoid" for *anti-* and "transoid" for *syn*-betaine), although the predicted energy differences between "cisoid" and "transoid" are lower in both cases and may play a role in the mechanism in some cases. The same interactions as for semi-stabilized ylides affect the addition step here. However, due to stabilization of the ylides the betaine formation step is now endothermic. In these cases, it is the ring-closure step which is the slow step, and both the betaine rotamer interconversion step and the initial addition step are reversible (the latter was demonstrated using cross-over experiments) [82]. The relative energies of the ring-closure transition states favor the *cis*-aziridine due to unfavorable steric interactions with the sulfonyl group in the transition state *trans*-aziridines. However, the energies of the transition states in the ring-closing steps leading to *cis-* and *trans*-aziridine are not very different and, indeed, low *cis*:*trans* ratios are observed experimentally.

To explain the enantioselectivity obtained with semi-stabilized ylides (e.g., benzyl-substituted ylides), the same factors as for the epoxidation reactions discussed earlier should be considered (see Section 10.2.1.10). The enantioselectivity is controlled in the initial, non-reversible, betaine formation step. As before, controlling which lone pair reacts with the metallocarbene and which conformer of the ylide forms are the first two requirements. The transition state for *anti*-betaine formation arises via a "head-on" or "*cisoid*" approach and, as in epoxidation, face selectivity is well controlled. The *syn*-betaine is predicted to be formed via a "head-to-tail" or "transoid" approach in which Coulombic interactions play no part. Enantioselectivity in *cis*-aziridine formation was more varied. Formation of the minor enantiomer in both cases is attributed to a lack of complete control of the conformation of the ylide rather than to poor facial control for imine approach. For stabilized ylides (e.g., ester-stabilized ylides), the enantioselectivity is controlled in the ring-closure step and moderate enantioselectivities have been achieved thus far. Due to differences in the stereocontrolling step for different types of ylides, it is likely that different sulfides will need to be designed to achieve high stereocontrol for the different types of ylides.

10.3.1.5 Use of S-Ylide Aziridinations in Synthesis

Aggarwal and Vasse have applied asymmetric ylide aziridination to the synthesis of the taxol side chain **26** [75]. By using stoichiometric quantities of chiral sulfide, **26** was obtained in six steps and 16% overall yield (see Scheme 10.19, route A), while the catalytic variant allowed access to **26** in seven steps and 20% overall yield (see Scheme 10.19, route B). Both methods gave the key aziridine intermediates in excellent enantiomeric purity. Unfortunately, the benzoyl imine **25** did not work well in the catalytic reaction, and so a slightly longer synthetic route

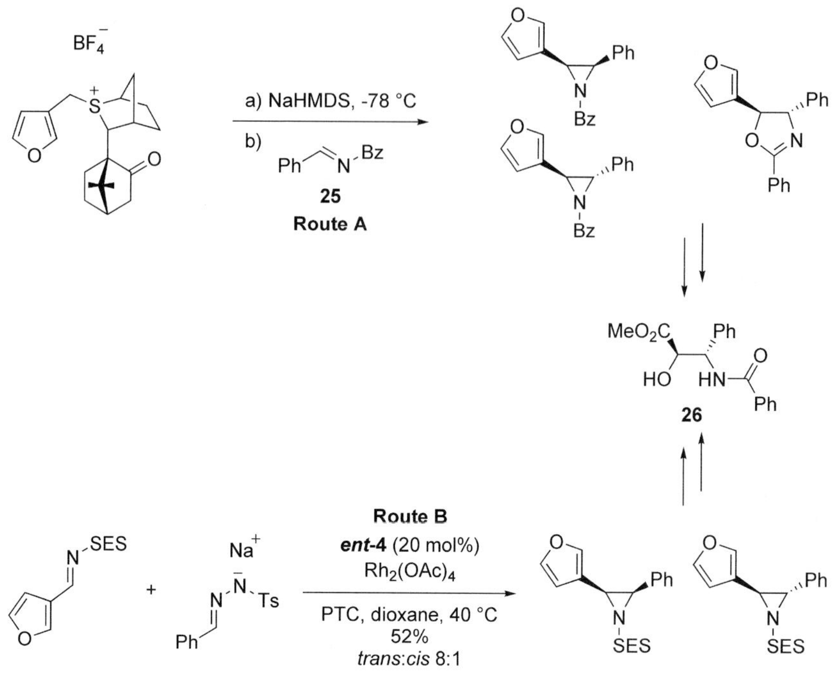

Scheme 10.19 Syntheses of taxol side-chain **26** via asymmetric ylide aziridinations. Overall yields: Route A 16% (six steps); Route B 20% (seven steps).

was necessary using the catalytic methodology. These syntheses demonstrate the usefulness of these aziridination reactions, although ultimately more efficient syntheses of **26** have been reported [84].

10.3.1.6 Summary of Ylide-Catalyzed Aziridination

A catalytic asymmetric aziridination methodology has been developed using chiral sulfur ylides as catalysts. The method is operationally simple and practical for the synthesis of aziridines bearing electron-withdrawing groups on nitrogen. In addition, NH-aziridines are easily accessible by the removal of protecting groups such as SES. Good to excellent yields and excellent enantioselectivities are obtainable, although further investigations are required to fully understand and improve the diastereoselectivity of this methodology. Although stoichiometric methods [16, 20, 22, 74, 75] are beyond the scope of this chapter, it is worth noting that in many cases the chiral sulfide can be recovered after the reaction, and in good yield. As such, these methods can constitute a viable alternative in some instances. Imine aziridinations using either chiral ylide or Lewis acid catalysts offer a complementary route to alkene-based routes [6, 14, 72, 73]. Further development of this methodology will establish its scope and usefulness in synthesis.

10.4
Asymmetric Cyclopropanation

Chiral non-racemic cyclopropanes are a common motif in natural and synthetic biologically active compounds [85]. They represent an important target in asymmetric synthesis, and a range of catalytic methods have been developed for their synthesis [6, 47, 86–90]. Many of the existing methods make use of a chiral metal complex as catalyst [6], but organocatalytic methods have also been developed. In this section we will review methods using a substoichiometric amount of a chiral ylide as a catalyst for cyclopropanation [91].

Stoichiometric ylide cyclopropanations have been known for some time, with asymmetric variants using aminosulfoxonium ylides having been reported as early as 1968 [27]. Since then, procedures using stoichiometric amounts of sulfur, nitrogen, and tellurium ylides to achieve asymmetric cyclopropanations have been reported [16, 22, 86, 92–94]. The catalytic analogues of these reactions are discussed in the following sections.

10.4.1
S-Ylide-Mediated Cyclopropanation

10.4.1.1 Catalytic Cycle Based on Carbene Generation of Ylide

The first catalytic asymmetric cyclopropanation using an ylide as catalyst was reported by Aggarwal et al. in 1997 [95, 96]. Phenyl diazomethane was added slowly to a mixture containing sulfide **12**, an enone and $Rh_2(OAc)_4$ (1 mol%). A sulfur ylide was generated *in situ* from the sulfide and phenyl diazomethane in the presence of the transition-metal catalyst (see Scheme 10.20), as in the epoxidations discussed earlier (see Section 10.2.1.3).

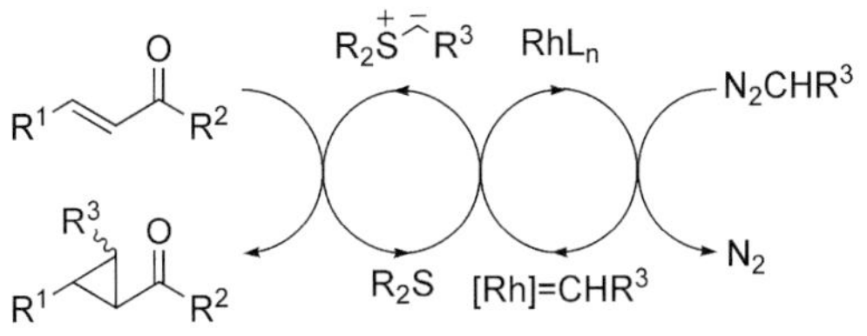

Scheme 10.20 Catalytic process for cyclopropanation.

The ylide then reacted with enones to give the corresponding cyclopropanes in excellent enantioselectivity and moderate relative stereoselectivity (see Table 10.3, conditions A). The *ee*-values were the same using 20 mol% and 100 mol% loadings of sulfide, and no cyclopropanation occurred in the absence of sulfide, demonstrating that there was no background cyclopropanation. Although turnover was observed, catalytic amounts of **12** (20 mol%) gave low yields when compared to pentamethylene sulfide **27** or stoichiometric loadings of either sulfide. Replac-

Table 10.3 Results obtained for the catalytic asymmetric cyclopropanation of α,β-unsaturated alkenes (according to Scheme 10.21 [79, 95, 96]).

Conditions[a]	Sulfide	Substrate	Product	Yield (%)	dr[b]	ee (%)
A	12	29a	30a	38	4:1	97
A		29b	30b	14	4:1	>98
B	4	29a	30a	30[c]	5:1	89
B	28	29a	30a	73	4:1	91
		29c	30c	50	4:1	90
		29b	30b	5	–	–
		29d	30d/31d/32	12	1:1:1	–
		29e	30e	10	7:1	–
		33	35	55	1:7	91
		34	36	72	1:6	92

[a] Reaction conditions: A: PhCHN$_2$ (1 equiv.), toluene, rt, 12 h.
B: PhClINNTsNa (1.5 equiv.), BnEt$_3$NCl (20 mol%), dioxane, 40 °C.
[b] 30:31.
[c] 50% starting material recovered.

ing phenyl diazomethane with ethyl diazoacetate as the source of carbene only gave cyclopropanes using achiral sulfides and Cu(acac)$_2$ as the transition-metal catalyst [97]. The major drawbacks of this method are that phenyl diazomethane is potentially hazardous, the reactions cannot be easily scaled up, and the sulfide cannot be fully recovered. Therefore, a modified procedure was required.

27   **28**

10.4.1.2 Catalysis via *In Situ* Generation of Diazo Compounds

In 2001, a modified procedure for sulfur ylide-catalyzed epoxidation, aziridination and cyclopropanation was introduced by Aggarwal and co-workers that utilized the generation of the diazo compounds *in situ* from tosyl hydrazone salts at 40 °C in the presence of a phase-transfer catalyst [46, 79]. (For experimental details see Chapter 14.12.1). Using this modified protocol, sulfide **4** was shown to be effective for epoxidation and aziridination (see Sections 10.2.1.4 and 10.3), but was not an effective cyclopropanation catalyst (see Table 10.3). Sulfide **28** was tried instead as it had been shown in achiral studies [96] that six-membered sulfides were more effective than five-membered analogues. This change gave rise to a more effective catalyst which was then screened against a range of α,β-unsaturated compounds (see Table 10.3). Good yields and high enantioselectivities were obtained for phenyl ketones and α-amino-substituted acrylates (see Table 10.3). Diastereoselectivities ranged from 4:1 to 7:1 for these substrates, and the sulfide could be recovered quantitatively. Although the enantioselectivities were not as high as when **12** was used, the improvement in yield makes this a better process. Methyl ketones (**29b,d**) and unsubstituted acrylates (such as **29e**) were not good substrates. Overall, the demonstrated scope of this reaction under catalytic conditions remains limited.

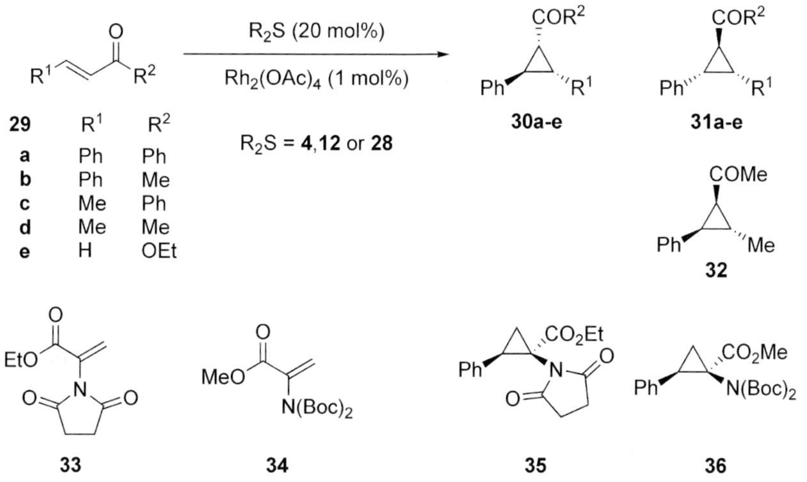

Scheme 10.21 Catalytic asymmetric cyclopropanation using chiral sulfides.

10.4.1.3 Control of Selectivity Using Carbene-Derived Ylides

To explain the observed selectivities, a similar model was proposed as for the epoxidation reactions (see Section 10.2.1.10) [96]. In this model, the ylide conformation is controlled as before, and the alkene selectively attacks one face in an analogous manner to the aldehyde in the epoxidation reactions. It was noted that in this case betaine formation was non-reversible so the diastereoselectivity is controlled by non-bonding interactions during the betaine-formation step [79].

10.4.1.4 Catalytic Cycle Based on Sulfide Alkylation and Sulfonium Salt Deprotonation

Recently, Tang, Wu and co-workers have reported the synthesis of vinylcyclopropanes using an alternative catalytic cycle for sulfur ylide-catalyzed cyclopropanation (see Scheme 10.22) [98]. Sulfonium salt **41a** or **41b** was deprotonated by Cs_2CO_3 to form an ylide which then reacted with chalcones **37** to form cyclopropanes and a sulfide. The sulfonium salt was regenerated *in situ* through reaction

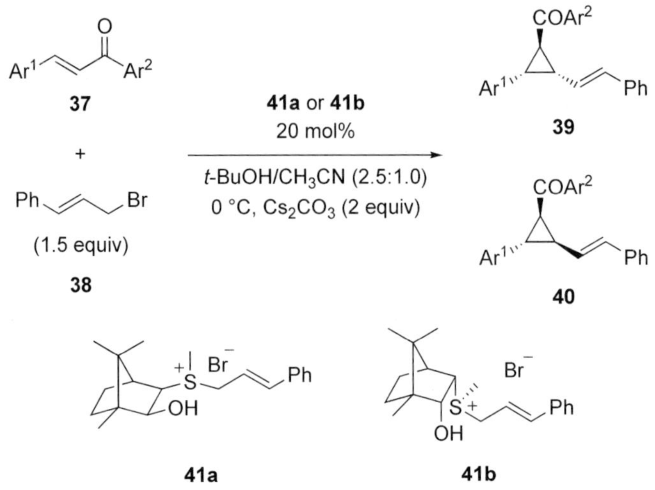

Scheme 10.22 Catalytic asymmetric cyclopropanation using sulfur ylides via an alkylation/deprotonation route.

with allylic bromide **38**. The cycle is analogous to those reported for epoxidation (see Scheme 10.3) and aziridination.

10.4.1.5 Scope and Limitations of the Alkylation/Deprotonation Method

A range of vinylcyclopropanes were synthesized in good to excellent yield (see Table 10.4). Good enantioselectivities (77–88% *ee*) and moderate diastereoselectiv-

Table 10.4 Catalytic asymmetric cyclopropanation with **41a** using sulfide alkylation/deprotonation route (according to Scheme 10.22).

Ar¹	Ar²	Time (h)	Yield (%)	dr[a]	ee (%)[b]
Ph	Ph	36	92	86:14	82
Ph	Ph	36	85	87:13	78[c]
p-Br-C$_6$H$_4$	Ph	30	89	75:25	77
p-Me-C$_6$H$_4$	Ph	54	80	86:14	81
p-MeO-C$_6$H$_4$	Ph	80	66 (90)[d]	86:14	80
o-Br-C$_6$H$_4$	Ph	20	90	77:13	88
o-MeO-C$_6$H$_4$	Ph	60	86 (92)[d]	67:33	78
Ph	p-Me-C$_6$H$_4$	57	87	87:13	81

[a] 39:40.
[b] *ee* of **39**.
[c] **41b** was used, producing the enantiomer of **39**.
[d] Conversion.

ities (3:1–7:1) were obtained at 0 °C in a mixture of tBuOH and CH$_3$CN in 20 to 80 h. Similar results were reported starting from the corresponding sulfides rather than sulfonium salts. Interestingly, switching from **41a** to **41b** resulted in the formation of the opposite enantiomer of the cyclopropane. Higher enantio- and diastereoselectivities were obtained using stoichiometric amounts of sulfonium salt with KOtBu in THF at −78 °C, but presumably these conditions were too slow for catalytic loadings. Only reactions with chalcones were successful under catalytic conditions, although α,β-unsaturated esters, amides and nitriles were suitable substrates under stoichiometric conditions.

10.4.1.6 Control of Selectivity in the Alkylation/Deprotonation Method

The side-arm alcohol plays a crucial role in controlling the reactivity and selectivity of the reaction. No reaction was observed if tetrahydrothiophene-derived sulfonium salts were used, or if the alcohol was protected as a methoxy group. The substrate is both activated and orientated by a hydrogen bond to its carbonyl group. No reaction intermediates were found in computational studies, so selectivity is determined in a single transition state. As before, controlling the ylide conformation and substrate approach determines the selectivity. The substrate approaches the ylide from the side bearing the alcohol in all cases, but **41a** and **41b** are predicted to present a different face of the ylide and so the opposite enantiomer is formed.

10.4.1.7 Use of S-Ylide-Catalyzed Cyclopropanations in Synthesis

The products of cyclopropanation have been further elaborated to cyclopropane-containing amino acids (**42**) and Boc-protected amino esters (**43**) (see Scheme 10.23) [79]. A single recrystallization either before or after deprotection afforded enantiomerically pure material. As the acrylate precursors are made in one step, this sequence provides an efficient route to this important class of conformationally locked amino acids.

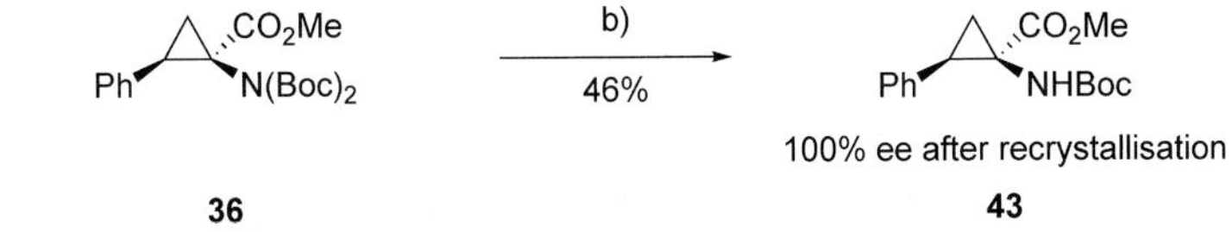

Scheme 10.23 Synthetic elaborations of cyclopropanes. Reagents and conditions: (a) 6 N HCl, reflux, 4 h; (b) CF$_3$COOH, Et$_3$SiH, CH$_2$Cl$_2$, 0 °C, 2 h.

10.4.2
Tellurium Ylide-Mediated Cyclopropanation

Tang and co-workers have reported the use of chiral allylic telluronium ylides derived from telluronium salts (e.g., **44**) in the cyclopropanation of α,β-unsaturated carbonyl compounds such as **29** to give vinylcyclopropanes **46**, **47** (see Scheme 10.24) [16, 99]. (For experimental details see Chapter 14.11.3). The allylic Te-ylides are more reactive than the corresponding S-ylides, and good yields were obtained using the tellurium analogue of Metzner's sulfide. The diastereoselectivity of the stoichiometric reaction could be switched by appropriate choice of reagents (LiTMP/HMPA or LDA/LiBr), and high enantioselectivities were achieved. A catalytic version of the reaction (20 mol%) was also demonstrated, although only two examples were given (see Scheme 10.24). Good yields, enantioselectivities and diastereoselectivities were obtained using α,β-unsaturated ketones **29a,f** and allylic bromide **45** in refluxing THF. Future reports will presumably provide more information on the scope and usefulness of this reaction.

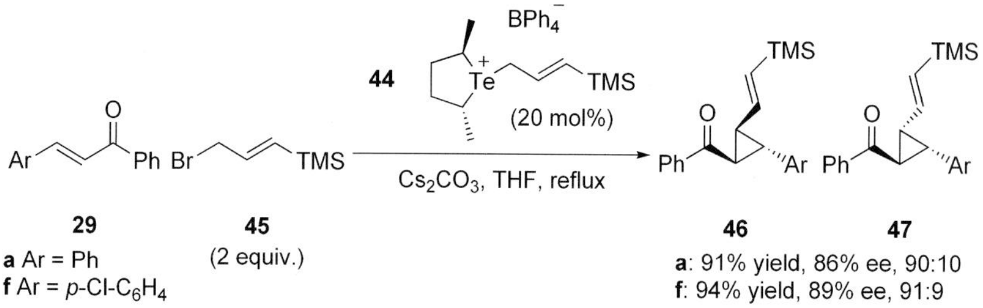

Scheme 10.24 Catalytic asymmetric cyclopropanation using telluronium salts.

10.4.3
Nitrogen Ylide-Mediated Cyclopropanation

10.4.3.1 Development of a Catalytic Method
Gaunt and co-workers have reported the cyclopropanation of α,β-unsaturated carbonyl compounds using nitrogen ylides generated *in situ* from tertiary amines [100–102]. Initial studies [102] indicated that the ammonium salt **49** could be deprotonated to form an ylide which reacted with acrylate **48** at 80 °C to give cyclopropane **52** in good yield and high *trans*-selectivity after a simple aqueous work-up (see Scheme 10.25). In addition, the salt could be generated *in situ* from DABCO **50** and phenacyl chloride **51**. In general, α,β-unsaturated alkenes yielded *trans*-substituted cyclopropanes in good to excellent isolated yields after chromatography. Catalytic turnover was demonstrated using 20 mol% of DABCO and 1.5–3 equivalents of electron-deficient alkene. An intramolecular catalytic version was later demonstrated yielding [n.1.0]bicycloalkanes in moderate to excellent yield, with excellent diastereoselectivity [101].

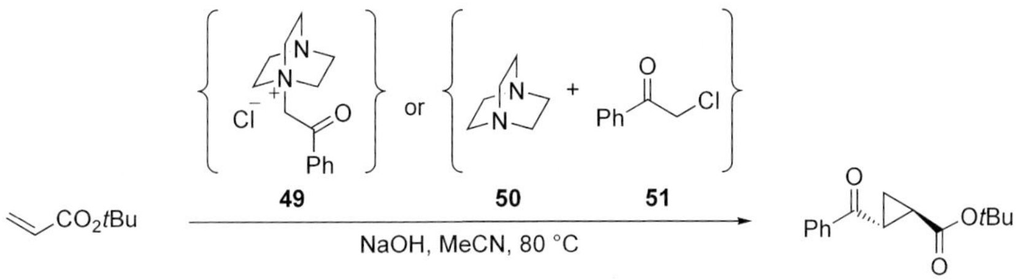

48 52

Scheme 10.25 Cyclopropanation of an α,β-unsaturated ester using a N-ylide.

10.4.3.2 **Enantioselective Cyclopropanation**

Enantioselective variants of these cyclopropanation reactions were achieved by replacing DABCO with cinchona alkaloid catalysts **53a–d** [100–102].

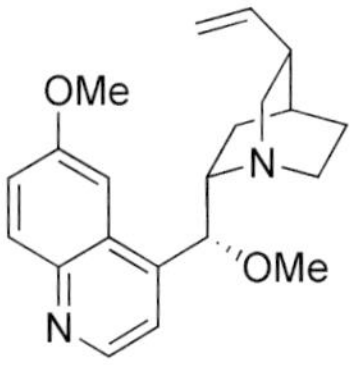

53a

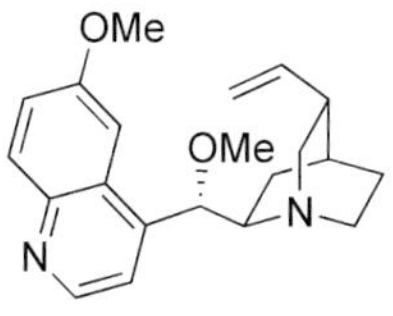

53b

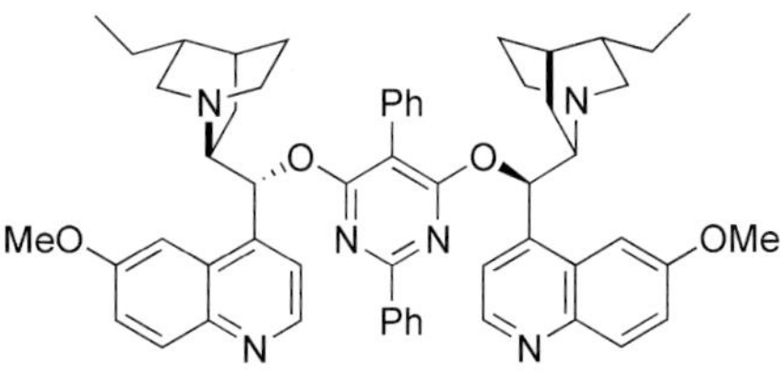

53c

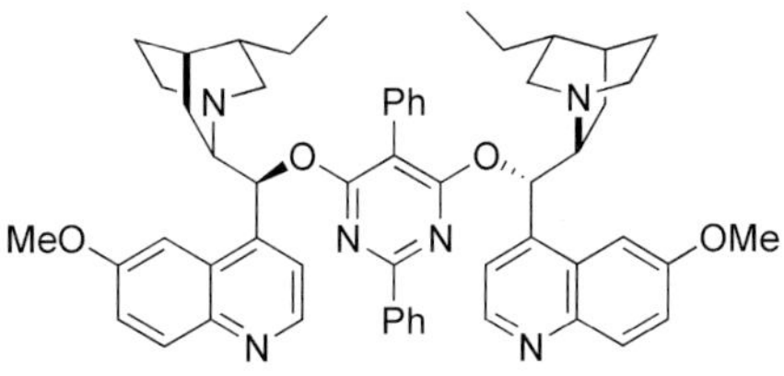

53d

Intramolecular cyclization of **54** catalyzed by **53a** produced (+)-55 in 94% *ee* and 61% yield (see Scheme 10.26) [101]. If **53b** was used instead, the opposite enantiomer was produced in 94% *ee* and 48% yield. The addition of 40 mol% NaBr or NaI improved reaction times from 4–5 days to 1 day, and was necessary to obtain high enantioselectivities.

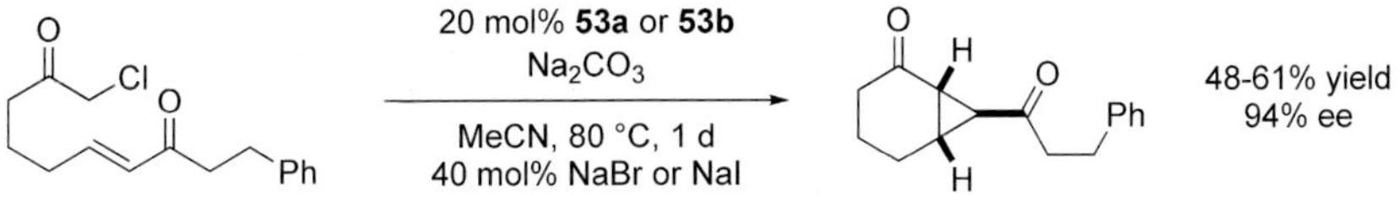

54 **55**

Scheme 10.26 N-ylide-catalyzed asymmetric intramolecular cyclopropanation.

In intermolecular cyclopropanations [100], it was found better to use α-bromoesters and amides as ylide precursors and α,β-unsaturated ketones and esters as electron-deficient alkenes – rather than using α-haloketones as the ylide precursor. (For experimental details see Chapter 14.11.4). The reaction gives access to a range of 1,2-dicarbonyl-substituted cyclopropanes (see Fig. 10.5). The alkene could have an aryl-, alkyl- or indole-substituted ketone, and α-substitution was also tolerated. Notably, Weinreb amides could be used as the ylide precursor and the product subsequently transformed into a diketocyclopropane. Both enan-

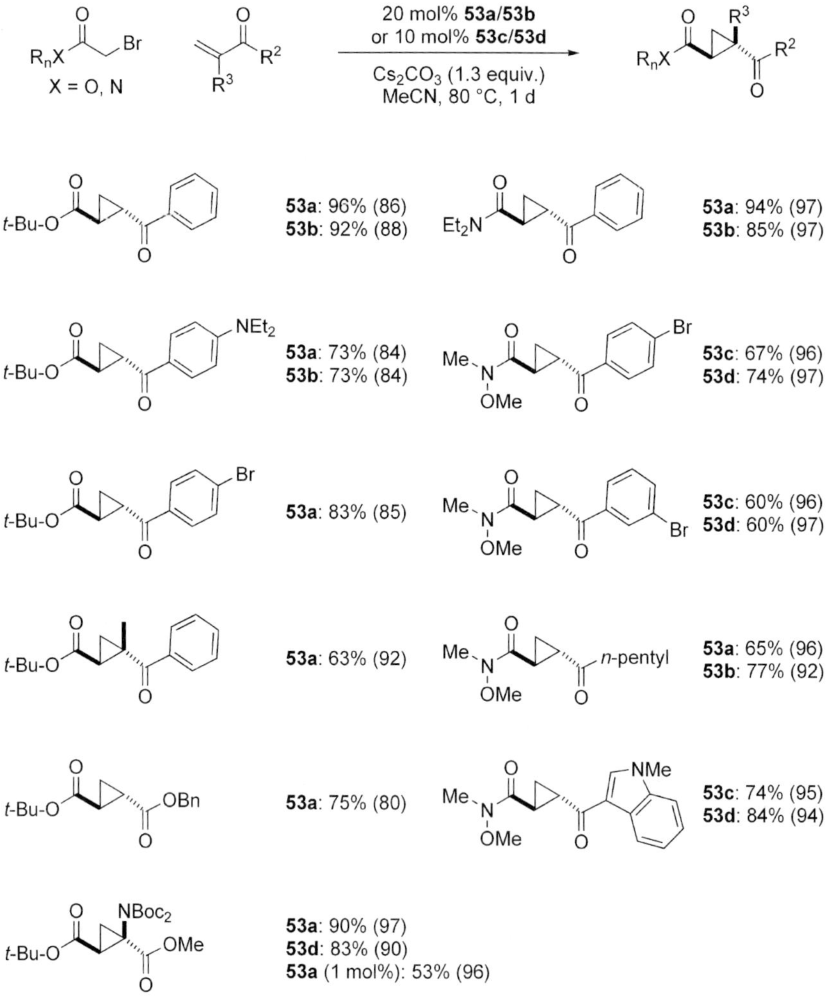

Fig. 10.5 Catalytic asymmetric cyclopropanation using N-ylides derived from **53a–d**. Catalyst, yield (% *ee*) are shown for each product. **53a,c** favored the (+)-enantiomer, while **53b,d** favored the (−)-enantiomer in all cases.

tiomers of the products could be accessed by appropriate choice of the pseudo-enantiomeric amines.

10.4.3.3 Mechanism of N-Ylide-Catalyzed Cyclopropanation

The proposed mechanism for the reaction is shown in Scheme 10.27. The amine salt is deprotonated to form the ylide, which undergoes conjugate addition followed by cyclization to produce the cyclopropane and regenerate the catalyst. Cs_2CO_3 was found to provide better yields than Na, K or Rb carbonates, although a slightly lower enantioselectivity was obtained. By using 10–20 mol% of tertiary amines **53a–d**, *trans*-substituted cyclopropanes were obtained in moderate to excellent yield with high to excellent enantioselectivity (see Fig. 10.5). As in the intramolecular reaction, both enantiomers of the cyclopropanes could be obtained in high enantioselectivity by using the pseudoenantiomeric amines. Slow addition of the α-bromo-carbonyl compound improved yields in some cases. The moderate catalyst loading is required for reasonable reaction times, with loadings of 1 mol% leading to slower reactions but without any change in selectivity being observed.

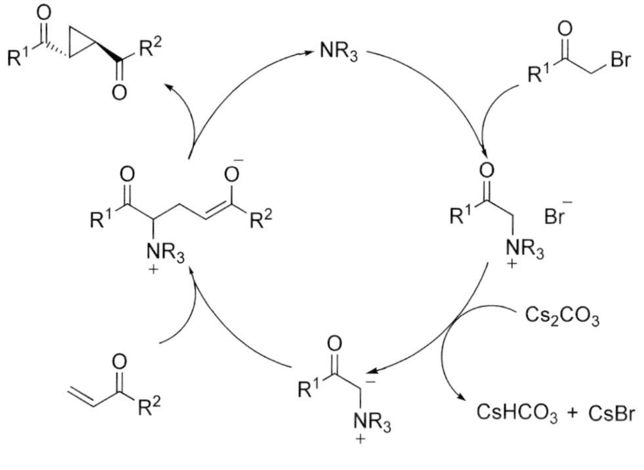

Scheme 10.27 Proposed mechanism for N-ylide-mediated catalytic cyclopropanation.

10.4.4 Summary of Ylide-Catalyzed Cyclopropanation

Sulfur, tellurium and nitrogen ylide-catalyzed cyclopropanations have been developed. Enantioenriched cyclopropanes can be synthesized in good yield and moderate to excellent diastereoselectivity. These catalysts are complementary to metal-based cyclopropanation catalysts, and have the potential for further development.

Sulfur-ylide-catalyzed cyclopropanation gives access to a range of cyclopropanes in good yield, moderate diastereoselectivity, and good to excellent enantioselectivity. Two complementary catalytic cycles have been used for ylide regeneration *in situ*: (1) reacting the sulfide with a metallocarbene; or (2) reacting the sulfide with an allylic bromide followed by deprotonation of the resulting sulfonium salt. The reported scope of the carbene methodology is wider, while the alkylation/deprotonation route gives access to vinylcyclopropanes, which are an important subclass of cyclopropanes.

More recently, nitrogen and tellurium ylide-catalyzed cyclopropanations have been developed with catalytic cycles based on ylide formation via alkylation and subsequent deprotonation. Both of these reactions give better diastereoselectivity than the corresponding sulfur-ylide reactions. The N-ylide reaction gives access to cyclopropanes in high yield, enantio- and diastereoselectivity. This N-ylide-catalyzed asymmetric cyclopropanation has three major advantages when compared to other ylide cyclopropanations: (1) the amine catalysts are commercially available; (2) the starting materials are readily available; and (3) the products have two carbonyl groups which allows a wide range of transformations to be used in further elaborations. Its usefulness in organic synthesis will, no doubt, be demonstrated further in the future.

10.5
Summary of Ylide-Catalyzed Asymmetric Reactions

Ylides of sulfur, nitrogen, tellurium and selenium have been used as highly enantioselective organocatalysts in epoxidation, aziridination, and cyclopropanation reactions. The sulfur ylide-mediated epoxidation is the best developed methodology and can be incorporated in synthetic planning with confidence. In contrast, ylide-mediated aziridination requires more development but is beginning to emerge as a useful methodology. Sulfur, nitrogen, and tellurium ylide-mediated cyclopropanations have advanced considerably in recent years, and the ammonium ylide methodology is particularly attractive because of the commercial availability of the catalysts and the potential of the products for further elaboration. Overall, ylide catalysts are useful components in the synthetic chemist's toolbox, and can be expected to feature in a growing number of total syntheses in the future.

References

1 R. Robiette, M. Conza, V.K. Aggarwal, *Org. Biomol. Chem.* **2006**, *4*, 621–623.
2 V.K. Aggarwal, J.N. Harvey, R. Robiette, *Angew. Chem. Int. Ed.* **2005**, *44*, 5468–5471.
3 A.K. Yudin, *Aziridines and Epoxides in Organic Synthesis*. Wiley-VCH, Weinheim, **2006**.
4 E.M. McGarrigle, D.G. Gilheany, *Chem. Rev.* **2005**, *105*, 1563–1602.
5 Q.-H. Xia, H.-Q. Ge, C.-P. Ye, Z.-M. Liu, K.-X. Su, *Chem. Rev.* **2005**, *105*, 1603–1662.
6 T. Katsuki, *Comprehensive Coordination Chemistry II* **2004**, *9*, 207–264.
7 Y. Shi, *Acc. Chem. Res.* **2004**, *37*, 488–496.
8 D. Yang, *Acc. Chem. Res.* **2004**, *37*, 497–505.

9 T. Katsuki, *Synlett* **2003**, 281–297.
10 E.N. Jacobsen, *Acc. Chem. Res.* **2000**, 33, 421–431.
11 R.A. Johnson, K.B. Sharpless, in: I. Ojima (Ed.), *Catalytic Asymmetric Synthesis*, 2nd edn. Wiley-VCH, New York, **2000**, pp. 231–280.
12 E.N. Jacobsen, M.H. Wu, in: E.N. Jacobsen, A. Pfaltz, H. Yamamoto (Eds.), *Comprehensive Asymmetric Catalysis I–III, Vol. 2*. Springer, New York, **1999**, pp. 649–677.
13 For a recent example involving a one-pot, two-step olefination/catalytic epoxidation procedure, see, S. Matsunaga, T. Kinoshita, S. Okada, S. Harada, M. Shibasaki, *J. Am. Chem. Soc.* **2004**, 126, 7559–7570.
14 V.K. Aggarwal, D.M. Badine, V.A. Moorthie, in: A.K. Yudin (Ed.), *Aziridines and Epoxides in Organic Synthesis*. Wiley-VCH, Weinheim, **2006**, pp. 1–35.
15 V. Blot, J.-F. Briere, M. Davoust, S. Miniere, V. Reboul, P. Metzner, *Phosphorus, Sulfur Silicon Relat. Elem.* **2005**, 180, 1171–1182.
16 Y. Tang, S. Ye, X.-L. Sun, *Synlett* **2005**, 2720–2730.
17 V.K. Aggarwal, C.L. Winn, *Acc. Chem. Res.* **2004**, 37, 611–620.
18 V. Aggarwal, J. Richardson, in: A. Padwa, D. Bellus (Eds.), *Science of Synthesis, Vol. 27*. George Thieme Verlag, Stuttgart, **2004**, pp. 21–104.
19 V.K. Aggarwal, in: E.N. Jacobsen, A. Pfaltz, H. Yamamoto (Eds.), *Comprehensive Asymmetric Catalysis I–III, Vol. 2*. Springer, New York, **1999**, pp. 679–693.
20 L.-X. Dai, X.-L. Hou, Y.-G. Zhou, *Pure Appl. Chem.* **1999**, 71, 369–376.
21 V.K. Aggarwal, *Synlett* **1998**, 329–336.
22 A.-H. Li, L.-X. Dai, V.K. Aggarwal, *Chem. Rev.* **1997**, 97, 2341–2372.
23 A.W. Johnson, R.B. LaCount, *Chem. & Ind. (London)* **1958**, 1440–1441.
24 E.J. Corey, M. Chaykovsky, *J. Am. Chem. Soc.* **1965**, 87, 1353–1364.
25 Y.G. Gololobov, V.P. Lysenko, I.E. Boldeskul, *Tetrahedron* **1987**, 43, 2609–2651.
26 V.K. Aggarwal, C.L. Winn, in: *Encyclopedia of Reagents for Organic Synthesis Online*. www.mrw.interscience.wiley.com/eros/, accessed 24 March **2006**. DOI:10.1002/047084289X.rd372, John Wiley & Sons Ltd., Chichester, **2006**.
27 C.R. Johnson, C.W. Schroeck, *J. Am. Chem. Soc.* **1968**, 90, 6852–6854.
28 C.R. Johnson, C.W. Schroeck, *J. Am. Chem. Soc.* **1973**, 95, 7418–7431.
29 T. Hiyama, T. Mishima, H. Sawada, H. Nozaki, *J. Am. Chem. Soc.* **1975**, 97, 1626–1627.
30 For leading references on stoichiometric asymmetric ylide epoxidations, see Refs. [14–22, 31, 32].
31 V.K. Aggarwal, I. Bae, H.-Y. Lee, J. Richardson, D.T. Williams, *Angew. Chem. Int. Ed.* **2003**, 42, 3274–3278.
32 A. Solladie-Cavallo, M. Roje, T. Isarno, V. Sunjic, V. Vinkovic, *Eur. J. Org. Chem.* **2000**, 1077–1080.
33 N. Furukawa, Y. Sugihara, H. Fujihara, *J. Org. Chem.* **1989**, 54, 4222–4224.
34 M. Davoust, J.-F. Briere, P.-A. Jaffres, P. Metzner, *J. Org. Chem.* **2005**, 70, 4166–4169.
35 J. Zanardi, C. Leriverend, D. Aubert, K. Julienne, P. Metzner, *J. Org. Chem.* **2001**, 66, 5620–5623.
36 C.L. Winn, B.R. Bellenie, J.M. Goodman, *Tetrahedron Lett.* **2002**, 43, 5427–5430.
37 K. Julienne, P. Metzner, V. Henryon, A. Greiner, *J. Org. Chem.* **1998**, 63, 4532–4534.
38 K. Julienne, P. Metzner, V. Henryon, *J. Chem. Soc., Perkin Trans. 1* **1999**, 731–736.
39 J. Zanardi, D. Lamazure, S. Miniere, V. Reboul, P. Metzner, *J. Org. Chem.* **2002**, 67, 9083–9086.
40 K. Li, X.-M. Deng, Y. Tang, *Chem. Commun.* **2003**, 2074–2075.
41 V.K. Aggarwal, H. Abdel-Rahman, R.V.H. Jones, H.Y. Lee, B.D. Reid, *J. Am. Chem. Soc.* **1994**, 116, 5973–5974.
42 V.K. Aggarwal, H. Abdel-Rahman, R.V.H. Jones, M.C.H. Standen, *Tetrahedron Lett.* **1995**, 36, 1731–1732.
43 V.K. Aggarwal, H. Abdel-Rahman, L. Fan, R.V.H. Jones, M.C.H. Standen, *Chem. Eur. J.* **1996**, 2, 1024–1030.

44 V.K. Aggarwal, A. Thompson, R.V.H. Jones, M. Standen, *Tetrahedron: Asymm.* **1995**, *6*, 2557–2564.

45 V.K. Aggarwal, J.G. Ford, A. Thompson, R.V.H. Jones, M.C.H. Standen, *J. Am. Chem. Soc.* **1996**, *118*, 7004–7005.

46 V.K. Aggarwal, E. Alonso, G. Hynd, K.M. Lydon, M.J. Palmer, M. Porcelloni, J.R. Studley, *Angew. Chem. Int. Ed.* **2001**, *40*, 1430–1433.

47 V.K. Aggarwal, E. Alonso, I. Bae, G. Hynd, K.M. Lydon, M.J. Palmer, M. Patel, M. Porcelloni, J. Richardson, R.A. Stenson, J.R. Studley, J.-L. Vasse, C.L. Winn, *J. Am. Chem. Soc.* **2003**, *125*, 10926–10940.

48 V.K. Aggarwal, C. Aragoncillo, C.L. Winn, *Synthesis* **2005**, 1378–1382.

49 V.K. Aggarwal, A. Ali, M.P. Coogan, *J. Org. Chem.* **1997**, *62*, 8628–8629.

50 V.K. Aggarwal, M.P. Coogan, R.A. Stenson, R.V.H. Jones, R. Fieldhouse, J. Blacker, *Eur. J. Org. Chem.* **2002**, 319–326.

51 B.R. Bellenie, J.M. Goodman, *Chem. Commun.* **2004**, 1076–1077.

52 Stoichiometric sulfur ylide epoxidation of paraformaldehyde gives terminal epoxides in good yield and selectivity, A. Solladie-Cavallo, A. Diep-Vohuule, *J. Org. Chem.* **1995**, *60*, 3494–3498.

53 V.K. Aggarwal, P. Blackburn, R. Fieldhouse, R.V.H. Jones, *Tetrahedron Lett.* **1998**, *39*, 8517–8520.

54 R. Imashiro, T. Yamanaka, M. Seki, *Tetrahedron: Asymm.* **1999**, *10*, 2845–2851.

55 V.K. Aggarwal, J.P.H. Charmant, D. Fuentes, J.N. Harvey, G. Hynd, D. Ohara, W. Picoul, R. Robiette, C. Smith, J.-L. Vasse, C.L. Winn, *J. Am. Chem. Soc.* **2006**, *128*, 2105–2114.

56 V.K. Aggarwal, G.Y. Fang, C.G. Kokotos, J. Richardson, M.G. Unthank, *Tetrahedron* **2006** (in press).

57 V.K. Aggarwal, S. Calamai, J.G. Ford, *J. Chem. Soc., Perkin Trans. 1* **1997**, 593–599.

58 V.K. Aggarwal, J.N. Harvey, J. Richardson, *J. Am. Chem. Soc.* **2002**, *124*, 5747–5756.

59 S. Miniere, V. Reboul, R.G. Arrayas, P. Metzner, J.C. Carretero, *Synthesis* **2003**, 2249–2254.

60 V.K. Aggarwal, J. Charmant, L. Dudin, M. Porcelloni, J. Richardson, *Proc. Natl. Acad. Sci. USA* **2004**, *101*, 5467–5471.

61 V.K. Aggarwal, J.G. Ford, S. Fonquerna, H. Adams, R.V.H. Jones, R. Fieldhouse, *J. Am. Chem. Soc.* **1998**, *120*, 8328–8339.

62 V.K. Aggarwal, S. Schade, B. Taylor, *J. Chem. Soc., Perkin Trans. 1* **1997**, 2811–2813.

63 M.A. Silva, B.R. Bellenie, J.M. Goodman, *Org. Lett.* **2004**, *6*, 2559–2562.

64 V.K. Aggarwal, J. Richardson, *Chem. Commun.* **2003**, 2644–2651.

65 V.K. Aggarwal, J. Bi, *Beilstein J. Org. Chem.* **2005**, *1*, 4.

66 M. Catasus, A. Moyano, V.K. Aggarwal, *Tetrahedron Lett.* **2002**, *43*, 3475–3479.

67 V.K. Aggarwal, I. Bae, H.-Y. Lee, *Tetrahedron* **2004**, *60*, 9725–9733.

68 J.-F. Briere, H. Takada, P. Metzner, *Phosphorus, Sulfur Silicon Relat. Elem.* **2005**, *180*, 965–968.

69 W.-H. Ou, Z.-Z. Huang, *Synthesis* **2005**, 2857–2860.

70 H. Takada, P. Metzner, C. Philouze, *Chem. Commun.* **2001**, 2350–2351.

71 W. McCoull, F.A. Davis, *Synthesis* **2000**, 1347–1365.

72 E.N. Jacobsen, in: E.N. Jacobsen, A. Pfaltz, H. Yamamoto (Eds.), *Comprehensive Asymmetric Catalysis I–III, Vol. 2.* Springer, New York, **1999**, pp. 607–618.

73 P. Mueller, C. Fruit, *Chem. Rev.* **2003**, *103*, 2905–2919.

74 K. Hada, T. Watanabe, T. Isobe, T. Ishikawa, *J. Am. Chem. Soc.* **2001**, *123*, 7705–7706.

75 V.K. Aggarwal, J.-L. Vasse, *Org. Lett.* **2003**, *5*, 3987–3990.

76 A.-H. Li, L.-X. Dai, X.-L. Hou, *J. Chem. Soc., Perkin Trans. 1* **1996**, 867–869.

77 V.K. Aggarwal, A. Thompson, R.V.H. Jones, M.C.H. Standen, *J. Org. Chem.* **1996**, *61*, 8368–8369.

78 V.K. Aggarwal, M. Ferrara, C.J. O'Brien, A. Thompson, R.V.H. Jones, R. Fieldhouse, *J. Chem. Soc., Perkin Trans. 1* **2001**, 1635–1643.

79 V.K. Aggarwal, E. Alonso, G. Fang, M. Ferrara, G. Hynd, M. Porcelloni, *Angew. Chem. Int. Ed.* **2001**, *40*, 1433–1436.

80 J.R. Fulton, V.K. Aggarwal, J. de Vicente, *Eur. J. Org. Chem.* **2005**, 1479–1492.

81 T. Saito, M. Sakairi, D. Akiba, *Tetrahedron Lett.* **2001**, *42*, 5451–5454.

82 V.K. Aggarwal, J.P.H. Charmant, C. Ciampi, J.M. Hornby, C.J. O'Brien, G. Hynd, R. Parsons, *J. Chem. Soc., Perkin Trans. 1* **2001**, 3159–3166.

83 R. Robiette, *J. Org. Chem.* **2006**, *71*, 2726–2734.

84 For example see, G. Li, H.-T. Chang, K.B. Sharpless, *Angew. Chem., Int. Ed. Engl.* **1996**, *35*, 451–454.

85 A. de Meijere, Guest Editor, *Chem. Rev.* **2003**, *103*, 931–1648.

86 H. Lebel, J.-F. Marcoux, C. Molinaro, A.B. Charette, *Chem. Rev.* **2003**, *103*, 977–1050.

87 A. Pfaltz, in: E.N. Jacobsen, A. Pfaltz, H. Yamamoto (Eds.), *Comprehensive Asymmetric Catalysis I–III, Vol. 2.* Springer, New York, **1999**, pp. 513–538.

88 K.M. Lydon, M.A. McKervey, in: E.N. Jacobsen, A. Pfaltz, H. Yamamoto (Eds.), *Comprehensive Asymmetric Catalysis I–III, Vol. 2.* Springer, New York, **1999**, pp. 539–580.

89 T. Aratani, in: E.N. Jacobsen, A. Pfaltz, H. Yamamoto (Eds.), *Comprehensive Asymmetric Catalysis I–III, Vol. 3.* Springer, New York, **1999**, pp. 1451–1460.

90 M.P. Doyle, in: I. Ojima (Ed.), *Catalytic Asymmetric Synthesis*, 2nd edn. Wiley-VCH, New York, **2000**, pp. 191–228, Chapter 5.

91 For a recent example of an organocatalytic cyclopropanation using a stoichiometric amount of ylide and an amine catalyst, see R.K. Kunz, D.W.C. MacMillan, *J. Am. Chem. Soc.* **2005**, *127*, 3240–3241.

92 A. Solladie-Cavallo, A. Diep-Vohuule, T. Isarno, *Angew. Chem. Int. Ed.* **1998**, *37*, 1689–1691.

93 S. Kojima, K. Fujitomo, Y. Shinohara, M. Shimizu, K. Ohkata, *Tetrahedron Lett.* **2000**, *41*, 9847–9851.

94 K. Huang, Z.-Z. Huang, *Synlett* **2005**, 1621–1623.

95 V.K. Aggarwal, H.W. Smith, R.V.H. Jones, R. Fieldhouse, *Chem. Commun.* **1997**, 1785–1786.

96 V.K. Aggarwal, H.W. Smith, G. Hynd, R.V.H. Jones, R. Fieldhouse, S.E. Spey, *Perkin 1* **2000**, 3267–3276.

97 For stoichiometric asymmetric cyclopropanations using ethyl diazoacetate-derived sulfur ylides and a unified model for cyclic substrates, see V.K. Aggarwal, E. Grange, *Chem. Eur. J.* **2006**, *12*, 568–575.

98 X.-M. Deng, P. Cai, S. Ye, X.-L. Sun, W.-W. Liao, K. Li, Y. Tang, Y.-D. Wu, L.-X. Dai, *J. Am. Chem. Soc.* **2006**, *128*, 9730–9740.

99 W.-W. Liao, K. Li, Y. Tang, *J. Am. Chem. Soc.* **2003**, *125*, 13030–13031.

100 C.D. Papageorgiou, M.A. Cubillo de Dios, S.V. Ley, M.J. Gaunt, *Angew. Chem. Int. Ed.* **2004**, *43*, 4641–4644.

101 N. Bremeyer, S.C. Smith, S. Ley, M.J. Gaunt, *Angew. Chem. Int. Ed.* **2004**, *43*, 2681–2684.

102 C.D. Papageorgiou, S.V. Ley, M.J. Gaunt, *Angew. Chem. Int. Ed.* **2003**, *42*, 828–831.

11
Organocatalytic Enantioselective Reduction of Olefins, Ketones, and Imines

Henri B. Kagan

11.1
Introduction

Although non-enzymatic asymmetric catalysis was first utilized during the early 1900s, unfortunately it afforded enantioselectivities that were much inferior to those achieved with enzymatic catalysis. For a review on the history of asymmetric catalysis, see Ref. [1]. The first reports of this approach involved catalysts derived from natural products, such as alkaloids. This typical organocatalysis approach was surpassed by the veritable "explosion" of enantioselective homogeneous organometallic catalysis during the early 1970s.

Subsequent major events, up until the early 1980s, have been reviewed [2], with one of the major reactions involved being that of asymmetric hydrogenation, which is especially useful and efficient. This was first developed using rhodium complexes equipped with chiral mono- or diphosphines [3–6], though many other types of reaction (e.g., hydroformylation, Diels–Alder reaction) are now well controlled in the presence of chiral organometallic catalysts. Over the past few years there has been a clear renewal of interest for organocatalysis [7], and consequently this chapter will review the specific and unusual case of the catalytic enantioselective reduction of C=C, C=O, and C=N double bonds.

11.2
The General Concept

The principles of the basic processes of reduction are indicated in Scheme 11.1. The fixation of 2 H on an organic substrate can formally involve several combinations of sequential events: hydride transfer followed by a protonation, two H additions, combinations of electron transfers, and protonation or H transfers. One may also envisage an indirect process, where first a X-Y or H-Y reagent (e.g., a hydrosilane) is added before the subsequent replacements of X and Y groups into H. Very few processes of Scheme 11.1 are known where a metal is *not*

Enantioselective Organocatalysis: Reactions and Experimental Procedures
Edited by Peter I. Dalko
Copyright © 2007 WILEY-VCH Verlag GmbH & Co. KGaA, Weinheim
ISBN: 978-3-527-31522-2

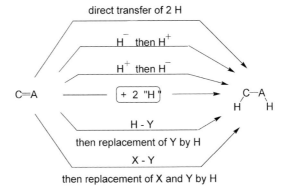

Scheme 11.1 Reduction of a C=A double bond, where A is C, O or N: some schematic processes are indicated (except electron-transfer reactions).

involved. Meerwein–Ponndorf–Verley reductions are hydrogen transfers from isopropanol to a ketone, which requires the presence of a sodium alcoholate or a ruthenium complex as catalyst. One reaction which does seem possible in absence of a metal is the hydride transfer from dihydropyridines to ketones. The hydrogenation of double bonds catalyzed by a base is unusual, although it has been described with sodium or potassium alcoholates at high temperature [8]. To date, there have been no reports of non-metallic catalysts for hydrogenation.

Another approach for the reduction of double bonds is to use hydride reagents H-Y, such as HBR_2 or $HSiR_3$. Once the adduct has been formed, the task is to replace Y by H. This is straightforward when Y is fixed to an oxygen or nitrogen, as acidic hydrolysis will cleave the O–Y or N–Y bond. This is illustrated in Scheme 11.2 for the reduction of ketones and conjugated ketones.

Scheme 11.2 The reduction of conjugated systems.

The ionic hydrogenation of alkenes is a process where a proton is transferred before a hydride, through the combined use of a strong acid (CF_3CO_2H) and a silane ($HSiR_3$) [9]. This reaction is different from a C=C hydrosilylation, which involves C–H and C–Si bond formations. The final transformation of C–Si into C–H requires an oxidation into C–OH (this is known to occur with retention of configuration), followed by the replacement of OH by H after some additional steps. Hydrogen transfer of dihydroaromatic compounds to C=C double bonds may occur, the driving force being the aromatization of the H donor. For example, cyclohexadiene acts as H donor for various substrates in the presence of metallic catalysts [10]. A final reducing system which might be of potential interest is based on dihydropyridines. In biological systems, dihydronicotinamide adenine dinucleotide (NADH) is a cofactor in enzyme-catalyzed hydride reductions. Biomimetic reductions are possible with structural analogues of NADH such as Hantzsch esters, and this opens a route to organocatalytic reductions (*vide infra*). Dihydroflavin adenine dinucleotide (FADH) is another organic cofactor in some enzyme-catalyzed reductions; analogues have been used with some asymmetric complexes, but these have not yet found application in organocatalysis. Diimide (HN=NH) is a good hydrogen donor to C=C with the generation of nitrogen, and can be prepared *in situ* by oxidative processes. A flavin-catalyzed generation of diimide has been recently discovered, allowing the smooth aerobic hydrogenation of olefins [11] whereby the terminal double bonds were preferentially reduced. The stereochemistry of the reaction has been investigated by deuterium labeling of hydrazine, establishing a *cis*-addition on the double bond of a norbornene derivative. The reaction was performed with NH_2NH_2, H_2O (1–2 eq.) and oxygen (1 atm) in the presence of 1 mol% catalyst (see below). A mechanism has been proposed which accounts for the production of 2 moles of reduced product per mole of oxygen. A chiral version of such a system has not been reported, however.

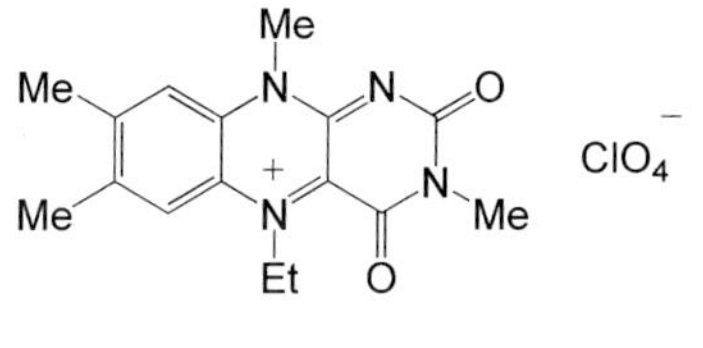

Scheme 11.3 A flavin-catalyst for the aerobic generation of diimide from hydrazine.

11.3
Asymmetric Organocatalytic Reduction of Olefins

This area has undergone very recent development, with List et al. first reporting the possibility of using ammonium salts as catalysts for the reduction of α,β-unsaturated aldehyde in 2004 [12]. These authors used a Hantzsch ester **1** (commercially available) as the hydride source, and preliminary screening showed that several ammonium salts were able to catalyze the reduction in an efficient manner. Some typical examples are indicated in Scheme 11.4, where salt **2** serves as the catalyst.

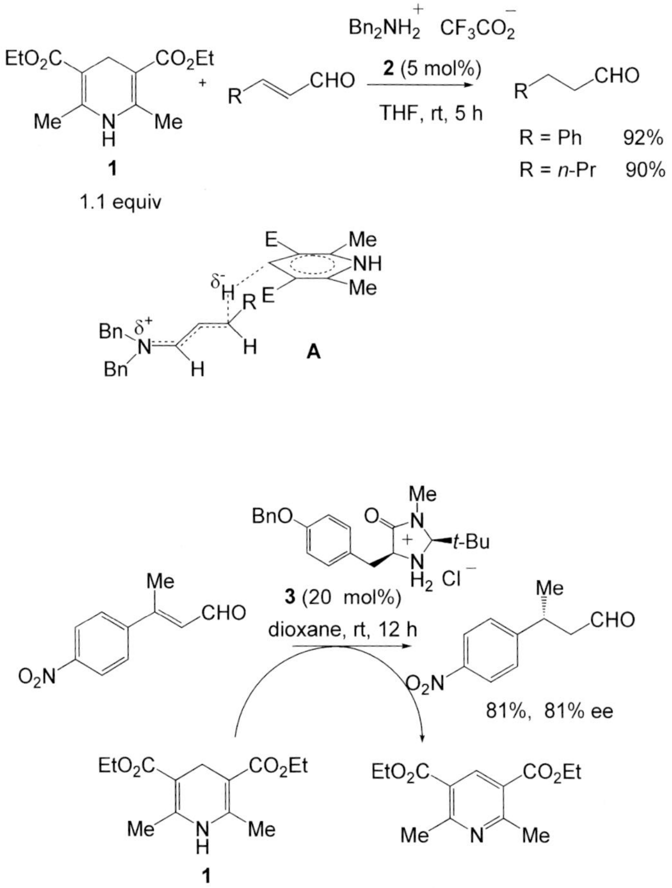

Scheme 11.4 Catalytic reduction of conjugated aldehydes with ethyl Hantzsch ester [12].

The reversible formation of a N,N-dibenzyl iminium intermediate, which is reduced by hydride capture from the Hantzsch ester **1** was proposed. Subsequent hydrolysis regenerates catalyst **2** and releases the saturated aldehyde. The transition state A has been suggested for the hydride transfer. An example of the asymmetric version of this reaction was also realized, by using a chiral imidazolidinone catalyst (the McMillan imidazolidinium salt **3** [13]) (see Scheme 11.4).

In 2005, the groups of List and McMillan simultaneously described excellent results in the asymmetric reduction of α,β-unsaturated aldehydes with a prochiral center in the β position [14, 15]. (For experimental details see Chapters 14.22.1 and 14.22.2). In both cases the catalyst used was a chiral imidazolidinone (**6** or **8**), and some representative examples are listed in Tables 11.1 and 11.2. The reactions were run at 10–20 mol% of catalyst, at moderate temperature (13 °C or 4 °C) over several hours. The hydride source (Hantzsch ester) was utilized in stoichiometric quantities, and the chemical yields and enantiomeric excesses proved to be

11.3 Asymmetric Organocatalytic Reduction of Olefins

Table 11.1 Organocatalytic reduction of conjugated aldehydes by Hantzsch ester **4** [14].

Ar	Yield (%)	ee (%)
Ph	77	90
p-CN-C$_6$H$_4$	89	96
p-NO$_2$-C$_6$H$_4$[a]	80	94
2-Napht	86	92

[a] Configuration: E or Z or E / Z = 1:1.

Table 11.2 Organocatalytic reduction of conjugated aldehydes by Hantzsch ester **1** [15].

7	T (°C)	Time (h)	Yield (%)	ee (%)
R^1 = Ph, R^2 = Me (E / Z > 20:1)	−30	23	91	93
R^1 = Cy, R^2 = Me (E / Z = 5:1)	−50	10	91	96
R^1 = Ph, R^2 = Et (E / Z > 20:1)	−30	16	74	94
R^1 = CO$_2$Me, R^2 = Me (E / Z > 20:1)[a]	−50	26	83	91
R^1 = t-Bu, R^2 = Me (E / Z > 20:1)	23	0.5	95	97

[a] 5 mol% catalyst at 23°C.

good. Indeed, in many examples enantioselectivities of >90% enantiomeric excess (*ee*) could be achieved. Interestingly, a mixture of *E/Z* enals provided the saturated aldehyde with high *ee*-values, a situation which was explained by a rapid $E \leftrightarrows Z$ equilibrium catalyzed by **6** or **8**, followed by asymmetric reduction of the *E* isomer [15]. McMillan et al. conducted a survey of reaction media and concluded that chloroform was superior to toluene in this respect [15].

An enamine intermediate has been proposed as being formed by hydride reduction of a transient iminium ion [14, 15]. The electrophilic capture of the enamine is possible by a Michael acceptor; thus, reductive Michael cyclizations of enal enones such as **9** or **11** were described in many cases (intramolecular reactions) (Scheme 11.5) [16].

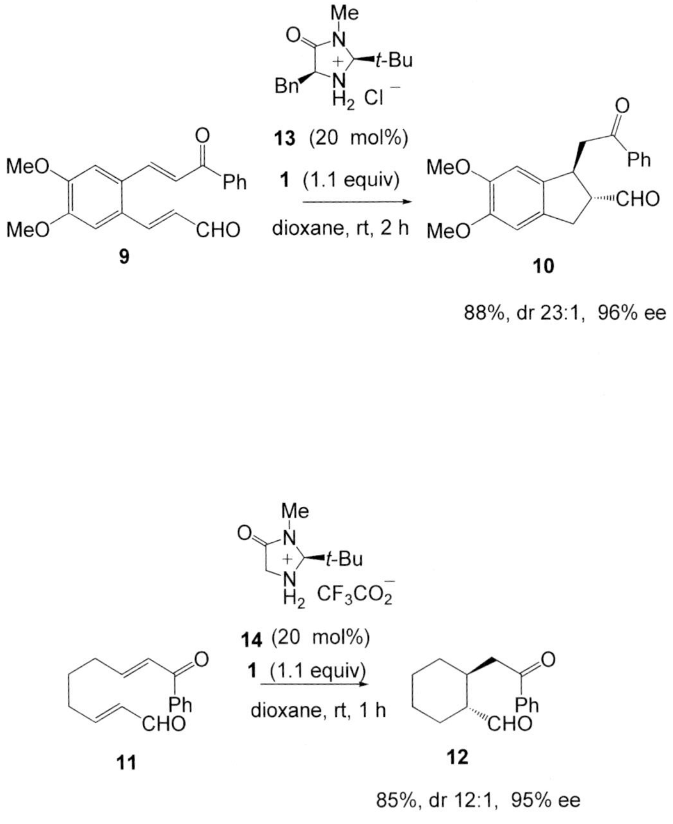

Scheme 11.5 Asymmetric reductive Michael cyclization [16].

The *anti*-diastereoselectivity in products **10** or **12** is very good (typically from 12:1 to 50:1, according to the substrate). This reaction is of synthetic use and provides clear confirmation of the enamine intermediate postulated in the reactions shown in Tables 11.1 and 11.2.

11.4
Reduction of Ketones or Ketimines

11.4.1
Hantzsch Ester/Chiral Brønsted Acids

In 2005, both Rueping et al. and List et al. reported the first transfer hydrogenation with Hantzsch ester **1** of several N-protected ketimines catalyzed by chiral Brønsted acids derived from 1,1′-binaphthol [17, 18]. The reaction typically requires 1 to 20 mol% of catalyst, is performed in benzene at 60 °C, and enantioselectivities of up to 90% are obtained. The chiral Brønsted acid protonates the ketimine at nitrogen, giving an ion-pair which is reduced by Hantzsch ester **1**. (For experimental details see Chapter 14.21.2). A preferred transition state has

Table 11.3 Transfer hydrogenation of ketimines catalyzed by chiral Brønsted acids.

15a Ar = 3,5-(CF_3)-Phenyl
15b Ar = 2,4,6-(i-Pr)-Phenyl

Catalyst 15	Ketimine 16	Amine 17		Ref.
		Yield (%)	ee (%)	
15a (20 mol%)	R^1 = 2-Napht, R = Ph	69	68	16
15a (20 mol%)	R^1 = 2-Napht, R = PMP	82	70	16
15a (20 mol%)	R^1 = Ph, R = Ph	71	72	16
15a (20 mol%)	R^1 = Ph, R = PMP	76	74	16
15a (20 mol%)	R^1 = 2-F-C_6H_4, R = PMP	82	84	16
15b (1 mol%)	R^1 = Ph, R = PMP	96	88	16
15b (1 mol%)	R^1 = 4-CN-C_6H_4, R = PMP	87	80	17
15b (1 mol%)	R^1 = 2-Napht, R = PMP	85	84	17
15b (1 mol%)	R^1 = i-Pr, R = PMP	80	90	17

For easier comparison, the configuration of catalyst **15** has been kept constant in this table, although the configuration of **15** (and product **17**) are opposite in Ref. [17].

PMP = p-methoxyphenyl.

been proposed by the authors. The key feature of the catalyst is the need for bulky substituents in the 3,3′ positions of the 1,1′-binaphthyl system. In Table 11.3, examples of the catalysts **15a** and **15b** have been selected, with results provided for the transfer hydrogenation of various ketimines.

To date, there have been no examples of asymmetric reductions of ketones by Hantzsch esters and a chiral organocatalyst.

11.4.2
Silanes/Phase-Transfer Catalysts

The asymmetric organocatalytic transformation of a ketone into an alcohol may be realized with the combination achiral silane:chiral phase-transfer catalyst, such a quaternary ammonium salt. The final alcohol is then recovered by an additional hydrolytic step. The asymmetric reduction of aryl alkyl ketones with silanes has been reported (*ee*-values up to 70%), the catalysts utilized being ammonium fluorides prepared from the quinine/quinidine series (e.g., **18** in Scheme 11.6) [19]. (For experimental details see Chapter 14.21.1). The more appropriated silanes were $(Me_3SiO)_3SiH$ or $(MeO)_3SiH$ (some examples are

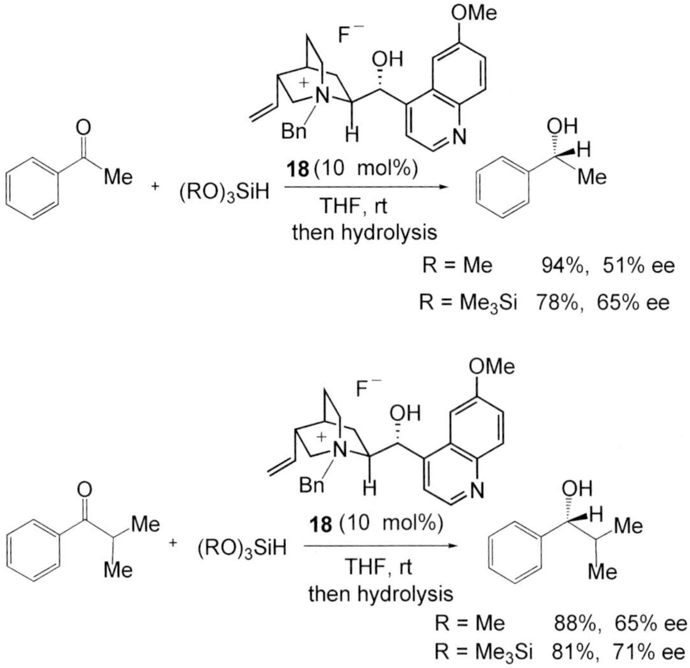

Scheme 11.6 Asymmetric reduction of ketones catalyzed by chiral quaternary ammonium salts [19].

shown in Scheme 11.6). Catalytic activation of the silane was achieved by a nucleophilic addition of fluoride on the silicon hydride, with the generation of a chiral ion pair. Polymethylhydroxysiloxane (PHMS) provided a very active system, but with inferior enantioselectivities. Electrochemical reduction with quaternary ammonium salts as the supporting electrolytes [20, 21] is not considered in this chapter.

11.4.3
Silanes/Formamide Catalysts

The nucleophilic activation of hydrosilanes as $HSi(OR)_3$ offers an opportunity to transfer one hydride on the carbon of ketones or imines [22]. The enantioselective organocatalytic hydrosilylation of ketones was first reported in 1999 by Matsumura et al. [23], the catalyst employed being a proline derivative **19** (Scheme 11.7). Amide **20** was also able to catalyze the hydrosilylation of ketimines, as indicated in Scheme 11.7 [24]. Improved results were recently reported by Kocovsky and Malkov [25], who prepared from valine some acyclic analogues of prolina-

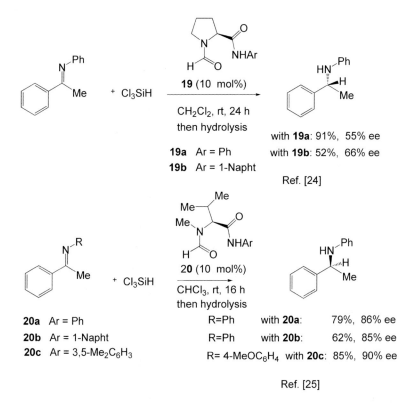

Scheme 11.7 Asymmetric hydrosilylation of imines catalyzed by chiral formamides.

mide **19**. These gave increased enantioselectivities, but of the opposite configuration to that obtained with the proline-derived catalyst. A careful optimization of the parameters of the reaction (solvent, substituents on the aromatic ring of the catalyst) allowed *ee*-values to reach 90% (Scheme 11.7).

A transition state B (Scheme 11.8) has been proposed where the catalyst coordinates to silicon as a bidentate ligand through carbonyls and where H-bonding and π ... π interactions are considered. A recent detailed report provided examples of the modifications of the structure of catalysts **19** and **20**, the product configuration being controlled by the nature of the amino acid side chain. The replacement of *i*-Pr in **19** by Me leads to a reversal of configuration for the product **20**, but this is obtained with a modest *ee*-value [26]. (For experimental details see Chapter 14.21.3).

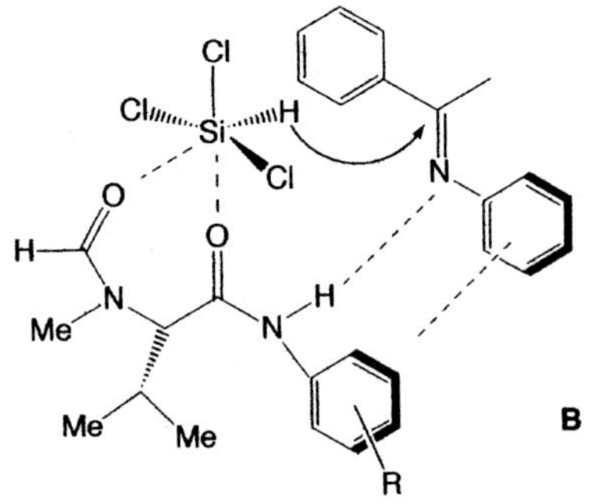

Scheme 11.8 Preferred transition state proposed for reaction IV-1 [25, 26].

11.5
Conclusions

The organocatalytic enantioselective reduction of C=C, C=O, and C=N double bonds is a relatively young area for which many new and exciting developments can be expected in the near future. Hantzsch esters are useful organic hydrides, and a recent review has summarized the results obtained to date in organocatalysis [27]. The case of silicon hydrides is convenient for imine or ketone reductions, as a chiral base can act as an organic catalyst. The asymmetric reductions of ketones catalyzed by oxazaborolidines and pioneered by Itsuno [28] and Corey [29] could not be included in this chapter.

Although the turnover frequency (TOF) and turnover number (TON) of the various above-described organocatalysts remain modest when compared to metal catalysis, the chiral organocatalysts are of relatively simple structure, and this permits easy tuning of the reaction. Clearly, one can predict many improvements in this area in the future, and undoubtedly new organocatalytic reactions relating to reduction will be introduced, using some of the basic processes described in Section 11.1.

Acknowledgments

The author thanks the Université Paris-Sud and CNRS for financial support, and the Department of Organic Chemistry of the Indian Institute of Science (Bangalore) which hosted him in 2006 as Visiting Professor, and provided the facilities to write this chapter.

References

1 H.B. Kagan, in: E.N. Jacobsen, A. Pfaltz, H. Yamamoto (Eds.), *Comprehensive Asymmetric Catalysis*. Springer-Verlag, Berlin, **1999**, Vol. 1, pp. 101–118.
2 H.B. Kagan, in: G. Wilkinson (Ed.), *Comprehensive Organometallic Chemistry*. Pergamon Press, **1982**, Vol. 8, pp. 483–493.
3 W.S. Knowles, M.S. Sabacky, *J. Chem. Soc. Chem. Commun.* **1968**, 1445.
4 L. Horner, H. Siegel, H. Büthe, *Angew. Chem. Int. Ed.* **1968**, 7, 942.
5 T.P. Dang, H.B. Kagan, *Chem. Commun.* **1971**, 481–482.
6 A. Miashita, A. Yasuda, H. Takaya, K. Torikumi, T. Ito, T. Souchi, R. Noyori, *J. Am. Chem. Soc.* **1980**, 102, 7932–7933.
7 P. Dalko, L. Moisan, *Angew. Chem. Int. Ed.* **2001**, 40, 3726–3748.
8 A. Berkessel, T.J.J. Schubert, T.N. Müller, *J. Am. Chem. Soc.* **2002**, 124, 8693–8698.
9 D.N. Kursanov, Z.N. Parnes, N.M. Loim, *Synthesis* **1974**, 633–651.
10 G. Brieger, T.J. Nestrick, *Chem. Rev.* **1974**, 74, 567–580.
11 Y. Imada, H. Iida, T. Nata, *J. Am. Chem. Soc.* **2005**, 127, 14544–14545.
12 J.W. Yang, M.T. Hechavarria Fonseca, B. List, *Angew. Chem. Int. Ed.* **2004**, 43, 6660–6662.
13 K.A. Ahrendt, C.J. Borths, D.W.C. McMillan, *J. Am. Chem. Soc.* **2000**, 122, 4243–4244.
14 J.W. Yang, M.T. Hechavarria Fonseca, B. List, *Angew. Chem. Int. Ed.* **2005**, 44, 108–110.
15 S.G. Ouellet, J.B. Tuttle, D.W.C. MacMillan, *J. Am. Chem. Soc.* **2005**, 127, 32–33.
16 J.W. Yang, M.T. Hechavarria Fonseca, B. List, *J. Am. Chem. Soc.* **2005**, 127, 15036–15037.
17 M. Rueping, E. Sugimo, C. Azap, T. Theismann, M. Bolte, *Organic Lett.* **2005**, 7, 3781–3783.
18 S. Hoffmann, A.M. Seagal, B. List, *Angew. Chem. Int. Ed.* **2005**, 44, 7424–7427.
19 M.D. Drew, N.J. Lawrence, W. Watson, S.A. Bowles, *Tetrahedron Lett.* **1997**, 38, 5857–5860.
20 L. Horner, W. Brich, *Liebigs Ann. Chem.* **1978**, 710–716.
21 A. Tallec, *Bull. Soc. Chim. Fr.* **1985**, 743–761.
22 S.E. Denmark, J. Fu, *Chem. Rev.* **2003**, 103, 2763–2794.
23 F. Iwasaki, O. Onomura, K. Mishima, T. Maki, Y. Matsumura, *Tetrahedron Lett.* **1999**, 40, 7507–7510.
24 F. Iwasaki, O. Onomura, K. Mishima, T. Kanematsu, T. Maki, Y. Matsumura, *Tetrahedron Lett.* **2001**, 42, 2525–2528.
25 A.V. Malkov, A. Mariani, K.N. MacDougall, P. Kocovsky, *Organic Lett.* **2004**, 6, 2253–2256.
26 A.V. Malkov, S. Stoncius, K.N. MacDougall, A. Mariani, G.D. MacGeod, P. Kocovsky, *Tetrahedron* **2006**, 6, 264–2256.
27 H. Adolfson, *Angew. Chem. Int. Ed.* **2005**, 44, 3340–3342.
28 S. Itsuno, in: E.N. Jacobsen, A. Pfaltz, H. Yamamoto (Eds.), *Comprehensive Asymmetric Catalysis*. Springer-Verlag, Berlin, **1999**, Vol. 1, pp. 289–315.
29 E.J. Corey, R.K. Bakshi, S. Shibata, *J. Am. Chem. Soc.* **1987**, 109, 7926–7926.

12
Oxidation Reactions

Alan Armstrong

Oxidation reactions – notably alkene epoxidations – were some of the first asymmetric organocatalytic processes to develop into generally useful synthetic methods applicable to a range of substrates [1]. This chapter surveys these reactions, with emphasis placed on the most practical and general. Some recent, very useful oxidation reactions involving α-oxidation of carbonyl compounds are covered elsewhere (see Chapter 2).

12.1
Epoxidation of Alkenes [2]

The discovery of the Sharpless Ti/tartrate-catalyzed asymmetric epoxidation was a major landmark in the field of asymmetric synthesis [3]. Although an exceptionally valuable process, it is largely limited to allylic alcohols as substrates. For the more challenging epoxidation of "unsubstituted" alkenes – that is, those lacking co-ordinating groups in the substrate – the Jacobsen/Katsuki Mn-salen complexes represented another major breakthrough [4]. Although these have again proved to be highly useful, they often give poor results for *trans*-disubstituted and terminal alkenes. Organocatalytic alkene epoxidation processes now offer highly complementary alternatives. The most productive processes discovered so far have involved the activation of the triple salt Oxone® ($2KHSO_5 \cdot KHSO_4 \cdot K_2SO_4$) of which the active constituent is $KHSO_5$ [5]. Three major classes of organocatalyst have been found for Oxone-mediated epoxidation, which will be considered in turn: ketones; iminium salts; and secondary amines or ammonium salts [6]. The former two classes are believed to operate in mechanistically similar ways (Scheme 12.1): interaction of the Oxone with the ketone or iminium salt precursor produces the reactive three-membered heterocyclic intermediate, the dioxirane **1** or oxaziridinium salt **2**, respectively. These species transfer oxygen to alkenes in a concerted (though possibly asynchronous) manner, resulting in stereospecific epoxidation (i.e., retention of the geometry of the starting alkene), which is a major synthetic advantage. Importantly, the starting ketone or iminium salt is regenerated, and so

Enantioselective Organocatalysis: Reactions and Experimental Procedures
Edited by Peter I. Dalko
Copyright © 2007 WILEY-VCH Verlag GmbH & Co. KGaA, Weinheim
ISBN: 978-3-527-31522-2

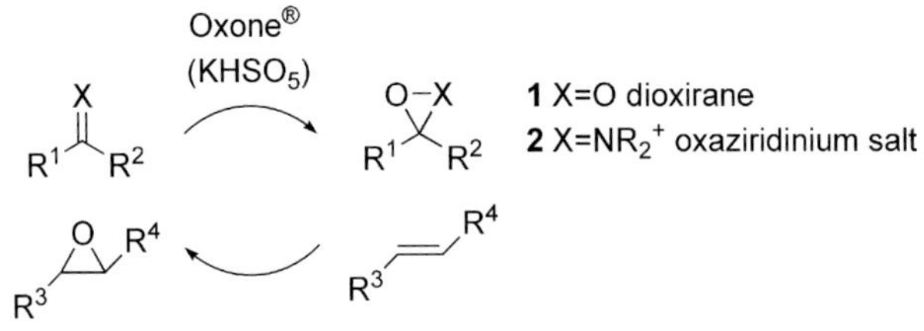

Scheme 12.1 Ketone- and iminium salt-catalyzed alkene epoxidation.

in principle can be used in substoichiometric quantities. Our understanding of the factors influencing the reactivity of the ketone or iminium catalyst, as well as the factors allowing efficient asymmetric induction and hence catalytic enantioselective epoxidation, have increased greatly over the past decade.

12.1.1
Ketone-Mediated Epoxidation

Kinetic and labeling studies conducted during the 1970s established that interaction of ketones with Oxone led to the formation of a dioxirane intermediate **1**, and the discovery that dimethyl dioxirane could be distilled from acetone/Oxone solutions provided a remarkably versatile new oxidant for organic synthesis [7]. However, the potential for ketone catalysis was recognized at an early stage and offered even more exciting possibilities. In a pioneering early attempt at asymmetric catalysis, Curci showed that chiral ketones, including (+)-isopinocamphone **3** (Fig. 12.1), would effect asymmetric epoxidation in a two-phase (organic solvent/water) system [8]. However, reaction rates and enantioselectivities were low (<15% ee).

Mechanistic studies of this two-phase reaction system showed that the solubility properties of both ketone and alkene, as well as the reaction pH, were important factors [9]. In this context, the report by Yang in 1995 of a monophasic

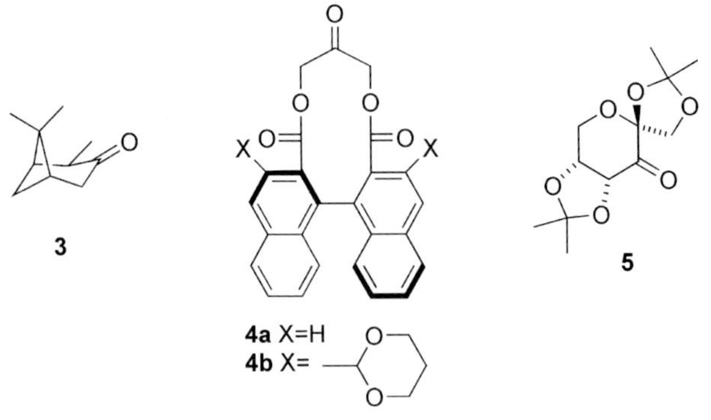

Fig. 12.1 Chiral ketone catalysts.

(CH₃CN/H₂O), self-buffering (pH 7–7.5) oxidation system proved the key to allowing the screening of a wider range of novel ketone catalysts [10]. Substituents α- to the ketone were found to have considerable influence on the reaction rate: electron-withdrawing groups (e.g., F, Cl, OAc) are beneficial, whilst steric hindrance reduces catalyst efficiency. For improvement of asymmetric induction, Yang's strategy was to examine C_2-symmetrical ketones, in which the approach on the two oxygens of the resulting dioxirane is stereochemically equivalent. This concept led to design of the new ketone family **4**, prepared from 1,1′-binaphthyl-2,2′-dicarboxylic acid and dihydroxyacetone, which gave markedly improved enantiomeric excesses (*ee*-values) [11, 12] compared to the earlier studies conducted by Curci [8]. While the parent ketone **4a** gave moderate enantioselectivities for epoxidation (47% *ee* for epoxidation of *E*-stilbene), tailoring of the substrate (by addition of para-substituents on the aromatic rings) or the catalyst (by introduction of "steric sensors", for example as in **4b**) afforded higher *ee*-values (up to 87%) (Scheme 12.2). However, *cis*- and terminal alkenes gave low enantioselectivities.

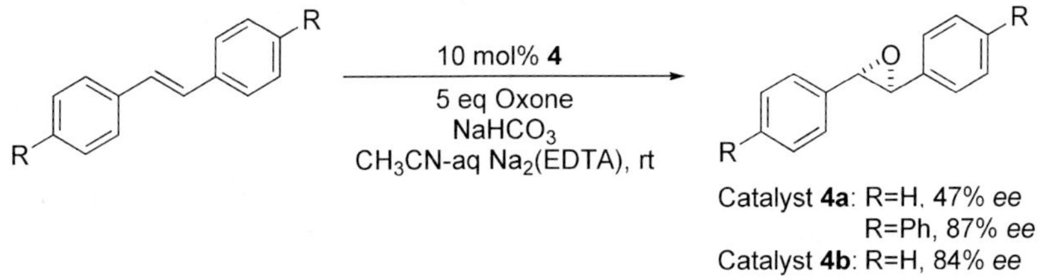

Scheme 12.2 Asymmetric epoxidation catalyzed by C_2-symmetrical chiral ketones **4**. (For experimental details see Chapter 14.13.4).

In 1996, Shi reported what is still the most generally useful chiral ketone catalyst for asymmetric epoxidation: ketone **5** [13–15]. Although now commercially available, this ketone may be obtained in two steps from D-fructose [16]; the other enantiomer may be prepared in five steps from L-sorbose. Initial studies using this catalyst employed the above Yang reaction conditions and, while high enantioselectivities were observed, a large excess (ca. 3 eq.) of the ketone was required [13]. This was due to catalyst decomposition by the Baeyer–Villiger reaction. Shi reasoned that this process could be avoided if the reaction were carried out at higher pH [17], which would favor formation of the anion **7** leading to the formation of dioxirane **8** (Scheme 12.3).

The optimum pH proved to be ca. 10.5, which could be maintained by adding either KOH or K_2CO_3 as the reaction proceeded. These modified conditions allowed substoichiometric (20–30 mol%) quantities of the ketone to be employed (Scheme 12.4). (For experimental details see Chapter 14.13.5).

A 1:2 mixture of CH₃CN and dimethoxymethane (DMM) as organic solvent often gives better results than CH₃CN alone [14]. This catalytic system provides

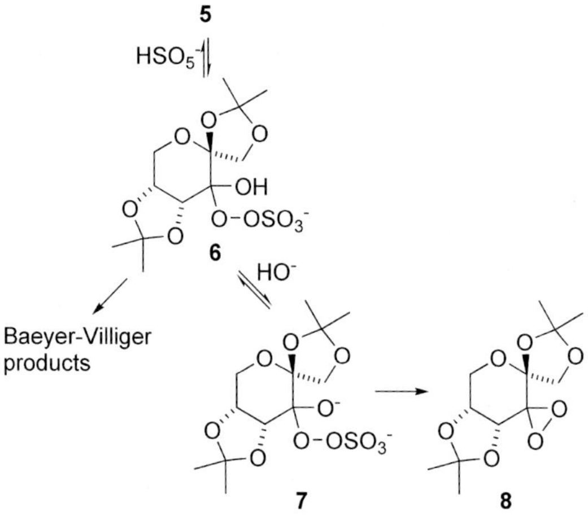

Scheme 12.3 Effect of pH on the Baeyer–Villiger reaction of ketone **5**.

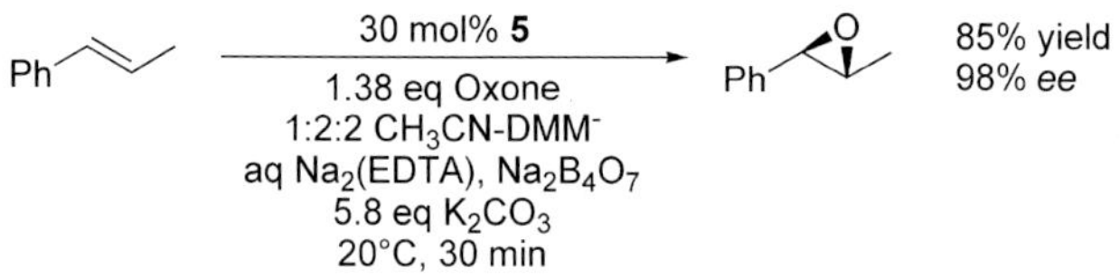

Scheme 12.4 Shi epoxidation with ketone **5**.

excellent enantioselectivities for trisubstituted and *trans*-1,2-disubstituted alkenes: representative examples are shown in Figure 12.2. The reaction is stereospecific (i.e., the alkene geometry is maintained in the product epoxide). Conjugated dienes [18] provide unsaturated monoepoxides, while epoxidation of conjugated

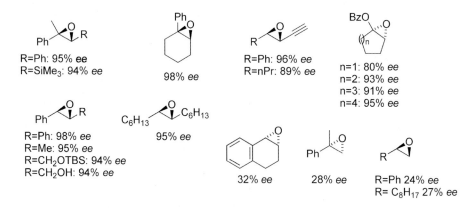

Fig. 12.2 The scope of Shi epoxidation with ketone **5**.

enynes [19] results in chemoselective epoxidation of the alkene. Kinetic resolution and desymmetrization of cyclic trisubstituted alkenes can also be accomplished [20]. Although the epoxidation of silyl enol ethers provides moderate enantioselectivities, enol benzoates also work well: the resulting epoxides can be isolated and converted to α-acyloxycarbonyl compounds with either retention (protic acid conditions) or with inversion when Lewis acids are used, allowing synthesis of either enantiomer of α-acyloxycarbonyl compound from the same epoxide precursor (Scheme 12.5) [21].

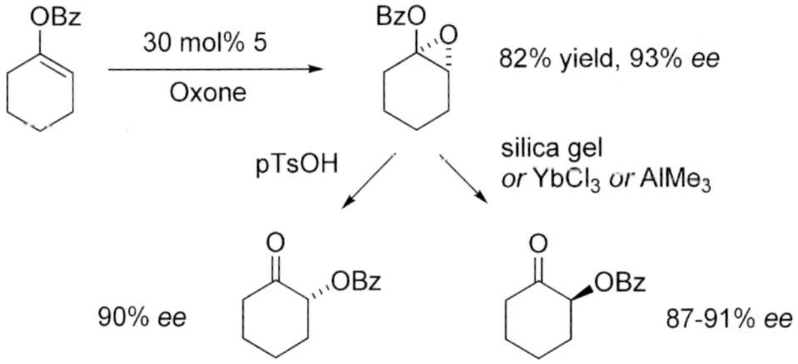

Scheme 12.5 Epoxidation and rearrangement of enol benzoates. (For experimental details see Chapter 14.13.5).

The Shi epoxidation has found several applications in total synthesis [15]. Particularly attractive are examples in which it has been used to establish the stereochemistry of polyepoxides which can undergo cascade cyclizations to polyether products, mimicking possible biosynthetic pathways. An example is the construction of the tetahydrofuran rings of the natural product glabrescol via highly stereoselective formation of the tetraepoxide **10** from the polyene **9** (Scheme 12.6) [22].

The high salt waste associated with Oxone means that its replacement with alternative co-oxidants would be attractive. Hydrogen peroxide has the desirable features of high oxygen content and a benign byproduct (water), and this can be

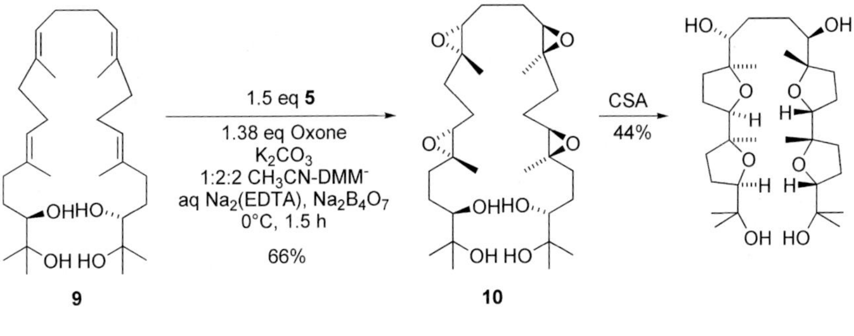

Scheme 12.6 Shi epoxidation in the total synthesis of glabrescol (right).

used in place of Oxone, provided that the reaction is performed in the presence of a nitrile such as CH_3CN or EtCN [23, 24]. (For experimental details see Chapter 14.13.6). A perimidic acid (from addition of hydrogen peroxide to the nitrile) is believed to be formed, which can then convert the ketone to the active dioxirane intermediate. The reaction pH is a crucial parameter for both conversion and *ee*-value, with pH 11 being optimal. Whilst the procedure is attractive, these authors mentioned that it is somewhat sensitive to the reactivity and solubility of olefins, and that optimization (catalyst loading, time, temperature, amount of H_2O_2 and solvent) may be required for different substrates [25].

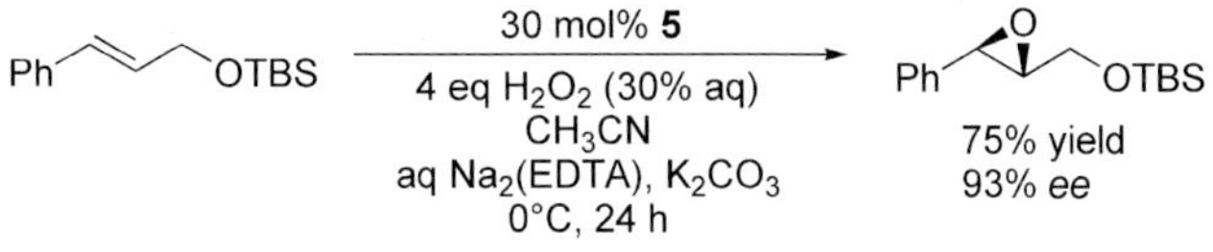

Ph〜〜OTBS → 30 mol% **5**, 4 eq H_2O_2 (30% aq), CH_3CN, aq Na_2(EDTA), K_2CO_3, 0°C, 24 h → Ph–epoxide–OTBS, 75% yield, 93% ee

Scheme 12.7 Shi epoxidation using hydrogen peroxide as co-oxidant. (For experimental details see Chapter 14.13.6).

One highly attractive feature of ketone-catalyzed epoxidation via chiral dioxiranes is that reliable models can be developed to rationalize the observed enantioselectivities. For the reaction of a dioxirane with an alkene, two extreme transition states can be envisaged: the so-called spiro and planar modes (Fig. 12.3).

Both computational studies and the large body of experimental evidence on observed enantioselectivities support a preference for the spiro mode (ca. 7.4 kcal according to B3LYP/6-31G* calculations) [26]. The spiro mode is preferred as this allows a stabilizing orbital interaction between the alkene π^*-orbital and a lone pair on the dioxirane oxygen that is being transferred to the alkene. This interaction is not possible in the planar mode. In the dioxirane derived from the Shi ketone **5**, the major epoxide enantiomer is then explained by the TS model of Figure 12.4, with attack on the less-hindered equatorial dioxirane oxygen in a spiro-

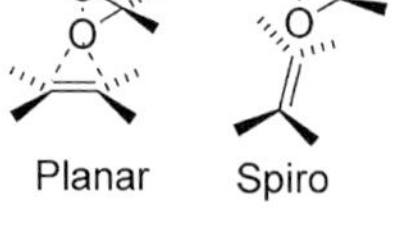

Planar   Spiro

Fig. 12.3 Spiro and planar transition states.

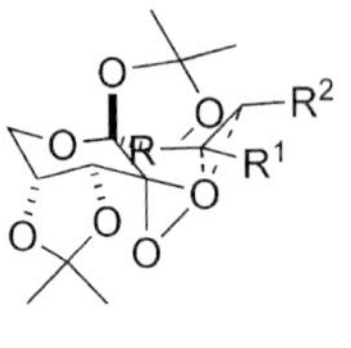

Fig. 12.4 Model for enantioselectivity with dioxirane from ketone **5**.

mode, with an alkene hydrogen substituent in the region of the axial oxygen of the spirocyclic acetonide motif. Recently, kinetic isotope effect studies along with calculations have supported the essential features of this model, whilst identifying that the transition state may be significantly asynchronous for certain alkenes [27].

Detailed studies of a range of alkenes and modified catalysts have led Shi to suggest that the minor enantiomer in the epoxidation of trisubstituted alkenes arises from a competing planar approach [14, 15]. The model also explains why *cis*- and terminal alkenes (R = H) give poor results – either enantiotopic face of the alkene can approach this dioxirane oxygen with a hydrogen placed in the spirocyclic acetonide region (R^1 and R^2 interchanged in Fig. 12.4). Shi reasoned that modification of the spirocyclic substituent may lead to the development of catalysts capable of differentiating between R^1 and R^2. This idea led to the discovery of the novel catalyst **11**, which can be made in six steps from D-glucose [28]. This catalyst gives good results for *cis*-alkenes, and promising results for terminal ones [29]. The major enantiomer can be predicted by the TS in Figure 12.5, where the oxazolinone spirocycle is suggested to have an attractive interaction with the π-substituent on the alkene [30]. Good enantioselectivities are obtained with *cis*-disubstituted alkenes, making this chemistry complementary to Jacobsen–Katsuki Mn(salen) catalysis. Very promising results have also been obtained for styrenes, with 1,1-disubstituted alkenes remaining a challenge (Fig. 12.5).

The Shi catalysts described above are the most practical to date in terms of high enantioselectivities and (particularly for **5**) ease of catalyst synthesis. However, it should be noted that as well as extensive studies by Shi himself [15], several other groups have made major contributions to the design and understanding of new ketone catalysts [31]. Examples are given in Figure 12.6, which illustrates just some of the diverse catalyst structures that have been evaluated. Bicyclooctanone **12b** affords 93% *ee* for the epoxidation of *E*-stilbene [32]; fluoroketone **13** gives up to 89% *ee* [33], while the arabinose-derived ketone **14** affords a very encouraging

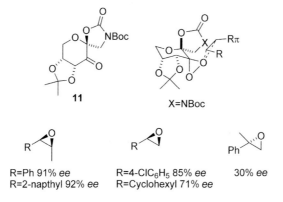

Fig. 12.5 Epoxidation of *cis*- and terminal alkenes with ketone **11**.

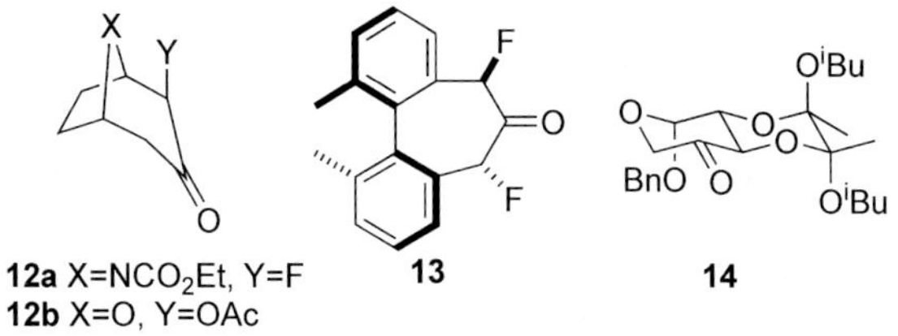

12a X=NCO$_2$Et, Y=F
12b X=O, Y=OAc

13

14

Fig. 12.6 Selected chiral ketone catalysts.

93% *ee* for a *trans*-aliphatic alkene [34, 35]. Some of these ketones show improved stability towards the Baeyer–Villiger reaction and can be used at lower catalyst loadings under the experimentally more convenient Yang pH 7–7.5 reaction conditions. For example, the stability of fluorotropinone **12a** [36] allowed the development of the first example of a supported, recyclable, chiral ketone catalyst [37].

12.1.2
Iminium Salt-Catalyzed Epoxidation

Closely related to the ketone/Oxone epoxidation system is the use of iminium salts as promoters. As isolated oxaziridinium salts are known to effect alkene epoxidation [38], these are presumed to be the reactive intermediates in this catalytic system (see Scheme 12.1; X = NR$_2^+$). The first asymmetric example used the dihydroisoquinolinium-based system **15** (Fig. 12.7), which afforded 33% *ee* for the epoxidation of *E*-stilbene [39].

In 1996, Aggarwal reported a chiral binaphthyl iminium **16** which gave 31% *ee* for the epoxidation of *E*-stilbene and 71% *ee* for the epoxidation of 1-phenylcyclohexene [40]. Several other classes of chiral iminium salt have since

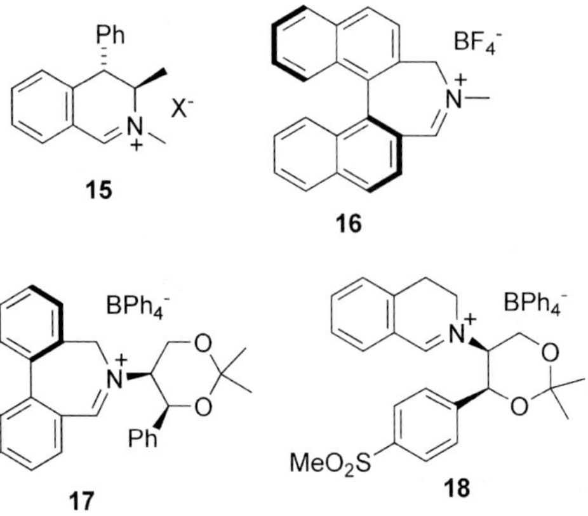

Fig. 12.7 Chiral iminium salt catalysts.

been tested [41], and the *in-situ* preparation of iminium salts from amines and aldehydes has been described which potentially allows easier catalyst structural variation [42]. However, enantioselectivities have until recently generally remained moderate, and only in 2004 did Page report the first high levels of enantiocontrol. The biaryl **17**, bearing exocyclic chirality in addition to Aggarwal's chiral biaryl motif, affords up to 95% *ee* for the epoxidation of cyclic aromatic alkenes (Scheme 12.8) [43]. Page has also developed non-aqueous reaction conditions employing tetraphenylphosphonium monoperoxysulfate which allow epoxidations to be performed at lower temperatures [44]. These conditions, along with catalyst **18**, resulted in the highly asymmetric epoxidation of alkene **19**, allowing enantioselective synthesis of the antihypertensive agent levcromakalim **21** (Scheme 12.9) [45].

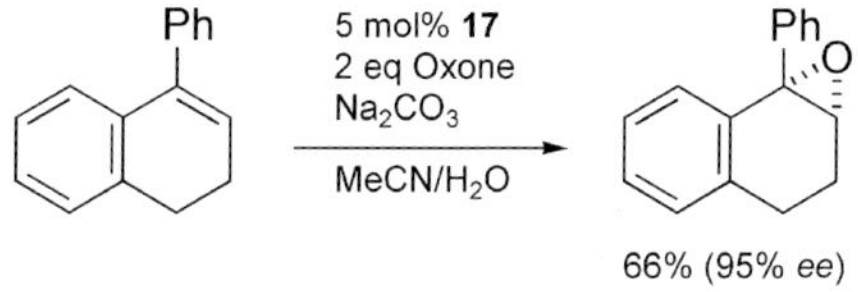

Scheme 12.8 Asymmetric epoxidation with iminium salt **17**.

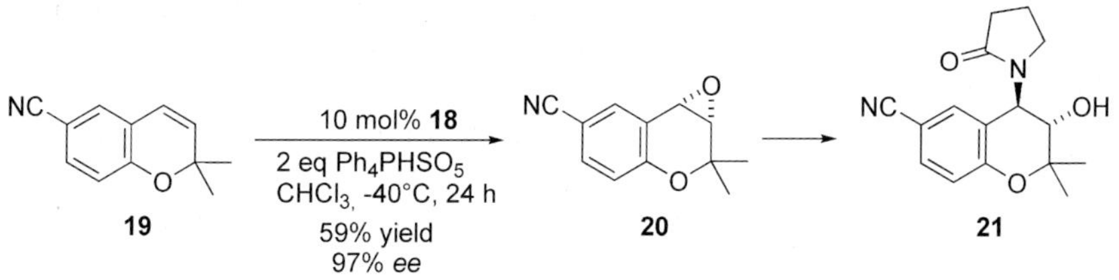

Scheme 12.9 Synthesis of levcromakalim **21** via iminium salt-catalyzed epoxidation. (For experimental details see Chapter 14.13.7).

Calculations [46] and studies of intramolecular oxaziridinium epoxidations [47] suggest that, like their dioxirane counterparts, these epoxidation processes proceed via spiro-transition states. However, the iminium epoxidations are generally more substrate-specific than those using dioxiranes, and models to explain the observed trends in stereocontrol have proved more difficult to construct. One complication is the possibility of formation of diastereomeric oxaziridinium salts from most of the iminium catalysts. Houk has rationalized computationally the observed enantioselectivity with Aggarwal's catalyst **16** [46]. The results of a recent study by Breslow suggest that hydrophobic interactions are important in these processes [48], and aromatic–aromatic interactions between catalyst and substrate may also play a role.

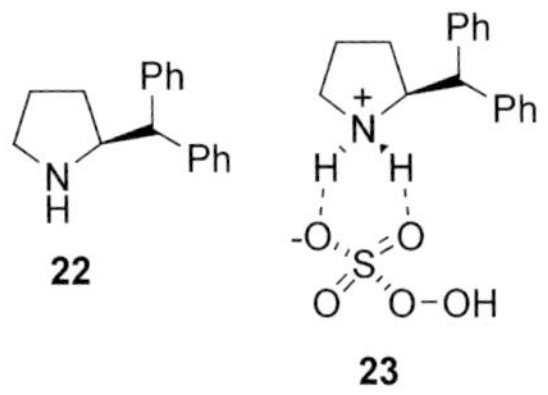

Fig. 12.8 Ammonium-activation of Oxone.

12.1.3
Amine-Catalyzed Epoxidation

During attempts to prepare iminium salts *in situ* from amines and carbonyl compounds, Aggarwal made the remarkable discovery that certain amines were capable of activating Oxone towards alkene epoxidation [49, 50]. Chiral amines such as **22** (Fig. 12.8) effected enantioselective epoxidation, albeit with moderate *ee*-values. After extensive optimization studies [50], it was found that the monohydrochloride salts of secondary amines gave reproducible results, and that the presence of pyridine was beneficial to obviate epoxide hydrolysis (Scheme 12.10). Aggarwal proposed a novel mode of activation of Oxone, involving hydrogen bonding to the ammonium salt (e.g., **23**, Fig. 12.8). The fact that this interaction can occur in several ways with differing orientations of the ammonium salt and HSO_5^- is likely to be one reason for the modest epoxidation enantioselectivities (66% *ee* for the best substrate and amine combination, 1-phenylcyclohexene with the naphthyl analogue of **22**·HCl). In general, trisubstituted alkenes give good conversion, but terminal alkenes (apart from styrene) do not. Disubstituted alkenes give inconsistent results, with *E*-stilbene being unreactive due to its low solubility. Yang has reported a further detailed study on amine-catalyzed Oxone epoxidation, using fluorinated amines and employing slightly acidic reaction conditions which mean that amines can be used directly, rather than as their hydrochloride salts [51]. These studies support a dual role of the amine as Oxone activator and phase-transfer catalyst, and *ee*-values of up to 61% were obtained.

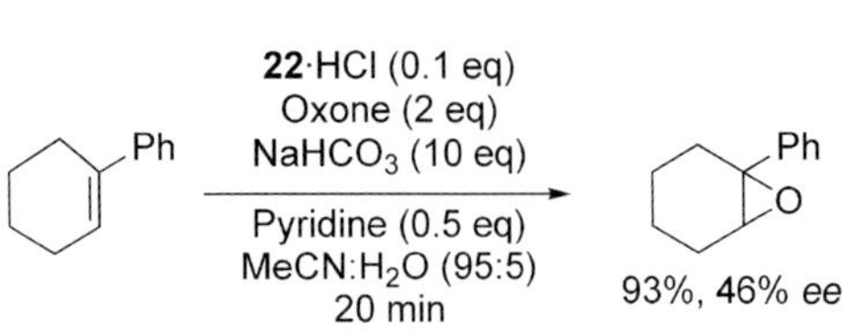

Scheme 12.10 Ammonium-catalyzed asymmetric alkene epoxidation.

12.2
Epoxidation of Enoates, Enones, and Enals [52]

12.2.1
Epoxidation of Enoates

The dioxirane and oxaziridinium reagents described in the previous section are electrophilic oxidants, and thus react relatively slowly with electron-poor alkenes (e.g., enoates and enones). There are therefore relatively few examples of ketone-catalyzed epoxidations of these substrate types. Often, the slow reaction times allow decomposition of the catalyst or of the Oxone co-oxidant to compete. For example, the Shi ketone **5** is not effective as it undergoes decomposition by Baeyer–Villiger reaction. A more reactive catalyst is required. Some fluoroketones epoxidize E-cinnamate esters with moderate reactivity and selectivity [36, 53], and catalysts related to **14** (see Fig. 12.6) have given up to 81% ee for E-cinnamate epoxidation [35]. Better enantioselectivities have been obtained by Shi, whose strategy was to increase catalyst reactivity by replacing the acetonide unit in **5** with more electron-withdrawing ester groups. This led to the discovery of bisacetate catalyst **24** (Fig. 12.9) [54], which can be prepared in two steps (deketalization, acetylation) from **5** [55]. Ketone **24** gives very good enantioselectivities for trisubstituted α,β-unsaturated esters and for *trans*-cinnamate esters (Scheme 12.11) [54], and its hydrate is equally effective [55]. However, **24** does not provide good ee-values for *trans*-aliphatic α,β-unsaturated esters. *cis*-Cinnamates also give low ee-values, in line with the trends seen with **5**. Imashiro and Seki have pointed

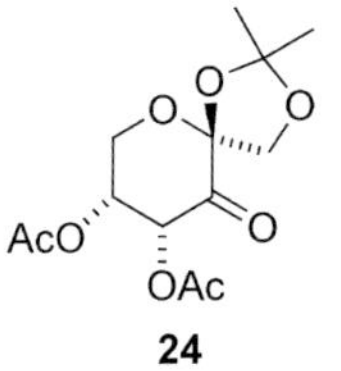

Fig. 12.9 Ketone **24** for epoxidation of unsaturated esters.

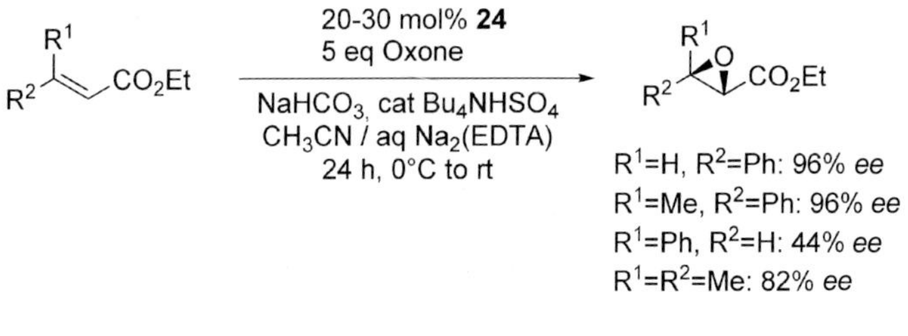

$R^1=H, R^2=Ph$: 96% ee
$R^1=Me, R^2=Ph$: 96% ee
$R^1=Ph, R^2=H$: 44% ee
$R^1=R^2=Me$: 82% ee

Scheme 12.11 Epoxidation of enoates catalyzed by ketone **24**.

out that the catalyst is used in relatively high loading (20–30 mol%) and that 5 equivalents of Oxone are required. They therefore optimized a system using the Yang catalyst **4a** for the epoxidation of *E*-methyl cinnamates, and obtained good yields using 5 mol% of **4** and 2 equivalents of Oxone [56]. Moreover, *ee*-values of up to 92% were obtained when the cinnamate possessed bulky groups in the 4-position of the aryl ring, and 74% *ee* was observed for *E*-methyl cinnamate itself. Cinnamic acid derivatives have been epoxidized with up to 95% *ee* by using Oxone promoted by dihydrocholic acid derivatives, but these chiral ketones are used in stoichiometric quantities [57].

12.2.2
Epoxidation of Enones

Other electron-poor alkenes generally require nucleophilic epoxidation conditions. These reactions usually proceed via non-concerted pathways (nucleophilic addition followed by epoxide ring closure), and so do not have the advantage of retaining the alkene geometry. Nevertheless, for the *trans*-epoxide, which is usually the predominant product, several methods exist that afford excellent levels of enantioselectivity.

12.2.2.1 Phase-Transfer Catalysis
Pioneering studies by Wynberg during the 1970s established that cinchona alkaloid-derived chiral quaternary ammonium salts could be used to effect asymmetric phase-transfer-catalyzed nucleophilic epoxidation of enones [58]. Diastereomeric catalysts derived from quinidine and quinuclidine produce opposite epoxide enantiomers. Apart from some examples of epoxidation of benzoquinone monoketals [59] and isoflavones [60], enantioselectivities with these early catalysts were generally low. However, contributions from several groups during the late 1990s led to dramatic improvements. Lygo reported enone epoxidation using the *N*-anthracenyl catalyst **25** (Fig. 12.10), along with NaOCl as oxidant (Scheme 12.12) [61, 62]. Typically, *ee*-values in the range 83–89% were obtained; a range of

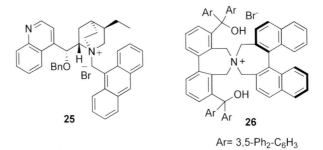

Fig. 12.10 Chiral phase-transfer catalysts for enone epoxidation.

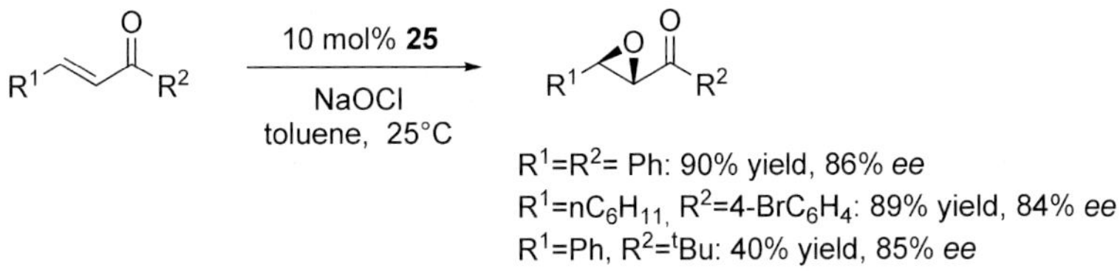

Scheme 12.12 Phase-transfer-catalyzed asymmetric enone epoxidation.

R^1=R^2= Ph: 90% yield, 86% ee
R^1=nC$_6$H$_{11}$, R^2=4-BrC$_6$H$_4$: 89% yield, 84% ee
R^1=Ph, R^2=^{t}Bu: 40% yield, 85% ee

substituents R^1 are tolerated, but substrates with R^2 = alkyl usually give slower rates. Corey independently developed the same catalyst and reported a similar epoxidation system using KOCl at −40 °C, giving slightly improved *ee*-values [63]. Corey also proposed a model to explain the observed product enantiomer in terms of binding of the enone and the hypochlorite to the chiral ammonium catalyst. Arai described the epoxidation of chalcone using hydrogen peroxide and catalysts similar to **25** in which the *N*-anthracenyl substituent is replaced by a 4-iodobenzyl group but, intriguingly, reported formation of the opposite epoxide enantiomer [64]. Lygo has recently reported optimized reaction conditions for chalcone epoxidations that allow lower catalyst loadings [65], and has also coupled the epoxidation step to the *in-situ* oxidation of allylic alcohols to enones (Scheme 12.13) [66].

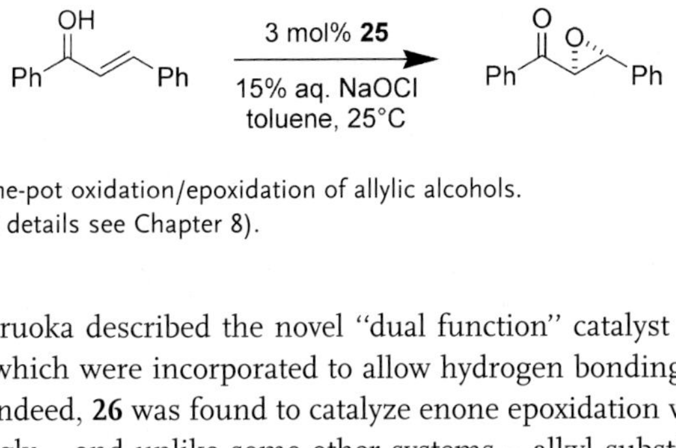

Scheme 12.13 One-pot oxidation/epoxidation of allylic alcohols. (For experimental details see Chapter 8).

Recently, Maruoka described the novel "dual function" catalyst **26** bearing hydroxyl groups which were incorporated to allow hydrogen bonding to the enolate intermediate. Indeed, **26** was found to catalyze enone epoxidation with 89–99% *ee* [67]. Interestingly – and unlike some other systems – alkyl substitution is tolerated (Scheme 12.14).

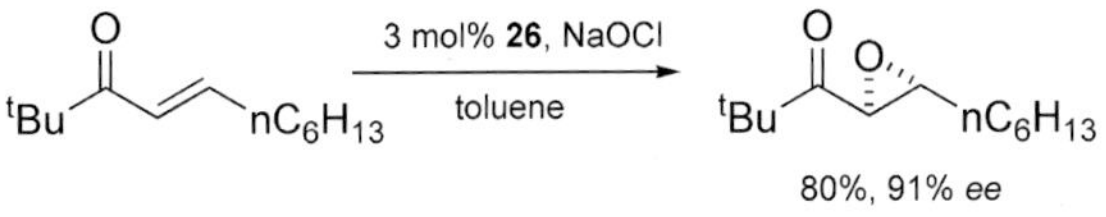

80%, 91% ee

Scheme 12.14 Enone epoxidation with "dual function" phase-transfer catalyst **26**.

12.2.2.2 Polyleucine-Mediated Epoxidation

Another fascinating system for enone epoxidation uses polyleucine as catalyst. This intriguing chemistry was first reported in 1980 by Julia and Colonna [68], who effected asymmetric epoxidation of chalcone in a triphasic system using an aqueous solution of H_2O_2 and NaOH, a solution of chalcone in an organic solvent, and the insoluble polyleucine catalyst. This system can give good enantioselectivities, but the reaction times are often long, and enolizable substrates in particular can be problematic. Extensive studies by Roberts and co-workers have resulted in several improvements to the chemistry [69]. This includes biphasic reaction systems (e.g., the use of DBU in THF along with urea-H_2O_2 as oxygen source) which can give shorter reaction times and improved scope [70]. Most of the enones used in this chemistry have aryl, heteroaryl or alkenyl substituents in the $β$-position, but alkyl substituents can be tolerated in the $α'$-position. $α$-Substituted enones are generally unreactive, with the exception of some examples where the alkene is exocyclic to a cyclic ketone, which is suggested to be due to the need for an s-cis enone conformation for reaction to occur. Most of the enantioselectivities obtained are excellent (>90% ee). The synthetic utility of the products has been demonstrated by application to target synthesis; the preparation of the blood pressure-lowering agent diltiazem **30** (Scheme 12.15) [71] illustrates a common tactic in that epoxidation of enone **27** is followed by Baeyer–Villiger reaction to give the desired epoxy ester **29**.

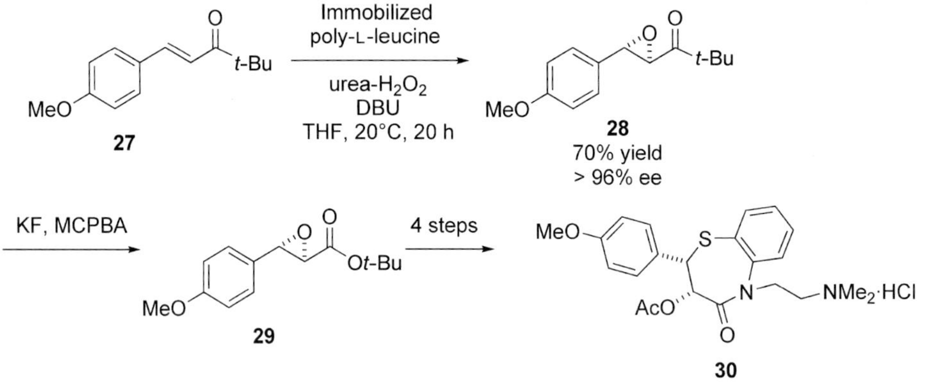

Scheme 12.15 Polyleucine-mediated epoxidation in the synthesis of diltiazem **30**. This procedure utilizes immobilized (polystyrene) poly-L-leucine, which is more easily separated from the products. (For experimental details see Chapter 14.13.9).

Although they can extend the substrate scope, these modified systems use more expensive bases and oxidants than the original conditions. In 2004, the triphasic system saw renewed interest when Geller and co-workers found that the addition of a phase-transfer catalyst allows much faster reactions without the induction period otherwise needed for catalyst activation, and far lower quantities of base and

H_2O_2 are required [72]. These authors also investigated the effect of the method used for the preparation of polyleucine, and found that material made by high-temperature polymerization gave the best results. Much lower loadings of polyleucine can now be used (down to 0.5 mol% for chalcone epoxidation). The same group has carried out the process on the 100-g scale [73]. Although these conditions have not yet been tested as widely as the Roberts biphasic ones, a non-chalcone substrate was reported (Scheme 12.16).

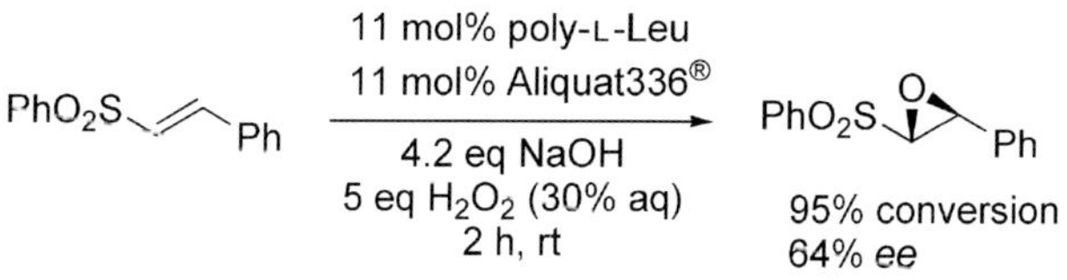

Scheme 12.16 Triphasic polyleucine-mediated epoxidation in the presence of a phase-transfer catalyst.

The mechanism of the polyleucine-catalyzed epoxidation is still under investigation [74]. Kinetic studies indicate that the reaction proceeds via the reversible addition of chalcone to a polyleucine-bound hydroperoxide [75]. Recent discussions have included studies of asymmetric amplification: polyleucine derived from non-enantiopure amino acid shows highly amplified epoxide enantiomeric excess, and the results fit a mathematical model requiring the active catalyst to have five terminal homochiral residues, as rationalized by molecular modeling studies [76].

12.2.3
Epoxidation of Enals

An impressive addition to the now extensive repertoire of amine-catalyzed reactions is the epoxidation of α-β-unsaturated aldehydes described by Jorgensen and co-workers, catalyzed by proline-derivative **31** (Fig. 12.11) [77, 78]. As with related processes, this chemistry likely proceeds via activation of the unsaturated alde-

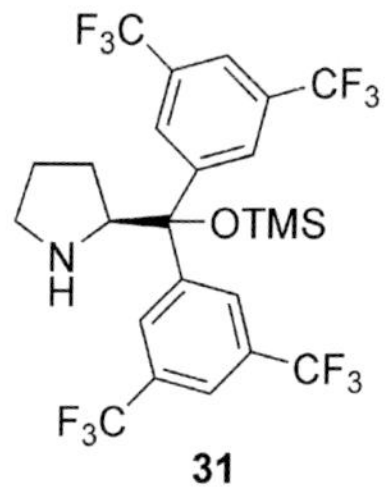

Fig. 12.11 Organocatalyst **31**.

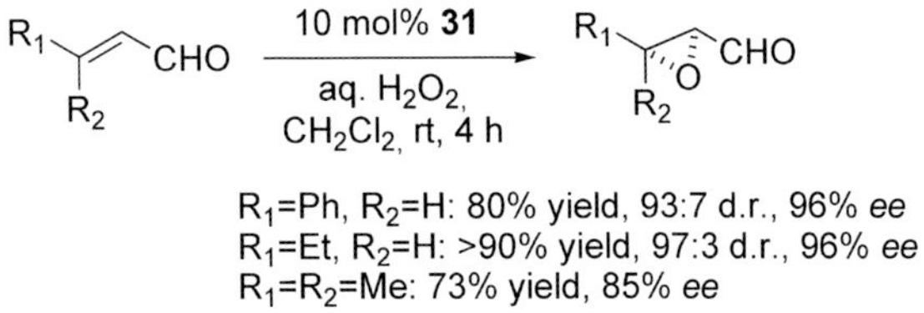
10 mol% 31
aq. H_2O_2,
CH_2Cl_2, rt, 4 h
R_1=Ph, R_2=H: 80% yield, 93:7 d.r., 96% ee
R_1=Et, R_2=H: >90% yield, 97:3 d.r., 96% ee
R_1=R_2=Me: 73% yield, 85% ee

Scheme 12.17 Organocatalyzed enal epoxidation.

hyde via iminium ion formation with the organocatalyst **31**. The reaction uses an environmentally benign oxidant (aqueous hydrogen peroxide). Excellent enantioselectivities and high diastereoselectivities are observed for *trans*-aliphatic and aromatic enals, with a slightly lower *ee*-value for a β,β-disubstituted example (Scheme 12.17). Nucleophilic epoxidation of the unsaturated aldehyde can be performed in the presence of an isolated alkene (Scheme 12.18). The chemistry can also be performed in aqueous alcohol solvent mixtures [79].

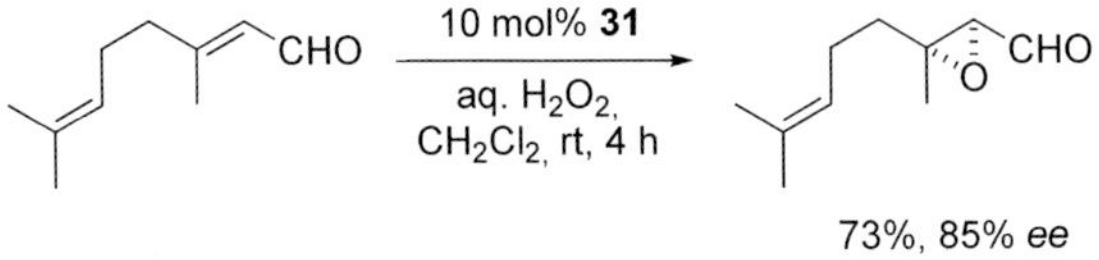
10 mol% 31
aq. H_2O_2,
CH_2Cl_2, rt, 4 h
73%, 85% ee

Scheme 12.18 Selective enal epoxidation.
(For experimental details see Chapter 14.13.10).

12.3 Asymmetric Baeyer–Villiger Reaction

Although the use of asymmetric Baeyer–Villiger reactions for desymmetrization of several types of prochiral ketone using enzymes or metal-catalyzed systems is well established [81], few studies have been conducted on organocatalytic variants. An interesting organocatalytic Baeyer–Villiger reaction has been reported in which the planar-chiral bisflavin **32** (Fig. 12.12) promotes desymmetrization

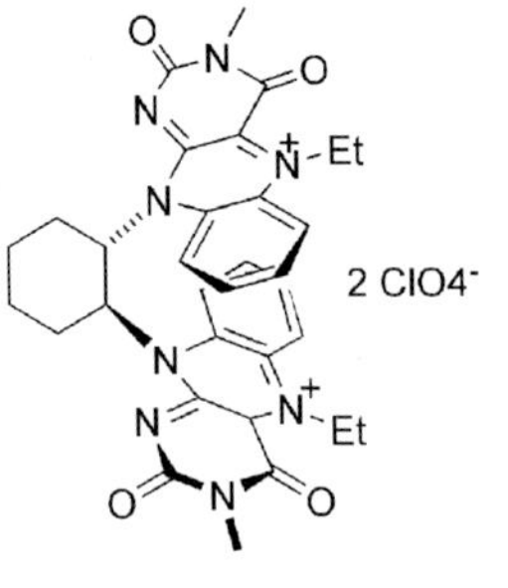
2 ClO4-

32

Fig. 12.12 Chiral bisflavin catalyst **32**.

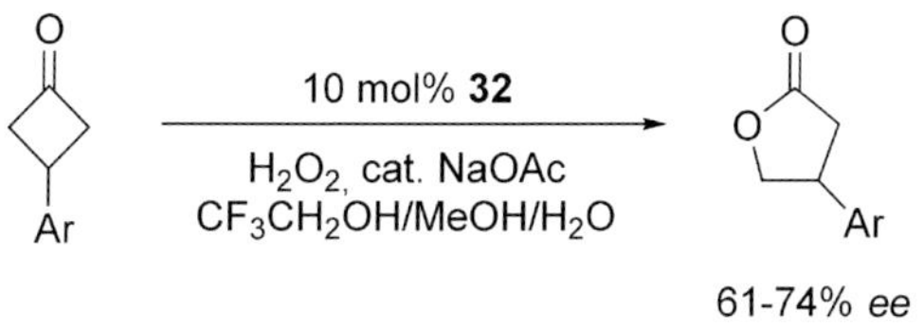

Scheme 12.19 Organocatalytic enantioselective Baeyer–Villiger reaction.

of 4-arylcyclobutanones (Scheme 12.19) [82]. Mechanistically, it is proposed that hydrogen peroxide is activated upon nucleophilic addition to the iminium group of **32**. The use of a protic solvent is essential for enantioselectivity; this is suggested to indicate that hydrophobic π-π stacking between catalyst and substrate plays a role in the enantiocontrol.

12.4
Oxidation of Thioethers

The organocatalyzed oxidation of sulfur appears to lag behind its metal-catalyzed counterpart in terms of enantioselectivities [83]. Although stoichiometric chiral N-sulfonyloxaziridines will effect sulfoxidation with high ee-values, and this class of reagent can be generated *in situ* from an N-sulfonylimine, hydrogen peroxide and DBU, enantioselectivities are low with substoichiometric imine [84]. Sulfoxidation of *p*-tolyl methylsulfide with hydrogen peroxide catalyzed by chiral iminium salt **15** (see Fig. 12.7) has been reported [39] (32% *ee*), while the same transformation has been reported in 65% *ee* with a chiral flavinium salt catalyst [85]. Another interesting organocatalytic system involves catalysis in reversed micelles in the presence of dibenzoyl tartrate **33** (Fig. 12.13) as chiral additive (Scheme 12.20) [86]. Enantioselectivities of up to 72% *ee* were observed; the product *ee*-

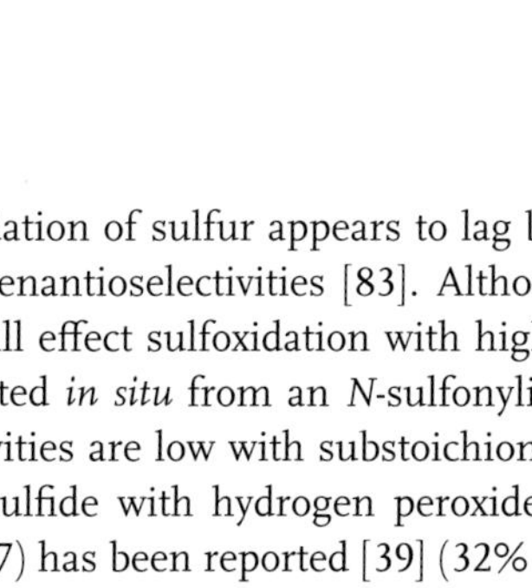

Fig. 12.13 Dibenzoyl tartrate additive for asymmetric sulfoxidation.

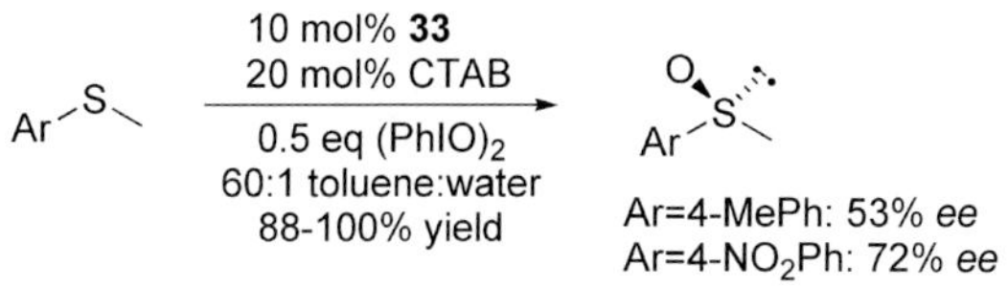

Scheme 12.20 Asymmetric sulfoxidation in reversed micelles.

value was found to increase as the reaction proceeds, which was attributed to the *in-situ* formation of a chiral hypervalent iodine reagent.

12.5 Resolution of Alcohols by Oxidation

An interesting feature of dioxirane chemistry is the possibility of oxidation of C–H bonds, including the oxidation of alcohols. Adam has shown that Shi's catalyst **5** can be used for asymmetric alcohol oxidation, although the *ee*-values are moderate [87, 88]. For example, desymmetrization of *meso*-stilbene diol or kinetic resolution of the DL-isomer can be effected, with the two substrates giving opposite product enantiomers (Scheme 12.21). Electronic effects operate, with higher levels of enantiocontrol being observed when there are electron-withdrawing groups on the aromatic rings [88].

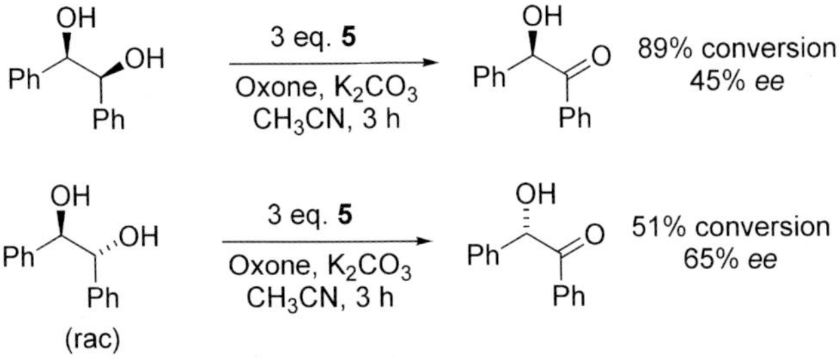

Scheme 12.21 Ketone-catalyzed asymmetric alcohol oxidation.

In an alternative application of asymmetric alcohol oxidation, Rychnovsky has reported the use of the chiral nitroxyl radical **34** (Fig. 12.14) along with bleach, allowing kinetic resolution of secondary alcohols [89]. The best substrates were simple benzylic alcohols, for which S factors (= k_S/k_R) were in the range 3.9 to 7.1 (Scheme 12.22). Other chiral C_2-symmetric nitroxyl radicals reported recently give lower selectivities [90].

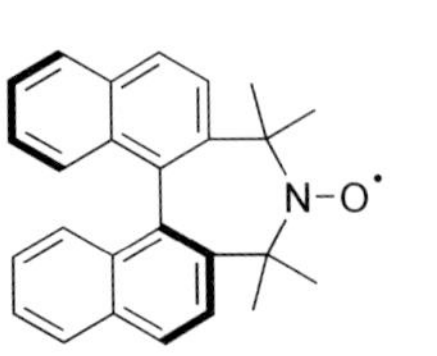

34

Fig. 12.14 Chiral nitroxyl radical **34** for oxidative kinetic resolution of alcohols.

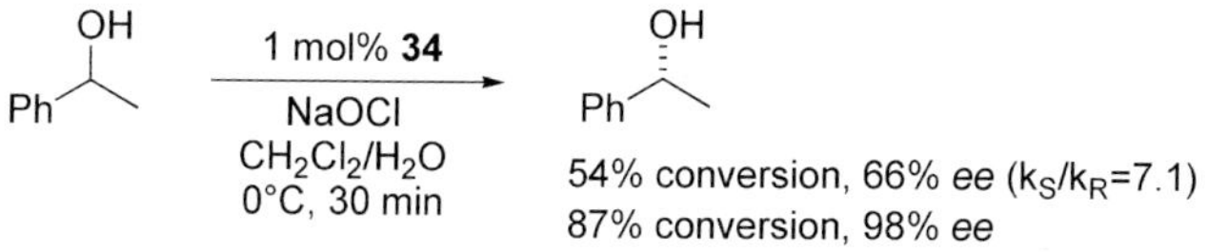

Scheme 12.22 Oxidative kinetic resolution of secondary alcohols.

Ph-CH(OH)-CH₃ with 1 mol% **34**, NaOCl, CH$_2$Cl$_2$/H$_2$O, 0 °C, 30 min → Ph-CH(OH)-CH₃
54% conversion, 66% ee (k_S/k_R=7.1)
87% conversion, 98% ee

12.6
Conclusions

The information contained in this chapter has demonstrated the enormous power and versatility of organocatalytic oxidation processes. However, challenges still remain: in addition to the substrate limitations highlighted for several of the transformations and moderate current levels of enantioselectivity for some of the processes, catalyst loadings are often high compared to metal-catalyzed reactions, and non-environmentally friendly oxidants are sometimes required. Thus, further progress in the development of more selective and more active catalysts is eagerly anticipated.

References

1 For a review of organocatalyzed oxidations, see: W. Adam, C.R. Saha-Moller, P.A. Ganeshpure, *Chem. Rev.* **2001**, *101*, 3499–3548.
2 For a review of catalytic asymmetric epoxidation, see: Q.H. Xia, H.Q. Ge, C.P. Ye, Z.M. Liu, K.X. Su, *Chem. Rev.* **2005**, *105*, 1603–1662.
3 T. Katsuki, V.S. Martin, *Org. React.* **1996**, *48*, 1–300.
4 E.M. McGarrigle, D.G. Gilheany, *Chem. Rev.* **2005**, *105*, 1563–1602.
5 For information and technical details on Oxone, see: http://www.dupont.com/oxone/
6 A. Armstrong, *Angew. Chem., Int. Ed.* **2004**, *43*, 1460–1462.
7 R.W. Murray, *Chem. Rev.* **1989**, *89*, 1187–1201.
8 R. Curci, M. Fiorentino, M.R. Serio, *J. Chem. Soc., Chem. Commun.* **1984**, 155–156.
9 S.E. Denmark, D.C. Forbes, D.S. Hays, J.S. Depue, R.G. Wilde, *J. Org. Chem.* **1995**, *60*, 1391–1407.
10 D. Yang, M.K. Wong, Y.C. Yip, *J. Org. Chem.* **1995**, *60*, 3887–3889.
11 D. Yang, Y.C. Yip, M.W. Tang, M.K. Wong, J.H. Zheng, K.K. Cheung, *J. Am. Chem. Soc.* **1996**, *118*, 491–492; D. Yang, X.C. Wang, M.K. Wong, Y.C. Yip, M.W. Tang, *J. Am. Chem. Soc.* **1996**, *118*, 11311–11312; D. Yang, *Acc. Chem. Res.* **2004**, *37*, 497–505.
12 D. Yang, M.K. Wong, Y.C. Yip, X.C. Wang, M.W. Tang, J.H. Zheng, K.K. Cheung, *J. Am. Chem. Soc.* **1998**, *120*, 5943–5952.
13 Y. Tu, Z.X. Wang, Y. Shi, *J. Am. Chem. Soc.* **1996**, *118*, 9806–9807.
14 Z.X. Wang, Y. Tu, M. Frohn, J.R. Zhang, Y. Shi, *J. Am. Chem. Soc.* **1997**, *119*, 11224–11235.
15 Y. Shi, *Acc. Chem. Res.* **2004**, *37*, 488–496.
16 Y. Tu, M. Frohn, Z.-X. Wang, Y. Shi, *Organic Syntheses* **2003**, *80*, 1–4.
17 Z.X. Wang, Y. Tu, M. Frohn, Y. Shi, *J. Org. Chem.* **1997**, *62*, 2328–2329.
18 M. Frohn, M. Dalkiewicz, Y. Tu, Z.X. Wang, Y. Shi, *J. Org. Chem.* **1998**, *63*, 2948–2953.
19 G.A. Cao, Z.X. Wang, Y. Tu, Y. Shi, *Tetrahedron Lett.* **1998**, *39*, 4425–4428.

20 J.C. Lorenz, M. Frohn, X.M. Zhou, J.R. Zhang, Y. Tang, C. Burke, Y. Shi, *J. Org. Chem.* **2005**, *70*, 2904–2911.
21 Y.M. Zhu, L.H. Shu, Y. Tu, Y. Shi, *J. Org. Chem.* **2001**, *66*, 1818–1826; Y.M. Zhu, K.J. Manske, Y. Shi, *J. Am. Chem. Soc.* **1999**, *121*, 4080–4081.
22 Z.M. Xiong, E.J. Corey, *J. Am. Chem. Soc.* **2000**, *122*, 9328–9329; Z.M. Xiong, E.J. Corey, *J. Am. Chem. Soc.* **2000**, *122*, 4831–4832.
23 L.H. Shu, Y. Shi, *Tetrahedron* **2001**, *57*, 5213–5218.
24 L.H. Shu, Y. Shi, *Tetrahedron Lett.* **1999**, *40*, 8721–8724.
25 Z.-X. Wang, L. Shu, Y. Tu, Z.-X. Wang, Y. Shi, *Organic Syntheses* **2003**, *80*, 9.
26 K.N. Houk, J. Liu, N.C. DeMello, K.R. Condroski, *J. Am. Chem. Soc.* **1997**, *119*, 10147–10152; A. Armstrong, I. Washington, K.N. Houk, *J. Am. Chem. Soc.* **2000**, *122*, 6297–6298.
27 D.A. Singleton, Z.H. Wang, *J. Am. Chem. Soc.* **2005**, *127*, 6679–6685.
28 L.H. Shu, Y.M. Shen, C. Burke, D. Goeddel, Y. Shi, *J. Org. Chem.* **2003**, *68*, 4963–4965.
29 H.Q. Tian, X.G. She, H.W. Yu, L.H. Shu, Y. Shi, *J. Org. Chem.* **2002**, *67*, 2435–2446; H.Q. Tian, X.G. She, J.X. Xu, Y. Shi, *Org. Lett.* **2001**, *3*, 1929–1931; H.Q. Tian, X.G. She, L.H. Shu, H.W. Yu, Y. Shi, *J. Am. Chem. Soc.* **2000**, *122*, 11551–11552.
30 M. Hickey, D. Goeddel, Z. Crane, Y. Shi, *Proc. Natl. Acad. Sci. USA* **2004**, *101*, 5794–5798; L.H. Shu, P.Z. Wang, Y.H. Gan, Y. Shi, *Org. Lett.* **2003**, *5*, 293–296; L.H. Shu, Y. Shi, *Tetrahedron Lett.* **2004**, *45*, 8115–8117; Z. Crane, D. Goeddel, Y.H. Gan, Y. Shi, *Tetrahedron* **2005**, *61*, 6409–6417.
31 For example: W. Adam, C.R. Saha-Moller, C.G. Zhao, *Tetrahedron: Asymm.* **1999**, *10*, 2749–2755; S.E. Denmark, Z.C. Wu, *Synlett* **1999**, 847–859; A. Solladie-Cavallo, L. Bouerat, L. Jierry, *Eur. J. Org. Chem.* **2001**, 4557–4560; A. Solladie-Cavallo, L. Jierry, A. Klein, *C. R. Chim.* **2003**, *6*, 603–606; A. Solladie-Cavallo, L. Jierry, H. Norouzi-Arasi, D. Tahmassebi, *J. Fluor. Chem.* **2004**, *125*, 1371–1377; A. Solladie-Cavallo, L. Jerry, A. Klein, M. Schmitt, R. Welter, *Tetrahedron: Asymm.* **2004**, *15*, 3891–3898; C.E. Song, Y.H. Kim, K.C. Lee, S. Lee, B.W. Jin, *Tetrahedron: Asymm.* **1997**, *8*, 2921–2926; C.J. Stearman, V. Behar, *Tetrahedron Lett.* **2002**, *43*, 1943–1946; G. Bez, C.G. Zhao, *Tetrahedron Lett.* **2003**, *44*, 7403–7406; A. Armstrong, T. Tsuchiya, *Tetrahedron* **2006**, *62*, 257–263.
32 A. Armstrong, B.R. Hayter, *Tetrahedron* **1999**, *55*, 11119–11126; A. Armstrong, W.O. Moss, J.R. Reeves, *Tetrahedron: Asymm.* **2001**, *12*, 2779–2781.
33 S.E. Denmark, H. Matsuhashi, *J. Org. Chem.* **2002**, *67*, 3479–3486.
34 T.K.M. Shing, G.Y.C. Leung, K.W. Yeung, *Tetrahedron Lett.* **2003**, *44*, 9225–9228.
35 T.K.M. Shing, G.Y.C. Leung, T. Luk, *J. Org. Chem.* **2005**, *70*, 7279–7289.
36 A. Armstrong, G. Ahmed, B. Dominguez-Fernandez, B.R. Hayter, J.S. Wailes, *J. Org. Chem.* **2002**, *67*, 8610–8617; A. Armstrong, B.R. Hayter, *Chem. Commun.* **1998**, 621–622.
37 G. Sartori, A. Armstrong, R. Maggi, A. Mazzacani, R. Sartorio, F. Bigi, B. Dominguez-Fernandez, *J. Org. Chem.* **2003**, *68*, 3232–3237.
38 G. Hanquet, X. Lusinchi, P. Milliet, *Tetrahedron Lett.* **1988**, *29*, 3941–3944.
39 L. Bohe, G. Hanquet, M. Lusinchi, X. Lusinchi, *Tetrahedron Lett.* **1993**, *34*, 7271–7274.
40 V.K. Aggarwal, M.F. Wang, *Chem. Commun.* **1996**, 191–192.
41 J. Vachon, C. Perollier, D. Monchaud, C. Marsol, K. Ditrich, J. Lacour, *J. Org. Chem.* **2005**, *70*, 5903–5911.
42 M.K. Wong, L.M. Ho, Y.S. Zheng, C.Y. Ho, D. Yang, *Org. Lett.* **2001**, *3*, 2587–2590.
43 P.C.B. Page, B.R. Buckley, A.J. Blacker, *Org. Lett.* **2004**, *6*, 1543–1546.
44 P.C.B. Page, D. Barros, B.R. Buckley, A. Ardakani, B.A. Marples, *J. Org. Chem.* **2004**, *69*, 3595–3597.
45 P.C.B. Page, B.R. Buckley, H. Heaney, A.J. Blacker, *Org. Lett.* **2005**, *7*, 375–377.

46 I. Washington, K.N. Houk, *J. Am. Chem. Soc.* **2000**, *122*, 2948–2949.
47 A. Armstrong, A.G. Draffan, *J. Chem. Soc., Perkin Trans. 1* **2001**, 2861–2873; A. Armstrong, A.G. Draffan, *Tetrahedron Lett.* **1999**, *40*, 4453–4456.
48 M.R. Biscoe, R. Breslow, *J. Am. Chem. Soc.* **2005**, *127*, 10812–10813.
49 M.F.A. Adamo, V.K. Aggarwal, M.A. Sage, *J. Am. Chem. Soc.* **2002**, *124*, 11223–11223; M.F.A. Adamo, V.K. Aggarwal, M.A. Sage, *J. Am. Chem. Soc.* **2000**, *122*, 8317–8318.
50 V.K. Aggarwal, C. Lopin, F. Sandrinelli, *J. Am. Chem. Soc.* **2003**, *125*, 7596–7601.
51 C.Y. Ho, Y.C. Chen, M.K. Wong, D. Yang, *J. Org. Chem.* **2005**, *70*, 898–906.
52 M.J. Porter, J. Skidmore, *Chem. Commun.* **2000**, 1215–1225.
53 A. Solladie-Cavallo, L. Bouerat, *Org. Lett.* **2000**, *2*, 3531–3534.
54 X.Y. Wu, X.G. She, Y.A. Shi, *J. Am. Chem. Soc.* **2002**, *124*, 8792–8793.
55 N. Nieto, P. Molas, J. Benet-Buchholz, A. Vidal-Ferran, *J. Org. Chem.* **2005**, *70*, 10143–10146.
56 R. Imashiro, M. Seki, *J. Org. Chem.* **2004**, *69*, 4216–4226.
57 O. Bortolini, M. Fogagnolo, G. Fantin, S. Maietti, A. Medici, *Tetrahedron: Asymm.* **2001**, *12*, 1113–1115; O. Bortolini, G. Fantin, M. Fogagnolo, R. Forlani, S. Maietti, P. Pedrini, *J. Org. Chem.* **2002**, *67*, 5802–5806; O. Bortolini, G. Fantin, M. Fogagnolo, L. Mari, *Tetrahedron: Asymm.* **2004**, *15*, 3831–3833.
58 R. Helder, J.C. Hummelen, R. Laane, J.S. Wiering, H. Wynberg, *Tetrahedron Lett.* **1976**, 1831–1834; H. Wynberg, B. Marsman, *J. Org. Chem.* **1980**, *45*, 158–161; H. Pluim, H. Wynberg, *J. Org. Chem.* **1980**, *45*, 2498–2502; J.C. Hummelen, H. Wynberg, *Tetrahedron Lett.* **1978**, 1089–1092.
59 G. Macdonald, L. Alcaraz, N.J. Lewis, R.J.K. Taylor, *Tetrahedron Lett.* **1998**, *39*, 5433–5436; L. Alcaraz, G. Macdonald, J.P. Ragot, N. Lewis, R.J.K. Taylor, *J. Org. Chem.* **1998**, *63*, 3526–3527; A.G.M. Barrett, F. Blaney, A.D. Campbell, D. Hamprecht, T. Meyer, A.J.P. White, D. Witty, D.J. Williams, *J. Org. Chem.* **2002**, *67*, 2735–2750.
60 W. Adam, P.B. Rao, H.G. Degen, A. Levai, T. Patonay, C.R. Saha-Moller, *J. Org. Chem.* **2002**, *67*, 259–264; W. Adam, P.B. Rao, H.G. Degen, C.R. Saha-Moller, *Tetrahedron: Asymm.* **2001**, *12*, 121–125.
61 B. Lygo, P.G. Wainwright, *Tetrahedron Lett.* **1998**, *39*, 1599–1602.
62 B. Lygo, P.G. Wainwright, *Tetrahedron* **1999**, *55*, 6289–6300.
63 E.J. Corey, F.Y. Zhang, *Org. Lett.* **1999**, *1*, 1287–1290.
64 S. Arai, H. Tsuge, T. Shioiri, *Tetrahedron Lett.* **1998**, *39*, 7563–7566; S. Arai, H. Tsuge, M. Oku, M. Miura, T. Shioiri, *Tetrahedron* **2002**, *58*, 1623–1630.
65 B. Lygo, D.C.M. To, *Tetrahedron Lett.* **2001**, *42*, 1343–1346.
66 B. Lygo, D.C.M. To, *Chem. Commun.* **2002**, 2360–2361.
67 T. Ooi, D. Ohara, M. Tamura, K. Maruoka, *J. Am. Chem. Soc.* **2004**, *126*, 6844–6845.
68 S. Julia, J. Masana, J.C. Vega, *Angew. Chem., Int. Ed. Engl.* **1980**, *19*, 929–931.
69 M.J. Porter, S.M. Roberts, J. Skidmore, *Bioorg. Med. Chem.* **1999**, *7*, 2145–2156.
70 B.M. Adger, J.V. Barkley, S. Bergeron, M.W. Cappi, B.E. Flowerdew, M.P. Jackson, R. McCague, T.C. Nugent, S.M. Roberts, *J. Chem. Soc., Perkin Trans. 1* **1997**, 3501–3507.
71 S. Itsuno, M. Sakakura, K. Ito, *J. Org. Chem.* **1990**, *55*, 6047–6049.
72 T. Geller, A. Gerlach, C.M. Kruger, H.C. Militzer, *Tetrahedron Lett.* **2004**, *45*, 5065–5067; T. Geller, C.M. Kruger, H.C. Militzer, *Tetrahedron Lett.* **2004**, *45*, 5069–5071.
73 A. Gerlach, T. Geller, *Adv. Synth. Catal.* **2004**, *346*, 1247–1249.
74 A. Berkessel, N. Gasch, K. Glaubitz, C. Koch, *Org. Lett.* **2001**, *3*, 3839–3842.
75 S.P. Mathew, S. Gunathilagan, S.M. Roberts, D.G. Blackmond, *Org. Lett.* **2005**, *7*, 4847–4850.

76 D.R. Kelly, A. Meek, S.M. Roberts, *Chem. Commun.* **2004**, 2021–2022; D.R. Kelly, S.M. Roberts, *Chem. Commun.* **2004**, 2018–2020; D.R. Kelly, E. Caroff, R.W. Flood, W. Heal, S.M. Roberts, *Chem. Commun.* **2004**, 2016–2017.

77 M. Marigo, J. Franzen, T.B. Poulsen, W. Zhuang, K.A. Jorgensen, *J. Am. Chem. Soc.* **2005**, *127*, 6964–6965.

78 For related work, see: H. Sunden, I. Ibrahem, A. Cordova, *Tetrahedron Lett.* **2006**, *47*, 99–103.

79 W. Zhuang, M. Marigo, K.A. Jorgensen, *Org. Biomol. Chem.* **2005**, *3*, 3883–3885.

80 M. Marigo, T.C. Wabnitz, D. Fielenbach, K.A. Jorgensen, *Angew. Chem., Int. Ed.* **2005**, *44*, 794–797.

81 M.D. Mihovilovic, F. Rudroff, B. Grotzl, *Curr. Org. Chem.* **2004**, *8*, 1057–1069.

82 S.I. Murahashi, S. Ono, Y. Imada, *Angew. Chem., Int. Ed.* **2002**, *41*, 2366–2368.

83 I. Fernandez, N. Khiar, *Chem. Rev.* **2003**, *103*, 3651–3705.

84 D. Bethell, P.C.B. Page, H. Vahedi, *J. Org. Chem.* **2000**, *65*, 6756–6760.

85 S. Shinkai, T. Yamaguchi, O. Manabe, F. Toda, *J. Chem. Soc., Chem. Commun.* **1988**, 1399–1401.

86 H. Tohma, S. Takizawa, H. Watanabe, Y. Fukuoka, T. Maegawa, Y. Kita, *J. Org. Chem.* **1999**, *64*, 3519–3523.

87 W. Adam, C.R. Saha-Moller, C.G. Zhao, *Tetrahedron: Asymm.* **1998**, *9*, 4117–4122.

88 W. Adam, C.R. Saha-Moller, C.G. Zhao, *J. Org. Chem.* **1999**, *64*, 7492–7497.

89 S.D. Rychnovsky, T.L. McLernon, H. Rajapakse, *J. Org. Chem.* **1996**, *61*, 1194–1195.

90 B. Graetz, S. Rychnovsky, W.H. Leu, P. Farmer, R. Lin, *Tetrahedron: Asymm.* **2005**, *16*, 3584–3598.

13
Shape- and Site-Selective Asymmetric Reactions

Nicolas Bogliotti and Peter I. Dalko

13.1
Introduction

Catalysts may be able to select substrates depending on physical criteria via dynamic molecular recognition through multivalent interactions [1, 2]. These molecules, which mimic enzyme functions, yet have entirely different functions, are called "synzymes". In this chapter, we will review accounts of man-made catalysts that have no metal elements [3].

Synzymes may achieve rate enhancements via binding and proximity effects. Such effects can occur when two reactive partners are bound within the same nanospace, thus increasing their relative encounter frequencies (i.e., the effective concentrations) (Fig. 13.1a) [4]. Some synzymes participate as reagents, and these catalysts feature structurally distinct substrate-binding site(s), together with a catalytically effective site(s) (Fig. 13.1b).

The concept, that enzymes provide binding by complementing the shapes and characteristics of transition states – that is, the so-called "Pauling paradigm" [5, 6] – is a major guiding principle in artificial catalyst design. However, the rational design of structures and functions is problematic, with difficulties arising from a lack of applicable theory for modeling transition states involved in enzymes

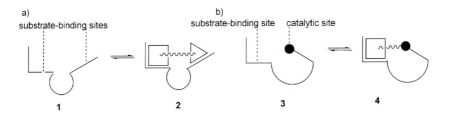

Fig. 13.1 Synzymes may catalyze reactions by binding two (or more) reagents within the same nanospace (a), or they may perform the transformation with the help of catalytically effective site(s) (b).

Enantioselective Organocatalysis: Reactions and Experimental Procedures
Edited by Peter I. Dalko
Copyright © 2007 WILEY-VCH Verlag GmbH & Co. KGaA, Weinheim
ISBN: 978-3-527-31522-2

operating. Also, our actual understanding of the catalytic acceleration of enzymes supposes that the venerable transition state theory is far from complete. Transition state stabilization is accompanied – or even superseded by – phenomena such as electrostatics, quantum mechanical tunneling, coupled protein motion, low-barrier hydrogen bonds and near-attack conformations, all of which enable enzymes to enhance reaction rates by up to 10^{20}-fold [7]. Further problems are related to the lack of dynamic response of the catalyst, notably when the structure of the starting material is similar to that of the product(s). There is now ample evidence for the occurrence of protein domain motions that result in active site opening and closing events as integral features of enzyme-catalyzed reactions [8]. The lack of such a dynamic response in actual artificial enzymes has consequences both in overall efficiency and also, in the fact that the formed host–guest complex is often too stable and results in product inhibition. Whilst enzyme catalysts are capable of adopting conformational changes in order to favor product release, the rigidity of the artificial catalysts does not allow efficient catalytic turnover. Thus, the search for a dynamic response represents a new and promising area for further developments.

The design of shape- and site-selective catalysts based on host–guest interactions is undoubtedly an innovative approach in organic synthesis, which foresees major perspectives [9]. As an immediate application, such catalysts may render unnecessary the use of protecting or directing groups in multi-step synthesis. In particular, both Yoshida [10] and Vasella [11] have made important contributions in this area, studying a set of catalysts for the selective functionalization of a range of minimally protected carbohydrate monomers.

Although this chapter describes advances in this field, it does not exclusively select enantioselective transformations. Although some of the reactions presented here feature poor catalytic turnover, and require stoichiometric (or even excess) quantities of catalyst, we chose to include such examples because of the scope of the concept, the possible recovery of the catalyst, and the hope that further optimization would allow the use of these systems in substoichiometric catalyst loadings.

13.2
Acylation Reactions with Oligopeptide-Based Catalysts

13.2.1
Regioselective Acetylation of Glucose Derivatives

In a series of elegant studies, Miller and colleagues demonstrated that small peptides containing modified histidine residues (π-methyl-histidine) were effective catalysts for enantioselective acylation and phosphorylation reactions [12, 13]. As peptides in this family are able to transfer stereochemical information, these catalysts were also tested to carry out site- and regio-selective reactions, which take place against kinetic expectations. The screening of a small library (36 members) showed that some peptides were indeed capable of perturbing the inherent selectivities of the acylation reactivity of a glucosamine derivative. Catalyst **5** allowed regioselective conversion of the glucose derivative **6** into the monoacetate **10** in 58% isolated yield (Scheme 13.1) [12]. In a control experiment, when glucoside **6** was treated in the presence of N-methylimidazole (NMI), the primary acetate was predominantly formed, albeit with a low conversion.

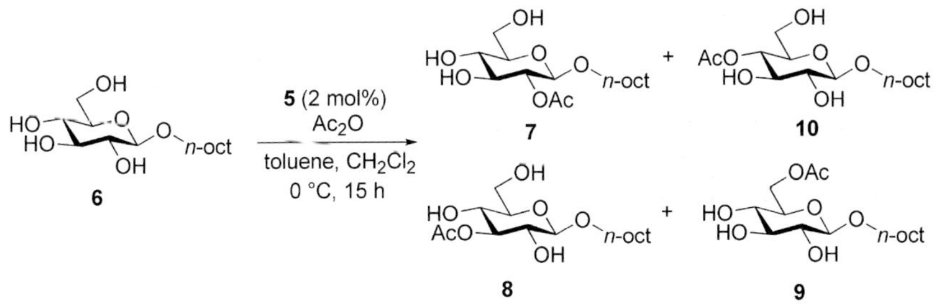

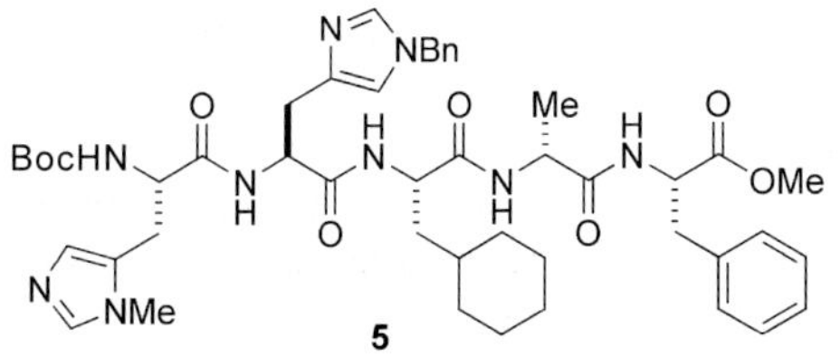

catalyst	conversion (%)	ratio			
		7	8	9	10
NMI	14	0	20	64	16
5	100	9	11	22	58

Scheme 13.1 The regioselective acylation of a glucose derivative **6**.

13.2.2
Regioselective Phosphorylation

As in acylation reactions, the histidine residue can serve as the catalytically active site in phosphate transfer [13], this reaction being reminiscent of that seen in certain classes of kinases. Just as enzymes rely on the structure of their binding pockets to determine the binding orientation of their substrate, Miller et al. induced enantioselectivity in phosphate transfer catalysts by modulating their secondary structure (Scheme 13.2). Enantiodivergent catalysts for the synthesis of D-myo-inositol-1-phosphate and D-myo-inositol-3-phosphate were obtained from randomized combinatorial libraries. Both catalysts contain a terminal L-(π-methyl-histidine) residue, but differ in their secondary structure. While catalyst **11** was picked from a library of β-turn-containing peptides, catalyst **12** was chosen from a randomized library. Optimization of the structure of catalysts **11** and **12** allowed a complete inversion of regioselectivity, yielding compounds **14** and **15**

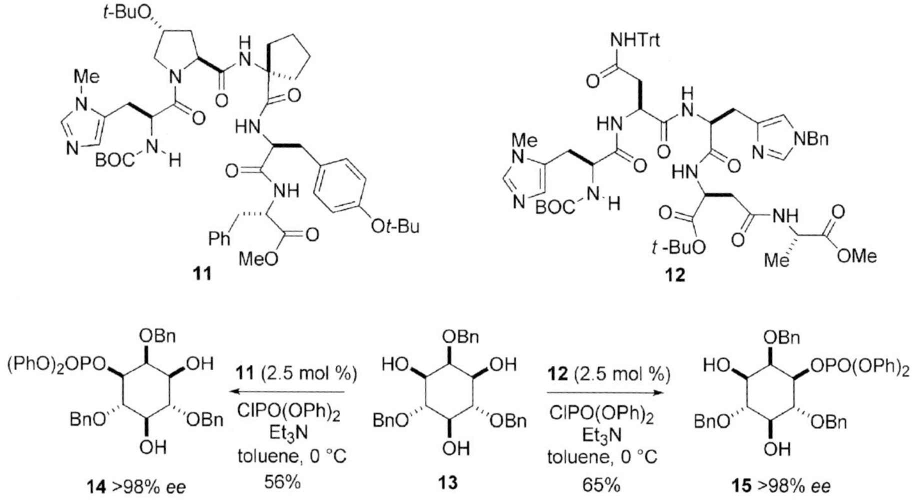

Scheme 13.2

with high enantioselectivities (>98% ee in both cases). (For experimental details see Chapter 14.17.14). In comparison to small-molecule catalysts such as DMAP and NMI, catalysts **11** and **12** were more active, indicating that the presence of a secondary structure not only provides site selectivity but also enhances the rate of the reaction.

13.3
Sequence-Specific RNA-Cleaving Agents [14]

The development of sequence-specific RNA-cleaving agents based on complementary base pairing is of major interest, not only in the mimicking of natural restriction enzymes for sequence-selective manipulation of RNA molecules, but also for applications as catalytic antisense oligonucleotides in chemotherapy. Under biological conditions, ribonucleases (RNases) catalyze cleavage of the 3′ O–P phosphodiester bond within a polynucleotide chain. This process plays a fundamental role in diverse eukaryotic functions, including viral defense, chromatin remodeling, genome rearrangement, developmental timing, brain morphogenesis, and stem cell maintenance. Under physiological conditions, RNA cleavage involves an attack of the 2′-oxyanion on the phosphorus atom, resulting in formation of the unstable cyclic dianion, which undergoes a rate-limiting fragmentation by departure of the 5′-linked nucleoside as an oxyanion (Scheme 13.3). It should be noted that this intramolecular mechanism requires some molecular (structural) flexibility. For example, when the structure is too rigid to allow intrastrand cooperation, the phosphodiester bond can be cleaved via an intermolecular attack by water, or by other nucleophiles.

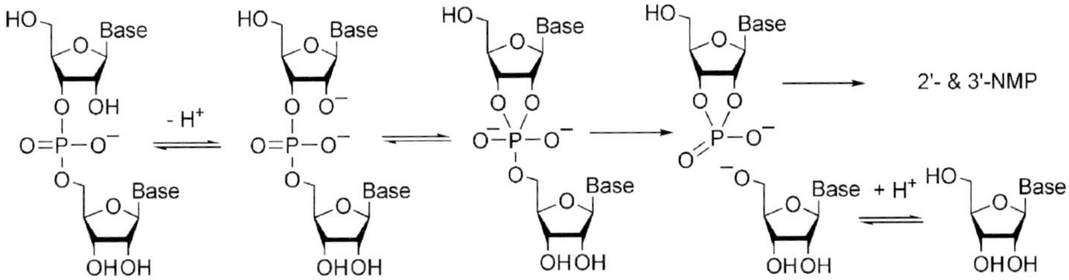

Scheme 13.3

Under biological conditions, RNases may require a divalent metal ion such as Mg^{2+} for activity [15, 16]. Such a co-catalyst may first, enhance the electrophilicity of the phosphorus atom and neutralize the negative charge of the pentacoordinate intermediate, and second, protonate or otherwise increase the acidity of the 3′-oxygen leaving group.

It was observed, under *in-vitro* conditions, that nucleotides located in an unpaired region were more prone to cleavage than those implied in intra-strand base stacking. It is suggested that RNA hydrolysis, which proceeds via an in-line attack mechanism on the phosphorus atom, is favored if the attacking 2′-oxyanion and the 5′-linked leaving group adopt an apical orientation during the transesterification reaction. It has also been indicated that the backbone geometry of base-paired regions prohibits suitable arrangement for the in-line attack, while the unpaired sites are preferentially hydrolyzed [17].

13.3.1
RNA Hydrolysis by DNA-DETA Adduct

Sequence-selective RNA scission was accomplished by the attachment of oligoamines to synthetic DNA oligomers [18]. The cleaving agent **16**, composed of diethylenetriamine (DETA) linked to 19-mer DNA, hydrolyzed linear 30-mer RNA strand **17** in a sequence-selective manner (Scheme 13.4). Hence, treatment of the linear RNA in the presence of DNA–DETA complex **16** and EDTA at pH 8 for 4 h at 50 °C, furnished, with good selectivity, the scission products **18** and **19**, albeit with a low conversion (10%) of the starting material.

The first event in this process is the formation of a RNA–DNA hetero-duplex between the complementary 19-mer sequences of **16** and **17**. This allows the catalytic moiety to be correctly placed near the selective scission site. The cleavage proved to be most efficient at pH 8, where the ethylenediamine exists mostly as a monocation (pK_a values are 9.2 and 6.5). This observation is consistent with a mechanism involving an intramolecular acid–base cooperation between the neutral amine and the ammonium ion of the DETA moiety, as depicted in Scheme 13.5. The basic amino moiety pulls a proton from the 2′-hydroxyl group of the ribose to promote its attack toward the phosphorus atom, while the protonated ammonium residue provides Lewis acid assistance to the reaction [7d, 19]. This

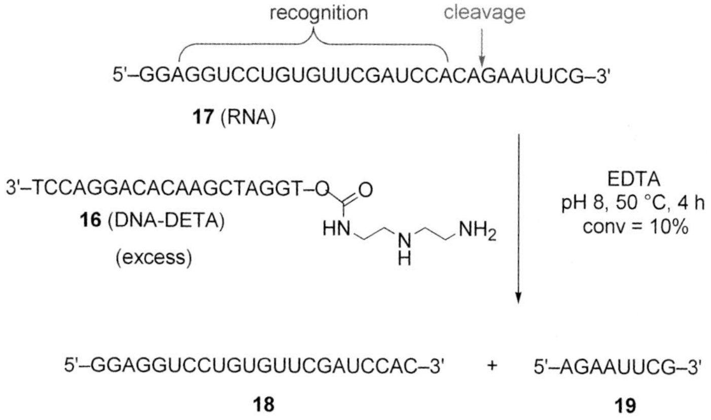

```
                    recognition        cleavage
            ┌─────────────────────┐ ┌─────────┐
            5'–GGAGGUCCUGUGUUCGAUCCACAGAAUUCG–3'
                        17 (RNA)
```

```
   3'–TCCAGGACACAAGCTAGGT–O   O                    EDTA
          16 (DNA-DETA)       ‖              pH 8, 50 °C, 4 h
            (excess)      HN     N     NH₂      conv = 10%
                                 H
```

5'–GGAGGUCCUGUGUUCGAUCCAC–3' + 5'–AGAAUUCG–3'
                    18   19

Scheme 13.4

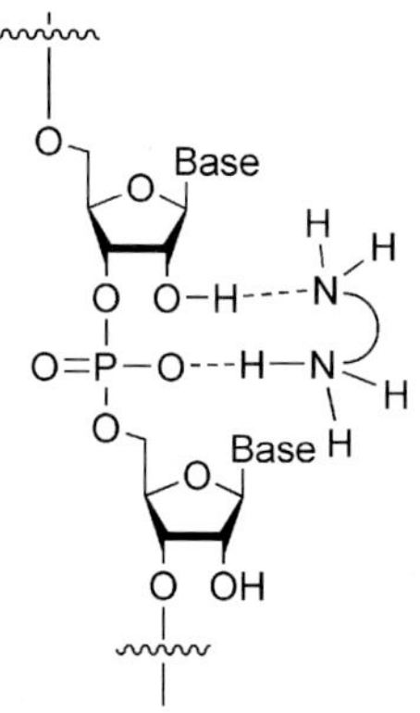

Scheme 13.5

cooperativity between two types of catalysts can proceed either simultaneously or in a stepwise manner, and the mechanism is related to the bifunctional acid–base cooperation between two imidazole moieties of histidine in ribonuclease A [7d].

13.3.2
RNA Hydrolysis by PNA-DETA Adduct

The affinity towards a complementary RNA strand can be improved by using peptide nucleic acids (PNAs) [20]. These DNA mimics, in which the nucleic acid bases are linked to a polyamide backbone [21], usually exhibit improved binding properties with a complementary DNA or RNA strand [22]. The use of an excess PNA-DETA adduct **20** in a Tris-HCl buffer (pH 7), in the presence of NaCl and EDTA at 40 °C, resulted in a sequence-selective cleavage of the RNA strand **21**

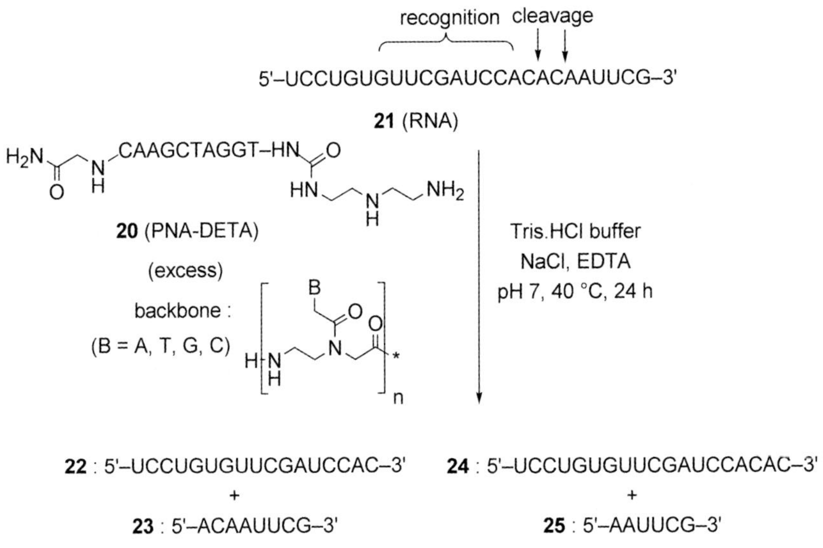

13.3 Sequence-Specific RNA-Cleaving Agents

recognition cleavage

5'-UCCUGUGUUCGAUCCACACAAUUCG-3'

21 (RNA)

20 (PNA-DETA) (excess)

backbone: (B = A, T, G, C)

Tris.HCl buffer
NaCl, EDTA
pH 7, 40 °C, 24 h

22 : 5'-UCCUGUGUUCGAUCCAC-3'
+
23 : 5'-ACAAUUCG-3'

24 : 5'-UCCUGUGUUCGAUCCACAC-3'
+
25 : 5'-AAUUCG-3'

Scheme 13.6

after 24 h (Scheme 13.6). The improved binding affinity of PNAs compared to DNAs toward a RNA strand allowed the use of a shorter recognition sequence (10-mer versus 19-mer). However, cleavage of the phosphodiester bond is less site-selective, as two major different C–A scissions were observed, resulting in the formation of compounds **22–25**. Since a slight modification in the RNA sequence occurred around the cleavage site compared to the previous example (replacement of G by C), it would be difficult to determine whether the intrinsic properties of the catalyst or a modification of the cleaving ability of the RNA strand are responsible for this decrease in site selectivity.

13.3.3
Methyl Phosphonate Oligonucleotides as RNA-Cleaving Agents

Methyl phosphonate oligonucleotides (MPOs) can be obtained by replacing the phosphodiester group of the parent oligonucleotide by a methyl phosphonate unit (Scheme 13.7). These compounds were used as site-specific RNA-cleaving agents [23]. Since RNA regions involved in base stacking are usually not prone to hydrolysis of the phosphodiester bond (*vide supra*), the catalytic site of **26–28** was built with a non-complementary, non-nucleotide-based linker **L** (Scheme 13.7). Treatment of the RNA strand **29** with an excess of **26** or **27**, functionalized with two proton donor/acceptor moieties, in the presence of phosphate buffer at 25 °C, resulted in a sequence-specific RNA cleavage, leading to compounds **30** and **31**, albeit with low conversion (<10% after 5 days). This sluggish catalytic activity is believed to be a consequence of the low binding affinity of these modi-

MPOs : 5'-TAGCTTCCTT**L**GCTCCTG-3'

backbone: Me-P=O structure

26, L =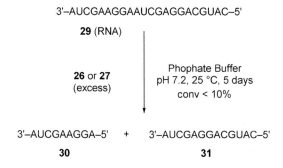

27, L = (structure)

28, L = (structure)

3'-AUCGAAGGAAUCGAGGACGUAC-5'
29 (RNA)

26 or 27 (excess) → Phophate Buffer pH 7.2, 25 °C, 5 days, conv < 10%

3'-AUCGAAGGA-5'   +   3'-AUCGAGGACGUAC-5'
30   **31**

Scheme 13.7

fied MPOs with the complementary RNA strand. Interestingly, treatment of the RNA strand **29** in the presence of **28** under the same conditions resulted in no reaction, thus confirming the role of the two basic **L** residues of **26** and **27** in the cleavage process. However, **28** could act as a protecting group. Hence, it promoted the site-selective RNA cleavage when incubated in the presence of an exogenous 1 M solution of ethylenediamine, while other regions remained intact. In this intermolecular reaction, the region facing the non-base-stacked site **L** site of **28** became more prone to hydrolytic cleavage than the hydrogen-bonded (protected) domains.

13.4
Chiral Cavitands: Calixarenes and Cyclodextrins

Calixarenes and cyclodextrins (CDs) are the simplest combination of organic compounds able to form molecular cavities. The relative ease of functionalization makes them versatile enzymatic binding site mimics [24]. These cavitands can afford inclusion complexes with a variety of guests based on size, structure, and polarity criteria [25, 26]. In this class of substances, the cavity adopts the shape of a truncated cone, featuring two inequivalent entrances. Ideally, the whole complex could function as a funnel able to select substrates according to their size and shape before they undergo a catalytic transformation in the confined space. The position of the substrate is expected to be controlled by weak forces within the cavity walls. Cavitands provided with catalytic site and appended groups engendering water solubility are of particular interest, as they constitute potential microreactors that would allow a catalytic reaction to take place in an aqueous medium (in reality, within the hydrophobic space of the interior).

13.4.1
Calixarenes [27]

Water-soluble calixarenes of the p-sulfonato type have been under intense investigation for catalytic, biomimetic, sensor, and separation applications [28–30]. For p-sulfonatocalix[n]arene (CXn), a particularly common derivative, electrostatic and cationic interactions are presumed to be the dominant driving force for complexation; this is reflected by the high binding constants of CX4 with organic (ammonium) cations (10^3 to 10^5 M^{-1}) [28]. The binding of non-charged organic guest molecules by CX4 has also been intensively studied, owing to, among others, the potential of water-soluble calixarenes to serve as inverse phase-transfer catalysts [29a]. The binding constants, however, were consistently found to be undesirably low (10^1 to 10^2 M^{-1}) [29, 30], as were selectivities [31], implying a very weak hydrophobic driving force for complexation. p-Sulfonatocalix[4]arene has been shown to display a relatively moderate guest size selectivity but a strong guest shape selectivity, namely a preference for inclusion of bicyclic (spherical) guests [32].

13.4.1.1 Specific Esterification of *N*-Acetyl-L-Amino Acids

p-Sulfonatocalix[6]arene **32** catalyzes the specific esterification of *N*-acetyl-L-amino acids bearing basic residues. Indeed, the reaction rates of His **33**, Lys **34** and Arg **35** were markedly enhanced in the presence of p-sulfonatocalix[6]arenes compared to Phe (Scheme 13.8) [33]. The reaction rates were measured in the presence of a large excess of calixarenes (>10 equiv.) and were compared to those obtained in the presence of the non-cyclic p-hydroxybenzenesulfonic acid (pHBS).

The size of the cavity plays a crucial role on the selectivity of the reaction. For example, when the esterification was performed with p-sulfonatocalix[4]arenes, the $k_{\text{calixarene}}/k_{p\text{HBS}}$ value for histidine was indeed increased from 24 to 86. ^{1}H NMR studies supported the formation of an inclusion complex of calixarenes with basic amino acids, and the reaction followed Michaelis–Menten kinetics. The specific rate enhancement observed for basic amino acids His **33**, Lys **34** and Arg **35** is the result of a stabilization by the anionic sulfonate groups of the cationic intermediate, which can undergo esterification (see Scheme 13.8). In contrast, the formation of Phe **36**, might proceed via simple acid catalysis.

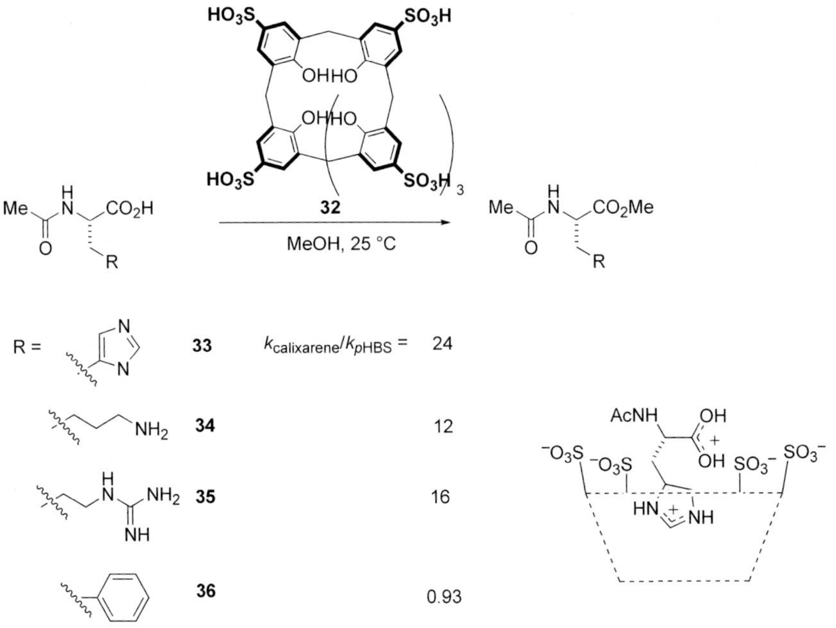

Scheme 13.8

13.4.2
Cyclodextrins

Cyclodextrins (CDs, Schardinger dextrins) are naturally occurring toroidal molecules consisting of six, seven, eight, nine or ten α-1,4-linked D-glucose units joined head-to-tail in a ring (α-, β-, γ-, δ- and ε-cyclodextrins, respectively). They may be synthesized from starch by the cyclomaltodextrin glucanotransferase from *Bacillus macerans*. Such a spatial arrangement results in the formation of hydrophobic pockets, due to the glycosidic oxygen atoms and inwards-facing C–H groups, which can differ in the diameter of the cavities (by about 0.5–1 nm) but remain about 0.7 nm deep. All of the C-6 hydroxyl groups project to one end, and all the C-2 and C-3 hydroxyl groups to the other. Although cyclodextrins are water-soluble, the presence of their hydrophobic pocket enables them to bind hydrophobic molecules of an appropriate size. The vast majority of complexes involving cyclodextrins and neutral organic molecules have average dissociation constants in the decimolar (10^{-1} M) to hundred micromolar (10^{-4} M) range, corresponding to the weakest energy host–guest interactions. Cyclodextrins can be derivatized in order to introduce catalytically relevant groups. Although their chiral discriminating properties have long been recognized [34], and amply exploited in the chromatographic separations of enantiomeric compounds, little has been done to develop enzyme-like catalysts of these substrates. While enantioselective reactions are known, few of them are amenable to catalytic transformations.

13.4.2.1 Aryl Glycoside Hydrolysis

Functionalized CDs may catalyze the hydrolysis of various aryl glycosides [35, 36]. Cyclodextrin diacid **37** is an artificial glycosidase with a k_{cat}/k_{uncat} of 35 for the hydrolysis of 4-nitrophenyl-β-D-glucoside (**40**) (Scheme 13.9) [35c]. The dicyanohydrin **38** [37] is a significantly better artificial enzyme, achieving k_{cat}/k_{uncat} of 200–2000 for the hydrolysis of aryl glycosides. While no marked difference in the hydrolysis of nitrophenyl α-glucoside versus β-glucoside was found using the latter catalyst, the related α-manno and α-galacto substrates were hydrolyzed at a slower rate, compared to the β-isomers. The catalysis can be inhibited by the addition of cyclopropanol, thus confirming that the CD cavity is involved in the catalytic process. The corresponding dinitrile-derivative (i.e. without the α-OH) showed an approximately 250-fold lower catalytic activity than **38**, whereas the dialdehyde dihydrate **39** exhibited a catalytic efficacy 20-fold lower than **38**. Furthermore, no reaction was observed neither with β-CD nor with mandelonitrile, showing that the supramolecular positioning of the binding cavity and the cyanohydrin group is essential for catalytic activity. An important feature of catalyst **38** is that the 6-OH groups are fixed in the *gt* conformation pointing toward the binding site. The proposed mechanism suggests the formation of an inclusion complex involving the aromatic moiety of the molecule, followed by activation of the glycosidic linkage by the cyanohydrin OH group.

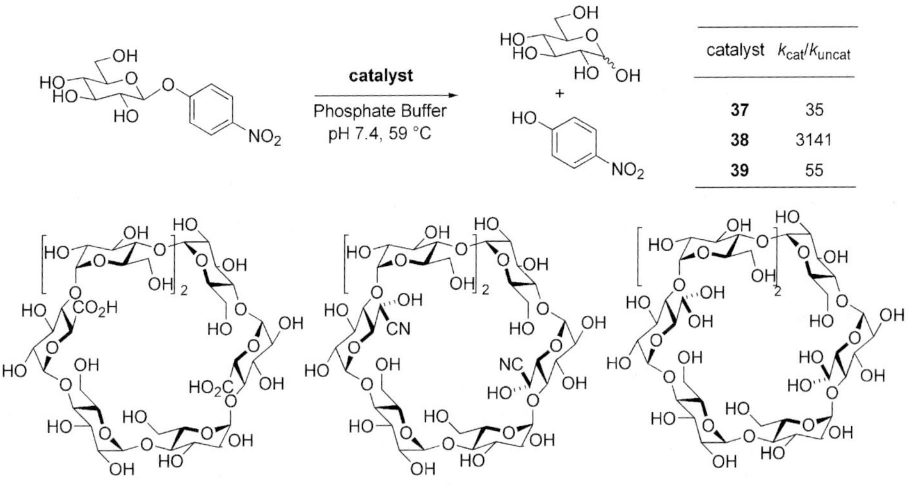

catalyst	k_{cat}/k_{uncat}
37	35
38	3141
39	55

Scheme 13.9

13.4.2.2 Phosphate Hydrolysis

Regioselective phosphate hydrolysis of **41** and **42** was realized with bifunctional catalysts **43–45**, in which one imidazole acts as a base, while the other protonated species acts as an acid (Scheme 13.10) [38]. Whereas all three catalysts were able to hydrolyze **41**, only **44** and **45** could cleave **42**, and in all cases the maximum

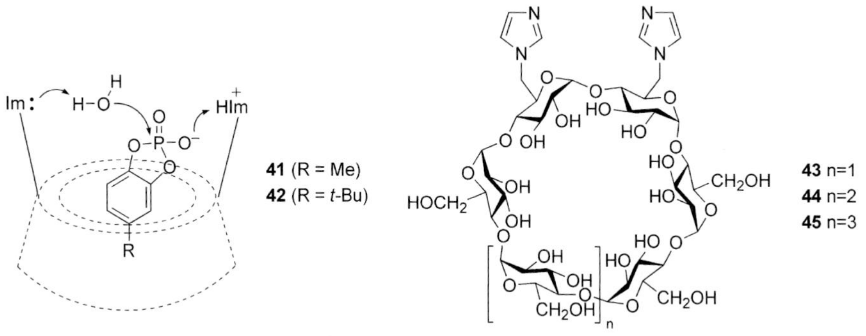

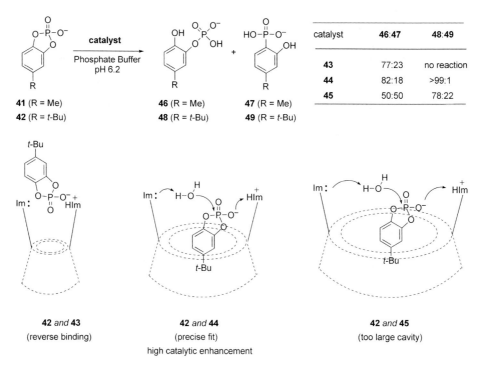

Scheme 13.10

rate was observed at ca. pH 6. In these reactions the size of the cavity might influence not only the catalyst-substrate binding affinity, but also the catalytic efficiency, and it is interesting to compare the substrate/catalyst binding forces to the rate constants of the reactions. Indeed, the affinity of compound **41** towards the catalysts (**44** > **43** > **45**) is not directly correlated to the rate of hydrolysis

catalyst	46:47	48:49
43	77:23	no reaction
44	82:18	>99:1
45	50:50	78:22

42 and **43** (reverse binding)

42 and **44** (precise fit) high catalytic enhancement

42 and **45** (too large cavity)

Scheme 13.11

(**43** > **44** > **45**). The contrasts are enhanced in the case of **42**, as no catalysis is observed with **43**, whereas both k_{cat} and binding efficiency are in the order **44** > **45**.

The anionic cyclic phosphates bind both to the cavity via their hydrophobic core and to the cationic imidazole groups via their polar head (Scheme 13.11). The precise fit of the *t*-Bu group of **42** inside the cavity results not only in an increased binding affinity, but also in a considerable increase in the catalytic rate constant and regioselectivity. In contrast, when **42** binds **45**, with its larger cavity, the advantages are less pronounced. The lack of reactivity of compound **42** in the presence of **43** was attributed to a reverse binding mode, due to impossibility for the bulky *t*-Bu group to fit inside the cavity.

13.4.2.3 Transamination Reactions

The conversion of α-keto acids to the corresponding α-amino acids is catalyzed under biological conditions by amino transferases, as depicted in Eq. (1):

$$\text{L-Glu} + \alpha\text{-keto acid} = \alpha\text{-keto glutaric acid} + \text{L-}\alpha\text{-amino acid} \quad (1)$$

In this transformation, L-Glu can be replaced by L-Asp, L-Orn, or by other amino acids; the reaction requires the presence of pyridoxal-5-phosphate (vitamin B_6), or, thiamine (vitamin B_1) [39, 40], as coenzyme.

In mimicking transaminases, the C-6 hydroxyl group of β-CD was covalently derivatized by an activated pyridoxal coenzyme [41]. In order to reproduce the catalytic role of the basic residue at the active site of transaminase enzymes, imidazoles were incorporated into the CD framework. Two isomeric compounds, **50** and **51**, carrying adjacent pyridoxamine and imidazole units on the C-6 position of the glucose backbone, have been synthesized (Scheme 13.12) [42]. These catalysts were tested in the transamination reaction of phenylpyruvic acid **52** producing phenylalanine (Phe). Interestingly, the compound **50** afforded a mixture of L-Phe and D-Phe in a 5:1 ratio, whereas the regioisomer **51** afforded the same amino acid, albeit with an opposite stereochemical preference (1:5 ratio). The examination of molecular models of **50** and **51** indicated that, during formation of the new amino acid, the proton was added on the face *away* from the imidazole group. Thus, the imidazole ring probably serves to prevent proton addition to one face of the imine intermediate, by steric blocking, rather than protonating the transamination intermediates.

13.4.2.4 Amidation Reactions

The substrate-specific condensation of aromatic carboxylates **52** and **53** having strong affinity for the CD cavity of **54** was realized with *in-situ* activation by formation of an acyloxytriazine, which undergoes aminolysis to give an amide [43]. Competitive experiments between **52** and **53** showed high selectivity, resulting in the major formation of compound **55** over compound **56** (Scheme 13.13). The observed selectivity can be attributed to differences in the affinity of the substrates for the CD cavity (Scheme 13.13). Most importantly, the CD-based catalyst **54**

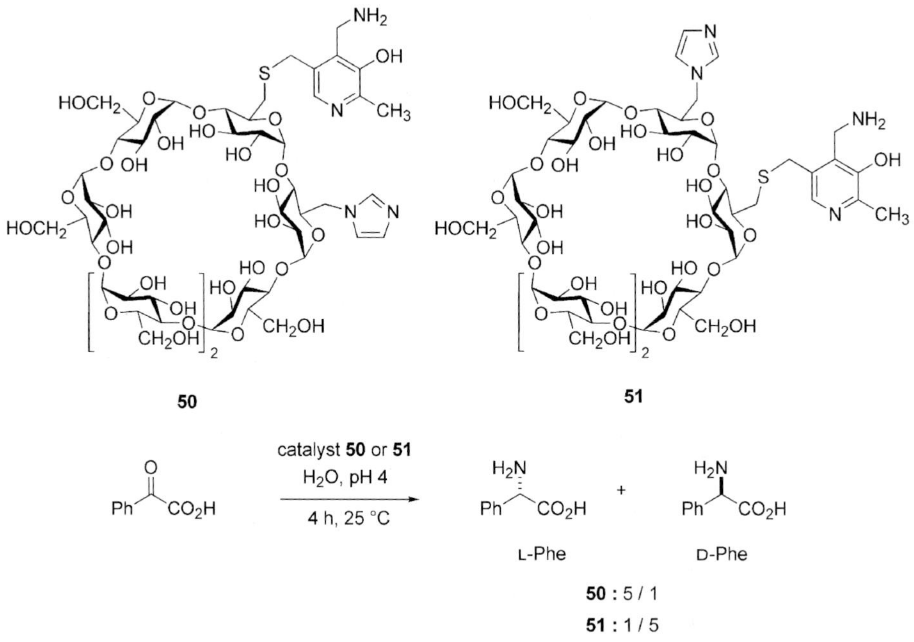

Scheme 13.12

accelerates the condensation of **52** by a factor of 13, compared to *N,N*-dimethyl-glycine ethyl ester. Product inhibition resulting from competitive binding of the hydrophobic part of the product to the cavity of **54** is overcome by precipitation of the products under the reaction conditions.

13.4.2.5 Epoxidation of Alkenes

Cyclodextrins have been covalently modified for catalytic oxidation, such as compounds **57**, **62–65** (Schemes 3.14 to 3.16) [44, 45]. Enantioselective epoxidation of styrene derivatives, and carene using 20–100 mol% of the CD-ketoester **57** has been achieved. The inclusion-complex formation was confirmed by ^{1}H NMR titration experiments, confirming the 1:1 substrate:catalyst stoichiometry under the reaction conditions. In the oxidation of carene, NOE and ROESY experiments showed different behavior according to the size of the R group (Scheme 13.14). Evidence was found for the formation of inclusion complexes with compounds **58** and **59**. On the other hand, compounds **60** and **61** proved to interact with the catalyst via a "tail" inclusion (*vide infra*). The increased diastereoselectivity observed with compounds **58** and **59** might be explained by a closer proximity to the covalently linked dioxirane.

This reaction has also been successfully applied to the enantioselective epoxidation of various styrenes, and furnished the corresponding epoxides in up to 99% chemical yield, and *ee*-values up to 40% (Scheme 13.15). It should be noted that

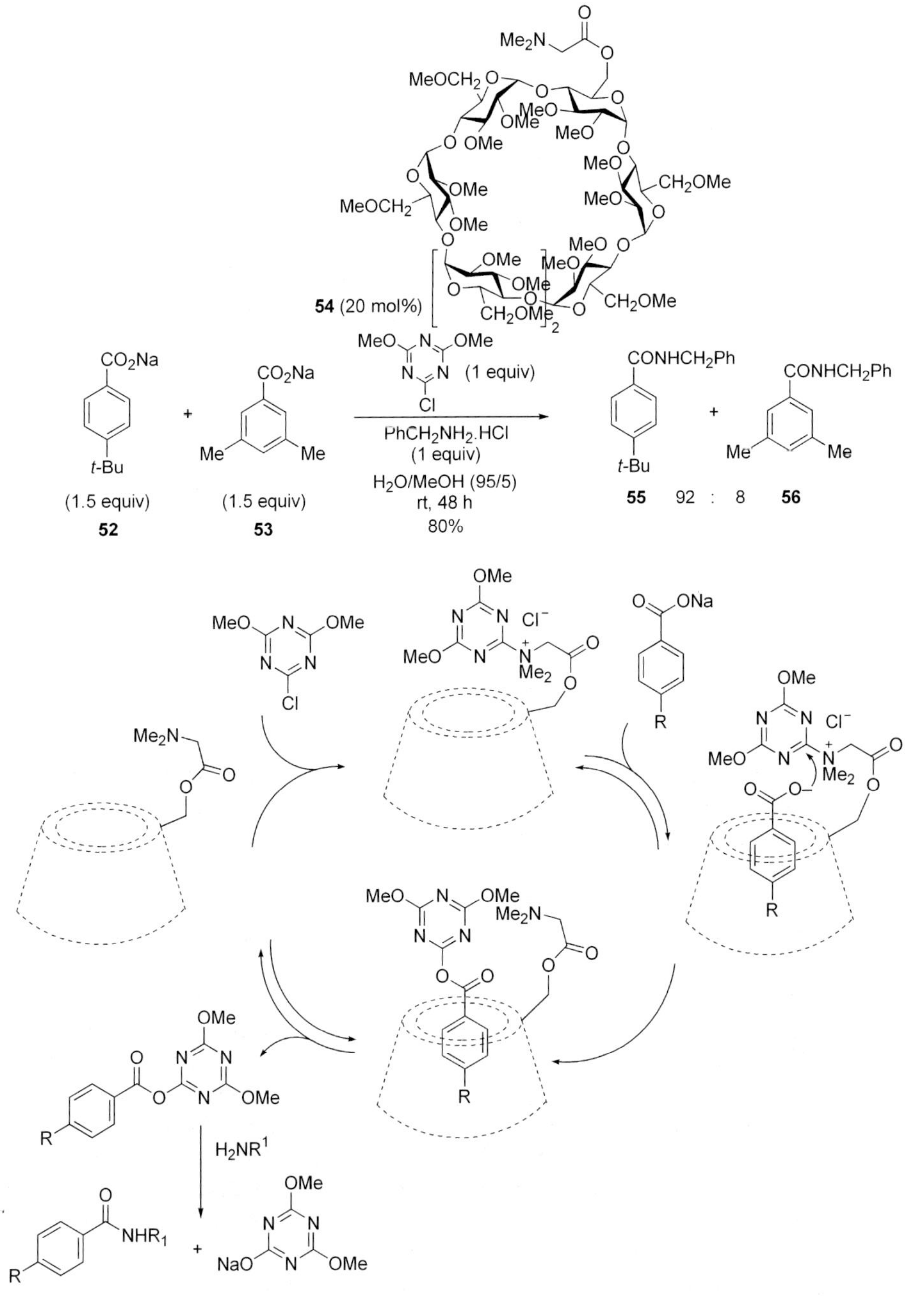

13.4 Chiral Cavitands: Calixarenes and Cyclodextrins

Scheme 13.13

440 | 13 Shape- and Site-Selective Asymmetric Reactions

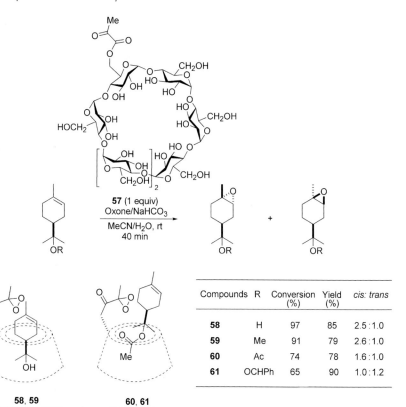

Compounds	R	Conversion (%)	Yield (%)	cis: trans
58	H	97	85	2.5:1.0
59	Me	91	79	2.6:1.0
60	Ac	74	78	1.6:1.0
61	OCHPh	65	90	1.0:1.2

58, 59 complete inclusion

60, 61 "tail" inclusion

Scheme 13.14

Ph–CH=CH–Ph + **57** (20 mol%), Oxone, NaHCO$_3$, Na$_2$EDTA (aq), CH$_3$CN, rt, 2.5h (88%) → Ph-epoxide-Ph

ee = 29%

Scheme 13.15

the catalyst loading could be decreased to 20 mol%, without any significant decrease in enantioselectivity.

Another strategy for positioning a catalytic center across the entrance of a conical cavity is to employ a cavitand functionalized at one entrance by a pendent chelate arm (Scheme 13.16). Enantioselective epoxidations of aromatic alkenes was realized with catalysts **62**, **63**, and **64**, **65**, although the enantioselectivity remained modest [46]. (For experimental details see Chapter 14.13.11). The reaction requires the slow addition (over 1 h) of a solution of alkene **66** and Oxone to a solution of the catalyst. Both the size of the cavity and the structure of the bridged ketone influenced the reactivity. Hence, whilst the formation of the diol **68** was observed when **62** and **63** were used, the presence of **64** and **65** resulted only in the formation of epoxide **67**.

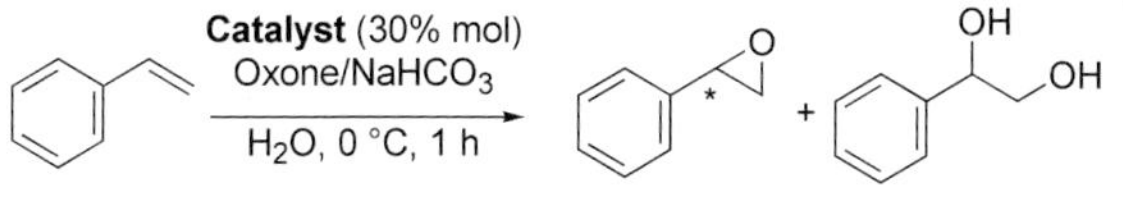

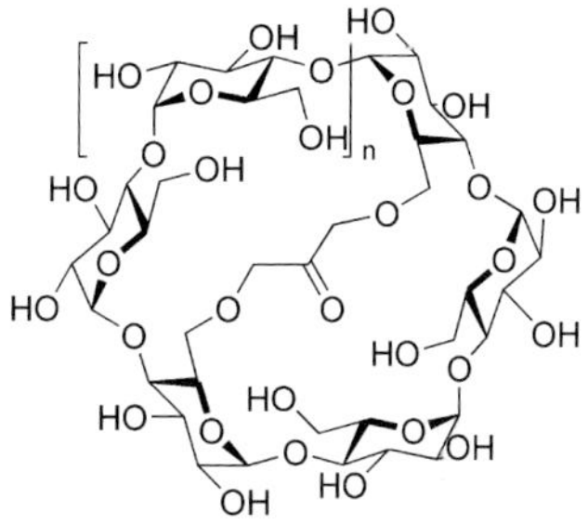

Catalyst	Yield (%) 67	ee (%) 67	Yield (%) 68
62	50	12	50
63	19	0	44
64	70	<5	-
65	quant	45	-

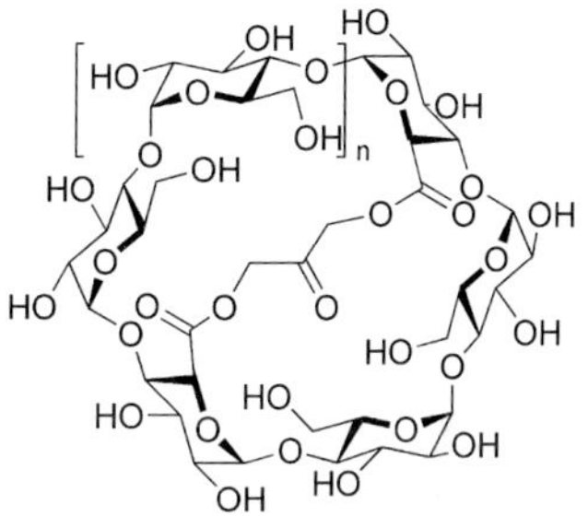

Scheme 13.16

The epoxide formation is thought to proceed via Oxone-mediated dioxirane formation and inclusion of the aromatic moiety inside the cavity. The influence of the substrate binding inside the CD cavity was evaluated by inhibitory experiments with naphthalene-2-sulfonate, which is known to bind both α- and β-CD with good affinity. As expected, the addition of 2 equivalents of inhibitor to the reaction mixture resulted in a significant decrease in reaction rate.

13.5
Molecular Imprinting

Molecular imprinting involves the formation of a substance, most often a polymer, around a molecular template, such as **74** (Scheme 13.17) [25g, 47, 48]. The functional monomers interact with a template through covalent or non-covalent interactions. Once the polymer is formed, the template is removed by extraction, leaving a catalytic cavity (i.e., a selective binding site), which is characterized both by the shape and the spatial arrangement of the functional groups. In order to exhibit selectivity, the molecular imprinting polymer (MIP) cavity needs to provide a well-defined (geometric) orientation of the binding and catalytic sites. Usually, the shape of the cavity corresponds to the shape of the substrate, or better, of the transition state of the given reaction [6]. The shape-based molecular selection is the result of a combination of covalent, ion-pairing and hydrogen bonds. The

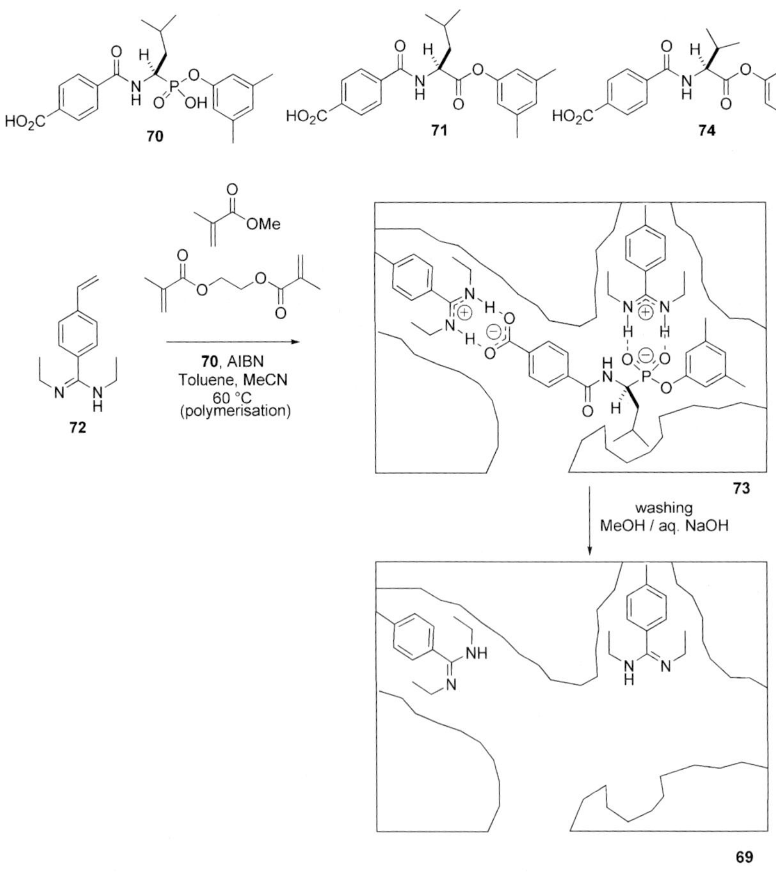

Scheme 13.17

cavity can be endowed with a variety of engineered properties, ranging from selective binding of specific molecules (molecular recognition) to enzyme-like catalysis. Such a simple and appealing concept has for more than 50 years attracted chemists from several disciplines, and actually promises a breakthrough in enantioselective catalysis. Highly cross-linked network polymers were developed which exhibited excellent enantiomer recognition properties for a highly interesting range of substances such as carbohydrates, esters, amino acids and peptides, steroids, and nucleotide bases. In parallel, innovative receptor design approaches based on dynamic combinatorial libraries are emerging, and new applications in sensors and catalysis are being developed in addition to traditional applications in chromatography [49]. While the concept is appealing, this seemingly "simple" approach is not without difficulties however. Problems arise, for example, from the proper technique used for tailoring catalytically active polymers: either due to microstructural factors of the catalyst such as accessibility, local solvation, non-homogeneity of the catalytic centers, etc., or to the rigidity of the structure [50]. Although shape-selective transformation usually infers enantiodiscrimination, truly asymmetric transformations using imprinted polymers remain scarce [51].

13.5.1
Enantioselective Ester Hydrolysis

Optically active phosphine monoester **69** was prepared as a stable transition state analogue of the hydrolysis of **70**, and used as a template (Scheme 13.17) [52]. When heated in the presence of amidine **72**, methyl methacrylate, ethylene glycol dimethacrylate and AIBN, polymerization took place and offered the polymer **73**, in which the template **69** is attached to the binding sites by a combination of hydrogen bonds and electrostatic interactions. After washing in the presence of MeOH and aqueous NaOH, the template was removed from the cavity, thus leading to a catalytically active MIP.

The MIP-mediated hydrolysis of ester **70** in HEPES buffer (0.15 M, pH 7.3) and MeCN at 20 °C showed Michaelis–Menten kinetics and a 325-fold rate enhancement compared to the reaction performed in solution. By comparison, the hydrolysis of ent-**70** showed a 234-fold enhancement under the same conditions, resulting in a pronounced enantioselectivity. Interestingly, the apparently slight difference between **70** and **71** resulted in a significant decrease in catalyst activity, as the rate enhancement compared to the reaction performed in solution was "only" 103-fold.

The mechanism of the ester hydrolysis or transesterification is thought to proceed via a pre-equilibrium state involving binding between the ester **70** and the MIP catalyst **74**, followed by an activation of the carbonyl group by one amidine group, and attack of a water molecule (Scheme 13.18). The reaction is believed to involve a high-energy tetrahedral oxanion intermediate (transition state-like). Rate enhancements caused by esterase enzymes can, to a large part, be ascribed to the electrostatic stabilization of this structure [53]. The transition states and the inter-

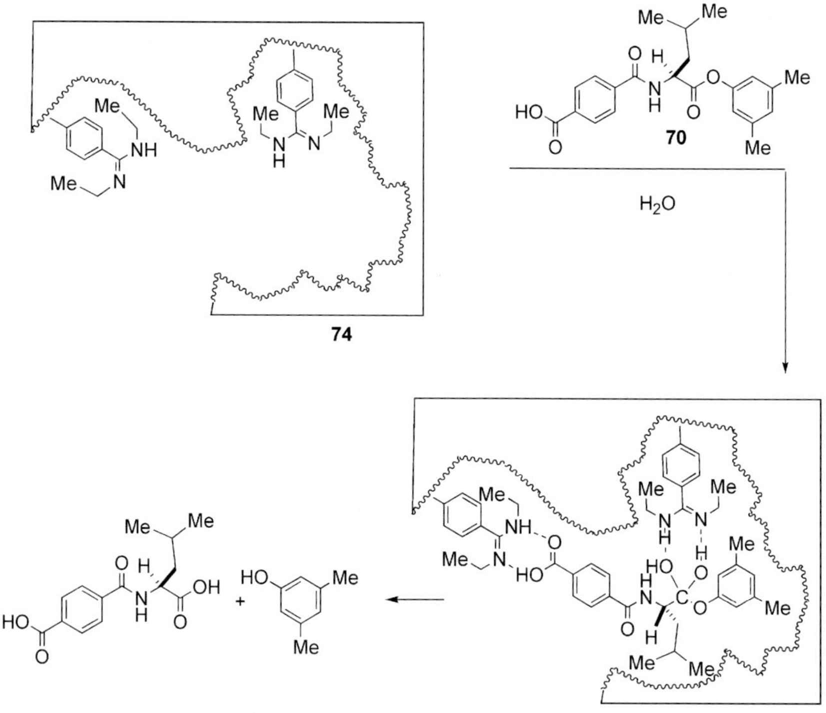

Scheme 13.18

mediate **I** show small energy differences due to similar structures. This justifies why an intermediate analogue and not a transition state analogue could be used as a template during synthesis of the MIP catalyst **74**.

A similar concept was used in the development of artificial chymotrypsin mimics [54]. The esterase-site was modeled by using the phosphonate template **75** as a stable transition state analogue (Scheme 13.19). The catalytic triad of the active site of chymotrypsin – that is, serine, histidine and aspartic acid (carboxylate anion) – was mimicked by imidazole, phenolic hydroxy and carboxyl groups, respectively. The catalytically active MIP catalyst **76** was prepared using free radical polymerization, in the presence of the phosphonate template **75**, methacrylic acid, ethylene glycol dimethacrylate and AIBN. The template removal conditions had a decisive influence on the efficiency of the polymer-mediated catalysis, and best results were obtained with aqueous Na_2CO_3.

A relatively low pH (pH 4.5) was necessary to obtain a selective MIP-catalyzed hydrolysis of ester **77** (Scheme 13.19); this was most likely due to the involvement of the carboxylic acid moieties in substrate binding. In addition, at this pH, the

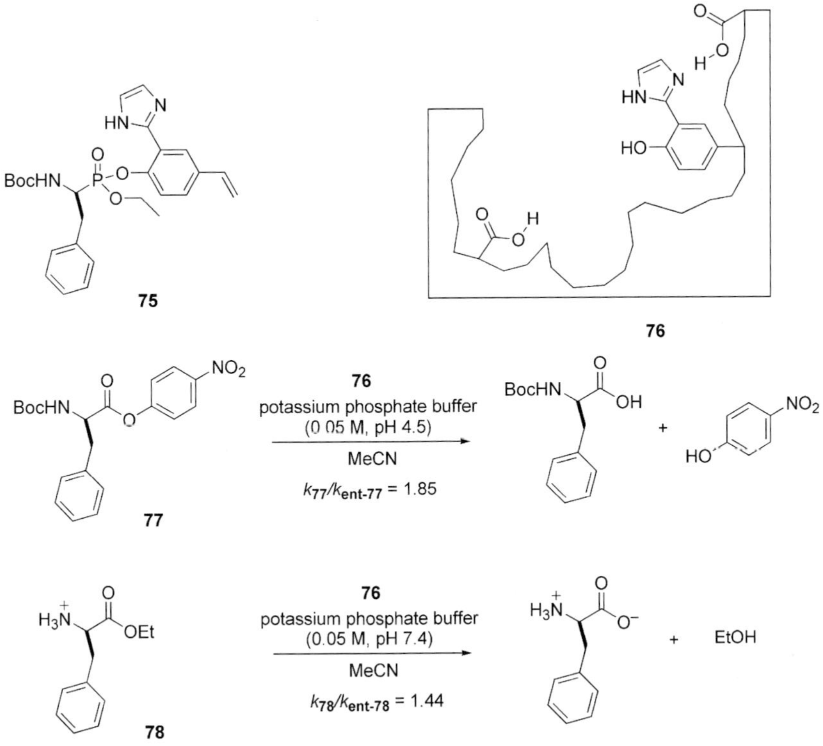

Scheme 13.19

imidazole moiety should exist as the protonated form, and act as a general acid catalyst, allowing a water molecule to attack the substrate.

MIP catalyst **76** also performs enantioselective hydrolysis of non-activated D- and L-phenylalanine ethyl ester **78** at pH 7.4, with characteristic saturation kinetics (Scheme 13.19). While this catalyst does not have the precise shape complementarity to ester **78**, a threefold enhancement compared to the blank reaction was observed. Furthermore, enantiodiscrimination was obtained, as compound **78** was hydrolyzed 1.44-fold faster than ent–**78**. Although the reported rate enhancement was modest compared to the standards in asymmetric synthesis, it can be considered as a first step in the development of imprinted polymers for enantioselective catalysts.

Enantioselective hydrolysis of *p*-nitrophenyl-*N*-(benzylocycarbonyl)-L-leucinate compared to its D-isomer could also be performed with a MIP catalyst built up with a racemic template [55]. In that case, the enantiodiscrimination is insured by the presence of L-His monomer during the polymerization process, performed under finely tuned conditions. The random distribution of quaternary trimethylammonium groups through the polymer framework makes this MIP very soluble in water. The MIP-catalyzed ester hydrolysis was performed in a mixture of

MeCN and Tris buffer (pH 7.15) at 30 °C, and showed a predominant reaction of Boc-L-Leu-PNP compared to Boc-D-Leu-PNP ($k_L/k_D = 8.41$).

13.6
Conclusion

The utilization of artificial enzyme-like catalysts, which are able to select substrates based on size or site/sequence criteria, is in its infancy. This field combines receptor-like mid-size or macromolecular structures forming either nanoscale organic vessels, or complementary receptor-like structures, and a catalytic site, which are operating typically as bifunctional catalysts. The design of shape- and site-selective catalysts based on host–guest interactions is undoubtedly an innovative approach in organic chemistry, which foresees major perspectives. Moreover, the scope in organic synthesis is notable, rendering protecting-directing groups unnecessary. In addition, these molecules can be used as tools in molecular biology, and as catalysts for organic chemistry, notably for hydrolytic reactions which mimic EC 3.1-type enzymes. In these catalysts the catalytic (nucleophilic) site adopts a spatial orientation suitable for proton transfer, this being usually insured by the chiral template. Although these catalysts perform efficient size or shape discrimination for non-asymmetric reactions, only a few enantioselective transformations have been identified, and these remain to be improved for synthetic applications. Our growing understanding of shape- and site-selective catalysts provides new insight into old problems of organocatalysis and newer problems such as enzyme design.

References

1 (a) R. Breslow (Ed.), *Artificial Enzymes*. Wiley-VCH, Weinheim, **2005**; (b) R. Breslow, *Science* **1982**, *218*, 532–537.

2 (a) M. Mammen, S.-K. Choi, G.M. Whitesides, *Angew. Chem. Int. Ed.* **1998**, *37*, 2754–2794; (b) L.L. Kiessling, J.E. Gestwicki, L.E. Strong, *Curr. Opin. Chem. Biol.* **2000**, *4*, 696–703; (c) C.W. Lim, B.J. Ravoo, D.N. Reinhoudt, *Chem. Commun.* **2005**, 5627–5629; (d) See also: T. Bando, H. Sugiyama, *Acc. Chem. Res.* **2006**, *39*, 935–744.

3 (a) For related antibody-derived catalysts (abzymes), see: J.D. Stewart, L.J. Liotta, S.J. Benkovic, *Acc. Chem. Res.* **1993**, *26*, 396–404; (b) F. Tanaka, *Chem. Rev.* **2002**, *102*, 4885–4906; (c) P.G. Schultz, J. Yin, R.A. Lerner, *Angew. Chem. Int. Ed.* **2002**, *41*, 4427–4437, and also Ref. [1a].

4 D.M. Vriezema, M. Comellas Aragones, J.A.A.W. Elemans, J.J.L.M. Cornelissen, A.E. Rowan, R.J.M. Nolte, *Chem. Rev.* **2005**, *105*, 1445–1490.

5 (a) H.F. Fisher, *Acc. Chem. Res.* **2005**, *38*, 157–166; (b) J. Kraut, *Science* **1988**, *242*, 533–540.

6 (a) L. Pauling, *Nature* **1948**, *161*, 707–709; (b) L. Pauling, *Chem. Eng. News* **1946**, *24*, 1375–1380.

7 (a) M. Garcia-Viloca, J. Gao, M. Karplus, D.G. Truhlar, *Science* **2004**, *303*, 186–195; (b) S.J. Benkovic, S.

Hammes-Schiffer, *Science* **2003**, *301*, 1196–1202. See also: (c) R. Wolfenden, M.J. Snider, *Acc. Chem. Res.* **2001**, *34*, 938–945; (d) H. Dugas (Ed.), *Bioorganic Chemistry*, 3rd edn. Springer-Verlag, New York, **1996**; (e) S. Borman, *Chem. Eng. News* **2004**, *82*, 35–39; (f) D. Fiedler, D.H. Leung, R.G. Bergman, K.N. Raymond, *Acc. Chem. Res.* **2005**, *38*, 349–358.

8 T.J. Stillman, P.J. Baker, K.L. Britton, D.W. Rice, *J. Mol. Biol.* **1993**, *234*, 1131–1139.

9 (a) C. Jeunesse, D. Armspach, D. Matt, *Chem. Commun.* **2005**, *45*, 5603–5614; (b) H.K.A.C. Coolen, P.W.N.M. van Leeuwen, R.J.M. Nolte, *Angew. Chem. Int. Ed. Engl.* **1992**, *31*, 905–907; (c) H.K.A.C. Coolen, J.H.N. Reek, J.M. Ernsting, P.W.N.M. van Leeuwen, R.J.M. Nolte, *Recl. Trav. Chim. Pays-Bas* **1995**, *114*, 381–386; (d) C. Loeber, C. Wieser, D. Matt, A. De Cian, J. Fischer, L. Toupet, *Bull. Soc. Chim. Fr.* **1995**, *132*, 166; (e) M.T. Reetz, S.R. Waldvogel, *Angew. Chem. Int. Ed. Engl.* **1997**, *36*, 865–867; (f) D. Armspach, D. Matt, *Chem. Commun.* **1999**, 1073–1074; (g) M. Vézina, J. Gagnon, K. Villeneuve, M. Drouin, P.D. Harvey, *Organometallics* **2001**, *20*, 273–281; (h) B. Kersting, *Anorg. Allg. Chem.* **2004**, *630*, 765–780.

10 (a) T. Kurahashi, T. Mizutani, J. Yoshida, *Tetrahedron* **2002**, *58*, 8669–8677; (b) T. Kurahashi, T. Mizutani, J. Yoshida, *J. Chem. Soc., Perkin Trans. 1* **1999**, 465–474.

11 G. Hu, A. Vasella, *Helv. Chim. Acta* **2002**, *85*, 4369–4391.

12 K.S. Griswold, S.J. Miller, *Tetrahedron* **2003**, *59*, 8869–8875.

13 (a) B.R. Sculimbrene, S.J. Miller, *J. Am. Chem. Soc.* **2001**, *123*, 10125–10126; (b) B.R. Sculimbrene, A.J. Morgan, S.J. Miller, *J. Am. Chem. Soc.* **2002**, *124*, 11653–11656; (c) B.R. Sculimbrene, A.J. Morgan, S.J. Miller, *Chem. Commun.* **2003**, 1781–1785; (d) A.J. Morgan, S. Komiya, Y. Xu, S.J. Miller, *J. Org. Chem.* **2006**, *71*, 6923–6931.

14 Reviews on sequence-selective artificial ribonucleases: (a) A. De Mesmaeker, R. Häner, P. Martin, H.E. Moser, *Acc. Chem. Res.* **1995**, *28*, 366–374; (b) M. Komiyama, *J. Biochem.* **1995**, *118*, 665–670; (c) B.N. Trawick, A.T. Daniher, J.K. Bashkin, *Chem. Rev.* **1998**, *98*, 939–960; (d) T. Niittymäki, H. Lönnberg, *Org. Biomol. Chem.* **2006**, *4*, 15–25.

15 W. Sun, A. Pertzev, A.W. Nicholson, *Nucleic Acids Res.* **2005**, *33*, 807–815.

16 I.J. MacRae, K. Zhou, F. Li, A. Repic, A.N. Brooks, W.Z. Cande, P.D. Adams, J.A. Doudna, *Science* **2006**, *311*, 195–198.

17 (a) G.A. Soukup, R.R. Breaker, *RNA* **1999**, *5*, 1308–1325; (b) V. Tereshko, S.T. Wallace, N. Usman, F.E. Wincott, M. Egli, *RNA* **2001**, *7*, 405–420; (c) U. Kaukinen, L. Bielecki, S. Mikkola, R.W. Adamiak, H. Lönnberg, *J. Chem. Soc., Perkin Trans. 2* **2001**, 1024–1031.

18 M. Komiyama, T. Inokawa, K. Yoshinari, *J. Chem. Soc., Chem. Commun.* **1995**, 77–78.

19 (a) K. Yoshinari, K. Yamazaki, M. Komiyama, *J. Am. Chem. Soc.* **1991**, *113*, 5899–5901; (b) M. Komiyama, K. Yoshinari, *J. Org. Chem.* **1997**, *62*, 2155–2160.

20 J.C. Verheijen, B.A.L.M. Deiman, E. Yeheskiely, G.A. van der Marel, J.H. van Boom, *Angew. Chem. Int. Ed.* **2000**, *39*, 369–372.

21 P.E. Nielsen, M. Egholm, R.H. Berg, O. Buchardt, *Science* **1991**, *254*, 1497–1500.

22 M. Egholm, O. Buchardt, L. Christensen, C. Behrens, S.M. Freier, D.A. Driver, R.H. Berg, S.K. Kim, B. Norden, P.E. Nielsen, *Nature* **1993**, *365*, 566–568.

23 M.A. Reynolds, T.A. Beck, P.B. Say, D.A. Schwartz, B.P. Dwyer, W.J. Daily, M.M. Vaghefi, M.D. Metzler, R.E. Klem, L.J. Arnold, Jr., *Nucleic Acids Res.* **1996**, *24*, 760–765.

24 (a) I. Tabushi, *Acc. Chem. Res.* **1982**, *15*, 66–72; (b) R. Breslow, *Acc. Chem. Res.* **1995**, *28*, 146–153; (c) M. Komiyama, H. Shigekawa, in: J.-M. Lehn, J.L. Atwood, J.E.D. Davies, D.D. MacNicol, F. Vögtle (Eds.), *Comprehensive Supramolecular Chemistry*.

Pergamon, Oxford, **1996**, Vol. 3, pp. 401–422.
25 Reviews: (a) R. Breslow, *Acc. Chem. Res.* **1995**, *28*, 146–153; (b) R. Breslow, S.D. Dong, *Chem. Rev.* **1998**, *98*, 1997–2012; (c) K. Takahashi, *Chem. Rev.* **1998**, *98*, 2013–2033; (d) E. Engeldinger, D. Armspach, D. Matt, *Chem. Rev.* **2003**, *103*, 4147–4173. On enzyme mimics, see also: (e) A.J. Kirby, *Angew. Chem. Int. Ed.* **1994**, *33*, 551–553; (f) Y. Murakami, J.-I. Kikuchi, Y. Hisaeda, O. Hayashida, *Chem. Rev.* **1996**, *96*, 721–758; (g) W.B. Motherwell, M.J. Bingham, Y. Six, *Tetrahedron* **2001**, *57*, 4663–4686; (h) R. Breslow, C. Schmuck, *J. Am. Chem. Soc.* **1996**, *118*, 6601–6605; (i) R. Breslow, B. Zhang, *J. Am. Chem. Soc.* **1992**, *114*, 5882–5883; (j) R. Breslow, B. Zhang, *J. Am. Chem. Soc.* **1994**, *116*, 7893–7894; (k) T. Akiike, Y. Nagano, Y. Yamamoto, A. Nakamura, H. Ikeda, A. Veno, F. Toda, *Chem. Lett.* **1994**, 1089–1092; (l) N.M. Milovic, J.D. Badjic, N.M. Kostic, *J. Am. Chem. Soc.* **2004**, *126*, 696–697.
26 Review on calyxarenes: P. Lhoták, *Eur. J. Org. Chem.* **2004**, 1675–1692.
27 K. Goto, Y. Yano, E. Okada, C.-W. Liu, K. Yamamoto, R. Ueoka, *J. Org. Chem.* **2003**, *68*, 865–870.
28 (a) S. Shinkai, K. Araki, T. Matsuda, N. Nishiyama, H. Ikeda, I. Takasu, M. Iwamoto, *J. Am. Chem. Soc.* **1990**, *112*, 9053–9058; (b) G. Arena, A. Casnati, A. Contino, F.G. Gulino, D. Sciotto, R. Ungaro, *J. Chem. Soc., Perkin Trans. 2* **2000**, 419–423; (c) W. Abraham, *J. Inclusion Phenom. Macrocyclic Chem.* **2002**, *43*, 159–174; (d) J.-M. Lehn, R. Meric, J.-P. Vigneron, M. Cesario, J. Guilhem, C. Pascard, Z. Asfari, J. Vicens, *Supramol. Chem.* **1995**, *5*, 97–103.
29 (a) M. Baur, M. Frank, J. Schatz, F. Schildbach, *Tetrahedron* **2001**, *57*, 6985–6991; (b) G. Arena, A. Contino, F.G. Gulino, A. Magrì, D. Sciotto, R. Ungaro, *Tetrahedron Lett.* **2000**, *41*, 9327–9330; (c) N. Kon, N. Iki, S. Miyano, *Org. Biomol. Chem.* **2003**, *1*, 751–755; (c) F. Perret, J.-P. Morel, N. Morel-Desrosiers, *Supramol. Chem.* **2003**, *15*, 199–206.
30 S. Kunsági-Máté, I. Bitter, A. Grün, G. Nagy, L. Kollár, *Anal. Chim. Acta* **2002**, *461*, 273–279.
31 (a) S. Shinkai, K. Araki, O. Manabe, *J. Chem. Soc., Chem. Commun.* **1988**, 187–189; (b) S. Shinkai, in: J. Vicens, V. Böhmer (Eds.), *Calixarenes*. Kluwer Academic Publishers, Dordrecht, **1990**, pp. 173–198; (c) P. Ballester, A. Shivanyuk, A.R. Far, J. Rebek, Jr. *J. Am. Chem. Soc.* **2002**, *124*, 14014–14016.
32 H. Bakirci, A.L. Koner, W.M. Nau, *J. Org. Chem.* **2005**, *70*, 9960–9966.
33 K. Goto, Y. Yano, E. Okada, C.-W. Liu, K. Yamamoto, R. Ueoka, *J. Org. Chem.* **2003**, *68*, 865–870.
34 (a) F. Cramer, W. Dietsche, *Chem. Ber.* **1959**, *92*, 1739–1747; (b) for a review, see: C.J. Easton, S.F. Lincoln, *Chem. Soc. Rev.* **1996**, *25*, 163–170.
35 (a) F. Ortega-Caballero, C. Rousseau, B. Christensen, T.E. Petersen, M. Bols, *J. Am. Chem. Soc.* **2005**, *127*, 3238–3239; (b) F. Ortega-Caballero, J. Bjerre, L.S. Lausten, M. Bols, *J. Org. Chem.* **2005**, *70*, 7217–7226; (c) C. Rousseau, N. Nielsen, M. Bols, *Tetrahedron Lett.* **2004**, *45*, 8709–8711.
36 For earlier works on glycosidase mimics, see: (a) D.T.H. Chou, J. Zhu, X. Huang, A.J. Bennet, *J. Chem. Soc., Perkin Trans. 2* **2001**, 83–89; (b) T. Ohe, Y. Kajiwara, T. Kida, W. Zhang, Y. Nakatsuji, I. Ikeda, *Chem. Lett.* **1999**, 921–922.
37 Compound 38 can be prepared from cyclodextrin dialdehyde via cyanohydrin synthesis according to Ref. [35a]. This reaction affords a single diastereomer as product.
38 R. Breslow, C. Schmuck, *J. Am. Chem. Soc.* **1996**, *118*, 6601–6605.
39 R. Breslow, *J. Am. Chem. Soc.* **1958**, *80*, 3719–3726.
40 J.A. Murry, D.E. Frantz, A. Soheili, R. Tillyer, E.J.J. Grabowski, P.J. Reider, *J. Am. Chem. Soc.* **2001**, *123*, 9696–9697.
41 (a) R. Breslow, M. Hammond, M. Lauer, *J. Am. Chem. Soc.* **1980**, *102*, 421–422; (b) R. Breslow, A.W.

Czarnik, M. Lauer, R. Leppkes, J. Winkler, S. Zimmerman, *J. Am. Chem. Soc.* **1986**, *108*, 1969–1979; (c) R. Breslow, J. Chmielewski, D. Foley, B. Johnson, N. Kumabe, M. Varney, R. Mehra, *Tetrahedron* **1988**, *44*, 5515–5524; (d) R. Breslow, J.W. Canary, M. Varney, S.T. Waddell, D. Yang, *J. Am. Chem. Soc.* **1990**, *112*, 5212–5219.

42 E. Fasella, S.D. Dong, R. Breslow, *Bioorg. Med. Chem.* **1999**, *7*, 709–714.

43 M. Kunishima, K. Yoshimura, H. Morigaki, R. Kawamata, K. Terao, S. Tani, *J. Am. Chem. Soc.* **2001**, *123*, 10760–10761.

44 W.-K. Chan, W.-Y. Yu, C.-M. Che, M.-K. Wong, *J. Org. Chem.* **2003**, *68*, 6576–6582.

45 For related cyclodextrin-based transformations see (a) R. Breslow, J. Yang, J. Yan, *Tetrahedron* **2002**, *58*, 653–659; (b) J. Yang, B. Gabriele, S. Belvedere, Y. Huang, R. Breslow, *J. Org. Chem.* **2002**, *67*, 5057–5067, and references cited therein.

46 (a) C. Rousseau, B. Christensen, T.E. Petersen, M. Bols, *Org. Biomol. Chem.* **2004**, *2*, 3476–3482; (b) C. Rousseau, B. Christensen, M. Bols, *Eur. J. Org. Chem.* **2005**, 2734–2739.

47 (a) R.A. Bartsch, M. Maeda (Eds.), *Molecular and Ionic Recognition with Imprinted Polymers*. American Chemical Society, **1998**; (b) B. Sellergren (Ed.), *Molecularly Imprinted Polymers: Man-Made Mimics of Antibodies and their Application in Analytical Chemistry*. Elsevier, **2000**; (c) K.J. Shea, M. Yan, M.J. Roberts (Eds.), *Molecularly Imprinted Materials – Sensors and Other Devices*. Materials Research Society, Symposia Held April 2–5, **2002**, San Francisco; (d) M. Komiyama, T. Takeuchi, T. Mukawa, H. Asanuma (Eds.), *Molecular Imprinting: From Fundamentals to Applications*. Wiley-VCH, Weinheim, **2003**.

48 Reviews: (a) C. Alexander, L. Davidson, W. Hayes, *Tetrahedron* **2003**, *59*, 2025–2056; (b) G. Wulff, *Chem. Rev.* **2002**, *102*, 1–28; (c) B. Clapham, T.S. Reger, K.D. Janda, *Tetrahedron* **2001**, *57*, 4637–4662; (d) B. Sellergren, *Angew. Chem. Int. Ed.* **2000**, *39*, 1031–1037.

49 (a) S. Otto, R.L.E. Furlan, J.K.M. Sanders, *Curr. Opin. Chem. Biol.* **2002**, *6*, 321–327; (b) R.L.E. Furlan, S. Otto, J.K.M. Sanders, *Proc. Natl. Acad. Sci. USA* **2002**, *99*, 4801–4804; (c) S. Otto, R.L.E. Furlan, J.K.M. Sanders, *Science* **2002**, *297*, 590–593; (d) S. Otto, R.L.E. Furlan, J.K.M. Sanders, *Drug Discovery Today* **2002**, *7*, 117–125; (e) S.J. Rowan, S.J. Cantrill, G.R.L. Cousins, J.K.M. Sanders, J.F. Stoddart, *Angew. Chem. Int. Ed.* **2002**, *41*, 898–952; (f) G.R.L. Cousins, S.-A. Poulsen, J.K.M. Sanders, *Curr. Opin. Chem. Biol.* **2000**, *4*, 270–279.

50 W.B. Motherwell, M.J. Bingham, J. Pothier, Y. Six, *Tetrahedron* **2004**, *60*, 3231–3241.

51 See for example: (a) A. Biffis, G. Wulff, *New J. Chem.* **2001**, *25*, 1537–1542; (b) J.-M. Kim, K.-D. Ahn, G. Wulff, *Macromol. Chem. Phys.* **2001**, *202*, 1105–1108; (c) A.G. Strikovsky, D. Kasper, M. Grun, B.S. Green, J. Hradil, G. Wulff, *J. Am. Chem. Soc.* **2000**, *122*, 6295–6296 and references cited therein.

52 M. Emgenbroich, G. Wulff, *Chem. Eur. J.* **2003**, *9*, 4106–4117.

53 (a) A.R. Fersht (Ed.), *Enzyme Structure and mechanisms*. Freeman, New York, **1985**; (b) T. Bruice, F.C. Lightstone, *Acc. Chem. Res.* **1999**, *32*, 127–136.

54 (a) B. Sellergren, K.J. Shea, *Tetrahedron: Asymm.* **1994**, *5*, 1403–1406; (b) B. Sellergren, R.N. Karmalkar, K.J. Shea, *J. Org. Chem.* **2000**, *65*, 4009–4027.

55 K. Ohkubo, K. Sawakuma, T. Sagawa, *Polymer* **2001**, *42*, 2263–2266.

14
Appendix I: Reaction Procedures

> This chapter provides sample procedures of asymmetric organocatalytic transformations organized according to reaction types.

14.1
Aldol Reactions

14.1.1
General Procedures for (S)-Proline-Catalyzed Cross-Aldol Reactions of Aldehyde Donors (p. 28)

Method A [1a]:

anti : syn = 3 : 1
anti = 97% ee

A solution of donor aldehyde (2.0 mmol, 2 equiv) in DMF (500 µL) was added slowly over the course of 2.5 h to a stirring mixture of acceptor aldehyde (1.0 mmol, 1 equiv) and (S)-proline (0.10 mmol, 0.1 equiv) in DMF (500 µL) at 4 °C. The resulting mixture was stirred for 16 h at the same temperature.

Enantioselective Organocatalysis: Reactions and Experimental Procedures
Edited by Peter I. Dalko
Copyright © 2007 WILEY-VCH Verlag GmbH & Co. KGaA, Weinheim
ISBN: 978-3-527-31522-2

Method B [1b]:

A mixture of donor aldehyde (2 mmol, 1 equiv), acceptor aldehyde (20 mmol, 10 equiv), and (S)-proline (0.6 mmol, 0.3 equiv) in N-methylpyrrolidine (NMP) (1 mL) was stirred at 4 °C for 36 h.

14.1.2
General Procedure for O-tert-Butyl-L-Threonine Catalyzed Cross-Aldol Reactions of Ketone Donors and Aldehyde Acceptors [2] (p. 23)

To a solution of ketone (1.0 mmol, 2 equiv) and aldehyde (0.50 mmol, 1 equiv) in NMP (0.50 mL) was added O-*tert*-butyl-L-threonine (0.10 mmol, 0.2 equiv) and the resulting mixture was stirred at 4 °C until the aldehyde was consumed as monitored by TLC.

14.1.3
Intramolecular Aldol Reactions: The Robinson Annulation [3] (p. 31)

(quant.), 93% ee

The triketone (1.82 g, 10 mmol) and (S)-(−)-proline (34.5 mg, 0.3 mmol) were stirred in anhydrous N,N-DMF (distilled from calcium hydride) under argon for 20 h. The brown-colored reaction mixture was filtered, and the filtrate evaporated under high vacuum at 22 °C (bath temperature) to give 2.4 g of an oil. This was dissolved in 10 mL of ethyl acetate and filtered through 8.0 g of silica gel. The adsorbent was eluted with 150 mL of ethyl acetate, and the solvent evaporated in vacuo to give 2.0 g of an oil, which crystallized upon seeding. The crystalline mass was broken up and placed under high vacuum at 55 °C (bath temperature) for 1 h to remove traces of DMF to give 1.82 g (quant.) of crude product as a tan-colored solid, $[\alpha]_D + 56.1$ (c 1.0, $CHCl_3$).

14.1.4
Quaternary Ammonium Fluoride-Mediated Mukaiyama-Type Asymmetric Aldol Reaction [4] (p. 122)

The catalyst can be prepared according to Ref. [4]. To a solution of the catalyst (49 mg, 0.12 mmol) in tetrahydrofuran (THF) (8 mL) were successively added benzaldehyde (110 μL, 1.1 mmol) and a solution of silyl enol ether (233 mg, 1.0 mmol) in THF (2 mL) at −78 °C under argon atmosphere. After stirring for 6 h, water (2 mL) was added and the mixture warmed to r.t. for 2 h. Usual work-up followed by chromatographic purification gave the aldol adduct (198 mg, 74%) with an *erythro:threo* ratio of 7:3.

14.1.5
Synthesis of *tert*-Butyl (2S,3R)-3-Hydroxyleucinate [5] (p. 126)

To a solution of silyl ketene acetal [5] (248 mg, 0.676 mmol) in CH_2Cl_2 (4.0 mL) and hexanes (14.4 mL) were added isobutyraldehyde (310 μL, 3.38 mmol) and catalyst (40 mg, 16.9 µmol) in CH$_2$Cl$_2$ (0.8 mL) at −78 °C under nitrogen atmosphere. The solution was stirred for 7 h at −78 °C and then treated with saturated aqueous NH$_4$Cl and ether. The ethereal solution was extracted with water and brine, dried over MgSO$_4$, filtered, and concentrated *in vacuo*. To a solution of the crude reaction product in THF (8.0 mL) was added 0.5 M citric acid aqueous solution (5.0 mL, 2.5 mmol) at 23 °C and the solution was stirred at 23 °C for 15 h. After removal of the THF *in vacuo* at 20 °C or below, the aqueous solution was extracted twice with ether, neutralized with NaHCO$_3$, and then saturated with NaCl and Rochelle salt. The mixture was extracted three times with CH$_2$Cl$_2$. The extract was dried over MgSO$_4$ and concentrated *in vacuo*. Column chromatography on silica gel (CH$_2$Cl$_2$:methanol, 96:4 as eluent) afforded *syn*-C (79 mg, 61%) and *anti*-C (12 mg, 9%). To a solution of *syn*-C (10.9 mg, 57.0 µmol) in CH$_2$Cl$_2$ (2.0 mL) was added thiocarbonyl bisimidazole (10.2 mg, 57.0 µmol). After stirring at 23 °C for 2 h, the solvent was removed and the residue purified by column chromatography (hexane:EtOAc, 3:2 as eluent) to give the corresponding cyclic thiocarbamate (11.8 mg, 48.1 µmol, 84%), the enantiomeric excess of which was determined as 95% *ee* by HPLC analysis (DAICEL Chiralpak AD, hexanes:2-propanol, 9:1).

14.1.6
Quaternary Ammonium Salt-Mediated Asymmetric Direct Aldol Reaction of Glycinate Benzophenone Schiff Base with 3-Phenylpropanal Under Phase-Transfer Conditions [6] (p. 145)

To a mixture of *tert*-butyl glycinate benzophenone Schiff base (88.6 mg, 0.3 mmol), catalyst (9.9 mg, 2 mol%) and NH$_4$Cl (1.6 mg, 0.03 mmol) in toluene (1.5 mL) was added 1% NaOH aqueous solution (180 µL, 0.045 mmol) at 0 °C under argon atmosphere, after which 3-phenylpropanal (79.0 µL, 0.6 mmol) was introduced dropwise. After stirring the whole mixture at 0 °C for 1.5 h, saturated NH$_4$Cl and ether were added sequentially. The ethereal phase was separated, washed with brine, and dried over Na$_2$SO$_4$. Evaporation of solvents gave the crude product, which was dissolved in THF (8.0 mL); 1 M HCl (1.0 mL) was then added at 0 °C. After stirring for 1 h, the THF was removed *in vacuo*. The resulting aqueous solution was washed three times with ether and neutralized with NaHCO$_3$. The mixture was then extracted three times with CH$_2$Cl$_2$. The combined extracts were dried over Na$_2$SO$_4$ and concentrated. Purification of the residue by column chromatography on silica gel (MeOH:CH$_2$Cl$_2$, 1:15 as eluent) afforded *tert*-butyl 2-amino-3-hydroxy-5-phenylpentanoate as a mixture of diastereomers (65.6 mg, 0.247 mmol; 82%, *anti/syn* = 96:4). The diastereomeric ratio was determined by ^{1}H NMR analysis. The enantiomeric excess of the *anti*-product was determined as 98% *ee*, after conversion to the corresponding oxazoline-2-thione [thiocarbonyl diimidazole (1.0 equiv), CH$_2$Cl$_2$], by HPLC analysis [DAICEL Chiralpak AD-H, hexane:2-propanol = 10:1, flow rate = 0.5 mL min^{-1}, retention time (*cis* isomer from *anti*) = 18.9 min (major) and 28.7 min (minor)].

14.1.7
Taddol-Catalyzed Mukaiyama Aldol Reactions of Aldehydes [7] (p. 245)

To a solution of A (18 mg, 25 μmol, 10 mol%) in toluene (500 μL) at r.t. was added 4-(trifluoromethyl)benzaldehyde (67 μL, 0.5 mmol, 2 equiv) in one portion under an argon atmosphere. The solution was cooled to −78 °C and treated with silyl ketene acetal (54 mg, 0.25 mmol, 1 equiv). The reaction temperature was maintained at −78 °C for 48 h prior to quenching with HF (500 μL, 5% solution in CH$_3$CN). The mixture was warmed to r.t. and stirring continued until complete consumption of the silyl ether was observed (TLC-monitored). The reaction mixture was diluted with CH$_2$Cl$_2$ (4 mL) and washed with saturated aqueous NaHCO$_3$ (5 mL). The aqueous layer was extracted with CH$_2$Cl$_2$ (3 × 5 mL) and the combined organic extracts were dried over anhydrous Na$_2$SO$_4$ prior to *in-vacuo* solvent removal. The product diastereomer ratio (>25:1 *syn:anti*) was determined by ^{1}H NMR spectroscopy using the crude material. The latter was purified by flash chromatography (FC) on silica gel (ethyl acetate:hexanes, 2:1) to give the *syn* product (58 mg, 84% yield, 96% *ee*) as a white solid.

14.1.8
(S)-(−)-4-Hydroxy-4-Phenyl-2-Butanone [8] (p. 267)

Trichloro[(1-methylethenyl)oxy]silane (421.3 mg, 2.2 mmol, 1.1 equiv) was added quickly to a cold (−74 °C) solution of catalyst (37.3 mg, 0.1 mmol, 0.05 equiv) in CH_2Cl_2 (2 mL). A cold solution (−78 °C) of benzaldehyde (203 μL, 2.0 mmol) in CH_2Cl_2 (2 mL) was then added quickly via a short cannula; during the addition, the temperature rose to −67 °C. The reaction mixture was stirred at −75 °C for 2 h. The mixture was then quickly poured into a cold (0 °C) saturated aqueous solution of $NaHCO_3$ and the slurry was stirred for 15 min. The two-phase mixture was filtered through Celite, the phases separated, and the aqueous layer was extracted with CH_2Cl_2 (3 × 50 mL). The organic phases were combined, dried over Na_2SO_4, and concentrated under reduced pressure. The residue was purified by chromatography on a column of silica gel with a mixture of pentane and ether (1:1) to furnish 321.6 mg product (98%, 87% *ee*) as a clear oil: $[\alpha]_D$ − 51.4 (*c* 1.47, $CHCl_3$).

14.1.9
(S)-(−)-Methyl 3-Hydroxy-3-Phenylbutanoate [9] (p. 269)

Trichlorosilyl ketene acetal (380 μL, 2.4 mmol, 1.2 equiv) was added to a solution of acetophenone (240 μL, 2.0 mmol) and the bipyridine bis-*N*-oxide catalyst (101 mg, 0.2 mmol, 0.1 equiv) in CH_2Cl_2 (10 mL) at −20 °C under nitrogen in a flame-dried, round-bottomed flask with a magnetic stirrer. The mixture was stirred at −20 °C for 12 h and then transferred dropwise to an ice-cold saturated aqueous solution of $NaHCO_3$ (20 mL) with vigorous stirring. The mixture was further stirred at r.t. for 30 min. The silicate precipitate was filtered off through Celite and the filtrate was extracted with CH_2Cl_2 (4 × 20 mL). The combined organic layers were dried over MgSO$_4$ and concentrated under reduced pressure. The crude aldol product was separated from the catalyst by distillation and the further purified by silica gel chromatography to afford the product. Analytically pure sample (364 mg, 94%, 82% *ee*) was obtained as a colorless liquid via Kugelrohr distillation: b.p. 110 °C ABT (1 mmHg); [α]$_D$ − 5.62 (*c* 1.09, EtOH).

14.2 Mannich Reaction

14.2.1 General Procedure for Direct Mannich Reaction of Aldehydes and Preformed Imines [10] (p. 39)

Aldehyde (0.75 mmol, 1.5 equiv), N-PMP-protected α-imino ethyl glyoxylate (0.5 mmol, 1 equiv), (*S*)-proline (0.025 mmol, 0.05 equiv), dioxane (5 mL), r.t., 2–24 h.

14.2.2 General Procedures for Three-Component Mannich Reactions (p. 46)

Method A: To a mixture of ArCHO (0.5 mmol), 4-methoxyaniline (0.5 mmol), and (*S*)-proline (0.15 mmol) in DMF (3 mL), donor aldehyde (5.0 mmol) in DMF (2 mL) was slowly added (0.2 mL min^{-1}) at −20 °C. The mixture was stirred at the same temperature for 4–10 h. The mixture was diluted with Et$_2$O and reduction performed by the addition of NaBH$_4$.

Method B: After stirring a solution of ArCHO (1.0 mmol), 4-methoxyaniline (1.1 mmol), and (*S*)-proline (0.1 mmol) in N-methyl-2-pyrrolidinone (1.0 mL) for 2 h at r.t., propionaldehyde (3.0 mmol) was added to the mixture at −20 °C, and stirring continued for 20 h at the same temperature. The reaction was worked-up and reduction with NaBH$_4$ was performed without purification.

14.2.3
Catalytic Asymmetric Mannich-Type Reaction under Solid/Liquid Phase Transfer Conditions [11] (p. 146)

To a solution of benzaldehyde, N-(*tert*-butoxycarbonyl)imine (10.7 mg, 52.3 μmol), *tert*-butyl glycinate benzophenone Schiff base (14.8 mg, 50.1 μmol) and catalyst (5.0 mg, 5.0 μmol) in fluorobenzene (0.5 mL) was added Cs_2CO_3 (32.6 mg, 100.1 μmol) at −30 °C. After stirring for 19 h, the reaction was quenched by the addition of water. The water layer was extracted three times with EtOAc. The combined organic layer was dried over $MgSO_4$ and concentrated. The residue was purified by column chromatography on silica gel (hexane:ether, 10:1 as eluent) to give the product (24.5 mg, 98%).

14.2.4
Thiourea-Catalyzed Mukaiyama-Mannich Reactions of Aldimines [12] (p. 215)

A 5-mL flask was charged sequentially with catalyst (15 mg, 25 μmol, 0.05 equiv) and anhydrous toluene (250 μL). 3-Pyridylaldehyde N-Boc imine (0.53 mmol, 1 equiv) was added in one portion with stirring. When the solution was homogeneous, the flask was immersed in a dry ice/acetone bath and cooled to −30 °C. Silyl ketene acetal (216 mg, 1.0 mmol, 2 equiv) was then added slowly along the flask wall over 10 min. The flask was sealed under a nitrogen atmosphere and stirred at −30 °C for 48 h. Excess silyl ketene acetal was quenched at −30 °C via the rapid addition of a 3 M solution of trifluoroacetic acid in toluene (500 μL; cooled to −20 °C prior to addition). The reaction was allowed to warm to ∼5 °C, and then partitioned between saturated aqueous Na$_2$CO$_3$ and CH$_2$Cl$_2$ (1:1, 2 mL). The aqueous layer was extracted with CH$_2$Cl$_2$ (3 × 2 mL) and the combined organic extracts were dried over Na$_2$SO$_4$, filtered, and concentrated *in vacuo*. The resulting residue was purified via FC on silica gel (2.5–5% MeOH in CH$_2$Cl$_2$) to give the product (162 mg, 99% yield, 98% *ee*) as a pale, yellow oil.

14.3
Pictet–Spengler Reaction

14.3.1
Thiourea-Catalyzed Acyl-Pictet–Spengler Reaction of Imines [13] (p. 223)

A flame-dried flask maintained under nitrogen at 23 °C was charged with molecular sieves (250 mg, 3 Å spherical, flame-dried *in vacuo*), dichloromethane (DCM) (1.25 mL), and 6-methoxytryptamine (51 mg, 0.25 mmol, 1 equiv). 2-Ethylbutanal (32 µl, 0.275 mmol, 1 equiv) was added dropwise, and the suspension allowed to stand at 23 °C for 7.5 h, with occasional swirling to ensure content mixing. The resulting mixture was filtered by cannula transfer to a flame-dried round-bottomed flask. The desiccant was rinsed twice with DCM (5 mL) and the rinses were combined with the filtrate by cannula transfer. The solution was concentrated *in vacuo* to yield the imine as an oil. After dissolution of the imine in anhydrous diethyl ether (5.0 mL), catalyst (13.5 mg, 25 µmol, 10 mol%) was added. The resulting solution was cooled to −78 °C in a dry ice/acetone bath and treated with 2,6-lutidine (29 µL, 0.25 mmol, 1 equiv), followed by dropwise addition of acetyl chloride (18 µL, 1 equiv). The mixture was stirred at −78 °C for 5 min, then warmed to −50 °C and stirred for 65 h. The heterogeneous mixture was allowed to warm to 23 °C, and then concentrated *in vacuo*. Purification of the residue by FC on silica gel (CH$_2$Cl$_2$:ethyl acetate, 9:1) afforded the product (60 mg, 76% yield, 86% *ee*) as a white solid.

14.4
Nitroaldol (Henry) Reaction

14.4.1
Typical Procedure for the Quaternary Ammonium Salt-Mediated Addition of Nitromethane to Aldehydes [14] (p. 123)

A well-stirred mixture of catalyst (12 mg, 0.02 mmol), potassium fluoride (KF, 145 mg, 2.5 mmol) and nitromethane (27 μL, 0.5 mmol) in THF (0.3 mL) was cooled to −10 °C and treated with a solution of (S)-N,N-dibenzylphenylalaninal (66 mg, 0.2 mmol) in THF (0.4 mL). After stirring for 6 h at −10 °C, the mixture was filtered to remove KF, and concentrated. The solid quaternary ammonium salt of the catalyst was precipitated with diethyl ether:hexane (7:3) and the product-containing soluble fraction evaporated *in vacuo* to yield an oil, which was purified by FC on silica gel (hexane:EtOAc, 10:1 as eluent) to afford the 2R,3S nitro alcohol (66 mg) and the 2S diastereomer (4 mg).

14.4.2
Aza-Henry Reaction Catalyzed by Polar Ionic Hydrogen Bonding [15] (p. 221)

A flame-dried flask containing nitroethane (0.4 mL) and catalyst (10.4 mg, 20 μmol, 0.1 equiv) was cooled to −20 °C, and then treated with the imine (50.0 mg, 200 μmol, 1.0 equiv). Stirring was continued at −20 °C and the reaction monitored by gas chromatography (GC). On complete consumption of the imine, the solution was concentrated *in vacuo* and purified by FC on silica gel (hexanes:ethyl acetate, 4:1) to afford the product (38.9 mg, 60% yield, 7:1 *syn:anti*, 90% *ee syn*) as a yellow oil.

14.5
Hydrocyanation and Cyanosilylation Reactions

14.5.1
Synthetic Peptide-Promoted Hydrocyanation of Benzaldehyde [16] (p. 208)

To a flask charged with catalyst (2.8 mg, 10 µmol, 2 mol%) in toluene (1 mL) was added benzaldehyde (51 µL, 0.5 mmol, 1.0 equiv) under a nitrogen atmosphere. After cooling the mixture to −20 °C, HCN (40 µL, 1.0 mmol, 2 equiv) was added dropwise via a pre-cooled syringe. Stirring was continued at −20 °C for 8 h. The reaction mixture was quenched with 0.1 M methanolic HCl (0.5 mL) and the remaining HCN removed under reduced pressure. The reaction mixture was washed with 2 M HCl and the resulting aqueous layer extracted twice with ether. The combined organic layers were dried over Na_2SO_4 and concentrated *in vacuo* to give a crude oil that was subjected to FC on silica gel (hexanes:ethyl acetate, 5:1) to provide product (64.6 mg, 0.49 mmol, 97% yield, 97% *ee*).

14.5.2
Strecker Reactions of *N*-Allyl Aldimines [17] (p. 214)

To a clear solution of *N*-allylbenzaldimine (34 µL, 0.2 mmol, 1 equiv) and catalyst (21 mg, 20 µmol, 0.1 equiv) in CH_2Cl_2 (1 mL) under nitrogen was added liquid HCN (15 µL, 0.4 mmol, 1 equiv) by precooled syringe at −70 °C. The resulting homogeneous solution was stirred at −70 °C until the aldimine was consumed, as monitored by TLC (36 h). The reaction mixture was treated with trifluoroacetic anhydride (TFAA) and flushed with a stream of nitrogen that was then passed through bleach, to remove excess HCN. After concentration, the residue was taken up in a minimum amount of CH_2Cl_2 and purified by FC on silica gel (1–5% ethyl acetate in hexanes) to give the product (51.0 mg, 0.19 mmol, 95% yield, 92% ee) and recovered catalyst (80–95% recovered yield).

14.5.3
Thiourea-Catalyzed Cyanosilylations of Ketones [18] (p. 214)

A flame-dried 5-mL flask was charged with catalyst (19.2 mg, 50 μmol, 5 mol%), 2-bromo-2-cyclohexen-1-one (109 μL, 1 mmol, 1 equiv), trimethylsilyl cyanide (TMSCN; 0.294 mL, 2.2 mmol, 2.2 equiv) and CH_2Cl_2 (2.0 mL). The flask was sealed with a rubber septum and Parafilm®, and the reaction cooled to −78 °C and stirred for 15 min. 2,2,2-Trifluoroethanol (73 μL, 1 mmol, 1.0 equiv) was then added via syringe and the reaction stirred at −78 °C for 12 h, after which the contents of the flask were placed under high vacuum at −78 °C for 5 min to remove excess HCN (*CAUTION*: HCN is highly toxic). After warming to r.t., the entire reaction mixture was loaded onto a silica gel column for FC. Elution (hexanes:ethyl acetate, 20:1) gave the expected product as a white solid (261 mg, 95% yield, 97% ee).

14.5.4
Cyanosilylation of Aromatic Aldehydes with (R)-BINOL/i-PrOLi [19] (p. 273)

A mixture of (R)-BINOL (28.6 mg, 0.1 mmol) and i-PrOLi (6.6 mg, 0.1 mmol) in toluene (2 mL) was stirred at r.t. for 20 min under a nitrogen atmosphere. Benzaldehyde (101.6 μL, 1.0 mmol) was then added and the mixture stirred at r.t. for 20 min. This pale yellow solution was cooled to −78 °C and stirred for 10 min at that temperature. Trimethylsilyl cyanide [*CAUTION!*] (133.1 μL, 1.0 mmol) was added dropwise, and the mixture stirred at −78 °C for 1 h. The mixture was then diluted with a 10% (w/w) solution of HCl in MeOH (1 mL) and stirred at −78 °C for 10 min. Water (10 mL) and ethyl acetate (10 mL) were then added and the product was extracted with ethyl acetate (2 × 10 mL), the combined organic phase was washed with brine (10 mL), dried over MgSO$_4$, and evaporated under reduced pressure to give the crude cyanohydrin. The latter product was dissolved in DCM (2 mL), cooled to −78 °C, and pyridine (0.16 mL, 2 mmol) and acetic anhydride (0.14 mL, 1.5 mmol) were added. The mixture was stirred at r.t. for 30 min and then diluted with ethyl acetate (20 mL) and 1 M HCl (10 mL). The organic phase was separated, washed with water (10 mL), a saturated aqueous solution of NaHCO$_3$ (10 mL), and brine (10 mL), and then dried over MgSO$_4$ and evaporated under reduced pressure to give the cyanohydrin acetate. This was purified by neutral silica gel column chromatography (hexane:AcOEt, 10:1) to afford the pure cyanohydrin acetate as a colorless oil (175.2 mg, >99%, 95% *ee*).

14.5.5
(S)-(+)-N-*tert*-Butyl-2-Hydroxy-2-(2-Naphthyl)acetamide [20] (p. 273)

A solution of *tert*-butyl isocyanide (125 μL, *d* = 0.735, 1.2 mmol, 1.2 equiv) in CH$_2$Cl$_2$ (1.0 mL) was added via a syringe pump over 4 h to a cold (−74 °C, internal temperature) solution of 2-naphthaldehyde (156 mg, 1.0 mmol), (*R*,*R*)-catalyst (43 mg, 0.05 mmol, 0.05 equiv), SiCl$_4$ (125 μL, *d* = 1.483, 1.1 mmol, 1.1 equiv), and diisopropylethylamine (18 μL, *d* = 0.742, 0.1 mmol, 0.1 equiv) in CH$_2$Cl$_2$ (1.0 mL) in a two-necked, round-bottomed flask fitted with an argon inlet adapter and thermocouple. The mixture was stirred further at −74 °C for 4 h. The reaction mixture was then transferred dropwise to a vigorously stirred, ice-cold, saturated aqueous solution of NaHCO$_3$ (20 mL) and the resulting mixture stirred at r.t. for 2 h. The white precipitate was filtered off through Celite and the filtrate extracted with CH$_2$Cl$_2$ (4 × 20 mL). The combined organic layers were dried (MgSO$_4$) and concentrated under reduced pressure. The crude product was purified by chromatography on silica gel (20-mm column, 60 g SiO$_2$, R_f 0.17, hexane:AcOEt, 3:1) to afford the (*S*)-(+) product (242 mg, 93%, >99 *ee*) as a white solid. An analytically pure sample was obtained after sublimation (90 °C at 0.5 mmHg): mp 94–96 °C (sublimed); $[\alpha]_D^{24}$ +44.9 (*c* 0.96, MeOH).

14.6
Alkylation of α-Diazoesters

14.6.1
Brønsted Acid-Catalyzed Direct Alkylations of α-Diazoesters [21] (p. 228)

A dry test tube containing catalyst (1.4 mg, 2 µmol, 2 mol%) and the imine (44.4 mg, 0.15 mmol, 1.5 equiv) was purged with nitrogen and then charged with toluene (1 mL). The resulting solution was treated with neat *tert*-butyl diazoacetate (14.2 mg, 0.1 mmol, 1 equiv) at r.t. After stirring for 24 h at ambient temperature, the mixture was subjected directly to FC on silica gel (hexanes:ethyl acetate, 10:1 to 3:1) to give the product (32.9 mg, 75 µmol, 75% yield, 95% *ee*) as an oil.

14.7
Heteroatom Addition to Imines

14.7.1
Hydrophosphonylation of Benzaldimines [22] (p. 225)

An oven-dried 4-mL vial equipped with a magnetic stirrer bar was charged with catalyst (28.8 mg, 50.1 µmol, 10 mol%), di-(2-nitrobenzyl) phosphite (177 mg,

501 μmol, 1.0 equiv) and diethyl ether (1.2 mL). The vial was sealed with a plastic cap and cooled to 0 °C in an ice-water bath, with stirring. To this suspension was added the imine (97.9 mg, 501 μmol, 1.0 equiv), and the reaction mixture was warmed to 4 °C. After 72 h, THF (0.5 mL) and an aqueous solution of HCl (1 M, 0.5 mL) were added, followed by additional stirring for 1 h at 4 °C. The reaction was partitioned between an aqueous solution of HCl (1 M) and CH_2Cl_2 (1:1, 40 mL). The aqueous layer was separated and extracted with CH_2Cl_2 (5 mL). The combined organic solutions were washed with a saturated aqueous Na_2CO_3 solution (20 mL). The separated Na_2CO_3 phase was back-extracted with CH_2Cl_2 (2 × 5 mL), and the combined organic extracts were dried over Na_2SO_4, filtered, and concentrated *in vacuo*. The resulting residue was purified via FC on silica gel (hexanes:ethyl acetate, 3:2) to provide the product as a very pale yellow solid (0.239 g, 0.436 mmol, 87% yield, 98% *ee*).

14.7.2
Phosphoric Acid-Catalyzed Imine Amidation [23] (p. 228)

A dry test tube was charged with Boc-imine (119.9 mg, 0.5 mmol, 2.0 equiv), sulfonamide (42.8 mg, 0.25 mmol, 1.0 equiv) and catalyst (15.0 mg, 25 μmol, 10 mol%). The tube was fitted with a septum, evacuated, and back-filled with argon (cycle repeated three times), and charged via a syringe with anhydrous toluene (1–2 mL). The reaction mixture was stirred at ambient temperature for 17 h, with reaction progress being monitored by TLC. On completion, acetone (2–3 mL) was added, followed by a scoop of silica gel. The solvent was removed *in vacuo* and the crude solid mixture subjected to FC on silica (hexanes:acetone, 4:1) to give the product (90.0 mg, 0.22 mmol, 88% yield, 95% *ee*) as a white solid.

14.8
α-Heteroatom Functionalization

14.8.1
Synthesis of α-Hydrazino Aldehydes and Subsequent Conversion to the Corresponding Boc-Protected Amino Acid Methyl Ester [24] (p. 60)

a) $KMnO_4$
b) $TMSCHN_2$
c) TFA
d) H_2/Ra-Ni
e) $(Boc)_2O$-DMAP

To a stirred solution of the azodicarboxylate (1.0 mmol) and aldehyde (1.5 mmol) in the solvent (3 mL), L-proline was added and stirred at r.t. The reaction mixture was quenched with H_2O (5 mL), extracted with Et_2O, and dried over anhydrous Na_2SO_4. The solvent and the excess of aldehyde were removed by evaporation.

To a solution of α-hydrazino aldehyde (568 mg, 1.8 mmol) in tert-butyl alcohol (10 mL) were added successively 1 M NaH_2PO_4 (10 mL) and 1 M $KMnO_4$ (10 mL). After 1 min vigorous stirring, saturated $NaHSO_3$ (10 mL) was added and at 0 °C the resulting pH was adjusted to 3 with 1 M HCl to dissolve the colloidal MnO_2. The resulting mixture was extracted with EtOAc, washed with water and brine, and dried over $MgSO_4$. The combined organic layers were concentrated, the residual acid was dissolved in the mixture of toluene (10 mL) and MeOH (4 mL), and $TMSCHN_2$ (1.8 mL, 2 M in hexane solution) was added dropwise. The solution was stirred for 10 min and the excess $TMSCHN_2$ decomposed by careful addition of a few drops of AcOH. Solvents were removed in vacuo and without further purification, the crude methyl ester product was dissolved in TFA/CH_2Cl_2 (1:1, 3 mL) and stirred for 10 min. After evaporation of the solvent, the residue and ca. 200 mg of Raney nickel (washed twice with MeOH) was suspended in 5 mL MeOH. The reaction mixture was stirred under 30 bar of H_2 for 10 h, followed by the addition of 20 mg DMAP and 500 mg di-tert-butyldicarbonate. The solution was stirred at r.t. overnight, and then filtered to remove Raney nickel; the mixture was concentrated in vacuo.

14.8.2
General Procedure for the Organocatalytic α-Chlorination of Aldehydes, and Subsequent Conversion to the Corresponding 2-Chloro Esters [25] (p. 70)

R = alkyl

The catalyst (0.05 mmol, 10 mol%) was added to a stirred, ice-cooled (0 °C) solution of the aldehyde (0.5 mmol) in CH_2Cl_2 (1.0 mL), followed by the addition of NCS (87 mg, 0.65 mmol, 1.3 equiv). After 1 h, the reaction mixture was allowed to warm to ambient temperature, and then stirred until the aldehyde was completely consumed (as determined by ^1H NMR spectroscopy of the reaction mixture, and confirmed by GC analysis). Pentane was added to the reaction mixture and the precipitated NCS, succinimide and the catalyst were filtered off. After removal of the solvents the pure products could be extracted into pentane, and isolated.

The reaction mixture containing the 2-chloro aldehyde was diluted with 3.0 mL t-BuOH, after which 3.0 mL 1 M NaH_2PO_4 (aq.) and 3.0 mL 1 M $KMnO_4$ (aq.) were added successively. After 1 min of vigorous stirring, 3.0 mL saturated $NaHSO_3$ (aq.) was added and the pH adjusted to ca. 3 with 1 M HCl at 0 °C. The resulting mixture was extracted with EtOAc, washed with water and brine, and dried over Na_2SO_4. The organic layer was concentrated and the residual acid dissolved in toluene (2.0 mL) and MeOH (5.0 mL); $TMSCHN_2$ (2.0 M in hexane) was then added drop-wise until the yellow color persisted. The solution was stirred for an additional 5 min and quenched with one drop of AcOH. The solvents were removed and the crude product was purified by FC to afford the pure product.

14.9
1,4-Additions

14.9.1
General Procedure for Conjugate Addition of Nitroalkanes to α,β-Unsaturated Enones [26a–c] (p. 109)

The desired nitroalkane (1 mL) was placed in an ordinary test tube equipped with a magnetic stirring bar. Enone (0.5 mmol) and catalyst (20 mol%) were then added, and the tube was closed with a robber stopper. The mixture was stirred at r.t. until completion of the reaction. After evaporation of excess nitroalkane, the crude reaction mixture was purified by FC to afford title compound.

14.9.2
General Procedure for the Michael Addition of Cyclic Ketones to Nitroalkenes [27] (p. 80)

To a solution of the catalyst (0.1 mmol), *p*-toluene sulfonic acid monohydrate (0.1 mmol) and the nitroalkene (0.5 mmol) in DMF (1 mL) was added the ketone (0.25 mmol). The solution was stirred at ambient temperature under reduced pressure, and the residue purified by FC on silica gel. Alternatively, ethyl acetate was added, and the solution washed with water, 1 M HCl, dried (Na$_2$SO$_4$), and concentrated to give the crude product which was purified by FC on silica gel.

14.9.3
Diamine-Catalyzed Addition of Unmodified Ketones to Nitroolefins [28] (p. 81)

To a solution of pyrrolidine catalyst (0.05 mmol, 15 mol%) in CHCl$_3$ (3 mL) was added at r.t. the ketone (3.35 mmol) and the nitroolefin (0.335 mmol). The evolution of the reaction was monitored by TLC. The solution was then hydrolyzed with 1 M HCl (2 mL). The layers were separated and the aqueous phase was extracted with CH$_2$Cl$_2$ (2 × 3 mL). The combined organic phases were dried over MgSO$_4$, filtered, concentrated and purified by flash column chromatography on silica gel using a mixture of cyclohexane and ethyl acetate as eluent.

14.9.4
Thiourea-Mediated Michael Addition of Ketones to Nitroolefins [29] (pp. 82–83)

To a stirred solution of catalyst (0.15 equiv) in toluene (0.5 mL) and ketone (10 equiv) at r.t., was added water (2 equiv) and, after 5 min, nitroolefin (1 equiv). The reaction mixture was stirred at r.t. for the appropriate time. The solvent was evaporated and the residue purified by TLC or chromatography on silica gel (hexane:ethyl acetate, 1:1) to afford the desired product.

14.9.5
Thiourea-Amine-Catalyzed Malononitrile Enolate Additions to α,β-Unsaturated Imide [30] (pp. 196–197)

A solution of imide (43.1 mg, 0.2 mmol, 1.0 equiv), malononitrile (226.4 mg, 0.4 mmol, 2.0 equiv), and thiourea catalyst (8.2 mg, 20 µmol, 10 mol%) in dry toluene was stirred for 2.5 days at r.t. The reaction was concentrated *in vacuo* and the residue purified by FC on silica gel (Et$_2$O), to afford the product as a colorless amorphous solid (52.3 mg, 93% yield, 94% *ee*).

14.9.6
Thiourea-Amine-Catalyzed Michael Additions of 2,4-Pentanedione [31] (pp. 205–206)

Catalyst (1 mg, 1.7 μmol, 1 mol%) was added to a vial containing 2,4-pentanedione (38 μL, 0.34 mmol, 2.0 equiv) and *trans*-β-nitrostyrene (26 mg, 0.17 mmol, 1.0 equiv) in Et$_2$O (1 mL) at r.t. After stirring for 28 h, the reaction was concentrated *in vacuo*. The residue was purified by FC on silica gel (hexanes:ethyl acetate, 10:1 to 3:1) to afford the product as a white solid (37 mg, 87% yield, 95% *ee*).

14.9.7
General Procedure for the Pyrrolidine-Mediated 1,4-Addition of Aldehydes to Nitroolefins [32] (p. 86)

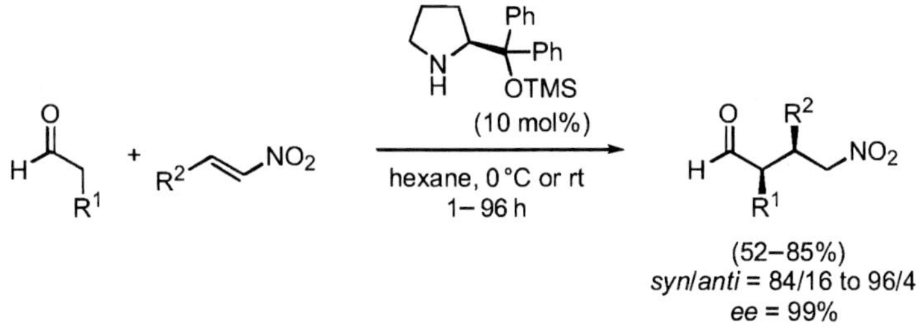

(52–85%)
syn/anti = 84/16 to 96/4
ee = 99%

To a hexane solution of the nitroolefin (1.0 mmol) and catalyst (34 mg, 0.1 mmol) was added an aldehyde (10 mmol) at 0 °C. The mixture was stirred until reaction completion, and then quenched by the addition of aq. 1 M HCl. Organic materials were extracted three times with ethyl acetate. The combined organic phases were dried over Na$_2$SO$_4$, filtered, concentrated, and purified by preparative TLC (chloroform) to afford the Michael adduct.

14.9.8
General Procedure for the Intramolecular Michael Addition with Chiral Imidazolidinone Salt [33] (p. 90)

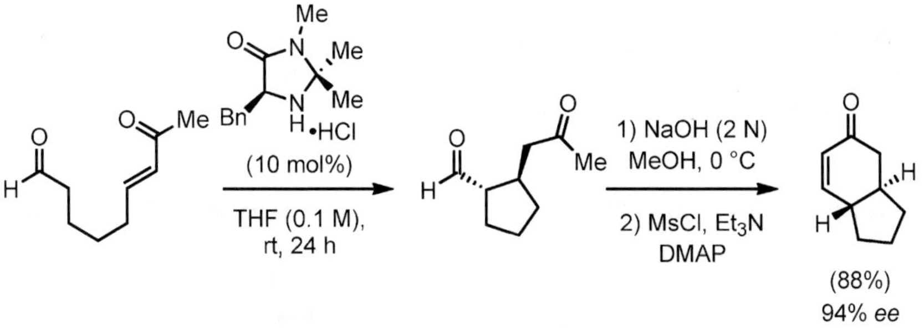

(88%)
94% ee

Ketoaldehyde (123 mg, 0.8 mmol) was dissolved in dry THF (8 mL) and treated with catalyst (20.4 mg, 0.08 mmol, 10%). The resulting mixture was stirred under argon at r.t. until the starting material had disappeared (15 h). The solution was then concentrated and filtered through silica. The product was dissolved in methanol (4 mL), cooled to 0 °C, and then treated with NaOH (2 M, 4 mL). TLC analysis indicated completion of the reaction after 5 min. The mixture was poured into saturated aqueous ammonium chloride solution, the resulting mixture was extracted with ether, and the extracts were dried, filtered, and concentrated to give the crude aldol product. This material was dissolved in CH_2Cl_2 (5.2 mL) and stirred with methanesulfonyl chloride (0.16 mL), DMAP (35.6 mg), and triethylamine (0.39 mL) under argon for 30 min. After an aqueous work-up and purification by chromatography on silica gel (20% ether/pentane), enone was obtained as a colorless oil (95.2 mg, 0.7 mmol, 88%).

14.9.9
Addition of Silyl Nitronates to Enals Catalyzed by Quaternary Ammonium Bifluoride; General Procedure for the Isolation of Enol Silyl Ether [34] (p. 128)

The catalyst can be prepared by using the modified Shioiri method according to Ref. [35].

(a) To a solution of chiral quaternary ammonium bifluoride (R,R)-catalyst (5.6 mg, 6 µmol) in toluene (3 mL) was added *trans*-cinnamaldehyde (38 µL, 0.3 mmol) at r.t. and the mixture was cooled to −78 °C under an argon atmosphere. Silyl nitronate (60 µL, 0.36 mmol) was then introduced. After stirring for 1 h at that temperature, 1 M HCl (2 mL) and THF (6 mL) were added. The mixture was stirred vigorously at 0 °C for 30 min, after which the whole mixture was extracted with ether. The combined organic extracts were washed with brine and dried over Na_2SO_4. Removal of solvents and purification of the residue by column chromatography on silica gel (ether:hexane, 1:3) gave γ-nitro aldehyde [62 mg, 0.3 mmol, 99% yield, syn:anti = 81:19, 97% ee (*syn* isomer), 79% ee (*anti* isomer)].

(b) The Michael addition was performed in a similar manner as described above. When completion of the reaction had been confirmed by TLC analysis, the resulting mixture was purified directly by column chromatography on silica gel 60 silanized (ether:hexane as eluent) to afford the corresponding enol silyl ethers.

14.9.10
1,4-Addition/Lactonization of Silyl Ketene Acetals to Enones [36] (p. 124)

The catalyst can be prepared according to Ref. [36].

To a stirred solution of catalyst (13.5 mg, 0.015 mmol) in THF (1.0 mL) were successively added a solution of chalcone (62.5 mg, 0.3 mmol) in THF (1.4 mL) and a solution of ketene silyl acetal (113 mg, 0.48 mmol) in THF (0.6 mL) at −78 °C. The reaction mixture was stirred for 1 h at the same temperature, then quenched with 1 M HCl aqueous solution and extracted with EtOAc. The organic layer was washed with brine, dried over anhydrous Na_2SO_4, and evaporated. The crude product was purified by preparative TLC (82.0 mg, 98%, 90% ee).

14.9.11
Michael Addition of Benzhydryl Glycinate Benzophenone Schiff Base to Methyl Vinyl Ketone [37] (p. 143)

84% yield, 94% ee

A solution of benzhydryl glycinate benzophenone Schiff base (0.12 mmol) and catalyst (1.2 μmol, 1 mol%) in diisopropyl ether (4 mL) was degassed with argon and then cooled to 0 °C. Methyl vinyl ketone (MVK, 0.24 mmol) was added, followed by anhydrous Cs_2CO_3 (0.06 mmol), and the mixture was stirred at 1500 r.p.m. After complete consumption of starting imine (ca. 2 h), the reaction mixture was warmed to r.t., filtered, and concentrated under reduced pressure. The crude product could be purified by FC on silica gel (petroleum ether:EtOAc:Et_3N = 94:5:1 as eluent) to give the corresponding Michael adduct as a colorless oil (84% yield, 94% ee).

14.9.12
Cinchona-Derivative-Catalyzed Conjugate Additions to Vinyl Sulfones [38] (p. 206)

To a solution of α-phenyl α-cyanoacetate (113 mg, 0.6 mmol, 3.0 equiv) in 0.4 mL dry toluene was added catalyst (19 mg, 40 μmol, 20 mol%) and vinyl phenyl sulfone (34 mg, 0.2 mmol, 1.0 equiv) at −25 °C. After 72 h, the reaction mixture was subjected directly to purification by FC on silica gel (hexanes:ethyl acetate, 8:1 to 6:1) to afford the 1,4-adduct as a colorless oil (64 mg, 89% yield, 95% ee). Catalyst was recovered in >95% yield on washing the column with methanol.

14.9.13
General Procedure for Organocatalytic Friedel–Crafts Alkylation of Pyrroles [39] (p. 106)

68–90%, 87–97% ee

A 10-mL round-bottomed flask equipped with a magnetic stirrer bar was charged with catalyst·HX (20 mol%), THF and H_2O, and then cooled to the desired temperature. The solution/suspension was stirred for 5 min before the pyrrole was added. At this time, the α,β-unsaturated aldehyde was added over 1 min with swirling of the reaction mixture by hand. The resulting suspension was stirred at constant temperature until complete consumption of the α,β-unsaturated aldehyde, as determined by TLC or GC. The resulting solution was then transferred to a flask containing an excess of NaBH$_4$ and an equal volume of EtOH. After 15 min, the resulting mixture was treated with saturated aqueous NaHCO$_3$ and the organics were extracted with CH$_2$Cl$_2$. The organic layers were combined and washed successively with saturated aqueous NaHCO$_3$ and brine. The organics were dried (Na$_2$SO$_4$) and concentrated. The resulting residue was purified by silica gel chromatography (15–50% AcOEt/hexane) to afford substituted 3-(2-pyrrolyl)-propanols.

14.9.14
Organocatalytic Pyrroloindoline Construction [40] (p. 111)

An amber 2-dram vial equipped with a magnetic stirrer bar, containing tryptamine substrate (1 equiv) and catalyst·p-TSA (20 mol%) was charged with CH$_2$Cl$_2$ (0.15 M), and then placed in a bath at the appropriate temperature. The solution was stirred for 5 min before addition of the α,β-unsaturated aldehyde (4 equiv). The resulting suspension was stirred at constant temperature until complete consumption of the starting material. To the reaction mixture was added pH 7.0 buffer, and the whole was extracted with Et$_2$O, and concentrated *in vacuo*. The resulting residue was purified by column chromatography to afford the title compound.

14.9.15
1,2-Aza-Friedel–Crafts Reactions of 2-Methoxyfuran and Aldimines [41] (p. 224)

A dry test tube containing catalyst (1.95 mg, 2 μmol, 2 mol%) was purged with nitrogen and charged with 1,2-dichloroethane (1 mL). The solution was cooled to −35 °C prior to sequential addition of neat imine (24.0 mg, 0.1 mmol, 1 equiv) and 2-methoxyfuran (11.1 μL, 0.12 mmol, 1.2 equiv). The resulting solution was stirred at −35 °C for 24 h prior to loading directly onto a column for purification by FC on silica gel (ethyl acetate:hexanes, 4:1) to give the product (29.7 mg, 0.09 mmol, 88% yield, 97% ee) as a solid.

14.9.16
Michael Reaction of α,β-Unsaturated Enones with Cyclic 1,3-Dicarbonyl Compounds [26d] (p. 109)

To a stirred solution of α,β-unsaturated enone (0.5 mmol) and Michael donor (1.05 equiv) in 1 mL solvent at rt was added catalyst (10 mol%) and the mixture was stirred until consumption of starting material. The crude reaction mixture was purified by FC to yield the pure product.

14.10
Morita–Baylis–Hillman (MBH) Reactions

14.10.1
General Procedure for Asymmetric MBH Reaction of Aldehydes and HFIPA Catalyzed by β-ICD [42] (pp. 160 and 232)

To a solution of aldehyde (1.0 mmol) and β-ICD catalyst (0.1 mmol), in DMF (2 mL) at −55 °C was added HFIPA (1.3 mmol). The solution was stirred at r.t. until total consumption of the starting material. The reaction was quenched with HCl (3 mL, 0.1 M) and extracted with ethyl acetate. The organic layer was washed with saturated NaHCO$_3$ and brine, dried, and concentrated. The crude material was purified by column chromatography (EtOAc/hexane).

14.10.2
Typical Procedure for the Triethyl Phosphine and Chiral Binaphthol-Derived Brønsted Acid Co-Catalyzed Asymmetric MBH Reaction [43] (pp. 173 and 232)

An oven-dried 10-mL flask was charged with (R)-3,3′-bis-(3,5-bis-trifluoromethylphenyl)-5,6,7,8,5′,6′,7′,8′-octahydro-[1,1′]binaphthalenyl-2,2′-diol catalyst (72 mg, 0.1 mmol) under an argon atmosphere. The catalyst was dissolved in THF (1.0 mL), and the solution cooled to −78 °C. 2-Cyclohexen-1-one (190 mg, 2.0 mmol), triethylphosphine (240 mg, 2.0 mmol), and 3-phenylpropionaldehyde (134 mg, 1.0 mmol) were added successively at −78 °C. The flask was placed in a −10 °C bath and stirred for 48 h. The reaction mixture was subjected directly to FC on silica gel, and eluted with hexanes and ethyl acetate (3:1 to 1:1) to afford (S)-2-(1-hydroxy-3-phenylpropyl)cyclohex-2-enone (202 mg, 0.88 mmol, 88% yield) as a colorless oil $[\alpha]^{21}_D - 37.4$ (c 1.05, CHCl$_3$).

14.10.3
General Procedure for the β-ICD-Mediated Asymmetric Aza-MBH Reaction of Tosyl Aryl Aldimines and Methyl Acrylate [44] (pp. 177 and 233)

To a solution of imine (1.0 mmol) and β-ICD catalyst (10 mol%), in CH$_2$Cl$_2$ (2 mL) at 0 °C was added methyl acrylate (2 mmol). The solution was stirred at the same temperature until total consumption of starting material. Solvents were evaporated under reduced pressure and the crude material purified chromatographically (EtOAc/hexane).

14.10.4
Typical Procedure for the Bifunctional BINOL-Promoted Aza-MBH Reactions of Aldimines [45] (pp. 179 and 234)

A solution of catalyst (2.2 mg, 5 µmol, 10 mol%) in CPME:toluene (9:1, 0.1 mL, both distilled from CaH_2 prior to use) was treated with methyl vinyl ketone (12.5 µL, 0.15 mmol, 3 equiv) and imine (11.8 µL, 50 µmol, 1 equiv) at −15 °C. The stirred solution was kept at −15 °C for 60 h until TLC indicated complete consumption of the imine. The mixture was directly purified by FC on silica gel (hexanes:ethyl acetate, 2:1) to give the product (17.5 mg, 48 µmol, 96% yield, 95% *ee*) as a white solid.

14.11
Cyclopropanation

14.11.1
General Procedure of Cyclopropanation of Enals with Sulfonium Ylides [46] (p. 104)

A 50-mL round-bottomed flask equipped with a magnetic stirrer bar was charged with aldehyde (0.5 mmol, 4 equiv) and $CHCl_3$ (21 mL), then cooled to −10 °C. The mixture was stirred for 20 min before the catalyst (20 mol%) and 2-(dimethyl-λ^4-sulfanylidene)-1-phenyl-ethanone (0.127 mmol, 1 equiv) were added. The homogeneous solution was stirred at a constant temperature for 24–48 h until complete consumption of the starting material was observed. The cold reaction mixture was then eluted with Et_2O through a short silica gel plug and the filtrate concentrated *in vacuo*. The resulting yellow residue was purified by silica gel chromatography to provide the title compounds.

14.11.2
General Procedure for S-Ylide-Catalyzed Cyclopropanation [49] (pp. 371 and 372)

To a 10-mL round-bottomed flask fitted with a nitrogen balloon was added sulfide catalyst (0.2 equiv), anhydrous dioxane (4.0 mL), rhodium(II) acetate dimer (2 mg, 0.01 equiv), substrate (0.5 mmol), benzyl triethylammonium chloride (23 mg, 0.2 equiv), and tosylhydrazone sodium salt (1.5 equiv). The reaction mixture was stirred vigorously at 40 °C for 48 h. Work-up consisted of the sequential addition to the reaction mixture of water (5 mL) and ethyl acetate (5 mL). The aqueous layer was washed with ethyl acetate (2.5 mL) and the combined organic phases dried (MgSO$_4$), filtered, and concentrated *in vacuo*. The crude products were analyzed by ^{1}H NMR to determine the diastereomeric ratio, and then purified by FC to afford the corresponding cyclopropane.

14.11.3
General Procedure for Te-Ylide-Catalyzed Cyclopropanation [47] (p. 382)

Ar = Ph: 91% yield, 86% ee, 90:10
Ar = p-Cl-C$_6$H$_4$: 94% yield, 89% ee, 91:9

To a mixture of α,β-unsaturated compound (0.25 mmol), Cs$_2$CO$_3$ (0.27 mmol), and telluronium salt catalyst (0.05 mmol) in THF (0.5 mL) was added trimethylsilyl allylic bromide (0.5 mmol) in portions over 8 h at reflux. When the reaction was complete (monitored by GC), the resulting mixture was filtered through short silica gel column and washed with ethyl acetate. The filtrate was concentrated, and the residue was purified by FC to afford the desired product.

14.11.4
General Procedure for N-Ylide-Catalyzed Cyclopropanation [48] (p. 384)

One-pot Method: The catalyst **A**, **B** (0.2 equiv) or **C**, **D** (0.1 equiv) was added to a solution of the α-bromo carbonyl compound (1.0 equiv), the alkene (1.2 equiv), and Cs$_2$CO$_3$ (1.2 equiv) in MeCN (0.25 M) and stirred at 80 °C for 24 h. The reaction was quenched with aqueous HCl (1 M) and extracted three times with Et$_2$O or EtOAc. The combined organic phases were washed with a saturated aqueous solution of NaHCO$_3$, dried (MgSO$_4$), and concentrated under reduced pressure. The residue was purified by flash column chromatography.

Slow Addition Method: A solution of the α-bromo carbonyl compound (1.0 equiv) and the alkene (1.2 equiv) in MeCN (0.25 M with respect to the α-bromo carbonyl compound) was added to a solution of the catalyst **A**, **B** (0.2 equiv) or **C**, **D** (0.1 equiv) and Cs$_2$CO$_3$ (1.2 equiv) in MeCN (0.25 M) at 80 °C over 20 h by means of a syringe pump. The syringe was rinsed with MeCN and the reaction mixture stirred for a further 4 h. The reaction was quenched with aqueous HCl (1 M) and extracted three times with Et$_2$O or EtOAc. The combined organic phases were washed with a saturated aqueous solution of NaHCO$_3$, dried (MgSO$_4$), and concentrated under reduced pressure. The residue was purified by flash column chromatography.

14.12
Aziridination

14.12.1
General Procedure for Aziridination of Imines using Chiral Sulfide Catalyst [49] (p. 378)

To a 5-mL round-bottomed flask fitted with a nitrogen balloon was added sulfide catalyst (16 mg, 64 mmol), anhydrous solvent (1.0 mL), rhodium(II) acetate dimer (1.5 mg, 3.3 mmol), benzyl triethylammonium chloride (7.5 mg, 33 mmol), imine (0.33 mmol) and tosylhydrazone sodium salt (146.5 mg, 0.495 mmol). The reaction mixture was stirred vigorously at 40 °C for 24–48 h. Work-up consisted of the sequential addition to the reaction mixture of water (5 mL) and ethyl acetate (5 mL). The aqueous layer was washed with ethyl acetate (2.5 mL) and the combined organic phases dried (MgSO$_4$), filtered, and concentrated *in vacuo*. The crude products were analyzed by ^{1}H NMR to determine the diastereomeric ratio, and then purified by FC to afford the corresponding aziridine.

14.13
Epoxide Formation

14.13.1
General Procedure for Chiral Amine-Catalyzed Epoxidations of Enals with H$_2$O$_2$ [50] (p. 104)

The catalyst (10 mol%) was added at r.t. to a solution of the aldehyde (1 equiv) in CH$_2$Cl$_2$ (0.5 M), followed by the addition of 35% aqueous H$_2$O$_2$ (1.3 equiv). After 4 h of reaction time the crude reaction mixture was passed through a silica gel FC column (Et$_2$O/pentane) to produce oxirane-2-carbaldehyde.

14.13.2
Epoxide Synthesis via Sulfide Alkylation/Deprotonation Procedure [51] (p. 360)

PhCHO + BnBr (2 equiv) → [catalyst (20 mol%), n-Bu$_4$NI (1 equiv), catechol (0.4 mol%), t-BuOH/H$_2$O, NaOH (2 equiv), 1 d] → Ph-epoxide-Ph
88% yield
dr = 83:17
ee = 96% (R,R)

To a solution of chiral sulfide catalyst (9.5 mg, 0.05 mmol) and catechol (1 mg, 1 μmol) in t-BuOH (450 μL) was added the benzaldehyde (26 μL, 0.25 mmol), benzyl bromide (60 μL, 0.5 mmol) and n-Bu$_4$NI (92 mg, 0.25 mmol). The resulting mixture was vigorously stirred at r.t. for 15 min (20–24 °C), and an aqueous solution of NaOH 28.5% (52 μL, 0.5 mmol, giving a 90:10 ratio of t-BuOH:water) was subsequently added. After 24 h of reaction, water (6 mL) was added, the two phases were separated, and the aqueous layer was extracted with diethyl ether (2 × 5 mL). The combined organic layers were dried over MgSO$_4$ and concentrated *in vacuo*. Purification by silica gel column chromatography afforded the two diastereoisomeric epoxides.

14.13.3
Typical Procedure for Sulfur Ylide Epoxidation via *In Situ* Diazo Compound Generation [52] (p. 361)

PhCHO + Ph-CH=N-N(Na$^+$)-Ts (2 equiv) → [catalyst (5 mol%), Rh$_2$(OAc)$_4$ (1 mol%), BnEt$_3$NCl (10 mol%), 40 °C, CH$_3$CN, 2 d] → Ph-epoxide-Ph
78% yield
dr = 98:2
ee = 94% (R,R)

Benzaldehyde tosylhydrazone sodium salt (5.9 g, 20.0 mmol), benzyl triethylammonium chloride (460 mg, 2.0 mmol), Rh$_2$(OAc)$_4$ (88 mg, 0.2 mmol), sulfide catalyst (250 mg, 1.0 mmol), benzaldehyde (2.0 mL, 20 mmol) and anhydrous acetonitrile (35 mL) were added sequentially to a two-necked flask equipped with a nitrogen balloon and a mechanical stirrer. After stirring vigorously for 10 min at r.t., the reaction was heated at 40 °C for 24 h. A second equivalent of tosyl hydrazone salt (5.9 g, 20.0 mmol) was added and stirring continued for a further 24 h at 40 °C. The reaction was quenched by the addition of a saturated solution of NH_4Cl (20 mL) and EtOAc (20 mL). The aqueous layer was washed with EtOAc (20 mL) and the combined organic phases were dried over $MgSO_4$, filtered, and concentrated *in vacuo*. Flash column chromatography afforded the epoxide in 78% yield (3.9 g, 98:2 *trans:cis*, 94% *ee*). The sulfide could be recovered in 70% yield by chromatography and used in further reactions.

14.13.4
Typical Procedure for Epoxidation of Stilbenes under Yang Conditions with CH_3CN-H_2O Solvent System [53] (p. 405)

E-Stilbene (18 mg, 0.1 mmol) and ketone (3.8 mg, 0.01 mmol) were dissolved in CH_3CN (1.5 mL) at r.t. An aqueous $Na_2(EDTA)$ solution (1 mL, 4×10^{-4} M) was added. To the stirred mixture was added in portions a mixture of Oxone (307 mg, 0.5 mmol) and sodium bicarbonate (130 mg, 1.55 mmol). On completion of the reaction according to TLC analysis (20 min), the reaction mixture was poured into water (20 mL) and extracted with CH_2Cl_2 (3 × 20 mL). The combined organic layers were dried over anhydrous Na_2SO_4. After removal of the solvent under reduced pressure, the residue was purified by flash column chromatography on silica gel (hexane, then 95:5 hexanes:ethyl acetate) to give *trans*-stilbene epoxide (19.4 mg, 99% yield) in 47% *ee*.

14.13.5
Typical Procedure for Epoxidation of Stilbenes under Shi Conditions (pH 10.5) [54] (p. 405)

Shi advises that all glassware is carefully washed prior to use in order to remove trace metals that can cause the decomposition of Oxone.

E-Stilbene (0.181 g, 1 mmol) was dissolved in 1:2 acetonitrile:DMM (15 mL). To this solution were added buffer (10 mL, 0.05 M solution of $Na_2B_4O_7 \cdot 10H_2O$ in 4×10^{-4} M aqueous Na_2(EDTA)), tetrabutylammonium hydrogen sulfate (0.015 g, 0.04 mmol), and ketone catalyst (77.4 mg, 0.3 mmol). The mixture was cooled in an ice bath. A solution of Oxone (0.85 g, 1.38 mmol) in aqueous Na_2(EDTA) (4×10^{-4} M, 6.5 mL) and a solution of K_2CO_3 (0.8 g, 5.8 mmol) in water (6.5 mL) were added dropwise separately over a period of 1.5 h (via syringe pumps or addition funnels). The best results were obtained if these solutions were added in a steady, uniform manner. After 2 h, the reaction was diluted with water (30 mL), and extracted with hexanes (4×40 mL). The combined extracts were washed with brine, dried (Na_2SO_4), filtered, concentrated, and purified by FC on silica gel (previously buffered with 1% triethylamine solution in hexane) using 1:0 to 50:1 hexane:ether as eluent. This provided *trans*-stilbene oxide (0.153 g, 78%) with 98.9% ee.

14.13.6
Asymmetric Epoxidation of E-β-Methylstyrene Catalyzed by Shi Ketone Using H_2O_2 as Co-Oxidant [55] (p. 408)

To a solution of *trans-β*-methylstyrene (0.118 g, 1.0 mmol) and ketone catalyst (38.7 mg, 0.15 mmol) in CH_3CN (1.5 mL) was added a solution of 2.0 M K_2CO_3 in 4×10^{-4} M aqueous $Na_2(EDTA)$ (1.5 mL), followed by H_2O_2 (30% aqueous, 0.4 mL, 4 mmol) at 0 °C. After stirring at 0 °C for 12 h, the reaction mixture was extracted with hexane, washed with 1 M aqueous $Na_2S_2O_3$ and brine, dried (Na_2SO_4), filtered, concentrated, and purified by chromatography on silica gel (previously buffered with 1% triethylamine solution in hexane) using 1:0 to 50:1 hexane:ether as eluent, to afford the epoxide product as a colorless oil (0.124 g, 93%) in 92% ee.

14.13.7
Iminium Salt-Catalyzed Epoxidation Under Non-Aqueous Conditions [56] (p. 411)

Tetraphenylphosphonium monoperoxysulfate can be prepared from Oxone and tetraphenyphosphonium chloride [57]. Synthesis of the iminium salt catalyst is described in the source paper (see [57]).

Tetraphenylphosphonium monoperoxysulfate (0.9 g, 2.0 mmol) was dissolved in chloroform (18 mL), and the solution cooled to −40 °C. Iminium salt (72 mg, 0.1 mmol) was dissolved in chloroform (4.5 mL) and the solution cooled to −40 °C. The solution of iminium salt was added dropwise to the solution of oxidant over 15–20 min; the temperature was monitored during the addition, and any increase in temperature minimized by controlling the rate of addition. A solution of 6-cyanobenzopyran (185 mg, 1.0 mmol) in chloroform (4.5 mL) was added in the same manner. The mixture was stirred at −40 °C for 24 h. Diethyl ether (180 mL, pre-cooled to −40 °C) was added to induce precipitation of the remaining oxidant. The mixture was filtered through Celite, and the solvent removed under reduced pressure. Column chromatography, eluting with ethyl acetate:light petroleum (1:99), afforded the product as a colorless oil which solidified on standing (119 mg, 59%, 97% ee).

14.13.8
Phase-Transfer-Catalyzed Asymmetric Epoxidation of *E*-Chalcone [58] (p. 415)

A solution of *E*-chalcone (708 mg, 3.4 mmol) and catalyst [59] (20 mg, 0.03 mmol) in toluene (10 mL) in a 25-mL round-bottomed flask was treated with 15% aqueous sodium hypochlorite solution (6.8 mmol), and the resulting mixture was stirred vigorously (magnetic stirrer at ca. 1000 r.p.m.) at 25 °C for 12–24 h. Water (20 mL) was then added and the layers separated. The aqueous layer was further extracted with ethyl acetate (30 mL), and the combined organic extracts dried (Na_2SO_4). Concentration under reduced pressure gave the crude epoxide. Chromatography on silica gel (petroleum ether:ethyl acetate, 9:1) gave the epoxide (747 mg, 98%) in >95% *de* and 86% *ee*. Alternatively, the product can be purified by recrystallization (EtOAc:petrol); here, the epoxide is produced in 72% yield and 98% *ee*.

14.13.9
Typical Procedure for Poly-L-Leucine-Catalyzed Epoxidation of Enones [60] (p. 416)

This procedure utilizes immobilized (polystyrene) poly-L-leucine, which is more easily separated from the reaction products, and can be prepared according to a published procedure [61].

To immobilized poly-L-leucine (7.0 g) was added THF (50 mL), urea hydrogen peroxide (2.07 g, 22 mmol) and DBU (4.11 mL, 27.5 mmol). This mixture was stirred for 3–5 min, after which the enone (4.01 g, 18.4 mmol) in THF (10 mL) was added. After a further 3 h, additional urea hydrogen peroxide (1.06 g, 11.3 mmol) and DBU (2.5 mL, 16.1 mmol) were added. After 28 h, the reaction was filtered to remove the poly-L-leucine. The filtrate was added to saturated aqueous ammonium chloride solution and extracted with ethyl acetate. The combined organic layers were dried ($MgSO_4$) and concentrated *in vacuo*. The acid-sensitive epoxide product was sufficiently pure for use in the next step, but could be purified by column chromatography on silica using eluent (2–12% ethyl acetate in hexanes) doped with 1% triethylamine. This gave the epoxide product (3.0 g, 70%) in >96% *ee*.

14.13.10
Amine-Catalyzed Epoxidation of *E*-Cinnamaldehyde with Hydrogen Peroxide [62] (p. 418)

Amine catalyst can be prepared in three steps from *N*-Boc-L-proline methyl ester [63].

E-Cinnamaldehyde (66 mg, 0.5 mmol) was dissolved in CH_2Cl_2 (1 mL). To this solution at ambient temperature was added the catalyst (30 mg, 0.05 mmol) followed by 35% aqueous hydrogen peroxide (63 mg, 0.65 mmol). After 4 h, the mixture was passed down a silica gel column, eluting with ether:pentane, to give the epoxide (59 mg, 80%), in a 93:7 *dr* and 96% *ee*.

14.13.11
General Procedure for Selective Epoxidation of Styrene [64] (p. 441)

To a solution of styrene (0.5 mmol) and catalyst (0.15 mmol) in water (5 mL) at 0 °C were added Oxone (153 mg) and $NaHCO_3$ (84 mg) at 10-min intervals over a period of 1 h. Water was then added, the aqueous layer was extracted with CH_2Cl_2 and the organic phase was dried over $MgSO_4$, and the solvents were evaporated.

14.14
Epoxide Opening

14.14.1
Desymmetrization of *meso*-Epoxides [65] (p. 281)

Tetrachlorosilane (86 μL, 0.75 mmol) was added dropwise to a stirred solution of cyclooctene oxide (63 mg, 0.50 mmol) and catalyst (18.0 mg, 0.05 mmol) in anhydrous CH_2Cl_2 (5 mL) at −90 °C. The mixture was then stirred at this temperature for 48 h under an argon atmosphere. The reaction was then quenched with a saturated aqueous solution of $NaHCO_3$ (10 mL) and the product extracted with CH_2Cl_2 (3 × 30 mL). The extract was washed with brine, dried over anhydrous $MgSO_4$, and the solvent evaporated. Purification using column chromatography on silica gel with a petroleum ether:ethyl acetate mixture (9:1) afforded the product (74 mg, 91%, 87% *ee*) as an oil: $[\alpha]_D$ − 24.4 (*c* 1.5, $CHCl_3$). For the *ee* determination, chlorohydrin was converted into the trifluoroacetyl derivative by stirring with trifluoroacetic anhydride (10 equiv) in CH_2Cl_2 for 3 h. Removal of the solvent, followed by purification using column chromatography on silica gel with a petroleum ether:ethyl acetate mixture (9:1), afforded (−)*trans*-1-chloro-2-trifluoroacetoxy-cyclooctane as a colorless oil.

14.15
Nucleophilic Substitution Reactions

14.15.1
Benzylation of *tert*-Butyl Alaninate P-Chlorobenzaldehyde Schiff Base Under Phase-Transfer Conditions [66] (p. 139)

14.15 Nucleophilic Substitution Reactions

To a mixture of Schiff base (134 mg, 0.5 mmol), (S,S)-catalyst (4.6 mg, 5 μmol) and benzyl bromide (72.8 μL, 0.6 mmol) in toluene (2 mL) was added CsOH·H$_2$O (420 mg, 2.5 mmol) at 0 °C, and the reaction mixture was stirred vigorously for 30 min. After consumption of the starting material, 1 M KHSO$_4$ (5 mL) and THF (5 mL) were added to this mixture, and the biphasic solution was stirred at r.t. to neutralize the base and hydrolyze the Schiff base moiety. After stirring for 1 h, the resulting organic wastes were extracted with hexane. The aqueous phase was then basified with NaHCO$_3$ and the target molecule extracted with ether, dried over Na$_2$SO$_4$, and concentrated. Purification of the crude product by column chromatography on silica gel (hexane:EtOAc, 1:2 as eluent) gave the alkylation (R) product (100 mg, 0.425 mmol, 85%, 98% ee) as a colorless oil.

14.15.2
Typical Procedure for the Catalytic Asymmetric Benzylation of Phenyl Oxazoline Derivatives [67] (p. 144)

RX = PhCH$_2$Br : 98%, >99% ee
CH$_2$=CHCH$_2$Br : 87%, 97% ee
EtI : 48%, 93% ee

(S,S)-catalyst
Ar = 3,4,5-F$_3$-C$_6$H$_2$

Benzyl bromide (120 μL, 1.0 mmol) was added to a solution of oxazoline (50.0 mg, 0.2 mmol), (S,S)-cat (9.55 mg, 5 μmol) and KOH (56.1 mg, 1.0 mmol) in toluene (0.8 mL) at 0 °C. The reaction mixture was stirred for 5 h. Upon completion of the alkylation, the mixture was diluted with EtOAc (20 mL), washed with water (2 × 5 mL), dried over MgSO$_4$, filtered, and concentrated *in vacuo*. The residue was purified by column chromatography on silica gel (hexanes:EtOAc, 20:1 as eluent) to afford the product (65 mg, 98%) as a pale yellow oil.

14.15.3
Hydrolysis of Oxazoline [67] (p. 141)

98% (R = CH$_2$Ph)

HCl (6 M; 1.5 mL) was added to a solution of oxazoline (500 mg, 1.48 mmol) in ethanol (1.5 mL), and the mixture heated at reflux for 24 h. After the solvent was removed *in vacuo*, the residue was purified by column chromatography (15% aqueous NH$_4$OH) through an ion-exchange resin (Dowex 50WX8-100) to give (S)-(+)-α-benzyl serine as a white solid (365 mg, 98%).

14.16
Allylation Reactions

14.16.1
Catalytic Asymmetric Allylic Alkylation [68] (p. 138)

To a suspension of glycinate Schiff base (50.0 mg, 0.169 mmol), catalyst (10.6 mg, 16.9 µmol), [Pd(C$_3$H$_5$)Cl]$_2$ (5.4 mg, 0.015 mmol) and triphenyl phosphite (21.0 mg, 67.7 µmol) in toluene (0.28 mL) were successively added a solution of *p*-chlorocinnamyl acetate (35.7 mg, 0.169 mmol) in toluene (0.56 mL) and 50% KOH aqueous solution (66.4 mg, 0.591 mmol) at 0 °C under an argon atmosphere. After stirring vigorously at 0 °C for 7 h, the mixture was diluted with ether (15 mL). The organic phase was washed with saturated NaHCO$_3$ aqueous solution (3 × 5 mL) and brine (5 mL). The extract was dried over Na$_2$SO$_4$, filtered, and concentrated *in vacuo*. The crude product was purified by column chromatography on basic silica gel (hexane:EtOAc, 300:1 as eluent) to give the allylated product (66.1 mg, 85%) as colorless crystals.

14.16.2
Reaction of Allyl Trichlorosilane with Benzaldehyde [69] (p. 262)

Allyl trichlorosilane (0.47 mmol) was added to a solution of the METHOX catalyst (0.02 mmol), diisopropylethylamine (2 mmol) and aldehyde (0.4 mmol) in acetonitrile (2 mL) under nitrogen at −40 °C. The mixture was stirred at the same temperature for 18 h, after which the reaction was quenched with aqueous saturated NaHCO$_3$ (1 mL). The aqueous layer was extracted with ethyl acetate (3 × 10 mL) and the combined organic extracts were washed with brine and dried over Na$_2$SO$_4$. The solvent was removed *in vacuo* and the residue purified by FC on a silica gel column (15 cm × 1 cm) with a petroleum ether-ethyl acetate mixture (95:5) to produce (*S*)-(−)-1-phenyl-but-3-en-1-ol (95%, 96% *ee*).

14.17
Acylation

14.17.1
PBO-Catalyzed KR of an Aryl Alkyl sec-Alcohol: KR of (±)-1-(2-Methylphenyl)ethanol (Preparative Scale) [70] (p. 292)

A solution of phosphine catalyst (16 mg, 0.045 mmol; 99.7% *ee*) in deoxygenated *n*-heptane (74 mL) was added to a N$_2$-purged flask containing (±)-1-(2-methylphenyl)ethanol (1.02 g, 7.5 mmol). After cooling the mixture to −40 °C, (*i*-PrCO)$_2$O (3.05 mL, 18.4 mmol) was added via a syringe. After stirring for 14 h at −40 °C, the mixture was quenched by the addition of isopropylamine (4 mL, 47 mmol). The solution was stirred at −40 °C for 10 min, after which the flask was allowed to warm to r.t. (ca. 1 h). After removal of the solvent *in vacuo*, the residue was purified by FC on silica gel (CH$_2$Cl$_2$:hexanes, 2:3 → CH$_2$Cl$_2$) to yield the ester as an oil [725 mg, 48%, 95.7% *ee* by chiral-HPLC on the alcohol obtained by hydrolysis of an aliquot using NaOH:MeOH, 1:19], and the alcohol as an oil [482 mg, 46%, 90.2% *ee* by chiral-HPLC following additional purification by FC on silica gel (EtOAc:hexanes, 1:5)]. The calculated selectivity value at 48.5% conversion was *s* = 142.

14.17.2
KR of an Aryl Alkyl sec-Alcohol with Ferrocenyl DMAP Catalyst: KR of (±)-1-(2-Methylphenyl)ethanol (Preparative Scale) [71] (p. 294)

In a glove-box, 1-(2-methylphenyl)ethanol (1.11 g, 8.14 mmol), *t*-amyl alcohol (16 mL), and Et$_3$N (0.67 mL, 4.8 mmol) were added to a flask containing catalyst (27.7 mg, 0.0419 mmol). A septum was added and the flask removed from the glove box. After gentle heating to dissolve the catalyst, the flask was cooled to 0 °C. Ac$_2$O (0.46 mL, 4.9 mmol) was added dropwise, and after 25.5 h the reaction was quenched with MeOH (5 mL). The mixture was passed through a short plug of silica gel to separate the catalyst from the alcohol:acetate mixture (EtOAc:hexanes, 1:1 to 3:1, then Et$_3$N:EtOAc, 1:9). The solution of alcohol and acetate was concentrated *in vacuo* and the residue purified by FC on silica gel (Et$_2$O:pentane, 1:20 to 1:4) to afford the (*R*)-acetate (639 mg, 44%, 90.2% *ee* by chiral-GC on the alcohol obtained by reduction using LiAlH$_4$), and the (*S*)-alcohol (517 mg, 47%, 92.9% *ee* by chiral-GC). The calculated selectivity value at 50.7% conversion was $s = 65.9$. The recovered catalyst was purified by FC on silica gel (EtOAc:hexanes, 1:1–EtOAc:hexanes:Et$_3$N, 9:9:2), which provided 24.9 mg of the pure catalyst (90%).

14.17.3
KR of (±)-1-(1-Naphthyl)-1-Ethanol [72] (p. 297)

A solution of (±)-1-(1-naphthyl)-1-ethanol (2.416 g, 14.0 mmol), DIPEA (1.93 mL, 10.5 mmol) and catalyst (74 mg, 0.28 mmol) in CHCl$_3$ (14 mL) was stirred at 0 °C for 15 min, and then treated with (*n*-PrO)$_2$O (1.35 mL, 10.5 mmol). The mixture was stirred at 0 °C for 10 h, at which time it was quenched with MeOH (10 mL), allowed to warm slowly, and left at r.t. for 1 h. The reaction mixture was diluted with CH$_2$Cl$_2$, washed twice with 1 M HCl, then twice with saturated aqueous NaHCO$_3$, and then dried (NaSO$_4$). The solution was concentrated *in vacuo* and purified by FC on silica gel (Et$_2$O:hexanes, 1:19 to 1:4) to produce the ester (1.672 g, 52%, 82.5% *ee* by chiral-HPLC), and the alcohol (1.091 g, 45%, 98.8% *ee* by chiral-HPLC). The calculated selectivity value at 54.5% conversion was $s = 52.3$. The aqueous phase obtained during the work-up was basified with 0.5 M NaOH and repeatedly extracted with CH$_2$Cl$_2$ (until the aqueous phase was pale-yellow), the extract was dried (Na$_2$SO$_4$), concentrated *in vacuo*, and purified by FC on silica gel (*i*-PrOH:hexanes, 1:19 to 1:9) to provide 50 mg of recovered catalyst (68%).

14.17.4
KR of an Aryl Alkyl *sec*-Alcohol Using Chiral Carbene Catalyst:
KR of (±)-1-Phenylethanol (Analytical Scale) [73] (p. 299)

To a solution of chiral carbene catalyst (0.03 mmol, 5 mol%) and 1-phenylethanol (73.2 mg, 72 μL, 0.6 mmol) in THF (2 mL) was added vinyl diphenylacetate (117.5 mg, 0.45 mmol) dropwise at −78 °C. The reaction mixture was stirred at this temperature for 3 h, and then treated with 0.1 M HCl and extracted with ether. The combined organic layers were washed with brine, dried (Na_2SO_4), and concentrated *in vacuo*. The residue was purified by FC on silica gel (ether:hexanes, 1:50 to 1:2) to afford (*R*)-1-phenylethyl diphenylacetate (65 mg, 32%, 96% *ee* by chiral-HPLC) and (*S*)-1-phenylethanol (44.6 mg, 61%, 52% *ee* by chiral HPLC). The calculated selectivity value at 34% conversion was $s = 80$.

14.17.5
KR of an Allylic *sec*-Alcohol Using Vedej's PBO Catalyst:
KR of (±)-1-(3,4-Dihydronaphthalen-1-yl)-ethanol (Analytical Scale) [74] (p. 299)

1-(3,4-Dihydronaphthalen-1-yl)-ethanol (21 mg, 0.12 mmol) was added to a solution of phosphine catalyst (1.97 mg, 6 μmol) in toluene (1.2 mL). The solution was cooled to −40 °C, and (*i*-PrCO)$_2$O (50 mL, 0.3 mmol) was added via a syringe. The reaction was stirred for 72 h, followed by quenching with *i*-PrNH$_2$

(120 mL, 1.4 mmol). After stirring for 15 min at −40 °C the mixture was warmed to r.t. and concentrated *in vacuo*. ^{1}H-NMR (d_6-acetone) revealed that 51% conversion to the ester had occurred, and confirmed that 5 mol% of catalyst had been used. Purification by FC on silica gel (CH_2Cl_2:hexanes, 6:1) gave the ester (86.7% *ee* by chiral-HPLC on the alcohol obtained by hydrolysis using NaOH:MeOH, 1:19) and the alcohol (96.1% *ee* by chiral-HPLC). The calculated selectivity value at 51% conversion was $s = 55$.

14.17.6
KR of an Allylic Alcohol (Preparative Scale) [75] (p. 301)

In the air, *t*-amyl alcohol (8.75 mL) and Et_3N (0.36 mL, 2.6 mmol) were added to a vial containing (±)-alcohol (1.16 g, 4.42 mmol) and catalyst (29.0 mg, 43.9 µmol). The vial was closed with a Teflon-lined cap and sonicated to assist catalyst dissolution. The reaction mixture was cooled to 0 °C, and Ac_2O (0.25 mL, 2.6 mmol) was added. After 42.5 h, the reaction was quenched with MeOH (0.25 mL). The mixture was passed through a pad of silica gel (EtOAc:hexanes, 1:5–EtOAc–Et_3N:EtOAc, 1:1) to separate the catalyst (27.6 mg, 95%) from the alcohol:acetate mixture. The solution of alcohol and acetate was concentrated *in vacuo* and the residue purified by FC on silica gel (EtOAc:hexanes, 1:9 to 1:4) to afford the (−)-acetate (0.7 g, 52%, 91.8% *ee* by chiral-HPLC) and the (+)-alcohol (0.55 g, 47%, 98.0% *ee* by chiral-HPLC). The calculated selectivity value at 51.6% conversion was $s = 107$.

14.17.7
Procedure for KR of a Propargylic *sec*-Alcohol: KR of (±)-4-Phenyl-3-butyn-2-ol (Analytical Scale) [76] (p. 302)

A vial containing (±)-4-phenyl-3-butyn-2-ol (73.0 mg, 0.5 mmol) and catalyst (3.3 mg, 5.0 μmol) in *t*-amyl alcohol (1.0 mL) was capped with a septum and sonicated to assist catalyst dissolution. The resulting purple solution was cooled to 0 °C, and Ac$_2$O (35.4 μL, 0.375 mmol) was added via a syringe. After 49 h, the reaction mixture was quenched by the addition of a large excess of MeOH. After concentration *in vacuo*, the residue was purified by FC on silica gel (EtOAc:hexanes, 1:9 to 1:1, then EtOAc:hexanes:Et$_3$N, 9:9:2) to afford the (*R*)-acetate (68.6% *ee* by chiral-GC) and the (*S*)-alcohol (96.0% *ee* by chiral-GC on the acetate obtained following esterification). The calculated selectivity value at 58.3% conversion was $s = 20.2$.

14.17.8
KR of a Monoprotected-1,2-diol: KR of (±)-*cis*-N-(2-Hydroxycyclohexanoxycarbonyl)pyrrolidine (Analytical Scale) [77] (p. 306)

To a solution of (±)-cis-N-(2-hydroxycyclohexanoxycarbonyl)pyrrolidine (0.25 mmol) and catalyst (12.5 μmol) in CCl4 (2.5 mL) was added i-Pr$_2$NEt (21.8 μL, 125 μmol) and (i-PrCO)$_2$O (20.7 μL, 125 μmol). The reaction mixture was stirred at 0 °C for 3 h, and then treated with 0.1 M aqueous HCl and extracted with EtOAc. The organic layer was washed with saturated aqueous NaHCO$_3$, dried (Na$_2$SO$_4$), and concentrated to provide a crude mixture of the unreacted alcohol (97% ee by chiral-HPLC) and acylated product (90% ee by chiral HPLC). The calculated selectivity value at 51.9% conversion was $s = 87$.

14.17.9
KR of an α-Chiral Primary Amine: KR of (±)-1-Phenylethylamine (Analytical Scale) [78] (p. 308)

Catalyst (5.2 mg, 14 μmol), (±)-1-phenylethylamine (17.0 mg, 140 μmol) and CHCl$_3$ (2.5 mL) were added to a Schlenk flask under an argon atmosphere. The resulting purple solution was cooled to −50 °C and a solution of O-carbonyloxyazlactone (13.5 mg, 42.0 μmol) in CHCl$_3$ (0.15 mL) was added via a syringe. After 4 h, additional O-carbonyloxyazlactone (13.5 mg, 42.0 μmol) in CHCl$_3$ (15 μL) was added. After 24 h in total, the solution was concentrated *in vacuo* and the residue purified by FC on silica gel (EtOAc:hexanes, 1:4) to afford the carbamate (7.3 mg, 29%, 79% ee by chiral-HPLC) and the amine, which was immediately acylated (Et$_3$N, Ac$_2$O, CH$_2$Cl$_2$, r.t.) and then purified by FC on silica gel (EtOAc) to afford the acetamide (11.4 mg, 50%, 42% ee by chiral-GC). The calculated selectivity value at 35% conversion was $s = 13$.

14.17.10
ASD of a *meso*-1,2-Diol: ASD of *cis*-1,2-Cyclohexanediol (Preparative Scale) [79] (p. 308)

Substrate	*cis*-cyclohexanediol	cyclohexenediol	tetrahydronaphthalenediol	Ph/Ph diol	Me/Me diol
% ee	96% ee	90% ee	66% ee	60% ee	94% ee
yield	83%	81%	89%	80%	85%

To 4 Å MS (400 mg) was added a solution of (*S*)-catalyst (3.3 mg, 15.1 μmol) in CH_2Cl_2 (2.5 mL), and the resulting reaction mixture was cooled to −78 °C. A solution of Et_3N (306 mg, 3.02 mmol) in CH_2Cl_2 (2.5 mL), a solution of *cis*-1,2-cyclohexanediol (351 mg, 3.02 mmol) in CH_2Cl_2 (20 mL), and a solution of BzCl (636 mg, 4.52 mmol) in CH_2Cl_2 (2.5 mL) were then added sequentially. After 3 h at −78 °C the reaction was quenched by the addition of a phosphate buffer (pH 7), and extracted with Et_2O. The combined organic extracts were dried (Na_2SO_4), and concentrated *in vacuo*. The residue was purified by FC on silica gel (EtOAc:hexanes, 1:15) to afford *cis*-benzoyloxy-1-cyclohexanol (554 mg, 83%, 96% ee by chiral-HPLC).

14.17.11
Procedure for ASD of a Cyclic *meso*-Anhydride Using (DHQD)$_2$AQN as Catalyst: ASD of *cis*-Cyclopentane-1,2-Dicarboxylic Acid Anhydride (Analytical Scale) [80] (p. 316)

Dry MeOH (32 mg, 40 μL, 1.0 mmol) was added dropwise to a stirred solution of cis-cyclopentane-1,2-dicarboxylic acid anhydride (14 mg, 0.1 mmol) and (DHQD)$_2$AQN (95%, 72.2 mg, 0.08 mmol) in dry Et$_2$O (5 mL) under an argon atmosphere at −30 °C. The reaction mixture was stirred at −30 °C until the starting material was consumed (TLC-monitored, 71 h). The reaction was quenched by the addition of HCl (1 M, 3 mL) in one portion. The aqueous phase was extracted with EtOAc (2 × 10 mL), and the combined organic phases were dried (MgSO$_4$) and concentrated *in vacuo* to afford the hemiester as a clear oil (17 mg, 99%, 95% *ee* by ^{1}H NMR on the diastereomeric amides formed by coupling the hemiesters to (R)-1-naphthalen-1-yl-ethylamine). The (DHQD)$_2$AQN catalyst was recovered quantitatively by basification (pH 11) of the aqueous phase with aqueous KOH (1 M), extraction with Et$_2$O, drying of the Et$_2$O extracts (MgSO$_4$), and concentration *in vacuo*.

14.17.12
PKR of a Monosubstituted Succinic Anhydride Using (DHQD)$_2$AQN as Catalyst: PKR of (±)-2-Methysuccinic Anhydride (Preparative Scale) [81] (p. 317)

2,2,2-Trifluoroethanol (0.73 mL, 10 mmol) was added to a solution of 2-methylsuccinic anhydride (114 mg, 1.0 mmol) and (DHQD)$_2$AQN (95%, 180 mg, 0.2 mmol) in Et$_2$O (50.0 mL) at −24 °C. The resulting reaction mixture was stirred at this temperature until the anhydride was consumed (TLC-monitored, 50 h). The reaction mixture was washed with aqueous HCl (1 M, 3 × 10 mL). The aqueous phase was extracted with Et$_2$O (3 × 20 mL), and the combined organic phases were dried (MgSO$_4$) and then concentrated *in vacuo*. The residue was purified by FC on silica gel (cyclohexane:butyl acetate:acetic acid, 50:1:1) to afford the α-methyl hemiester (77 mg, 36%, 93% *ee* by chiral-HPLC on the diastereomeric amides formed by coupling the hemiesters to (R)-1-naphthalen-1-yl-ethylamine) and the β-methyl hemiester (88 mg, 41%, 80% *ee* by chiral-HPLC on the diastereomeric amides formed by coupling the hemiesters to (R)-1-naphthalen-1-yl-ethylamine). The (DHQD)$_2$AQN catalyst was recovered quantitatively by basification (pH 11) of the aqueous phase with aqueous KOH (2 M), extraction with EtOAc (3 × 15 mL), drying of the EtOAc extracts (MgSO$_4$), and concentration *in vacuo*.

14.17.13
DKR of a UNCA using (DHQD)$_2$AQN as Catalyst: DKR of (±)-2,5-Dioxo-4-Phenyl-3-Oxazolidine Carboxylic Acid Phenylmethyl Ester (Analytical Scale) [82] (p. 318)

A mixture of UNCA (62.2 mg, 0.2 mmol) and 4 Å MS (20 mg) in anhydrous Et$_2$O (14.0 mL) was stirred at r.t. for 10 min and then warmed to 34 °C, after which (DHQD)$_2$AQN (95%, 36.1 mg, 0.04 mmol) was added. The resulting mixture was stirred for a further 5 min, after which a solution of allyl alcohol in Et$_2$O (1:99, 0.24 mmol) was introduced dropwise via a syringe over a period of 1 h. The resulting reaction mixture was stirred for 1 h at 34 °C, washed with aqueous HCl (2 M, 2.0–3.0 mL) and brine (3.0 mL), dried (Na$_2$SO$_4$), and concentrated to provide a light yellow solid. Purification by FC on silica gel (EtOAc:hexanes, 1:9) gave (R)-allyl-(N-benzyloxycarbonyl)phenylglycinate as a white solid (63 mg, 97%, 91% ee by chiral-HPLC). The (DHQD)$_2$AQN catalyst was recovered quantitatively by washing the combined aqueous extracts with Et$_2$O (2 × 2.0 mL), and then basifying first with KOH (pH ∼ 4) and then with Na$_2$CO$_3$ (pH ∼ 11). The resulting solution was extracted with EtOAc (2 × 5.0 mL) and the combined organic extracts washed with brine (2.0 mL), dried (Na$_2$SO$_4$), and concentrated *in vacuo*.

14.17.14
Procedure for Regioselective Phosphorylation [83] (p. 428)

Triol (0.5 g, 1.11 mmol) was dissolved in 30 mL of toluene and an aliquot of peptide catalyst (27.6 μmol, 2.5 mol%) in CH$_2$Cl$_2$ (1.0 mL) was delivered. The mixture was then cooled to 0 °C, and triethylamine (0.325 mL, 2.33 mmol) was added. Diphenyl chlorophosphate (0.46 mL, 2.22 mmol) was then added dropwise. After 4 h at 0 °C, the reaction was filtered to remove triethylamine salts, quenched with 4 mL of methanol, and concentrated under reduced pressure. The crude mixture was purified using silica gel FC, eluting with a gradient of 0 to 40% ethyl acetate:hexanes to yield diol (0.422 g, 56% yield), which was found to exhibit > 98% ee.

14.18
[4+2] Cycloadditions

14.18.1
General Procedure for the Diels–Alder Reaction of α,β-Unsaturated Aldehydes [84] (p. 98)

72–99%, 83–94% ee
exo:endo 1.3:1 – 1:14

To a solution of the catalyst·HCl (5–20 mol%) in wet MeOH (5% H_2O) was added the α,β-unsaturated aldehyde (1 M). After stirring for 1–2 min, diene (3 equiv) was added to the solution. Upon consumption of the limiting reagent (as judged by TLC, ^{1}H-NMR, or GLC analysis), the reaction mixture was diluted into Et_2O and washed successively with H_2O and brine. The aqueous layer was extracted twice with Et_2O, and the combined organic layers were dried (Na_2SO_4), filtered, and concentrated. The resulting dimethy acetal was then rapidly stirred in TFA:H_2O:$CHCl_3$ (1:1:2, 4 mL) for 2 h at r.t., followed by neutralization with saturated aqueous $NaHCO_3$ solution and extraction with Et_2O. Purification of the Diels–Alder adduct was performed using silica gel chromatography.

14.18.2
General Procedure for Asymmetric Cycloaddition of Dienes with α-Acyloxyacroleins [85] (p. 100)

81–>99%, 79–92% ee
exo:endo 6.7:1–99:1

To a solution of catalyst (0.08 mmol) and pentafluorobenzenesulfonic acid (0.22 mmol) in nitroethane (0.125 mL) or THF (0.25 mL) was added the diene (1.6 mmol). After stirring at 0 °C or r.t., α-substituted acrolein (0.8 mmol) was added in one portion. Upon consumption of α-substituted acrolein, the reaction was quenched with saturated aqueous $NaHCO_3$, and the reaction mixture was diluted with pentane and washed successively with H_2O and brine. The organic layer was dried ($MgSO_4$), filtered, and concentrated under reduced pressure. Purification of the Diels–Alder adduct was accomplished by silica gel chromatography, eluting with pentane:ether.

14.18.3
General Procedure for Intramolecular Diels–Alder Cycloaddition [86] (p. 100)

71% yield
>20:1 dr, 90% ee

A solution of the catalyst·HX (20 mol%) in wet solvent (2% H_2O, 0.1 M) was cooled to the desired temperature (−20 to +4 °C). The trienal (100 mol%) was then added and the mixture stirred until completion of the reaction, while the temperature warmed to r.t. (from 16 h to 6 days). The crude reaction mixture was loaded directly onto a silica column, and purified.

14.18.4
Taddol-Catalyzed Hetero-Diels–Alder Reaction of 1-Aldimine-3-Siloxy Dienes [87] (p. 236)

A solution of catalyst (67 mg, 0.1 mmol, 20 mol%) and benzaldehyde (61 µL, 0.6 mmol, 1.2 equiv) in toluene (0.5 mL) was cooled to −78 °C. Diene (114 mg, 0.5 mmol, 1.0 equiv) was then added slowly, and the resulting slightly yellow solution was maintained at −78 °C for 2 days. The reaction was then diluted with CH_2Cl_2 (2 mL), followed by slow addition of freshly distilled acetyl chloride (71 µL, 1.0 mmol, 2.0 equiv). The mixture was stirred at −78 °C for 30 min and transferred directly to a column for FC purification on silica gel (hexanes:ethyl acetate, 7:3), which afforded pure dihydropyranone (61 mg, 70% yield, >98% *ee*) as an oil.

14.18.5
Taddol-Catalyzed Diels–Alder Reactions of 1-Amino-3-Siloxydienes [88] (p. 242)

A solution of taddol catalyst (66.7 mg, 0.1 mmol, 10 mol%) and methacrolein (41.5 µL, 0.5 mmol, 1.0 equiv) in 0.75 mL of toluene at −80 °C was treated with aminosiloxydiene (260 µL, 1.0 mmol, 2.0 equiv). The reaction was stirred for 48 h and treated with $LiAlH_4$ (1 M in Et_2O, 2.0 mL, 2 mmol, 4.0 equiv) at −80 °C. The reaction mixture was stirred for 0.5 h at −80 °C and then for an additional 1.5 h at r.t. After cooling to 0 °C, excess $LiAlH_4$ was quenched with water (0.5 mL), and the solids were removed by filtration. The solids were rinsed with Et_2O (5 × 3.0 mL) and the combined filtrate was concentrated *in vacuo*. The resulting oil was taken up in CH_3CN (2 mL), cooled in an ice bath, and treated with HF (5% solution in CH_3CN, 3.0 mL). After stirring for 0.5 h at r.t., the volatiles were removed *in vacuo* and the residue was subjected to FC on silica gel (ethyl acetate: hexanes, 7:3) to afford the product as a colorless oil (116 mg, 83% yield, 91% *ee*).

14.19
[3+2]-Cycloadditions

14.19.1
[3+2]-Cycloaddition of Nitrones to α,β-Unsaturated Aldehydes [89] (p. 102)

66–98%, 90–99% ee
exo:endo 4:1–>99:1

A flask containing nitrone (1 equiv) and imidazolidinone catalyst·HX (20 mol%) was charged with MeNO$_2$ (0.1 M), and then with the appropriate amount of H$_2$O (3 equiv). After cooling the solution to the desired temperature, α,β-unsaturated aldehyde (4 equiv) was added dropwise to the flask. Additional aldehyde (3 equiv) was added to the reaction mixture at 24-h intervals until the reaction was complete (96–160 h). The resulting solution was passed through a silica gel column with AcOEt. The removal of volatiles afforded an oily residue that was purified by silica gel chromatography to afford the title compound.

14.19.2
General Procedure for the Catalytic Annulation of Enals and Aldehydes [90] (p. 334)

R^1 = Ph, p-MeO-Ph, 1-Naphth, TIPS—≡

R^2 = Ph, p-Br/Cl-Ph, p-MeCO$_2$-Ph, p-CF$_3$-Ph, m-F/Cl-Ph, o-Cl-Ph

dr (cis:trans) = 3:1–5:1

Into an oven-dried 20-mL vial were weighed both aldehydes (2.0 mmol each), and catalyst (28.4 mg, 0.083 mmol). The vial was crimped, evacuated, and back-filled with argon. To this mixture were added 2 mL of a 10:1 THF:t-BuOH solution, followed by DBU (11.0 µL, 0.074 mmol). The resulting yellow-orange solution was stirred for 15 h at r.t. The reaction mixture was concentrated under reduced pressure, and the residue purified by FC (*n*-hexane:ethyl acetate, 4:1) to afford the corresponding lactone products as colorless solids.

14.20
Acyloin and Benzoin Condensations

14.20.1
General Procedure for the Asymmetric Synthesis of Acyloins [91] (p. 337)

The precatalyst triazolium salt can be prepared according to Ref. [92].
In a Schlenk tube with an argon atmosphere at r.t., precatalyst (20.6 mg, 0.055 mmol, 20 mol%) was suspended with absolute THF (1.7 mL). A solution of freshly sublimed KO*t*-Bu (5.9 mg, 0.052 mmol, 19 mol%) in absolute THF (0.6 mL) was added slowly, and the solution stirred for 5 min. An aldehyde ketone (0.275 mmol) was dissolved in absolute THF (0.5 mL) and added to the carbene solution. The reaction mixture was stirred for 48 h, diluted with DCM, quenched with water, extracted twice with DCM, and dried over $MgSO_4$. The solvent was evaporated and the crude product purified by FC on silica gel (*n*-pentane:DCM, 1:2) to yield the corresponding acyloin.

14.20.2
Asymmetric Benzoin Condensation [93] (p. 335)

The triazolium precatalyst can be prepared according to Ref. [93].
Under an argon atmosphere, 10 mmol of the aromatic aldehyde were added at r.t. to a solution of 331 mg (1.0 mmol, 10 mol%) of the triazolium salt in 7 mL absolute THF. KO*t*-Bu (112 mg, 1.0 mmol, 10 mol%) in 4 mL absolute THF was then added dropwise. The reaction was stirred at r.t. for 16 h, and then poured into distilled water. After extraction with DCM the organic phase was dried over

MgSO₄. The solvent was removed *in vacuo*. After recrystallization or column chromatography, the benzoins were obtained as colorless crystals or pale yellow oils.

14.21
Reduction of Ketones and Imines

14.21.1
General Procedure for the Reduction of Aromatic Ketones with Silanes Under Phase-Transfer Conditions [94] (p. 398)

To a stirred solution of ketone (1 mmol) and fluoride salt (0.02 mmol) in dry THF was added the silane (1.5 mmol). The mixture was stirred at r.t. until the reaction was complete (TLC-monitored). NaOH (5 mL of a 3 M solution) was added dropwise. After stirring vigorously overnight, the solution was extracted with diethyl ether (3 × 15 mL). The combined organic extracts were washed with water, dried (MgSO₄), and evaporated *in vacuo*. The residue was purified by chromatography (SiO₂) or distillation if necessary. Enantiomeric excess was measured by chiral stationary phase GC-analysis (Chiraldex G-TA column).

14.21.2
Phosphoric Acid-Catalyzed Transfer Hydrogenations of Imines [95] (pp. 230 and 397)

A mixture of the imine (90 mg, 0.4 mmol, 1.0 equiv), Hantzsch ester (142 mg, 0.56 mmol, 1.4 equiv), and catalyst (3 mg, 4 μmol, 1 mol%) in toluene (4 mL) was stirred at 35 °C under an argon atmosphere for 45 h. The solvent was evaporated at reduced pressure, and the products were isolated by FC on silica gel (3% ethyl acetate in hexanes) to give the product (87 mg, 96% yield, 88% ee).

14.21.3
Typical Procedure for the Catalytic Hydrosilylation of Imines [96] (pp. 275 and 400)

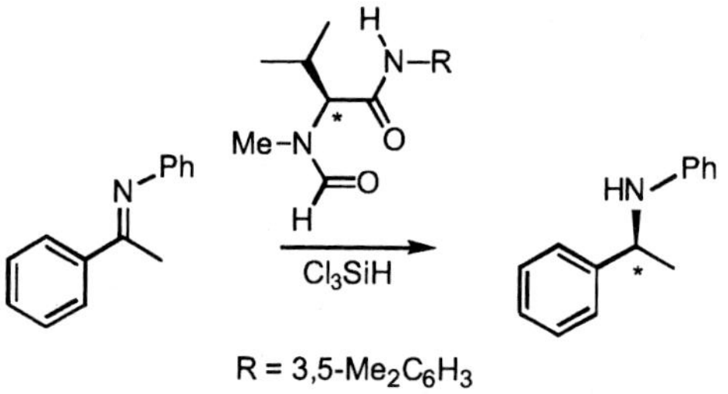

R = 3,5-Me$_2$C$_6$H$_3$

Trichlorosilane (77 µL, 0.77 mmol) was added dropwise to a stirred solution of the imine (100 mg, 0.51 mmol) and catalyst (13.5 mg, 0.051 mmol) in anhydrous toluene (2 mL) at 0 °C, and the mixture was stirred at r.t. overnight under an argon atmosphere. The reaction was quenched with a saturated solution of NaHCO$_3$ (10 mL) and the product was extracted with ethyl acetate (3 × 30 mL). The extract was washed with brine, dried over anhydrous MgSO$_4$, and the solvent was evaporated. Purification using column chromatography on silica gel with a petroleum ether-ethyl acetate mixture (24:1) afforded the (S)-(+)-amine (81 mg, 81%, 92% ee) as an oil: $[\alpha]_D$ + 16.8 (c 0.5, MeOH).

14.22
Hydrogenation of Olefins

14.22.1
General Procedure for Transfer Hydrogenation of Enals Using Symmetric Hantzsch Ester [97] (pp. 110 and 394)

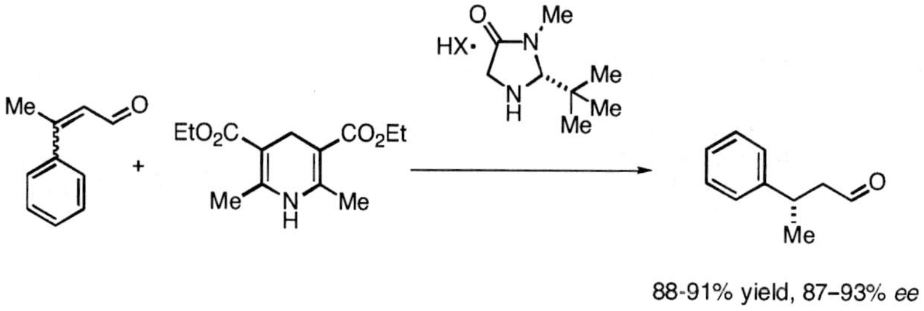

88–91% yield, 87–93% ee

A solution of β,β-disubstituted enal (1 mmol) in CHCl$_3$ (0.2 M) was cooled to the desired temperature. Catalyst·TCA (20 mol%) and Hantzsch ester (1.2 equiv) were added, and the mixture was stirred at the same temperature until the reaction was determined to be complete (by TLC). The reaction mixture was then diluted with Et$_2$O and passed through a short pad of silica gel. The resulting solution was concentrated *in vacuo* and purified by FC to provide the title compound.

14.22.2
Asymmetric Reduction of α,β-Unsaturated Aldehydes Using Symmetric Hantzsch Ester [98] (p. 394)

To a stirred solution of α,β-unsaturated aldehyde (0.5 mmol) in dioxane (7 mL) at 13 °C was added catalyst (20.4 mg, 0.05 mmol, 10 mol%) and, after a further 5 min, crystalline dihydropyridine (Hantzsch ester, 129.2 mg, 0.51 mmol). After a reaction time of 48 h the mixture was poured into distilled water (20 mL) and extracted with DCM (2 × 125 mL). The combined organic layers were dried (MgSO$_4$), filtered, and concentrated. The product was isolated by FC (SiO$_2$, ethyl acetate/hexane) to give the saturated aldehyde. Enantiomeric excess was measured by chiral stationary phase GC-analysis.

14.22.3
Procedure for the Selective Amidation of Carboxylic Acids [99] (p. 437)

To a stirred solution of sodium benzoate derivatives (45.1 mg, 0.225 mmol, and 38.7 mg, 0.225 mmol), PhCH$_2$NH$_2$·HCl (21.5 mg, 0.15 mmol) and catalyst

(45.0 mg, 0.03 mmol) in water (9.5 mL) was added 2-chloro-4,6-dimethoxy-1,3,5-triazine (26.3 mg, 0.15 mmol) in MeOH (0.5 mL), at r.t. The solution was stirred for 48 h at r.t. until a white precipitate was produced. The mixture was extracted with Et$_2$O, and the organic phase washed successively with 1 M NaOH, 1 M HCl, and brine. The crude mixture was purified by preparative TLC (hexane:AcOEt, 70:30) to give 31.6 mg of a mixture of both amides, of which the yield and ratio was determined by NMR analysis.

14.23
Multicomponent Domino Reactions

14.23.1
The Organocatalytic Mukaiyama–Michael Reaction [100] (p. 108)

To a 2-dram vial equipped with a magnetic stirrer bar and charged with catalyst·HX was added solvent; the vial was then placed in a bath at the appropriate temperature. The solution was stirred for 10 min before addition of the silyloxyfuran substrate in one portion. The resulting solution was stirred at constant temperature until reaction was determined to be complete (by GC conversion assay). The reaction mixture was then transferred cold through a silica gel plug with ether into a flask and carefully concentrated *in vacuo*. The resulting residue was purified by silica gel chromatography to provide the title compound.

14.23.2
General Procedure for Cascade Organocatalytic 1,4-Addition–Chlorination Sequence of Enals [101] (p. 114)

To a 2-dram vial equipped with a magnetic stirrer bar and charged with catalyst (0.05 mmol, 20 mol%) was added AcOEt (0.25 mL) and TFA (20 mol%); the vial was then placed in a bath at the desired temperature. The solution was stirred for 5 min before the addition of aldehyde (3 equiv), followed by the nucleophile (1 equiv) and chlorinating agent (2 equiv). The resulting mixture was stirred at constant temperature until the reaction was determined to be complete (by TLC or GTLC analysis). The reaction was quenched by filtration through Iatrobeads, eluting with Et_2O, and concentrated *in vacuo* under an ice bath. The resulting residue was purified by Iatrobead chromatography and fractions carefully concentrated *in vacuo* under an ice bath to provide the title compounds.

14.23.3
General Procedure for Organocatalytic Sulfenylation–Amination Reaction Sequence [102] (p. 114)

The aldehyde (0.38 mmol, 1.5 equiv), the catalyst (10 mol%), and benzoic acid (10 mol%) were stirred in toluene and cooled to −15 °C. The thiol (1 equiv) was added and the reaction mixture stirred for 30 min. Diazo dicarboxylate derivative (1.3 equiv) was added and the reaction progress monitored by TLC (3–24 h). The crude reaction mixture was diluted with MeOH (2 mL) and cooled to 0 °C, followed by the addition of $NaBH_4$ (2 equiv). After 10 min, 2 M NaOH (2 mL) and THF (2 mL) were added and the crude reaction mixture was stirred for 2 h. After standard aqueous work-up, the product was purified by FC on silica gel.

References

1 (a) A.B. Northrup, D.W.C. MacMillan, *J. Am. Chem. Soc.* **2002**, *124*, 6798; (b) R. Thayumanavan, F. Tanaka, C.F. Barbas, III, *Org. Lett.* **2004**, *6*, 3541.
2 S.S.V. Ramasastry, H. Zhang, F. Tanaka, and C.F. Barbas, III, *J. Am. Chem. Soc.* **2006**, *128*, ASAP; DOI: 10.1021/ja0677012.
3 Z.G. Hajos, D.R. Parrish, *J. Org. Chem.* **1974**, *39*, 1615. See also: U. Eder, G. Sauer, R. Wiechert, *Angew. Chem., Int. Ed.* **1971**, *10*, 496.
4 A. Ando, T. Miura, T. Tatematsu, T. Shioiri, *Tetrahedron Lett.* **1993**, *34*, 1507.
5 M. Horikawa, J. Busch-Petersen, E.J. Corey, *Tetrahedron Lett.* **1999**, *40*, 3843.
6 T. Ooi, M. Kameda, M. Taniguchi, K. Maruoka, *J. Am. Chem. Soc.* **2004**, *126*, 9685.
7 J.D. McGilvra, A.K. Unni, K. Modi, V.H. Rawal, *Angew. Chem., Int. Ed.* **2006**, *45*, 6130.
8 S.E. Denmark, R.A. Stavenger, S.B.D. Winter, K.-T. Wong, P.A. Barsanti, *J. Org. Chem.* **1998**, *63*, 9517.
9 S.E. Denmark, Y. Fan, *J. Am. Chem. Soc.* **2002**, *124*, 4233.
10 (a) A. Cordova, S. Watanabe, F. Tanaka, W. Notz, C.F. Barbas, III. *J. Am. Chem. Soc.* **2002**, *124*, 1866; (b) W. Notz, F. Tanaka, S. Watanabe, N.S. Chowdari, J.T. Turner, R. Thayumanavan, C.F. Barbas, III. *J. Org. Chem.* **2003**, *68*, 9624; (c) A. Cordova, C.F. Barbas, III. *Tetrahedron Lett.* **2003**, *44*, 1923; (d) N. Chowdari, D.B. Ramachary, A. Cordova, C.F. Barbas, III. *Synlett* **2003**, 1906.
11 A. Okada, T. Shibuguchi, T. Ohshima, H. Masu, K. Yamaguchi, M. Shibasaki, *Angew. Chem. Int. Ed.* **2005**, *44*, 4564.
12 A.G. Wenzel, E.N. Jacobsen, *J. Am. Chem. Soc.* **2002**, *124*, 12964.
13 M.S. Taylor, E.N. Jacobsen, *J. Am. Chem. Soc.* **2004**, *126*, 10558.
14 E.J. Corey, F.-Y. Zhang, *Angew. Chem. Int. Ed.* **1999**, *38*, 1931.
15 B.M. Nugent, R.A. Yoder, J.N. Johnston, *J. Am. Chem. Soc.* **2004**, *126*, 3418.
16 K. Tanaka, A. Mori, S. Inoue, *J. Org. Chem.* **1990**, *55*, 181.
17 J.K. Huang, E.J. Corey, *Org. Lett.* **2004**, *6*, 5027.
18 D.E. Fuerst, E.N. Jacobsen, *J. Am. Chem. Soc.* **2005**, *127*, 8964.
19 M. Hatano, T. Ikeno, T. Miyamoto, K. Ishihara, *J. Am. Chem. Soc.* **2005**, *127*, 10776.
20 S.E. Denmark, Y. Fan, *J. Org. Chem.* **2005**, *70*, 9667.
21 D. Uraguchi, K. Sorimachi, M. Terada, *J. Am. Chem. Soc.* **2005**, *127*, 9360.
22 G.D. Joly, E.N. Jacobsen, *J. Am. Chem. Soc.* **2004**, *126*, 4102.
23 G.B. Rowland, H.L. Zhang, E.B. Rowland, S. Chennamadhavuni, Y. Wang, J.C. Antilla, *J. Am. Chem. Soc.* **2005**, *127*, 15696.
24 A. Bøgevig, K. Juhl, N. Kumaragurubaran, W. Zhuang, K.A. Jørgensen, *Angew. Chem. Int. Ed.* **2002**, *41*, 1790.
25 N. Halland, A. Braunton, S. Bachmann, M. Marigo, K.A. Jørgensen, *J. Am. Chem. Soc.* **2004**, *126*, 4790.
26 (a) N. Halland, R.G. Hazell, K.A. Jørgensen, *J. Org. Chem.* **2002**, *67*, 8331; (b) A. Prieto, N. Halland, K.A. Jørgensen, *Org. Lett.* **2005**, *7*, 3897; (c) C.E.T. Mitchell, S.E. Brenner, S.V. Ley, *Chem. Commun.* **2005**, 5346; (d) N. Halland, T. Hansen, K.A. Jørgensen, *Angew. Chem. Int. Ed.* **2003**, *42*, 4955.
27 S.V. Pansare, K. Pandya, *J. Am. Chem. Soc.* **2006**, *128*, 9624.
28 O. Andrey, A. Alexakis, G. Bernardinelli, *Org. Lett.* **2003**, *5*, 2559.
29 S.B. Tsogoeva, S. Wei, *Chem. Commun.* **2006**, 1451.
30 Y. Hoashi, T. Okino, Y. Takemoto, *Angew. Chem., Int. Ed.* **2005**, *44*, 4032.
31 J. Wang, H. Li, W.H. Duan, L.S. Zu, W. Wang, *Org. Lett.* **2005**, *7*, 4713.
32 Y. Hayashi, H. Gotoh, T. Hayashi, M. Shoji, *Angew. Chem. Int. Ed.* **2005**, *44*, 4212.

33 M.T. Hechavarria Fonseca, B. List, *Angew. Chem. Int. Ed.* **2004**, *43*, 3958.
34 T. Ooi, K. Doda, K. Maruoka, *J. Am. Chem. Soc.* **2003**, *125*, 9022.
35 T. Ooi, K. Doda, K. Maruoka, *J. Am. Chem. Soc.* **2003**, *125*, 2054.
36 T. Tozawa, Y. Yamane, T. Mukaiyama, *Chem. Lett.* **2006**, *35*, 56.
37 B. Lygo, B. Allbutt, E.H.M. Kirton, *Tetrahedron Lett.* **2005**, *46*, 4461.
38 H. Li, J. Song, X. Liu, L. Deng, *J. Am. Chem. Soc.* **2005**, *127*, 8948.
39 N.A. Paras, D.W.C. MacMillan, *J. Am. Chem. Soc.* **2001**, *123*, 4370.
40 J.F. Austin, S.-G. Kim, C.J. Sinz, W.-J. Xiao, D.W.C. MacMillan, *Proc. Natl. Acad. Sci. USA* **2004**, *101*, 5482.
41 D. Uraguchi, K. Sorimachi, M. Terada, *J. Am. Chem. Soc.* **2004**, *126*, 11804.
42 (a) Y. Wabuchi, M. Nakatani, S. Yokoyama, S. Hatakeyama. *J. Am. Chem. Soc.* **1999**, *121*, 10219; See also in: (b) A. Nakano, S. Kawahara, K. Morokuma, M. Nakatami, Y. Iwabuchi, K. Takahashi, J. Ishihara, S. Hatakeyama, *Tetrahedron* **2006**, *62*, 381.
43 N.T. McDougal, S.E. Schaus, *J. Am. Chem. Soc.* **2003**, *125*, 12094.
44 (a) M. Shi, Y.M. Xu, *Angew. Chem. Int. Ed. Engl.* **2002**, *41*, 4507; (b) M. Shi, Y.-M. Xu, Y.-L. Shi, *Chem. Eur. J.* **2005**, *11*, 1794.
45 K. Matsui, S. Takizawa, H. Sasai, *J. Am. Chem. Soc.* **2005**, *127*, 3680.
46 R.K. Kunz, D.W.C. MacMillan, *J. Am. Chem. Soc.* **2005**, *127*, 3240.
47 W.-W. Liao, K. Li, Y. Tang, *J. Am. Chem. Soc.* **2003**, *125*, 13030.
48 C.D. Papageorgiou, M.A. Cubillo de Dios, S.V. Ley, M.J. Gaunt, *Angew. Chem. Int. Ed.* **2004**, *43*, 4641.
49 V.K. Aggarwal, E. Alonso, G. Fang, M. Ferrara, G. Hynd, M. Porcelloni, *Angew. Chem. Int. Ed.* **2001**, *40*, 1433.
50 (a) M. Marigo, J. Franzén, T.B. Poulsen, W. Zhuang, K.A. Jørgensen, *J. Am. Chem. Soc.* **2005**, *127*, 6964; (b) W. Zhuang, M. Marigo, K.A. Jørgensen, *Org. Biomol. Chem.* **2005**, 3883.
51 M. Davoust, J.-F. Briere, P.-A. Jaffres, P. Metzner, *J. Org. Chem.* **2005**, *70*, 4166.
52 V.K. Aggarwal, C. Aragoncillo, C.L. Winn, *Synthesis* **2005**, 1378.
53 D. Yang, M.K. Wong, Y.C. Yip, X.C. Wang, M.W. Tang, J.H. Zheng, K.K. Cheung, *J. Am. Chem. Soc.* **1998**, *120*, 5943.
54 Z.X. Wang, Y. Tu, M. Frohn, J.R. Zhang, Y. Shi, *J. Am. Chem. Soc.* **1997**, *119*, 11224.
55 L.H. Shu, Y. Shi, *Tetrahedron* **2001**, *57*, 5213.
56 P.C.B. Page, B.R. Buckley, H. Heaney, A.J. Blacker, *Org. Lett.* **2005**, *7*, 375.
57 P.C.B. Page, D. Barros, B.R. Buckley, A. Ardakani, B.A. Marples, *J. Org. Chem.* **2004**, *69*, 3595.
58 B. Lygo, D.C.M. To, *Tetrahedron Lett.* **2001**, *42*, 1343.
59 B. Lygo, P.G. Wainwright, *Tetrahedron* **1999**, *55*, 6289.
60 B.M. Adger, J.V. Barkley, S. Bergeron, M.W. Cappi, B.E. Flowerdew, M.P. Jackson, R. McCague, T.C. Nugent, S.M. Roberts, *J. Chem. Soc., Perkin Trans. 1* **1997**, 3501.
61 S. Itsuno, M. Sakakura, K. Ito, *J. Org. Chem.* **1990**, *55*, 6047.
62 M. Marigo, J. Franzen, T.B. Poulsen, W. Zhuang, K.A. Jørgensen, *J. Am. Chem. Soc.* **2005**, *127*, 6964.
63 M. Marigo, T.C. Wabnitz, D. Fielenbach, K.A. Jørgensen, *Angew. Chem., Int. Ed.* **2005**, *44*, 794.
64 C. Rousseau, B. Christensen, M. Bols, *Eur. J. Org. Chem.* **2005**, 2734.
65 A.V. Malkov, P. Kocovsky, unpublished results.
66 T. Ooi, M. Takeuchi, M. Kameda, K. Maruoka, *J. Am. Chem. Soc.* **2000**, *122*, 5228.
67 S.-S. Jew, Y.-J. Lee, J. Lee, M.J. Kang, B.-S. Jeong, J.-H. Lee, M.-S. Yoo, M.-J. Kim, S.-H. Choi, J.-M. Ku, H.-G. Park, *Angew. Chem. Int. Ed.*, **2004**, *43*, 2382.
68 (a) M. Nakoji, T. Kanayama, T. Okino, Y. Takemoto, *Org. Lett.* **2001**, *3*, 3329; (b) M. Nakoji, T. Kanayama, T. Okino, Y. Takemoto, *J. Org. Chem.* **2002**, *67*, 7418.
69 A.V. Malkov, M. Bell, F. Castelluzzo, P. Kocovsky, *Org. Lett.* **2005**, *7*, 3219.

70 E. Vedejs, O. Daugulis, *J. Am. Chem. Soc.* **1999**, *121*, 5813.
71 G.C. Fu, J.C. Ruble, J. Tweddell, *J. Org. Chem.* **1998**, *63*, 2794.
72 V.B. Birman, E.W. Uffman, H. Jiang, X. Li, C.J. Kilbane, *J. Am. Chem. Soc.* **2004**, *126*, 12226.
73 T. Kano, K. Sasaki, K. Maruoka, *Org. Lett.* **2005**, *7*, 1347.
74 E. Vedejs, J.A. MacKay, *Org. Lett.* **2001**, *3*, 535.
75 S. Bellemin-Laponnaz, J. Tweddell, J.C. Ruble, F.M. Breitling, G.C. Fu, *Chem. Commun.* **2000**, 1009.
76 B. Tao, J.C. Ruble, D.A. Hoic, G.C. Fu, *J. Am. Chem. Soc.* **1999**, *121*, 5091, and erratum p. 10452.
77 K. Ishihara, Y. Kosugi, M. Akakura, *J. Am. Chem. Soc.* **2004**, *126*, 12212.
78 S. Arai, S. Bellemin-Laponnaz, G.C. Fu, *Angew. Chem. Int. Ed.* **2001**, *40*, 234.
79 T. Oriyama, K. Imai, T. Sano, T. Hosoya, *Tetrahedron Lett.* **1998**, *39*, 3529.
80 Y.G. Chen, S.-K. Tian, L. Deng, *J. Am. Chem. Soc.* **2000**, *122*, 9542.
81 Y.G. Chen, L. Deng, *J. Am. Chem. Soc.* **2001**, *123*, 11302.
82 J. Hang, H. Li, L. Deng, *Org. Lett.* **2002**, *4*, 3321.
83 B.R. Sculimbrene, A.J. Morgan, S.J. Miller, *J. Am. Chem. Soc.* **2002**, *124*, 11653.
84 K.A. Ahrendt, C.J. Borths, D.W.C. MacMillan, *J. Am. Chem. Soc.* **2000**, *122*, 4243.
85 K. Ishihara, K. Nakano, *J. Am. Chem. Soc.* **2005**, *127*, 10504.
86 (a) R.M. Wilson, W.S. Jen, D.W.C. MacMillan, *J. Am. Chem. Soc.* **2005**, *127*, 11616; (b) S.A. Selkälä, A.M.P. Koskinen, *Eur. J. Org. Chem.* **2005**, 1620.
87 (a) Y. Huang, A.K. Unni, A.N. Thadani, V.H. Rawal, *Nature (London)* **2003**, *424*, 146; (b) A.K. Unni, *Asymmetric Catalysis of Hetero-Diels-Alder Reactions Through Hydrogen Bonding*. PhD Thesis, The University of Chicago, Chicago, IL, June 2005; (c) Y. Huang, *Exploring the New Horizon of Diels-Alder Reactions: Asymmetric Catalysis*. PhD Thesis, The University of Chicago, Chicago, IL, August 2002.
88 A.N. Thadani, A.R. Stankovic, V.H. Rawal, *Proc. Natl. Acad. Sci. USA* **2004**, *101*, 5846.
89 W.S. Jen, J.J.M. Wiener, D.W.C. MacMillan, *J. Am. Chem. Soc.* **2000**, *122*, 9874.
90 S.S. Sohn, E.V. Rosen, J.W. Bode, *J. Am. Chem. Soc.* **2004**, *126*, 14370.
91 D. Enders, O. Niemeier, T. Balensiefer, *Angew. Chem. Int. Ed.* **2006**, *45*, 1463.
92 (a) D. Enders, O. Niemeier, T. Balensiefer, *Angew. Chem. Int. Ed.* **2006**, *45*, 1463; (b) M.D. Ennis, R.L. Hoffman, N.B. Ghazal, D.W. Old, P.A. Mooney, *J. Org. Chem.* **1996**, *61*, 5813; (c) J.A. Nieman, M.D. Ennis, *Org. Lett.* **2000**, *2*, 1395; (d) Y. Suzuki, K. Yamauchi, K. Muramatsu, M. Sato, *Chem. Commun.* **2004**, 2770.
93 D. Enders, U. Kallfass, *Angew. Chem. Int. Ed. Engl.* **2002**, *41*, 1743.
94 M.D. Drew, N.J. Lawrence, W. Watson, S.A. Bowles, *Tetrahedron Lett.*, **1997**, *38*, 5857.
95 S. Hoffmann, A.M. Seayad, B. List, *Angew. Chem., Int. Ed.* **2005**, *44*, 7424.
96 A.V. Malkov, S. Stončius, K.N. MacDougall, A. Mariani, G.D. McGeoch, P. Kocovsky, *Tetrahedron* **2006**, *62*, 264.
97 S.G. Ouellet, J.B. Tuttle, D.W.C. MacMillan, *J. Am. Chem. Soc.* **2005**, *127*, 32.
98 J.W. Yang, M.T. Hechavarria Fonseca, B. List, *Angew. Chem. Int. Ed.*, **2005**, *44*, 108.
99 M. Kunishima, K. Yoshimura, H. Morigaki, R. Kawamata, K. Terao, S. Tani, *J. Am. Chem. Soc.* **2001**, *123*, 10760.
100 S.P. Brown, N.C. Goodwin, D.W.C. MacMillan, *J. Am. Chem. Soc.* **2003**, *125*, 1192.
101 Y. Huang, A.M. Walji, C.H. Larsen, D.W.C. MacMillan, *J. Am. Chem. Soc.* **2005**, *127*, 15051.
102 M. Marigo, T. Schulte, J. Franzén, K.A. Jørgensen, *J. Am. Chem. Soc.* **2005**, *127*, 15710.

15
Appendix II: Catalysts

A selection of chiral organocatalysts is presented in this chapter. The compounds are classed according to the chiral inductor classes preset in their structure. Bold numbers below the molecules refer to the corresponding chapter where further information on typical reaction scope can be found.

Enantioselective Organocatalysis: Reactions and Experimental Procedures
Edited by Peter I. Dalko
Copyright © 2007 WILEY-VCH Verlag GmbH & Co. KGaA, Weinheim
ISBN: 978-3-527-31522-2

Aminoacid-derived catalysts

Proline derivatives and proline analogs:

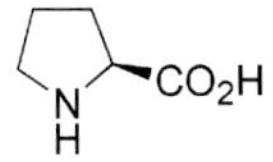

2.1
2.2
3
5

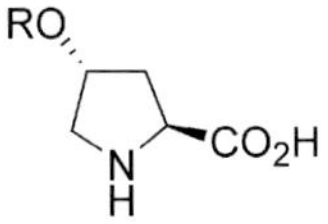

R = H; **2.2**
R = TMS; **2.2**
R = t-Bu; **2.1**
R = TBDPS; **2.1**

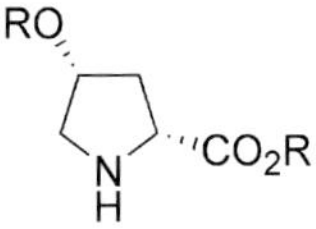

R = TBDPS; **2.1**

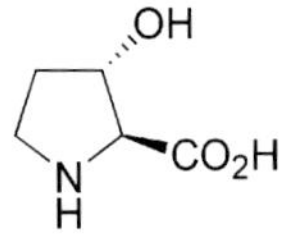

2.1

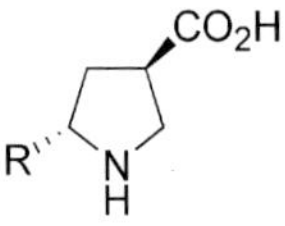

R = H, Me; **2.1**

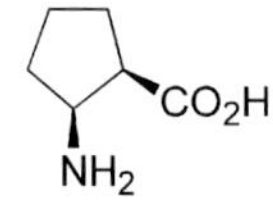

2.1

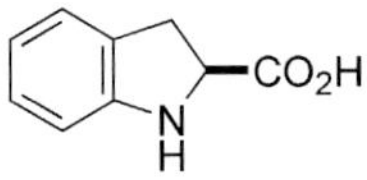

3

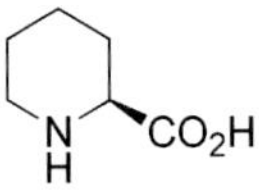

2.1
5

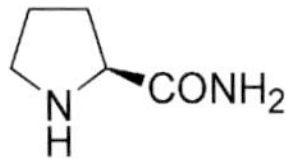

2.2

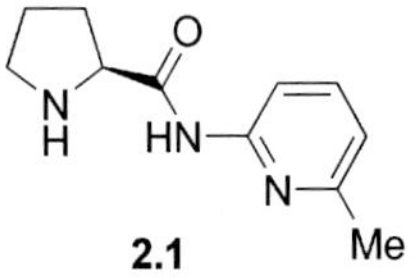

2.1

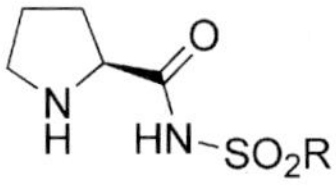

2.1
2.2

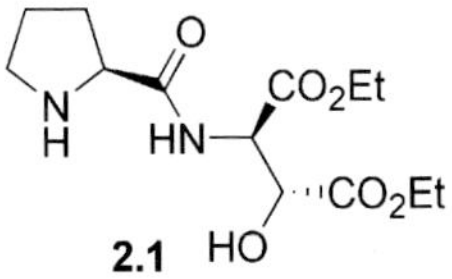

2.1

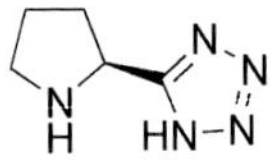

2.1
2.2
2.3
3

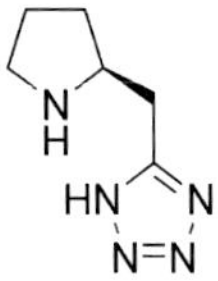

2.3

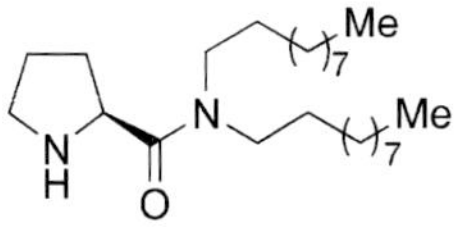

2.1

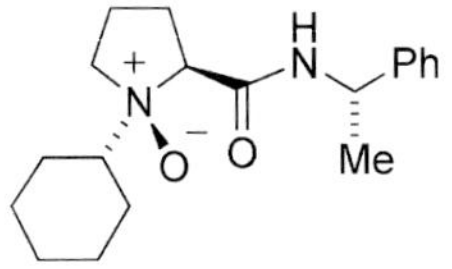

7

Imidazolidines:

Miscellaneous aminoacid-derived catalysts:

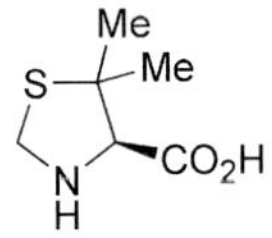

2.1

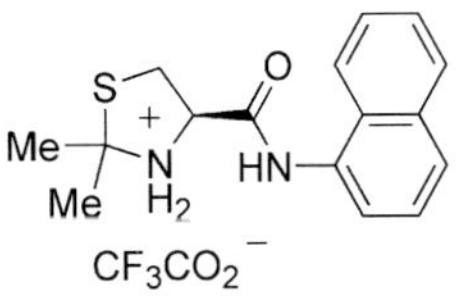

2.3

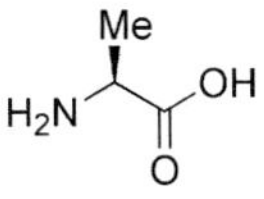

2.3

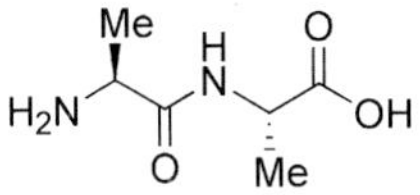

2.3

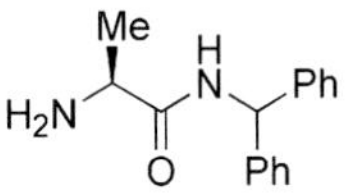

2.3

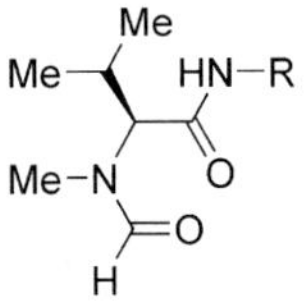

R = Ph, 4-MeOC₆H₄, 3,5-Me₂C₆H₃; **7**

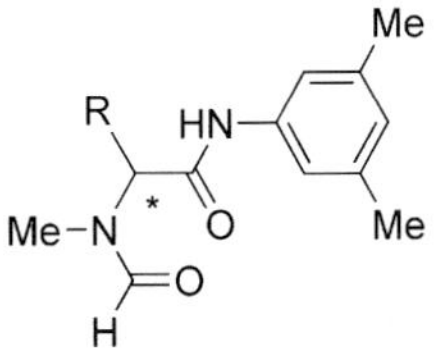

R = c-hexyl, t-Bu, Ph, Bn, Me; **7**

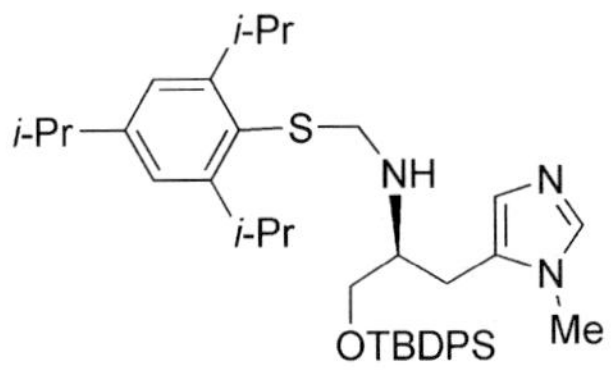

8

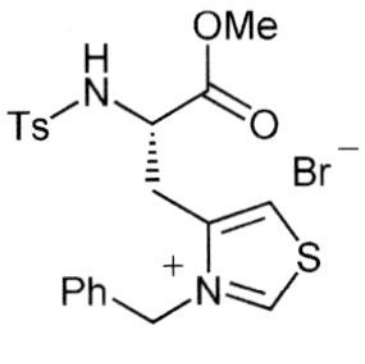

9

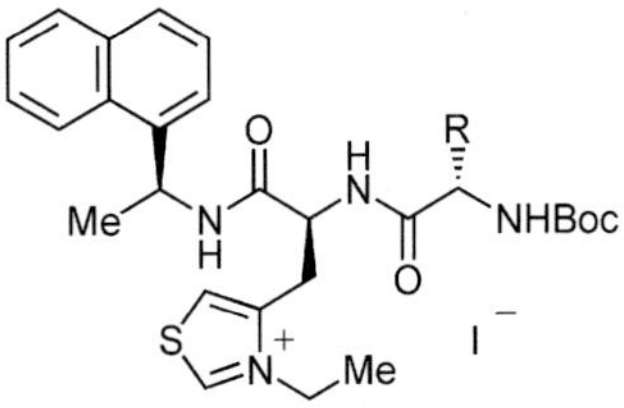

9   R = L-Thr (Bn)

Cyclic peptides:

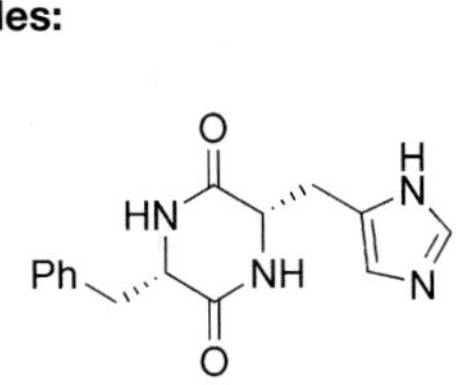

6

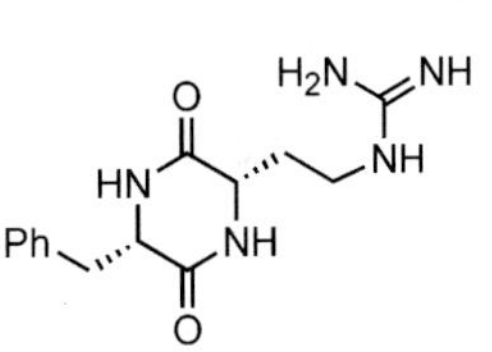

6

Oligopeptides:

1,2-Diamines

1,2-Diaminocyclohexane-derived catalysts:

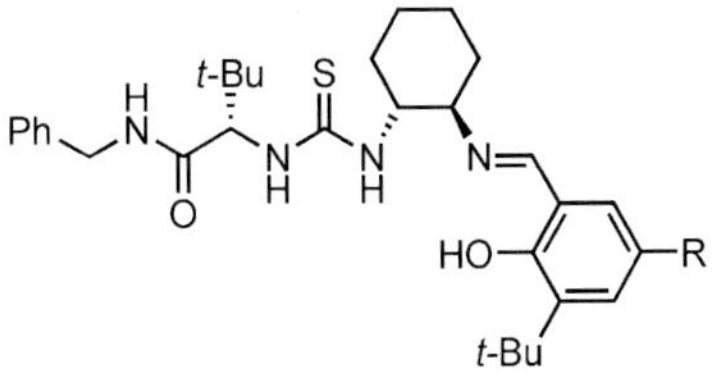

R = OMe; **6**
R = t-Bu; **5**, **6**

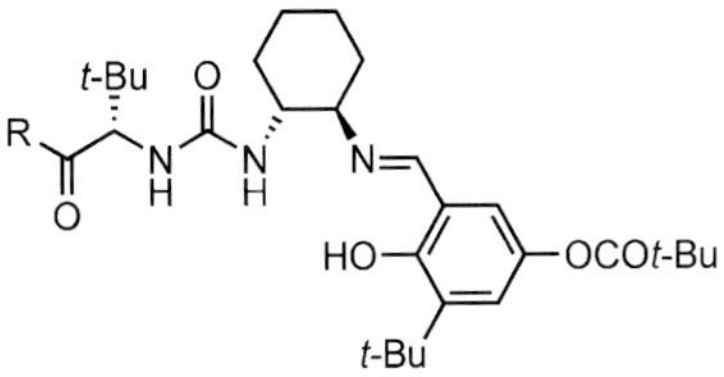

R = BnNH; **6**
R = Me₂N; **5**, **6**

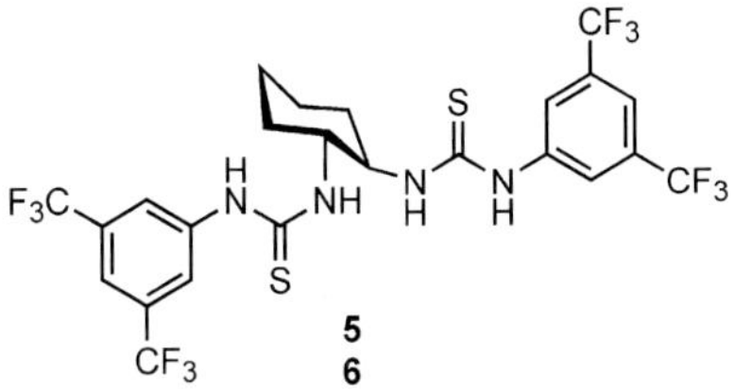

5
6

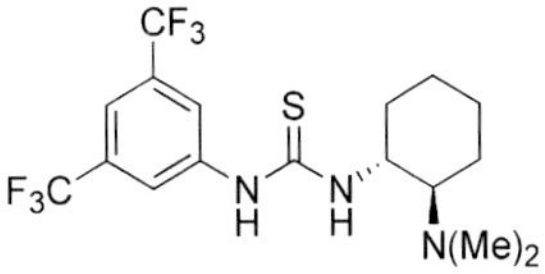

5
6

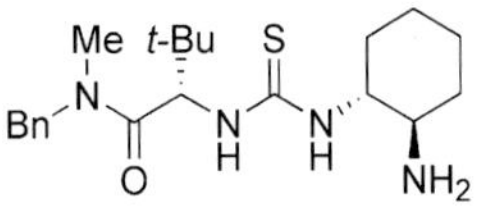

2.3

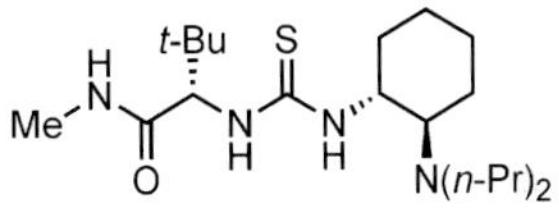

6

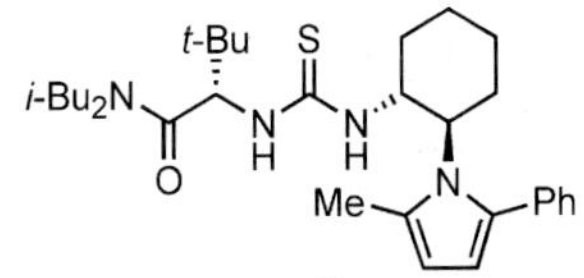

6

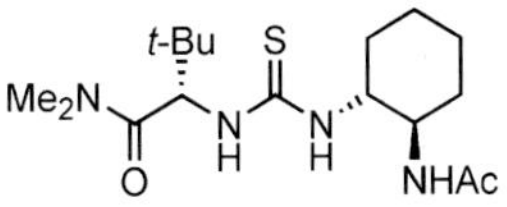

6

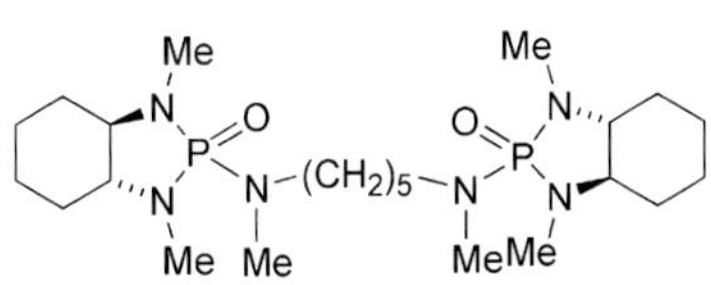

7

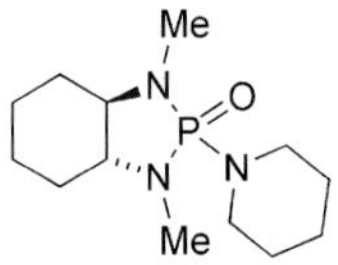

7

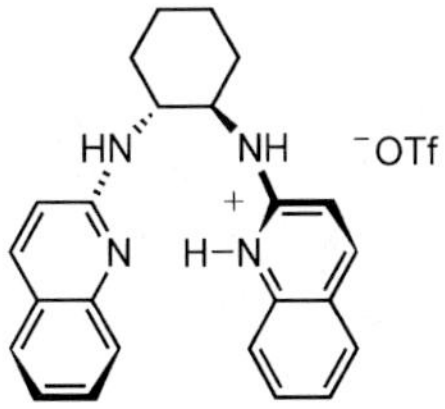

6

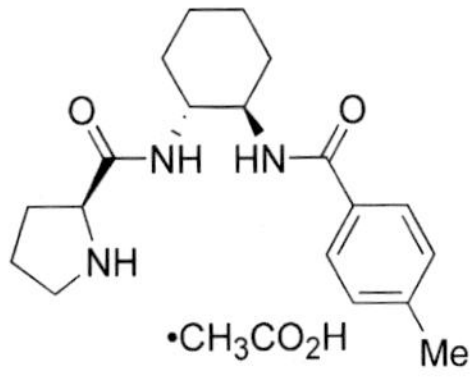

2.1

1,2-Diphenylethylenediamine:

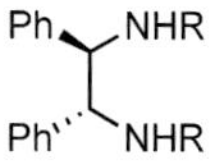

R = Tf; **6**
R = Nf; **6**

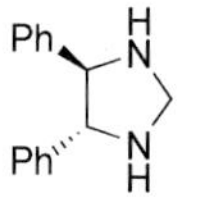

2.2

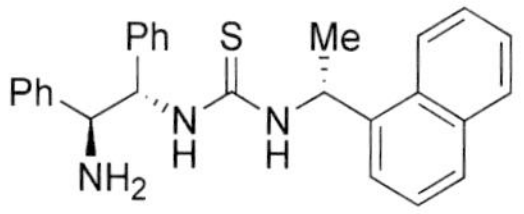

2.3

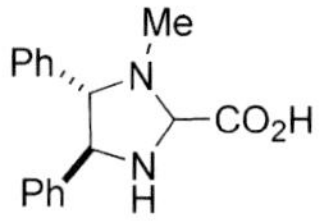

3

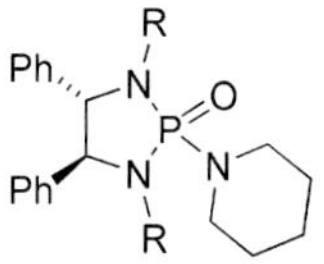

7

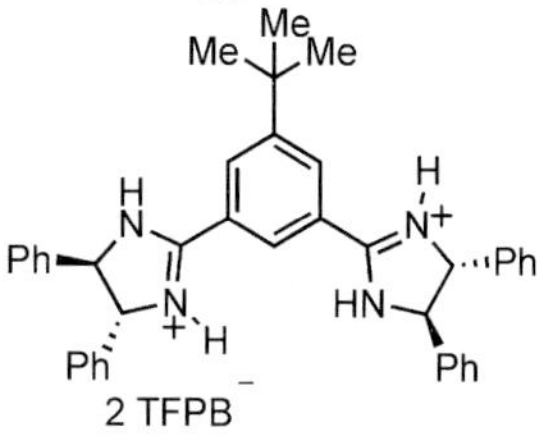

6
2 TFPB⁻

Miscellaneous 1,2-diamines:

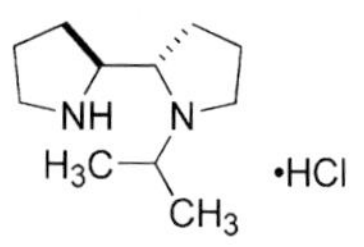

2.3 ·HCl

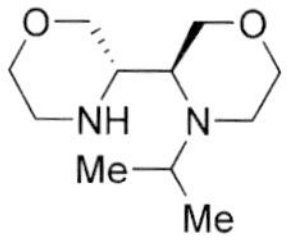

2.3

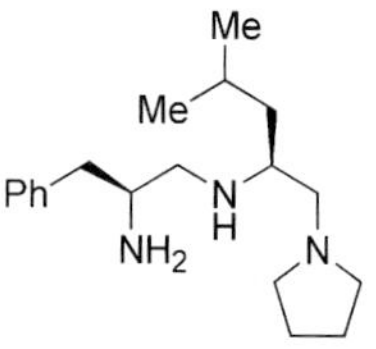

3

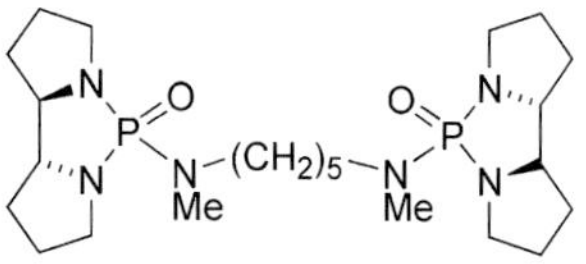

7

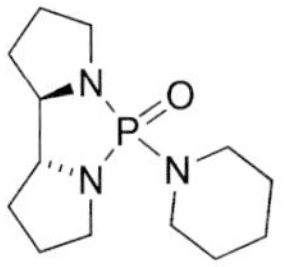

7

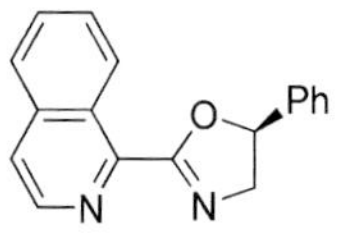

7

Phenethylamine derivatives

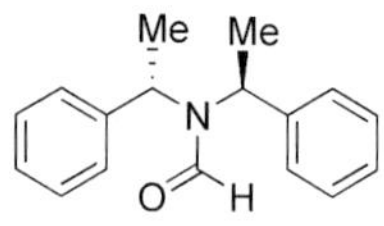

7

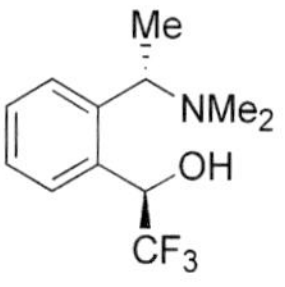

8

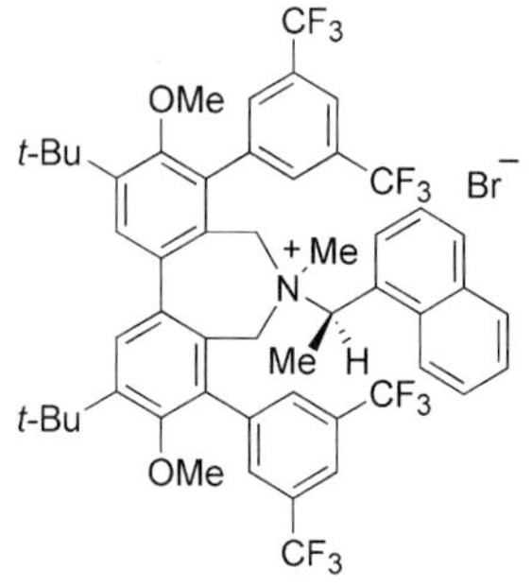

4

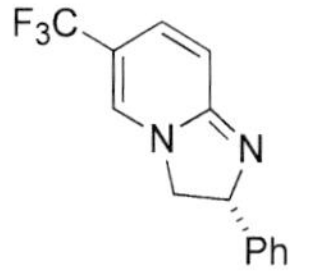

8

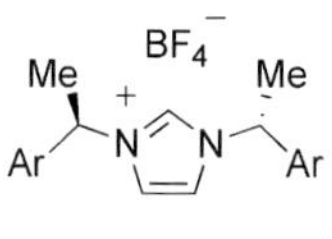

9

Aminopyridines

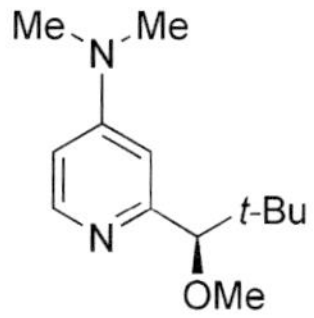

8 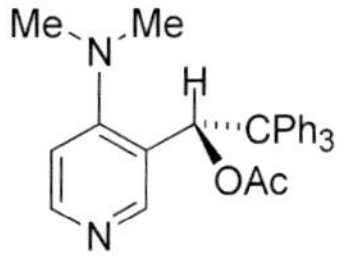 **8** 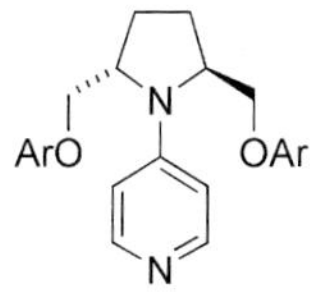 **8**

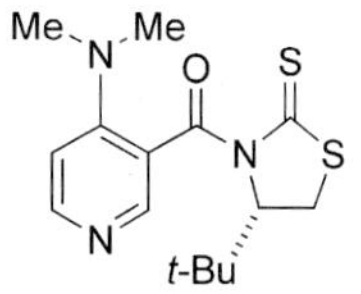

8 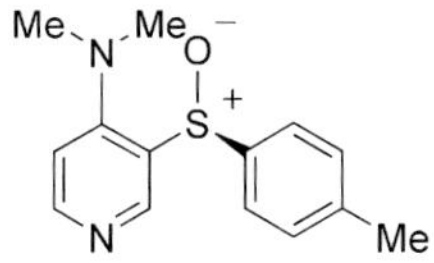 **8** 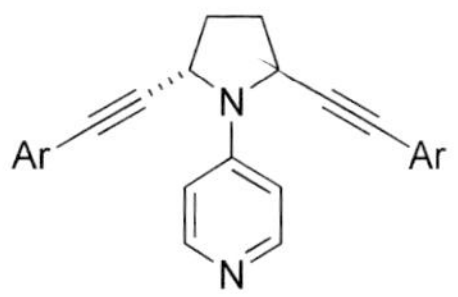 **8**

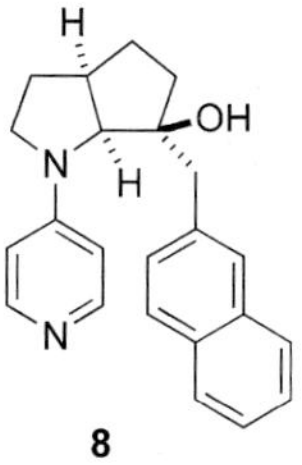

8 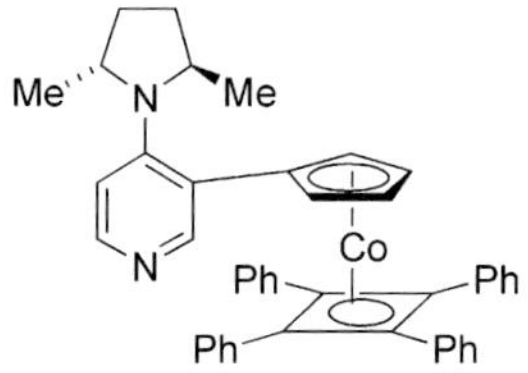 **8** 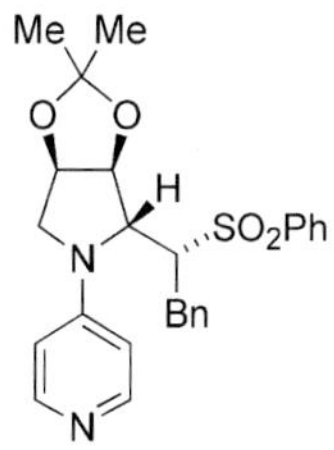 **8**

Triazoles

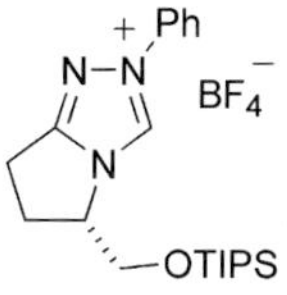

9 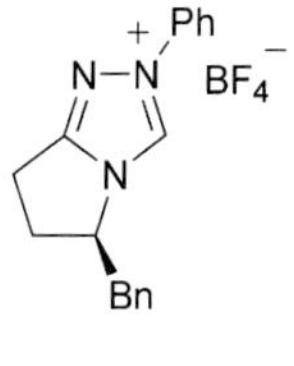 **9** 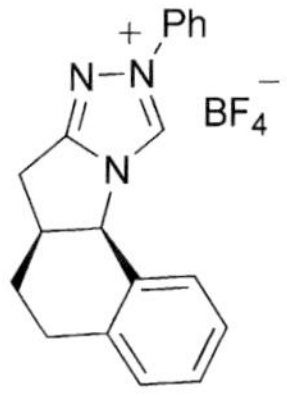 **9** 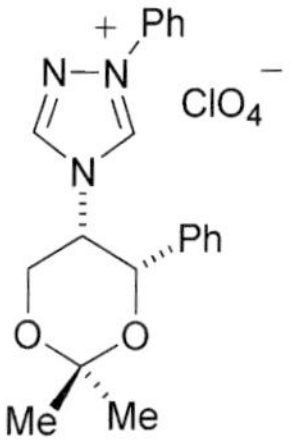 **9**

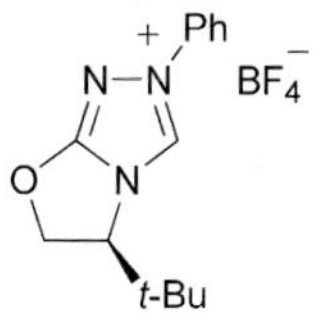

9 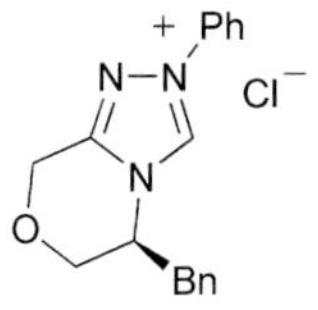 **9**

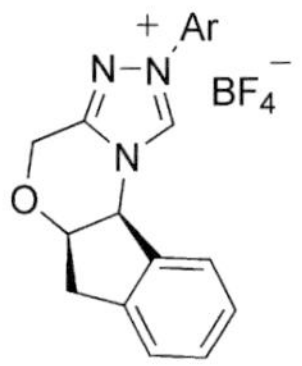

Ar = pentafluorophenyl; **9**

Miscellaneous catalysts

Cinchona-derived catalysts

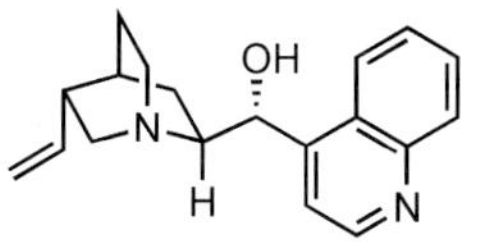

cinchonidine; **5**

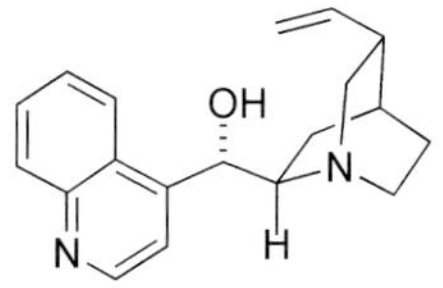

cinchonine; **5**

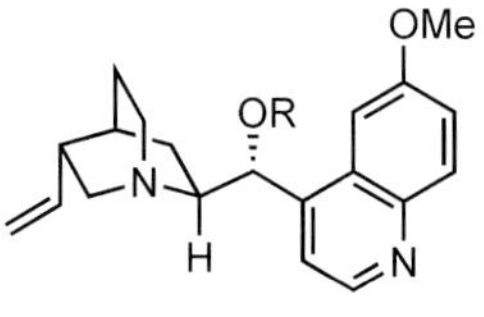

R = H quinine; **5**, **2.2**
R = Me; **7**

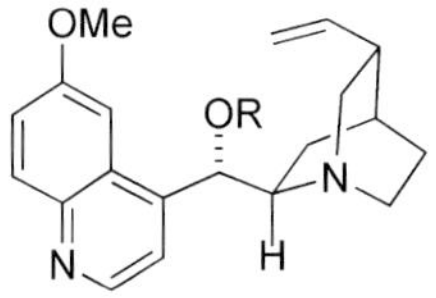

R = H quinidine; **5**, **8**
R = Me; **7**
R = Bz; **2.2**

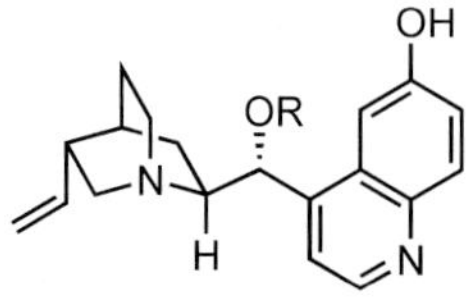

R = H, Bn, 9-phenanthrene; **2.3**

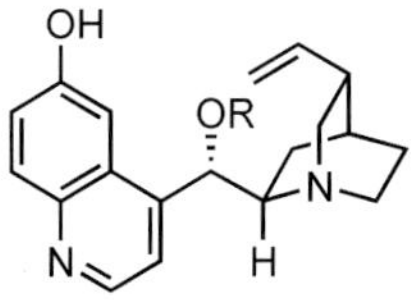

R = H, Bn, 9-phenanthrene; **2.3**

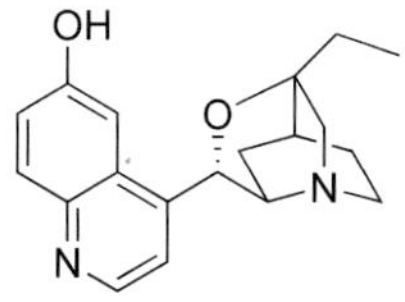

2.2
5
6

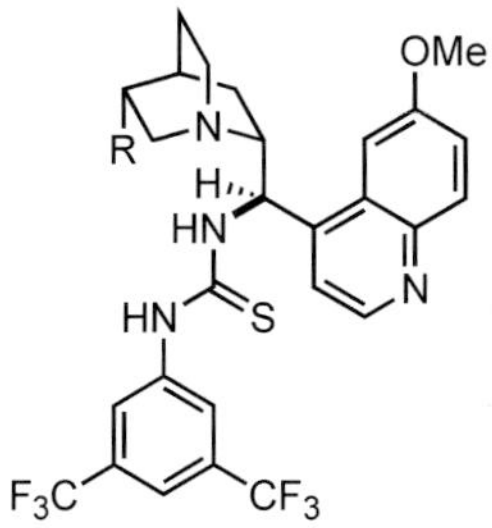

R = Et, vinyl; **6**

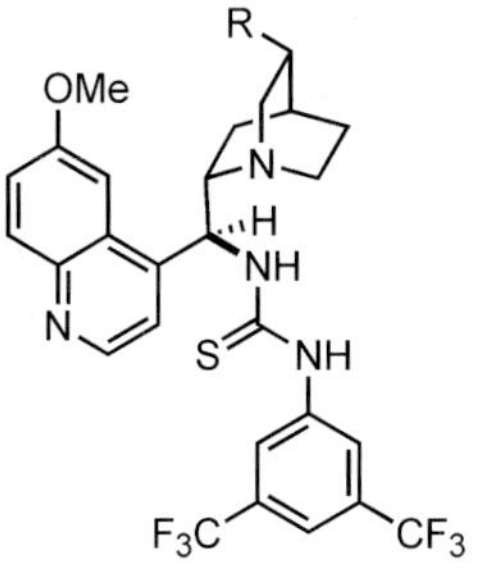

R = Et, vinyl; **6**

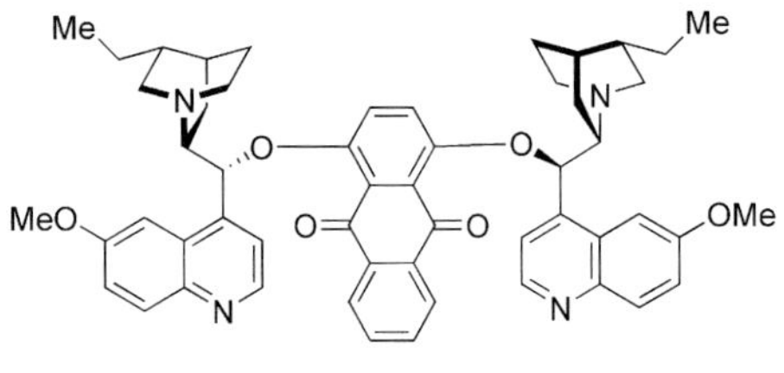

(DHQ)₂AQN

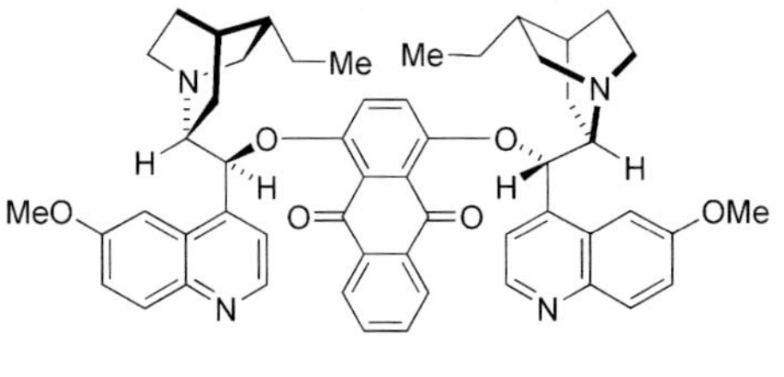

(DHQD)₂AQN

5
8

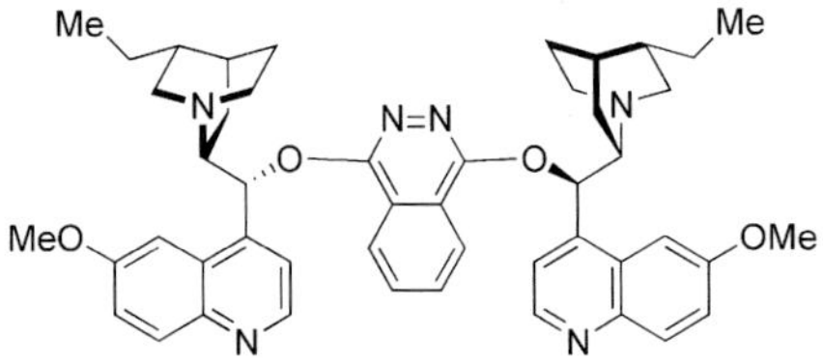

(DHQ)₂PHAL

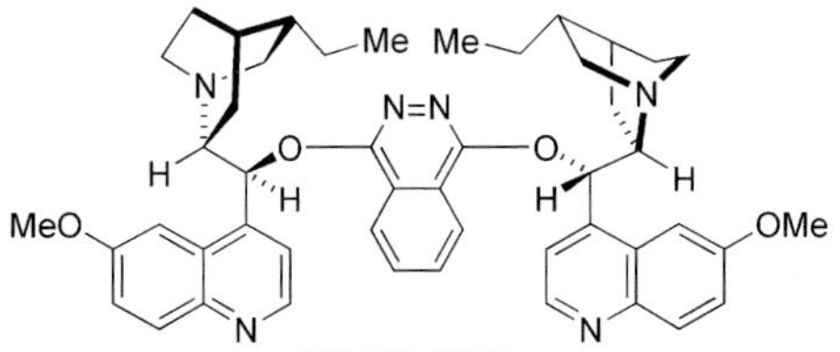

(DHQD)₂PHAL

5
8

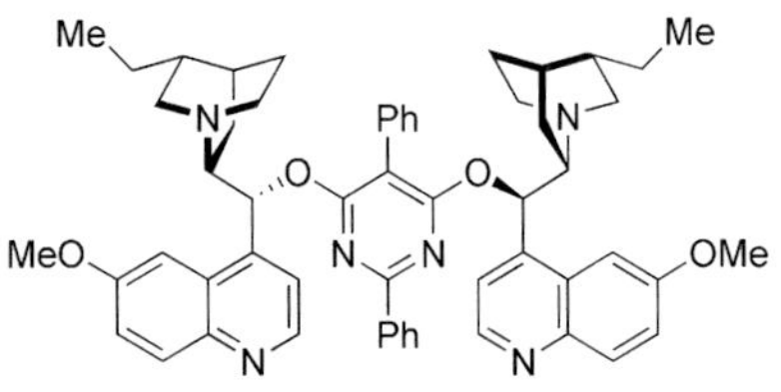

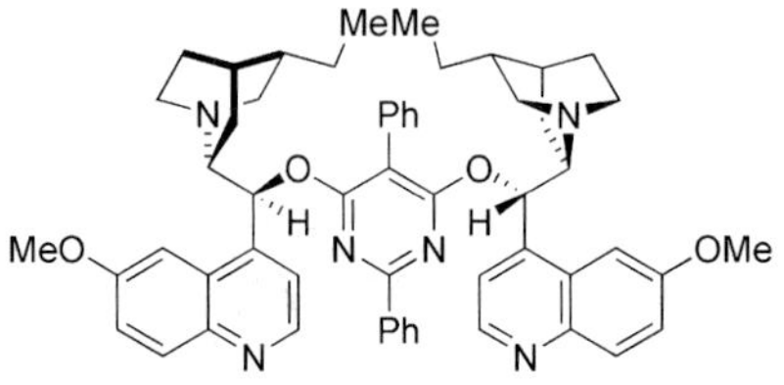

2.2
10

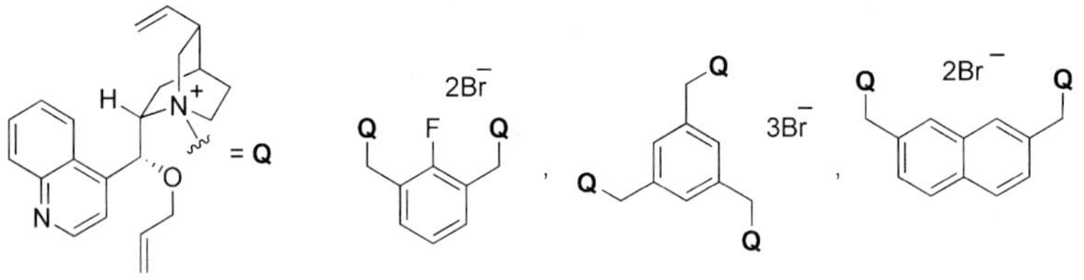

= Q

2Br⁻

3Br⁻

2Br⁻

4

·CF₃CO₂H

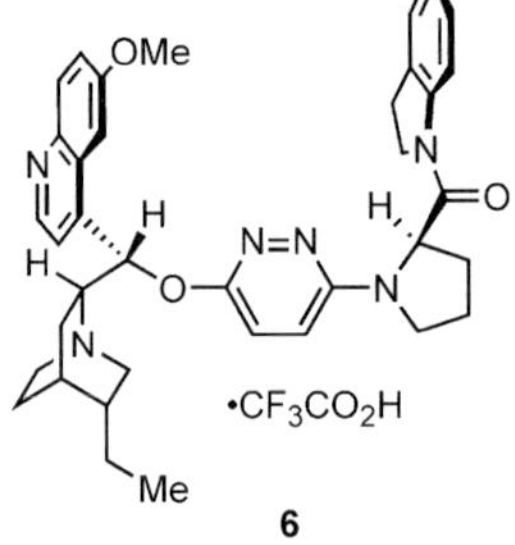

6

Terpene-derived catalysts

Biaryl catalysts

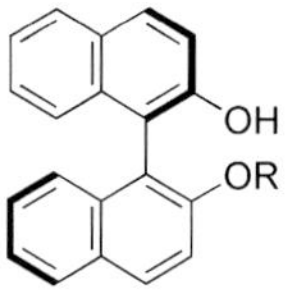

R = H (BINOL); **5, 7**
R = Me; **5**

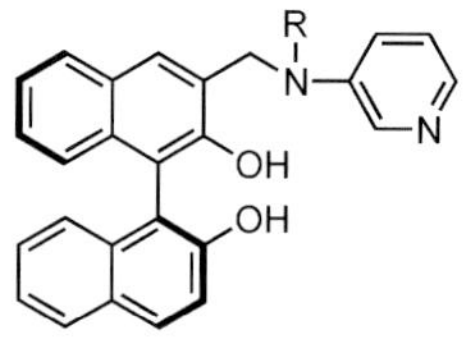

R = H, Et, *t*-Bu, Bn, *i*-Pr; **5, 6**

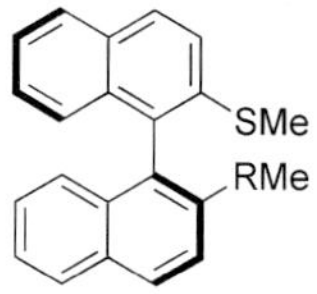

R = S, O; **10**

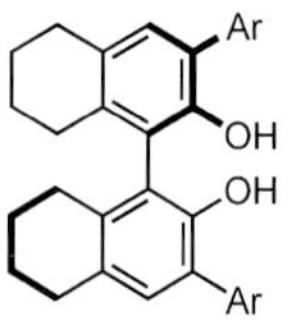

Ar = phenyl, 3,5-bis(CF₃)phenyl, 3,5-dimethylphenyl; **5, 6**

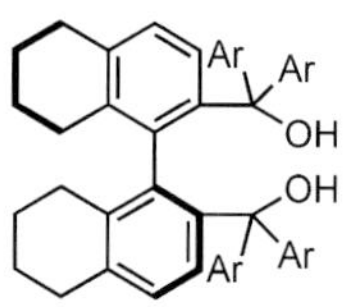

Ar = 4-F-3,5-Me₂C₆H₂, 4-F-3,5-Et₂C₆H₂; **6**

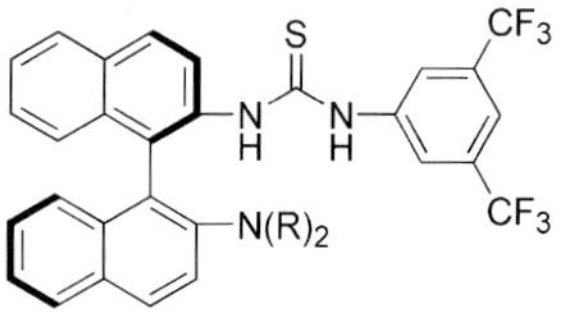

R = Me, Et; **5, 6**

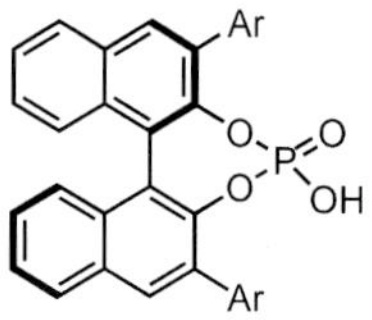

Ar = 9-anthryl, 4-(β-naphthyl)-C₆H₄, 3,5-dimesitylphenyl; **6**
Ar = 3,5-bis(CF₃)phenyl, 2,4,6-(*i*-Pr)-phenyl; **6, 11**

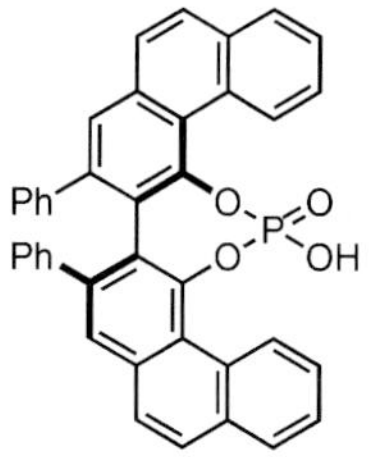

6

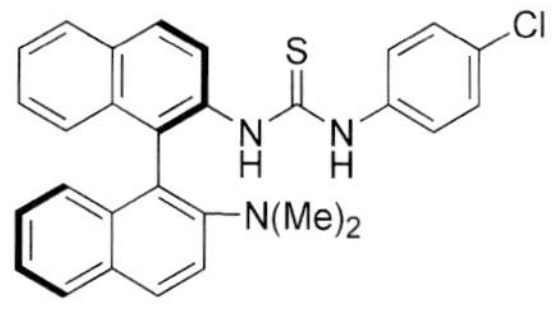

5

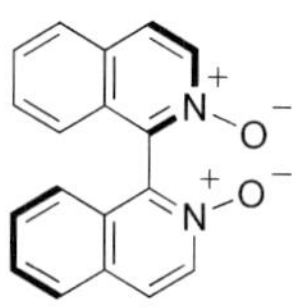

7

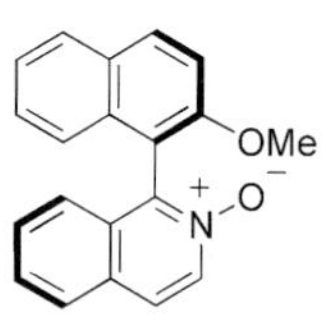

7

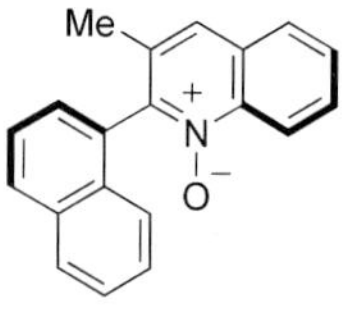

7

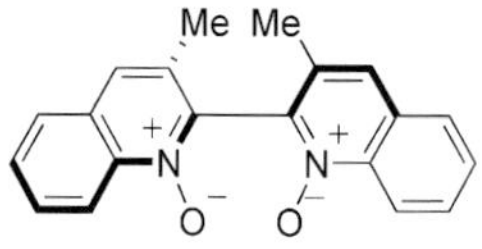

7

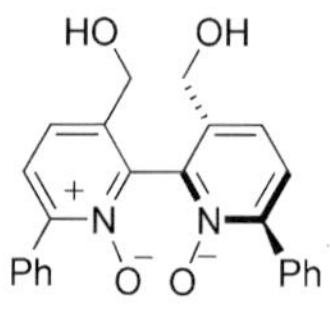

7

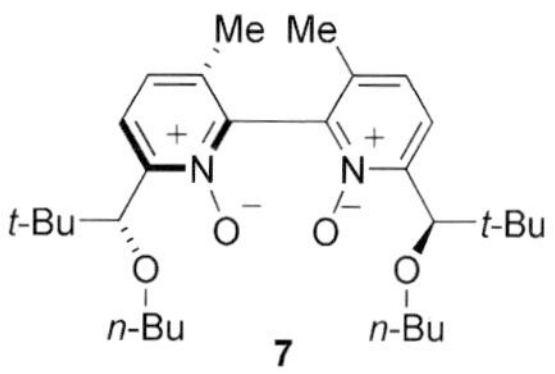

7

15 Appendix II: Catalysts

Ferrocenes

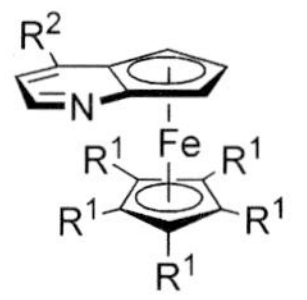

R¹ = Me, Ph;
R² = N(Me)₂, pyrrolidine; **8**

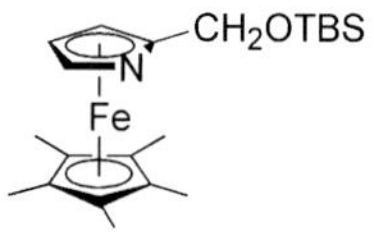

8

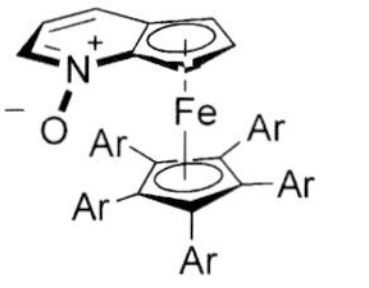

7

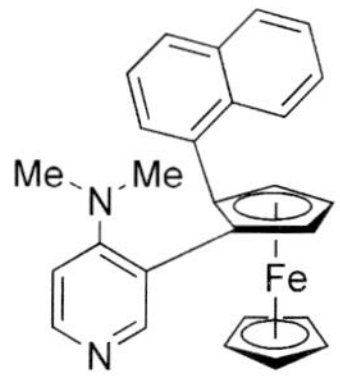

8

Tartaric acid-derived catalysts

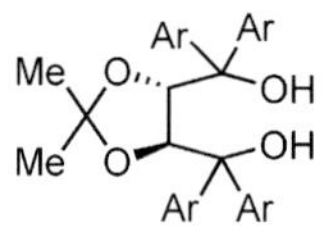

6

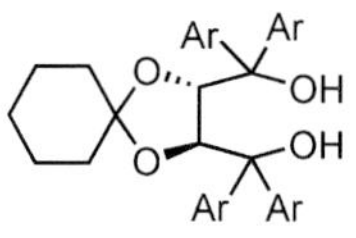

Ar = 1-Nap; **6**

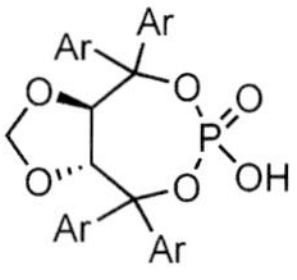

6

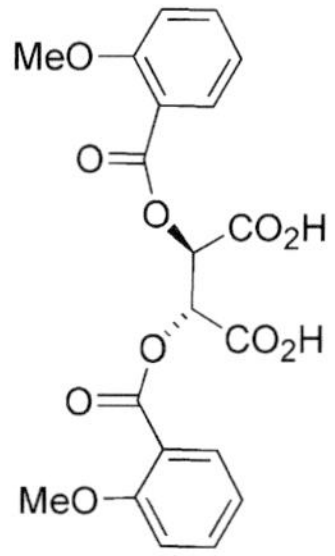

12

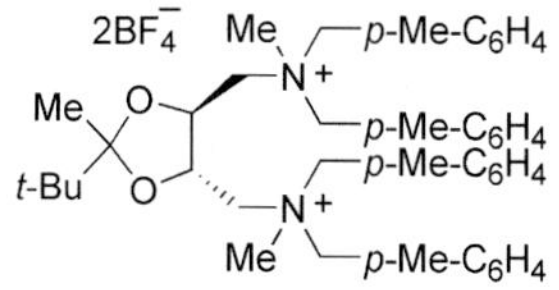

4

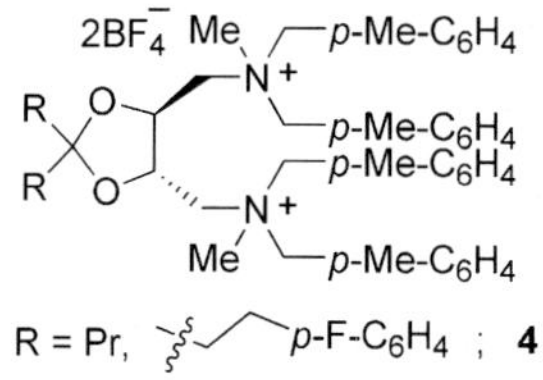

R = Pr, ⤳⌒p-F-C₆H₄ ; **4**

12

Mandelic acid

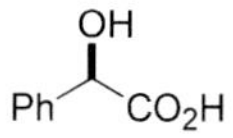

6

Sugar-derived catalysts

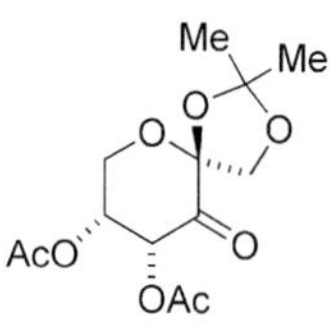

12

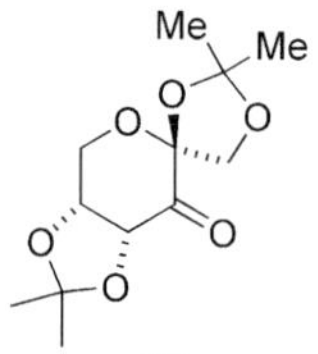

12

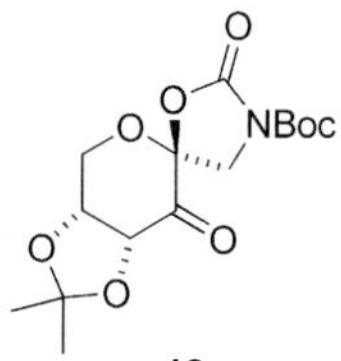

12

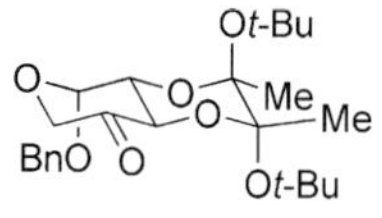

12

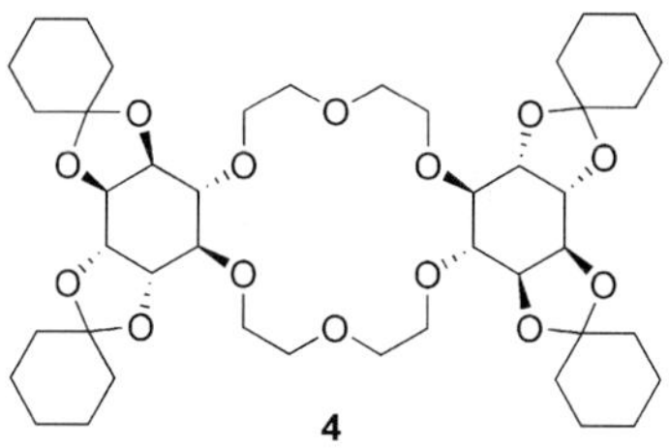

4

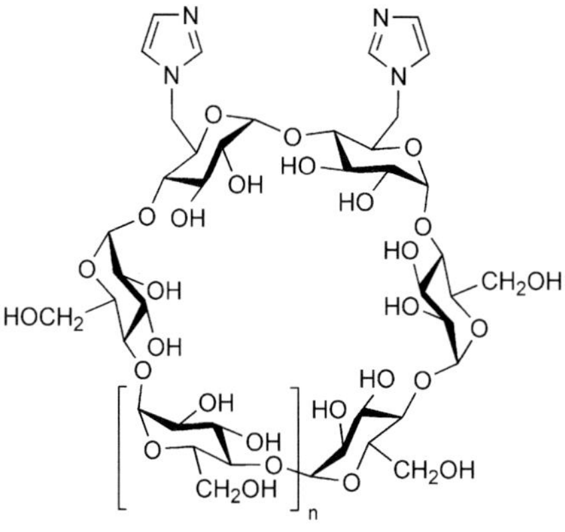

n = 1, 2, 3; **13**

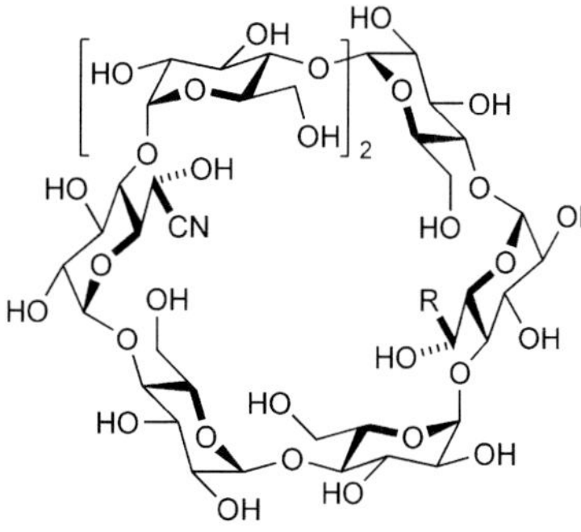

R = CN; OH; =O; **13**

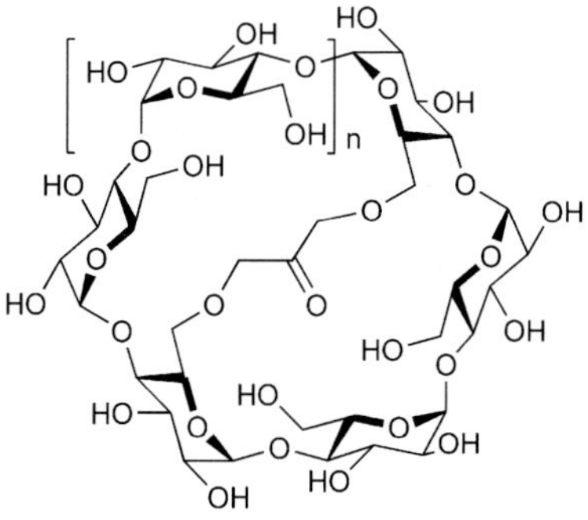

n = 1, 2; **13**

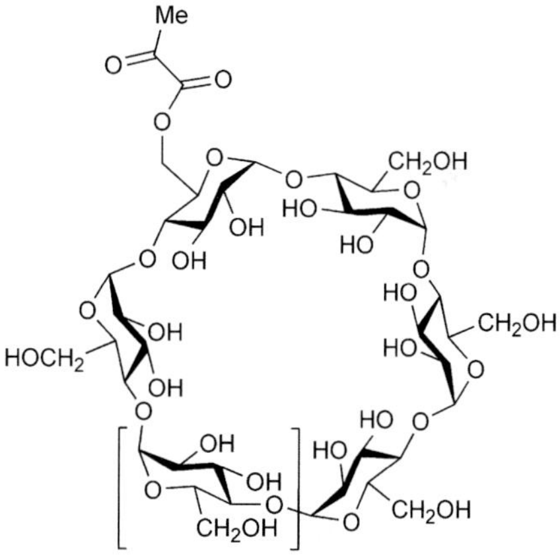

13

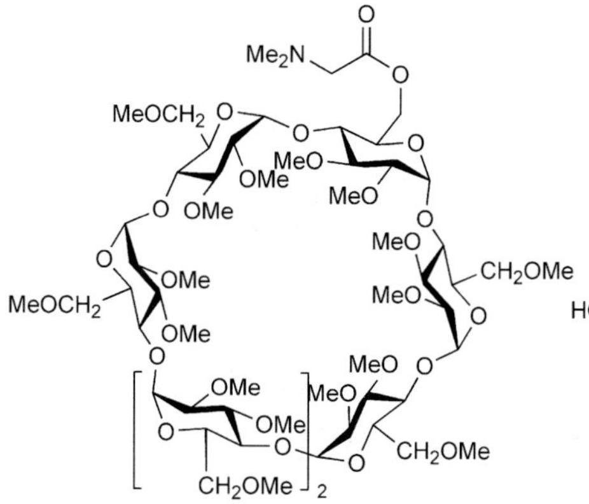

13

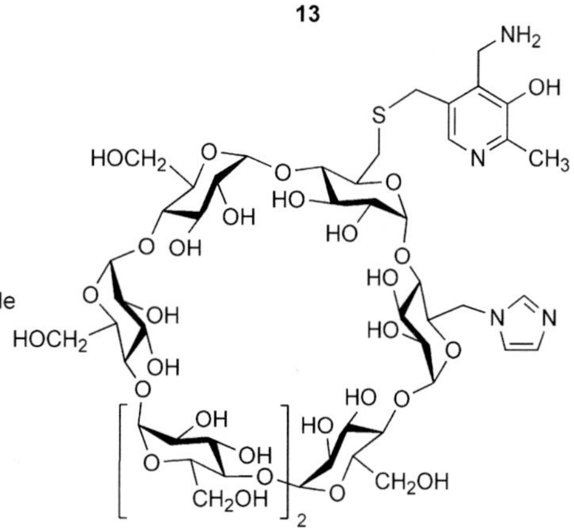

13

Subject Index

a

acyl Pictet-Spengler reaction   222 ff.
– hydrogen bond activation   223
– iminium ions   222 f.
– tetrahydro-β-carboline systems   222
acyl transfer reactions   287 ff.
– asymmetric desymmetrization   287
– biomimetic catalysts   297
– ketenes   321 ff.
– kinetic resolution   287
– π-nucleophiles   309 ff.
– Steglich rearrangements   309 ff.
– Type I   290 ff.
– Type II   311 ff., 321 ff.
aldehyde donors   25 ff.
– cross-aldol reactions   26, 28
– Mannich-type reactions   38, 45 f.
– self-aldolization   25
aldimines   46, 139, 152, 174 ff., 209 ff., 234 f., 250, 265 f., 274
– alkynylation   139 f., 266
– aza Baylis–Hillmann   174
– aza-Friedel–Crafts reaction   224 f.
– hydrophosphonylation   225 f.
aldol cycloisomerization   165
aldol reactions   2, 5 f., 19 ff., 88, 121 f., 125 ff., 142, 244 ff., 267 ff.
– aldehyde donors   25 f.
– alkyl ketone donors   20 ff.
– chiral hydrogen bond donors   244
– chiral N-oxides   268
– ketone acceptors   30 ff.
– intramolecular aldol reaction   31
– Lewis bases   272
– Mukaiyama aldol reaction   28
– nitrosobenzene   244 ff.
– (S)-proline   21
– α-oxyketone donors   23 f.
– scope   269
– silyl ketene acetals   267

– TADDOL-catalyzed   245
– tandem Michael aldol   20
– transition states   31 ff.
– water   36 f.
aldolase antibodies   20
allylation   256 ff.
– aldehydes   264
– competing processes   263
– electronic effects   262
– imines   264
– phosphoramides   257 f.
– pyridine-type N-oxides   259 ff.
allylic alkylation   133
amidation   351
– mechanism   351
α-amidation   59 ff.
– aldehydes   59
– α-cyanoacetates   64 f.
– β-dicarbonyl compounds   64
– ketones   59 f.
– mechanisms   63
– proline-catalyzed   60 f.
– quaternary stereocenter   61 f.
– transition states   63
aminocatalysis   7
ammonium bifluorides   125 ff.
– aldol reactions   125 ff.
– Michael reactions   127 ff.
ammonium ions   121 ff.
– nitroaldol reaction   122, 125 f.
ammonium phenoxides   129
amprenavir   123
ASD   see asymmetric desymmetrization
asymmetric desymmetrization   287, 307 ff.
– achiral/meso-diols   307 ff.
– Sharpless ligands   315
– Type I   307
– Type II   311 ff.
asymmetric proton catalysis   189 ff.
– conjugate additions   193 ff.

Enantioselective Organocatalysis: Reactions and Experimental Procedures
Edited by Peter I. Dalko
Copyright © 2007 WILEY-VCH Verlag GmbH & Co. KGaA, Weinheim
ISBN: 978-3-527-31522-2

aza-Darzens reaction 227
aza-Friedel–Crafts reaction 224 ff.
– aldimines 224 ff.
– chiral phosphoric acid catalysts 224
– hydrogen bond interaction 225
aza-Henry reaction 166, 220 ff.
– aldimines 220 ff.
– polar ionic hydrogen-bonding 221
– quinolinium salt catalyst 221
– thiourea catalysts 220 f.
aza-Morita–Baylis–Hillman 174 ff.
– chiral amine catalysis 175 ff.
– conditions 174
– mechanism 175
– phosphine catalysts 180 f.
aziridination 370 ff.
– carbene-based cycle 371
– ring expansions 372
– scope 375 f.
– selectivity 374 f.
– sulfonium salt formation 373 f.
– sulfur ylide-catalyzed 370 f.

b

Bacillus macerans 434
Baeyer–Villiger reaction 418 ff.
– bisflavin catalyst 418
benzoin condensation 331 ff.
– catalytic cycle 332
– chiral thiazolium salts 333 f.
– crossed benzoin condensations 337 f.
– "formoin" reaction 334
– stable carbenes 333
– stereoselectivity 335
– transition states 336
benzoylformate decarboxylase 336
binaphthol (BINOL) derivatives 11 f.
biodegradable polyesters 352
α-bromination 72 f.

c

calixarenes 432 ff.
– esterification 433
cascade reactions 31, 111 ff.
    see also domino reaction
– addition-chlorination cascade 113
– addition-cyclization reaction 111
– enamine activations 113
– iminium catalysis 113
– pyrroloindoline catalytic cycle 112
catalyst recycling 38 f., 103
catalytic alkylations 133, 227 f.
– Brønsted acid catalyzed 227
– α-diazoesters 227 f.

cell-adhesion inhibitor BIRT-377 61 f.
chiral diketopiperazines 6
chiral Lewis bases 255 ff.
– phosphoramides 257 f.
– pyridine N-oxides 259 ff.
α-chlorination 69 ff.
– aldehydes 69 f.
– β-ketoesters 72 f.
chymotrypsin mimetics 444
cinchona alkaloids 10
cinchona catalyst 5, 10 f., 59, 64, 67, 69, 72 f., 129 ff., 175 f., 193 ff., 231, 249 f., 312 ff., 414
conjugate addition (*see also* Michael addition) 72 ff., 97 f., 109, 114, 155, 193 ff., 197 ff., 249, 339, 341 ff., 392 ff.
– alkylidene malonates 79, 84 f.
– bifunctional cinchona catalysts 193 ff., 199
– α-cyanoacetates 206 f.
– enamine rotamers 78
– enones 87 f.
– intramolecular 90
– Michael addition 86 ff., 196 f.
– nitroolefins 79, 84 f., 197 ff.
– prochiral imide 195
– selectivity model 195
– stereoselectivity 77 f.
– thiourea catalysts 197
– transition state 78 f.
– α,β-unsaturated carbonyl compounds 193
– vinyl sulfones 206 f.
covalent calalysis 12
(+)-curcuphenol 107
[3+2]-cycloaddition 102 ff.
– imidazolidinone catalysis 102
– scope 103
[4+2]-cycloaddition 65
[4+3]-cycloadditions 105 f.
– imidazolidinone-catalyzed 105 f.
cyclodextrins 432, 434 ff.
– alkene epoxidation 438
– amidation 437
– aryl glycoside hydrolysis 435
– phosphate hydrolysis 435
– transamination 437
cyclomaltodextrin glucanotransferase 434
cyclopropanations 103 f., 377 ff.
– catalytic cycle 379 f.
– diazo compounds 378 f.
– direct electrostatic activation 104
– enantioselective 383 f.
– iminium-catalysis 103
– Nitrogen ylide catalyzed 382 f.

– reaction mechanism   385
– scope   380 f.
– selectivity   379, 381
– Tellurium ylide catalyzed   382
– S-ylide catalyzed   377 f.

d

Dakin–West reaction   317
Deng's selectivity model   201 f.
Diels–Alder reaction   98 f., 234, 241 ff.
– amine-catalyzed   99
– bis-amidinium salts   241 f.
– chiral Lewis-acids   100
– donor-acceptor "π-stacking"   242
– iminium-activated   99
– intramolecular   100 f.
– TADDOL-catalyzed   242 f.
dihydroflavine adenine dinucleotide   393
dihydronicotinamide adenine dinucleotide   393
domino-type mulitstep reaction   2, 8, 110
   see also cascade reactions

e

enal epoxidation   417
enamine activation   77 ff., 92
enamine catalysis   19 ff.
enamine mechanism   31
enoates   413 f.
– epoxidation   413
enone epoxidation   414 ff.
– cinchona alkaloids   414
– phase-transfer catalysis   414 f.
– substrate scope   416
enzyme inhibitors   43
enzyme-like catalysts   446
(−)-epibatidine   199
epopromycin B   162
epothilone A   301
epoxidations   104 f., 357 ff.
– alkenes   403
– alkylidenation   357
– amine catalyzed   412
– betaine intermediate   358
– carbene source   360 f., 364
– catalysis   359
– cyclodextrins   438, 441
– diastereoselectivity   365
– enantioselectivity   366 f.
– enoates   413 f.
– enole benzoates   407
– glycidic amides   364
– iminium salt-catalyzed   410 ff.
– in situ diazo compound generation   361

– ketone catalysts   404 ff.
– oxone-mediated   403
– scope   362, 367
– Se-ylide catalysis   369
– Shi epoxidations   406, 408
– Simmons-Smith reagent   364
– sulfur ylide epoxidation   358 f.
– Te-ylide catalysis   369
– transition states   408, 411
– α,β-unsaturated aldehydes   104
epoxide opening   278 f.
– chiral Lewis bases   280 f.
– chlorosilanes   278
– mechanism   279
esterases   289

f

FADH   see dihydroflavine adenine dinucleotide
(−)-flustramine   113
Friedel–Crafts alkylation   106 f., 201 ff., 224, 250
– amine-catalyzed   106
– imidazolidinone catalysts   106
– nucleophiles   107
– α,β-unsaturated carbonyl compounds   106

g

glabrescol   407
G-protein-coupled receptors   62

h

Hajos–Parish-type reaction   5 f., 31
α-halogenation   68 ff.
– α-bromination   72
– chiral amines   70
– cinchona alkaloids   69
– α-chlorination   69 f.
– α-fluorination   68 f.
– N-fluorobenzenesulfonimide   68
– β-ketoesters   69
– α-selenation   74
– α-sulfenylation   73 f.
Hantzsch dihydropyridine   91, 110, 229, 278, 393 ff.
Heck-type cyclisation   349
Hetero-Diels–Alder reactions   235 ff.
– Danishefsky diene   239 f., 250
– hydrogen bond donor catalysts   236
– oxazoline   239
– TADDOLs   236 ff.
α-heteroatom functionalization   56 ff.
– electronic interaction   57 f.

– face-selectivity   58
– steric interaction   57 f., 60
heterogeneous catalysis   129 ff.
– aldol reaction   142, 144
– α-amino acids   130 f.
– catalysts   134 ff.
– cinchona alkaloids   130 ff.
– dialkylation   138 ff.
– Mannich reaction   145 ff.
– Michael reaction   141 ff.
– Schiff bases   130 ff.
HIV protease inhibitor   123
hydrocyanations   207 ff., 273 ff.
– aldehydes   207 ff., 273
– α-amino nitrils   209
– N-benzhydryl aldimines   212
– guanidine catalysts   212
– hydrogen bonding activation   213
– Inoue catalyst   208 f.
– ketoimines   211
– ketones   214 ff.
– Schiff base catalysts   210
– Strecker reaction   209 ff., 274
– thiourea catalysts   214 f.
hydrogen bond   35, 161, 189 ff., 381, 415, 426, 432, 442 ff.
– acceptor   190 f.
– donor   190 f.
– electronegativities   190
– Etter's concept   192
– frontier molecular orbitals   192
– solvent effects   191
hydrolase enzymes   289
hydrophosphonylation   225 ff.
– chiral phosphoric acid catalyst   226
– imines   226
– reaction mechanism   226
– selectivity model   226
– thiourea catalysts   225
hydrosilylation   124, 392, 399
– chiral ammonium fluoride   124

i
imidazolidine catalysts   96 ff.
– aza-Diels–Alder reaction   96
– Diels–Alder reaction   101
– first generation   96 f.
– reductive amination   96
imines   229 f.
– chiral Lewis bases   276 f.
– Kwon annulation   266
– phosphoric acid catalysis   230
– pipecolinic acid   275
– reduction   275 ff., 391

– transfer hydrogenations   229 f.
imine amidation   228
– unsymmetrical aminals   228
iminium activation strategy   96 f.
iminium catalysis   95 ff.
– acyl Pictet–Spengler reaction   222 ff.
– catalysis concept   95 f.
– geometry control   97
– imidazolidinone catalysts   96 ff.
– lone pair activation   101
– phase-transfer catalysis
– scope   97
ionic hydrogenation   392
ion-pairing mechanism   6

k
ketone epoxidations   403 ff.
– transition states   408
(−)-ketorolac   106 f.
kinetic resolution   287, 291 ff.
– allylic sec-alcohols   299 f.
– α-chiral primary amines   306 f.
– nucleophilic N-heterocyclic carbenes   298
– planar chiral pyrrole catalysts   293 ff.
– propargylic sec-alcohols   302
– Type I acylative   291
– Type II alcoholative   316 ff.
– Vedejs' PBO catalyst   291 f, 298 f.
KR   see kinetic resolution

l
Lewis acid, Lewis base activation   8
Lewis bases   255 ff.
lipases   289

m
MacMillan's organocatalyst   90 ff.
Mannich reactions   38 ff., 142 ff., 215 ff.
– aldehyde donors   38 ff.
– allylation   44
– anti-selective reactions   50
– cyanation   44
– glyoxylate imines   38 f.
– heterogeneous catalysis   145 f.
– ketone donors   47 f.
– (S)-proline-catalysis   40 ff.
– proton catalysis   215 ff.
– self-Mannich reactions   46
– three-component reaction   45, 48
– transition states   51
Meerwein–Ponndorf–Verley reduction   392
metabotropic glutamate receptors   62
Michael addition   25, 31, 87 ff., 108 ff., 114, 127 ff., 141 ff., 154, 174, 196 ff., 205 f., 249